Lecture Notes of the Institute for Computer Sciences, Social Informatics and Telecommunications Engineering   156

More information about this series at http://www.springer.com/series/8197

Mark Weichold · Mounir Hamdi
Muhammad Zeeshan Shakir · Mohamed Abdallah
George K. Karagiannidis · Muhammad Ismail (Eds.)

Cognitive Radio Oriented Wireless Networks

10th International Conference, CROWNCOM 2015
Doha, Qatar, April 21–23, 2015
Revised Selected Papers

 Springer

Editors
Mark Weichold
Texas A&M University at Qatar
Doha
Qatar

Mohamed Abdallah
Texas A&M University at Qatar
Doha
Qatar

Mounir Hamdi
Hamad Bin Khalifa University
Doha
Qatar

George K. Karagiannidis
Aristotle University of Thessaloniki
Greece and Khalifa University
United Arab Emirates

Muhammad Zeeshan Shakir
Texas A&M University of Qatar
Doha
Qatar

Muhammad Ismail
Texas A&M University at Qatar
Doha
Qatar

ISSN 1867-8211   ISSN 1867-822X (electronic)
Lecture Notes of the Institute for Computer Sciences, Social Informatics
and Telecommunications Engineering
ISBN 978-3-319-24539-3   ISBN 978-3-319-24540-9 (eBook)
DOI 10.1007/978-3-319-24540-9

Library of Congress Control Number: 2015950861

Springer Cham Heidelberg New York Dordrecht London

Printed on acid-free paper

Springer International Publishing AG Switzerland is part of Springer Science+Business Media
(www.springer.com)

CROWNCOM 2015

Preface

2015 marks the 10th anniversary of the International Conference on Cognitive Radio-Oriented Wireless Networks (Crowncom). Crowncom 2015 was jointly hosted by Texas A&M University at Qatar and Hamad Bin Khalifa University in Doha, Qatar, April 21–23, 2015. The event was a special occasion to look back at the contribution of Crowncom toward the advancements of cognitive radio technology since its inaugural conference in 2006 in Mykonos, Greece, as well as to look forward to the decades ahead, the ways that cognitive radio technology would like to evolve, and the ways its emerging applications and services can ensure everyone is connected everywhere.

Evolution of cognitive radio technology pertaining to 5G networks was the theme of the 2015 edition of Crowncom. The technical program of Crowncom 2015 was structured to bring academic and industrial researchers together to identify and discuss recent developments, highlight the challenging gaps, and forecast the future trends of cognitive radio technology toward its integration with the 5G network deployment. One of the key topics of the conference was cognition and self-organization in the future networks, which are now widely considered as a striking solution to cope with the future ever-increasing spectra demands. Going beyond the theoretical development and investigation, further practical advances and standardization developments in this technology could provide potential dynamic solutions to cellular traffic congestion problems by exploiting new and underutilized spectral resources. One of the challenging issues that Crowncom 2015 brought forward was to facilitate the heterogeneous demands of users in heterogeneous-type environments — particularly in the 5G network paradigm, where the networks are anticipated to incorporate the provision of high-quality services to users with extremely low delays and consider these requirements without explicit demand from users. Machine-type communications and Internet of Everything are now representing emerging use cases of such ubiquitous connectivity over limited spectra.

Crowncom 2015 strongly advocated that the research community, practitioners, standardization bodies, and developers should collaborate on their research efforts to further align the development initiatives toward the evolution of emerging highly dynamic spectrum access frameworks. The biggest challenge is to design unified cross-layer new network architectures for successful aggregation of licensed and unlicensed spectra, addressing the spectrum scarcity problem for ubiquitous connectivity and preparing the ground for "The Age of the ZetaByte."

Crowncom 2015 received a large number of submissions, and it was a challenging task to select the best and most relevant meritorious papers to reflect the theme of the 2015 edition of Crowncom. All submissions received high-quality reviews from the Technical Program Committee (TPC) members/reviewers and eventually 66 technical papers (with an acceptance ratio of 56 %) were selected for the technical program of the conference. The technical program of Crowncom 2015 is the result of the tireless efforts of 14 track chairs, and more than 200 TPC members and reviewers. We are grateful to the track chairs for handling the paper review process and their outstanding efforts, and to the reviewers/TPC for their high-quality evaluations. We offer our sincere gratitude to the Advisory Committee, local Organizing Committee (especially colleagues at Texas A&M University at Qatar), and the Steering Committee members for their insightful guidance. We would like to acknowledge the invaluable support from European Alliance for Innovation and the Qatar National Research Fund for the success of Crowncom 2015.

2015

Mark Weichold
Mounir Hamdi
Muhammad Zeeshan Shakir
Mohamed Abdallah
George K. Karagiannidis
Muhammad Ismail

Organization

General Chair

Mark Weichold   Texas A&M University at Qatar, Qatar
Mounir Hamdi   Hamad Bin Khalifa University, Qatar

Technical Program Chair

Muhammad Zeeshan Shakir   Texas A&M University at Qatar, Qatar
Mohamed Abdallah   Texas A&M University at Qatar, Qatar
George K. Karagiannidis   Aristotle University of Thessaloniki, Greece,
                                 and Khalifa University, UAE

Advisory Board

Athanasios V. Vasilakos   Kuwait University, Kuwait
Khalid A. Qaraqe   Texas A&M University at Qatar, Qatar
Jinsong Wu   Bell Labs, China
David Grace   University of York, UK
Naofal Al-Dhahir   University of Texas, Dallas, USA
Kaushik Chowdhury   Northeastern University, USA

Special Session Chair

Alhussein Abouzeid   Rensselaer Polytechnic Institute, USA

Panel Chair

Maziar Nekovee   Samsung, UK

Publication Chair

Muhammad Ismail   Texas A&M University at Qatar, Qatar

Tutorial Chair

Mohamed Nafie   Nile University, Egypt

Exhibitions and Demos Chair

Majid Butt   Qatar University, Qatar

Web Chair

İslam Şafak Bayram   Qatar Environment and Energy Research Institute,
                                     Qatar

Local Arrangements

Carol Nader   Texas A&M University at Qatar, Qatar
Mohamed Kashef   Texas A&M University at Qatar, Qatar

Track Chairs

Track 1: Dynamic Spectrum Access/Management

Mohammad Shaqfeh   Texas A&M University at Qatar, Qatar

Track 2: Networking Protocols for CR

Tamer Khattab   Qatar University, Qatar
Amr Mohamed   Qatar University, Qatar

Track 3: Modeling and Theory

Zouheir Rezki   King Abdullah University of Science and Technology,
                                     Saudi Arabia
Syed Ali Raza Zaidi   University of Leeds, UK

Track 4: HW Architecture and Implementations

Ahmed El-Tawil   University of California, Irvine, USA
Fadi Kurdahi   University of California, Irvine, USA

Track 5: Next Generation of Cognitive Networks

Muhammad Ali Imran   CCSR/5G Innovation Centre University of Surrey, UK
Richard Demo Souza   Federal University of Technology - Paraná (UTFPR),
                                     Curitiba - PR - Brazil

Track 6: Standards and Business Models

Stanislav Fillin   National Institute of Information and Communications
                                     Technology (NICT), Japan
Stephen J. Shellhammer   Qualcomm Technologies, Inc., USA
Markus Dominik Mueck   INTEL Mobile Communications, Germany

Track 7: Emerging Applications for Cognitive Networks

Octavia A. Dobre   Memorial University, Canada
Hai Lin   Osaka Prefecture University, Japan

Contents

Dynamic Spectrum Access/Management

Fractional Low Order Cyclostationary-Based Spectrum Sensing in Cognitive Radio Networks

Hadi Hashemi[1], Sina Mohammadi Fard[1], Abbas Taherpour[1],
and Tamer Khattab[2]([✉])

[1] Department of Electrical Engineering, Imam Khomeini International University,
Qazvin, Iran
h.hashemi@edu.ikiu.ac.ir
[2] Electrical Engineering, Qatar University, Doha, Qatar
tkhattab@ieee.org

Abstract. In this paper, we study the problem of cyclostationary spectrum sensing in cognitive radio networks based on cyclic properties of linear modulations. For this purpose, we use fractional order of observations in cyclic autocorrelation function (CAF). We derive the generalized likelihood ratio (GLR) for designing the detector. Therefore, the performance of this detector has been improved compared to previous detectors. We also find optimum value of the fractional order of observations in additive Gaussian noise. The exact performance of the GLR detector is derived analytically as well. The simulation results are presented to evaluate the performance of the proposed detector and compare its performance with their counterpart, so to illustrate the impact of the optimum value of fractional order over performance improvement of these detectors.

Keywords: Cognitive radio · Spectrum sensing · Cyclostationary signal · Fractional low order

1 Introduction

Increasing need for bandwidth in telecommunication and limited environmental resources lead us to take advantage of other system's spectrum. In spectrum sensing, cognitive radio networks monitor the status of the frequency spectrum by observing their surroundings to exploit the unused frequency bands. There are several methods of spectrum sensing which need different and extra information about the primary user (PU) signal, such as accuracy and implementation complexity [1]. The most important methods are matched filter, energy detection, eigenvalues-based detection, detection based on the covariance matrix and cyclostationary based detection.

Among those, cyclostationary-based detector is one of the best way of spectrum sensing in terms of performance and robustness against environmental parameters like ambient noise. In the context of cyclostationary-based spectrum

© Institute for Computer Sciences, Social Informatics and Telecommunications Engineering 2015
M. Weichold et al. (Eds.): CROWNCOM 2015, LNICST 156, pp. 3–16, 2015.
DOI: 10.1007/978-3-319-24540-9_1

sensing, in [2,3], this detector has been investigated for one specific cyclic frequency. The authors in [2] have reviewed collaborative case and have demonstrated channel fading effects in its performance. The authors in [4–6] have used multiple cyclic frequencies for detection of PU signal and improvement the detection performance has been shown. Furthermore, several research such as [2,7,8] have been conducted where the benefit of using cyclostationary-based detectors in the collaborative systems are investigated. It is known that cyclostationary-based detectors have poor performance for situations where the environment is impulsive noisy and to compensate, the CAF with fractional order of observations are used [9–11]. In these works, the problem of fractional order of observations, is investigated in Alpha stable noisy environment.

In this paper, we provide a spectrum sensing method which benefits of PU signal's cyclostationary property and improve performance of cyclostationary-based detector in different practical cases and noise models. We suggest using fractional order of observed signals. We assume an additive Gaussian noise, thought the results could be extended for the other model of ambient noises. For this purpose, we formulate the spectrum sensing as a binary hypothesis testing problem and then derive the corresponding GLR detectors for the different practical scenarios. Then we investigate the optimum value of fractional order which results in best performance in related cases.

The remaining of the paper is organized as follows. In Section 2, we introduce the system model and the assumptions. In Section 3, we derive cyclostationary-based detectors in different scenarios for signal and noise prameters. In Section 4, we study the performance of the proposed detectors. The optimization of the performance of the proposed detectors is presented in Section 6. The simulation results are provided in Section 7 and finally Section 8 summarizes the conclusions.

Notation: Lightface letters denote scalars. Boldface lower-and upper-case letters denote column vectors and matrices, respectively. x(.) is the entries and $\mathbf{x}_i$ is sub-vector of vector $\mathbf{x}$. The inverse of matrix $\mathbf{A}$ is $\mathbf{A}^{-1}$. The $M \times M$ identity matrix is $\mathbf{I}_M$. Superscripts $*$, T and H are the complex conjugate, transpose and Hermitian (conjugate transpose), respectively. $\mathbb{E}[.]$ is the statistical expectation. $\mathcal{N}(\mathbf{m}, \mathbf{P})$ denotes Gaussian distribution with mean $\mathbf{m}$ and covariance matrix $\mathbf{P}$. $Q(x)$ is Q-function $Q(x) = \frac{1}{\sqrt{2\pi}} \int_x^\infty exp\left(\frac{-u^2}{2}\right) du$.

2   System Model

Suppose a cognitive radio network in which PU and secondary user (SU) equipped with a single antenna. For presentation, it's assumed that the PU signal is transmitted with linear modulation such that

$$s(t) = \sum_{i=-\infty}^{\infty} d_i p(t - iT_{\mathrm{P}}), \tag{1}$$

where d_i is the PU data and $p(t)$ is shaping pulse in the PU transmitter. We suppose PU data, d_i, is a random variable with zero-mean Gaussian distribution, $\mathcal{N}(0, \sigma_s^2)$. For the shaping pulse, a rectangular pulse with unit amplitude and time spread T_P is assumed. Received signal in SU has been sampled with sampling rate of $f_s = \frac{1}{T_s}$. The wireless channel between PU transmitter and SU is assumed to be a flat fading channel with additive Gaussian noise and the channel gain. The random variable $w(n) \sim \mathcal{N}(0, \sigma_w^2)$ denotes noise samples and we assume noise and PU signal samples are mutually independent. Therefore observed signal samples in SU under two hypotheses can be shown as follows,

$$\begin{cases} \mathcal{H}_0: & x(n) = w(n), \\ \mathcal{H}_1: & x(n) = hs(n) + w(n), \end{cases} \tag{2}$$

where h is channel gain between the PU and SU antennas. It is assumed that the channel gain is constant during the sensing time. CAF for the SU observed signal samples is defined based on the correlation between samples and their complex conjugate with lag time $\tau_i < T_P$. The CAF for fractional order is defined as,

$$R_{xx^*}^\alpha(\tau_i) = \frac{1}{N} \sum_{n=0}^{N-1} x^p(n) x^{*p}(n + \tau_i) e^{-j2\pi\alpha n}, \tag{3}$$

where p is fractional order $0 < p < 1$, $\alpha \in \{\frac{k}{T_P}, k = 1, 2, ...\}$ is cyclic frequency for linear modulation which is assumed to be known to SU and $\tau_i, i = 1, \ldots, M$s is M lag times where the CAF is calculated.

We introduce vector $\mathbf{r}_{xx^*}^\alpha$ consisting of CAF real parts for M different lag times as,

$$\mathbf{r}_{xx^*}^\alpha = [Re(R_{xx^*}^\alpha(\tau_1)), ..., Re(R_{xx^*}^\alpha(\tau_M))]^T. \tag{4}$$

By considering central limit theorem (CLT), since the CAF is summation of N random variables, according to [12], for sufficiently large number of observation samples, each member of vector $\mathbf{r}_{xx^*}^\alpha$ has Gaussian distribution . Thus, we have,

$$\mathbf{r}_{xx^*}^\alpha \sim \begin{cases} \mathcal{N}(\boldsymbol{\mu}_0, \boldsymbol{\Sigma}_0) & \text{for} \quad \mathcal{H}_0, \\ \mathcal{N}(\boldsymbol{\mu}_1, \boldsymbol{\Sigma}_1) & \text{for} \quad \mathcal{H}_1. \end{cases} \tag{5}$$

where $\boldsymbol{\mu}_0$ and $\boldsymbol{\mu}_1$ can be calculated for any given p. In Section 5 for the known noise and signal variance these values are computed.

3 Cyclostationary-Based Detectors

SUs use different detection methods in spectrum sensing to make decision about PU's presence. In this section, we assume SU determines PUs situation based on cyclostationary properties of PU signal in which the SU has knowledge about cyclic frequency of observation signal by consideration of different scenarios. These scenarios are investigated in following subsections.

3.1 Known Signal and Noise Variance

Since in (5) covariance matrices under two hypotheses are unknown, we have to use their estimations to construct the likelihood ratio (LR) function which results in a GLR detector. Covariance matrices estimation have been calculated under two hypotheses in Appendix. It has been shown that both of the covariance matrices have same estimation. Thus, $\widehat{\Sigma_0} = \widehat{\Sigma_1} = \Sigma$. Now for the LR function, we have,

$$LR(\mathbf{r}^{\alpha}_{xx^*}) = \exp\{\boldsymbol{\mu}_0^T \Sigma^{-1} \boldsymbol{\mu}_0 - \boldsymbol{\mu}_1^T \Sigma^{-1} \boldsymbol{\mu}_1 + 2\mathbf{r}^{\alpha T}_{xx^*} \Sigma^{-1}(\boldsymbol{\mu}_1 - \boldsymbol{\mu}_0)\} \underset{\mathcal{H}_0}{\overset{\mathcal{H}_1}{\gtrless}} \eta. \quad (6)$$

By incorporating the constant terms into threshold and taking logarithm in (6), we obtain,

$$\mathrm{T}_{sub1} = \mathbf{r}^{\alpha T}_{xx^*} \Sigma^{-1}(\boldsymbol{\mu}_1 - \boldsymbol{\mu}_0) \underset{\mathcal{H}_0}{\overset{\mathcal{H}_1}{\gtrless}} \eta_1, \quad (7)$$

where $\boldsymbol{\mu}_0$ and $\boldsymbol{\mu}_1$ can be calculated. It can be seen that detector is the weighted summation of CAF real part for different lag times $\tau_i, i = 1, 2, ..., M$.

3.2 Known Noise Variance, Unknown Signal Variance

The mean of (4), when SU has just knowledge about noise variance, can be derived under null hypothesis according to section 5.1. But as mentioned, signal variance is unknown and thus, mean of the CAF real parts under alternative hypothesis cannot be calculated. In this situation, we can use Hotelling-test [13,17], because we definitely know that the mean under two hypotheses are different. Suppose, $L > M + 1$ given vector $\mathbf{r}^{\alpha}_{xx^*}$ in a vector are considered together, $\mathbf{r} = [\mathbf{r}^{\alpha}_{xx^*}(1), \mathbf{r}^{\alpha}_{xx^*}(2), ..., \mathbf{r}^{\alpha}_{xx^*}(L)]$. Statistical distribution of this vector under hypothesis $\mathcal{H}_j, j = 0, 1$ can be written in the form below,

$$f(\mathbf{r}|\mathcal{H}_j) = \frac{\exp\left\{-\frac{1}{2}tr([\frac{1}{L}\boldsymbol{\Psi} + (\bar{\mathbf{r}} - \boldsymbol{\mu}_j)(\bar{\mathbf{r}} - \boldsymbol{\mu}_j)^T]\Sigma_j^{-1})\right\}}{(2\pi)^{\frac{LM}{2}}|\Sigma_j|^{\frac{L}{2}}}, \quad (8)$$

where $\bar{\mathbf{r}} = \frac{1}{L}\sum_{i=1}^{L}\mathbf{r}^{\alpha}_{xx^*}(i)$ and $\boldsymbol{\Psi} = \sum_{i=1}^{L}(\mathbf{r}^{\alpha}_{xx^*}(i) - \bar{\mathbf{r}})(\mathbf{r}^{\alpha}_{xx^*}(i) - \bar{\mathbf{r}})^T$, under alternative hypothesis, $\bar{\mathbf{r}}$ is estimate of $\boldsymbol{\mu}_1$ and the statement inside the bracket of function $tr(.)$ is the estimation covariance matrix under two hypotheses. Thus after eliminating the constants we have,

$$\Lambda = \frac{|\frac{1}{L}(\boldsymbol{\Psi} + L(\bar{\mathbf{r}} - \boldsymbol{\mu}_0)(\bar{\mathbf{r}} - \boldsymbol{\mu}_0)^T)|^{\frac{L}{2}}}{|\frac{1}{L}\boldsymbol{\Psi}|^{\frac{L}{2}}} = |\mathbf{I} + L\boldsymbol{\Psi}^{-1}(\bar{\mathbf{r}} - \boldsymbol{\mu}_0)(\bar{\mathbf{r}} - \boldsymbol{\mu}_0)^T|^{\frac{L}{2}}. \quad (9)$$

By using the matrix determinant lemma that computes the determinant of the sum of an invertible matrix $\mathbf{I}$ and the dyadic product, $\boldsymbol{\Psi}^{-1}(\bar{\mathbf{r}} - \boldsymbol{\mu}_0)(\bar{\mathbf{r}} - \boldsymbol{\mu}_0)^T$,

$$\Lambda = \left(1 + L(\bar{\mathbf{r}} - \boldsymbol{\mu}_0)^T \boldsymbol{\Psi}^{-1}(\bar{\mathbf{r}} - \boldsymbol{\mu}_0)\right)^{\frac{L}{2}} = (1 + \mathrm{T}_{sub2})^{\frac{L}{2}}. \quad (10)$$

Since Λ is the strictly ascending function of T_{sub2}, therefore, T_{sub2} can be considered as a statistic.

$$T_{sub2} = L(\bar{\mathbf{r}} - \boldsymbol{\mu}_0)^T \boldsymbol{\Psi}^{-1}(\bar{\mathbf{r}} - \boldsymbol{\mu}_0) \tag{11}$$

3.3   Unknown Signal and Noise Variance

In this situation, by considering covariance matrices estimation as (A-4), we have two Gaussian distribution by same covariance matrices and different mean under two hypotheses. If estimation is used for means of CAF real parts under both hypotheses, due to equality of estimation under two hypotheses the result of GLR test does not give any information to make decision. Thus, mean of CAFs for various lag time is considered as statistic and compared with a proper threshold.

$$T_{sub3} = \frac{1}{M} \sum_{m=1}^{M} Re(R^\alpha_{xx*}(\tau_m)) \underset{\mathcal{H}_0}{\overset{\mathcal{H}_1}{\gtrless}} \eta_3. \tag{12}$$

4   Analytical Performance

In this section, we evaluate the performance of our proposed cyclostationary-based detectors in terms of detection and false alarm probabilities, P_{d} and P_{fa}, respectively.

4.1   Analytical Performance of T_{sub1}

We should derive statistical distribution of (7) under two hypotheses. We can rewrite (7) as follows,

$$T_{sub1} = (\mathbf{r}^{\alpha T}_{xx*} \boldsymbol{\Sigma}^{-\frac{1}{2}})(\boldsymbol{\Sigma}^{-\frac{1}{2}}(\boldsymbol{\mu}_1 - \boldsymbol{\mu}_0)) = \tilde{\mathbf{r}}^{\alpha T}_{xx*} \mathbf{w} \underset{\mathcal{H}_0}{\overset{\mathcal{H}_1}{\gtrless}} \eta_1, \tag{13}$$

where $\mathbf{w} = \boldsymbol{\Sigma}^{-\frac{1}{2}}(\boldsymbol{\mu}_1 - \boldsymbol{\mu}_0)$ and $\tilde{\mathbf{r}}^\alpha_{xx*} = \boldsymbol{\Sigma}^{-\frac{1}{2}}\mathbf{r}^\alpha_{xx*}$ which is distributed as Gaussian under two hypotheses, i.e.,

$$\tilde{\mathbf{r}}^\alpha_{xx*}|\mathcal{H}_\nu \sim \mathcal{N}(\mathbf{m}_\nu, \mathbf{I}_M), \quad \nu = 0, 1, \tag{14}$$

where $\mathbf{m}_\nu = \boldsymbol{\Sigma}^{-\frac{1}{2}}\boldsymbol{\mu}_\nu$. As we can see in (13), our detector is a linear combination of independent Gaussian random variables mentioned in (14). Therefore, mean of statistic is,

$$\mu_{T_{sub1}|\mathcal{H}_\nu} = \sum_{i=1}^{M} m_\nu(i)w(i), \quad \nu = 0, 1. \tag{15}$$

And similarly variance has been derived,

$$\sigma^2_{\mathrm{T}_{sub1}|\mathcal{H}_\nu} = \sum_{i=1}^{M} w^2(i), \ \nu = 0, 1. \tag{16}$$

Then, the false alarm and detection probabilities can be calculated.

$$P_{\mathrm{fa}} = P\left[\mathrm{T}_{sub1} > \eta_1|\mathcal{H}_0\right] = Q\left(\frac{\eta_1 - \mu_{\mathrm{T}_{sub1}|\mathcal{H}_0}}{\sigma_{\mathrm{T}_{sub1}|\mathcal{H}_0}}\right) \tag{17}$$

If β is maximum acceptable probability false alarm, then threshold of detector can be set, $\eta_1 = F^{-1}_{\mathrm{T}_{sub1}|\mathcal{H}_0}(\beta) = Q^{-1}(\beta) \times \sigma_{\mathrm{T}_{sub1}|\mathcal{H}_0} + \mu_{\mathrm{T}_{sub1}|\mathcal{H}_0}$. Similarly for probability of detection, we have,

$$P_{\mathrm{d}} = P\left[\mathrm{T}_{sub1} > \eta_1|\mathcal{H}_1\right] = Q\left(\frac{\eta_1 - \mu_{\mathrm{T}_{sub1}|\mathcal{H}_1}}{\sigma_{\mathrm{T}_{sub1}|\mathcal{H}_1}}\right). \tag{18}$$

4.2 Analytical Performance of T_{sub2}

We should derive statistical distribution of (11) under two hypotheses. According to [13], the asymptotic distribution of (11) under null hypothesis is central chi-squared with M degrees of freedom. Thus, probability of false alarm is as follows,

$$P_{\mathrm{fa}} = P\left[\mathrm{T}_{sub2} > \eta_2|\mathcal{H}_0\right] = 1 - \frac{\gamma\left(\frac{M}{2}, \frac{\eta_2}{2}\right)}{\Gamma\left(\frac{M}{2}\right)}, \tag{19}$$

where $\Gamma(.)$ and $\gamma(.,.)$ are Gamma and lower incomplete Gamma function, respectively. The asymptotic distribution of (11) under alternative hypothesis is noncentral chi-squared with noncentrality parameter, λ. Probability of detection is as follows,

$$P_{\mathrm{d}} = P\left[\mathrm{T}_{sub2} > \eta_2|\mathcal{H}_1\right] = Q_{\frac{M}{2}}(\sqrt{\lambda}, \sqrt{\eta_2}), \tag{20}$$

where $Q(.,.)$ is Marcum Q-function and non-centrality parameter is, $\lambda = \frac{L}{2}(\boldsymbol{\mu}_1 - \boldsymbol{\mu}_0)^T \boldsymbol{\Sigma}_1^{-1}(\boldsymbol{\mu}_1 - \boldsymbol{\mu}_0)$.

4.3 Analytical Performance of T_{sub3}

Because (12) is a linear combination of Gaussian random variables, therefore, T_{sub3} distribution is Gaussian under two hypotheses. According to Appendix 8, mean and variance of (12) can be calculated. Thus, probability of false alarm and detection are as follow,

$$P_{\mathrm{fa}} = Q\left(\frac{\eta_3 - \mu_{\mathrm{T}_{sub3}|\mathcal{H}_0}}{\sigma_{\mathrm{T}_{sub3}|\mathcal{H}_0}}\right), \tag{21}$$

$$P_{\mathrm{d}} = Q\left(\frac{\eta_3 - \mu_{\mathrm{T}_{sub3}|\mathcal{H}_1}}{\sigma_{\mathrm{T}_{sub3}|\mathcal{H}_1}}\right). \tag{22}$$

5 Calculation of $r_{xx^*}^\alpha$ Means

In this section, we have provided computations for expectation of $r_{xx^*}^\alpha$ under two hypotheses when all variables are known.

5.1 Null Hypothesis

In this subsection, we investigate mean of $r_{xx^*}^\alpha$ under null hypothesis. By consideration of noise samples independency, expectation of (3) can be easily derived for ith lag time as follows,

$$\mathbb{E}[R_{xx^*}^\alpha(\tau_i)|\mathcal{H}_0] = \frac{1}{N}\sum_{n=0}^{N-1} \mathbb{E}[w^p(n)]\mathbb{E}[w^{*p}(n+\tau_i)]e^{-j2\pi\alpha n}. \tag{23}$$

pth moment of Gaussian random variable has been calculated in Appendix, since $w(n)$ is zero mean Gaussian random variable, therefore,

$$\mathbb{E}[R_{xx^*}^\alpha(\tau)|\mathcal{H}_0] = \frac{e^{-j\pi\alpha(N-1)}}{N}\frac{sin(\pi\alpha N)}{sin(\pi\alpha)}\frac{(-2)^p\pi\sigma_n^{2p}}{\Gamma^2\left(\frac{1-p}{2}\right)}. \tag{24}$$

Mean of (4) for $i = 1,..,M$,

$$\mu_0(i) = \frac{sin(\pi\alpha N)}{N sin(\pi\alpha)}\frac{\pi(2\sigma_n^2)^p}{\Gamma^2\left(\frac{1-p}{2}\right)}cos(\pi(\alpha(1-N)+p)). \tag{25}$$

5.2 Alternative Hypothesis

As mentioned earlier, each of the observation samples at SU is distributed as,

$$X = x(n) \sim \mathcal{N}(0, h^2p^2\sigma_s^2 + \sigma_n^2) \triangleq \mathcal{N}(0,\sigma_1^2). \tag{26}$$

Now, we assume random variable Y to be the ith lag time of observation samples which is distributed same as X, i.e., $Y = x(n+\tau_i)$. It can be easily demonstrated that correlation coefficient between X and Y is,

$$r = \frac{\mathbb{E}(XY) - \mathbb{E}(X)\mathbb{E}(Y)}{\sigma_1 \times \sigma_1} = \frac{h^2}{\sigma_1^2}\mathbb{E}[s(t)s(t+\tau_i)] = \frac{h^2p^2\sigma_s^2}{\sigma_1^2}, \tag{27}$$

which reveals that X and Y are correlated. Thus, X and Y have joint Gaussian distribution, $\mathcal{N}(0,0,\sigma_1^2,\sigma_1^2,r)$. To determine the mean of CAF under alternative hypothesis, we need to calculate $\mathbb{E}[X^pY^p] = \mathbb{E}[Z^p] = \mathbb{E}[T]$. First we must derive probability density function (PDF) of Z which is product X and Y. i.e.,

$$f_Z(z) = \int_0^\infty \frac{1}{x}f_{XY}(x,\frac{z}{x})dx - \int_{-\infty}^0 \frac{1}{x}f_{XY}(x,\frac{z}{x})dx. \tag{28}$$

$$\frac{(x\sigma_1\sqrt{2(1-r^2)})^p}{j^p\sqrt{2\sigma_1^2}} e^{\left(-\frac{x^2}{2\sigma_1^2(1-r^2)}\right)} \sum_{k=0}^{\infty} \left[\frac{\left(\frac{-p}{2}\right)^{\overline{k}}}{\Gamma\left(\frac{1-p}{2}\right)\left(\frac{1}{2}\right)^{\overline{k}} k!} - \frac{\sqrt{2}jrx\left(\frac{1-p}{2}\right)^{\overline{k}}}{\Gamma\left(-\frac{p}{2}\right)\sigma_1\sqrt{1-r^2}\left(\frac{3}{2}\right)^{\overline{k}} k!}\right] \times$$

$$\left(-\frac{r^2x^2}{2\sigma_1^2(1-r^2)}\right)^k = \sum_{k=0}^{\infty} \left[A(r,\sigma_1,k,p)x^{2k+p} - B(r,\sigma_1,k,p)x^{2k+p+1}\right] e^{\left(-\frac{x^2}{2\sigma_1^2(1-r^2)}\right)}$$

$$(33)$$

In second step, we can declare distribution of T as function of Z PDF, as follows,

$$f_T(t) = \frac{1}{p}t^{\frac{1}{p}-1}f_Z(t^{\frac{1}{p}}).$$
$$(29)$$

And thus, for computation of T mean, we have,

$$\mathbb{E}[T] = \int_0^{\infty}\int_{-\infty}^{\infty}\frac{t^{\frac{1}{p}}}{px}f_{XY}(x,\frac{t^{\frac{1}{p}}}{x})dtdx - \int_{-\infty}^0\int_{-\infty}^{\infty}\frac{t^{\frac{1}{p}}}{px}f_{XY}(x,\frac{t^{\frac{1}{p}}}{x})dtdx. \quad (30)$$

Common part of above equation is derived in following expression,

$$\int_{-\infty}^{\infty}\frac{t^{\frac{1}{p}}}{px}f_{XY}(x,\frac{t^{\frac{1}{p}}}{x})dt = \frac{\exp\left(-\frac{x^2}{2\sigma_1^2}\right)}{px2\pi\sigma_1^2\sqrt{1-r^2}}\int_{-\infty}^{\infty}t^{\frac{1}{p}}\exp\left\{-\frac{\left(t^{\frac{1}{p}}-rx^2\right)^2}{2x^2\sigma_1^2(1-r^2)}\right\}dt.$$
$$(31)$$

Integral expression in equation (31) is in the form of p-th moment of Gaussian random variable with respectively mean and variance rx^2 and $x^2\sigma_1^2(1-r^2)$ that is calculated in Appendix. Therefore,

$$\int_{-\infty}^{\infty}\frac{t^{\frac{1}{p}}}{px}f_{XY}(x,\frac{t^{\frac{1}{p}}}{x})dt = \frac{(x\sigma_1\sqrt{(1-r^2)})^p}{j^p\sqrt{2\pi\sigma_1^2}}\exp\left(-\frac{(2-r^2)x^2}{4\sigma_1^2(1-r^2)}\right)D_p\left(\frac{jrx}{\sigma_1\sqrt{1-r^2}}\right).$$
$$(32)$$

Result of replacement Apendix equations in (32) also some calculations and simplifications, has led to (33), which is at the top of next page. In (33),

$$A(r,\sigma_1,k,p) = \frac{\left(\frac{-p}{2}\right)^{\overline{k}}(\sigma_1\sqrt{2(1-r^2)})^p r^{2k}}{\Gamma\left(\frac{1-p}{2}\right)\left(\frac{1}{2}\right)^{\overline{k}} k!\sqrt{2\sigma_1^2}j^p(2\sigma_1^2(r^2-1))^k}, \quad (34)$$

$$B(r,\sigma_1,k,p) = \frac{\sqrt{2^{p+1}}\left(\frac{1-p}{2}\right)^{\overline{k}}(\sigma_1\sqrt{(1-r^2)})^{p-2k-1}r^{2k+1}}{\Gamma\left(-\frac{p}{2}\right)\left(\frac{3}{2}\right)^{\overline{k}} k!j^{p-1}(-2)^k}. \quad (35)$$

Finally, from (36) and according to [14], mean of T is derived in the next page. Therefore, ith member of $\boldsymbol{\mu}_1$ for $i = 1, ..., M$ is,

$$\mu_1(i) = \frac{sin(\pi\alpha N)}{N sin(\pi\alpha)}cos(\pi\alpha(N-1))\mathbb{E}[T] \quad (37)$$

$$\mathbb{E}[T] = \sum_{k=0}^{\infty}(1+(-1)^p)\left[A(r,\sigma_1,k,p)2^{2k+p+1}(2\sigma_1^2(1-r^2))^{\frac{2k+p+1}{2}}\frac{\Gamma(2k+p+1)\sqrt{\pi}}{\Gamma(\frac{2k+p+2}{2})}\right.$$

$$\left. - B(r,\sigma_1,k,p)2^{2k+p+2}(2\sigma_1^2(1-r^2))^{\frac{2k+p+2}{2}}\frac{\Gamma(2k+p+2)\sqrt{\pi}}{\Gamma(\frac{2k+p+3}{2})}\right] \quad (36)$$

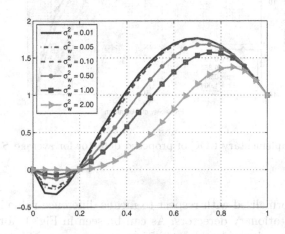

Fig. 1. Normalized difference of means for ith lag time

6 Performance Optimization

To optimize the performance of proposed detector and obtain an appropriate threshold by using the Neyman-Pearson criterion, we have to maximize the probability of detection respect to fractional order of observations, p. The difference between the null and alternative is just in the mean value while their covariance matrix is estimated to be similar. Therefore, since $\mathbf{r}_{xx^*}^{\alpha}$ has Gaussian distribution, for maximizing the probability of detection, statistical means difference between two hypotheses should be maximized.

$$p = \arg\max_{0<p<1}\{\mu_1(i) - \mu_0(i)\}, \quad (38)$$

where i denotes ith lag time.

Therefore, for a specific value of p, if the difference between the means of null and alternative hypotheses is maximized, it can be concluded that the performance has improved. Due to complex relations obtained for the means in (25) and (37), differentiation and solve the result of its equation for this purpose is not possible, however, with the help of numerical results, we can obtain the optimal amount of fractional order, p.

In Fig. 1, difference of means under two hypotheses for a certain lag time is plotted versus changes of p for various value of noise variance, σ_w^2. In this figure,

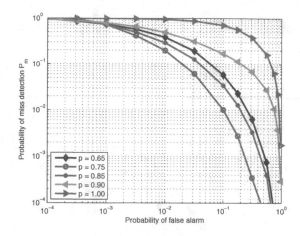

Fig. 2. The complementary ROC of proposed detector for average $SNR = -3dB$.

the values are normalized with respect to means difference value in $p = 1$ which is used in cyclostationary detectors. As can be seen in Fig. 1, for example, the difference of means increases about 0.75 percent in $p = 0.75$ for $\sigma_w^2 = 1$ and also for other value of noise variance, we can found specific p that improves the detector performance.

7   Simulation Results

In this section, we provide simulation results of cyclostationary-based detectors performance in fractional order of observations Monte Carlo simulation and we compared it with other detectors. For this purpose, we assume a linear modulation for PU signals which its pulse width for outgoing data is $1ms$. This signal has Gaussian distribution with unit variance which has been sampled in receiver. To detect these signals that affected by environmental additive Gaussian noise, we have used cyclostationary detector in fractional order of observations. Also, we assume the number of lag times is 16.

In Fig. 4, performance of this detector has been investigated in orders of $p = 0.65, 0.75, 0.85, 0.9$ and 1, with the probability of detection P_d versus SNR with assumption $\sigma_w^2 = 1$ and fixed probability of false alarm 0.01. As can be seen, by changing the fractional orders, the detector performance will changes and when the value get close to 0.75, detector performance improves approximately $3dB$ compared to $p = 1$ has been used used in previous detectors. This change and improvement is due to an increase in mean difference of observations under the two hypotheses.

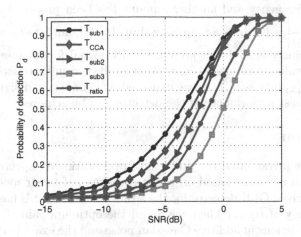

Fig. 3. The probability of detection of different detectors versus SNR for $P_{fa} = 0.01$.

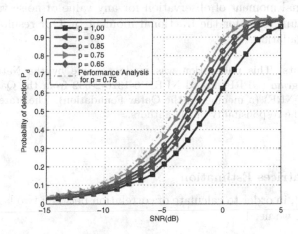

Fig. 4. The probability of detection of T_{sub1} versus SNR for $P_{fa} = 0.01$ and various fractional moment with assumption $\sigma_w^2 = 1$.

Fig. 2 depicts the receiver operating characteristics (ROC) curve of proposed cyclostationary detector for different fractional order of observations. This figure reveals of the detector behavior for different values of the false alarm probability P_{fa}.

In Fig. 3, performance of detectors has been investigated by the probability of detection P_d versus SNR with assumption $\sigma_w^2 = 1$ and fixed probability of false alarm 0.01. This figure compares performance of obtained GLR-based detectors with detectors that are mentioned in [15, 16]. In [15], the ratio of CAF absolute value in cyclic frequency and another amount has been proposed as detector, $\text{T}_{ratio} = \left| \frac{R_{xx*}^{\alpha}(\tau)}{R_{xx*}^{\alpha+\delta}(\tau_i)} \right|$, where δ is a frequency shift. In [16], authors by using canonical correlation analysis to detect presence of PU signal for M antennas SU. If λ_m is mth eigenvalue of canonical correlation analysis result, statistic is defined as, $\text{T}_{CCA} = \sum_{m=1}^{M} \ln(1 - \lambda_m^2)$. As we expected, when noise and signal variance are known, the best performance of the detector can be achieved.

8   Conclusion

In this paper, we investigated the problem of cyclostationary spectrum sensing in cognitive radio networks based on cyclic properties of linear modulated signal. First, we derived GLR detector for the situation in which SU has knowledge of cyclic frequency of signal. Then, we found the optimum value for fractional moment of observations in additive Gaussian noise and the exact performance of the GLR detector is evaluated analytically. Finally, we simulated and derived the GLR detector performance for various values of fractional moment of observations. We revealed that GLR detector performance improves for Gaussian noise if we use fractional moment of observation for any value of noise variance. We found the optimum value for the fractional moment, p. Our results have been confirmed by simulation.

Acknowledgments. This publication was made possible by the National Priorities Research Program (NPRP) award NPRP 6-1326-2-532 from the Qatar National Research Fund (QNRF) (a member of the Qatar Foundation). The statements made herein are solely the responsibility of the authors.

Appendix

Covariance Matrices Estimation

According to [14], in order to calculate of correlation between two lag times mth and nth of CAF, we need,

$$S_{x_{\tau_m} x_{\tau_n}}(2\alpha, \alpha) = \frac{1}{T} \sum_{s=-\frac{T-1}{2}}^{\frac{T-1}{2}} W(s) F_{\tau_n}\left(\alpha - \frac{2\pi s}{N}\right) F_{\tau_m}\left(\alpha + \frac{2\pi s}{N}\right), \qquad (A\text{-}1)$$

$$S_{x_{\tau_m} x_{\tau_n}}^*(0, -\alpha) = \frac{1}{T} \sum_{s=-\frac{T-1}{2}}^{\frac{T-1}{2}} W(s) F_{\tau_n}^*\left(\alpha + \frac{2\pi s}{N}\right) F_{\tau_m}\left(\alpha + \frac{2\pi s}{N}\right). \qquad (A\text{-}2)$$

Where $S_{x_{\tau_m} x_{\tau_n}}(2\alpha, \alpha)$ and $S_{x_{\tau_m} x_{\tau_n}}^*(0, -\alpha)$, respectively are unconjugated and conjugated cyclic-spectrum of observations and

$$F_{\tau}(\omega) = \frac{1}{\sqrt{N}} \sum_{n=0}^{N-1} x^p(n) x^{*p}(n + \tau) e^{-j\omega n}. \qquad (A\text{-}3)$$

Thus, covariance matrix estimation of vector $\mathbf{r}_{xx^*}^\alpha$ can be calculated as,

$$[\mathbf{\Sigma}]_{i,j} = Re\{\frac{S_{x_{\tau_i} x_{\tau_j}}(2\alpha, \alpha) + S^*_{x_{\tau_i} x_{\tau_j}}(0, -\alpha)}{2}\}, i,j = 1, 2, ..., M. \qquad \text{(A-4)}$$

pth Moment of Gaussian Random Variable

Suppose N is a Gaussian random variable with mean μ and variance σ_n^2. Thus,

$$\mathbb{E}[N^p] = \frac{2^{\frac{p}{2}} \sigma_n^p e^{-\frac{\mu^2}{2\sigma_n^2}}}{\sqrt{\pi} j^p} \int (jt)^p e^{\left(-t^2 - j\frac{\sqrt{2}\mu j}{\sigma_n} t\right)} dt. \qquad \text{(B-1)}$$

By assumption of $\beta^2 = 1$ and $q = \frac{\sqrt{2}\mu j}{\sigma_n}$ in section 3.462 of [16], (B-1) has been calculated for $p > -1$ as follows,

$$\mathbb{E}[N^p] = \frac{2^{\frac{p}{2}} \sigma_n^p e^{-\frac{\mu^2}{2\sigma_n^2}}}{\sqrt{\pi} j^p} \left[2^{-\frac{p}{2}} \sqrt{\pi} e^{\frac{\mu^2}{4\sigma_n^2}} D_p\left(\frac{j\mu}{\sigma_n}\right)\right] = \frac{\sigma_n^p e^{-\frac{\mu^2}{4\sigma_n^2}}}{j^p} D_p\left(\frac{j\mu}{\sigma_n}\right), \qquad \text{(B-2)}$$

where $D_p(.)$ is parabolic cylinder function,

$$D_p(z) = 2^{\frac{p}{2}} e^{\frac{-z^2}{4}} \left[\frac{\sqrt{\pi}}{\Gamma\left(\frac{1-p}{2}\right)} \Phi\left(-\frac{p}{2}, \frac{1}{2}; \frac{z^2}{2}\right) - \frac{\sqrt{2\pi} z}{\Gamma\left(-\frac{p}{2}\right)} \Phi\left(\frac{1-p}{2}, \frac{3}{2}; \frac{z^2}{2}\right)\right], \qquad \text{(B-3)}$$

and also $\Phi(.,.;.)$ is Kummer confluent hypergeometric function, $\Phi(a, b; c) = \sum_{k=0}^{\infty} \frac{a^{\overline{k}}}{b^{\overline{k}}} \frac{c^k}{k!}$. Where, $a^{\overline{k}}$ is rising factorial function, $a^{\overline{k}} = \frac{\Gamma(a+k)}{\Gamma(a)}$.

Mean and Variance of (12)

Mean of (12) under two hypotheses is,

$$\mu_{T_{sub3}|\mathcal{H}_\nu} = \frac{1}{M} \sum_{m=1}^{M} \mu_\nu(m), \nu = 0, 1, \qquad \text{(C-1)}$$

and variance of (12) can be calculated as follows,

$$\sigma_{T_{sub3}|\mathcal{H}_\nu}^2 = \frac{1}{M^2} \sum_{m_1=1}^{M} \sum_{m_2=1}^{M} \mathbb{E}[r_{xx^*}^\alpha(m_1) r_{xx^*}^\alpha(m_2)|\mathcal{H}_\nu] - \mu_\nu(m_1)\mu_\nu(m_2). \qquad \text{(C-2)}$$

Therefore, variance of (12) is sum of (A-4) entries.

References

1. Taherpour, A., Nasiri-Kenari, M., Gazor, S.: Multiple antenna spectrum sensing in cognitive radios. IEEE Transactions on Wireless Communications **9**(2), 814–823 (2010)

2. Sadeghi, H., Azmi, P.: A cyclic correlation-based cooperative spectrum sensing method for OFDM signals. In: 2013 21st Iranian Conference on Electrical Engineering (ICEE), pp. 1–5, May 2013

3. An, J., Yang, M., Bu, X.: Spectrum sensing for OFDM systems based on cyclostationary statistical test. In: 2010 6th International Conference on Wireless Communications Networking and Mobile Computing (WiCOM), pp. 1–4, September 2010

4. Tani, A., Fantacci, R.: A low-complexity cyclostationary-based spectrum sensing for UWB and WiMAX coexistence with noise uncertainty. IEEE Transactions on Vehicular Technology 59(6), 2940–2950 (2010)

5. Sedighi, S., Taherpour, A., Khattab, T., Hasna, M.O.: Multiple antenna cyclostationary-based detection of primary users with multiple cyclic frequency in cognitive radios, pp. 799–804, December 2014

6. Ali, O., Nasir, F., Tahir, A.: Analysis of OFDM parameters using cyclostationary spectrum sensing in cognitive radio. In: 2011 IEEE 14th International Multitopic Conference (INMIC), pp. 301–305, December 2011

7. Chaudhari, S., Kosunen, M., Makinen, S., Cardenas-Gonzales, A., Koivunen, V., Ryynanen, J., Laatta, M., Valkama, M.: Measurement campaign for collaborative sensing using cyclostationary based mobile sensors. In: 2014 IEEE International Symposium on Dynamic Spectrum Access Networks (DYSPAN), pp. 283–290, April 2014

8. Derakhshani, M., Nasiri-Kenari, M., Le-Ngoc, T.: Cooperative cyclostationary spectrum sensing in cognitive radios at low SNR regimes. In: 2010 IEEE International Conference on Communications (ICC), pp. 1–5, May 2010

9. hong You, G., shuang Qiu, T., Min Songi, A.: Novel direction findings for cyclostationary signals in impulsive noise environments, vol. 32, May 2013

10. Zha, D., Zheng, Z., Gao, X.: Robust time delay estimation method based on fractional lower order cyclic statistics, pp. 1304–1307, September 2007

11. Ma, S., Zhao, C., Wang, Y.: Fractional low order cyclostationary spectrum sensing based on eigenvalue matrix in alpha-stable distribution noise, pp. 500–503, September 2010

12. Dandawate, A., Giannakis, G.: Statistical tests for presence of cyclostationarity. IEEE Transactions on Signal Processing 42(9), 2355–2369 (1994)

13. Hotelling, H.: The generalization of student's ratio. Ann. Math. Statist. 2(3), 360–378 (1931)

14. Gradshteyn, I.S., Ryzhik, I.M.: Table of integrals, series, and products, 7th edn. Elsevier/Academic Press, Amsterdam (2007)

15. Urriza, P., Rebeiz, E., Cabric, D.: Multiple antenna cyclostationary spectrum sensing based on the cyclic correlation significance test. IEEE Journal on Selected Areas in Communications 31(11), 2185–2195 (2013)

16. Gradshteyn, I.S., Ryzhik, I.M.: Table of integrals, series, and products, 7th edn. Elsevier/Academic Press, Amsterdam (2007)

17. Urriza,P., Rebeiz,E., Cabric, D.: Multiple antenna cyclostationary spectrum sensing based on the cyclic correlation significance test. IEEE Journal on Selected Areas in Communications 31(11), 2185–2195 (2013)

Achievable Rate of Multi-relay Cognitive Radio MIMO Channel with Space Alignment

Lokman Sboui[✉], Hakim Ghazzai, Zouheir Rezki, and Mohamed-Slim Alouini

Computer, Electrical and Mathematical Sciences and Engineering (CEMSE) Division,
King Abdullah University of Science and Technology (KAUST),
Thuwal, Makkah Province, Saudi Arabia
{lokman.sboui,hakim.ghazzai,zouheir.rezki,slim.alouini}@kaust.edu.sa

Abstract. We study the impact of multiple relays on the primary user (PU) and secondary user (SU) rates of underlay MIMO cognitive radio. Both users exploit amplify-and-forward relays to communicate with the destination. A space alignment technique and a special linear precoding and decoding scheme are applied to allow the SU to use the resulting free eigenmodes. In addition, the SU can communicate over the used eigenmodes under the condition of respecting an interference constraint tolerated by the PU. At the destination, a successive interference cancellation (SIC) is performed to estimate the secondary signal. We present the explicit expressions of the optimal PU and SU powers that maximize their achievable rates. In the numerical results, we show that our scheme provides cognitive rate gain even in absence of tolerated interference. In addition, we show that increasing the number of relays enhances the PU and SU rates at low power regime and/or when the relays power is sufficiently high.

Keywords: Underlay cognitive radio · MIMO space alignment · Amplify-and-forward multiple-relay

1 Introduction

In order to cope with the continuous growth of wireless networks, new emerging systems need to offer higher data rate and to overcome bandwidth shortage. Consequently, many techniques have been presented to enhance the network performances and spectrum scarcity [1]. From one side, the cognitive radio (CR) paradigm was introduced to avoid spectrum inefficient allocation. In this paradigm, unlicensed secondary users (SU's) are allowed to share the spectrum with licensed primary users (PU's) under the condition of maintaining the PU quality of service (QoS). One of the CR modes is the underlay mode in which the PU tolerates a certain level of interference coming from the SU [2]. From the other side, relay-assisted communications [3], was introduced as a solution to considerably enhance distant and non-line of sight communications. The relaying was first intended to enhance single-antenna communications. Nevertheless, relaying in MIMO systems was shown to improve performances as well [4]. In addition, adopting MIMO power allocation within a CR framework has been

© Institute for Computer Sciences, Social Informatics and Telecommunications Engineering 2015
M. Weichold et al. (Eds.): CROWNCOM 2015, LNICST 156, pp. 17–29, 2015.
DOI: 10.1007/978-3-319-24540-9_2

studied previously in, e.g., [5–7]. In [5], MIMO space alignment was adopted but without relaying. In [6], the space alignment (SA) technique was introduced to mitigate SU interference by exploiting the free eigenmodes of MIMO systems. In [7], the authors present the CR rate after optimizing the power under interference and budget power constraints. From another side, CR with multi-relays networks was studied in [8]. To the best of our knowledge, sharing multiple-relays with the PU in a CR setting was not studied before. In [8], the authors only consider the SU transmission and respecting only interference constraints. In[9], the multiple-relays CR with interference constraint was studied. However, the analyzed performance metric was the outage probability. In [10], the authors consider a multiple relay CR without considering the PU. In addition, only the interference from the relays is considered, and the SU interference was not analyzed. From another side, communicating to the same destination in CR context was studied in previously, i.e. [11–13], but with no multiple-relaying.

In this paper, we study a multi-relay CR system with a proposed linear precoding and decoding that simplify the derivation of the optimal power. Our objective is to maximize the achievable rate of both the primary and the cognitive users, as well as the effect of the number of relays on these rates. The motivation of this study is to investigate the eventual gain of the cognitive users when sharing, in addition to the spectrum, the multiple relays with the PU's. Hence, we are interested in analyzing the effect of the number of relays on the rates. In our setting, after a particular precoding at the PU transmitter, the set of PU-relays channels is transformed into parallel channels with some free eigenmodes that can be freely exploited by the SU. Nevertheless, the SU also transmits through the used eigenmodes but respecting an interference constraint tolerated by the PU. At the destination, the PU and the SU signals are decoded using a Successive Interference Cancellation (SIC) decoder [14].

The rest of this paper is organized as follows. In Section 2, the system model is presented. Section 3 describes the PU precoding under space alignment. SU achievable rate expressions are derived for various SIC accuracies in Section 4. Numerical results are presented in Section 5. Finally, the paper is concluded in Section 6.

2   System Model

In our system model, we study an uplink communication scenario as depicted in Fig.1 where the "PU" and the "SU" are transmitting their signals simultaneously to a common destination "D". The destination could be seen as a base station to which the SU is trying to communicate under the underlay CR concept. We assume that there is no direct link between the transmitters and the common receiver. Instead, there are L relays, $R_1, \cdots, R_L$, that can receive and amplify the PU and SU signals and forward the amplified to the destination D. As a licensed user, the PU is free to exploit the channel. Meanwhile, the SU, as an unlicensed user, can share opportunistically the spectrum under some constraints that preserve a certain Quality of Service (QoS) of the PU communication.

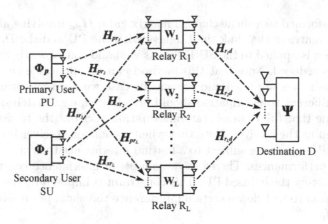

Fig. 1. Uplink spectrum sharing communication with multiple relays.

We assume that each node has N antennas, and the channel gain matrices representing the links between the PU and R_l (PU-R_l) between SU and R_l (SU-R_l), and between R_l and D (R_l-D) are denoted by $\boldsymbol{H_{pr_l}}$, $\boldsymbol{H_{sr_l}}$, and $\boldsymbol{H_{rd_l}}$, respectively, $l = 1, \ldots, L.$. All channel matrices are assumed to be independent. In the first time slot, the transmitters transmit simultaneously their signal to the relays where the complex received vector at each relay $R_l, l = 1, \ldots, L$, is given by:

$$\boldsymbol{y_{R_l}} = \boldsymbol{H_{pr_l}} \boldsymbol{\Phi_p} \boldsymbol{s_p} + \boldsymbol{H_{sr_l}} \boldsymbol{\Phi_s} \boldsymbol{s_s} + \boldsymbol{z_{R_l}}, \qquad (1)$$

where $\boldsymbol{\Phi_p}$ and $\boldsymbol{\Phi_s}$ are the linear precoding matrices applied at the PU and SU, and $\boldsymbol{s_p}$ and $\boldsymbol{s_s}$ are the transmitted signals by PU and SU, respectively, assumed to be independent and identically distributed (i.i.d) complex Gaussian. The covariance matrix of the vector $\boldsymbol{s_i}$, $i \in \{p, s\}$, are $\boldsymbol{P_i} = \mathbb{E}[\boldsymbol{s_i} \boldsymbol{s_i}^h]$, where $\mathbb{E}[\cdot]$ is the expectation operator over all channel realizations and $.^h$ is the transpose conjugate operator. This covariance matrix is constrained by a power constraint $Tr\left(\boldsymbol{\Phi_i} \boldsymbol{P_i} \boldsymbol{\Phi_i}^h\right) \le P_{tot}$ where $Tr\left(\boldsymbol{A}\right) = \sum_j A(j, j)$ is the trace of the matrix $\boldsymbol{A}$, and P_{tot} is the total power budget considered, without loss of generality, to be the same for both users. The noise $\boldsymbol{z_{R_l}}$, $l = 1, \ldots, L$, is a zero mean additive white Gaussian noise (AWGN) vector at the relay $R_l, l = 1, \ldots, L$, with an identity covariance matrix, $\boldsymbol{I_N}$.

In the second time slot, each relay $R_l, l = 1, \ldots, L$, amplifies the signal $\boldsymbol{y_{R_l}}$ through a gain matrix denoted $\boldsymbol{W_l}$ before retransmitting the signal to D. We denote by P_{R_l} the budget power of each relay R_l. The total received signal $\boldsymbol{y_D}$ at the receiver D can be written as follows

$$\boldsymbol{y_D} = \boldsymbol{H_{pd}} \boldsymbol{\Phi_p} \boldsymbol{s_p} + \boldsymbol{H_{sd}} \boldsymbol{\Phi_s} \boldsymbol{s_s} + \boldsymbol{z}, \qquad (2)$$

where $\boldsymbol{H_{pd}} = \sum_{l=1}^{L} \boldsymbol{H_{r_l d}} \boldsymbol{W_l} \boldsymbol{H_{pr_l}}$, $\boldsymbol{H_{sd}} = \sum_{l=1}^{L} \boldsymbol{H_{r_l d}} \boldsymbol{W_l} \boldsymbol{H_{sr_l}}$ and $\boldsymbol{z} = \boldsymbol{z_D} + \sum_{l=1}^{L} \boldsymbol{H_{r_l d}} \boldsymbol{W_l} \boldsymbol{z_{R_l}}$. The noise $\boldsymbol{z_D}$ is a AWGN vector at the destination D with an identity covariance matrix, $\boldsymbol{I_N}$. Consequently, the link between the PU and the D is transformed to a single channel matrix gain, $\boldsymbol{H_{pd}}$ involving all the $2 \times L$ channel gain matrices that link the L relays with the PU and the D. The same transformation is applied to the SU-D link as well and consequently the problem complexity is reduced. Note that this method can be applicable since the gain matrix at the relays are fixed and known. In case we need to optimize theses matrices, a different transformation should be adopted, e.g., matched filter [15].

We assume that full channel state information (CSI) at the transmitters, at the relays and at the receiver. Note that when a common receiver is considered, the PU and SU signals are subject to a mutual interference that may affect both PU and SU performances. Therefore, we adopt an interference constraint [16], in order to protect the licensed PU. This constraint is imposed by the PU on the SU transmission to be below a certain interference threshold per receive antenna denoted by I_{th}.

3   Primary User Precoding with Space Alignment

We propose a linear precoding and decoding matrices used to maximize the both PU and SU rates while respecting the PU's QoS. In this scheme the space alignment technique [6] is adopted. This technique allows the SU to transmit through the unused primary eigenmodes. By having perfect CSI as well as the relay gain matrices, the PU performs an optimal power allocation that maximizes its rate by applying a Singular Value Decomposition (SVD) to $\boldsymbol{H_{pd}}$ denoted $\boldsymbol{H_{pd}} = \boldsymbol{U \Lambda V}^h$ where $\boldsymbol{U}$ and $\boldsymbol{V}$ are two unitary matrices and $\boldsymbol{\Lambda}$ is a diagonal matrix that contains the ordered singular values of $\boldsymbol{H_{pd}}$ denoted as $\lambda_1 \geq \lambda_2 \geq \cdots \geq \lambda_N$. Thus, the PU transmits through parallel channels associated to their eigenmodes. Afterwards, in order to transform the PU channel to N parallel channels, we employ the linear precoding at the PU transmitter $\boldsymbol{\Phi_p}$ such as $\boldsymbol{\Phi_p} = \boldsymbol{V}$ and the decoding $\boldsymbol{\Psi}$ at the destination such as $\boldsymbol{\Psi} = \boldsymbol{U}$. Consequently, the received signal after decoding is given by

$$r = \boldsymbol{\Psi}^h \boldsymbol{y_D} = \boldsymbol{\Lambda} \boldsymbol{s_p} + \boldsymbol{U}^h \boldsymbol{H_{sd}} \boldsymbol{\Phi_s} \boldsymbol{s_s} + \tilde{z}, \tag{3}$$

where $\tilde{z} = \boldsymbol{U}^h z$ is a zero mean AWGN with a N-by-N covariance matrix $\boldsymbol{Q_{\tilde{z}}} = \boldsymbol{I_N} + \boldsymbol{U}^h \boldsymbol{H_{rd}} \boldsymbol{W} \boldsymbol{W}^h \boldsymbol{H_{rd}}^h \boldsymbol{U}$.

Meanwhile, the PU communication is protected by forcing the coming interference to be below a certain threshold denoted I_{th}. Let s be the received signal related to the SU transmission, i.e., $s = \boldsymbol{U}^h \boldsymbol{H_{sd}} \boldsymbol{\Phi_s} \boldsymbol{s_s}$. Let also $\boldsymbol{Q_s}$ to be its covariance matrix. Respecting the interference constraint means that, for each antenna $j, j = 1, \ldots, N$, we have $Q_s(j,j) \leq I_{th}$. In our study, the PU considers the SU interference to be I_{th} in each antenna so that the power allocation is performed. This study presents a lower bound of the PU performance since the threshold I_{th} is not always reached by the SU. The PU rate expression can be written as

$$R_p = \log_2 \left(\det \left[\boldsymbol{I_N} + (\boldsymbol{\Lambda s_p})(\boldsymbol{\Lambda s_p})^h (I_{th} \boldsymbol{I_N} + \boldsymbol{Q_{\tilde{z}}})^{-1} \right] \right), \tag{4}$$

where $\det[\cdot]$ is the determinant operator. Since all the matrices are diagonal, this rate can be simply written as

$$R_p = \sum_{j=1}^{N} \log_2 \left(1 + \frac{P_p(j,j)\lambda_j^2}{I_{th} + Q_{\tilde{z}}(j,j)} \right). \tag{5}$$

Meanwhile, the PU power must respect two types of constraints: budget power constraint and the relays constraints. As mentioned in Section 2, the budget power constraint is written as $Tr\left(\boldsymbol{\Phi_p} \boldsymbol{P_p} \boldsymbol{\Phi_p}^h\right) \leq P_{tot}$. By using the invariance of the *Trace* operator under the cyclic permutation and the unitarity of the matrix $\boldsymbol{\Phi_p}$, this constraint becomes $Tr\left(\boldsymbol{P_p}\right) \leq P_{tot}$. From another side, the relays constraints reflect the fact that for a given relay, $R_l, l = 1, \ldots, L$, the transmit a signal power cannot exceed its own budget P_{R_l} which can be written as:

$$Tr\left(\boldsymbol{W}_{l y R_l} \left(\boldsymbol{W}_{l y R_l}\right)^h\right) \leq P_{R_l}. \tag{6}$$

However, since the SU must respect the interference threshold, the SU power is considered to be this threshold so that the PU can allocate its power without the need to know the exact SU power. Then, the PU actual achieved rate is greater or equal to this lower bound and is mainly derived by considering the actual SU interference instead of I_{th}. By denoting $\boldsymbol{H}_{p_l} = \boldsymbol{W}_l \boldsymbol{H}_{pr_l} \boldsymbol{\Phi_p}$ and $\boldsymbol{H}_{s_l} = \boldsymbol{W}_l \boldsymbol{H}_{sr_l} \boldsymbol{\Phi_s}$, the optimal PU power and the rate lower

$$\underset{\boldsymbol{P_p}}{\text{maximize }} R_p = \sum_{j=1}^{N} \log_2 \left(1 + \frac{P_p(j,j)\lambda_j^2}{I_{th} + Q_{\tilde{z}}(j,j)} \right) \tag{7}$$

$$\text{s.t. } Tr\left(\boldsymbol{P_p}\right) \leq P_{tot}, \tag{8}$$

$$Tr\left(\boldsymbol{H}_{p_l} \boldsymbol{P_p} \boldsymbol{H}_{p_l}^h + I_{th} \boldsymbol{H}_{s_l} \boldsymbol{H}_{s_l}^h + \boldsymbol{W}_l \boldsymbol{W}_l^h\right) \leq P_{R_l}, \forall \, l = 1, \ldots, L, \tag{9}$$

Since the objective function (7) is convex and the constraints are linear, this optimization problem is convex [17]. Consequently, we use the Lagrangian method to solve this problem. We first compute the Lagrangian function and then find its derivative with regards to each $P_p(j,j)$. The optimal power is given such as the derivative is equal to zero and is given, $\forall j = 1, \ldots, N$, by:

$$P_p^*(j,j) = \left[\frac{1}{\mu_p + \sum_{l=1}^{L} \left(\eta_{p_l} \sum_{i=1}^{N} |H_{p_l}(j,i)|^2 \right)} - \frac{I_{th} + Q_{\tilde{z}}(j,j)}{\lambda_j^2} \right]^+, \tag{10}$$

where $[.]^+ = \max(0,.)$. μ_p and $\eta_{p_l}, l = 1, \ldots, L$, are the Lagrangian multipliers corresponding to the power budget constraint and the relays power constraints expressed in (8) and (9), respectively. The optimal power allocation in (10) is similar to the water-filling power expression. Note also that when the channel gain is low, i.e., λ_j's have small values, the PU is using fewer eigenmodes than the number of antennas N which gives the opportunity to the SU to exploit more free eigenmodes.

4   Secondary User Achievable Rate

In this section, we investigate the achievable rate of SU using the proposed strategy described in Section 3 depending on the SIC performance. First, we derive the SU optimal power allocation assuming a perfect SIC (a sort of genie SIC). Then, we investigate the gain in performance with an imperfect SIC (i.e., totally erroneous SIC). We introduce a parameter α ($0 \leq \alpha \leq 1$) that corresponds to the probability of detecting the PU signal s_p correctly before applying the SIC. Let n ($0 \leq n < N$) be the number of unused eigenmodes. Then, there are $N - n$ eigenmodes used by the PU and n unused eigenmodes that can be freely exploited by the SU. In order to totally eliminate the effect of interference, an appropriate choice of Φ_s has been proposed in [5] for a Line-of-Sight channel without relaying scheme where the SU is allowed to transmit in all the eigenmodes by respecting a certain interference temperature threshold I_{th} when sharing the used eigenmodes. $(H_{sd})^{-1} U \bar{P}_p$, where $\bar{P}_p$ is a diagonal matrix with the following entries:

$$\bar{P}_p(j,j) = \begin{cases} 1 & \text{if } P_p(j,j) = 0 \\ 0 & \text{if } P_p(j,j) \neq 0, \end{cases} \quad \text{for } j = 1 \dots N \quad (11)$$

In order to allow the SU to transmit in all the eigenmodes by respecting a certain interference temperature threshold I_{th} when sharing the used eigenmodes, we choose Φ_s as follows:

$$\Phi_s = (H_{sd})^{-1} U. \quad (12)$$

without loss of generality we assume that H_{sd} is invertible otherwise $(H_{sd})^{-1}$ can be taken as the pseudo-inverse of H_{sd}. In addition, since the SU knows the PU CSI and the relay gain matrices, (i.e., H_{pr_l}, $H_{r_l d}$ and W_l), the unitary matrix U can be computed at the SU transmitter. We assume here that there is a feedback through which the receiver can broadcast this information to the cognitive user. Consequently,

the received signal at the D is expressed in the two following sets depending on the number of free eigenmodes

$$r_{Dj} = \lambda_j s_{p_j} + s_{sj} + \tilde{z}_j, \ \forall j = 1, \dots, N - n,$$
$$r_{Dj} = s_{sj} + \tilde{z}_j, \ \forall j = N - n + 1, \dots, N. \quad (13)$$

Since the SU power is constrained by I_{th}, a SIC is performed at the D to decode the SU signal and to remove the effect s_p from the received signal. Meanwhile, the SU signal transmitted over the n free eigenmodes is only constrained by the budget and relays constraints.

4.1   Perfect SIC

A perfect SIC is reached when the PU signal is always decoded perfectly, i.e., $\hat{s}_{p_j} = s_{p_j}, \forall j = 1, \dots, N - n$, where $\hat{s}_{p_j}$ is the estimated PU signal at the jth receive antenna. Consequently, the cancellation of the PU effect on the SU signal is performed correctly ($\alpha = 1$) and corresponding received signal after the SIC decoding, $\tilde{r}$, is written as

$$\tilde{r} = r - \Lambda \hat{s}_p = s_s + \tilde{z}. \tag{14}$$

Hence, the corresponding SU rate is given by solving the following optimization problem

$$\max_{P_s} R_s^{(1)} = \sum_{j=1}^{N} \log_2 \left(1 + \frac{P_s(j,j)}{Q_{\tilde{z}}(j,j)} \right) \tag{15}$$

$$\text{s.t. } Tr(\boldsymbol{\Phi_s} \boldsymbol{P_s} \boldsymbol{\Phi_s}^h) \leq P_{tot}, \tag{16}$$

$$Tr\left(\boldsymbol{H_{p_l}} \boldsymbol{P_p^*} \boldsymbol{H_{p_l}}^h + \boldsymbol{H_{s_l}} \boldsymbol{P_s} \boldsymbol{H_{s_l}}^h + \boldsymbol{W_l} \boldsymbol{W_l}^h \right) \leq P_{R_l}, \forall\, l = 1, \ldots, L, \tag{17}$$

$$P_s(j,j) \leq I_{th}, \forall j = 1, \ldots, N - n, \tag{18}$$

where $\boldsymbol{P_p^*}$ is the optimal PU power obtained after solving the optimization problem given in (7)-(9). The problem (15)-(18) is a convex problem as the objective function is convex and the constraints are linear. The constraint (16) can be written as $Tr(\boldsymbol{\Phi_s}^h \boldsymbol{\Phi_s} \boldsymbol{P_s}) \leq P_{tot}$ after using the invariance of the *Trace* operator under the cyclic permutation. Let the matrix $\boldsymbol{A_s} = \boldsymbol{\Phi_s}^h \boldsymbol{\Phi_s}$, then (16) becomes $Tr(\boldsymbol{A_s} \boldsymbol{P_s}) \leq P_{tot}$. Now, since the constraint (18) is a peak constraint, we solve this problem by solving two subproblems with the same objective function. The first subproblem has the constraints (16), (17) whereas the second has the constraint (18). Afterward, the solution of the main problem is given by taking minimum between the two solutions [18]. The first subproblem is solved by using the Lagrange method [17], and an optimal solution similar to (10) is found. In the second subproblem, the optimal solution is simply $I_{th}\ \forall j = 1, \ldots, N - n$. Consequently, the resulting power profile is given as follows:

$$P_s^*(j,j) =$$

$$\begin{cases} \min\left\{ \left[\frac{1}{\mu_s A_s(j,j) + \sum_{l=1}^{L} \left(\eta_{s_l} \sum_{i=1}^{N} |H_{s_l}(j,i)|^2 \right)} \right. \right. \\ \left. \left. - Q_{\tilde{z}}(j,j) \right]^+, I_{th} \right\}, \forall j = 1, \ldots, N - n, \\ \left[\frac{1}{\mu_s A_s(j,j) + \sum_{l=1}^{L} \left(\eta_{s_l} \sum_{i=1}^{N} |H_{s_l}(j,i)|^2 \right)} - Q_{\tilde{z}}(j,j) \right]^+, \\ \forall j = N - n + 1, \ldots, N, \end{cases} \tag{19}$$

where μ_s and η_{s_l}, $l = 1, \ldots, L$, are the Lagrange multipliers associated to the budget power and the relays constraints, respectively. The optimal SU power in (19) does not involve directly the PU power allocation. However, it is affected by the number of free eigenmodes. Moreover, even in the case where the PU does not tolerate any interference, i.e. $I_{th} = 0$, the SU is able to transmit through the free eigenmodes and the corresponding rate is called the free eigenmodes (FE) rate.

4.2 Imperfect SIC

We previously analyzed the case where capacity achieving codes are employed by the PU transmitter. In this subsection, instead of using capacity achieving codes, the PU uses more practical coding schemes that may lead to unavoidable decoding errors. To this extent, we have introduced the parameter α the represents the accuracy of the SIC. We now investigate the case of $\alpha = 0$, when an imperfect SIC is employed. In this case, the interference power at each antenna is equal to $\mathbb{E}\left[\left|\tilde{\lambda}_j\left(s_{p_j} - \hat{s}_{p_j}\right)\right|^2\right] = 2P_p^*(j,j)\lambda_j^2$. The corresponding SU achievable rate is obtained by solving the following optimization problem:

$$\max_{P_s} R_s^{(0)} = \sum_{j=1}^{N-n} \log_2\left(1 + \frac{P_s(j,j)}{Q_{\tilde{z}}(j,j) + 2P_p^*(j,j)\lambda_j^2}\right)$$

$$+ \sum_{j=N-n+1}^{N} \log_2\left(1 + \frac{P_s(j,j)}{Q_{\tilde{z}}(j,j)}\right) \tag{20}$$

$$\text{s.t. } Tr(A_s P_s) \leq P_{tot}, \tag{21}$$

$$Tr\left(H_{p_l} P_p^* H_{p_l}{}^h + H_{s_l} P_s H_{s_l}{}^h + W_l W_l{}^h\right) \leq P_{R_l}, \forall\, l = 1, \dots, L, \tag{22}$$

$$P_s(j,j) \leq I_{th}, \forall j = 1, \dots, N-n, \tag{23}$$

Using the convexity of this problem, the optimal power is computed by using the Lagrange method, similarly to the perfect SIC case. The resulting solution is

$$P_s^*(j,j) =$$

$$\begin{cases} \min\left\{\left[\frac{1}{\mu_s A_s(j,j)+\sum_{l=1}^{L}\left(\eta_{s_l}\sum_{i=1}^{N}|H_{s_l}(j,i)|^2\right)}\right.\right. \\ \left.\left. - \left(Q_{\tilde{z}}(j,j) + 2P_p^*(j,j)\lambda_j^2\right)\right]^+, I_{th}\right\}, \forall j = 1, \dots, N-n, \\[2mm] \left[\frac{1}{\mu_s A_s(j,j)+\sum_{l=1}^{L}\left(\eta_{s_l}\sum_{i=1}^{N}|H_{s_l}(j,i)|^2\right)} - Q_{\tilde{z}}(j,j)\right]^+, \\ \hspace{3cm} \forall j = N-n+1, \dots, N. \end{cases} \tag{24}$$

We notice, here, that the optimal power involves directly the primary power and eigenmodes. Consequently, the SU power allocation is more sensitive tot eh PU channel variation than in the case of perfect SIC.

   We adopt a Rayleigh fading channel in which the channel gains are complex Gaussian random variables with zero mean and unit variance. We choose $N = 4$ antennas, and the rates expressed in bits per channel use (BPCU). We consider the same budget power at the PU and the SU transmitters, i.e., $P_{tot,p} = P_{tot,s} = P_{tot}$. For simplicity, we assume that the relays amplification matrices are diagonal and equal and are given by: $W = w \times I_N$ where w is a positive scalar and I_N is the N-dimension identity matrix. We also take an equal power budget at all the relays, i.e., $P_{R_1} = \cdot = P_{R_L} = P_R$. Note that the proposed scheme can be

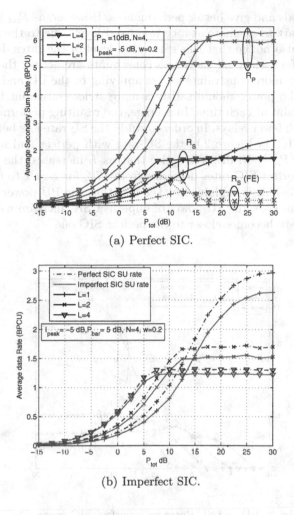

(a) Perfect SIC.

(b) Imperfect SIC.

Fig. 2. PU and SU Rates versus P_{tot}.

applied to any fixed amplification gain matrix. The optimization of W is left to a future extension of this work.

In Figure 2.a, we plot the PU and SU rates as a function of P_{tot} for $P_R = 10$ dB and $w = 0.4$ with perfect SIC ($\alpha = 0$) and with various number of relays, $L = 1, 2, 4$. We show that the space alignment technique allows the SU to achieve a free eigenmodes rate $R_S(FE)$, i.e. there is no tolerated interference from the PU, up to 0.5 BPCU for $L = 1$ and 1.1 BPCU for $L = 4$. However, this rate becomes zero when P_{tot} exceeds 17 dB for $L = 1$ and becomes constant for $L > 1$ since, in this regime, the PU is using most of the eigenmodes. We also show that at low values of P_{tot}, increasing the number of relays enhances both PU and SU rates. In fact, in this range, the relays are not saturated, i.e. the relays constraints are not active. That is, adding more relays will further amplify the

PU and SU signals and give better performances. However, as P_{tot} increases, the performances start to stagnate at a certain fixed levels due to the saturation of the relays. We also notice that this saturation level of the rates decrease when L increases. In fact, since all the relays constraints are active, the more relays are available, the more constraints we are applying to the PU and SU transmit power. Hence, the power should satisfy a more strict constraint by respecting the lower constraint at each time. In average, the resulting performance is lower than the one with fewer relays. In order to study the SU rate loss between perfect and imperfect SIC, in Figure 2.b, the SU rate with perfect and imperfect SIC is presented for $P_R = 10$ dB. We notice that as L increases, the gap between perfect and imperfect SIC rates decreases from 17% for $L = 1$ to 6% for $L = 4$ for $P_{tot} = 20$ dB. This is explained by the fact that the PU power decrees with L and hence from (24), the SU power of imperfect SIC increase with L and the imperfect SIC rate becomes closer to the perfect SIC one.

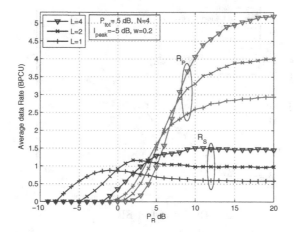

Fig. 3. PU and SU Rates with perfect SIC versus P_R.

Figure 3 shows the effect of the relay's power, P_R, on the PU and SU rates with different values of L. First, we notice that when P_R is low, the SU reaches its maximum rate before starting to slightly decrease. Meanwhile, the primary rate is very reduced since its power is limited, in (9), by the low relay's power and the terms involving $I_{th}\boldsymbol{H}_{\boldsymbol{s}_l}\boldsymbol{H}_{\boldsymbol{s}_l}{}^h$ which is independent of P_R. Hence the optimal PU power, P_p^*, is limited and close to zero. Meanwhile, the SU power in (17) is limited by the relays power P_R, in addition, $\boldsymbol{H}_{\boldsymbol{p}_l}\boldsymbol{P}_{\boldsymbol{p}}^*\boldsymbol{H}_{\boldsymbol{p}_l}{}^h$ which is already very low, consequently the power budgets of the relays are, in this regime, entirely dedicated to the SU. However, when P_R becomes greater, the cognitive rate stagnates or decreases while the primary rate increases remarkably to the no cognition upper bound. Hence, the choice of P_R is critical to the PU since the SU rate is almost the same.

In Figure 4, we highlight the effect of the relay amplification matrix gain of the relays W on PU and SU rates for different values of L. Recall that, we considered all the gain matrices to be equal to $W = w \times I_N$, which is not necessarily the optimal choice but is a simple one to quantify the effect of this matrix on the system performance. We notice that the PU and SU rates reach a maximum for a particular value of w before decreasing to zero as w increases. The reason behind this rate shape is that increasing w enhances the power as the relays constraints are not active. When reached, i.e., the values of w are large, the transmit power should be small in order to respect the constraint and as w increases further, the power should be near zero which applies for both PU and SU rates. Besides, the optimal w giving the maximum rate is slightly

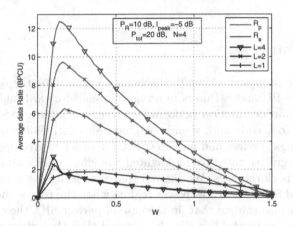

Fig. 4. PU and SU rates with perfect SIC versus w.

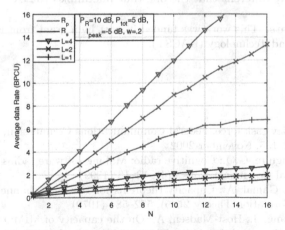

Fig. 5. PU and SU rates with perfect SIC versus N.

different for PU and SU and can favor one over the other as shown in Figure 4. By measuring the maxima rate increase between $L = 1$ and $L = 4$, we find and increase of 97% for the PU and 105% for the SU. Meanwhile, we notice that this maximum is independent from the number of relays, in fact the optimal w for $L = 1$ is the same for $L = 2$ and $L = 3$ which means that finding the optimal w is important since any additional relays to the system should adopt this values in order to give maximum performance.

In Figure 5, we study the effect of the number of antennas on the PU and SU rates with perfect SIC and different number of relays L. We first notice that the increasing slope of the PU and SU rates with N is almost linear except for the PU rate when $L = 1$. However, increasing the number of relays enhances considerably both PU and SU rates, e.g. for $N = 8$, the PU and SU rate increases by 110% and 106%, respectively between $L = 1$ and $L = 4$.

5   Conclusion

In this paper, we proposed a simplified scheme to determine the optimal power allocation for the PU and SU users in an amplify-and-forward multi-relays network. The common destination performs a successive interference cancellation (SIC) technique to decode both signals. We have also derived the optimal power in different settings (perfect and imperfect SIC) in order to give upper and lower bounds of the cognitive rate. We highlighted the effect of the number of relays on the system's performances. We showed that increasing the number of relays enhances PU and SU rates at low power regime when the relays budget power is not attained. We also showed that, in the case of perfect SIC, the corresponding SU rate drops by at most 17%. We also showed that the relays gain matrices considerably affect PU and SU rates and that the relay gain that maximizes the PU rate is slightly different that the one that maximizes the SU rate.

Acknowledgments. This work was funded in part by a grant from King Abdulaziz City of Sciences and Technology (KACST).

References

1. Spectrum policy task force. Federal Communications Commission, Tech. Rep. ET Docket no. 02–135, November 2002
2. Mitola, J., Maguire, G.Q.: Cognitive radio: Making software radios more personal. IEEE. Personal Communications **6**(4), 13–18 (1999)
3. Cover, T., EL Gamal, A.: Capacity theorems for the relay channel. IEEE Transactions on Information Theory **25**(5), 572–584 (1979)
4. Wang, B., Zhang, J., Host-Madsen, A.: On the capacity of MIMO relay channels. IEEE Transactions on Information Theory **51**(1), 29–43 (2005)

5. Sboui, L., Ghazzai, H., Rezki, Z., Alouini, M.-S.: Achievable rate of cognitive radio spectrum sharing MIMO channel with space alignment and interference temperature precoding. In: Proc. of the IEEE International Conference on Communications (ICC 2013), Budapest, Hungary, pp. 2656–2660, June 2013
6. Perlaza, S., Debbah, M., Lasaulce, S., Chaufray, J.-M.: Opportunistic interference alignment in MIMO interference channels. In: Proc. of the 19th IEEE International Symposium on Personal, Indoor and Mobile Radio Communications (PIMRC 2008), Cannes, France, September 2008
7. Kang, X., Liang, Y.-C., Nallanathan, A.: Optimal power allocation for fading channels in cognitive radio networks under transmit and interference power constraints. In: Proc. of the IEEE International Conference on Communications (ICC 2008), Beijing, China, pp. 3568–3572, May 2008
8. Choi, M., Park, J., Choi, S.: Simplified power allocation scheme for cognitive multi-node relay networks. IEEE Transactions on Wireless Communications 11(6), 2008–2012 (2012)
9. Lee, J., Wang, H., Andrews, J.G., Hong, D.: Outage probability of cognitive relay networks with interference constraints. IEEE Transactions on Wireless Communications 10(2), 390–395 (2011)
10. Naeem, M., Lee, D., Pareek, U.: An efficient multiple relay selection scheme for cognitive radio systems. In: IEEE International Conference on Communications Workshops (ICC), Cape Town, South Africa, pp. 1–5 (2010)
11. Sboui, L., Ghazzai, H., Rezki, Z., Alouini, M.-S.: On the throughput of a Relay-Assisted cognitive radio MIMO channel with space alignment. In: 12th International Symposium and Workshops on Modeling and Optimization in Mobile, Ad Hoc and Wireless Networks (WiOpt 2014), Hammamet, Tunisia, pp. 317–323, May 2014
12. Krikidis, I.: A SVD-based location coding for cognitive radio in MIMO uplink channels. IEEE Communications Letters 14(10), 912–914 (2010)
13. Sboui, L., Ghazzai, H., Rezki, Z., Alouini, M.-S.: Achievable rate of a cognitive MIMO multiple access channel with multi-secondary users. IEEE Communications Letters 19(3), 403–406 (2015)
14. Popovski, P., Yomo, H., Nishimori, K., Di Taranto, R., Prasad, R.: Opportunistic interference cancellation in cognitive radio systems. In: Proc. of the 2nd IEEE International Symposium on New Frontiers in Dynamic Spectrum Access Networks (DySPAN 2007), Dublin, Ireland, pp. 472–475, April 2007
15. Qingyu, M., Osseiran, A., Gan, J.: MIMO amplify-and-forward relaying: spatial gain and filter matrix design. In: Proc. IEEE International Conference on Communications Workshops (ICC Workshops 2008), Beijing, China (2008)
16. Haykin, S.: Cognitive radio: Brain-empowered wireless communications. IEEE Journal on Selected Areas in Communications 23(2), 201–220 (2005)
17. Boyd, S., Vandenberghe, L.: Convex Optimization. Cambridge University Press (2004)
18. Sboui, L., Rezki, Z., Alouini, M.-S.: A unified framework for the ergodic capacity of spectrum sharing cognitive radio systems. IEEE Transactions on Wireless Communications 12(2), 877–887 (2013)

Effective Capacity and Delay Optimization in Cognitive Radio Networks

Mai Abdel-Malek[1]([✉]), Karim Seddik[2], Tamer ElBatt[1,3],
and Yahya Mohasseb[1,4]

[1] Wireless Intelligent Networks Center (WINC), Nile University, Giza, Egypt
m.elkady@nileu.edu.eg, {telbatt, mohasseb}@ieee.org
[2] ECNG Department, American University in Cairo, New Cairo 11835, Egypt
kseddik@aucegypt.edu
[3] Department of EECE, Faculty of Engineering, Cairo University, Giza, Egypt
[4] Department of Communications, The Military Technical College, Cairo, Egypt

Abstract. In this paper, we study the fundamental trade-off between delay-constrained primary and secondary users in cognitive radio networks. In particular, we characterize and optimize the trade-off between the secondary user (SU) effective capacity and the primary user (PU) average packet delay. Towards this objective, we employ Markov chain models to quantify the SU effective capacity and average packet delay in the PU queue. Afterwards, we formulate two constrained optimization problems to maximize the SU effective capacity subject to an average PU delay constraint. In the first problem, we use the spectrum sensing energy detection threshold as the optimization variable. In the second problem, we extend the problem and optimize also over the transmission powers of the SU. Interestingly, these complex non-linear problems are proven to be quasi-convex and, hence, can be solved efficiently using standard optimization tools. The numerical results reveal interesting insights about the optimal performance compared to the unconstrained PU delay baseline system.

Keywords: Cognitive radios · Effective capacity · Delay constraints · Optimization · Quality of service (QoS)

1 Introduction

The rapid evolution and ubiquity of wireless connectivity, as well as the wide proliferation of smartphones and powerful hand-held devices, mandate the handling of a huge amount of data in applications such as wireless multimedia. These applications are characterized by high bandwidth requirements and relatively stringent delay constraints. The limited wireless spectrum presents a major challenge and adds to the complexity of the problem.

This publication was made possible by NPRP 4-1034-2-385, NPRP 09-1168-2-455 and NPRP 5-782-2-322 from the Qatar National Research Fund (a member of Qatar Foundation). The statements made herein are solely the responsibility of the authors.

© Institute for Computer Sciences, Social Informatics and Telecommunications Engineering 2015
M. Weichold et al. (Eds.): CROWNCOM 2015, LNICST 156, pp. 30–42, 2015.
DOI: 10.1007/978-3-319-24540-9_3

The concept of cognitive radios was originally introduced by J. Mitola III in 1999 as a paradigm shift due to the severe under-utilization of the spectrum in some bands [1]. Cognitive radios enable opportunistic, or secondary users (SUs), to share part of the spectrum with licensed spectrum owners, referred to as the primary users (PUs), and possibly coexist with them in some paradigms.

Striking a balance between improving the performance of the opportunistic SUs and maintaining QoS requirements for the PUs is crucial for cognitive radio systems. This is particularly true for PUs running multimedia applications with stringent QoS constraints, which are particularly sensitive to potential performance degradation caused by the SUs. A significant portion of research in the cognitive radios arena has focused on improving the ability of the SUs to communicate over their, inherently, unreliable links. This is achieved either through improvements in SU data encoding schemes, e.g., [2], [3] or by proper adaptation of the SU power and rate in response to the time-varying channel conditions to achieve SU QoS requirements, e.g., [4], [5], [6].

Physical-layer channel models cannot be easily linked to QoS metrics. Therefore, in [7], the notion of "Effective Capacity" was originally introduced to express the maximum constant arrival rate that can be supported by a given channel service process while satisfying a statistical QoS requirement as specified by the QoS exponent, θ. The effective capacity may be thought of as the dual wireless concept to the "Effective Bandwidth" notion originally introduced in [8].

Applied to cognitive radio systems, the effective capacity has been employed to characterize the performance of the SU in [9]. The authors derived expressions for the effective capacity of the SU for combinations of, both, fixed and adaptive power and rate scenarios. To enhance the prediction of the PU channel state, the authors in [10] proposed a feedback model, where the SU leverages the overheard primary ARQ message and uses information gleaned from the PU feedback channel to improve its sensing. It was shown, analytically, and using simulations that such side information can potentially increase the effective capacity of the SU.

In essence, the scheme in [10] minimizes the probability of PU *re-transmission* failure. However, this does not automatically guarantee the PU's ability to satisfy delay constraints. In this paper, we incorporate an explicit QoS constraint on the PU, namely an average delay constraint, which is more relevant to users engaged in interactive or multimedia sessions. Thus, in this paper our prime objective is to maximize the SU effective capacity under an average delay constraint on the PU packets.

Our main contribution in this paper is two-fold. First, we develop a mathematical model that incorporates delay constraints into the optimization of the primary users and secondary users performance in cognitive radio networks. Second, we formulate, establish quasi-convexity and efficiently solve two optimization problems for maximizing the SU effective capacity subject to a constraint on the PU average packet delay. Towards this objective, we analyze the Markov chain models capturing the SU channel sensing model and the PU queue dynamics. Next, we formulate, assess complexity (establish quasi-convexity) and solve our first (basic) optimization problem which decides the optimal spectrum sensing energy detection threshold to maximize the effective capacity subject to the average PU delay constraint. Afterwards, we generalize the problem to jointly optimize over the energy detection threshold along with the SU transmission powers, yielding superior performance. Finally, we solve the problems numerically and demonstrate promising performance results for plausible scenarios.

The rest of the paper is organized as follows. First, we introduce the system model, assumptions and secondary user channel access model in Section 2. Afterwards, the basic and generalized optimization problems are formulated, to maximize the SU effective capacity under an average PU delay constraint, and examined for convexity in Section 3. Performance results confirming the fundamental trade-off and the optimal solution are quantified under plausible scenarios in Section 4. Finally, conclusions are drawn and potential directions for future work are pointed out in Section 5.

2   System Model

We focus on a simple cognitive radio network with one PU and one SU for mathematical tractability of the proposed model. We consider a time slotted system with slots of equal duration, T seconds. All channels are assumed to experience Rayleigh block fading where the channel remains constant over a time slot and changes independently from one slot to another. The PU queue has a Bernoulli packet arrival process with rate $0 \leq \lambda_p \leq 1$ per slot and the SU is assumed to be fully backlogged (always has a packet to transmit) at the beginning of each time slot.

We assume a hybrid (underlay/interweave) cognitive radio model as introduced in [11], whereby the SU senses the channel at the beginning of the slot for N seconds, where $N < T$, in order to determine the mode of channel access, as in the interweave model. Subsequently, the SU accesses the channel with high power, P_i, if the primary user is idle. Otherwise, the SU accesses the channel with lower power $P_b < P_i$ and correspondingly lower rate (as in the underlay model). The discrete-time SU channel input-output relation in the ith symbol duration is given by

$$y(i) = \begin{cases} \text{PU idle}: & h_s(i)x(i) + n(i), & i = 1, 2, \cdots \\ \text{PU active}: h_s(i)x(i) + s_p(i) + n(i), & i = 1, 2, \cdots, \end{cases} \quad (1)$$

where $x(i)$ is the complex channel input, $y(i)$ denotes the complex channel output and $h_s(i)$ is the channel fading coefficient between the secondary transmitter and receiver and $|h_s(i)|^2 = z_s(i)$. The primary transmitted signal, as perceived by the secondary receiver, is denoted $s_p(i)$, and $n(i)$ denotes the additive white Gaussian noise at the secondary receiver, with zero mean and variance of σ_n^2.

We adopt a simple energy detection spectrum sensing mechanism whereby the RF energy measured at the secondary transmitter is compared to an energy detection threshold, η, to decide whether the PU is active or idle. The channel

sensing problem is known to be modeled as a binary hypothesis testing problem
[12]. Under these assumptions, the optimal Neyman-Pearson detector is [9]:

$$Y = \frac{1}{NB} \Sigma_{i=0}^{NB} |y(i)|^2 \lessgtr_{\mathcal{H}_1}^{\mathcal{H}_0} \eta, \tag{2}$$

where NB denotes the number of complex symbols in the N (in seconds) sensing
duration with B (in Hz) channel bandwidth. The test statistic Y is chi-square
distributed with $2NB$ degrees of freedom. In this case, the probability of detection can be derived as follows

$$P_d = \mathcal{P}rob\{Y > \eta | \mathcal{H}_1\} = 1 - \frac{\gamma\left(\frac{NB\eta}{(\sigma_n^2 + \sigma_{sp}^2)}, NB\right)}{\Gamma(NB)}, \tag{3}$$

where $\gamma(x, s)$ is the lower incomplete gamma function and $\Gamma(x)$ is the Gamma
function. The probability of false alarm can be written as follows

$$P_f = \mathcal{P}rob\{Y > \eta | \mathcal{H}_0\} = 1 - \frac{\gamma\left(\frac{NB\eta}{\sigma_n^2}, NB\right)}{\Gamma(NB)}. \tag{4}$$

2.1 The Secondary User Access Model

We employ a classic SU access model adopted earlier in the literature, e.g., [1,11],
which is represented by the Markov chain in Fig.1. The prime objective of this
model is to capture the secondary user access decision and transmission parameters, namely, power and rate, depending on the spectrum sensing outcome and
the instantaneous capacity of the secondary user channel. This is instrumental
in characterizing the secondary user effective capacity for a statistical QoS constraint, θ, as shown later in the sequel. Thus, the secondary user transmission
parameters have four cases, depending on the sensing outcome:

1. The PU channel is busy and the SU detects it as such: SU transmits with
   the lowest acceptable power P_b and rate r_b.
2. The PU channel is idle while the SU detects it busy (false alarm (FA)): the
   SU sends its packet with power P_b and rate r_b as in case (1).
3. The PU channel is busy while the SU detects it idle (mis-detection (MD)):
   the SU sends with power P_i and rate r_i, where $P_i > P_b$ and $r_i > r_b$.
4. The PU channel is idle and the SU detects it as such: the SU sends with
   power P_i and rate r_i.

On the other hand, the secondary user channel has two states, OFF and
ON, depending on whether the secondary user transmission rate exceeds the
instantaneous channel capacity or not, respectively. This is caused by the time-
varying fluctuations in the SU channel.

Next, we present the SU effective capacity (EC) subject to statistical QoS
constraints. It has been established in [7] that the EC for a given QoS exponent,
θ, is given by

$$EC = -\lim_{t \to \infty} \frac{1}{\theta t} \log_e \mathbb{E}\left\{e^{-\theta S(t)}\right\} = -\frac{\Lambda(-\theta)}{\theta}, \tag{5}$$

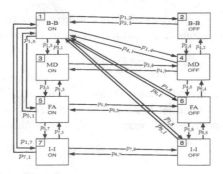

Fig. 1. SU access and transmission model.

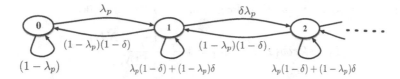

Fig. 2. A discrete-time Markov chain modeling the PU queue evolution.

where $S(t) = \sum_{k=1}^{t} r(k)$ represents the time accumulated service process and $\{r(k), k= 1, 2, ...\}$ is the discrete and ergodic stochastic service process.

For Markov modulated processes, like the model in Fig.1 , it has been shown in [13] that the effective capacity can be reduced to

$$EC = \frac{-1}{\theta TB} \log_e \left(\mathbf{sp}(\mathbf{\Phi}(-\theta)\mathbf{R})\right), \qquad (6)$$

where $\mathbf{sp}(.)$ denotes the spectral radius of a matrix, $\mathbf{R}$ represents the transition probability matrix of the secondary user Markov chain and θ is the delay exponent. $\mathbf{\Phi}(\theta) = diag(\phi_1(\theta), \dots, \phi_M(\theta))$ is a diagonal matrix whose elements are the moment generating functions of the processes in the M states.

2.2   Modeling the Primary User Queue

The primary user queue evolution is modeled formally by the discrete-time Markov chain shown in Fig. 2 with Bernoulli arrival rate λ_p. The primary user service rate (i.e. queue length decreases by one), μ_p, depends only on two parameters, namely, the arrival rate and the primary channel outage probability, denoted δ, as follows,

$$\mu_p = (1 - \lambda_p)(1 - \delta). \qquad (7)$$

As the SU always uses the spectrum with different powers and rates depending on the PU activity, the primary channel outage may occur in two cases: first, if the SU attempts to transmit on the PU channel with power P_b and rate r_b, which in addition to the PU rate, r, brings the total rate on the channel above the instantaneous capacity. Alternatively, PU channel outage occurs when the SU attempts to transmit on the PU channel with power P_i on a busy channel due to mis-detection. Therefore, we can express the primary outage probability as follows:

$$\delta = P_d \mathcal{P}rob\{z_p(i) < \phi_1\} + (1 - P_d)\mathcal{P}rob\{z_p(i) < \phi_2\}, \tag{8}$$

where $z_p(i) = |h_p(i)|^2$ and $h_p(i)$ is the fading coefficient of the PU transmission channel, $\phi_1 = \frac{2^{\frac{r}{B}} - 1}{SINR_1}$ and $\phi_2 = \frac{2^{\frac{r}{B}} - 1}{SINR_2}$. $SINR_1$ and $SINR_2$ denote the average signal to interference plus noise ratios for the PU in each outage case.

Solving the Markov chain in Fig. 2 which is M/M/1 queue with steady-state probability of the PU queue having i packets, π_i, then using simple queuing analysis (see [14]), π_0 can be derived as

$$\pi_0 = \frac{\delta(\mu_p - \delta\lambda_p)}{\mu_p(1 + \delta) - \delta^2\lambda_p}, \tag{9}$$

and the utilization factor of the primary user queue, ρ, is

$$\rho = \frac{\delta\lambda_p}{\mu_p}, \tag{10}$$

and, finally, the expected delay for the PU packets can be derived as [14]

$$D_p = \frac{\rho(\pi_0 - \delta(1 - \rho)^2)}{\delta^2\lambda_p(1 - \rho)^2}. \tag{11}$$

3   SU Effective Capacity Optimization Under PU Average Delay Constraint

In this section, we study the problem of balancing the conflicting QoS requirements for the primary and secondary users in cognitive radio networks. In particular, we attempt to maximize the SU effective capacity (defined for a given statistical QoS exponent, θ) subject to an average PU delay constraint, denoted D_{max}. Towards this objective, we formulate two optimization problems. The first (basic) problem solves for the optimum energy detection threshold, η, that maximizes the SU effective capacity subject to the aforementioned PU constraint. The key insight guiding the choice of the energy detection threshold, η, is that it affects the power level used by the SU which, in turn, affects the channel outage probability. For this reason, increasing η always degrades the PU delay, yet, enhances the SU effective capacity up to some point beyond which it starts decreasing, as shown in Fig. 3. The explanation of this behavior is that when η exceeds the sensed PU RF energy, the SU will always detect an idle channel, and use high rate and power, hence, causing higher interference. On the other hand, when $0 \leqslant \eta \leqslant E_{s_p}(i)$ ($E_{s_p}(i)$ is the PU energy as perceived by the SU), the probability of miss-detection decreases with increasing η; as η gets closer to

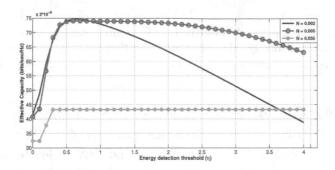

Fig. 3. The effect of the energy detection threshold, η, on the SU effective capacity for different sensing durations, N.

the sensed PU RF energy, the SU sensing is more reliable, hence, its effective capacity increases.

Fig. 3 shows the SU effective capacity for different values of the sensing duration, N. We notice that the relation between effective capacity and energy threshold can always be divided into the two modes discussed above; one increasing and the other is monotonically decreasing/non-increasing. We also notice that if we increase N from 0.002 sec to 0.005 sec, the SU effective capacity increases due to the more accurate estimate of the PU channel activity. On the other hand, when we increase N beyond 0.005 sec, the SU effective capacity decreases because the slot portion dedicated to data transmission, $T - N$, decreases linearly with N.

3.1   Basic Problem Formulation

In this section, we maximize the SU effective capacity with respect to the spectrum sensing energy detection threshold, η, subject to the PU average delay constraint. Thus, the basic optimization problem can be formulated as

$$\mathbf{P1} : \max_{\eta} \frac{-1}{\theta TB} \log_e(sp(\mathbf{\Phi}(-\theta)\mathbf{R}))$$

$$s.t. \ D_p \leq D_{max}, \tag{12}$$

where D_{max} is the maximum allowable average PU delay which represents the predefined QoS for the PU. Next, we establish an important complexity reduction result for **P1** which facilitates an efficient solution using standard optimization solvers.

Theorem 1. P1 *is a quasi-convex problem in* η.

Proof. In order to establish this result, we must assess the quasi-concavity of the SU effective capacity as well as the convexity of the delay constraint.

We first establish the convexity of the constraint with respect to the optimization variable, η. From (8-11), the expected PU delay, D_p, is given by

$$D_p = \frac{-1}{(1-\lambda_p)(1-\delta)} + \frac{(1-\lambda_p)(1-\delta)}{(1-\lambda_p-\delta^2)(1-\lambda_p-\delta)}. \tag{13}$$

Since D_p consists of two terms, it suffices to prove that each term is convex in η. This follows from the fact that the summation of convex functions is convex. This can be validated be examining the Hessian of each term. Performing partial differentiation of each term with respect to δ, then from (3) and (8), we get the partial derivative of δ in terms of η. The Hessian of the first term, denoted f, is given by:

$$f'' = \frac{qe^{-x}}{(1-\delta)^2} x^{(NB-x)} (1 - NB - x), \tag{14}$$

where $x = \frac{\eta NB}{\sigma_n^2 + \sigma_{sp}^2}$, q is a positive constant. It turns out from (14) that the Hessian is positive definite only under the condition that

$$\eta \leqslant (\sigma_n^2 + \sigma_{sp}^2)\left(1 - \frac{1}{NB}\right). \tag{15}$$

For a stable system ($0 < \rho < 1$ and $0 \leqslant P_d \leqslant 1$), η will never violate this condition, then the first term of the constraint function is convex over the optimization domain. Similarly, we can prove that the second term is also convex over the optimization domain, and, hence, the constraint in (12) is convex.

Second, we establish the quasi-concavity of the objective function, namely the SU effective capacity. To prove this, we must show that the function inside the logarithm (in (6)) is quasi-convex. This is based on the fact that if an arbitrary function U is quasi-convex and a function g is monotonically non-decreasing, then the function f defined as $f(x) = g(U(x))$ is also quasi convex [15]. Since the log function is non-decreasing, then it suffices to prove that its interior function is quasi-convex to establish the desired result. From (3), (4), (6) and ((28) in [9]), the objective function can be written as in (16) where $z_s = |h_s|^2$ and $\alpha_1, \alpha_2, \alpha_3$ and α_4 are the fading channel thresholds for no outage for the four SU sensing outcome cases stated in Section 2.1.

$$\begin{aligned} EC = \; & \frac{-1}{\theta TB} \log_e \left\{ e^{-\theta r_b(T-N)} \left[\rho(1-\kappa_2)\, \mathbf{p}(z_s > \alpha_1) + (1-\rho)(1-\kappa_1)\, \mathbf{p}(z_s > \alpha_3) \right] \right. \\ & + e^{-\theta r_i(T-N)} \left[\rho\kappa_2 \mathbf{p}(z_s > \alpha_2) + (1-\rho)\kappa_1 \mathbf{p}(z_s > \alpha_4) \right] + \rho(1-\kappa_2)\, \mathbf{p}(z_s < \alpha_1) \\ & \left. + (1-\rho)\kappa_2 \mathbf{p}(z_s < \alpha_2) + (1-\rho)(1-\kappa_1)\, \mathbf{p}(z_s < \alpha_3) + (1-\rho)\kappa_1 \mathbf{p}(z_s < \alpha_4) \right\} \end{aligned} \tag{16}$$

where $\kappa_1 = \gamma\left(\frac{\eta NB}{\sigma_n^2}, NB\right)/\Gamma(NB)$ and $\kappa_2 = \gamma\left(\frac{\eta NB}{\sigma_n^2 + \sigma_{sp}^2}, NB\right)/\Gamma(NB)$. From (16), we have eight terms that can classified into three categories. First category is

$$f_1 = q_1(1 - \kappa_1), \tag{17}$$

second category is of the form

$$f_2 = q_2\rho = q_2 \frac{c_1 - c_2 P_d}{k_1 + k_2 P_d}, \tag{18}$$

and the last category is of the form

$$f_3 = q_3 \rho P_d \text{ or } f_3 = q_4 \rho P_f, \tag{19}$$

where $q_1, q_2, c_1, c_2, k_1, k_2, q_3$ and q_4 are positive constants. For the first category

$$f_1'' = q \left(e^{-\frac{\eta N B}{\sigma_n^2}} \right) \left(\frac{\eta N B}{\sigma_n^2} \right)^{\left(N B - \frac{\eta N B}{\sigma_n^2} \right)} \left(1 - N B - \frac{\eta N B}{\sigma_n^2} \right), \tag{20}$$

which is convex only on the region indicated in (15), thus it is quasi-convex. Similarly, we can prove that all other terms are quasi-convex. Hence, their summation is quasi-convex. Thus, the effective capacity is quasi-concave in η since it is the negative of a quasi-convex function.

3.2   The Generalized Optimization Problem

In this section, we generalize the basic problem **P1** to optimize over the energy detection threshold, η, along with the SU transmission powers, P_b and P_i. As expected, the expanded policy space for the generalized problem yields noticeable performance improvement, as confirmed by the numerical results in Section 4, compared to the basic problem where the transmission powers are given and fixed throughout. Thus, the generalized optimization problem can be formulated as follows

$$\mathbf{P2}: \max_{\eta, P_b, P_i} \frac{-1}{\theta T B} \log_e (sp(\mathbf{\Phi}(-\theta)\mathbf{R}))$$

$$s.t. \ D_p \leq D_{max}$$
$$0 \leqslant P_i \leqslant P \tag{21}$$
$$0 \leqslant P_b \leqslant P,$$

where P is the maximum SU transmission power. In preparation for our main result in this section establishing the quasi-convexity of **P2**, we utilize the following two definitions from optimization theory.

Definition 1. *The Hessian of a multi-variate function $f(\mathbf{x})$ is:*

$$\mathbf{H}_n(\mathbf{x}) = \begin{bmatrix} f''(x)_{11} & \cdots & f''(x)_{1n} \\ \vdots & \vdots & \vdots \\ f''(x)_{n1} & \cdots & f''(x)_{nn} \end{bmatrix}, \tag{22}$$

where $f''(\mathbf{x})_{ij} = \frac{\partial^2 f}{\partial x_i \partial x_j}$. The function $f(\mathbf{x})$ is convex if its *Hessian* $\mathbf{H}_n(\mathbf{x})$ satisfies $\mathbf{H}_n(\mathbf{x}) \succeq 0$ which means that the Hessian matrix is positive semi-definite.

Definition 2. *The bordered Hessian of a multi-variate function $f(\mathbf{x})$, where $\mathbf{x} = (x_1, x_2, \ldots, x_n)$, is given by:*

$$\mathbf{H}_n^b(\mathbf{x}) = \begin{bmatrix} 0 & f'(x)_1 & \cdots & f'(x)_n \\ f'(x)_1 & f''(x)_{11} & \cdots & f''(x)_{1n} \\ \vdots & \vdots & \vdots & \vdots \\ f'(x)_n & f''(x)_{n1} & \cdots & f''(x)_{nn}, \end{bmatrix} \tag{23}$$

where $f'(\mathbf{x})_i = \frac{\partial f}{\partial x_i}$. If $|\mathbf{H}_n^b(x)| \leq 0$ for all n and x, then the function is quasi-convex (Note that the condition that $|\mathbf{H}_1^b| \leq 0$ is automatically satisfied).

Theorem 2. P2 *is a quasi-convex problem in the optimization variables* η, P_b, P_i

Proof. For problem **P2**, $n = 3$ as we have three optimization variables. We first start by establishing the convexity of the constraint with respect to the three optimization variables. Using (13) and (22), we can show that the delay constraint is convex and the other two constraints are affine in the optimization variables[1].

The only remaining step towards establishing the proof is to show the quasi-concavity of the objective function, EC, with respect to the three optimization variables. This involves characterizing the determinants of the bordered Hessian of the EC, defined in (23). Hence, from (23) and (16) we can simply show that $|\mathbf{H}_2^b| \leq 0$ and $|\mathbf{H}_3^b| \leq 0$ are satisfied and, hence, all terms are quasi-convex over the same domain 1. Thus, the effective capacity is quasi-concave in η, P_b and P_i since it is the negative of a quasi-convex function.

4   Numerical Results

In this section, we present the numerical results which confirm the fundamental trade-off under investigation, between the secondary user effective capacity and the primary user average packet delay. Moreover, we characterize the optimal solution using standard optimization tools for convex problems, e.g., CVX [15]. The system parameters used throughout this section are as follows: $r_b = 1.4$ Mbps, $r_i = 5.7$ Mbps, $P_b = 1$ unit power, $P_{max} = 2$ unit power, $P_i = 3$ unit

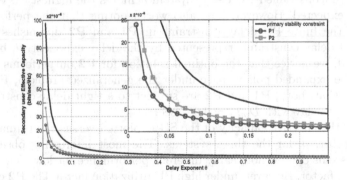

Fig. 4. Comparing the performance of the baseline (delay unconstrained system) to the optimal solutions of **P1** and **P2** ($D_{max} = 0.057$ sec)

---

[1] Details are omitted due to space limitations.

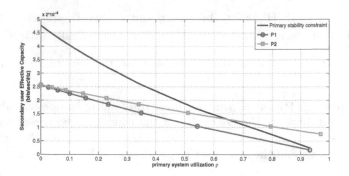

Fig. 5. The role of the primary user utilization, ρ.

power, $P = 3$ unit power, $B = 5$ MHz, $T = 0.1$ sec, $N = 0.002$ sec, $\lambda_p = 0.1$ and $\theta = 0.2$.

In Fig. 4, the effective capacity in (bit/sec/Hz) is plotted against the delay exponent, θ, for three different systems. The baseline is is a system with no constraints on the average PU packet delay and, hence in essence, it represents the case where the PU queue is stable "primary stability constraint". Thus the base line consider only maximizing the SU effective capacity which does not provide any delay guarantees for PU. On the other hand, **P1** represents the optimal solution for the basic problem which optimizes only over the energy detection threshold, η. **P2** represents the optimal solution for the generalized problem which optimizes over the three variables, η, P_b and P_i. A number of key observations are now in order. First, it is straightforward to observe that the SU effective capacity monotonically decreases under the three systems as the delay exponent increases, i.e. the statistical delay constraint becomes stricter. Second, the baseline (unconstrained PU delay) system achieves the highest SU effective capacity, as expected. However, it is also worth noting that the performance loss, due to the finite PU delay constraint in **P1** and **P2** diminishes as the SU statistical delay constraint, represented by the delay exponent, θ, becomes tighter. Third, we notice that the optimal solution for **P2** outperforms that of **P1** due to the expanded policy space under the generalized problem, **P2**. It is worth noting that, both, **P1** and **P2** are subject to a tight delay constraint on the PU average delay with $D_{max} = 0.057$ sec.

In Fig. 5, the effective capacity of the SU is plotted versus the primary user utilization factor, ρ, for the three systems. The most interesting observation is that all systems exhibit noticeably different performance for low to moderate PU utilization factor. However, under high PU utilization factor The **P2** exhibits superior performance to both the unconstrained system and **P1**. This behavior is attributed to the fact that as the PU utilization increases, the PU is using the channel more frequently and the probability of interference increases and, hence, the SU EC decreases. However, if we optimize over the transmission powers as in **P2**, this helps sustaining high SU EC. Also, we notice that the difference between the unconstrained system and **P1** diminishes as the PU utilization factor increases which is also attributed to higher interference.

5   Conclusions

In this paper we investigate a fundamental trade-off in delay-constrained cognitive radio networks. In particular, we characterize and optimize the trade-off between the secondary user effective capacity and the primary user average delay. Towards this objective, we employ Markov chain models to characterize the secondary user effective capacity and the average packet delay in the primary user queue. First, we formulate two constrained optimization problems to maximize the secondary user effective capacity subject to an average primary user delay constraint. Afterwards, the two formulated problems are proven to be quasi-convex and, hence, can be solved efficiently using standard techniques. Finally, the numerical results reveal interesting insights about the optimal performance compared to the unconstrained PU delay baseline system studied earlier in the literature.

References

1. Mitola, J.: Cognitive Radio: An Integrated Agent Architecture for Software Defined Radio. Ph.D. thesis, Royal Institute of Technology (KTH) (2000)
2. Chaoub, A., Ibn Elhaj, E., El Abbadi, J.: Multimedia traffic transmission over cognitive radio networks using multiple description coding. In: Advances in Computing and Communications, pp. 529–543. Springer, Heidelberg (2011)
3. Kushwaha, H., Xing, Y., Chandramouli, R., Heffes, H.: Reliable multimedia transmission over cognitive radio networks using fountain codes. Proc. of the IEEE Journal **96**(1), 155–165 (2007)
4. Liu, Q., Zhou, S., Giannakis, G.B.: Cross-layer modeling of adaptive wireless links for QoS support in multimedia networks. In: IEEE Quality of Service in Heterogeneous Wired/Wireless Networks Conf., pp. 68–75, October 2004
5. Tang, J., Zhang, X.: Quality-of-service driven power and rate adaptation for multichannel communications over wireless links. IEEE Trans. on Wireless Comm. 4349–4360 (2007)
6. Simeone, O., Bar-Ness, Y., Spagnolini, U.: Stable throughput of cognitive radios with and without relaying capability. IEEE Trans. on Comm. 2351–2360 (2007)
7. Wu, D., Negi, R.: Effective capacity: a wireless link model for support of quality of service. IEEE Trans. on Wireless Comm. **2**(4), 630–643 (2003)
8. Mohammadi, A., Kumar, S., Klymyshyn, D.: Characterization of effective bandwidth as a metric of quality of service for wired and wireless ATM networks. In: IEEE Int. Conf. on Comm. (ICC), pp. 1019–1024, June 1997
9. Akin, S., Gursoy, M.C.: Effective capacity analysis of cognitive radio channels for quality of service provisioning. IEEE Trans. on Wireless Comm. **9**(11), 3354–3364 (2010)
10. Anwar, A.H., Seddik, K.G., ElBatt, T., Zahran, A.H.: Effective capacity of delay constrained cognitive radio links exploiting primary feedback. In: Int. Symposium on Modeling Optimization in Mobile AdHoc Wireless Networks (WiOpt), pp. 412–419 (2013)

11. Chakravarthy, V., Xue, L., Zhiqiang, W.: Novel overlay/underlay cognitive radio waveforms using SD-SMSE framework to enhance spectrum efficiency- part I: theoretical framework and analysis in AWGN channel. IEEE Trans. on Comm. **57**(12), 3794–3804 (2009)
12. Ma, J., Li, Y.: Soft combination and detection for cooperative spectrum sensing in cognitive radio networks. In: IEEE Global Telecom. Conf., (GLOBECOM), pp. 3139–3143 (2007)
13. Chang, C.S.: Performance Guarantees in Communication Networks, chap. 7. Springer (1995)
14. Bertsekas, D.P., Gallager, R.G.: Data Networks, chap. 3. Prentice Hall (1992)
15. Boyd, S., Vandenberghe, L.: Convex Optimization. Cambridge Unive. Press (2004)

Auction Based Joint Resource Allocation with Flexible User Request in Cognitive Radio Networks

Wei Zhou$^{(\boxtimes)}$, Tao Jing, Yan Huo, Jin Qian, and Zhen Li

School of Electronic and Information Engineering, Beijing Jiaotong University,
Beijing, China
{11111032,tjing,yhuo,12111020,12111041}@bjtu.edu.cn

Abstract. Cognitive Radio (CR) has emerged as a promising technology to address the spectrum scarcity through encouraging the open access of licensed spectrum to unlicensed users. The incentives for licensed users and the resource allocation among unlicensed users are two main critical issues in practical implementation. Recently, auction has been introduced as an efficient tool to solve both incentive and allocation issues in cognitive radio networks. However, existing studies on auction are focusing on either channel allocation or power allocation. Few of them considers the channel and power allocation jointly. In addition, various transmission demands of unlicensed users push the need for flexible user request on spectrum resource. In this paper, we propose an auction scheme to study the joint resource allocation problem among unlicensed users and allow them to submit either range request or strict request according to their demands. To the best of our knowledge, we are the first to focus on this kind of problem. In the final, Theoretical analysis and numerical evaluations verify the truthfulness and efficiency of our scheme.

Keywords: Cognitive radio networks · Joint resource allocation · Auction theory

1 Introduction

Nowadays, the dramatic development of wireless devices and applications puts a growing demand on spectrum resource. The ever-increasing spectrum demand has posed a great challenge on current static spectrum allocation policy, in which the spectrum is allocated to licensed holders in long-term. Cognitive Radio (CR) [1], which utilizes the idle spectrum via opportunistic access, has emerged as a promising technology to alleviate the spectrum scarcity. There are two crucial issues in the adoption of cognitive radio technology: (1) Incentive problem: how to promote licensed holders to open the access of licensed spectrum; and (2) Allocation problem: how to allocate the spectrum resource among unlicensed users (i.e., Secondary Users, SUs).

Many economic tools have been introduced into cognitive radio networks to concurrently solve the incentive and allocation problem [2–4]. Among them,

© Institute for Computer Sciences, Social Informatics and Telecommunications Engineering 2015
M. Weichold et al. (Eds.): CROWNCOM 2015, LNICST 156, pp. 43–53, 2015.
DOI: 10.1007/978-3-319-24540-9_4

auction is preeminent due to its efficiency and fairness. However, prior works on auction have the following limitations: First, most of the studies only concentrate on the allocation of either spectrum channels [5–11] or transmitting power [12–14]. Joint channel and power allocation is rarely to see in existing studies. When adopting spectrum reuse in joint resource allocation, we need to consider not only which channel an SU is transmitting on, but also how much power the SU is transmitting with. Second, in existing auction schemes, user requests on spectrum resource are always assumed to be strict (i.e., an SU requests for a given amount of resource and accepts either all of the request or nothing) [5–7]. This assumption restricts the flexibility of user request and may compromise the efficiency of resource usage. The works in [10,11] introduce the concept of range request in the auction, in which an SU requests a given amount of spectrum resource and accepts any possible allocations, but they only consider range request on spectrum channels. Moreover, none of previous studies support both two types of user request in the auction.

In this paper, we allow SUs to bid for spectrum channel and transmitting power simultaneously. We model this joint resource allocation problem as an auction process. Moreover, we offer the SUs the flexibility on request format via allowing them to submit with either strict request or range request in the auction. The proposed auction scheme consists of two sequential sub-schemes, a multi-round auction for range request SUs and a greedy algorithm based auction for strict request SUs. Furthermore, a primary property of an auction scheme is truthfulness, since it makes the auction scheme invulnerable and keeps it from market manipulation. In the end of the paper, we theoretically analyze the truthfulness property of our auction scheme and conduct a numerical evaluation to verify the performance.

The rest of the paper is organized as follows. The network model, design goals and preliminary knowledge on auction are described in Section 2. Our proposed auction scheme and corresponding theoretical analysis are detailed in Section 3. The numerical evaluation is presented in Section 4 and the paper is concluded in Section 5.

2   Network Model and Preliminaries

2.1   Network Model

We consider a network model containing multiple SUs randomly distributed within a certain area. The set of SUs is denoted by $\mathcal{M}$ ($SU_i \in \mathcal{M}$). These SUs request spectrum channels and transmitting power on required channels to fulfill their transmission demands. There also exists a spectrum broker in the network who possesses a number of orthogonal channels and wants to lease out for additional profits. The set of channels is denoted by $\mathcal{C}$. A channel can be leased to multiple SUs as long as they are conflict-free, i.e., they are located out of the interference range from each other. Due to the power differentiation, the interference ranges of SUs are different. In this paper, we employ a conflict graph to reflect the interference relations among SUs. In the conflict graph, a vertex represents an SU and an edge exists between two vertices if they are conflict. We assume these channels are identical to SUs, which means SUs only care about the number of assigned channels and do not distinguish which channel.

We model the joint allocation problem as an auction process, wherein the spectrum broker is the seller and the SUs are buyers. Each SU submits a bid to the spectrum broker at the beginning of the auction. The bid of SU_i is denoted by $\mathcal{B}_i(x_i, n_i, P_i, \lambda_i)$, where x_i represents the type of user request, n_i represents the number of required channels, P_i represents the demand of transmitting power and λ_i represents the unit valuation (i.e., the valuation per channel per unit power). In this paper, we focus on two types of user request, strict request ($x_i = 1$) and range request ($x_i = 0$). In strict request, an SU only accepts the allocation of either transmitting on all n_i channels with power P_i or getting nothing. In range request, an SU is willing to accept any number of channels between 0 and n_i with any value of transmitting power less than P_i. The objective of the joint allocation problem can be formally written as

$$\max_{q_i, p_i, \mathcal{C}_i} \sum_{SU_i \in \mathcal{M}} \lambda_i \cdot q_i \cdot p_i$$

$$s.t. \ \mathcal{C}_i \subseteq \mathcal{C}, \ |\mathcal{C}_i| = q_i,$$
$$\mathcal{C}_i \cap \mathcal{C}_j = \emptyset, \ \forall SU_j \in N_i, \ SU_i \in \mathcal{M} \qquad (1)$$
$$q_i \in [0, n_i], \ 0 < p_i \le P_i, \ \forall SU_i \in \mathcal{M}, \ x_i = 0$$
$$q_i \in \{0, n_i\}, \ p_i \in \{0, P_i\}, \ \forall SU_i \in \mathcal{M}, \ x_i = 1.$$

q_i and p_i denote the number of allocated channels and the amount of allocated power, respectively. $\mathcal{C}_i$ represents the set of channels allocated to SU_i. N_i represents the set of conflicting neighbors of SU_i, i.e., they have common edges with SU_i in conflict graph. The second condition restricts that a channel cannot be reused among two conflicting SUs. The objective of the allocation is to maximize the social welfare, i.e., the sum of all winning SUs' valuations.

2.2 Truthfulness

In the auction, an SU's utility is determined by its valuation, final charge and the amount of obtained resource, which can be denoted as

$$u_i(\mathbf{B}) = \lambda_i \cdot q_i \cdot p_i - g_i, \qquad (2)$$

where $\mathbf{B} = \{\mathcal{B}_i\}, \forall SU_i \in \mathcal{M}$. The information of unit valuation is private, which means an SU may or may not report its true valuation in submitted bid. g_i is the price SU_i needs to pay for assigned resource. If an SU obtains nothing, its utility equals 0. Note that we assume the SUs only have incentives to lie about their unit valuations.

The performance of an auction design heavily depends on an economic property called truthfulness. The property requires that no SU_i can obtain a higher utility by submitting a false unit valuation $\tilde{\lambda}_i \neq \lambda_i$ in bid. In other words, revealing the true valuation is the dominant strategy for each SU in a truthful auction.

Let $\mathcal{B}_{-i}$ denote the bids submitted by all SUs other than SU_i, the truthfulness property can be formally written as

$$u_i(\mathcal{B}_i(\lambda_i), \mathcal{B}_{-i}) \geq \tilde{u}_i(\mathcal{B}_i(\tilde{\lambda}_i), \mathcal{B}_{-i}). \tag{3}$$

Guaranteeing the truthfulness keeps the auction scheme invulnerable and avoids market manipulation from SUs.

In general, the goal of this paper is to design an auction scheme to achieve the objective in (1) while satisfying the truthfulness property.

3   Auction Design Under Flexible Request

There are two challenges lying in the design of the auction scheme. First, how to charge the SUs with range requests. Due to the allocation is unfixed, it is difficult to directly apply the traditional pricing solution method which through finding the corresponding critical bids. Second, variable power requests may cause non-identical interference relations among SUs. The neighborhood of each SU varies with the allocated transmitting power. If we update the power allocation, we need to make sure whether preassigned channels are still available. With numerous SUs and continuous power region, the problem is more complicated.

3.1   Auction Design

We divide SUs into two sets $\mathcal{S}$ and $\mathcal{R}$, representing the set of strict request SUs and the set of range request SUs, respectively. We first conduct an auction among the SUs in $\mathcal{R}$. Taking account of the variability of SUs' resource requests, we design a multi-round auction scheme where all the channels are sequentially allocated. In each round, a single channel is allocated to a set of conflict-free SUs with relative higher raise on bid price. The power allocation of winning SUs is gradually updated with the interval δ to guarantee the free of conflicts.[1] Then, we propose a greedy algorithm based auction scheme among winning SUs in $\mathcal{R}$ and SUs in $\mathcal{S}$. The auction greedily assigns the channels and transmitting power to SUs in decreasing order of their bids as long as the allocation in feasible.

Multi-round Auction for SUs in $\mathcal{R}$: We start the auction with randomly distributing every SU_i into n_i different rounds. This ensures that each SU has no knowledge about other competitors, which is essential to keep the auction truthful. In each round, we use Φ_1 to denote the set of SUs that have not been assigned channels and Φ_2 to denote the left SUs. The details of the auction are shown in Algorithm 1.

Lines 1-5 describe the allocation among SUs in Φ_1. Specifically, we sort the SUs in decreasing order of their unit valuations and sacrifice the lowest-rank SU to determine other SUs' payments. In line 4, we gradually raise the power for

---

[1] δ is a small constant such that the assigned power is multiple times of δ.

Algorithm 1. Multi-round Auction for SUs in $\mathcal{R}$	

	for *each round l* **do**
	Input: Φ_1, Φ_2, $\{\mathcal{B}_i\}_{SU_i \in \Phi_1 \cup \Phi_2}$, $\mathcal{C}$, δ
1	$L = \{\lambda_i | SU_i \in \Phi_1\}$, $\phi_1 = \emptyset$;
2	Sort L in decreasing order and remove the last SU;
	for *each remaining SU_i in L* **do**
3	$\phi \leftarrow \{SU_i\}$;
4	Gradually raise the p_i with a step size of δ until it exceeds P_i or conflicts with others;
5	$q_i = 1$, $g_i^{q_i} = \lambda_{\underline{i}} \cdot p_i$; // $\lambda_{\underline{i}}$ *denotes the unit valuation of the removed SU*;
	end
6	$I = \{I_i = p_i \cdot \lambda_i | SU_i \in \Phi_2\}$, $\phi_2 = \emptyset$;
	while $I \neq \emptyset$ **do**
7	$SU_i = \arg\max\{I_i | I_i \in I\}$; $flag = 1$;
	for *each $SU_j \in \phi_2$* **do**
8	**if** $SU_j \in N_i^{\Phi_2}$ **then**
9	$g_j^{q_j} = \max(g_j^{q_j}, I_i)$; $flag = 0$;
10	break *for*;
	end
	end
11	**if** $flag == 1$ **then**
12	**if** $\sum_{SU_j \in N_i^{\phi_1}} p_j \cdot \lambda_j < I_i$ **then**
13	$\phi_2 \leftarrow \{SU_i\}$, $q_i = q_i + 1$;
14	$g_i^{q_i} = \sum_{SU_j \in N_i^{\phi_1}} p_j \cdot \lambda_j$;
15	$q_j = 0$, $\forall SU_j \in N_i^{\phi_1}$; $\phi_1 = \phi_1 - N_i^{\phi_1}$;
	end
	end
16	$I = I - \{I_i\}$;
	end
	end
	Output: $\{p_i, q_i, g_i^{q_i}, \mathcal{C}_i\}_{SU_i \in \mathcal{R}}$

each $SU_i \in \phi_1$ at a step size of δ until it conflicts with others or reaches the upper bound of power request. Once the power allocation of an SU is fixed, it would not change in following auction process. We could notice that, the power allocation only relates to SUs' locations and thus is independent of SUs' bid prices.

Lines 6-16 describe the final allocation among SUs in ϕ_1 and Φ_2. I_i denotes the increment on bid price for $SU_i \in \Phi_2$ if obtaining this channel. $N_i^{\phi_1}$ and $N_i^{\Phi_2}$ denote the set of conflicting SUs of SU_i in ϕ_1 and Φ_2, respectively. From lines 7 to 16, we check each element of I in decreasing order to see whether $SU_i \in \Phi_2$ satisfies the two conditions that enable the allocation: 1) do not conflict with granted SUs in ϕ_2 (lines 8-11); 2) able to cover the loss of social welfare caused by conflict (line 12). If so, SU_i obtains the channel and the conflicting SUs in ϕ_1

Algorithm 2. Auction for Winning SUs in $\mathcal{R}$ and SUs in $\mathcal{S}$

Input: $\{\mathcal{B}_i\}_{SU_i \in \mathcal{S}}$, $\{p_i, q_i, g_i^{q_i}, \mathcal{C}_i\}_{SU_i \in \mathcal{R}}$

1 $\mathcal{H} = \{H_i = n_i \cdot P_i \cdot \lambda_i | SU_i \in \mathcal{S}\}$; $\psi = \emptyset$;

   while $\mathcal{H} \neq \emptyset$ **do**

2      $SU_i = \arg\max\{H_i | H_i \in \mathcal{H}\}$; $flag = 1$;

3   **for** *each* $SU_j \in \psi$ **do**

4   **if** $SU_j \in N_i^S$ **then**

5              $g_i = \max(g_i, H_i)$; $flag = 0$;

6   break *for*;

        end

    end

7   **if** $flag == 1$ **then**

8          $\mathcal{C}_d = \bigcup_{SU_j \in N_i^R} \mathcal{C}_j$;

9   **if** $|\mathcal{C}| - |\mathcal{C}_d| \geq n_i$ **then**

10             $\psi \leftarrow \{SU_i\}$;

        else

11             $q_i = n_i - (|\mathcal{C}| - |\mathcal{C}_d|)$;

12             $\mathcal{T} = \{\sum_{ch_k \in \mathcal{C}_j, SU_j \in N_i^R} p_j \cdot \lambda_j | ch_k \in \mathcal{C}_d\}$;

13   Sort $\mathcal{T}$ in increasing order;

14   **if** $\sum_{k=1}^{q_i} T_k < H_i$ **then**

15                 $\psi \leftarrow \{SU_i\}$; $g_i = \sum_{k=1}^{q_i} T_k$;

16                 $g_j^{q_j} = 0$, $q_j = q_j - 1$, $\mathcal{C}_j = \mathcal{C}_j - \{ch_k\}$, $\forall SU_j \in N_i^R$, $ch_k \in \mathcal{C}_j$, $k = 1, 2, \cdots, q_i$;

            end

        end

    end

17     $\mathcal{H} = \mathcal{H} - \{H_i\}$;

   end

Output: ψ, $\{\mathcal{C}_i, g_i\}_{SU_i \in \psi}$, $\{p_i, q_i, g_i^{q_i}, \mathcal{C}_i\}_{SU_i \in \mathcal{R}}$

---

need to be eliminated (Line 15). Finally, the SUs in ϕ_1 and ϕ_2 win the channel in this round.

Auction for Winning SUs in $\mathcal{R}$ and SUs in $\mathcal{S}$: The proposed auction scheme is based on a greedy algorithm, the details of which are shown in Algorithm 2. We sort the SUs in $\mathcal{S}$ with decreasing order of their total valuations and examine each SU sequentially (lines 1-2). The allocation is feasible for $SU_i \in \mathcal{S}$ if and only if: 1) SU_i does not conflict with granted SUs in ψ (lines 3-6); 2) the available channels within its interference range can afford its demand (lines 8-10) or its valuation can cover the minimum loss on social welfare caused by its exclusive usage of channels within its interference range (lines 11-16). The minimum loss on social welfare is calculated by ranking the cumulative valuations of winning SUs in $\mathcal{R}$ on competitive channels in increasing order (lines 12-13) and selecting the q_i highest-rank channels (line 14).

Payment Calculation: The total payment of winning SUs in $\mathcal{R}$ is the sum of the payment on each assigned channel,

$$g_i = \sum_{k=1}^{q_i} g_i^k, \ \forall SU_i \in \mathcal{R}. \tag{4}$$

The payment on each assigned channel is obtained through finding out the critical user on this channel. On the first assigned channel, the critical user is the SU with minimum unit valuation, and thus we set the payment as $g_i^{q_i} = \lambda_{\underline{i}} \cdot p_i$. The critical user(s) on other assigned channels is either the set of conflicting SUs in ϕ_1 or the first SU in Φ_2 whose loss of the auction is caused by SU_i. Therefore, we set the payment equal to whichever is larger (line 9 in Algorithm 1).

The payment calculation for winning SU_i in $\mathcal{S}$ inherits the critical user based method. The critical user(s) of SU_i is either the set of conflicting winning SUs in $\mathcal{R}$ on competitive channels or the first SU whose loss of the auction is caused by SU_i. We set the payment in similar way as the last paragraph (line 5 in Algorithm 2).

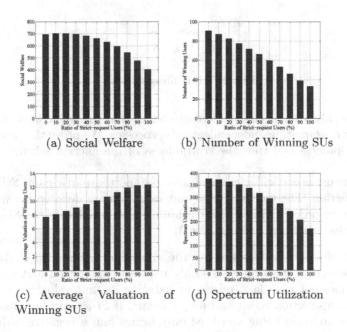

(a) Social Welfare

(b) Number of Winning SUs

(c) Average Valuation of Winning SUs

(d) Spectrum Utilization

Fig. 1. Impact of Request Type

3.2 Truthfulness Check

We analyze the truthfulness of SUs in $\mathcal{R}$ and $\mathcal{S}$ separately. We first consider the SUs in $\mathcal{R}$. In the auction, we randomly distribute the SUs into different rounds

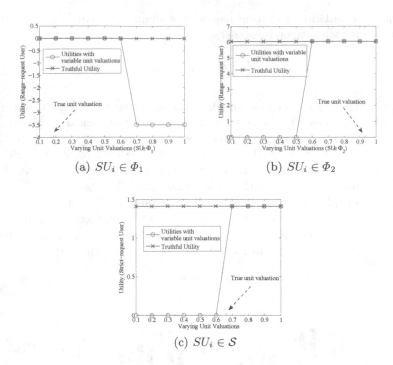

Fig. 2. Truthfulness Check

and allocate only a single channel in each round, thus the whole auction process can be viewed as multiple independent sub-processes. We prove the truthfulness of each sub-process and thus the truthfulness of the auction scheme could be proven.

We focus on a particular round l and assume the participant SU_i lies on its unit valuation. The true and false unit valuation are denoted by λ_i and $\tilde{\lambda}_i$, respectively. In order to prove the truthfulness, we need to show SU_i cannot obtain a higher utility by bidding $\tilde{\lambda}_i \neq \lambda_i$.

- If $SU_i \in \Phi_1$, it would participate the allocation in lines 1-5 of Algorithm 1. The selection of sacrificed SU only relates to submitted unit valuations. If SU_i does not rank last, raising or reducing its unit valuation would not change the selection result and its payment[2]. If SU_i ranks last, rasing unit valuation to avoid being sacrificed only brings him a negative utility since $\lambda_i \cdot p_i < g_i^1 < \tilde{\lambda}_i \cdot p_i$, g_i^1 is SU_i's payment when it lies.
- If $SU_i \in \Phi_2$, it only participates the allocation in lines 6-16. The allocation proceeds in a greedy fashion and the payment of each winning SU is set to an independent critical value below which the SU is unable to win the auction. If SU_i wins when bidding truthfully, raising or reducing the unit

---

[2] We have claimed above that the power allocation only depends on SUs' physical locations and is independent from SUs' unit valuations.

valuation cannot change the result and the payment. If SU_i loses when bidding truthfully, rasing the unit valuation to win the auction definitely generates a negative utility.

The proof of SUs in $\mathcal{R}$ ends. The proof of SUs in $\mathcal{S}$ is similar to that of SUs in Φ_2, so we omit it here.

4   Numerical Evaluation

In this section, we provide simulation results to evaluate the performance of our auction scheme. We stimulate a wireless cognitive radio networks in an area of $150 \times 150 \ m^2$, where a number of SUs are uniformly and randomly distributed. The relation between interference range and transmitting power is formulated as $Ir_i = \alpha \cdot \sqrt{p_i}$, based on free space propagation model. α is a systematic coefficient to match the parameter values[3]. The number of available channels is fixed to 10. The channel requests of SUs are randomly chosen from [1:1:5], the power requests are from $(0, 15]$ dBm and the unit valuations are from $(0, 1]$. The power interval δ is set to 0.2dBm. All simulation results are averaged over 200 runs to reduce randomness.

4.1   Impact of Request Type

To investigate the impact of request type, we fix the total number of SUs to 100 and vary the ratio of strict request SUs from 0 to 1 with a step size of 0.1. In Figure 1, we examine the performance of auction scheme in terms of four metrics: (1) **Social Welfare**, the sum of all winners' valuations; (2) **Number of Winning SUs**; (3) **Average Valuation of Winning SUs**; (4) **Spectrum Utilization**, is calculated based on Shannon's Theory:

$$StrUti = \sum_{SU_i \in \mathcal{RUS}} q_i \cdot \log(1 + p_i). \tag{5}$$

This metric can roughly quantify the achievable data throughput of secondary network.

Figure 1(a) depicts the result on social welfare. We see that, the social welfare declines as the number of strict request SUs increases. The strictness on request restricts the full allocation of network resource and thus discounts the social welfare. Range request SUs can provide adequate flexibility in resource distribution through accepting any possible allocations, which contributes to the increment on social welfare. The restriction of strict request could also be demonstrated in Figure 1(b) and Figure 1(d). Figure 1(b) shows that the number of winning SUs decreases as the ratio of strict request SUs grows. Figure 1(d) shows the result on spectrum utilization which verifies that range request can benefit the efficient usage of network resource.

---

[3] In practical implementation, the value of α can be set according to antenna gain, channel gain and SNR threshold.

In Figure 1(c), we present the result on average valuation of winning SUs. We can see that, the average valuation increases with the increment on ratio of strict request SUs. As illustrated before, range request can make the resource allocation more flexible by allowing more SUs to share the network resource. Although this could benefit the social welfare, there is a limitation that, a small number of range request SUs can obtain a relative large amount of resource, leading to a low individual valuation among SUs. On the contrary, strict request SUs selected from Algorithm 2 always own a higher individual valuation due to the strictness on request and greedy selection.

4.2   Truthfulness Check

Figure 2 examines the truthfulness of our auction scheme. We randomly select two SUs from $\mathcal{R}$ and $\mathcal{S}$ respectively, and check how their utilities change with variable unit valuation. The unit valuation varies from 0.1 to 1 at a step size of 0.1.

For the case when SU in $\mathcal{R}$, we further divide it into two subcases, SU in Φ_1 and SU in Φ_2. The results are shown in Figure 2(a) and 2(b). Figure 2(c) shows the result for the case when SU in $\mathcal{S}$. We can note that, no matter which type of request the SU bids, it cannot improve its utility by bidding untruthfully on its unit valuation.

5   Conclusion

In this work, we study the problem of joint channel and power allocation among multiple SUs in cognitive radio networks. We consider a mixed form of resource request, wherein the SUs can bid with either strict request or range request. To solve the problem, we propose an auction scheme consisting of two sequential sub-schemes, a multi-round auction for range request SUs and a greedy algorithm based auction for strict request SUs. The calculation of winners' payments is based on the corresponding critical value. We theoretical analyze the truthfulness of our auction scheme for both range request SUs and strict request SUs. The simulation results also evince the efficient performance of our auction scheme.

In our future work, we will investigate the power budget for SUs in power allocation which caused by PUs' interference constraints. Moreover, the heterogeneities among available channels and the truthfulness on other attributes in the demand are also worth exploiting.

Acknowledgments. The authors would like to thank the support from the National Natural Science Foundation of China (Grant No.61172074,61272505 and 61471028) and the Specialized Research Fund for the Doctoral Program of Higher Education (Grant No.20130009110015).

References

1. Xing, X., Jing, T., Cheng, W., Huo, Y., Cheng, X.: Spectrum prediction in cognitive radio networks. IEEE Wireless Communications **20**(2), 90–96 (2013)
2. Wang, F., Krunz, M., Cui, S.: Price-based spectrum management in cognitive radio networks. IEEE Journal on Selected Areas in Communications (JSAC) **2**(1), 74–87 (2008)
3. Zhang, T., Yu, X.: Spectrum sharing in cognitive radio using game theory-a survey. In: 6th IEEE International Conference on Wireless Communications, Networking and Mobile Computing, pp. 1–5 (2010)
4. Gao, L., Wang, X., Xu, Y., Zhang, Q.: Spectrum trading in cognitive radio networks: A contract-theoretic modeling approach. IEEE Journal on Selected Areas in Communications (JSAC) **29**(4), 843–855 (2011)
5. Jing, T., Zhao, C., Xing, X., Huo, Y., Li, W., Cheng, X.: A multi-unit truthful double auction framework for secondary market. In: IEEE International Conference on Communications, pp. 2817–2822 (2013)
6. Li, W., Wang, S., Cheng, X.: Truthful multi-attribute auction with discriminatory pricing in cognitive radio networks. In: 1st ACM Workshop on Cognitive Radio Architectures for Broadband, pp. 21–30 (2013)
7. Dong, M., Sun, G., Wang, X., Zhang, Q.: Combinatorial auction with time-frequency flexibility in cognitive radio networks. In: 31st IEEE International Conference on Computer Communications, pp. 2282–2290 (2012)
8. Feng, X., Chen, Y., Zhang, J., Zhang, Q., Li, B.: Tahes: Truthful double auction for heterogeneous spectrums. In: 31st IEEE International Conference on Computer Communications, pp. 3076–3080 (2012)
9. Chen, Y., Zhang, J., Wu, K., Zhang, Q.: Tames: a truthful auction mechanism for heterogeneous spectrum allocation. In: 32nd IEEE International Conference on Computer Communications, pp. 180–184 (2013)
10. Huang, H., Sun, Y., Xing, K., Xu, H., Xu, X., Huang, L.: Truthful multi-unit double auction for spectrum allocation in wireless communications. In: Wang, X., Zheng, R., Jing, T., Xing, K. (eds.) WASA 2012. LNCS, vol. 7405, pp. 248–257. Springer, Heidelberg (2012)
11. Zhou, X., Gandhi, S., Suri, S., Zheng, H.: ebay in the sky: strategy-proof wireless spectrum auctions. In: 14th Annual International Conference on Mobile Computing and Networking, pp. 2–13 (2008)
12. Chen, L., Iellamo, S., Coupechoux, M., Godlewski, P.: An auction framework for spectrum allocation with interference constraint in cognitive radio networks. In: 29th IEEE International Conference on Computer Communications, pp. 1–9 (2010)
13. Baidas, M., MacKenzie, A.: Auction-based power allocation for multi-source multi-relay cooperative wireless networks. In: IEEE Global Communications Conference, pp. 1–6 (2011)
14. Xu, H., Zou, J.: Auction-based power allocation for multiuser two-way relaying networks with network coding. In: IEEE Global Communications Conference, pp. 1–6 (2011)

Two-Stage Multiuser Access in 5G Cellular Using Massive MIMO and Beamforming

Hussein Seleem[✉], Abdullhameed Alsanie, and Ahmed Iyanda Sulyman

King Saud University, Riyadh, Saudi Arabia
{hseleem,sanie,asulyman}@ksu.edg.sa

Abstract. This paper explores the possibility of using multiuser massive MIMO and beamforming together as two-stage multiuser access methods in 5G cellular. Multi-carrier OFDM transmission as currently used in the 4G-LTE may be difficult to implement in 5G cellular because of the peculiarities of mmWave channels. Therefore, there is the need to analyze and propose suitable physical layer techniques suitable for 5G systems that are amenable to beamforming and/or massive MIMO. It turns out that both of these schemes are inherently multiuser access methods and can be used together in 5G cellular. Our results show that simple transmitter and receiver processing can be achieved when using the combined system. Moreover, the proposed approach will allow the system to accommodate more users at minor error rate degradation.

Keywords: Millimeter waves · 5G cellular networks · Massive MIMO · MU-MIMO · Beamforming

1 Introduction

Mobile communication has evolved significantly over the last four decades, from the early voice systems to today's highly sophisticated integrated platforms that provide numerous services, and support countless applications used by billions of people around the world [1]. The bandwidth-intensive media services that were earlier confined to wired transmission are now being used on mobile devices. The exponential growth of cellular data traffic and also the continued advances in computing and communications coupled with the emergence of new customer devices such as smart-phones are driving technology evolutions and moving the world toward a fully connected networked society where access to information and data sharing are possible anywhere, anytime, by anyone or anything [1,2].

This has created unprecedented challenges for wireless service providers. They need to overcome a global bandwidth shortage and support ever-growing data rate demands. Subsequently, an efficient radio access technology (RAT) combined with more spectrum availability is essential to meet this overwhelming challenges faced by wireless carrier [1–4]. Consequently, wireless carriers will

H. Seleem—Student Member IEEE.
A.I. Sulyman—Senior Member IEEE.

© Institute for Computer Sciences, Social Informatics and Telecommunications Engineering 2015
M. Weichold et al. (Eds.): CROWNCOM 2015, LNICST 156, pp. 54–65, 2015.
DOI: 10.1007/978-3-319-24540-9_5

be able to provide peak and cell edge rates higher than tens of Gb/s and 100 Mb/s respectively [2–4]. Also it will offer a minimum of 1 Gb/s data rate anywhere to provide a Gb/s experience to all end users and up to 5 and 50 Gb/s data rates for high mobility and pedestrian users, respectively [3]. In addition, they will be able to provide latency less than 1 -ms for local area networks (LANs) [4], ultra-low energy consumption, massive numbers of connected devices, very high volumes of data transfer, low cost devices, ultra-reliable connectivity with guaranteed availability for mission-critical machine type applications [1].

Recently the large amount of underutilized spectrum in the millimeter wave (mmWave) frequency bands has attracted researchers attentions as a potentially viable solution for achieving tens to hundreds of times more capacity compared to current fourth generation (4G) cellular networks [5–8]. Historically, mmWave bands were kept out for cellular usage mainly due to the high propagation loss and lack of cost-effective components. Accordingly, mmWave have mostly been utilized for outdoor point-to-point back-haul links or for carrying high-resolution multimedia streams in indoor applications, but not for cellular access links. In addition, mmWave frequencies can be severely vulnerable to shadowing, resulting in outages and irregular channel quality. Power consumption in mmWave to support large numbers of antennas with very wide bandwidths is also a key challenge. In order to reform these underutilized spectra for future outdoor cellular applications, two key hurdles must be overcome: sufficiently large geographical coverage and support for mobility even in non-line of sight (NLOS) environments where the direct communications path between the transmitter (TX) and the receiver (RX) is blocked by obstacles. Today, the cell sizes in urban environments are about 200 m, so it becomes clear that mmWave cellular can overcome these issues [7].

The next step is to develop an efficient RAT and core network technologies to efficiently utilize the abundant spectrum in the mmWave bands and achieve commercial viability. Due to the much smaller wavelength combined with advances in cost-effective CMOS-based low-power radio frequency (RF) circuits, enabling large numbers of miniaturized antennas (such as 64 elements), we may exploit polarization and new spatial processing techniques, such as multiuser multiple input multiple output (MU- MIMO) [9–11] and adaptive beamforming [12,13] using Massive MIMO [9,14–17], and highly directional antennas at both the user equipment (UE) and base station (BS). Consequently, the increase in omnidirectional path loss (PL) when using the higher frequencies can be completely compensated [18].

Using Adaptive Beamforming, one can focus the radio transmission from multiple antenna elements using very narrow beams, which results in reduced interference and improves overall system performance. In spatial multiplexing, we exploit the propagation properties to provide multiple data streams to one or more terminals simultaneously. These techniques are already integral parts of the long-term evolution (LTE) cellular system but their full potential remains to be released. They are now set to play an even bigger role in future cellular systems. MIMO is an advanced antenna solutions that include a substantial number of antenna elements that can be used to reduce the impact of RF imperfections and control the way interference is distributed in a network.

Therefore, the base station to device links, as well as back-haul links between base stations, will be able to handle much greater capacity than today's 4G networks in highly populated areas when deploying massive MIMO system. Also, as operators continue to reduce cell coverage areas to exploit spatial reuse, and implement new cooperative architectures such as cooperative MIMO, relays, and interference mitigation between BSs, the cost per BS will drop as they become more plentiful and more densely distributed in urban areas, making wireless back-haul essential for flexibility, quick deployment, and reduced ongoing operating costs.

This paper introduces a two-stage multiuser access in fifth generation (5G) cellular using baseband precoding and RF beamforming. The baseband (BB) precoding is done in digital domain to optimize capacity using various MIMO techniques. On the other hand, RF beamforming is executed in analog domain to overcome higher path loss with beamforming gain in mmWave bands. By implementing both schemes together, we could be able to reach high performance with low complexity for mmWave 5G cellular systems.

The rest of the paper is organized as follows. Section 2 describes the mmWave channel model. Section 3 introduces the system model for the two-stage multiuser access for 5G cellular system using massive MIMO and beamforming. The main contribution of the paper appear in this section. Section 4 and section 5 describe the multiuser massive MIMO downlink model and beamforming downlink model, respectively. Numerical results and discussions are presented in section 6, while the computational complexity analysis is given in section 7. Finally, the conclusions are presented in section 8.

2   Millimeter wave Channel Model

Over 90 percent of the allocated radio spectrum falls in the mmWave band (30 - 300 GHz). Industry has considered mmWave to be any frequency above 10 GHz. The most common misunderstanding of the propagation characteristics at higher frequencies is that they always suffer a much higher propagation loss even in free space compared to lower frequencies, and thus are not adequate for long-range communications. To clarify this misunderstanding, let us start with the Friis transmission equation, given by [19].

$$P_r = P_t G_t G_r (\frac{\lambda}{4\pi d})^2 = P_t G_t G_r \underbrace{(\frac{c^2}{4\pi f^2})}_{Aperture\ Size} \underbrace{(\frac{1}{4\pi d^2})}_{Shperical\ Area} \tag{1}$$

$$\underbrace{\phantom{P_t G_t G_r (\frac{\lambda}{4\pi d})^2}}_{PL}$$

The received power can be seen as inversely proportional to the frequency squared when an ideal isotropic radiator $G_t = 1$ and an ideal isotropic RX $G_r = 1$ are used at each end. By employing array of antennas at TX or/and RX with antenna gains of $G_t = A_{e,tx}(\frac{4\pi f^2}{c^2})$ or/and $G_r = A_{e,rx}(\frac{4\pi f^2}{c^2})$ respectively, which is greater than unity, the received power will be given as:

$$P_r = P_t . A_{e,tx} . A_{e,rx} (\frac{4\pi f^2}{c^2}) . (\frac{1}{4\pi d^2}) \tag{2}$$

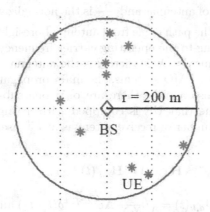

Fig. 1. Single cell scenario in mmWave massive MIMO, the BS is fixed and it is deployed with ULA antennas. The K UE are randomly uniform distributed and the BS communicates with all UE simultaneously.

Using antenna arrays at both TX and RX transmitting at higher frequencies, given the same physical aperture size, can allow us to send and receive more energy though narrower directed beams [3].

Consider a single cell massive MIMO transmission system working at mmWave frequencies and adopting time division duplex (TDD) transmission mode. BS is located at the cell center and deployed with N_t uniform linear array (ULA) antennas. A total number of K single antenna users are randomly distributed in the cell as shown in Fig. 1.

Using the spatial multi-path channel model (SCM) [12], the physical downlink (DL) channel Matrix for k^{th} user can be written as:

$$\bar{\mathbf{H}}_{d,k}(t) = \sum_{i=1}^{N_p} \sqrt{\frac{P_i}{N_t N_r}} \sum_{j=1}^{N_{sp}} \mathbf{\Lambda}_{ij} . [\mathbf{a}_r(\theta_{ij}^{AoA}) \mathbf{a}_t^T(\theta_{ij}^{AoD})] . e^{j2\pi f_{ij}^d t} \delta(t - t_i) \quad (3)$$

Where N_p is the total number of paths per channel link, N_{sp} is the total number of subpaths per path, P_i is the power of the ith path. N_r is the total number of antennas at receiver, $\mathbf{\Lambda}_{ij}$ is $N_r \times N_t$ initial phase matrix for the jth sub-path of the ith path. θ_{ij}^{AoA} and θ_{ij}^{AoD} is the angle of arrival (AoA) and angle of departure (AoD), respectively for the jth sub-path of the ith path. $\mathbf{a}_r(\theta_{ij}^{AoA})$ and $\mathbf{a}_t(\theta_{ij}^{AoD})$ is the array response vector of length $N_r \times 1$ and $N_t \times 1$ for the given AoA and AoD, respectively. f_{ij}^d is the Doppler frequency. Symbols $(.)$, $()^T$ and $()^H$ stands for element wise product, transpose, and Conjugate transpose operations, respectively.

For the ULA, the array response vector is given by:

$$\mathbf{a}(\theta) = \frac{1}{\sqrt{N}}[1 \quad e^{jkdsin(\theta)} \dots e^{jkd(N-1)sin(\theta)}]^T \quad (4)$$

Where N is the number of antennas and $\frac{1}{\sqrt{N}}$ is the normalization factor such that $\|\mathbf{a}(\theta)\|_2^2 = 1$, where θ is the path angle from antenna boresight and $\lambda = c/f_c$ is the wavelength corresponding to the operating carrier frequency f_c and $k = 2\pi/\lambda$ is the wave number. Assume $d = \lambda/2$ where d is the antenna spacing then $kd = \pi$.

In mmWave massive MIMO systems, the main propagation path is LOS, so $P_1 >> P_i$ for $i \geq 2$, also assume there are only one sub-path for simplicity. In our model we assume that UE is equipped with a single antenna element where $N_{r,k} = 1$ is the number of receive antennas at k^{th} user. Consequently, the physical channel becomes,

$$\bar{\mathbf{H}}_{d,k}(t) = \mathbf{H}_{d,k}(t)\mathbf{A}_{d,k}. \tag{5}$$

where $\bar{\mathbf{H}}_{d,k} \in \mathbb{C}^{1 \times N_t}$, $\mathbf{H}_{d,k}(t) = \sqrt{\frac{P_i}{N_t N_r}}\Lambda e^{j2\pi f^d t}\delta(t - t_1)$ and $\mathbf{A}_{d,k} = \mathbf{a}_t^T(\theta^{AOD})$

3   Two-Stage Multiuser Access Using Multiuser Massive MIMO and Beamforming

Consider the mmWave massive MIMO system shown in Fig. 2 in which a BS transmits to all users simultaneously. The received signal at k^{th} user is given by:

$$\bar{\mathbf{y}}_{d,k} = \bar{\mathbf{H}}_{d,k}\bar{\mathbf{W}}\bar{\mathbf{P}}\bar{\mathbf{s}} + \bar{\mathbf{n}}_{d,k} \tag{6}$$

Where $\bar{\mathbf{s}}$ is the $K \times 1$ vector of user symbols, $\bar{\mathbf{P}}$ is the $N_t^c \times K$ BB precoding matrix and it will used as the first stage multiuser access scheme. $\bar{\mathbf{W}}$ is the $N_t^c \times N_t$ RF beamformer weight matrix and it will be considered as the second stage

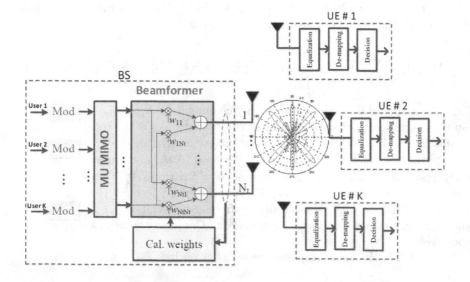

Fig. 2. System model for the mmWave multiuser massive MIMO system.

multiuser access scheme. N_t^c is the number of RF chains and this number we need to decrease it as much as possible to decrease the complexity. On the other hand by decreasing number of RF chains the performance will be worse. So that by using the RF beamforming in analog domain we can decrease complexity and increase the performance at the same time. The RF beamformer weights could be implemented using variable gain attenuators (VGAs) and phase shifters that are connected with the antennas, while BB precoder weights are processed in the digital domain as for the conventional MU-MIMO implementation. We consider a MIMO fading channel which yields a received signal $\bar{\mathbf{y}}_{d,k}$. $\bar{\mathbf{H}}_{d,k}$ is the channel matrix that can be analytically represented according to the SCM-based multipath channel model as described before in section 3. $\bar{\mathbf{n}}_{d,k} \in \mathbb{C}^{N_{r,k} \times 1}$ is the additive zero mean circular symmetric complex Gaussian noise at the kth user with the variance of σ^2.

4   Multiuser Massive MIMO Downlink System Model

MU-MIMO deal with multiple users who are sharing the same radio (time-frequency) resources. It has brought a great improvement in the system capacity by serving multiple users simultaneously. Multiple antennas allow the independent users to transmit their own data stream in the uplink (many-to-one) at the same time or the base station to transmit the multiple user data stream to be decoded by each user in the downlink (one-to-many). In the multiuser MIMO system, downlink and uplink channels are referred to as broadcast channel (BC) and multiple access channel (MAC), respectively. We are interested in the system where $N_t >> KN_r, k$, hence this system refers to a massive MU-MIMO.

We further assume that the channels will stay constant during a coherence interval of T symbols. The downlink transmission will occur in two phases: training phase and downlink data transmission phase, where τ is the coherence duration used for the training phase. In the training phase, the base station estimates the channel state information (CSI) from K users based on the received pilot sequences in the uplink. The base station uses this CSI and a precoding scheme to process the signals before transmitting them to K users. We assume that perfect CSI is available at the BS which can be achieved by simple TDD training method [20]. Also, this assumption is reasonable under the scenarios that the training power is large or the coherent interval is large (and hence, we can spend large τ for training).

In the case of downlink channel, Let $\mathbf{x} \in \mathbb{C}^{N_t \times N_r}$ is the transmit signal from the BS and can be written as $\bar{\mathbf{x}} = \bar{\mathbf{P}}\bar{\mathbf{s}}$. The average transmission power is constrained by $E_s = \mathbf{E}[\|\mathbf{x}_k\|_2^2] = tr(\mathbf{PP}^H)$. the received vector $\mathbf{y}_k \in \mathbb{C}^{N_{r,k} \times N_t}$ at the kth user, is expressed as: $\bar{\mathbf{y}}_{d,k} = \bar{\mathbf{H}}_{d,k}\bar{\mathbf{x}} + \bar{\mathbf{n}}_{k,d}$. Representing all user signals by a single vector, the overall system can be represented as: $\mathbf{y} = \mathbf{H}_d\mathbf{x} + \mathbf{n}$.

The linear precoding will be considered in this paper which include zero forcing (ZF), linear minimum mean squared error (LMMSE) and maximum ratio transmission (MRT) [21]. A simple way to deal with the inter user interference (IUI) is to set the constraint so to force all interference terms to zero, the method is known as the ZF. Since the transmitted power is limited by E_s, the precoding matrix has to be designed to satisfy the transmit power constraint, that is: $E_s = \mathbf{E}[\|(1/\beta)\bar{\mathbf{P}}\bar{\mathbf{s}}\|_2^2]$, where β is a factor to make sure that the transmitted signal power after precoding will not be changed. So that the estimation of the transmitted signal at the receiver side is given by: $\hat{\mathbf{s}} = \mathbf{s} + \beta_{CI}\mathbf{n}$.

Then, the ZF precoding design problem can be described as follows:

$$\underset{\mathbf{P},\beta}{\operatorname{argmin}} \ \mathbf{E}[\|\hat{\mathbf{s}} - \mathbf{s}\|_2^2]$$

$$\text{s. t.} \ \ \mathbf{E}[\|(1/\beta_{ZF})\mathbf{P}_{ZF}\mathbf{s}\|_2^2] = E_s$$

$$\hat{s}|_{n=0} = \mathbf{s} \qquad (7)$$

The solution of the above optimization problem is: $\mathbf{P}_{ZF} = \mathbf{H}_d^H(\mathbf{H}_d\mathbf{H}_d^H)^{-1}$. So, this precoding is assumed to implement a pseudo-inverse of the channel matrix. If the channel matrix is ill conditioned β_{ZF} will become larger and the performance of ZF precoding will be poor due to noise enhancement.

In order to reduce the noise enhancement and to maximize the signal to interference plus noise ratio (SINR), the second constraint in the optimization problem needs to be dropped. the resultant problem is the well known LMMSE constraint optimization problem that is average power at each transmitted antenna is constrained. Thus, LMMSE precoding technique is the optimal in MU-MIMO downlink system and its solution is given by:

$$\mathbf{P}_{LMMSE} = \mathbf{H}_d^H(\mathbf{H}_d\mathbf{H}_d^H + \sigma^2\mathbf{I})^{-1},$$

for MRT precoding, it is on of the common methods due to its simplicity, which maximizes the SNR. MRT works well in the MU-MIMO system where the base station radiates low signal power to the users. Hence precoder matrix can be expressed as:

$$\mathbf{P}_{MRT} = \mathbf{H}_d^H.$$

The normalization constants can be given by:

$$\beta = \sqrt{\frac{tr(\mathbf{P}\mathbf{P}^H)}{E_s}}$$

for all linear precoding techniques.

5   Beamforming for Downlink System Model

In this section, the proposed RF beamforming and equalization scheme is presented. This scheme consists of two stages. In the first stage, the beamforming at the base station is used to reduce the effect of the IUI. The second stage uses the equalization at the mobile unit to reduce the effect of the ISI and to provide a better estimate of the desired data. The uplink beamforming weights are calculated via the minimum variance distortion-less response (MVDR) algorithm [22]. This algorithm minimizes the total received power while maintaining a unity power gain towards the desired user with his N_p paths. Assuming reciprocity between the uplink and the downlink and thereby the calculated uplink beamforming weights are used in the downlink to mitigate the IUI. In our work, we assume for simplicity that all users send their signals at the same time in the uplink.

The uplink weights of all users with their N_p paths are calculated as follows; in uplink, all users send their signals synchronously after spreading and modulation:

$$\bar{\mathbf{y}}_{BS} = \mathbf{A}_u \mathbf{H}_u \bar{\mathbf{P}}_u \bar{\mathbf{s}}.$$

Where $\bar{\mathbf{y}}_{BS}$ is the received data matrix at the output of the antenna array at BS. $\mathbf{A}_u$ is the array response matrix of the antenna array for all active users with their paths. $\mathbf{H}_u$ is the uplink channel response matrix. $\bar{\mathbf{P}}_u$ is the precoding matrix that used as access scheme between users for uplink. The multi-path MVDR algorithm is applied for calculating the weights as follows:

$$\mathbf{w}_k = \frac{Cov^{-1}(\mathbf{R}_{yy})\,\mathbf{a}_k}{[\mathbf{a}_k^H]^{-1}Cov^{-1}(\mathbf{R}_{yy})\,\mathbf{a}_k} \tag{8}$$

Where $\mathbf{R}_{yy}$ is $N_t \times N_t$ covariance matrix of the received signal at antenna array and $\mathbf{w}_k$ are the weight vector of the kth user. These weights enable the antenna array to receive/transmit from/to a certain user in a multipath environment and to optimize the signal quality. The previous step is repeated K times to calculate all weights of the active users. Next, After calculating the weights of the active users at the uplink, we use these weights for downlink beamforming as: $\bar{\mathbf{y}}_{d,k} = \mathbf{H}_{d,k}\mathbf{A}_{d,k}\bar{\mathbf{W}}\bar{\mathbf{s}} + \bar{\mathbf{n}}_{d,k}$. Where $\bar{\mathbf{y}}_{d,k}$ is the received data vector at the desired UE. After that, the received signal is equalized to suppress the ISI as follows:

$$\hat{\mathbf{s}} = (\mathbf{H}_{d,k}^H \mathbf{H}_{d,k} + \alpha\mathbf{I})^{-1}\mathbf{H}_{d,k}^H \bar{\mathbf{y}}_{d,k}, \tag{9}$$

Finally, the decision process is performed to produce the estimate of the desired data. The optimum regularization factor α that minimizes the equalization error is 1/SNR, which is the MMSE weight.

6 Numerical Results and Discussion

Several simulations are carried out to test the performance of the proposed two-stage multiuser access scheme using MU-MIMO and beamforming. The simulation environment is based on the downlink MU-MIMO and beamforming system, in which each user transmits binary phase shift keying (BPSK) information symbols. The wireless channel model used in the simulation is the SCM model. It has N_p Raleigh fading taps. The fading is modeled as quasi-static (unchanged during the duration of a block).

In the first part of the simulations, a comparison study between the BB precoding, RF beamforming and both BB precoding combined with RF beamforming in single user (SU) case is conducted and the results, in terms of bit error rate (BER), are shown in Fig. 3. In BB precoding, the precoding is done

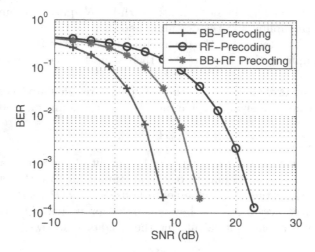

Fig. 3. Performance comparison for SU case between BB precoding with LMMSE precoder, RF beamforming with LMMSE equalization at the receiver side and BB with RF precoding using LMMSE precoder at $K = 1$, $N_t = 4$, $N_r = 1$, and BPSK

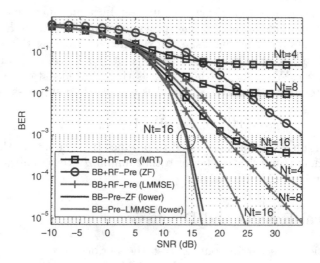

Fig. 4. Performance comparison for MU case between BB precoding and BB with RF precoding at $K = 4$, $N_t^c = 4$, $N_r = 1$, and BPSK

in the form of digital pre-processing that multiplies a particular coefficients with the modulated baseband signal per RF chain. For RF beamforming, the precoding is done in analog form where complex coefficients are applied to manipulate the RF beams by means of controlling the phase shifters and/or variable gain amplifiers (VGAs). As can be observed from Fig. 3, the BB precoding provides

Table 1. Computational complexity Comparison

Scheme	Complexity (flops)	for case: $N_t = 16$, $N_t^c = 4$, $K = 4$
BB Precoding (LMMSE)	$16N_t^3 + 3N_t^2$	66304
BB+RF precoding (LMMSE)	$16N_t^{c3} + 3N_t^{c2}$	1072

a higher degree of freedom as it has better performance, but at the expense of increased complexity. On the other hand, RF beamforming is a simple and effective method of generating high beamforming gains to user from a large number of antennas, but it has a degraded BER performance than BB precoding. Therefore, there is a tradeoff between performance and simplicity. Using a two-stage scheme as proposed, we can achieve good performance in between at reasonable complexity. This case is depicted by the middle curve in Fig. 3.

In the second part of the simulations, same comparison above is conducted for the case of multiuser system using ZF, LMMSE and MRT precoding for MU-MIMO as an access scheme. Fig. 4 shows the BER performance versus the SNR for that case. Note that in digital BB processing, the TX processing is complex while the RX processing is simple. In RF beamforming, the TX processing is simple while the RX processing is complex. The proposed two-stage approach will combine the benefits of both schemes achieving simple TX and RX processing. In addition, the results shows that the proposed two-stage scheme gives BER performance that is comparable to the digital BB processing only.

Moreover, the proposed two-stage scheme allows more users to be served since the BB MU-MIMO part will give access service to some users, while the RF BF part will also give access service to additional users.

7   Computational Complexity Analysis

In this section we assess the computational complexity of different multiuser access schemes. The computational load is primarily a function of the number of users K, the number of transmit antennas at BS Nt, the number of RF chains N_t^c and the number of receive antenna at each user $N_{r,k}$. The computational complexity considered here is expressed in terms of the total number of flops[1]. In real arithmetic, a multiplication followed by an addition needs 2 flops. With complex-valued quantities, a multiplication followed by an addition needs 8 flops. Thus, the complexity of a complex matrix multiplication is nearly 4 times its real counterpart. For a complex $m \times n$ matrix **A**, we summarize the total FLOPs needed for the matrix operations as shown below:

1. Multiplication of $m \times n$ and $n \times p$ complex matrices $= 8mnp - 2mp$.
2. Pseudo-inversion of an $m \times n$ $(m \leq n)$ complex matrix $= \frac{4}{3}m^3 + 7m^2n - m^2 - 2mn$.

---

[1] A flop stands for floating point operation. Operations such as addition, multiplication, subtraction, division and compare are considered as one flop.

The required number of flops for the BB precoding and BB with RF precoding are given in Table 1. The case given in the table with $N_t = 16$, $N_t^c = 4$, $K = 4$, it is worth noting that with the decrease in the number of RF chains, there is a considerable reduction in complexity. Therefore, the proposed BB with RF beamforming scheme has low complexity compared with the BB precoding only.

8   Conclusion

We have proposed a novel two-stage multiuser access scheme for 5G cellular using massive MIMO and Beamforming. The results show that the proposed two-stage scheme will combine the benefits of both BB and RF precoding, achieving simple TX and RX processing. For the multiuser case, the proposed approach will also allow the system to accommodate more users since BB MU-MIMO part will give access service to some users, while the RF beamforming part will also give access service to additional users at minor error rate degradation.

Acknowledgments. This work was supported by NSTIP strategic technologies programs (no 11-ELE1854-02) in the Kingdom of Saudi Arabia.

References

1. Dahilman, E., Mildh, G., Parkvall, S., Peisa, J., Sachs, J., Selen, Y.: 5G radio access. Ericsson Review (2014)
2. Rappaport, T.S., Shu, S., Mayzus, R., Hang, Z., Azar, Y., Wang, K., Wong, G.N., Schulz, J.K., Samimi, M., Gutierrez, F.: Millimeter Wave Mobile Communications for 5G Cellular: It Will Work!. IEEE Access **1**, 335–349 (2013b)
3. Roh, W., Seol, J.-Y., Park, J., Lee, B., Lee, J., Kim, Y., Cho, J., Cheun, K., Aryanfar, F.: Millimeter-Wave Beamforming as an Enabling Technology for 5G Cellular Communications: Theoretical Feasibility and Prototype Results. IEEE Communications Magazine **52**(2), 106–113 (2014)
4. Rangan, S., Rappaport, T.S., Erkip, E.: Millimeter wave cellular wireless networks: Potentials and challenges (2014). arXiv preprint arXiv:1401.2560
5. Rappaport, T.S., Gutierrez, F., Ben-Dor, E., Murdock, J.N., Qiao, Y., Tamir, J.I.: Broadband millimeter-wave propagation measurements and models using adaptive-beam antennas for outdoor urban cellular communications. IEEE Transactions on Antennas and Propagation **61**(4), 1850–1859 (2013a)
6. Rappaport, T.S., Roh, W., Cheun, K.: Mobile's millimeter-wave makeover. IEEE Spectrum **51**(9), 34–58 (2014)
7. Sulyman, A.I., Nassar, A.T., Samimi, M.K., Maccartney, G.R., Rappaport, T.S., Alsanie, A.: Radio propagation path loss models for 5G cellular networks in the 28 GHZ and 38 GHZ millimeter-wave bands. IEEE Communications Magazine **52**(9), 78–86 (2014)
8. Rappaport, T.S., Heath Jr., R.W., Daniels, R.C., Murdock, J.N.: Millimeter Wave Wireless Communications. Pearson Education (2015)
9. Hu, Y., Ji, B., Huang, Y., Yu, F., Yang, L.: Energy-Efficient Resource Allocation in Uplink Multiuser Massive MIMO Systems. International Journal of Antennas and Propagation (2014)

10. Yang, Y.-H., Lin, S.-C., Su, H.-J.: Multiuser MIMO downlink beamforming design based on group maximum SINR filtering. IEEE Transactions on Signal Processing **59**(4), 1746–1758 (2011)
11. Xiang, G., Edfors, O., Rusek, F., Tufvesson, F.: Linear pre-coding performance in measured very-large MIMO channels. In: 2011 IEEE Presented at Vehicular Technology Conference (VTC Fall) (2011)
12. Han, Y., Jin, S., Li, X., Huang, Y.: A joint SDMA and interference suppression multiuser transmission scheme for millimeter-wave massive MIMO systems. In: 2014 Sixth International Conference on Presented at Wireless Communications and Signal Processing (WCSP) (2014)
13. Sulyman, A.I., Hefnawi, M.: Adaptive MIMO Beamforming Algorithm Based on Gradient Search of the Channel Capacity in OFDM-SDMA Systems. IEEE Communications Letters **12**(9), 642–644 (2008)
14. Swindlehurst, A.L., Ayanoglu, E., Heydari, P., Capolino, F.: Millimeter-wave massive MIMO: the next wireless revolution? IEEE Communications Magazine **52**(9), 56–62 (2014)
15. Larsson, E., Edfors, O., Tufvesson, F., Marzetta, T.: Massive MIMO for next generation wireless systems. IEEE Communications Magazine **52**(2), 186–195 (2014)
16. Hoydis, J., ten Brink, S., Debbah, M.: Massive MIMO in the UL/DL of Cellular Networks: How Many Antennas Do We Need? IEEE Journal on Selected Areas in Communications **31**(2), 160–171 (2013)
17. Hoydis, J., ten Brink, S., Debbah, M.: Massive MIMO: How many antennas do we need? In: 2011 49th Annual Allerton Conference on Presented at Communication, Control, and Computing (Allerton) (2011)
18. Sun, S., Rappaport, T.S., Heath, R., Nix, A., Rangan, S.: MIMO for millimeter-wave wireless communications: beamforming, spatial multiplexing, or both? IEEE Communications Magazine **52**(12), 110–121 (2014)
19. Friis, H.T.: A note on a simple transmission formula. Proc. IRE **34**(5), 254–256 (1946)
20. Bogale, T.E., Le, L.B.: Pilot optimization and channel estimation for multiuser massive MIMO systems. In: 48th Annual Conference on Information Sciences and Systems (CISS), pp. 1–6 (2014)
21. Peel, C.B., Hochwald, B.M., Swindlehurst, A.L.: A vector-perturbation technique for near-capacity multiantenna multiuser communication-part I: channel inversion and regularization. IEEE Transactions on Communications **53**(1), 195–202 (2005)
22. Trees, H.V., Harry, L.: Optimum array processing, Part IV of detection, Estimation, and Modulation Theory. Wiley-Interscience, New York (2002)

Detection of Temporally Correlated Primary User Signal with Multiple Antennas

Hadi Hashemi[1], Sina Mohammadi Fard[1], Abbas Taherpour[1], Saeid Sedighi[1], and Tamer Khattab[2]([✉])

[1] Department of Electrical Engineering, Imam Khomeini International University, Qazvin, Iran
[2] Electrical Engineering, Qatar University, Doha, Qatar
tkhattab@ieee.org

Abstract. In this paper, we address the problem of multiple antenna spectrum sensing in cognitive radios (CRs) when the samples of the primary user (PU) signal as well as samples of noise are assumed to be temporally correlated. We model and formulate this multiple antenna spectrum sensing problem as a hypothesis testing problem. First, we derive the optimum Neyman-Pearson (NP) detector for the scenario in which the channel gains, the PU signal and noise correlation matrices are assumed to be known. Then, we derive the sub-optimum generalized likelihood ratio test (GLRT)-based detector for the case when the channel gains and aforementioned matrices are assumed to be unknown. Approximate analytical expressions for the false-alarm probabilities of the proposed detectors are given. Simulation results show that the proposed detectors outperform some recently-purposed algorithms for multiple antenna spectrum sensing.

1 Introduction

Using multiple antennas at the secondary user (SU) receiver is one efficient approach to overcome deleterious effects of noise uncertainty, fading and shadowing. Moreover, it improves the performance of spectrum sensing by exploiting available observations in the spatial domain [1–10]. [1] considers a blind spectrum sensing approach where the empirical characteristic function of the multiantenna samples is used in the formulation of the statistical test. In [2], the authors derive the optimum NP and sub-optimum GLRT-based multiantenna detectors of an orthogonal frequency division multiplexing (OFDM) signal with a cyclic prefix of known length. In [3–7], GLRT eigenvalues-based detectors of spatial rank-one PU signals robust to noise variance uncertainty are derived. In addition, some GLRT eigenvalue-based detectors for multiantenna spectrum sensing are proposed in [8,9], for PU signals with spatial rank larger than one. In CR networks, signals from far PUs arrive at the SU base station within a small beamwidth, which results in a high correlation between the channel gains of different antennas. The Roa test is applied to derive sub-optimum multiantenna detectors under the correlated receiving antennas model in [10].

© Institute for Computer Sciences, Social Informatics and Telecommunications Engineering 2015
M. Weichold et al. (Eds.): CROWNCOM 2015, LNICST 156, pp. 66–77, 2015.
DOI: 10.1007/978-3-319-24540-9_6

All the detectors proposed in [1–10] do not consider any temporal correlation between the samples of the received signal. Nevertheless, in practice, the PU signal samples as well as noise samples may be temporally correlated, which causes degradation in the performance of detectors proposed in [1–10]. In [11], the detection of temporally correlated signals over multipath fading channels is discussed and a modified energy detector (ED) is proposed for such a scenario. However, as known, the performance of the ED is susceptible to errors in the noise variance and it has been shown that in order for ED to achieve a desired probability of detection under noise (or in more general terms, under model uncertainties) the signal-to-noise ratio (SNR) must be above a certain threshold [12] (SNR wall).

In this paper, we consider multiple antenna spectrum sensing when there are temporal correlation between the PU signal samples in the presence of additive temporally correlated Gaussian noise. First, for benchmarking purposes, we obtain the optimum NP detector for the case when the SU receiver has complete knowledge about the channel gains, the PU signal and noise covariance matrices. Then, we derive the sub-optimum GLRT-based detector when the channel gains, the PU signal and noise covariance matrices are assumed to be unknown to the SU receiver. Approximate analytical expressions for the false-alarm probabilities of proposed detectors are also given. The simulation results are provided to evaluate the impact of the different parameters on the performance of the proposed detectors and, moreover, to compare the performance of the proposed detectors with some recently-proposed detectors.

The rest of the paper is organized as follows. In Section 2, we describe the system model and the basic assumptions about the PU signal and noise. In Section 3, we derived the optimum NP detector. The sub-optimum GLRT-based detector is obtained in Section 4. Asymptotic expressions for the false-alarm probabilities of the proposed detectors are evaluated in Sections 5. The simulation results and related discussions are given in Section 6. Finally, Section 7 concludes the paper.

Notation: Lightface letters denote scalars. Boldface lower-and upper-case letters denote column vectors and matrices, respectively. x[.] are the entries and $\mathbf{x}_i$ is sub-vector of vector $\mathbf{x}$. The determinant and inverse of matrix $\mathbf{A}$ are $|\mathbf{A}|$ and $\mathbf{A}^{-1}$, respectively. vec[$\mathbf{A}$] is the column-wise vectorization of matrix $\mathbf{A}$. The $M \times M$ identity matrix is $\mathbf{I}_M$ and the $M \times M$ zeros matrix is $\mathbf{0}_M$. Superscripts $*$, T and H are the complex conjugate, transpose and Hermitian (conjugate transpose) operations, respectively. $\mathbb{E}[\cdot]$ is the statistical expectation. $\mathbf{A} \otimes \mathbf{B}$ is kronecker product of matrices $\mathbf{A}$ and $\mathbf{B}$. $\mathcal{CN}(\mathbf{m}, \mathbf{P})$ denotes circularly symmetric complex Gaussian distribution with mean $\mathbf{m}$ and covariance matrix $\mathbf{P}$. $Q(x)$ is Q-function $Q(x) = \frac{1}{\sqrt{2\pi}} \int_x^{\infty} exp\left(\frac{-u^2}{2}\right) du$.

2   System Model

We consider a CR network including a single-antenna PU and a mutiantenna SU with M receiving antennas. We assume the received signal is downconverted to baseband and sampled at the frequency $f_s = \frac{1}{T_s}$ at each antenna to generate N consecutive time blocks, each of which contains L consecutive temporal samples. Define $\mathbf{y}_{i,j} \in \mathbb{C}^M$ as the vector of the received signal samples from M different antennas of the ith time block at the jth time instant. The observation vector $\mathbf{y}_{i,j}$ is given as

$$\mathbf{y}_{i,j} = \mathbf{h}_i s_j + \mathbf{n}_{i,j}, \ i = 1, \ldots, N; \ j = 1, \ldots, L \tag{1}$$

where $\mathbf{h}_i \in \mathbb{C}^M$ is the baseband equivalent of channel gains vector at ith time block, which is assumed to be constant during each time block. $s_j \in \mathbb{C}$ denotes the baseband samples of the PU signal, which is assumed to be distributed as a zero-mean complex Gaussian random variable with autocorrelation function $r_s[k] = \mathbb{E}[s_j s_{j-k}^*]$. We assume, in general, s_j exhibits temporal correlation, i.e. $r_s[k] \neq 0$ for $k \neq 0$. $\mathbf{n}_{i,j} \in \mathbb{C}^M$ denotes the baseband equivalent of noise samples which is assumed to be distributed as a zero-mean complex Gaussian random vector. In addition, $\mathbf{n}_{i,j}$ is assumed to be spatially uncorrelated but temporally correlated with autocorrelation function $r_n[k]$. We assume noise and the PU signal samples are mutually independent.

Let us define the matrix $\mathbf{Y}_i \doteq [\mathbf{y}_{i,1}, \ldots, \mathbf{y}_{i,L}] \in \mathbb{C}^{M \times L}$ containing L time samples of the ith time block. In addition, let us define $\mathbf{y}_i = \mathrm{vec}[\mathbf{Y}_i] \in \mathbb{C}^{LM \times 1}$ and $\mathbf{y} = \mathrm{vec}[\mathbf{y}_1, \ldots, \mathbf{y}_N] \in \mathbb{C}^{NLM \times 1}$. We denote the hypotheses of the presence and absence of the PU signal by $\mathcal{H}_1$ and $\mathcal{H}_0$, respectively. By defining the correlation matrix of $\mathbf{y}$: $\boldsymbol{\Sigma}_\nu = \mathbb{E}\{\mathbf{y}\mathbf{y}^H | \mathcal{H}_\nu\}$, $\nu = 0, 1$, it can be easily shown that,

$$\boldsymbol{\Sigma}_0 = \mathbf{I}_N \otimes \mathbf{I}_M \otimes \boldsymbol{\Sigma}_n = \mathbf{I}_{NM} \otimes \boldsymbol{\Sigma}_n, \tag{2}$$

where

$$\boldsymbol{\Sigma}_n \doteq \begin{pmatrix} r_n[0] & r_n^*[1] & \ldots r_n^*[L-1] \\ r_n[1] & r_n[0] & \ldots r_n^*[L-2] \\ \vdots & \vdots & \ddots & \vdots \\ r_n[L-1] & r_n[L-2] & \ldots & r_n[0] \end{pmatrix}, \tag{3}$$

is the temporal correlation matrix of noise, and

$$\boldsymbol{\Sigma}_1 \doteq \begin{pmatrix} \boldsymbol{\Sigma}_{1,1} & 0 & \ldots & 0 \\ 0 & \boldsymbol{\Sigma}_{2,2} & \ldots & 0 \\ \vdots & \vdots & \ddots & \vdots \\ 0 & \ldots & \ldots & \boldsymbol{\Sigma}_{N,N} \end{pmatrix}, \tag{4}$$

where

$$\boldsymbol{\Sigma}_{i,i} = \mathbf{h}_i \mathbf{h}_i^H \otimes \boldsymbol{\Sigma}_s + \mathbf{I}_M \otimes \boldsymbol{\Sigma}_n \tag{5}$$

with

$$\Sigma_s \doteq \begin{pmatrix} r_s[0] & r_s^*[1] & \cdots & r_s^*[L-] \\ r_s[1] & r_s[0] & \cdots r_s^*[L-2] \\ \vdots & \vdots & \ddots & \vdots \\ r_s[L-1] & r_s[L-2] & \cdots & r_s[0] \end{pmatrix}. \tag{6}$$

Accordingly, the distribution of observations under each hypothesis is given as

$$\begin{cases} \mathcal{H}_0 : & y \sim \mathcal{CN}(0_{NLM}, \Sigma_0) \\ \mathcal{H}_1 : & y \sim \mathcal{CN}(0_{NLM}, \Sigma_1). \end{cases} \tag{7}$$

3   Optimum Detector

In this section, we derive the NP detector for the case of completely known h_i, Σ_n and Σ_s. From (7) the probability density function (PDF) of the observations vector y under each hypothesis is given by,

$$\begin{aligned} f(y|\mathcal{H}_0, \Sigma_0) &= \frac{\exp\left\{-y^H \Sigma_0^{-1} y\right\}}{\pi^{NLM}|\Sigma_0|} \\ &= \frac{\exp\left\{-L\text{tr}(\Sigma_n^{-1} \sum_{i=1}^{N} \sum_{m=1}^{M} R_{i,mm})\right\}}{\pi^{NLM}|\Sigma_n|^{NM}}. \end{aligned} \tag{8}$$

and

$$\begin{aligned} f(y|\mathcal{H}_1, \Sigma_1) &= \frac{\exp\left\{-y^H \Sigma_1^{-1} y\right\}}{\pi^{NLM}|\Sigma_1|} \\ &= \frac{\exp\left\{-L\text{tr}(\sum_{i=1}^{N} R_i \Sigma_{i,i}^{-1})\right\}}{\pi^{NLM} \prod_{i=1}^{N} |\Sigma_{i,i}|}, \end{aligned} \tag{9}$$

where

$$R_i \doteq \frac{1}{L} y_i y_i^H = \begin{pmatrix} R_{i,11} & R_{i,12} & \cdots & R_{i,1M} \\ R_{i,21} & R_{i,22} & \cdots & R_{i,2M} \\ \vdots & \vdots & \ddots & \vdots \\ R_{i,M1} & \cdots & \cdots & R_{i,MM} \end{pmatrix}. \tag{10}$$

in which $R_{i,mk}$'s are the corresponding sub-matrices.

Taking logarithm from (8) and (9) and defining $\mathcal{L}_\nu(\mathbf{y}) \doteq \ln f(\mathbf{y}|\mathcal{H}_\nu, \boldsymbol{\Sigma}_\nu)$, $\nu = 0, 1$, we obtain,

$$\mathcal{L}_0(\mathbf{y}) = -L\mathrm{tr}(\boldsymbol{\Sigma}_n^{-1} \sum_{i=1}^{N} \sum_{m=1}^{M} \mathbf{R}_{i,mm}) - NLM \ln \pi - NM \ln |\boldsymbol{\Sigma}_n|, \qquad (11)$$

$$\mathcal{L}_1(\mathbf{y}) = -L\mathrm{tr}(\sum_{i=1}^{N} \mathbf{R}_i \boldsymbol{\Sigma}_{i,i}^{-1}) - NLM \ln \pi - \sum_{i=1}^{N} \ln |\boldsymbol{\Sigma}_{i,i}|. \qquad (12)$$

By constituting the logarithm of likelihood ratio (LLR) function from (11) and (12) and comparing it with a threshold, the optimum detector is given by

$$\mathrm{LLR} = L\mathrm{tr}(\sum_{i=1}^{N} \mathbf{R}_i[(\mathbf{I}_M \otimes \boldsymbol{\Sigma}_n)^{-1} - \boldsymbol{\Sigma}_{i,i}^{-1}]) \gtrless_{H_0}^{H_1} \tau', \qquad (13)$$

Now by using matrix inversion lemma, we have,

$$(\mathbf{h}_i \mathbf{h}_i^H \otimes \boldsymbol{\Sigma}_s + \mathbf{I}_M \otimes \boldsymbol{\Sigma}_n)^{-1} \qquad (14)$$
$$= (\mathbf{I}_M \otimes \boldsymbol{\Sigma}_n)^{-1} - (\mathbf{I}_M \otimes \boldsymbol{\Sigma}_n)^{-1}(\mathbf{h}_i \mathbf{h}_i^H \otimes \boldsymbol{\Sigma}_s)(\mathbf{h}_i \mathbf{h}_i^H \otimes \boldsymbol{\Sigma}_s + \mathbf{I}_M \otimes \boldsymbol{\Sigma}_n)^{-1}$$

Therefore, by substituting (14) in (13), we find,

$$T_{\mathrm{opt}} = L\mathrm{tr}(\sum_{i=1}^{N} \mathbf{R}_i \mathbf{C}_i^{-1}) \gtrless_{H_0}^{H_1} \tau, \qquad (15)$$

where $\mathbf{C}_i \doteq (\mathbf{h}_i \mathbf{h}_i^H \otimes \boldsymbol{\Sigma}_s + \mathbf{I}_M \otimes \boldsymbol{\Sigma}_n)(\mathbf{h}_i \mathbf{h}_i^H \otimes \boldsymbol{\Sigma}_s)^{-1}(\mathbf{I}_M \otimes \boldsymbol{\Sigma}_n)$.

In order to simplify the optimum detector more, we can use the singular value decomposition (SVD) of $\mathbf{C}_i$: $\mathbf{C}_i = \mathbf{U}\boldsymbol{\Lambda}_i\mathbf{U}^H$, where $\boldsymbol{\Lambda}_i = \mathrm{diag}\{\lambda_{i,1}, \lambda_{i,2}, ..., \lambda_{i,LM}\}$ contains eigenvalues of $\mathbf{C}_i$ and the columns of $\mathbf{U}$ its the corresponding eigenvectors. Therefore,

$$T_{\mathrm{opt}} = \mathrm{tr}(\sum_{i=1}^{N} \mathbf{y}_i \mathbf{y}_i^H \mathbf{U}\boldsymbol{\Lambda}_i^{-1}\mathbf{U}^H) = \sum_{i=1}^{N} \mathrm{tr}(\boldsymbol{\Lambda}_i^{-1}\mathbf{U}^H \mathbf{y}_i \mathbf{y}_i^H \mathbf{U}) \gtrless_{H_0}^{H_1} \tau. \qquad (16)$$

Let $\mathbf{w}_i \doteq \mathbf{U}^H \mathbf{y}_i$ and $w_i(n)$ be the nth elements of $\mathbf{w}_i$, then the optimum detector can be written as

$$T_{\mathrm{opt}} = \sum_{i=1}^{N} \sum_{n=1}^{LM} \lambda_{i,n}^{-1}|w_i(n)|^2 \gtrless_{H_0}^{H_1} \tau. \qquad (17)$$

4   Sub-Optimum GLRT-Based Detector

In this section, we assume the channel gains, the PU signal and noise covariance matrices are unknown to the SU receiver. We derive the GLRT-based detector for such a scenario. We first maximize (9) with respect to Σ_1 in order to compute the maximum likelihood estimate (MLE) of Σ_1. By setting the derivative of (9) with respect to Σ_1 equal to zero, we obtain

$$L(\Sigma_{i,i}^{-1}\mathbf{R}_i\Sigma_{i,i}^{-1})^T = (\Sigma_{i,i}^{-1})^T, \tag{18}$$

which results to $\hat{\Sigma}_{i,i} = L\mathbf{R}_i$.

In addition, in order to compute the MLE of Σ_n, we should take derivative of (8) with respect to Σ_n and set it equal to zero, which yields $\hat{\Sigma}_n = \frac{L}{NM}\sum_{i=1}^{N}\sum_{m=1}^{M}\mathbf{R}_{i,mm}$.

By constituting the LR function, the GLRT-based detector is obtained as,

$$T_{\text{sub}} = \frac{f(\mathbf{y}|\mathcal{H}_1,\hat{\Sigma}_1)}{f(\mathbf{y}|\mathcal{H}_0,\hat{\Sigma}_0)} = \frac{|\hat{\Sigma}_n|^{NM}}{\prod_{i=1}^{N}|\hat{\Sigma}_{i,i}|} = \frac{1}{(NM)^{NML}}\prod_{i=1}^{N}\frac{|\sum_{i=1}^{N}\sum_{m=1}^{M}\mathbf{R}_{i,mm}|^M}{|\mathbf{R}_i|}. \tag{19}$$

5 Analytical Performance

In the following section, we evaluate the performance of the proposed optimum and sub-optimum detectors in terms of the detection and false-alarm probabilities, i.e. P_d and P_{fa}, respectively.

5.1 Performance of the Optimum Detector

Performance of the optimum detector is evaluated in this sub-section. We can rewrite (17) in the null hypothesis as,

$$T_{\text{opt}|\mathcal{H}_0} = \sum_{i=1}^{N}\sum_{n=1}^{LM}\lambda_{i,n}^{-1}|w_i(n)|^2 = \sum_{i=1}^{N}\lambda_{i,1}^{-1}\lambda_{0_{i,1}}\frac{|w_i(1)|^2}{\lambda_{0_{i,1}}} + \dots$$

$$+ \lambda_{i,LM}^{-1}\lambda_{0_{i,LM}}\frac{|w_i(LM)|^2}{\lambda_{0_{i,LM}}}$$

$$= \sum_{i=1}^{N}\sum_{n=1}^{LM}\lambda_{i,n}^{-1}\lambda_{0_{i,n}}z_i(n), \tag{20}$$

where $\lambda_{0_{i,n}}$'s are eigenvalues of ith time block covariance matrix and $w_i(n)$ is a zero-mean Gaussian random variable with variance $\lambda_{0_{i,n}}$. Thus, from [13], $z_i(n) = \frac{|w_i(n)|^2}{\lambda_{0_{i,n}}}$ has the chi-squared distribution with one degree of freedom.

From central limit theorem (CLT), with NLM sufficiently large and, also, from [13], the distribution of the optimum detector under the null hypothesis is Gaussian. Hence, for evaluating performance of the optimum detector, we should compute its mean and variance as,

$$\mu_{T_{\text{opt}}|\mathcal{H}_0} = \sum_{i=1}^{N}\sum_{n=1}^{LM}\lambda_{i,n}^{-1}\lambda_{0_{i,n}}\mathbb{E}[z_i(n)|\mathcal{H}_0] = \sum_{i=1}^{N}\sum_{n=1}^{LM}\lambda_{i,n}^{-1}\lambda_{0_{i,n}}, \tag{21}$$

and

$$\sigma^2_{\mathrm{T_{opt}}|\mathcal{H}_0} = \sum_{i=1}^{N} \sum_{n=1}^{LM} \lambda_{i,n}^{-1} \lambda_{0_{i,n}} Var[z_i(n)|\mathcal{H}_0] = 2 \sum_{i=1}^{N} \sum_{n=1}^{LM} \lambda_{i,n}^{-1} \lambda_{0_{i,n}}. \qquad (22)$$

Similarly, under $\mathcal{H}_1$, $\mathrm{T_{opt}}$ has a Gaussian distribution with mean and variance as,

$$\mu_{\mathrm{T_{opt}}|\mathcal{H}_1} = \sum_{i=1}^{N} \sum_{n=1}^{LM} \lambda_{i,n}^{-1} \lambda_{1_{i,n}} \mathbb{E}[z_i(n)|\mathcal{H}_1] = \sum_{i=1}^{N} \sum_{n=1}^{LM} \lambda_{i,n}^{-1} \lambda_{1_{i,n}}, \qquad (23)$$

and

$$\sigma^2_{\mathrm{T_{opt}}|\mathcal{H}_1} = \sum_{i=1}^{N} \sum_{n=1}^{LM} \lambda_{i,n}^{-1} \lambda_{1_{i,n}} Var[z_i(n)|\mathcal{H}_1] = 2 \sum_{i=1}^{N} \sum_{n=1}^{LM} \lambda_{i,n}^{-1} \lambda_{1_{i,n}}, \qquad (24)$$

where $\lambda_{1_{i,n}}$'s are eigenvalues of ith time block covariance matrix under $\mathcal{H}_1$.

Thus, the false-alarm and detection probabilities can be calculated as,

$$P_{\mathrm{fa}} = P\{\mathrm{T_{opt}} > \tau | \mathcal{H}_0\} = Q\left(\frac{\tau - \mu_{\mathrm{T_{opt}}|\mathcal{H}_0}}{\sigma_{\mathrm{T_{opt}}|\mathcal{H}_0}} \right), \qquad (25)$$

$$P_d = P\{\mathrm{T_{opt}} > \tau | \mathcal{H}_1\} = Q\left(\frac{\tau - \mu_{\mathrm{T_{opt}}|\mathcal{H}_1}}{\sigma_{\mathrm{T_{opt}}|\mathcal{H}_1}} \right). \qquad (26)$$

5.2   Performance of the Sub-Optimum Detector

In this sub-section, we derive the asymptotic distribution of the proposed detectors under $\mathcal{H}_0$ by using the results existing for the asymptotic distribution of the GLRT.

Lemma 1. *Let $\boldsymbol{\Theta} = [\boldsymbol{\mu}_r, \boldsymbol{\mu}_s]^T$, with $\boldsymbol{\mu}_r \in \mathbb{R}^r$ and $\boldsymbol{\mu}_s \in \mathbb{R}^s$, be the set of unknown parameters under $\mathcal{H}_1$ and $\mathcal{H}_0$. For a composite hypothesis test of the form,*

$$\begin{cases} \mathcal{H}_0 : \boldsymbol{\mu}_r = \boldsymbol{\mu}_{r_0}, \boldsymbol{\mu}_s \\ \mathcal{H}_1 : \boldsymbol{\mu}_r \neq \boldsymbol{\mu}_{r_0}, \boldsymbol{\mu}_s \end{cases}, \qquad (27)$$

the asymptotic distribution of GLRT statistic, T_{GLRT}, under $\mathcal{H}_0$, as $N \to \infty$, is as

$$2 \ln T_{GLR} \sim \chi_r^2, \qquad (28)$$

where χ_n^2 denotes the central chi-squared distribution with n degrees of freedom.

Proof. See [14]

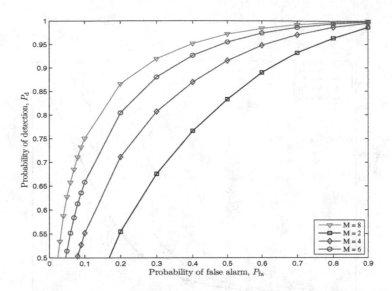

Fig. 1. The complementary ROC of the proposed GLRT-based detector for $SNR = -8$, $P_{fa} = 0.01$, $N = 32$, $L = 10$ and different value of M.

According to Lemma 1, as $N \to \infty$, we have

$$2 \ln T_{\text{sub}} \sim \chi_f^2, \tag{29}$$

where $f = NL^2(M^2 - 1)$. Thus, the false-alarm probability of Λ_{GLR1} can be obtained as

$$P_{\text{fa}} = \mathbb{P}[T_{\text{sub}} > \eta | \mathcal{H}_0] = \mathbb{P}[2 \ln T_{\text{sub}} > 2 \ln \eta | \mathcal{H}_0] = \frac{\gamma\left(\frac{1}{2}NL^2(M^2 - 1), \ln \eta\right)}{\Gamma\left(\frac{1}{2}NL^2(M^2 - 1)\right)}, \tag{30}$$

where $\gamma(k, z) \doteq \int_0^z t^{k-1} e^{-t} dt$ is the lower incomplete Gamma function.

6 Simulation Results

In this section, we provide simulations in order to evaluate the impact of the different parameters on the performance of the proposed detectors and, moreover, to compare the performance of the proposed detectors with other previously reported detectors used as a benchmark. Specifically, the benchmark detectors are: the AGM method [8, Eqn.(14)], the maximum eigenvalue to trace (MET) detector [3, Eqn.(39)], and maximum to minimum eigenvalue (MME) detector [15, Algorithm1].

The complementary ROC (receiver operating characteristics) of the proposed GLRT-based detector for $SNR = -8dB$, $P_{fa} = 0.01$, $N = 32$, $L = 10$ and

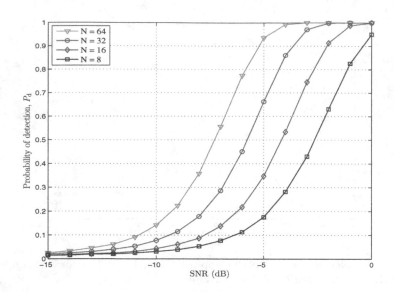

Fig. 2. The detection probability of the proposed GLRT-based detector versus SNR for $P_{fa} = 0.01$ and $M = 4$, $L = 10$ and different value of N.

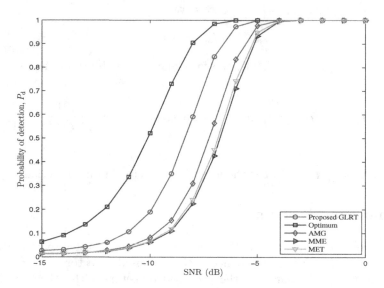

Fig. 3. The detection probability of different detectors versus SNR for $P_{fa} = 0.01$, $L = 10$, $N = 32$ and $M = 4$.

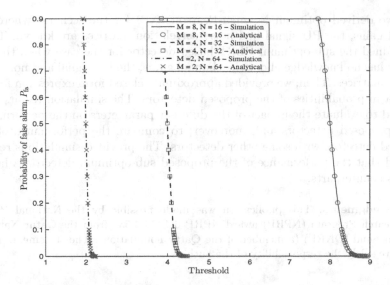

Fig. 4. The false-alarm probability versus threshold of the proposed GLRT-based detectors for $L = 10$ and different value of L and M.

different value of M is shown in Fig. 1. This figure shows that increase in the number of antennas is associated with improvement in performance, but the improvement declines when the number of antennas becomes larger.

Fig. 2 depicts the detection probability of the proposed GLRT-based detector versus SNR for $P_{fa} = 0.01$ and $M = 4$, $L = 10$ and different value of N. As expected, performance of the proposed GLRT-based detector improves by increasing the number of time blocks. Fig. 3 compares the performance of optimum NP detector and the proposed GLRT-based detectors with some other previously reported detectors used as a benchmark. Fig. 3 depicts detection probability of different detectors versus SNR for for $P_{fa} = 0.01$, $L = 10$, $N = 32$ and $M = 4$. As it can be seen, optimum detector has the best performance among all detectors and after that the proposed GLRT-based detector outperforms AGM, MET and MME. In addition, the proposed GLRT-based detector does not require to compute the eigenvalue of the sample covariance matrix in contrast to MET and MME. Hence, the proposed GLRT-based detector has lower computational complexity compare to MET and MME.

Finally, the validity of the approximate closed-form expression provided for proposed GLRT-based detectors is verified in Fig. 4. Fig. 4 shows that there is a good agreement between simulations and the approximate closed-form expression for different value of M and N.

7   Conclusion

In this paper, we investigated the multiple antenna spectrum sensing problem in CR networks when there are temporal correlation between received samples.

First, we derived optimum Neyman-Pearson (NP) for the scenario where the channel gains, the PU signal and noise correlation matrices are known. Then, we obtained the sub-optimum GLRT-based detector for the case when the PU receiver has no knowledge about the channel gains, the PU signal and noise correlation matrices. Then, we provided approximate closed-form expression for the false-alarm probabilities of the proposed detectors. The simulation results were provided to evaluate the impact of the different parameters on the performance of the proposed detectors and, moreover, to compare the performance of the proposed detectors with some other detectors. The provided simulations results revealed that the performance of the proposed sub-optimum detector is better than its counterparts.

Acknowledgments. This publication was made possible by the National Priorities Research Program (NPRP) award NPRP 6-1326-2-532 from the Qatar National Research Fund (QNRF) (a member of the Qatar Foundation). The statements made herein are solely the responsibility of the authors.

References

1. Shen, L., Wang, H., Zhang, W., Zhao, Z.: Multiple antennas assisted blind spectrum sensing in cognitive radio channels. IEEE Commun. Lett. **16**(1), 92–94 (2012)
2. Axell, E., Larsson, E.: Optimal and sub-optimal spectrum sensing of OFDM signals in known and unknown noise variance. IEEE J. Sel. Areas Commun. **29**(2), 290–304 (2011)
3. Taherpour, A., Nasiri-Kenari, M., Gazor, S.: Multiple antenna spectrum sensing in cognitive radios. IEEE Trans. Wireless Commun. **9**(2), 814–823 (2010)
4. Sedighi, S., Taherpour, A., Monfared, S.S.: Bayesian generalised likelihood ratio test-based multiple antenna spectrum sensing for cognitive radios. IET Communications **7**(18), 2151–2165 (2013)
5. Lopez-Valcarce, J.S.R., Vazquez-Vilar, G.: Multiantenna spectrum sensing for cognitive radio: overcoming noise uncertainty, pp. 310–315 (2010)
6. Bianchi, P., Debbah, M., Maïda, M., Najim, J.: Performance of statistical tests for single-source detection using random matrix theory. IEEE Trans. Inf. Theory **57**(4), 2400–2419 (2011)
7. Wang, P., Fang, J., Han, N., Li, H.: Multiantenna-assisted spectrum sensing for cognitive radio. IEEE Trans. Veh. Technol. **59**(4), 1791–1800 (2010)
8. Zhang, R., Lim, T., Liang, Y., Zeng, Y.: Multi-antenna based spectrum sensing for cognitive radios: A GLRT approach. IEEE Trans. Commun. **58**(1), 84–88 (2010)
9. Ramírcz, D., Vazquez-Vilar, G., López-Valcarce, R., Vía, J., Santamaría, I.: Detection of rank-p signals in cognitive radio networks with uncalibrated multiple antennas. IEEE Trans. Signal Process. **59**(8), 3764–3774 (2011)
10. Sedighi, S., Taherpour, A., Sala, J.: Spectrum sensing using correlated receiving multiple antennas in cognitive radios **12**(11) (November 2013)
11. Huang, Y., Huang, X.: Detection of temporally correlated signals over multipath fading channels. IEEE Trans. Wireless Commun. **12**(3), 1–10 (2013)

12. Tandra, R., Sahai, A.: SNR walls for signal detection. IEEE J. Sel. Topics Signal Process. **2**(1), 4–17 (2008)
13. Papoulis, A.: Probability, random variables and stochastic processes. McGraw-hill, New York (2002)
14. Kay, S.M.: Fundamentals of statistical signal processing, Volume II: Detection theory. Prentice Hall, Englewood Cliffs (1998)
15. Zeng, Y., Liang, Y.-C.: Maximum-minimum eigenvalue detection for cognitive radio, vol. 5, September 2007

Non-uniform Quantized Distributed Sensing in Practical Wireless Rayleigh Fading Channel

Sina Mohammadi Fard[1], Hadi Hashemi[1], Abbas Taherpour[1],
and Tamer Khattab[2]([✉])

[1] Department of Electrical Engineering, Imam Khomeini International University,
Qazvin, Iran
[2] Electrical Engineering, Qatar University, Doha, Qatar
tkhattab@ieee.org

Abstract. In this paper, we study non-uniform multilevel quantization problem in cognitive radio networks (CRNs). We consider a practical collaborative spectrum sensing (CSS) scenario in which secondary users (SUs) cooperate with each other to decide about the presence of the primary user (PU). We consider a cooperative parallel access channel (CPAC) scheme in reporting channels in which SUs transmit their quantized data to fusion center (FC) for the final decision. Also, we evaluate the final summation-based decision statistic and Kullback-Leibler (KL) divergence performance criterion in the Rayleigh fading channel and additive Gaussian noise. We compare the non-uniform quantization scheme performance with the uniform one and illustrate the sensitivity of the provided quantization scheme to average error probability of symbols. Furthermore, the effect of the collaboration in the CPAC scheme on performance of the distributed sensing compared with non-cooperative scheme is investigated.

Keywords: Collaborative spectrum sensing · Non-uniform quantization · Rayleigh fading channel

1 Introduction

Cooperative spectrum sensing (CSS) is one of the ways to improve the performance of spectrum sensing algorithms in the shadowing and channel fading situations [1–6]. In [1] the impact of sensors collaboration to achieve optimal performance is studied. Authors in[2], investigate a CSS problem in which the energy sensed by each SU is transmitted to the others and each SU attempts to decide based on its own information and received signals. In [3–5], cooperation in CRNs with assumption of independent channels is considered. The authors in [3] and [4] assume the same distribution for all users, while [5] assumes different distributions which is more realistic for practical scenarios. In [6], the authors propose a method which uses spatial diversity to deal with the devastating effects of a fading channel. However, due to the increasing wireless network users, the CSS algorithms are often faced with the problem of limited bandwidth. Thus, information received by the SUs must be properly quantized before transmitting

© Institute for Computer Sciences, Social Informatics and Telecommunications Engineering 2015
M. Weichold et al. (Eds.): CROWNCOM 2015, LNICST 156, pp. 78–91, 2015.
DOI: 10.1007/978-3-319-24540-9_7

to FC in order to occupy less bandwidth, while maintaining the accuracy of the observations [7–15]. In [7], the M-level quantization problem for a distributed detection system is investigated by assuming interfering nodes and Byzantine attacks. In [8] the problem of SUs binary decisions fusion in Rayleigh channels is studied. Also in [9], authors analyze quantizer design which is robust to link outages and/or sensor failure in a multi-user system. In [10], comparison of the performance of single-user and collaborative multi-user quantized system has been studied. In [11] and [12], authors review the performance of relay based collaborative distributed detection system for the quantize and forward scheme. The same problem with the assumption of orthogonal multiple access channels is considered in [12]. In [13], the authors provide a quantization system with multiple non-uniform threshold levels, while the impact of channel errors on the performance of the quantized detection system has been examined in [14]. In [15], a simulation-based investigation of energy quantization effect on proposed detector performance is provided.

In this paper, we investigate a multilevel quantization problem in a practical collaborative distributed detection system. We review an M-level non-uniform quantization procedure. In order to make the final decision, the SUs transmit their quantized data to FC in CPAC protocol. We assume a practical wireless report channel with Rayleigh fading and analytically evaluate the corresponding performances. A remarkable point of this paper is the comprehensive study of the SUs cooperation impact on practical wireless networks by deploying a non-uniform quantization technique. The rest of the paper is organized as follows. In Section 2, we introduce the system model and assumptions on SUs detector and applied transmission protocol. In Section 3, we provide boundaries calculations of uniform and non-uniform multilevel quantization scheme which we have used. Derivation and evaluation of final decision performance in FC and KL divergence performance criterion are investigated in Section 4. Simulation results are provided in Section 5 and finally, Section 6 summarizes the conclusions.

Notation: Lightface letters denote scalars. Boldface lower-and upper-case letters denote column vectors and matrices, respectively. x(.) is the entries and $\mathbf{x}_i$ is sub-vector of vector $\mathbf{x}$ and $[\mathbf{A}]_{.,.}$ is the entries of matrix $\mathbf{A}$. $\mathcal{N}(\mu, \sigma^2)$ denotes Gaussian distribution with mean μ and variance σ^2. Superscript T is transpose and $Q(x)$ is Q-function $Q(x) = \frac{1}{\sqrt{2\pi}} \int_x^\infty exp\left(\frac{-u^2}{2}\right) du$.

2   System Model

Suppose a CSS system in CRNs which K SUs detect presence or absence of PU signals in certain frequency range. We assume each SU has been equipped with a single antenna. Thus, the final goal of this spectrum sensing system is decision between the two following hypotheses,

$$y_i(t) = \begin{cases} w_i(t) & \mathcal{H}_0 \\ s(t) + w_i(t) & \mathcal{H}_1 \end{cases}, i = 1, 2, .., K, \tag{1}$$

where $y_i(t)$ is observed signal in ith SU, $s(t)$ is PU's transmitted signal and $w_i(t)$ is additive white Gaussian noise of the channel between the PU antenna and ith SU.

There are several methods for received signal detection in CRNs that each of them require different information about PU signal parameters. Generally, it is assumed that PU signal and priori probabilities of transmitted symbols are unknown for SU and the traditional and efficient detection method is energy detector (ED) in this situation. For this reason and also better expression, we use ED in this paper. Input band-pass filter of detector selects the center frequency f_c, and the bandwidth of interest, W. By assuming sampling time interval T, each SU takes $P = 2TW$ samples. Thus, ED for each user can be declared, as follows,

$$T_i = \sum_{t=1}^{P} |y_i(t)|^2. \tag{2}$$

Then, statistical distributions of the derived detector in each SU are,

$$T_i \sim \begin{cases} \chi_P^2 & \mathcal{H}_0 \\ \chi_P^2(\lambda_i) & \mathcal{H}_1 \end{cases} , i = 1, 2, \ldots, K, \tag{3}$$

where χ_P^2 and $\chi_P^2(\lambda_i)$ denote central and non-central chi squared distributions, respectively, each with P degrees of freedom and non-centrality parameter of λ_i for the latter distribution. λ_i is the instantaneous signal to noise ratio (SNR) of ith SU. To overcome the bandwidth constraint problem of reporting channels, which link SUs and FC, calculated statistic in each SU is quantized into M levels. The M quantized symbols are transmitted to FC over non-ideal channel. In different papers, cooperative and non-cooperative schemes are used for transmission scheme. The point we consider in this paper is that we benefit the whole capacity of the report channels to transmit the SUs data to FC. One scheme that has this feature is CPAC, which is described in the following. According to the results derived in [6], CPAC scheme has the best performance among the analogous ones. Thus, transmission scheme which is used in this network is CPAC, where sensors are assigned orthogonal channels for transmission.

CPAC transmitting protocol, as is shown in Fig. 1, involves K^2 phases that each SU transmits during K phases, despite the non-cooperative one, which is called PAC scheme, where each SU transmits data only in one phase. Thus, the symbol of each SU is transmitted over all of report channels between SUs and FC. Therefore, received vector in FC is,

$$\mathbf{y}_{FC} = \mathbf{H}_{eq}\mathbf{u}_{SU} + \mathbf{n}, \tag{4}$$

where $\mathbf{u}_{SU} = (u_{SU1}, u_{SU2}, \ldots, u_{SUK})^T$ is transmitted vector from SUs to FC, $\mathbf{n}$ is additive noise vector in which n_i, $i = 1, \ldots, K$ is a Gaussian random variable with distribution $\mathcal{N}(0, \sigma_n^2)$ and $\mathbf{H}_{eq}$ is the equivalent channel matrix as follows,

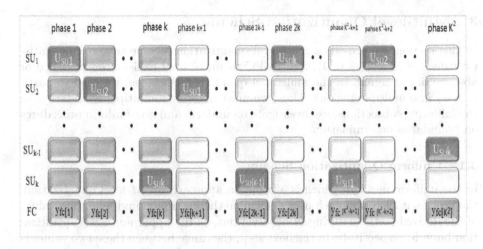

Fig. 1. Structure of CPAC scheme. White blocks represent inactive slots, blue blocks represent active slots (dark blue for transmitting and light blue for receiving).

$$\mathbf{H}_{eq} = \begin{pmatrix} h_1 & 0 & \dots & 0 \\ 0 & h_2 & \dots & 0 \\ \vdots & \vdots & \ddots & \vdots \\ 0 & 0 & \dots & h_K \\ h_2 & 0 & \dots & 0 \\ 0 & h_3 & \dots & 0 \\ \vdots & \vdots & \ddots & \vdots \\ 0 & 0 & \dots & h_1 \\ \vdots & \vdots & \ddots & \vdots \\ h_K & 0 & \dots & 0 \\ 0 & h_1 & \dots & 0 \\ \vdots & \vdots & \ddots & \vdots \\ 0 & 0 & \dots & h_{K-1} \end{pmatrix}, \qquad (5)$$

where h_i, $i = 1, ..., K$ is ith SU's channel fading gain which is assumed has Rayleigh distribution. Also channel gain and additive noise is assumed independent. Remarkable thing is that in this scheme the cooperative channels between sensors are assumed to be error-free. In other words, we assume that the symbols trasmitted from each SU, is received completely and correctly in others SUs.

3   Multilevel Quantization Scheme

In this section, we provide a multilevel quantization scheme to transmit statistic values of SUs $T_i, i = 1, ..., K$ to FC to make final decision on presence or absence of PU signal. In the multilevel quantization scheme, we need to partition T_i into multiple regions. Thus, we must determine multiple boundaries. In the following subsections, we investigate on uniform and non-uniform procedures on boundaries determination.

3.1   Uniform Quantization Scheme

In the uniform method, which is a common approach for M-level quantization, the process is such that each user determine the maximum and minimum values of the derived statistic during N observations, as the upper and lower quantization bounds, respectively. In the next step, the range between these two values is divided into M interval and finally quantized value for each interval is obtained, as follows,

$$\tau_{m,i} = \left(\frac{\theta_{max,i} - \theta_{min,i}}{M} \right) m + \theta_{min,i}, \quad m = 0, \ldots, M, \qquad (6)$$

where $\theta_{max,i}$ and $\theta_{min,i}$ are maximum and minimum of $T_i^{(j)}$, $j = 1, \ldots, N$, respectively. And thus, quantized values are,

$$\nu_{m,i} = \left(\frac{\tau_{m-1,i} + \tau_{m,i}}{2} \right), \quad m = 1, \ldots, M, \qquad (7)$$

3.2   Non-uniform Quantization Scheme

Since detection error usually occur in low SNR and nearby thresholds and in higher SNR detection of PU presence is much easier, it is more efficient that quantization boundaries density be higher nearby thresholds. In this study, we use a procedure to create non-uniform boundaries. In this procedure, quantization boundaries of each SU are derived based on the threshold value $\eta_i, i = 1, \ldots, K$ which has been calculated by consideration of Neyman-Pearson method for binary hypotheses and acceptable false alarm probability [13]. The M quantization levels of ith SU quantizer is represented by $U_{SU_i} = \{u_{1,i}, u_{2,i}, \ldots, u_{M,i}\}$ and its quantization boundaries are $\mathbf{t}_i = \{t_{0,i}, t_{1,i}, \ldots, t_{M,i}\}$. In fact, when $T_i \in (t_{m+1,i}, t_{m,i}]$ for $m = 1, \ldots, M$, quantizer decides $U_{SU_i} = u_{m,i}$. First, we determine an upper and lower bounds for $t_{m,i}$, in order to cover all possible values of each SU's statistic. Then, determination of boundaries values is in a way that number of boundaries near the binary hypothesis threshold η_i be greater. Therefore, boundaries are determined as follows,

$$\begin{cases} t_{M,i} = \mathcal{T}_i + \eta_i \\ t_{\frac{M}{2},i} = \eta_i \\ t_{0,i} = \eta_i - \mathcal{T}_i. \end{cases} \qquad (8)$$

where,

$$\mathcal{T}_i = \max_j |\mathrm{T}_i^{(j)} - \eta_i|, \ j = 1, \dots, N, \tag{9}$$

is the maximum distance of ith SU statistic from threshold in N observations and also $\mathrm{T}_i^{(j)}$ is ith SU statistic in jth observation. For middle values of boundaries, we have,

$$\begin{cases} t_{m,i} = \frac{t_{m+1,i} + \eta_i}{2}, & m > \frac{M}{2} \\ t_{n,i} = \frac{t_{n-1,i} + \eta_i}{2}, & n < \frac{M}{2}. \end{cases} \tag{10}$$

Therefore, quantized value of each level is defined as,

$$u_{m,i} = \frac{t_{m-1,i} + t_{m,i}}{2}, \ m = 1, \dots, M. \tag{11}$$

The probability mass function (PMF) of U_{SU_i}, which is the transmitted value to FC from ith SU, can be expressed as,

$$P(U_{\mathrm{SU}_i} = u_{m,i}|\mathcal{H}_j) = \int_{t_{m-1,i}}^{t_{m,i}} f(\mathrm{T}_i = t|\mathcal{H}_j)dt, \ j = 0, 1. \tag{12}$$

According to cumulative distribution function (CDF) of central and non-central chi-square random variables and (3), we have,

$$P(U_{\mathrm{SU}_i} = u_{m,i}|\mathcal{H}_0) = \frac{\gamma\left(\frac{N}{2}, \frac{t_{m,i}}{2}\right) - \gamma\left(\frac{N}{2}, \frac{t_{m-1,i}}{2}\right)}{\Gamma\left(\frac{N}{2}\right)}, \tag{13}$$

where $\Gamma(.)$ and $\gamma(.,.)$ are gamma and lower incomplete gamma function, respectively.

$$P(U_{\mathrm{SU}_i} = u_{m,i}|\mathcal{H}_1) = Q_{\frac{N}{2}}(\sqrt{\lambda_i}, \sqrt{t_{m-1,i}}) - Q_{\frac{N}{2}}(\sqrt{\lambda_i}, \sqrt{t_{m,i}}), \tag{14}$$

for $m = 1, \dots, M$, where $Q_{\frac{N}{2}}(.,.)$ is Marcum Q-function.

As mentioned before, the report channel is assumed fading channel with Rayleigh distribution and additive Gaussian noise. Therefore, according to (5) and Fig. 1, all of SUs quantized data are transmitted to FC over K different independent channels. i.e. for each SU's transmitted symbol u_{SU_i}, we have K observations in K different time phases. If variable l be index of observations time phases, it can be seen in Fig. 1 that transmitted symbol related to ith SU, u_{SU_i}, is received in $l = i + nK$, $n = 0, \dots, K - 1$, time phases. Thus, the received signal from ith SU in lth time phase is,

$$y_{\mathrm{FC}_i}[l] = u_{\mathrm{SU}_i}h_{i,l} + n_i = x_{\mathrm{SU}_{i,l}} + n_i, \tag{15}$$

where $h_{i,l} = [\mathbf{H}_{eq}]_{l,i}$ has Rayleigh distribution with unit power as follows,

$$f_H(h_{i,l}) = 2h_{i,l}e^{-h_{i,l}^2}, \quad h_{i,l} > 0. \tag{16}$$

For $x_{\mathrm{SU}_{i,l}}$ distribution, we have,

$$F_{X_{\mathrm{SU}_{i,l}}}(x_{\mathrm{SU}_{i,l}}) = P(X_{\mathrm{SU}_{i,l}} < x_{\mathrm{SU}_{i,l}}) = P(H_{i,l}U_{\mathrm{SU}_i} < x_{\mathrm{SU}_{i,l}})$$

$$= P\left(H_{i,l} < \frac{x_{\mathrm{SU}_{i,l}}}{u_{\mathrm{SU}_i}}\right) = F_H\left(\frac{x_{\mathrm{SU}_{i,l}}}{u_{\mathrm{SU}_i}}\right). \tag{17}$$

Then,

$$f_{X_{\mathrm{SU}_{i,l}}}(x_{\mathrm{SU}_{i,l}}) = \frac{1}{u_{\mathrm{SU}_i}} f_H\left(\frac{x_{\mathrm{SU}_{i,l}}}{u_{\mathrm{SU}_i}}\right) \tag{18}$$

$$= \frac{2}{u_{\mathrm{SU}_i}^2} x_{\mathrm{SU}_{i,l}} \exp\left(-\frac{x_{\mathrm{SU}_{i,l}}^2}{u_{\mathrm{SU}_i}^2}\right), \quad \frac{x_{\mathrm{SU}_{i,l}}}{u_{\mathrm{SU}_i}} > 0.$$

4   Decision in Fusion Center

In this section, we derive a posteriori probability of received symbols in FC from each user. Since, additive noise is assumed Gaussian, distribution of observed signal in FC from ith SU in lth time phase can be calculated as,

$$f(y_{\mathrm{FC}_i}[l]|u_{\mathrm{SU}_i}, \mathcal{H}_j) = f(y_{\mathrm{FC}_i}[l]|u_{\mathrm{SU}_i}) \tag{19}$$

$$= \int_{-\infty}^{+\infty} \frac{2x_{\mathrm{SU}_{i,l}}}{\sqrt{2\pi}u_{\mathrm{SU}_i}^2} e^{-\left(\frac{x_{\mathrm{SU}_{i,l}}^2}{u_{\mathrm{SU}_i}^2} + \frac{(y_{\mathrm{FC}_i}[l] - x_{\mathrm{SU}_{i,l}})^2}{2\sigma_n^2}\right)} dx_{\mathrm{SU}_{i,l}}.$$

By replacement $\tau_{\mathrm{SU}_{i,l}} = \frac{x_{\mathrm{SU}_{i,l}}}{u_{\mathrm{SU}_i}}$, and also from Section 3.462 in [16], (19) might be simplified as,

$$f(y_{\mathrm{FC}_i}[l]|u_{\mathrm{SU}_i}) = \frac{\sqrt{2}\sigma_n}{\sqrt{\pi}(u_{\mathrm{SU}_i}^2 + 2\sigma_n^2)} e^{\left(\frac{-y_{\mathrm{FC}_i}[l]^2}{2\sigma_n^2}\right)} \tag{20}$$

$$\times \left[1 + \frac{\sqrt{2\pi}u_{\mathrm{SU}_i}y_{\mathrm{FC}_i}[l]}{\sigma_n\sqrt{u_{\mathrm{SU}_i}^2 + 2\sigma_n^2}} e^{\left(\frac{u_{\mathrm{SU}_i}^2 y_{\mathrm{FC}_i}[l]^2}{2\sigma_n^2\left(u_{\mathrm{SU}_i}^2 + 2\sigma_n^2\right)}\right)} \left\{\frac{1}{2} - \frac{1}{2}\mathrm{erf}\left(\frac{1}{\sqrt{2}}\frac{-u_{\mathrm{SU}_i}y_{\mathrm{FC}_i}[l]}{\sigma_n\sqrt{u_{\mathrm{SU}_i}^2 + 2\sigma_n^2}}\right)\right\}\right].$$

In (20), $\mathrm{erf}(x) = \frac{2}{\sqrt{\pi}}\int_0^x e^{-t^2} dt$ is Gaussian error function of random variable x.

By assumption $A_{\mathrm{SU}_i} = \frac{1}{u_{\mathrm{SU}_i}^2 + 2\sigma_n^2}$, then (20) can be rewritten as follows,

$$f(y_{\mathrm{FC}_i}[l]|u_{\mathrm{SU}_i}) = \frac{\sqrt{2}}{\sqrt{\pi}}\sigma_n A_{\mathrm{SU}_i} e^{\left(\frac{-y_{\mathrm{FC}_i}[l]^2}{2\sigma_n^2}\right)}$$

$$+ 2u_{\mathrm{SU}_i} A_{\mathrm{SU}_i}^{\frac{3}{2}} y_{\mathrm{FC}_i}[l] e^{-A_{\mathrm{SU}_i}y_{\mathrm{FC}_i}[l]^2} Q\left(-\frac{u_{\mathrm{SU}_i}A_{\mathrm{SU}_i}^{\frac{1}{2}}}{\sigma_n}y_{\mathrm{FC}_i}[l]\right). \tag{21}$$

As we can see, by assumption of known noise variance, (21) is function of u_{SU_i} and $y_{\mathrm{FC}_i}[l]$ and is independent from instantaneous channel value. Since transmitted data from SUs are discrete random variables, we have to calculate PMF of received symbols as a posteriori probability of quantized data from distribution of received signals in FC. Thus,

$$P(U_{\mathrm{FC}_i}[l] = u_{m,i}|u_{\mathrm{SU}_i} = u_{k,i}) = \int_{\Omega_m} f(y_{\mathrm{FC}_i}[l] = t|u_{k,i})dt, \qquad (22)$$

where Ω_m is decision region of $u_{m,i}$ symbol. Calculation of integral in (22) is given in Appendix. Thus, from (12) and (A-4), a posteriori probability of received symbol from ith SU under both hypotheses can be represented as,

$$P(U_{\mathrm{FC}_i}[l] = u_{m,i}|\mathcal{H}_j) = \sum_{k=1}^{M} P(U_{\mathrm{FC}_i}[l] = u_{m,i}|u_{\mathrm{SU}_i} = u_{k,i})P(U_{\mathrm{SU}_i} = u_{k,i}|\mathcal{H}_j), \qquad (23)$$

for $j = 0, 1$. By averaging on K time phases in which ith SU's symbol is transmitted, we have, for $j = 0, 1$,

$$P(U_{\mathrm{FC}_i} = u_{m,i}|\mathcal{H}_j) = \frac{1}{K}\sum_{l=1}^{K} P(U_{\mathrm{FC}_i}[l] = u_{m,i}|\mathcal{H}_j). \qquad (24)$$

To determine transmitted symbol of the ith SU, maximum a posteriori probability (MAP) is used such that the received symbol with greater a posteriori probability is selected. Thus, for equal priori probabilities of both hypotheses, we have,

$$\widehat{u}_{\mathrm{SU}_i} = \max_{m=1,\ldots,M}\{u_{m,i} : P(U_{\mathrm{FC}_i} = u_{m,i})\}. \qquad (25)$$

4.1   Decision Rule Based on Received Symbol Summation

In this subsection, we consider summation based fusion scheme in FC in which FC sums the determined symbols of all SUs to make final decision. i.e.,

$$\mathrm{T_{FC}} = \sum_{i=1}^{K} \widehat{u}_{\mathrm{SU}_i} \underset{\gtrless \mathcal{H}_0}{\overset{\leq \mathcal{H}_1}{}} \eta_{\mathrm{FC}}. \qquad (26)$$

To evaluate performance of the expressed decision rule, we have to derive the PMF of $\mathrm{T_{FC}}$. Thus, From [17], we have,

$$p(\mathrm{T_{FC}}|\mathcal{H}_j) = p(\widehat{U}_{\mathrm{SU}_1}|\mathcal{H}_j) * \ldots * p(\widehat{U}_{\mathrm{SU}_K}|\mathcal{H}_j). \qquad (27)$$

Where $p(\widehat{U}_{\mathrm{SU}_i}|\mathcal{H}_j)$ denotes the PMF of ith received symbol estimation and $*$ denotes the convolution operator of discrete random variables. If $\mathcal{L}_{\mathrm{T_{FC}}}$, which has M^K elements, be the set of values which $\mathrm{T_{FC}}$ can take, then according to

Neyman-Pearson test for discrete random variables, decision rule at the FC is given by,

$$
\begin{cases}
T_{FC} < \eta_{_{FC}}, & \mathcal{H}_0 \\
T_{FC} = \eta_{_{FC}}, & \mathcal{H}_1 \text{ with probability } \gamma \\
T_{FC} > \eta_{_{FC}}, & \mathcal{H}_1,
\end{cases}
\tag{28}
$$

where for maximum acceptable false alarm probability α, $\eta_{_{FC}}$ is the threshold given by,

$$
\eta_{_{FC}} = \min_{t_{_{FC}} \in \mathcal{L}_{T_{FC}}} \{ t_{_{FC}} : P(T_{FC} > t_{_{FC}} | \mathcal{H}_0) < \alpha \}.
\tag{29}
$$

and γ is randomization parameter as follows,

$$
\gamma = \frac{\alpha - P(T_{FC} > \eta_{_{FC}} | \mathcal{H}_0)}{P(T_{FC} = \eta_{_{FC}} | \mathcal{H}_0)}.
\tag{30}
$$

Detection probability can be calculated as,

$$
P_{\mathrm{d}} = P(T_{FC} > \eta_{_{FC}} | \mathcal{H}_1) + \gamma P(T_{FC} = \eta_{_{FC}} | \mathcal{H}_1).
\tag{31}
$$

4.2 Kullback-Leibler Divergence Criterion in FC

In this subsection, we benefit KL divergence criterion for evaluation of detector performance in FC, which in probability, is a criterion of the difference between two probability distributions. The KL divergence is a fundamental equation to quantify the proximity of two probability distributions[18]. While, it may be used as criterion of divergence between the two hypotheses in detection, since it is the expected log-likelihood ratio[19]. In this paper, we express the KL divergence criterion in the following form,

$$
D_{KL_{FC}} = \sum_{\mathbf{m}} P(\widehat{\mathbf{u}} = \mathbf{m} | \mathcal{H}_1) \ln \frac{P(\widehat{\mathbf{u}} = \mathbf{m} | \mathcal{H}_1)}{P(\widehat{\mathbf{u}} = \mathbf{m} | \mathcal{H}_0)},
\tag{32}
$$

where $P(\widehat{\mathbf{u}} = \mathbf{m} | \mathcal{H}_j)$, $j = 0, 1$, is PMF of the received symbols estimations vector, $\widehat{\mathbf{u}} = \{\widehat{u}_{\mathrm{SU}_1}, \widehat{u}_{\mathrm{SU}_2}, ..., \widehat{u}_{\mathrm{SU}_K}\}$ and $\mathbf{m} \in \{u_{1,1}, ..., u_{M,1}\} \times ... \times \{u_{1,K}, ..., u_{M,K}\}$. As mentioned before, we assume that SUs are independent and the PMF of them are independent too. Therefore, (32) can be expressed as follows,

$$
D_{KL_{FC}} = \sum_{i=1}^{K} \sum_{m=1}^{M} P(\widehat{u}_{\mathrm{SU}_i} = u_{m,i} | \mathcal{H}_1) \ln \frac{P(\widehat{u}_{\mathrm{SU}_i} = u_{m,i} | \mathcal{H}_1)}{P(\widehat{u}_{\mathrm{SU}_i} = u_{m,i} | \mathcal{H}_0)}.
\tag{33}
$$

5   Simulation Results

In this section, we provide a comparative simulation-based performance of the system which is introduced in this paper, based on Monte-Carlo simulations.

Fig. 2 depicts comparison of the performance in FC for different quantization levels under uniform and non-uniform schemes based on detection probability P_{d} versus SNR at false alarm probability rate $P_{\mathrm{fa}} = 0.01$ and $K = 4$ in the

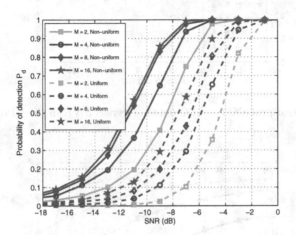

Fig. 2. Probability of detection P_{d} of the CPAC scheme versus SNR for $P_{\mathrm{fa}} = 0.01$, $K = 4$ and different uniform and non-uniform quantization levels.

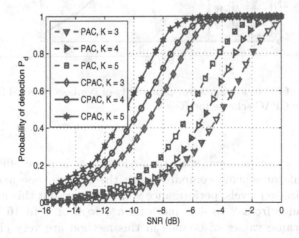

Fig. 3. Probability of detection P_{d} of the CPAC and PAC schemes versus SNR for $P_{\mathrm{fa}} = 0.01$, $M = 8$ and different number of users.

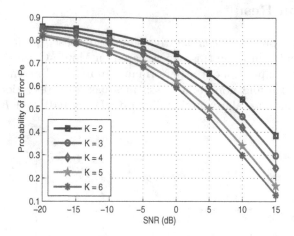

Fig. 4. Average error probability of FC detector versus SNR in the CPAC scheme for $M = 8$.

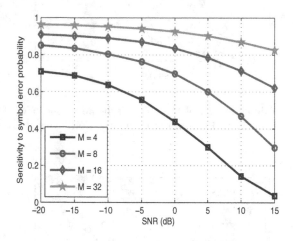

Fig. 5. sensitivity of FC to average error probability of symbols received from each SU versus SNR in the CPAC scheme for $K = 4$.

CPAC scheme. As can be seen, statistic performance is more efficient in the provided non-uniform scheme compared with the uniform one and also with increasing quantization levels, performance will be better, but this improvement in the levels change from 2 to 4 is more than 4 to 8 or 8 to 16 in the non-uniform one. Because values of bounds in this method are very close to each other near the threshold and so increasing of levels will not have noticeable effect. Therefore, non-uniform method does not require high quantization levels for better performance that will be effective for less occupied bandwidth.

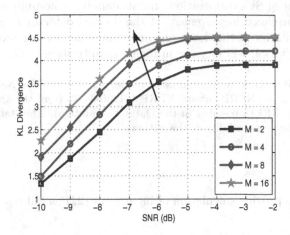

Fig. 6. KL divergence performance metric in FC versus SNR in CPAC scheme for $K = 2$.

In Fig. 3, we investigate the impact of CPAC protocol on performance of the FC statistic based on detection probability P_d versus SNR for $M = 8$ and $P_{fa} = 0.01$. As can be seen, in addition to better performance of the detector for more SUs, performance improvement in cooperative scheme compared with non-cooperative one, PAC, is also evident.

In Fig. 4, decreasing average error probability P_e of FC detector versus SNR in CPAC scheme for $M = 8$ and the higher number of SUs is depicted. Fig. 5 shows sensitivity of FC detector to average symbol error probability for $K = 4$ and cooperative scheme. In this figure, the maximum numerical value for sensitive criterion is equal 1, thus the values closer to one are more sensitive, which means in higher quantization levels, receiver is more sensitive to average symbol error and thus its decision is more accurate.

Finally, in Fig. 6 we provide simulation-based evaluation of the KL divergence criterion versus SNR for different quantization levels and $K = 2$ in CPAC scheme. Higher values of KL divergence which means more accurate decision can be observed in figure with increasing quantization levels.

6   Conclusion

In this paper, we investigated a practical CSS problem in which SUs use multi-level non-uniform quantization to transmit their data to FC based on cooperative scheme in Rayleigh fading wireless links. We detected symbols received in FC from each SU in different time phases of CPAC scheme based on MAP pattern. In addition, we provided summation-based final decision statistic and KL divergence metric to evaluate the system performance in FC. Simulation results revealed the effect of SUs cooperation and also applying non-uniform quantization on the performance improvement of the derived detector in FC. Also, we showed that sensitivity of the detector to average symbol error probability and KL divergence criterion will increase in higher quantization levels.

Acknowledgments. This publication was made possible by the National Priorities Research Program (NPRP) award NPRP 6-1326-2-532 from the Qatar National Research Fund (QNRF) (a member of the Qatar Foundation). The statements made herein are solely the responsibility of the authors.

Appendix

In this section we provide calculation of integral in (22) as follows,

$$\int_{\Omega_m} f(y_{\mathrm{FC}_i}[l] = t | u_{\mathrm{SU}i}) dt \tag{A-1}$$

$$= \int_{\Omega_m} \frac{\sqrt{2}\sigma_n A_{\mathrm{SU}i}}{\sqrt{\pi}} e^{\left(\frac{-t^2}{2\sigma_n^2}\right)} dt + \int_{\Omega_m} 2 u_{\mathrm{SU}i} A_{\mathrm{SU}i}^{\frac{3}{2}} t e^{-A_{\mathrm{SU}i} t^2} Q\left(-\frac{u_{\mathrm{SU}i}\sqrt{A_{\mathrm{SU}i}}}{\sigma_n} t\right) dt.$$

First and second parts of (A-1) can be calculated, respectively, as,

$$I: \quad \int \frac{\sqrt{2}\sigma_n A_{\mathrm{SU}i}}{\sqrt{\pi}} e^{\left(\frac{-t^2}{2\sigma_n^2}\right)} dt = \sigma_n^2 A_{\mathrm{SU}i} \, \mathrm{erf}\left(\frac{t}{\sqrt{2}\sigma_n}\right). \tag{A-2}$$

Also, by integration by parts method, for second part we have,

$$II: \quad \int 2 u_{\mathrm{SU}i} A_{\mathrm{SU}i}^{\frac{3}{2}} t e^{-A_{\mathrm{SU}i} t^2} Q\left(-\frac{u_{\mathrm{SU}i}\sqrt{A_{\mathrm{SU}i}}}{\sigma_n} t\right) dt$$

$$= \frac{\sqrt{\pi} u_{\mathrm{SU}i}^2 A_{\mathrm{SU}i}}{2} \mathrm{erf}\left(\frac{t}{\sigma_n}\right) - u_{\mathrm{SU}i}\sqrt{A_{\mathrm{SU}i}} Q\left(-\frac{u_{\mathrm{SU}i}\sqrt{A_{\mathrm{SU}i}}}{\sigma_n} t\right) e^{-A_{\mathrm{SU}i} t^2}. \tag{A-3}$$

Then, from (A-2), (A-3) and (22), we have,

$$P(U_{\mathrm{FC}_i}[l] = u_{m,i} | u_{\mathrm{SU}i} = u_{k,i}) \tag{A-4}$$

$$= \left[\sigma_n^2 A_{k,i} \mathrm{erf}\left(\frac{t}{\sqrt{2}\sigma_n}\right) + \frac{\sqrt{\pi} u_{k,i}^2 A_{k,i}}{2} \mathrm{erf}\left(\frac{t}{\sigma_n}\right) - u_{k,i}\sqrt{A_{k,i}} Q\left(-\frac{u_{k,i}\sqrt{A_{k,i}}}{\sigma_n} t\right) e^{-A_{k,i} t^2}\right]_{\Omega_m}.$$

References

1. Akyildiz, I., Lee, W., Vuran, M., Mohanty, S.: Next generation dynamic spectrum access cognitive radio wireless networks: A survey. Computer Networks **50**(13) (2006)

2. Taherpour, A., Nasiri-Kenari, M., Jamshidi, A.: Efficient cooperative spectrum sensing in cognitive radio networks. In: IEEE International Symposium on Personal, Indoor and Mobile Radio Communications (PIMRC) (2007)

3. Sun, C., Zhang, W., Letaief, K.: Cluster-based cooperative spectrum sensing in cognitive radio systems. In: Proc. of the IEEE International Conference on Communications (ICC), pp. 2511–2515, June 2007

4. Zhang, W., Mallik, R., Letaief, K.: Cooperative spectrum sensing optimization in cognitive radio networks. In: Proc. of the IEEE International Conference on Communications (ICC), pp. 3411–3415, May 2008

5. Rao, A., Alouini, M.-S.: Cooperative spectrum sensing over non-identical nakagami fading channels, pp. 1–4, May 2011

6. Salvo Rossi, P., Ciuonzo, D., Romano, G.: Orthogonality and cooperation in collaborative spectrum sensing through MIMO decision fusion. IEEE Transactions on Wireless Communications **12**(11), 5826–5836 (2013)

7. Nadendla, V., Han, Y., Varshney, P.: Distributed inference with M-Ary quantized data in the presence of byzantine attacks. IEEE Transactions on Signal Processing **62**(10), 2681–2695 (2014)

8. Niu, R., Chen, B., Varshney, P.: Fusion of decisions transmitted over Rayleigh fading channels in wireless sensor networks. IEEE Transactions on Signal Processing **54**(3), 1018–1027 (2006)

9. Lin, Y.: Quantization for distributed detection under link outages. In: 42nd Asilomar Conference on Signals, Systems and Computers, pp. 1948–1952, October 2008

10. Chen, H., Tse, C., Zhao, F.: Optimal quantization bit budget for a spectrum sensing scheme in bandwidth-constrained cognitive sensor networks. IET Wireless Sensor Systems **1**, 144–150 (2011)

11. Schwandter, S., Matz, G.: A practical forwarding scheme for wireless relay channels based on the quantization of log-likelihood ratios. In: IEEE International Conference on Acoustics Speech and Signal Processing (ICASSP), pp. 2502–2505, March 2010

12. Zeitler, G., Bauch, G., Widmer, J.: Quantize-and-forward schemes for the orthogonal multiple-access relay channel. IEEE Transactions on Communications **60**(4), 1148–1158 (2012)

13. Liang, H., Chen, Y., Li, S.: A log-likelihood ratio non-uniform quantization scheme for cooperative spectrum sensing. In: 8th International Conference on Wireless Communications, Networking and Mobile Computing (WiCOM), pp. 1–4, September 2012

14. Chaudhari, S., Koivunen, V.: Effect of quantization and channel errors on collaborative spectrum sensing. In: Conference Record of the Forty-Third Asilomar Conference on Signals, Systems and Computers, pp. 528–533, November 2009

15. Taherpour, A., Norouzi, Y., Nasiri-Kenari, M., Jamshidi, A., Zeinalpour-Yazdi, Z.: Asymptotically optimum detection of primary user in cognitive radio networks. IET Communication **1**, 1138–1145 (2007)

16. Gradshteyn, I., Ryzhik, I.: Table of integrals, series, and products

17. Evans, D., Leemis, L.: Algorithms for computing the distributions of sums of discrete random variables. Mathematical and Computer Modelling **40**, 1429–1452 (2014)

18. Shlens, J.: Notes on Kullback-Leibler divergence and likelihood (2014). CoRR, abs/1404.2000

19. Eguchi, S., Copas, J.: Interpreting Kullback-Leibler divergence with the Neyman-Pearson lemma. Journal of Multivariate Analysis **97**(9), 2034–2040 (2006)

Downlink Scheduling and Power Allocation in Cognitive Femtocell Networks

Hesham M. Elmaghraby$^{(\boxtimes)}$, Dongrun Qin, and Zhi Ding

University of California, Davis, CA 95616, USA
{hmelmaghraby,drqin,zding}@ucdavis.edu
http://web.ece.ucdavis.edu/~zding/

Abstract. We consider resource assignment and power allocation problem in femtocells under channel estimation errors. Our formulation is to maximize the throughput of femtocell users that share spectrum resources with macrocell base station (MBS) while limiting interference between macrocell and femtocells. Using cognitive capabilities, femtocell basestations (FBS) can acquire the needed information about the neighboring MBS users to reduce cross-tier interference between FBS and MBS users. We analyze the distributions of signal to interference and noise ratio (SINR) of MBS users and signal to interference ratio (SIR) of FBS users. Based on the analytical results, we present resource assignment and power allocation solutions to maximize the mean sum rate subject to SINR and SIR outage constraints, along with simulation verifications.

Keywords: Cognitive femtocell · Cross-tier interference · Resource assignment · Outage constraint · Power allocation

1 Introduction

For nearly a century, wireless capacity has doubled every 30 months. Capacity analysis shows that the capacity increased 25x due to wider spectrum, 5x from dividing spectrum into smaller portions, 5x from enhancements in modulation techniques, and 1600x through reducing the cell sizes and accordingly the communication distances [1]. Despite such high capacity growth, consumer demand for capacity rises even higher. Recent studies show that nearly 50% of voice traffic and 70% of data traffic take place from indoor consumers and it is predicted that this indoor traffic will increase to 60% and 90% for voice and data traffic respectively [1][2]. Femtocell is one promising solution to the traffic growth problem under limited spectrum. Femtocell basestation (FBS) is a short range, low-power and low-cost basestation, installed by users with internet connection, in order to provide better service for local or indoor users.

This material is based upon work supported by National Science Foundation under Grants ECCS-1307820, CNS-1443870, and CNS-1457060. The work of the 1st author is also supported by an Egyptian Government grant.

© Institute for Computer Sciences, Social Informatics and Telecommunications Engineering 2015
M. Weichold et al. (Eds.): CROWNCOM 2015, LNICST 156, pp. 92–105, 2015.
DOI: 10.1007/978-3-319-24540-9_8

A number of other existing works have focused on the interference problem that arises because of spectrum sharing between MBS and FBS [3]. Among various solutions, cognitive radio (CR) may effectively add the needed spectrum awareness functions to the FBS [4][5]. Such FBS with cognitive capabilities may obtain spectrum information needed to control interference level on the shared resources. The authors of [6] presented an algorithm for optimal power allocation in order to solve the downlink interference problem, requiring prior knowledge of all the system channel gains collected by a fusion center. In [7] the authors presented a decentralized interference management method for LTE-A femtocells by sharing measured pathloss information among neighboring femtocells. The authors in [8] formulated the optimization problem of the resource allocation as a Stackelberg game. They focused on the energy efficiency aspect of the shared spectrum in heterogeneous networks. Further, authors of [9] used game theory to model the resource allocation problem and introduced cognitive radio resource management and strategic game based radio resource management schemes to solve the given problem.

In this paper, we study the underlay femtocell scheduling and power assignment problem. Our main objective is to derive a decentralized technique for FBS resource scheduling so as to maximize the total capacity of home user equipments (HUEs) served by the FBS while keeping SINR of nearby macro-user equipments (MUEs) above a given threshold when sharing the same resources. We incorporate some sensing capabilities at the FBS for collecting needed information on the shared resources and for measuring the femtocell impact on nearby cochannel MUEs. Our main contribution is in considering an estimation error in FBS-to-HUE and MBS-to-HUE channel gains. By formulating the problem based on this channel uncertainty, we analyze the distributions of HUEs signal to interference ratio (SIR) and MUEs SINR. Another main contribution in this work is the analytical reduction of the given problem into a much simpler problem [10]. We further present two decentralized methods to solve the resulting channel assignment and power allocation problem based on the optimal Hungarian algorithm and a greedy suboptimal algorithm.

We organize our manuscript as follows: in section 2 we introduce the system model and our problem formulation to maximize the average sum rate subject to SINR and SIR outage constraints. section 3 provides the distribution analysis on SIR of HUEs and SINR of the MUEs. We present our proposed solution for the given optimization problem in section 4 and simulation results in section 5.

2   System Model

2.1   Network Architecture

Our underlying heterogeneous network consists of two tiers: a central macrocell and several femtocells. Each femtocell shares assigned bandwidth (BW) with the macrocell without intra-tier interference with other femtocells. This can be achieved by orthogonal bandwidth assignments for adjacent femtocells. The femtocells operate in closed access mode.

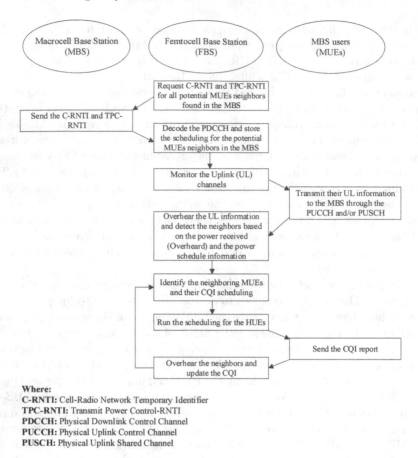

Fig. 1. The action sequence to identify the femtocell neighbors.

We assume cognitive capabilities in each FBS in order to assist in the scheduling and power assignment process. Given the cognitive capability plus indirect coordination of the macrocell base station (MBS), FBS can identify the neighboring MUEs as well as their power and channel assignments. Fig. 1 illustrates the actions of FBS, MBS, and MUEs in order for the FBS to acquire the needed information. Each FBS schedules its actions separately. In this paper, resource block (RB) and channel are synonymous. Before proceeding, here are some important notations we use:

- $\gamma_u^c(t)/\gamma_v^c(t)$: SINR of the HUE u / MUE v on channel c at time t.
- $CQI_v^c(t)$: The overheard channel quality information (CQI) report of the MUE v on channel c at time t.
- $\theta_u^c(t)$: SIR of the HUE u on channel c at time t.
- $\theta_{HUELB}/\gamma_{MUELB}$: The minimum SIR / SINR for the HUE / MUE that can guarantee reliable connection with the FBS / MBS.

- $H_{F-u}^c(t)/H_{M-u}^c(t)/H_{M-v}^c(t)/H_{F-v}^c(t)$: Complex channel gains for the FBS-HUE / MBS-HUE / MBS-MUE / FBS-MUE.
- $P_M^c(t)/P_F^c(t)$: MBS / FBS assigned power on channel c at time t.
- $P_{u,c}(t)$: Power assigned for the HUE u by the FBS on channel c at time t.
- $C_{out}(t)/C_{out}^I(t)/C_{out}^N(t)$: Normalized total capacity of HUEs at time t over all / overlapped / non-overlapped assigned resources.
- N_{HUE}/N_{MUE}: Number of HUEs / MUEs.
- α, β: Probability values from 0 to 1.
- $\mathbf{E}[g(t)]/\mathbf{V}[g(t)]$: Mean / Variance of $g(t)$.
- $\chi'^2(k,\lambda)$: Non-central chi square distribution with k degrees of freedom and non-centrality parameter λ.

2.2   Problem Formulation

Assume the MBS has N_{MUE} users (neighboring the FBS) and N_1 available RBs while the FBS has N_{HUE} users assigned with N_2 RBs ($N_2 < N_1$). Normally not all the N_2 RBs are occupied with neighboring MUEs. We can divide the total HUEs capacity according to

$$C_{out}(t) = C_{out}^I(t) + C_{out}^N(t), \tag{1}$$

where overlapped resources are RBs occupied by neighboring MUEs and assigned to HUE by the FBS for sharing whereas non-overlapped resources are either empty or occupied by far away MUEs.

In our problem formulation, we will consider maximizing HUEs capacity over shared RBs, in other words we are just considering optimizing the scheduling and power assignment over the interfered channels.

Let N_0 be the background noise power. Eqs. (2) and (3) define the SINR and SIR of HUE u respectively as

$$\gamma_u^c(t) = \frac{P_{u,c}(t)\left|H_{F-u}^c(t)\right|^2}{P_M^c(t)\left|H_{M-u}^c(t)\right|^2 + N_0}, \tag{2}$$

$$\theta_u^c(t) = \frac{P_{u,c}(t)\left|H_{F-u}^c(t)\right|^2}{P_M^c(t)\left|H_{M-u}^c(t)\right|^2}. \tag{3}$$

Because we only consider the overlapped channels, $P_M^c(t) > 0$ and consequently $N_0 \ll P_M^c(t)\left|H_{M-u}^c(t)\right|^2$, Hence $\gamma_u^c(t) \approx \theta_u^c(t)$, from which we can define the normalized capacity of the HUEs over the overlapped channels by

$$C_{out}^I(t) = \sum_{i=1}^{N_{HUE}} \left(\sum_{j=1}^{N_{Ch}} a_{i,j}(t) \log_2(1 + \theta_i^j(t)) \right) \tag{4}$$

where N_{Ch} is the number of overlapped channels and $a_{i,j}(t)$ is the action by the FBS such that:

$$a_{i,j}(t) = \begin{cases} 0, & \text{if Channel } j \text{ is not assigned to HUE } i \\ 1, & \text{if Channel } j \text{ is assigned to HUE } i \end{cases}.$$

Since channel gains are estimated by the HUE u before being sent to the FBS, according to estimation error model provided in [11], we can represent the channel gains as Random Variables (RVs) as shown in (5)

$$H_{F-u}^c(t) = \hat{H}_{F-u}^c(t) + \tilde{H}_{F-u}^c, \tag{5}$$

where $\hat{H}_{F-u}^c(t)$ is a constant complex value representing the estimated channel at time t and $\tilde{H}_{F-u}^c$ is a complex normal distributed RV represent the estimation error such that $\tilde{H}_{F-u}^c \sim \mathcal{CN}(0, 2\sigma_{F-u}^2)$. Therefore $H_{F-u}^c(t)$ can be modeled as a complex normally distributed RV such that $H_{F-u}^c(t) \sim \mathcal{CN}(\hat{H}_{F-u}^c(t), 2\sigma_{F-u}^2)$ Similarly we have

$$H_{M-u}^c(t) = \hat{H}_{M-u}^c(t) + \tilde{H}_{M-u}^c, \tag{6}$$

where $H_{M-u}^c(t)$ can be modeled as a complex normally distributed RV such that $H_{M-u}^c(t) \sim \mathcal{CN}(\hat{H}_{M-u}^c(t), 2\sigma_{M-u}^2)$. We also define the SINR of the MUE v

$$\gamma_v^c(t) = \frac{P_M^c(t)\left|H_{M-v}^c(t)\right|^2}{P_F^c(t)\left|H_{F-v}^c(t)\right|^2 + N_0}. \tag{7}$$

We assume that the MBS shares its power assignment information with the FBS in order to reduce the interference from FBS frequency reuse. In order to gain some information about $H_{M-v}^c(t)$ and $H_{F-v}^c(t)$, we assume that the channel is slow-fading channel, such that the channel gain is approximately constant in 3 consecutive time slots.

From this assumption, for $t - 2 \leq T \leq t$ we have

$$H_{M-v}^c(T) = H_{M-v}^c, \tag{8}$$

$$H_{F-v}^c(T) = H_{F-v}^c. \tag{9}$$

Therefore,

$$\gamma_v^c(t) = \frac{P_M^c(t)\left|H_{M-v}^c\right|^2}{P_F^c(t)\left|H_{F-v}^c\right|^2 + N_0}, \tag{10}$$

and

$$\gamma_v^c(t-1) = \frac{P_M^c(t-1)\left|H_{M-v}^c\right|^2}{P_F^c(t-1)\left|H_{F-v}^c\right|^2 + N_0}, \tag{11}$$

$$\gamma_v^c(t-2) = \frac{P_M^c(t-2)\left|H_{M-v}^c\right|^2}{P_F^c(t-2)\left|H_{F-v}^c\right|^2 + N_0}. \tag{12}$$

As a result of FBS's cognitive capabilities, the FBS can overhear $CQI_v^c(t-1)$ and $CQI_v^c(t-2)$, which represent quantized versions of $\gamma_v^c(t-1)$ and $\gamma_v^c(t-2)$, respectively. Therefore, $CQI_v^c(T)$ indicates the interval of $\gamma_v^c(T)$ such that

$$At\ CQI_v^c(T) = K \rightarrow \gamma_v^c(T) \in [a, b], \tag{13}$$

where K represent the overheard CQI value and a, b represent the interval boundaries corresponding to K that $\gamma_v^c(T)$ lies in. Therefore at time t, we can estimate H_{M-v}^c and H_{M-v}^c (assuming $CQI_v^c(t-1) \neq CQI_v^c(t-2)$) from Eqs. (11) and (12), thereby allowing us to find $P(\gamma_v^c(t)|CQI_v^c(t-1), CQI_v^c(t-2)) \geq \beta$.

We now formulate the maximization of HUE capacity:

$$\max_{P_{u,c}} \mathbf{E}[C_{out}^I(t)] = \max_{P_{u,c}} (\sum_{i=1}^{N_{HUE}} (\sum_{j=1}^{N_{Ch}} a_{i,j}(t)\mathbf{E}[\log_2(1 + \theta_i^j(t))])) \tag{14a}$$

$$s.t. \quad \sum_{i=1}^{N_{HUE}} a_{i,c}(t) \leq 1 \tag{14b}$$

$$\sum_{j=1}^{N_{Ch}} a_{u,j}(t) = 1 \tag{14c}$$

$$\mathbf{P}(\sum_{j=1}^{N_{Ch}} a_{u,j}(t)\theta_u^j(t) \geq \theta_{HUELB}) \geq \alpha \tag{14d}$$

$$\mathbf{P}(\sum_{j=1}^{N_{Ch}} \xi_{v,j}(t)\gamma_v^j(t) \geq \gamma_{MUELB} \Big| \sum_{j=1}^{N_{Ch}} \xi_{v,j}(t)CQI_v^j(t-1),$$

$$\sum_{j=1}^{N_{Ch}} \xi_{v,j}(t)CQI_v^j(t-2)) \geq \beta \tag{14e}$$

$$u = 1, 2...N_{HUE}, \quad v = 1, 2, ...N_{MUE}, \quad c = 1, 2, ...N_{Ch},$$

where $\xi_{v,j}(t)$ is the participation indicator at time t based on overheard scheduling information:

$$\xi_{v,j}(t) = \begin{cases} 0, & \text{if Channel } j \text{ is not scheduled to MUE } v \\ 1, & \text{if Channel } j \text{ is scheduled to MUE } v \end{cases}.$$

Note that [12] provides a Gaussian approximation for the objective function as:

$$\mathbf{E}[\log_2(1 + \theta_u^c(t))] \approx \log_2(1 + \mathbf{E}[\theta_u^c(t)]) - \frac{\mathbf{V}[\theta_u^c(t)]}{2(1 + \mathbf{E}[\theta_u^c(t)])^2}, \tag{15}$$

which provides our approximate objective function:

$$\max_{P_{u,c}} \mathbf{E}[C_{out}^I(t)] = \max_{P_{u,c}} (\sum_{i=1}^{N_{HUE}} (\sum_{j=1}^{N_{Ch}} a_{i,j}(t)(\log_2(1 + \mathbf{E}[\theta_u^c(t)]) - \frac{\mathbf{V}[\theta_u^c(t)]}{2(1 + \mathbf{E}[\theta_u^c(t)])^2}))) \tag{16}$$

The constraints shown in Eqs. (14b) and (14c) aim to ensure that each channel is occupied once and that each HUE gets only one channel respectively.

While the constraints in Eqs. (14d) and (14e) are to guarantee that there exists a minimum acceptable SIR and SINR levels for each HUE and MUE to sustain a reliable transmission with the FBS and MBS respectively.

In order to solve the shown problem, we need to analyze the distributions of $\gamma_v^c(t)$ and $\theta_u^c(t)$ as well as calculating the first order statistics of $\theta_u^c(t)$.

3 SIR and SINR Distribution Analysis

3.1 SIR Distribution Analysis

According to the channel model of (5), we define

$$\bar{H}_{F-u}^c(t) = H_{F-u}^c(t)/\sigma_{F-u}, \tag{17}$$

$$\bar{H}_{M-u}^c(t) = H_{M-u}^c(t)/\sigma_{M-u}, \tag{18}$$

Therefore

$$\theta_u^c(t) = \frac{P_{u,c}(t)\sigma_{F-u}^2\left|\bar{H}_{F-u}^c(t)\right|^2}{P_M^c(t)\sigma_{M-u}^2\left|\bar{H}_{M-u}^c(t)\right|^2}, \tag{19}$$

where the random variable $\bar{H}_{F-u}^c(t) \sim \mathcal{CN}(\frac{\hat{H}_{F-u}^c(t)}{\sigma_{F-u}}, 2)$ and accordingly

$$\left|\bar{H}_{F-u}^c(t)\right|^2 \sim \chi'^2(2, \left|\frac{\hat{H}_{F-u}^c(t)}{\sigma_{F-u}}\right|^2). \tag{20}$$

And similarly we have

$$\left|\bar{H}_{M-u}^c(t)\right|^2 \sim \chi'^2(2, \left|\frac{\hat{H}_{M-u}^c(t)}{\sigma_{M-u}}\right|^2). \tag{21}$$

Then we will have

$$\theta_u^c(t) = m\underbrace{\frac{\left|\bar{H}_{F-u}^c(t)\right|^2}{\left|\bar{H}_{M-u}^c(t)\right|^2}}_{\bar{H}}. \tag{22}$$

Therefore, from [13] we conclude that $\bar{H}$ has a doubly non-central F-Distribution with parameters $(2, 2, \left|\hat{H}_{F-u}^c(t)/\sigma_{F-u}\right|^2, \left|\hat{H}_{M-u}^c(t)/\sigma_{M-u}\right|^2)$, from which the probability density function (PDF) of $\theta_u^c(t)$ is also known.

3.2 SINR Distribution Analysis

In order to evaluate the constraint shown in equation (14e), we need to calculate the cumulative distribution function (CDF) of $\gamma_v^c(t)$. To do so we will start by substituting in Eq. (10) by Eqs. (11) and (12), to get the form in equation (23)

$$\gamma_v^c(t) = \frac{K_1\gamma_v^c(t-1)\gamma_v^c(t-2)}{K_2\gamma_v^c(t-1) + K_3\gamma_v^c(t-2)}, \tag{23}$$

where

$$K_1 = P_M^c(t)(P_F^c(t-2) - P_F^c(t-1))$$
$$K_2 = P_M^c(t-2)(P_F^c(t) - P_F^c(t-1))$$
$$K_3 = P_M^c(t-1)(P_F^c(t-2) - P_F^c(t))$$

and since we do not know the exact values of $\gamma_v^c(t-1)$ and $\gamma_v^c(t-2)$, we only can overhear their CQI level as we mentioned earlier. Therefore we can model $\gamma_v^c(t-1)$ and $\gamma_v^c(t-2)$ as random variables uniformly distributed within the known interval based on the CQI level.

Starting from equation (23), at $K_2 \neq 0$

$$\gamma_v^c(t) = (\frac{K_1}{K_2}) \frac{\gamma_v^c(t-1)\gamma_v^c(t-2)}{\gamma_v^c(t-1) + \frac{K_3}{K_2}\gamma_v^c(t-2)}, \tag{24}$$

$$\gamma_v^c(t) = (\frac{K_1}{K_2})S, \tag{25}$$

where

$$S = \frac{\gamma_v^c(t-1)\gamma_v^c(t-2)}{\gamma_v^c(t-1) + k\gamma_v^c(t-2)}, \tag{26}$$

and $k = K_3/K_2$. Applying Eq. (26) and PDFs of $\gamma_v^c(t-1)$ and $\gamma_v^c(t-2)$, we evaluated a closed-form PDF of S which is then used to determine $\gamma_v^c(t)$ PDF and CDF.

4   Proposed Solution

In this section we will introduce a solution to the given problem by first focusing on a power selection policy.

4.1   Optimum Power Level Selection

Considering the objective function (14a), Eqs. (14b) and (14c) guarantee that no channel assigned to more than one HUE and that every HUE gets only one channel, while equations (14d) and (14e) specify the minimum and maximum power limits respectively. Thus, for a valid assignment we can rewrite our problem as follows:

$$\max_{P_{u,c}} \mathbf{E}[C_{out}^I(t)] = \max_{P_{u,c}}(\sum_{i=1}^{N_{HUE}} (\sum_{j=1}^{N_{Ch}} a_{i,j}(t)\mathbf{E}[\log_2(1 + \theta_i^j(t))])) \tag{27a}$$

s.t.

$$P_{u,c}^{min} \leq P_{u,c} \leq P_{u,c}^{max} \tag{27b}$$

Lemma 1: For the objective function (27a) with any valid channel assignment, if there exists $P_{u,c}$ for HUE u to satisfy (27b), then its optimum power assignment equals $P_{u,c}^{max}$.

Proof: We can show that our objective function is monotonically increasing in $P_{u,c}$. The details are omitted here.

4.2   Main Structure of the Solution Algorithm

In order to explain the proposed solution, we will first describe the reduction/transformation used to transfer the given problem equivalently into an assignment problem. The term "problem reduction" is very popular in complexity theory. The main idea is in transform underlying problem from an unknown form (non-convex optimization problem) to a known one such that there exists an optimal and efficient algorithm to solve it. One common use of problem reduction is to show that a specific problem belongs to a certain class of complexity like P, NP and NP-complete. This reduction is based on the analytical results from the previous sections and it is described in the algorithm given below. These steps should be made regardless of the method we will use later to solve the assignment problem.

Algorithm: Optimum Channel Allocation (main structure)

---

1. Combining the calculated CDF of $\gamma_v^c(t)$ (section 3.2) and the constraint in (14e), we will be able to evaluate the maximum power $(P_{u,c}^{max})$ for all the available RBs.
2. The results of section 3.1, enable us to calculate the distribution of the random variable $\theta_{u,c}^{max}(t)$ as well as its first order statistics for each HUE at each RB.
3. Using the first order statistics of $\theta_{u,c}^{max}(t)$, we will be able to calculate the maximum capacity for each HUE on every channel.
4. In order to apply the constraint in equation (14d) we will use the $\theta_{u,c}^{max}(t)$ CDF to verify that all HUEs SIR exceeds θ_{HUELB}, otherwise exclude this channel assignment from the result.

Result: a lookup (rate) table $r(i,j)$ representing the maximum capacity for each HUE at each channel ($i = 1, ... N_{HUE}$ and $j = 1, ... N_{Ch}$)

---

where $\theta_{u,c}^{max}(t)$ is the maximum SIR of the HUE u on channel c at time t

$$\theta_{u,c}^{max}(t) = \frac{P_{u,c}^{max}(t)\left|H_{F-u}^c(t)\right|^2}{P_M^c(t)\left|H_{M-u}^c(t)\right|^2}. \tag{28}$$

Basically we start our solution by using the results in section 3 to calculate the lookup (rate) table or matrix $\mathbf{R}$ (where $\mathbf{R} = [r(i,j)]$ for $i = 1, ... N_{HUE}$ and $j = 1, .. N_{Ch}$). After completing the 4 steps, we will proceed to find the channel assignment to maximize the objective function.

4.3   Channel Assignment Algorithms

Given matrix $\mathbf{R}$ which represents the lookup table, we can also view this as the edge weight matrix of a bipartite graph. On one end of the bipartite graph are the user nodes, while on the opposite end of the bipartite graph are the available channels. To find the best pairing to maximum the sum rate, we can either use a simpler greedy algorithm or resort to the well known Hungarian Algorithm designed to solve such assignment problem optimally in shorter time.

Greedy Algorithm. The lookup table is a matrix $\mathbf{R}$ with HUEs as rows and channels as columns. We can determine the suboptimum channel assignment by applying a greedy algorithm to find the maximum pairing in each iteration. Let a matrix $\mathbf{P_1} = \mathbf{R}$. For the i-th iteration, our greedy algorithm find the maximum element in matrix $\mathbf{P_i}$ as a pairing choice before forming the next matrix $\mathbf{P_{i+1}}$ by removing the corresponding row and column of the maximum element from $\mathbf{P_i}$. We continue until all HUEs or channels are exhausted.

In the greedy algorithm, successful HUE acquires the maximum capacity from the available channels regardless the remaining HUEs. Although the complexity of greedy solution is very low and its time consumption grows linearly with increasing problem size, it is generally not optimal.

Hungarian Algorithm. Starting from the rate lookup table, our problem is viewed as an assignment problem in which the Hungarian algorithm has proven to solve optimally and in polynomial time [14][15]. The Hungarian algorithm is a combinatorial optimization algorithm first introduced in 1955 to solve an equivalent assignment problem [14].

In order to achieve the optimum channel assignment we add one more step on the algorithm main structure in section 4.2 by adopting the Hungarian algorithm. Unlike [16], we did not use the Hungarian algorithm to work on the original scheduling problem which may result near optimal solutions. Instead, we used Hungarian algorithm to determine the optimal combination from the rate lookup table.

5   Performance Evaluation

We will present our simulation results in three parts, in the first part we verify the analysis in section 3 with numerical examples. We will compare the two proposed solutions in the second part. The third part compares the results according to our channel gain estimation error assumption and the assumption of zero channel gain estimation error.

Fig. 2 shows both numerical and analytical distributions of MUE SINR, from which we can see excellent verification of the analytical results in section 3. We also presents both numerical and analytical distributions of HUE SIR, from which we also observe evident verification.

In Fig. 3, we present the maximum capacity against the estimation error standard deviation. We plot the Hungarian algorithm solution along with the solution from the greedy algorithm. Clearly, the Hungarian algorithm achieves optimal solution. The results from the greedy algorithm show sub-optimality but require lower complexity ($O(n)$) while the Hungarian algorithm requires $O(n^3)$ [17]. In Fig. 4 we compare two results: one from being ignorant of the channel estimation error existing in the channel gain by assigning resources based on purely the estimated channel (assuming estimation error $= 0$), another account for the channel estimation error and assign the resources based on this consideration as in our proposed algorithms.

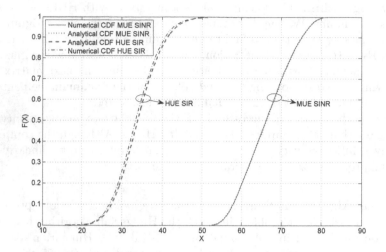

Fig. 2. The numerical and analytical distributions of the $\gamma_v^c(t)$ and $\theta_u^c(t)$.

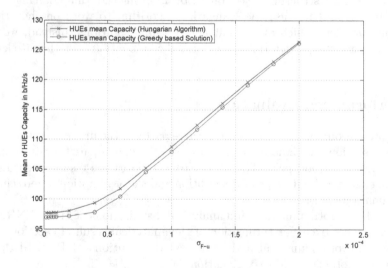

Fig. 3. The Hungarian and Greedy algorithms results.

For small variance in channel estimation error, the results of both cases are nearly the same. However, as channel estimation error variance grows, the first result starts to deteriorate to less than optimal while the second results remains optimum (as circled and diamond curves).

Moreover in the first case, the total capacity estimation is constant (Asterisk line) regardless of the value of the real capacity (diamond line) and the total

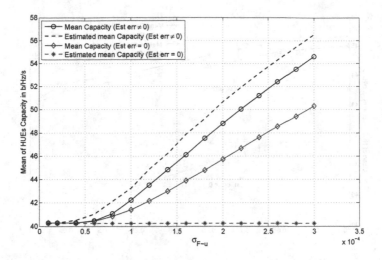

Fig. 4. Results due to zero and non-zero estimation error assumptions.

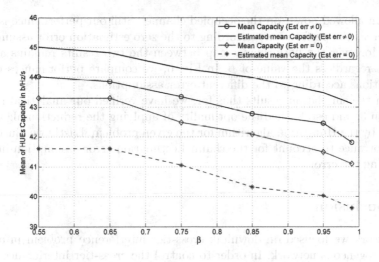

Fig. 5. Results according to error assumptions for different β.

capacity estimation error increase with growing channel error variance. On the other hand, for the second case, the total capacity estimation (dashed line) tracks the actual capacity (circled line).

Finally, in Fig. 5 and Fig. 6 we compare the performance of the two error assumptions illustrated earlier for different β. In Fig. 5, we can see that as β increases the total capacity decreases. This follows from (14e), as β affects the

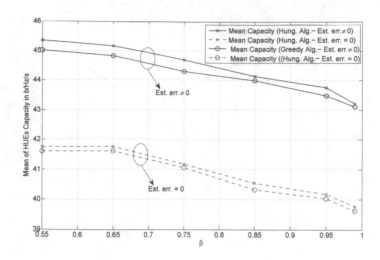

Fig. 6. Solutions results according to error assumptions for different β.

maximum allowed power on the occupied channel. Still our performance is optimal whereas the performance according to the zero estimation error assumption is not. Moreover the performance gap between the two results remains almost constant regardless the value of β. In Fig. 6, we compare performances of the two solutions according to the different error assumptions.

From the simulation results thus far, we have verified our analytical results in section 3, and established the optimality of applying the reduction algorithm followed by the Hungarian algorithm for the given problem. Lastly we illustrated the importance to account for the channel estimation error assumption instead of ignoring the error.

6   Conclusion

In this work we focused on downlink cross-tier interference problem in a two-tier heterogeneous network. In order to control the cross-tier interference while maximizing the femtocell capacity, we exploit the cognitive capabilities of FBS to acquire nearby MUE scheduling. Our problem formulation take into account channel estimation error and we provide full analysis for the distribution of the HUE SIR and MUE SINR. Both analytical results are verified via simulations. Based on our analysis we developed a problem reduction method for the given problem. We also suggested two solutions for the reduced problem based on the Hungarian and greedy algorithms, respectively, with demonstrated simulation results.

References

1. Chandrasekhar, V., Andrews, J., Gatherer, A.: Femtocell networks: a survey. IEEE Communications Magazine **46**(9), 59–67 (2008)
2. Al-Rubaye, S., Al-Dulaimi, A., Cosmas, J.: Cognitive femtocell. IEEE Vehicular Technology Magazine **6**(1), 44–51 (2011)
3. Andrews, J., Claussen, H., Dohler, M., Rangan, S., Reed, M.: Femtocells: Past, present, and future. IEEE Journal Selected Areas in Communications **30**(3), 497–508 (2012)
4. Wang, W., Yu, G., Huang, A.: Cognitive radio enhanced interference coordination for femtocell networks. IEEE Communications Magazine **51**(6), 37–43 (2013)
5. Oh, D., Lee, H., Lee, Y.: Cognitive radio based femtocell resource allocation. In: International Conference on Information and Communication Technology Convergence (ICTC), pp. 274–279, November 2010
6. Sun, D., Zhu, X., Zeng, Z., Wan, S.: Downlink power control in cognitive femtocell networks. In: International Conference on Wireless Communications and Signal Processing (WCSP), pp. 1–5, November 2011
7. Zhang, L., Yang, L., Yang, T.: Cognitive interference management for LTE-A femtocells with distributed carrier selection. In: IEEE 72nd Vehicular Technology Conference Fall (VTC 2010-Fall), pp. 1–5, September 2010
8. Xie, R., Yu, F., Ji, H.: Spectrum sharing and resource allocation for energy-efficient heterogeneous cognitive radio networks with femtocells. In: IEEE International Conference on Communications (ICC), pp. 1661–1665, June 2012
9. Lien, S., Lin, Y., Chen, K.: Cognitive and game-theoretical radio resource management for autonomous femtocells with QoS guarantees. IEEE Transactions on Wireless communication **10**(7), 2196–2206 (2011)
10. Karp, R.M.: Reducibility among combinatorial problems. In: A Symposium on the Complexity of Computer Computations, pp. 85–103 (1972)
11. Yoo, T., Goldsmith, A.: Capacity and power allocation for fading mimo channels with channel estimation error. IEEE Transactions on Information Theory **52**(5), 2203–2214 (2006)
12. Teh, Y., Newman, D., Welling, M.: A collapsed variational Bayesian inference algorithm for latent Dirichlet allocation. Advances in Neural Information Processing Systems **19**, 1353–1360 (2007)
13. Walck, C.: Handbook on Statistical Distributions for Experimentalists. University of Stockholm press, Sweden (2000)
14. Kuhn, H.W.: The hungarian method for the assignment problem. Naval Research Logistics **52**(1), 7–21 (2005)
15. Munkres, J.: Algorithms for the assignment and transportation problems. Society for Industrial and Applied Mathematics **5**(1), 32–38 (1957)
16. Tamura, S., Kodera, Y., Taniguchi, S., Yanase, T.: Feasiblity of hungarian algorithm based scheduling. In: IEEE International Conference on Systems Man and Cybernetics (SMC), pp. 1185–1190, October 2010
17. Bellur, U., Kulkarni, R.: Improved matchmaking algorithm for semantic web services based on bipartite graph matching. In: IEEE International Conference on Web Services, ICWS 2007, pp. 86–93, July 2007

Networking Protocols for CR

Optimization of Collaborative Spectrum Sensing with Limited Time Resource

Fariba Mohammadyan[1]([✉]), Zahra Pourgharehkhan[1], Abbas Taherpour[1], and Tamer Khattab[2]([✉])

[1] Department of Electrical Engineering, Imam Khomeini International University, Qazvin, Iran
{f_mohammadyan,pourgharehkhan}@edu.ikiu.ac.ir, taherpour@eng.ikiu.ac.ir
[2] Electrical Engineering, Qatar University, Doha, Qatar
tkhattab@ieee.org

Abstract. In this paper, Cognitive Radios (CRs) collaborate in spectrum sensing to detect random signals corrupted by Gaussian noise. Our analysis is based on a limited time resource assumption. This implies that the time resource dedicated for cooperative spectrum sensing process is constrained and shared between spectrum sensing time and results reporting time, which depends on the number of sensing users. We use common weighted gain combining detector to detect presence or absence of Primary User (PU). In order to find optimum gains, number of users and detection threshold, we maximize the achievable throughput with two approaches so that the predefined constraints on detection and false alarm probabilities are satisfied to protect the cooperative network performance quality. Analytical results in addition to simulation results show that the proposed schemes significantly outperform similar traditional detectors.

1 Introduction

The collaboration or cooperation among multiple Secondary Users (SUs) is one of the efficient approaches to make a reliable and accurate spectrum sensing in wireless channels where a single SU's sensing capability will be limited due to the deleterious channel effects such as shadowing [1–3]. Although, collaboration of SUs has a significant impact on decreasing the error probability of identifying the accurate status of the spectrum, it has some challenges: a large delay occurs for making final decision and CR network may be more affected by external attacks [4] and especially, the energy consumed in CR network is increased. Therefore, the analysis of the energy efficiency of cooperative spectrum sensing must be investigated before making any conclusions on the actual benefits of this approach. The energy efficiency of cooperative spectrum sensing has been investigated in many papers. However, the results available in the literature are not often directly comparable since the analysis is performed under different assumptions. Many works have investigated the optimization of the number of sensing users for several objectives. The problem was firstly formulated by [5],

© Institute for Computer Sciences, Social Informatics and Telecommunications Engineering 2015
M. Weichold et al. (Eds.): CROWNCOM 2015, LNICST 156, pp. 109–122, 2015.
DOI: 10.1007/978-3-319-24540-9_9

where the number of users is optimized to maximize a target function combining the detection performance and the usage efficiency of the resources. In [6] a new energy-efficient CSS scheme is investigated which implies that only a SU will broadcast its local decision among the whole network and other SUs will object to the fusion center, or agree with the announced decision. Using the proposed scheme, the broadcasting SU is selected so that to maximize energy efficiency. Several robust collaborative spectrum sensing schemes are presented in [7] wherein a trust value for each secondary user is obtained to reflect its suspicious level and mitigate its harmful effect on cooperative sensing. The motivation of the paper is to investigate the problem of the cooperative spectrum sensing considering limitation on time resources to make CR network more practically efficient. In order to have an efficient spectrum sensing with a controlled time, we suppose that a synchronous slotted communication protocol with duration T is employed by PU, in which SUs should perform sensing, result reporting and transmitting data operations according to PU's time slot. It is supposed that a fixed part of total time frame is dedicated for data transmission, while the rest is divided between local sensing and results reporting. The reporting channel between SUs and FC is considered Time Division Multiple Access (TDMA). So, the restriction on time duration of cooperative spectrum sensing causes relationship between number of SUs and their sample numbers. Unlike to the most of other works which assume a fixed sensing time and variable data/reporting times [8], our model does not affect data transmission and thus, makes cooperation a less ineffective process. In our approach, it is assumed that Fusion Center (FC) uses a useful and popular detector known as weighted gain combining (WGC)[9,10]. The WGC has better performance than the other energy detection-based detectors. In order to find optimum number of SUs, weighting gains vector and decision threshold, we maximize total achievable throughput which has an important role in efficiency of data transmission. Unlike to similar studied works available in literature, we analytically prove that our optimization problems are convex to make sure that derived optimal solutions are global. In studied optimization problems, we consider some constraints on number of SUs, predefined detection and false alarm probabilities to protect network requirements. We show that our proposed method outperforms the conventional WGC detectors in considered problems.

2   Basic Assumptions and System Model

Suppose there are N SUs available which are interested to detect presence or absence of the PU signal in a special frequency band and each SU receives M independent samples from the PU signal. A centralized topology is considered for secondary network in which the distance between users in secondary network is negligible compared to the distance between PU and SUs. Individual SUs use energy detector and send their sensing test statistics to FC through a control channel and in FC, final decision on presence or absence of PU is taken and then shared between SUs. We consider two basic assumptions for hypothesis testing problem and frame duration as follows:

Hypothesis Testing Problem. Regardless of any collaboration among the SUs, each SU has to decide individually based on its own received samples. In this case, the spectrum sensing for each SU in a wireless channel at m^{th} time instant can be modeled as a binary hypothesis testing problem as

$$\mathbf{y}_i(m) = \begin{cases} \mathbf{v}_i(m) & , \mathcal{H}_0 \\ h_i\,\mathbf{x}(m) + \mathbf{v}_i(m) & , \mathcal{H}_1 \end{cases} ; i = 1, 2, ..., N\,, \tag{1}$$

where $\mathbf{y}_i \in \mathbb{C}^M$ is the complex signal received by i^{th} SU, $\mathbf{h}_i$ is the channel gain between the PU and the ith SU which is assumed that changes slowly such that it can be considered to be constant during each operation period of interest [8]. Also, we assume that $\mathbf{x} \sim \mathcal{CN}(\mathbf{0}, \sigma_x^2 \mathbf{I}_M)$ is the vector of the PU signal samples, and $\mathbf{v}_i \sim \mathcal{CN}(\mathbf{0}, \sigma_v^2 \mathbf{I}_M)$ is the vector of additive noise samples at the ith SU. In order to do more accurate and faster spectrum sensing, the N SUs collaborate with each other by sharing information between themselves and after collaboration, the final collaborative decision about the absence or presence of the PU signal is made by the FC. Thus, for final collaborative decision at FC, we can write following binary hypothesis testing problem

$$\begin{cases} \mathcal{H}_0 : W < \eta, \text{PU is absent} \\ \mathcal{H}_1 : W > \eta, \text{PU is present.} \end{cases} \tag{2}$$

where η is the decision threshold and

$$W = \sum_{i=1}^{N} \mathbf{w}_i \mathbf{z}_i = \mathbf{w}^T \mathbf{z}. \tag{3}$$

is our total decision statistic at FC, where $\mathbf{w} = [\mathbf{w}_1, \mathbf{w}_2, ..., \mathbf{w}_N]^T$ is the combining coefficients vector and the elements of $\mathbf{z} = [\mathbf{z}_1, \mathbf{z}_2, ..., \mathbf{z}_N]^T$ are our local test statistics which are defined as

$$\mathbf{z}_i = \sum_{m=1}^{M} |\mathbf{y}_i(m)|^2 = \|\mathbf{y}_i\|^2 \tag{4}$$

Additionally, we assume that $\|\mathbf{w}\| = 1$. In accordance with [8] and [11], since the local test statistics ($\mathbf{z}_i$) are normally distributed, their linear combination would also be distributed normally. Consequently, for the performance of the proposed cooperative spectrum detection scheme at the FC, we have

$$P_{\text{fa}} = P[W > \eta \mid \mathcal{H}_0] = Q(\frac{\eta - M\sigma_v^2 \mathbf{w}^T \mathbf{1}}{\sqrt{2M\sigma_v^4 \mathbf{w}^T \mathbf{w}}}). \tag{5}$$

where $Q(x) = \frac{1}{\sqrt{2\pi}} \int_x^\infty \exp\left(-\frac{u^2}{2}\right) du$ is the tail probability of the standard normal distribution and $\mathbf{1}$ is a vector with all elements equal to one. In addition,

$$P_{\text{d}} = P[W > \eta \mid \mathcal{H}_1] = Q(\frac{\eta - M\mathbf{w}^T(\sigma_x^2 \mathbf{h} + \sigma_v^2 \mathbf{1})}{\sqrt{2M\sigma_v^4 \mathbf{w}^T \mathbf{C}\mathbf{w}}}). \tag{6}$$

where $\mathbf{h} = [|h_1|^2, |h_2|^2, ..., |h_N|^2]^T$ and $\mathbf{C} = diag\{[1 + 2\gamma]\}$ so that $\gamma = [\gamma_1^2, \gamma_2^2, ..., \gamma_N^2]^T$ and $\gamma_i^2 \triangleq \frac{|h_i|^2 \sigma_x^2}{\sigma_v^2}$ is the received Signal-to-Noise Ratio (SNR) at i^{th} SU.

Frame Duration Structure. The transmission is organized in frames of fixed time duration. The frame duration T is divided into three sub-frames: i) the sensing sub-frame of duration T_s, during which local sensing is performed; ii) the reporting sub-frame of duration T_r, where local results are reported to the FC; and iii) the data transmission sub-frame of duration T_t, where data transmission occurs if the channel is identified as free. As a consequence, $T = T_s + T_r + T_t$. We assume that T_t is given and fixed, while T_s and T_r are chosen in order to trade-off sensing and reporting reliability, respectively, such that T is kept fixed. The frame duration structure has been shown in Figure 1. If t_r is the time needed by each SU to report the sensed result to the FC, then $Tr = Nt_r$. It means that we have supposed the channel between SUs and the FC to be TDMA. Since T_t is assumed fixed, sensing duration can be expressed as

$$T_s = \underbrace{(T - T_t)}_{fixed} - T_r = T_{cte} - Nt_r \tag{7}$$

If $M = f_s T_s$ where f_s is sampling frequency, the number of sensing samples is expressed as a function of the number of SUs as follows

$$M = f_s(T_{cte} - Nt_r) \tag{8}$$

It can be observed that as N increases, sensing samples decreases.

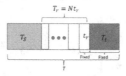

Fig. 1. The frame duration structure for cooperative spectrum sensing. Increasing the number of users yields decreasing sensing samples.

3   Optimization of Throughput

In this section, optimization of throughput function will be obtained with two distinct approaches. At first, we investigate joint optimization problem and then, we solve the optimization problem through optimizing a measure called *modified deflection coefficient*.

3.1 Joint Optimization

In order to create more chances for the SUs to send their data by higher rate when the frequency band is idle, the achievable throughput must be increased. The achievable throughput is defined here as [12]:

$$R(N, \mathbf{w}, \eta) = \pi_0 D_t T_t (1 - P_{\text{fa}}(N, \mathbf{w}, \eta)) \qquad (9)$$

where π_0 and D_t [bit/sec] are respectively the probability of the primary user being absent in the channel and the transmission rate. In addition, P_{fa} is the false alarm probability after replacing M from (8) which is shown as follow

$$P_{\text{fa}}(N, \mathbf{w}, \eta) = Q\left(\frac{\eta - f_s(T_{cte} - Nt_r)\sigma_v^2 \mathbf{w}^T \mathbf{1}}{\sqrt{2f_s(T_{cte} - Nt_r)\sigma_v^4 \mathbf{w}^T \mathbf{w}}}\right) \qquad (10)$$

Clearly, the higher throughput is achieved if the false alarm probability is decreased. On the other hand, more accurate and reliable cooperative spectrum sensing will be resulted when higher overall detection probability is provided. Therefore, we should make a compromise between the higher achievable throughput and more reliable sensing. From all above, the optimization problem can be defined as

$$\max_{\mathbf{w}, \eta, N} : \quad R(N, \mathbf{w}, \eta) \qquad (11a)$$

$$s.t. : \quad P_{\text{fa}} \leq \beta \qquad (11b)$$

$$P_{\text{d}} \geq \bar{P}_{\text{d}} \qquad (11c)$$

$$1 \leq N \leq N_{\max} \qquad (11d)$$

$$\|\mathbf{w}\| = 1 \qquad (11e)$$

$$\mathbf{w}, \eta > 0 \qquad (11f)$$

where P_{d} is the detection probability and $N_{\max} = \frac{T_{cte}}{t_r} - 1$ is obtained when we assum $T_s = 0$. Additionally, β and $\bar{P}_{\text{d}}$ are respectively the predefined constraints of the false alarm and detection probabilities to protect network performance quality, which are desired as $0 < \beta < \frac{1}{2}$ and $\frac{1}{2} < \bar{P}_{\text{d}} < 1$. From (8), it is obvious that the optimization of N is equal to optimization of M. Also, note that when transmission time T_t is constant, the minimization of the false alarm probability is equal to maximization of the achievable throughput. Thus, the optimization problem can be replaced by

$$\min_{\mathbf{w}, \eta, M} : \quad P_{\text{fa}}(M, \mathbf{w}, \eta) \qquad (12a)$$

$$s.t. : \quad P_{\text{fa}} \leq \beta \qquad (12b)$$

$$P_{\text{d}} \geq \bar{P}_{\text{d}} \qquad (12c)$$

$$t_r f_s \leq M \leq f_s(T_{cte} - t_r) \qquad (12d)$$

$$\|\mathbf{w}\| = 1 \qquad (12e)$$

$$\mathbf{w}, \eta > 0 \qquad (12f)$$

In order to solve the optimization problem (12), it is easily realized that the decision threshold should meet (12b) and (12c). Therefore, from (5) and (6) we have (13). On the other hand, P_{fa} and P_{d} are decreasing functions of η and so, to find the minimum value of false alarm probability, η should be maximized that causes reduction of P_{d}. As a consequence, the maximum value of the decision threshold which can satisfy $P_{\text{d}} = \bar{P}_{\text{d}}$ and minimize the objective function of problem (12) is

$$Q^{-1}(\beta)\sqrt{2M\sigma_v^4 \mathbf{w}^T\mathbf{w}} + M\sigma_v^2\mathbf{w}^T\mathbf{1} \leq \eta \leq \tag{13}$$
$$Q^{-1}(\bar{P}_{\text{d}})\sqrt{2M\sigma_v^4\mathbf{w}^T\mathbf{C}\mathbf{w}} + M\mathbf{w}^T(\sigma_x^2\mathbf{h} + \sigma_v^2\mathbf{1})$$

$$\eta_{\text{opt}} = Q^{-1}(\bar{P}_{\text{d}})\sqrt{2M\sigma_v^4\mathbf{w}^T\mathbf{C}\mathbf{w}} + M\mathbf{w}^T(\sigma_x^2\mathbf{h} + \sigma_v^2\mathbf{1}) \tag{14}$$

Thus, the optimization problem can be rewritten as

$$\min_{\mathbf{w},M} : \quad Q\left(\frac{Q^{-1}(\bar{P}_{\text{d}})\sqrt{2M\sigma_v^4\mathbf{w}^T\mathbf{C}\mathbf{w}} + M\sigma_x^2\mathbf{w}^T\mathbf{h}}{\sqrt{2M\sigma_v^4\mathbf{w}^T\mathbf{w}}}\right) \tag{15a}$$

$$s.t. : \quad P_{\text{fa}} \leq \beta \tag{15b}$$
$$t_r f_s \leq M \leq f_s(T_{cte} - t_r) \tag{15c}$$
$$\|\mathbf{w}\| = 1 \tag{15d}$$
$$\mathbf{w} > 0 \tag{15e}$$

To find the minimum value of objective function, one approach is to use convex optimization methods. Since we encounter with a complicated optimization problem, an efficient suboptimal method to solve (15) is to minimize the upper bound of its objective function. Using Rayleigh-Ritz theorem and (15d) and by noticing the fact that $Q^{-1}(\bar{P}_{\text{d}}) < 0$ (since $\bar{P}_{\text{d}} > \frac{1}{2}$), we have

$$Q\left(\frac{Q^{-1}(\bar{P}_{\text{d}})\sqrt{2M\sigma_v^4\mathbf{w}^T\mathbf{C}\mathbf{w}} + M\sigma_x^2\mathbf{w}^T\mathbf{h}}{\sqrt{2M\sigma_v^4\mathbf{w}^T\mathbf{w}}}\right)$$
$$= Q\left(Q^{-1}(\bar{P}_{\text{d}})\sqrt{\frac{\mathbf{w}^T\mathbf{C}\mathbf{w}}{\mathbf{w}^T\mathbf{w}}} + \frac{M\sigma_x^2\mathbf{w}^T\mathbf{h}}{\sqrt{2M\sigma_v^4\mathbf{w}^T\mathbf{w}}}\right)$$
$$\leq Q\left(Q^{-1}(\bar{P}_{\text{d}})\sqrt{\lambda_{\max}\mathbf{C}} + \frac{M'\mathbf{w}^T\boldsymbol{\gamma}}{\sqrt{\mathbf{w}^T\mathbf{w}}}\right)$$
$$= Q\left(Q^{-1}(\bar{P}_{\text{d}})\sqrt{\lambda_{\max}\mathbf{C}} + M'\mathbf{w}^T\boldsymbol{\gamma}\right) \tag{16}$$

where $M' \triangleq \sqrt{\frac{M}{2}}$ and $\lambda_{\max}\mathbf{C}$ denotes maximum eigenvalue of $\mathbf{C}$. Since matrix $\mathbf{C}$ is diagonal, the eigenvalues are simply recognizable on the diagonal of matrix, and when we assume the SNRs in descending order of their γ_i so that the 1st SU in the list has the highest received SNR ($\gamma_1 \geq \gamma_2 \geq ... \geq \gamma_N$), $\lambda_{\max}\mathbf{C}$ equals $1 + 2\gamma_1$. By minimizing the upper bound of the objective function, a good approximation to the optimal solution of the original problem is achieved.

Therefore, (15) can be reformulated into an equivalent form with an objective function upper bounded by a convex function as

$$\min_{\mathbf{w}, M'} : \quad f(M', \mathbf{w}) = Q\left(Q^{-1}(\bar{P}_d)\sqrt{\lambda_{\max}\mathbf{C}} + M'\mathbf{w}^T\boldsymbol{\gamma}\right) \tag{17a}$$

$$s.t. : \quad P_{fa} \leq \beta \tag{17b}$$

$$\sqrt{\frac{t_r f_s}{2}} \leq M' \leq \sqrt{\frac{f_s(T_{cte} - t_r)}{2}} \tag{17c}$$

$$\|\mathbf{w}\| = 1 \tag{17d}$$

$$\mathbf{w} > 0 \tag{17e}$$

Here, we have an optimization problem with $N + 1$ variables which is proved to be convex through following lemma:

Lemma 1. *The optimization problem (17) is convex with respect to M' and coefficient vector $\mathbf{w}$ if*

$$0 < \beta \leq Q\left(\frac{1}{-Q^{-1}(\bar{P}_d)\sqrt{\lambda_{\max}\mathbf{C}} + \sqrt{Q^{-1}(\bar{P}_d)^2\lambda_{\max}\mathbf{C} + 2}}\right)$$

Proof. See Appendix A.

Moreover, as mentioned before, in the throughput function all the elements are constant values, except P_{fa}. Thus, convexity of $P_{fa}(M', \mathbf{w})$, means concavity of $R(M', \mathbf{w})$ and so, the maximum value of throughput can be achieved easily. By applying the following proposed algorithm, the optimum values of $\mathbf{w}$ and N_{opt} are achieved. According to (8) and relation between M and M', there is

$$N_{opt} = \frac{T_{cte}}{t_r} - \frac{2M_{opt}'^2}{f_s t_r} \tag{18}$$

In this algorithm, the variable N is each time selected respectively from 1 to N_{max}, and every time for selected N, we have a vector variable with specified size, which is found from (17). Then, the objective function f is calculated every time and the values of N, $\mathbf{w}$ correspond to minimum one are interpreted as optimum values.

3.2 Optimization of Throughput by Maximizing Modified Deflection Coefficient

Here, we present an approach to solve optimization problem (11) via maximizing modified deflection coefficient. This measure is used for evaluating detection performance at the FC. When the test statistic W is normally distributed under both hypotheses, for a determined probability of false alarm, maximizing $d_N^2(\mathbf{w})$ leads to an increment of detection probability. Although the method incurs small performance degradation, due to its less computational complexity has been

---

Algorithm 1. Joint optimization algorithm for problem _____

1. Set $N_{max} = \frac{T_{cte}}{t_r} - 1$ and $f_0(M', \mathbf{w}) \triangleq \infty$
2. **for** $N = 1 : N_{max}$
3.   Find M_N from (8)
4.   Set $M'_N \triangleq \sqrt{\frac{M_N}{2}}$
5.   Find optimum N-dimensional vector $(\mathbf{w}_N^{opt})$ by (17)
6.   Put M'_N and $\mathbf{w}_N^{opt}$ in $f_N(M', \mathbf{w})$
7.   **if** $f_N(M', \mathbf{w}) > f_{N-1}(M', \mathbf{w})$ **then**
8. $f_{opt}(M', \mathbf{w}) = f_{N-1}(M', \mathbf{w})$,
            $\mathbf{w}_{opt} = \mathbf{w}_{N-1}^{opt}$, $M'_{opt} = M'_{N-1}$
9.   Using M'_{opt}, obtain N_{opt} from (18).
10.   **end if**
11. **end for**

---

interesting in the literature. This method is completely interpreted in [8] and [11]. By applying this method, we are able to find optimum weight vector value and replace it in the optimization problem. The modified deflection coefficient is defined as

$$d_N^2(\mathbf{w}) = \frac{(\mathbb{E}[W|\mathcal{H}_1] - \mathbb{E}[W|\mathcal{H}_0])^2}{Var[W|\mathcal{H}_1]} = \frac{f_s(T - Nt_r)(\mathbf{w}^T \gamma)^2}{2\mathbf{w}^T \mathbf{C} \mathbf{w}}$$

(19)

Hence, we have to maximize $d_N^2(\mathbf{w})$ while having a constraint on the weight vector to be on the unit-norm ball. So

$$\max_{\mathbf{w}} : \quad d_N^2(\mathbf{w}) \tag{20a}$$

$$s.t. : \quad \| \mathbf{w} \|_2 = 1 \tag{20b}$$

We can rewrite equation (19) to obtain

$$\frac{f_s(T - Nt_r)\mathbf{w}^T \gamma \gamma^T \mathbf{w}}{2\mathbf{w}^T \mathbf{C} \mathbf{w}} = \frac{f_s(T - Nt_r)\mathbf{w}'^T \mathbf{C}^{-\frac{T}{2}} \gamma \gamma^T \mathbf{C}^{-\frac{1}{2}} \mathbf{w}'}{2\mathbf{w}'^T \mathbf{w}'}$$

$$\leq \frac{f_s(T - Nt_r)}{2} \lambda_{max}(\mathbf{C}^{-\frac{T}{2}} \gamma \gamma^T \mathbf{C}^{-\frac{1}{2}}) \tag{21}$$

where, $\mathbf{w}'$ is defined as $\mathbf{w}' = \mathbf{C}^{\frac{1}{2}} \mathbf{w}$, and inequality results from Rayleigh-Ritz. Equality incurs when $\mathbf{w}'$ equals to eigenvector which is corresponded to maximum eigenvalue. Noting that $\mathbf{w}$ is a normalized vector, we can obtain it as

$$\mathbf{w}'_{opt} = \mathbf{C}^{-\frac{T}{2}} \gamma \rightarrow \mathbf{w}_{opt} = \frac{\mathbf{C}^{-1} \gamma}{\|\mathbf{C}^{-1} \gamma\|_2} \tag{22}$$

Now, we aim to solve problem (11). After we put $\mathbf{w}_{opt}$ in objective function, problem will change into

$$\max_{\eta,N} : \quad R(N,\eta) \tag{23a}$$

$$s.t. : \quad P_{\text{fa}} \le \beta \tag{23b}$$

$$P_{\text{d}} \ge \bar{P}_{\text{d}} \tag{23c}$$

$$1 \le N \le N_{\max} \tag{23d}$$

$$\eta > 0 \tag{23e}$$

As seen in Section A, we can minimize P_{fa} instead of maximizing throughput and so, the problem is equal to

$$\min_{\eta,N} : \quad Q\left(\frac{\eta\|\mathbf{C}^{-1}\gamma\| - f_s(T_{cte} - Nt_r)\sigma_v^2(\gamma^{\mathbf{T}}\mathbf{C}^{-T})\mathbf{1}}{\|\mathbf{C}^{-1}\gamma\|\sqrt{2f_s(T_{cte} - Nt_r)\sigma_v^4}}\right) \tag{24a}$$

$$s.t. : \quad P_{\text{fa}} \le \beta \tag{24b}$$

$$P_{\text{d}} \ge \bar{P}_{\text{d}} \tag{24c}$$

$$1 \le N \le N_{\max} \tag{24d}$$

$$\eta > 0 \tag{24e}$$

After rewriting constraint (24b) and (24c), we have (25) and (26).

$$Q^{-1}(\beta)\sqrt{2f_s(T_{cte} - Nt_r)\sigma_v^4} + f_s(T_{cte} - Nt_r)\sigma_v^2\frac{\gamma^{\mathbf{T}}\mathbf{C}^{-T}\mathbf{1}}{\|\mathbf{C}^{-1}\gamma\|} \le \eta \tag{25}$$

and

$$\eta < \frac{Q^{-1}(\bar{P}_{\text{d}})\sqrt{2f_s(T_{cte} - Nt_r)\sigma_v^4\gamma^{\mathbf{T}}\mathbf{C}^{-T}\gamma}}{\|\mathbf{C}^{-1}\gamma\|} + \frac{f_s(T_{cte} - Nt_r)\gamma^{\mathbf{T}}\mathbf{C}^{-T}(\sigma_x^2\mathbf{h} + \sigma_v^2\mathbf{1})}{\|\mathbf{C}^{-1}\gamma\|} \tag{26}$$

Similar to Section A, the optimum value of η occurs when η equals to its upper bound. Hence, by putting η in (24), the optimization problem turns into a single variable problem

$$\min_{N} : \quad Q\left(\frac{Q^{-1}(\bar{P}_{\text{d}})\sqrt{\gamma^{\mathbf{T}}\mathbf{C}^{-T}\gamma}}{\|\mathbf{C}^{-1}\gamma\|} + \frac{\sqrt{f_s(T_{cte} - Nt_r)}\gamma^{\mathbf{T}}\mathbf{C}^{-T}\gamma}{\sqrt{2}\|\mathbf{C}^{-1}\gamma\|}\right) \tag{27a}$$

$$s.t. : \quad 1 \le N \le N_{\max} \tag{27b}$$

Lemma 2. *The optimization problem (27) is convex in N and so, $R(N)$ is concave.*

To proof the lemma, the second derivative of objective function is here

$$\frac{\partial^2 P_{\text{fa}}}{\partial N^2} = \frac{f_s^2 t_r^2(\gamma^T\mathbf{C}^{-T}\gamma)^2(\rho)\exp\left(-\frac{\rho^2}{2}\right)}{8\sqrt{2\pi}f_s(T_{cte} - Nt_r)\|\mathbf{C}^{-1}\gamma\|^2} \ge 0 \tag{28}$$

where, $\rho = \dfrac{Q^{-1}(\bar{P}_\mathrm{d})\sqrt{\gamma^\mathbf{T}\mathbf{C}^{-T}\gamma}}{\|\mathbf{C}^{-1}\gamma\|} + \dfrac{\sqrt{f_s(T_{cte}-Nt_r)}\gamma^T\mathbf{C}^{-T}\gamma}{\sqrt{2}\|\mathbf{C}^{-1}\gamma\|}$

and also, $\dfrac{\partial^2 R(N)}{\partial N^2} = -\pi_0 D_t T_t \dfrac{\partial^2 P_\mathrm{fa}}{\partial N^2} \le 0$ which proves concavity of $R(N)$.

4   Numerical Results and Discussion

In this section, some simulation results are provided to evaluate the optimization problems and ensure the accuracy of the calculations. We have assumed that SNR of jth SU in dB domain equals to γ_j. The basic parameters which are determined fixed in simulation results are as :

1. Frequency of sampling in each SU: $f_s = 10\ kHz$
2. Time of reporting the results to the FC by each SU: $t_r = 0.2\ ms$
3. Time of transmission if frequency is detected as idle: $T_t = 3\ ms$

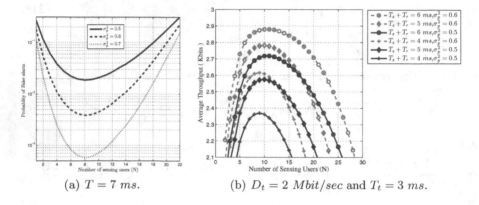

(a) $T = 7\ ms$.   (b) $D_t = 2\ Mbit/sec$ and $T_t = 3\ ms$.

Fig. 2. False alarm probability and Throughput with respect to number of SUs.

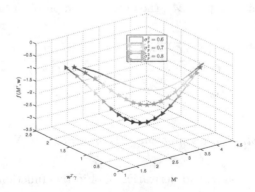

Fig. 3. Cost function of problem (17) versus M' and $\mathbf{w}^T\gamma$ in different values of σ_p^2 when $T = 7\ ms$ and $P_\mathrm{d} = 0.9$.

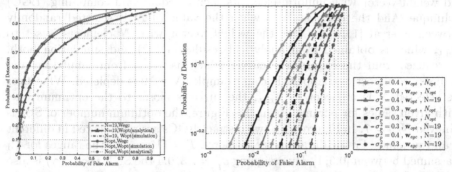

(a) Comparison between ROCs (b) Logarithmic ROCs with $N_{opt} = 10$, when $\sigma_p^2 = 0.3$ and $T = 9\ ms$. $T = 11\ ms$ and P_d changes between $(0.6, 1)$.

Fig. 4. Receiver Operation Characteristic (ROC)

4. Achievable throughput rate: $D_t = 2\ Mbit/sec$
5. Desired false alarm and detection probabilities to protect the QoS: $\beta = 0.1$ and $\bar{P}_d = 0.9$
6. The prior probabilities are both assumed: $\pi_1 = \pi_0 = 0.5$.
7. Finally, it is assumed that spectrum sensing performs in low SNR regime.

Channel gain between PU and SUs is assumed such a slow flat fading channel. We note that the number of SUs is limited by $N_{\max} = \frac{T_{cte}}{t_r} - 1$, that changes in different figures by changing the whole frame of time, So the optimum value of N is also changed. In Figure 2(a), using above algorithm, we depict the probability of false alarm versus number of SUs for three different cases of $\sigma_p^2 = \frac{\sigma_q^2}{\sigma_v^2} \triangleq$ 0.5, 0.6, 0.7 when we have set $T_{cte} = 4\ ms$. In each curve, the change in the number of users, causes variation in size of channel gain vector and as a result the SNR vector (γ) is also changed. So, the SNR of SUs are respectively arranged in these ranges:$(-12.5, -4.5)dB, (-12, -3.5)dB, (-11, -3)dB$ respectively. Now, looking at Figure 2(b), we can evaluate achievable throughput in three different cases where $T_s + T_r$ varies from $4\ ms$ to $6\ ms$ and as a result, the frame duration changes: $T = 7\ ms, T = 8\ ms, T = 9\ ms$. Also, it is supposed that $\sigma_p^2 = 0.8$. From this Figure, we find out that, increasing the value of sensing and reporting time, leads to higher achievable throughput. In Figure 3, using above algorithm, we depict the logarithmic cost function of problem (17) versus parameters M' and $\mathbf{w}^T \gamma$ for three different cases of $\sigma_p^2 = 0.6, 0.7, 0.8$ when we have set $T_{cte} = 4\ ms$ and $P_d = 0.9$. Having these parameters it is obvious that the minimum point is $(3.3, 1.57, -3.5)$ for $\sigma_p^2 = 0.8$ as an example. We can see that with a bit increase in the range of SNRs, $f(M', \mathbf{w})$ decreases significantly. We have illustrated the receiver operating characteristics (ROC) scheme of spectrum sensing in different states in Figure 4(a). It is assumed that $\sigma_p^2 = 0.3$ and $T = 9\ ms$. A comparison between the optimum state –which uses optimum values for variables N and $\mathbf{w}$– with two other cases is shown. One of them is obtained by allocating N_{opt}

and weight vector with uniform elements such as equal gain combining (EGC) technique, And the other state is when the value of N is selected randomly between interval $[1, N_{max}]$, but the weight vector with N elements is set to $\mathbf{w}_{opt}$, which is obtained from analytical results in (22) and simulation both. As realized from the Figure, the curves which use $\mathbf{w}_{opt}$ from simulation and analysis are almost overlapped. For an example, $N = 19$ is depicted. What ever the selected N is adjacent to N_{opt}, the curve is closer to the optimum one. With given values for parameters in this Figure, the optimum number of SUs is achieved $N_{opt} = 7$. In Figure 4(b), two distinct ROCs are represented in different N and $\mathbf{w}$ values. Frame duration is considered $T = 11$ ms and the ranges for P_d is assumed between $[0.6, 1]$. By assigning $\sigma_p^2 = 0.3, 0.4$ the SNR vector(γ) is also changed. In each curve, the change in the number of users, causes variation in size of channel gain vector and as a result the SNR vector (γ) is also changed. So, for curves with $N = 19$ and $\sigma_p^2 = 0.3$, the SNR elements are arranged in interval $[-31.5\ dB, -6.5\ dB]$, for $N = 19$ and $\sigma_p^2 = 0.4$: $[-30.5\ dB, -5.5\ dB]$, for curves with $N_{opt} = 10$, $\sigma_p^2 = 0.3$: $[-13.5\ dB, -6.5\ dB]$ and for $N_{opt} = 10$, $\sigma_p^2 = 0.4$: $[-12.5\ dB, -5.5\ dB]$. Looking at this Figure, it is obvious that increment of SNR ratio for all of the SUs leads to an enhancement in ROC as expected.

5   Conclusion

In this paper, we proved that to have an efficient cooperative spectrum sensing with limited time resource, while applying the optimum number and consuming lower energy resources, we would have a better performance such as higher achievable throughput. Moreover, by constraining the whole frame time of collaborative spectrum sensing, a relationship between number of samples and number of sensing users was obtained. From that, we could find the optimum number of samples too. Furthermore, through analytical results and also simulation results, it was shown that the upper bound of false alarm probability (and then the achievable throughput) is a convex (concave) function of SUs number and weighting vector with some constraints. So, the proposed scheme which uses jointly optimized number of SUs and weighting vector, outperforms significantly other traditional detectors that use one of the optimum values.

Acknowledgments. This publication was made possible by the National Priorities Research Program (NPRP) award NPRP 6-1326-2-532 from the Qatar National Research Fund (QNRF) (a member of the Qatar Foundation). The statements made herein are solely the responsibility of the authors.

A   Appendix

In order to show that cost function of the optimization problem is convex with respect to $\mathbf{w}$ and M', we should prove that its Hessian matrix is positive semi definite. From (17), if $f(M', \mathbf{w}) = Q(\xi)$ where $\xi \triangleq Q^{-1}(\bar{P}_\mathrm{d})\sqrt{\lambda_\mathrm{max}\mathbf{C}} + M'\mathbf{w}^T\boldsymbol{\gamma}$, we can obtain hessian matrix as (29).

$$\mathbf{H} = \nabla^2 f(M', \mathbf{w}) = \begin{pmatrix} a & \mathbf{b}^T \\ \mathbf{b} & \mathbf{D} \end{pmatrix} = \tag{29}$$

$$\frac{1}{\sqrt{2\pi}} \begin{pmatrix} (\mathbf{w}^T\boldsymbol{\gamma})^2 \xi \exp(-\xi^2/2) & -\boldsymbol{\gamma}^T \exp(-\xi^2/2)[1 - M'(\mathbf{w}^T\boldsymbol{\gamma})\xi] \\ -\boldsymbol{\gamma}\exp(-\xi^2/2)[1 - M'(\mathbf{w}^T\boldsymbol{\gamma})\xi] & M'^2 \xi \exp(-\xi^2/2)(\boldsymbol{\gamma}\boldsymbol{\gamma}^T) \end{pmatrix}$$

To prove positiveness we have: If $a \succ 0 \Longrightarrow$ then, $\mathbf{H} \succeq 0 \Leftrightarrow \mathbf{S} = \mathbf{D} - \mathbf{b}a^{-1}\mathbf{b}^T \succeq 0$. In other words , since a is positive, the Hessian matrix $\mathbf{H}$ is positive semi-definite if and only if its Schur complement is positive semi-definite. Its Schur complement is $\mathbf{S} = \frac{\mathbf{D}a - \mathbf{b}\mathbf{b}^T}{a}$. Thus, we should have

$$\mathbf{S} = \frac{M'^2 \boldsymbol{\gamma}\boldsymbol{\gamma}^T \xi \exp(-\xi^2/2)}{\sqrt{2\pi}} - \frac{\boldsymbol{\gamma}\boldsymbol{\gamma}^T \exp(-\xi^2)(M'\xi\mathbf{w}^T\boldsymbol{\gamma} - 1)^2}{\sqrt{2\pi}(\mathbf{w}^T\boldsymbol{\gamma})^2 \exp(-\xi^2/2)\xi} \succeq 0 \tag{30}$$

Therefore

$$\frac{(2M'\xi\mathbf{w}^T\boldsymbol{\gamma} - 1)\exp(-\xi^2/2)\boldsymbol{\gamma}\boldsymbol{\gamma}^T}{\sqrt{2\pi}(\mathbf{w}^T\boldsymbol{\gamma})^2\xi} \succeq 0 \tag{31}$$

The matrix $\boldsymbol{\gamma}\boldsymbol{\gamma}^T$ has rank 1 and all the eigenvalues are zero except its maximum eigenvalue which equals $\boldsymbol{\gamma}^T\boldsymbol{\gamma}$ and is positive. So $\boldsymbol{\gamma}\boldsymbol{\gamma}^T \succeq 0$. Thus $\mathbf{S} \succeq 0$ if $2M'\xi\mathbf{w}^T\boldsymbol{\gamma} - 1 \geq 0$. In fact, after manipulation the inequality, the condition $P_\mathrm{fa} \leq Q(\frac{1}{2M'\mathbf{w}^T\boldsymbol{\gamma}})$ should be satisfied. But $\Theta = Q(\frac{1}{2M'\mathbf{w}^T\boldsymbol{\gamma}})$ depends on $\mathbf{w}$. So, we should find a lower bound for Θ to be ensure that P_fa is lower than this term and condition is satisfied. To this end, we can see the condition above as

$$2M'\xi\mathbf{w}^T\boldsymbol{\gamma} - 1 \geq 0$$
$$2M'\mathbf{w}^T\boldsymbol{\gamma}\left(Q^{-1}(\bar{P}_\mathrm{d})\sqrt{\lambda_\mathrm{max}\mathbf{C}} + M'\mathbf{w}^T\boldsymbol{\gamma}\right) \geq 1 \tag{32}$$

So,

$$2Q^{-1}(\bar{P}_\mathrm{d})\sqrt{\lambda_\mathrm{max}\mathbf{C}}(M'\mathbf{w}^T\boldsymbol{\gamma}) + 2(M'\mathbf{w}^T\boldsymbol{\gamma})^2 - 1 \geq 0 \tag{33}$$

which is a quadratic function of $M'\mathbf{w}^T\boldsymbol{\gamma}$. It can be easily shown that for this quadratic function to be positive, just one of the answers is acceptable. So, we should have

$$2M'\mathbf{w}^T\boldsymbol{\gamma} \geq -Q^{-1}(\bar{P}_\mathrm{d})\sqrt{\lambda_\mathrm{max}\mathbf{C}} + \sqrt{(Q^{-1}(\bar{P}_\mathrm{d}))^2\lambda_\mathrm{max}\mathbf{C} + 2} \tag{34}$$

which is equal to

$$Q\left(\frac{1}{2M'\mathbf{w}^T\boldsymbol{\gamma}}\right) \geq Q\left(\frac{1}{-Q^{-1}(\bar{P}_\mathrm{d})\sqrt{\lambda_\mathrm{max}\mathbf{C}} + \sqrt{(Q^{-1}(\bar{P}_\mathrm{d}))^2\lambda_\mathrm{max}\mathbf{C} + 2}}\right)$$

Therefore, for satisfying condition $P_\mathrm{fa} \leq Q(\frac{1}{2M'\mathbf{w}^T\boldsymbol{\gamma}})$ from (35), we obtain the condition for convexity of cost function of the optimization problem as [13]:

$$P_\mathrm{fa} \leq Q\left(\frac{1}{-Q^{-1}(\bar{P}_\mathrm{d})\sqrt{\lambda_\mathrm{max}\mathbf{C}} + \sqrt{Q^{-1}(\bar{P}_\mathrm{d})^2\lambda_\mathrm{max}\mathbf{C} + 2}}\right) \tag{35}$$

References

1. Ahsant, B., Viswanathan, R.: A review of cooperative spectrum sensing in cognitive radios. In: Mukhopadhyay, S.C., Jayasundera, K.P., Fuchs, A. (eds.) Advancement in Sensing Technology. SSMI, vol. 1, pp. 69–80. Springer, Heidelberg (2013)
2. Sedighi, S., Pourgharehkhan, Z., Taherpour, A., Khattab, T.: Distributed spectrum sensing of correlated observations in cognitive radio networks. In: 7th IEEE GCC Conference and Exhibition (GCC), IEEE 2013, pp. 483–488 (2013)
3. Pourgharehkhan, Z., Sedighi, S., Taherpour, A. Uysal, M.: Spectrum sensing of correlated subbands with colored noise in cognitive radios. In: Wireless Communications and Networking Conference (WCNC). IEEE, pp. 1017–1022 (2012)
4. Akyildiz, I.F., Lo, B.F., Balakrishnan, R.: Cooperative spectrum sensing in cognitive radio networks: a survey. Physical Communication 4(1), 40–62 (2011)
5. Chen, Y.: Optimum number of secondary users in collaborative spectrum sensing considering resources usage efficiency. IEEE Communications Letters 12(12), 877–879 (2008)
6. Althunibat, S., Granelli, F.: An objection-based collaborative spectrum sensing for cognitive radio networks. IEEE Communications Letters 18(8), 1291–1294 (2014)
7. Li, H., Cheng, X., Li, K., Hu, C., Zhang, N., Xue, W.: Robust collaborative spectrum sensing schemes for cognitive radio networks. IEEE Transactions on Parallel andDistributed Systems 25(8), 2190–2200 (2014)
8. Quan, Z., Cui, S., Sayed, A.H.: Optimal linear cooperation for spectrum sensing in cognitive radio networks. IEEE Journal of Selected Topics in SignalProcessing 2(1), 28–40 (2008)
9. Quan, Z., Cui, S., Sayed, A.: An optimal strategy for cooperative spectrum sensing in cognitive radio networks. In: Global Telecommunications Conference, GLOBE-COM 2007. IEEE, pp. 2947–2951 (2007)
10. Peh, E., Liang, Y., Guan, Y., Zeng, Y.: Cooperative spectrum sensing in cognitive radio networks with weighted decision fusion schemes. IEEE Transactions on Wireless Communications 9(12), 3838–3847 (2010)
11. Quan, Z., Cui, S., Sayed, A.H., Poor, H.V.: Optimal multiband joint detection for spectrum sensing in cognitive radio networks. IEEE Transactions on SignalProcessing 57(3), 1128–1140 (2009)
12. Peh, E.C.Y., Liang, Y.-C., Guan, Y.L., Zeng, Y.: Optimization of cooperative sensing in cognitive radio networks: a sensing-throughput tradeoff view. IEEE Transactions on Vehicular Technology 58(9), 5294–5299 (2009)
13. Hoseini, P.P., Beaulieu, N.C.: An optimal algorithm for wideband spectrum sensing in cognitive radio systems. In: Communications (ICC). IEEE, pp. 1–6 (2010)

Stability and Delay Analysis for Cooperative Relaying with Multi-access Transmission

Mohamed Salman[1]([✉]), Amr El-Keyi[1], Mohammed Nafie[1], and Mazen Hasna[2]

[1] Wireless Intelligent Networks Center (WINC), Nile University, Giza, Egypt
Mohamed.Salman@nileu.edu.eg, {aelkeyi,mnafie}@nileuniversity.edu.eg
[2] College of Engineering, Qatar University, Doha, Qatar
hasna@qu.edu.qa

Abstract. We consider a cooperative relaying system with two source terminals, one full duplex relay, and a common destination. Each terminal has a local traffic queue while the relay has two relaying queues to store the relayed source packets. We assume that the source terminals transmit packets in orthogonal frequency bands. In contrast to previous work which assumes a time division multi-access cooperation strategy, we assume that the source terminals and the relay simultaneously transmit their packets to the common destination through a multi-access channel (MAC). A new cooperative MAC scheme for the described network is proposed. We drive an expression for the stable throughput and characterize the stability region of the network. Moreover, the fundamental trade-off between the delay and the stable throughput is studied. Numerical results reveal that the proposed protocol outperforms traditional time division multi-access strategies.

Keywords: Cooperative relaying · Multi-access channel · Stable throughput region · Queuing theory · Average delay

1 Introduction

In wireless networks, the transmission of a single node may successfully reach multiple nodes within its range, which is referred to as the wireless multicast advantage. As a result, intermediate nodes have the capability to capture the transmission and contribute to the communication by *cooperatively* relaying the data. This contribution enhances the aggregate throughput of the network and reduces the delay encountered by the packets of different nodes [1], [2]. Cooperative communication in wireless networks has been widely investigated. In [3], a time division multiple access (TDMA) policy is assumed, where a single relay cooperatively transmits the packets of the source nodes during idle time slots.

Recently, multi-packet reception (MPR) has received considerable attention in the literature. A generalized MPR model was first introduced in [4]. The number of successful transmissions in a time-slot was modelled as a random variable

This paper was made possible by NPRP grant # 4-1119-2-427 from the Qatar National Research Fund (a member of Qatar Foundation). The statements made herein are solely responsibility of the authors.

© Institute for Computer Sciences, Social Informatics and Telecommunications Engineering 2015
M. Weichold et al. (Eds.): CROWNCOM 2015, LNICST 156, pp. 123–134, 2015.
DOI: 10.1007/978-3-319-24540-9_10

which is a function of the number of attempted transmissions. Also, multi-access channel (MAC) systems have been addressed in the literature in different contexts, most of which do not deal with cognitive or cooperative systems [5], [6]. Nevertheless, in [7], a MAC network with two primary transmitters and a single secondary node was considered with a symmetric configuration. The primary users, simultaneously, access the channel to deliver their packets to a common destination. The cognitive node transmits during idle time slots. The impact of the cognitive node with and without relaying capability was studied.

Several metrics have been considered for evaluating the performance of cooperative networks and the average packet delay is one of these metrics. In [8], the delay analysis for a cognitive relaying scenario was presented, using the moment generation function approach, where a full priority is given to the relaying queue. However, in [9], the delay analysis for randomized cooperation policy was studied where the secondary user serves either its own data or the primary packets with certain service probabilities. This policy enhances the secondary user delay at the expense of a slight degradation in the primary user delay.

In this paper, we investigate a cooperative scenario with one full duplex relay and two source terminals. Unlike most of the existing work, e.g., [7], [3], we assume that the source terminals transmit their packets using two orthogonal frequency bands. We assume that the receivers have perfect CSI. In contrast with previous work in [10], a new MAC cooperation scheme is proposed. Under this scheme, the relay transmits only if the destination can decode the message of the relay by treating the source terminal message as noise. The relay may exploit one or both frequency bands for transmission. For comparison purpose, we introduce a TDMA cooperation scheme where the source terminals and the relay transmit their packets over disjoint fractions of time. The comparison between the two schemes shows that the proposed MAC scheme outperforms the conventional TDMA scheme.

The remainder of the paper is organized as follows. Section 2 introduces the system model and the proposed cooperative strategies. Section 3 presents the analysis of the stable throughput region. The average delay characterization is provided in Section 4. Numerical results are then presented in Section 5, followed by the conclusion in Section 6.

2   System Model

We assume a network consisting of two source terminals (s_1 and s_2), one common relay (r), and one common destination (d), as shown in Fig. 1. The source terminals transmit their signals to the common destination using two orthogonal frequency bands donated by w_1 and w_2 for s_1 and s_2, respectively. All wireless links are assumed to be stationary, frequency non-selective, and Rayleigh block fading. The fading coefficients, $h_{m,n}$, where $m \in \{s_1, s_2, r\}$ and $n \in \{r, d\}$, are assumed to be constant during one slot duration, but change independently from

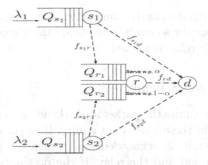

Fig. 1. System Model

one time slot to another according to a circularly symmetric complex Gaussian distribution with zero mean and variance $\rho_{m,n}^2$. All wireless links are corrupted by additive white Gaussian noise (AWGN) with zero mean and unit variance.

The ith source terminal, where $i \in \{1,2\}$, transmits with fixed power P_{s_i}. An outage occurs when the rate R is more than the instantaneous capacity of the link (m, n). Each link is characterized by the probability

$$f_{mn} = \mathbb{P}\{R < \log_2(1 + P_m|h_{m,n}|^2)\} = \exp\left(-\frac{2^R - 1}{P_m\rho_{m,n}^2}\right) \quad (1)$$

which denotes the probability that the link (m, n) is not in outage.

Time is slotted and the transmission of a packet takes exactly one slot duration. Each source terminal has an infinite queue to store its own incoming packets. Packet arrivals of both terminals are independent and stationary Bernoulli processes with means λ_1 and λ_2 (packets per slot) for s_1 and s_2, respectively.

The relay has two relaying queues (Q_{r_1} and Q_{r_2}) to store the packets of the source terminals that are not successfully decoded at the destination. Let Q_l^t denote the number of packets in the lth queue at the beginning of time slot t. The instantaneous evolution of the lth queue length is given by

$$Q_l^{t+1} = (Q_l^t - Y_l^t)^+ + X_l^t \quad (2)$$

where $l \in \{s_1, s_2, r_1, r_2\}$ and $(x)^+ = \max\{x, 0\}$. The binary random variables Y_l^t and X_l^t, denote the departures and arrivals of Q_l in time slot t, respectively, and their values are either 0 or 1.

We assume that the relay is full duplex, i.e., it can transmit and receive at the same time slot. In wireless networks when a node transmits and receives simultaneously on the same frequency, the problem of self-interference arises. Although there are some techniques that allow the possibility of perfect self-interference cancellation [11], in practice, there are currently several technological limitations and challenges that limit the accuracy and the effectiveness of self-interference cancellation [12]. Therefore, we assume that the relay can transmit and receive, simultaneously, over two *distinct* frequency bands. In addition, the relay also has the capability to receive or transmit packets on the two frequency bands simultaneously. It is worth noting that our network can be applied in the uplink of a cellular system where the source terminals are mobile nodes, the destination is a base station and the relay is a fixed node.

2.1 Cooperative MAC Scheme

Each source terminal transmits the packet at the head of its queue on its assigned frequency band whenever the queue is not empty. If the destination receives the packet successfully, it sends an acknowledgement message (ACK) which can be heard by both the terminal and the relay. If the destination does not succeed in receiving the packet correctly but the relay does, then the relay stores this packet at the end of its queue and sends an ACK to the source terminal. The source terminal drops the transmitted packet when it hears an ACK from the destination or the relay, otherwise, it retransmits the packet in the next time slot. The feedback messages are assumed to be error-free as short length packets and low rate codes can be employed in the feedback channel.

We assume that the destination knows the state of the channels from the sources and the relays, i.e., $h_{m,d}$, where $m \in \{s_1, s_2, r\}$. Note that this assumption is well-justified as the system can dedicate a small portion at the beginning of each time slot to transmit a short training sequence to the destination to be used for channel estimation[1]. We assume that the average channel gain between the source and the relay is higher than that between the source and the destination. In absence the of the relay, the source terminal wastes power when it transmits a packet that the destination can not successfully decode. However, the relay might still be able to decode that packet and this provides a diversity gain to the source terminals. According to the CSI, the destination decides the reception/transmission policy of the relay and sends it through a short error-free message to the relay at the beginning of each time slot[2]. The reception/transmission policy is described as follows

- The relay stores the transmitted packet from s_i, where $i \in \{1, 2\}$, if the destination can not decode this packet successfully.
- The relay transmits a packet on w_i, when it is not receiving packets on that band and the destination can decode the packet of the relay by treating s_i as noise.

If the relay transmits on both frequency bands simultaneously, a packet from each of the relaying queues is served. When the relay transmits on only one frequency band, a packet is served from Q_{r_1} with probability α or from Q_{r_2} with probability $1 - \alpha$. We do not assume that the destination controls the source terminals. We lose the diversity gain provided by the relay if the source transmits, only, according to the source-destination channel.

---

[1] The nodes re-transmit the training sequence whenever the channel changes.
[2] The side communication between the destination node and the relay and the amount of required training for CSI estimation is outside the scope of this paper.

Let $g_{r,d}^{s_i}$ denote the probability that the destination decodes the packet of the relay by treating s_i as noise. Therefore,

$$
\begin{aligned}
g_{rd}^{s_i} &= \mathbb{P}\left\{ R < \log_2\left(1 + \frac{P_r |h_{r,d}|^2}{P_s |h_{s_i,d}|^2 + 1} \right) \right\} \\
&= \exp\left(-\frac{2^R - 1}{\rho_{r,d}^2 P_r} \right) \frac{\rho_{r,d}^2 P_r}{\rho_{r,d}^2 P_r + (2^R - 1)\rho_{s_i,d}^2 P_s}
\end{aligned}
\tag{3}
$$

where P_r denotes the power transmitted by the relay per frequency band. Note that if the source terminal is not transmitting simultaneously with the relay, e.g., when the queue of the source is empty, the destination will be able to successfully decode the transmission of the relay with a higher probability than that in (3). It is obvious that there is an interaction between the queues of the source terminals and that of the relay because the probability of successful transmission of the relay depends on the queue state of the source terminals. Since, the analysis of the average delay of interacting queues is difficult [13], we resort to the use of a dominant system where s_i transmits dummy packet whenever the relay is transmitting on w_i [7]. In other words, s_1 transmits a dummy packet if the relay is transmitting on w_1 and Q_{s_1} is empty. Similarly, s_2 transmits a dummy packet if the relay is transmitting on w_2 and Q_{s_2} is empty. The dominant system decouples the interaction between the queues and provides an upper bound on the the the delay of the original system.

It is worth noting from the given description of the proposed policy that the system at hand is non work-conserving. A system is considered work-conserving if it is not idle whenever it has packets [14]. This condition is violated when the relay randomly selects to transmit a packet from a queue which is empty, while the other queue is non-empty. We resort to a non-conserving policy for its mathematical tractability.

2.2   Cooperative TDMA Scheme

The main difference between the two schemes is in the way the nodes utilize the available resources (time and frequency). Here, the cooperation policy depends on a TDMA frame work where s_1 and r_1 transmit their packets on w_1 only while s_2 and r_2 transmit using w_2. Each of s_i and r_i transmits in fixed fraction of time donated by m_{s_i} and m_{r_i} for s_i and r_i, respectively, where $m_{s_i} + m_{r_i} = 1$. Based on the cooperation policy described, there is no interaction between the queues because all nodes transmit over orthogonal resources.

3   Stable Throughput Region

A fundamental performance measure of a communication network is the stability of its queues. The stability of the overall system requires the stability of each individual queue. We can apply Loynes' theorem to check the stability of a queue [15]. Loynes' theorem states that if the arrival process and the service process of a queue are strictly stationary, then the queue is stable if and only if the average service rate is greater than the average arrival rate of the queue.

3.1   The Stability Analysis of MAC Scheme

A packet departs Q_{s_i} if it is successfully decoded by at least one node, i.e., the destination or the relay. Thus, the average service rate of Q_{s_i} is given by

$$\mu_i = f_{s_i d} + (1 - f_{s_i d}) f_{s_i r} \tag{4}$$

Thus, for stability of Q_{s_i}, the following condition must be satisfied

$$\lambda_i < f_{s_i d} + f_{s_i r}(1 - f_{s_i d}) \tag{5}$$

A packet arrives at Q_{r_1} if the following two conditions are met. First, if an outage occurs in the link between s_1 and the destination node while no outage occurs in the link between s_1 and the relay. Second, Q_{s_1} is not empty which has a probability of λ_1/μ_1. Thus, the average arrival rate of Q_{r_1} is given by

$$\lambda_{r_1} = (1 - f_{s_1 d}) f_{s_1 r} \frac{\lambda_1}{\mu_1} \tag{6}$$

A packet departs Q_{r_1} if the relay transmits on both frequency bands, simultaneously, which happens with probability $p_1 g_{rd}^{s_1} g_{rd}^{s_2}$ or the relay transmits on a single frequency which happens with probability $p_1(\overline{g_{sd}^{s_1}} g_{rd}^{s_2} + g_{rd}^{s_1} \overline{g_{rd}^{s_2}}) + p_2 g_{rd}^{s_2} + p_3 g_{rd}^{s_1}$ and Q_{r_1} is selected to transmit a packet which happens with probability α. Thus, the service rate of Q_{r_1} is given by

$$\mu_{r_1} = p_1 g_{rd}^{s_1} g_{rd}^{s_2} + \alpha(p_1 \overline{g_{sd}^{s_1}} g_{rd}^{s_2} + p_1 g_{rd}^{s_1} \overline{g_{rd}^{s_2}} + p_2 g_{rd}^{s_2} + p_3 g_{rd}^{s_1}) \tag{7}$$

where $p_1 = f_{s_1 d} f_{s_2 d}$, $p_2 = \overline{f_{s_1 d}} f_{s_2 d}$, $p_3 = f_{s_1 d} \overline{f_{s_2 d}}$, and $\overline{x} = 1 - x$.

For the stability of Q_{r_1}, the service rate must be higher than the arrival rate, i.e., $\lambda_{r_1} < \mu_{r_1}$, and hence, we have

$$\lambda_1 < \frac{\mu_{r_1}}{(1 - f_{s_1 d}) f_{s_1 r}} \mu_1 \tag{8}$$

Applying exactly the same analysis for Q_{r_2}, we get

$$\lambda_2 < \frac{\mu_{r_2}}{(1 - f_{s_2 d}) f_{s_2 r}} \mu_2 \tag{9}$$

where

$$\mu_{r_2} = p_1 g_{rd}^{s_1} g_{rd}^{s_2} + \overline{\alpha}(p_1 \overline{g_{sd}^{s_1}} g_{rd}^{s_2} + p_1 g_{rd}^{s_1} \overline{g_{rd}^{s_2}} + p_2 g_{rd}^{s_2} + p_3 g_{rd}^{s_1}) \tag{10}$$

From (5), (8) and (9), it is obvious that to guarantee the stability of the system the followings must be satisfied

$$\lambda_i < \min\{\mu_i, \mu_{u_i}\} \tag{11}$$

$$\text{where}\quad \mu_{u_i} = \frac{\mu_{r_i}}{(1 - f_{s_i d}) f_{s_i r}} \mu_i \quad i \in \{1, 2\} \tag{12}$$

The stable throughput of the system is constrained by the stability of the queues of the source terminals as long as $\mu_i \leq \mu_{u_i}$. The effect of α appears only when the value of μ_{u_i} is less than μ_i. Let α_1 denote the value of α that satisfies $\mu_{u_1} = \mu_1$

$$\alpha_1 = \min \left\{ 1, \frac{(1-f_{s_1d})f_{s_1r} - p_1 g_{rd}^{s_1} g_{rd}^{s_2}}{p_1(\overline{g_{rd}^{s_1}} g_{rd}^{s_2} + g_{rd}^{s_1} \overline{g_{rd}^{s_2}}) + p_2 g_{rd}^{s_2} + p_3 g_{rd}^{s_1}} \right\} \tag{13}$$

and α_2 denote the value of α that satisfy $\mu_{u_2} = \mu_2$

$$\alpha_2 = \max \left\{ 0, 1 - \frac{(1-f_{s_2d})f_{s_2r} - p_1 g_{rd}^{s_1} g_{rd}^{s_2}}{p_1(\overline{g_{rd}^{s_1}} g_{rd}^{s_2} + g_{rd}^{s_1} \overline{g_{rd}^{s_2}}) + p_2 g_{rd}^{s_2} + p_3 g_{rd}^{s_1}} \right\} \tag{14}$$

Therefore the interesting values of α are between α_2 and α_1 because above α_1 or below α_2 the stable throughput, λ_i, is constant and equals to μ_i.

3.2   The Stability Analysis of TDMA Scheme

We follow the same steps as those in the MAC scheme. A packet departs Q_{s_i} in the assigned time slot if it is successfully decoded by at least one node. Thus, the average service rate of Q_{s_i} is given by

$$\mu_i = m_{s_i}(f_{s_id} + f_{s_ir}(1 - f_{s_id})) \tag{15}$$

For Q_{s_i} stability, the following condition must be satisfied

$$\lambda_i < m_{s_i}(f_{s_id} + f_{s_ir}(1 - f_{s_id})) \tag{16}$$

A packet arrives at Q_{r_i} when s_i transmits on the assigned time slot and an outage occurs in the direct link from s_i to the destination node while no outage occurs in the link between s_i and the relay, yet, Q_{s_i} is not empty. Thus, the average arrival rate of Q_{r_i} is given by

$$\lambda_{r_i} = m_{s_i} \left(\frac{\lambda_i}{\mu_i}(1 - f_{s_id})f_{s_ir} \right) \tag{17}$$

A packet departs Q_{r_i} if there is no outage in the link between the relay and the destination. Thus, the average service rate of Q_{r_i} is given by

$$\mu_{r_i} = m_{r_i} f_{rd} \tag{18}$$

For stability of Q_{r_i}, $\lambda_{r_i} < \mu_{r_i}$, which yields

$$\lambda_i < \frac{\mu_{r_i}}{m_{s_i}(1 - f_{s_id})f_{s_ir}} \mu_i \tag{19}$$

From (16) and (19), the system is stable if

$$\lambda_i < \min\{\mu_i, \mu_{u_i}\} \tag{20}$$

$$\mu_{u_i} = \frac{\mu_{r_i}}{m_{s_i}(1 - f_{s_id})f_{s_ir}} \mu_i \quad i \in \{1, 2\} \tag{21}$$

It is obvious from (19) that the maximum stable throughput depends on m_{s_i}. We formulate an optimization problem to calculate the maximum achievable stable throughput for both source terminals. From (20), λ_i is a concave function in the parameter m_{s_i} as it is the minimum between two affine functions, μ_i and μ_{s_i} [16]. For the ith source terminal, the optimization problem is given by

$$\begin{array}{c} \underset{m_{s_i}}{\text{maximize}} \quad \min\{\mu_i, \mu_{u_i}\} \\ \text{subject to} \quad 0 \le m_{s_i} \le 1 \end{array} \tag{22}$$

The optimal solution of this problem, $m_{s_i}^*$, can be easily calculated because μ_i is monotonically increasing in m_{s_i}, while μ_{u_i} is monotonically decreasing in m_{s_i}. Therefore, $m_{s_i}^*$ is obtained at $\mu_i = \mu_{u_i}$ and is given by

$$m_{s_i}^* = \frac{f_{rd}}{f_{s_i r}(1 - f_{s_i d}) + f_{rd}} \tag{23}$$

4   Average Delay Characterization

In this section, we present the delay analysis for both cooperative schemes. Then, we investigate the fundamental trade-off between the average delay and the stable throughput for s_1 and s_2.

4.1   Delay Analysis of MAC Scheme

If a packet is directly delivered to the destination then this packet experiences a queueing delay at the source only. This event occurs for the ith source with probability $\epsilon_i = \frac{f_{s_i d}}{f_{s_i d} + f_{s_i r} - f_{s_i d} f_{s_i r}}$, which is the the probability that the packet is successfully decoded by the destination given that it is dropped from Q_{s_i}. If the first successful transmission for this packet is not to the destination, then the packet experiences two delays; a queuing delay at Q_{s_i} in addition to the queuing delay at Q_{r_i}. This event occurs with probability $1 - \epsilon_i$. Therefore, the average delay is given by

$$D_i = T_{s_i} + (1 - \epsilon_i) T_{r_i} \tag{24}$$

where T_{s_i} and T_{r_i} denote the average queueing delays at s_i and r_i, respectively. Since the arrival rates at Q_{s_i} and Q_{r_i} are given by λ_i and λ_{r_i}, respectively, then applying Little's law yields

$$T_{s_i} = N_i / \lambda_i, \quad T_{r_i} = N_{r_i} / \lambda_{r_i} \tag{25}$$

where N_i and N_{r_i} denote the average queue size of Q_{s_i} and Q_{r_i}, respectively. The dominant system, described before, de-couples the interaction between the queues. Thus, we can easily calculate N_{s_i} and N_{r_i} by observing that Q_{s_i} and Q_{r_i} are discrete-time M/M/1 queues with Bernoulli arrivals and geometrically distributed service rates. Then, by applying the Pollaczek-Khinchine formula [17], we obtain N_i and N_{r_i} as

$$N_i = \frac{-\lambda_i^2 + \lambda_i}{\mu_i - \lambda_i}, \quad N_{r_i} = \frac{-\lambda_{r_i}^2 + \lambda_{r_i}}{\mu_{r_i} - \lambda_{r_i}} \tag{26}$$

Substituting (25) and (26) in (24), we can write the average queueing delay for the ith source terminal as

$$D_i = \frac{1 - \lambda_i}{\mu_i - \lambda_i} + \frac{f_{s_i r}(1 - f_{s_i d})}{f_{s_i d} + f_{s_i r} - f_{s_i d} f_{s_i r}} \frac{1 - \lambda_{r_i}}{\mu_{r_i} - \lambda_{r_i}} \tag{27}$$

4.2   Delay Analysis of TDMA Scheme

Since the source terminals and the relay transmit their packets over orthogonal resources, there is no interaction between the queues. Using exactly the same analysis as that used in the MAC scheme, we can write the average queueing delay for the i-th source terminal as

$$D_i = \frac{1 - \lambda_i}{\mu_i - \lambda_i} + \frac{f_{s_i r}(1 - f_{s_i d})}{f_{s_i d} + f_{s_i r} - f_{s_i d} f_{s_i r}} \frac{1 - \lambda_{r_i}}{\mu_{r_i} - \lambda_{r_i}} \tag{28}$$

$$\text{where } \mu_i = m_{s_i}(f_{s_i d} + f_{s_i r}(1 - f_{s_i d})) \tag{29}$$

$$\lambda_{r_i} = m_{s_i}\left(\frac{\lambda_i}{\mu_i}(1 - f_{s_i d})f_{s_i r}\right) \tag{30}$$

$$\mu_{r_i} = m_{r_i} f_{rd}, \quad i \in \{1, 2\} \tag{31}$$

5   Numerical Results

In this section, we investigate the performance of the proposed cooperative schemes. First, we show the effect of changing the direct link channel gain, between the sources and the destination, on the stability region. Next, we demonstrate the effect of varying α on the maximum stable throughput for the MAC scheme. Furthermore, we characterize the fundamental trade-off between the average delay and the stable throughput for both source terminals. Finally, we demonstrate effect of varying α on the delay experienced by the packets of s_1 and s_2 and validate our results via queue simulation.

In Fig. 2a, we plot the stable throughput region of the studied schemes for different direct link channel conditions. Hereafter, the system parameters are chosen as follows: $P_r = P_{s_i} = 6$, $R = 1$, $\rho_{s_1,r}^2 = 0.8$, $\rho_{s_2,r}^2 = 0.86$, and we define four different sets each contains a channel condition for the direct links according to the following: $S_1 = \{\rho_{s_1,d}^2 = 0.14, \rho_{s_2,d}^2 = 0.1\}$, $S_2 = \{\rho_{s_1,d}^2 = 0.2, \rho_{s_2,d}^2 = 0.16\}$, $S_3 = \{\rho_{s_1,d}^2 = 0.27, \rho_{s_2,d}^2 = 0.2\}$, and $S_4 = \{\rho_{s_1,d}^2 = 0.32, \rho_{s_2,d}^2 = 0.28\}$. In Fig. 2a, it is obvious that the MAC scheme provides the worst performance for both terminals for low direct channel gains. The poor direct link causes slow emptying

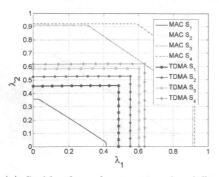 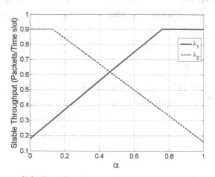

(a) Stable throughput region for different direct link channels

(b) Stable throughput for s_1 and s_2

Fig. 2. Stable Throughputs

of the source queues and a very few relay transmission opportunities. In this case, it is more efficient to use the TDMA scheme to achieve higher throughput. As the direct link channel gain increases, the performance of the MAC scheme improves and outperforms the TDMA scheme which becomes inefficient due to the division of the available degrees of freedom.

Next, we show the effect of varying α on the stability of the MAC scheme. We use S_2 for the direct channel condition. In Fig. 2b, we plot the stable throughput versus α for s_1 and s_2. Increasing the value of α increases the maximum stable arrival rate at s_1 while decreasing α increases the maximum stable arrival rate at s_2. This result is intuitive, since increasing the value of α gives more chance for transmitting the packets of s_1 at the cooperative queues and this reduces the amount of cooperation that the s_2 experiences from the relay.

Using (13) and (14), we can compute the values of α_1 and α_2 as $\alpha_1=0.76$, $\alpha_2=0.13$. It is clear from the figure that the stable throughput for s_1, λ_1, becomes constant when the value of α exceed α_1 because μ_{u_1} becomes greater than μ_1 which is constant and does not depend on α and this emphasize the results obtained in (11) and (13). It is exactly the same for s_2 when the system operates with value of α below α_2.

Next, we characterize a fundamental trade-off that arises between the average delay and the stable throughput for s_1 and s_2. Given that the system is stable, the throughput of any node is equal to its packet arrival rate. Thus, increasing the throughput means injecting more packets into the system which yields a higher delay. In Fig. 3a, we illustrate the delay throughput trade-off for s_2. We plot the average delay versus the stable throughput for the proposed cooperative schemes. The system parameters are chosen as follows: $P_r=P_{s_i}=6, R=1$, $\rho_{s_1,r}^2=0.8$, $\rho_{s_2,r}^2=0.86$, S_3, and for a fair comparison we choose $\lambda_1=0.6$, which is the maximum stable stable throughput for the TDMA scheme at S_3. The trade-off is obvious where as the throughput increases the delay also increases.

The MAC scheme sustains the stability of the system up to $\lambda_2 \approx 0.65$, while in the TDMA scheme the system is unstable with $\lambda_2 \approx 0.58$. These values appear clearly in Fig. 2a where at $\lambda_1 = 0.6$ the maximum stable throughput for s_2 in the MAC scheme is $\lambda_2 \approx 0.65$ while in the TDMA scheme is $\lambda_2 \approx 0.58$.

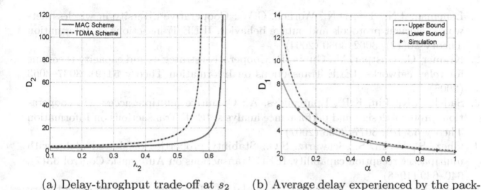

(a) Delay-throghput trade-off at s_2 (b) Average delay experienced by the packets of s_1

Fig. 3. Average Delay

Finally, we demonstrate effect of varying α on the delay of the packets of s_1 in the MAC scheme. In Fig. 3b, it is clear that the results obtained through simulations are close to the expressions derived in (27). The gap between the upper bound and the queue simulation emerges due to the dominant system where the nodes transmit dummy packets which affect the average delay experienced by the packets. We also introduce a new dominant where the relay transmits its packet over new frequency bands. This dominant system provides a lower bound on the delay of the original system and we can calculate the delay for this dominant system by substituting in (27) by

$$\mu_{r_1} = p_1 f_{rd} + \alpha(p_2 f_{rd} + p_3 f_{rd}) \tag{32}$$
$$\mu_{r_2} = p_1 f_{rd} + \overline{\alpha}(p_2 f_{rd} + p_3 f_{rd}) \tag{33}$$

Moreover, given λ_1 and λ_2, it is clear that as α increases D_1 decreases and this matches the result stated in (27).

6   Conclusion

In this paper, we have proposed a novel randomized MAC cooperative policy where a full duplex relay can efficiently transmit the packets of two source terminals. We have characterized the stable throughput region for the proposed cooperative scheme in addition to a TDMA scheme. The results indicate that the MAC scheme can provide significant gain over the TDMA scheme in the case of high direct channel gain. Moreover, we have also addressed the throughput delay trade-off. The results show that the MAC scheme achieves higher stable throughput than that of the TDMA scheme.

References

1. Laneman, J.N., Tse, D.N., Wornell, G.W.: Cooperative diversity in wireless networks: efficient protocols and outage behavior. IEEE Transactions on Information Theory **50**(12), 3062–3080 (2004)
2. Kramer, G., Gastpar, M., Gupta, P.: Cooperative strategies and capacity theorems for relay networks. IEEE Transactions on Information Theory **51**(9), 3037–3063 (2005)
3. Sadek, A.K., Liu, K.R., Ephremides, A.: Cognitive multiple access via cooperation: protocol design and performance analysis. IEEE Transactions on Information Theory **53**(10), 3677–3696 (2007)
4. Ghez, S., Verdu, S., Schwartz, S.C.: Stability properties of slotted aloha with multipacket reception capability. IEEE Transactions on Automatic Control **33**(7), 640–649 (1988)
5. ParandehGheibi, A., Médard, M., Ozdaglar, A., Eryilmaz, A.: Information theory vs. queueing theory for resource allocation in multiple access channels. In: IEEE Personal Indoor and Mobile Radio Communications, pp. 1–5, Septemper 2008
6. Naware, V., Mergen, G., Tong, L.: Stability and delay of finite-user slotted aloha with multipacket reception. IEEE Transactions on Information Theory **51**(7), 2636–2656 (2005)
7. Krikidis, I., Devroye, N., Thompson, J.S.: Stability analysis for cognitive radio with multi-access primary transmission. IEEE Transactions on Wireless Communications **9**(1), 72–77 (2010)
8. Rong, B., Ephremides, A.: Cooperative access in wireless networks: stable throughput and delay. IEEE Transactions on Information Theory **58**(9), 5890–5907 (2012)
9. Ashour, M., El-Sherif, A.A., ElBatt, T., Mohamed, A.: Cooperative access in cognitive radio networks: stable throughput and delay tradeoffs (2013). arXiv preprint arXiv:1309.1200
10. Bao, X., Martins, P., Song, T., Shen, L.: Stable throughput analysis of multi-user cognitive cooperative systems. In: IEEE Global Telecommunications Conference, pp. 1–5, December 2010
11. Cover, T.M., Thomas, J.A.: Elements of information theory. John Wiley & Sons (2012)
12. Eliezer, O.E., Staszewski, R.B., Bashir, I., Bhatara, S., Balsara, P.T.: A phase domain approach for mitigation of self-interference in wireless transceivers. IEEE Transactions on Solid-State Circuits **44**(5), 1436–1453 (2009)
13. Rao, R.R., Ephremides, A.: On the stability of interacting queues in a multiple-access system. IEEE Transactions on Information Theory **34**(5), 918–930 (1988)
14. Wolff, R.W.: Stochastic modeling and the theory of queues. Prentice hall Englewood Cliffs, vol. 14, NJ (1989)
15. Loynes, R.: The stability of a queue with non-independent inter-arrival and service times, vol. 58, no. 3, pp. 497–520. Cambridge Univ Press (1962)
16. Boyd, S.P., Vandenberghe, L.: Convex optimization. Cambridge University Press (2004)
17. Kleinrock, L.: Queueing systems. volume 1: Theory (1975)

An Efficient Switching Threshold-Based Scheduling Protocol for Multiuser Cognitive AF Relay Networks with Primary Users Using Orthogonal Spectrums

Anas M. Salhab[1]($\boxtimes$), Fawaz Al-Qahtani[2], Salam A. Zummo[1], and Hussein Alnuweiri[2]

[1] Electrical Engineering Department,
King Fahd University of Petroleum & Minerals, Dhahran, Saudi Arabia
{salhab,zummo}@kfupm.edu.sa
[2] Electrical and Computer Engineering Program,
Texas A&M University at Qatar, Doha, Qatar
{fawaz.al-qahtani,hussein.alnuweiri}@qatar.tamu.edu

Abstract. In this paper, we propose and evaluate the performance of multiuser switched diversity (MUSwiD) cognitive amplify-and-forward (AF) relay networks with multiple primary receivers using orthogonal spectrums. Using orthogonal spectrum bands aims to mitigate the interference between users in wireless networks. The spectrum of primary receiver whose channel results in the best performance for the secondary system is shared with secondary users. To reduce the channel estimation load in the secondary cell, the MUSwiD selection scheme is used to select among secondary users. In this scheme, the user whose end-to-end (e2e) signal-to-noise ratio (SNR) satisfies a predetermined switching threshold is scheduled to receive data from the source instead of the best user. In the analysis, an upper bound on the e2e SNR of a user is used in deriving of analytical approximations of the outage probability and average symbol error probability (ASEP). The performance is also studied at the high SNR regime where the diversity order and coding gain are derived. The derived expressions are verified by Monte-Carlo simulations. Results illustrate that the diversity order of the studied MUSwiD cognitive AF relaying network is the same as its non-cognitive counterpart. Unlike the existing papers where the same spectrum band is assumed to be shared by the primary receivers, our findings demonstrate that increasing the number of primary receivers in the proposed scenario enhances the system performancevia improving the coding gain.

Keywords: Amplify-and-forward · Multiuser cognitive relay network · Switching threshold · Orthogonal spectrums

1 Introduction

Cognitive radio is an important tool used to improve the spectrum resource utilization efficiency in wireless networks [1]. Several cognitive radio paradigms have

© Institute for Computer Sciences, Social Informatics and Telecommunications Engineering 2015
M. Weichold et al. (Eds.): CROWNCOM 2015, LNICST 156, pp. 135–148, 2015.
DOI: 10.1007/978-3-319-24540-9_11

been proposed in [2], among which is the underlay scheme. This scheme allows users in a secondary cell to utilize the frequency bands of users in a primary cell only if the interference is below a certain threshold. Beside the cognitive radio networks, a lot of research has been recently done on relay network which is used to deal with the multipath fading problem in wireless systems [3].

In the area of decode-and-forward (DF) cognitive relay networks (CRNs), closed-form expressions were derived in [4] for the outage and error probabilities of DF CRNs considering various relay selection scenarios. The outage and symbol error probabilities of amplify-and-forward (AF) CRNs with opportunistic and partial-relay selection schemes were evaluated in [5]. In [6], the error rate performance of an AF CRN was studied using the partial-relay selection scheme. The outage performance of an AF CRN with multiple primary users was recently studied in [7]. In addition to deriving the ergodic channel capacity, Bao *et al.* evaluated in [8] some lower bounds for the outage and error rate probabilities of AF CRNs assuming Rayleigh fading channels. Recently, the outage performance of opportunistic AF and DF CRNs with multiple secondary users and direct link was studied in [9].

Currently, the performance of CRNs with multiple secondary users is attracting a lot of researchers to work on such important topic. In [10], the secondary user was selected to achieve the largest secondary rate while satisfying primary rate target. In [11], the outage performance of AF and DF CRNs was studied assuming multiple secondary sources, single secondary relay and destination, and multiple primary receivers. The secondary source which maximizes the SNR at the destination combiner output considering the presence of the direct link is selected to send its message.

As can be seen, the only considered scenario in CRNs is the one where the multiple primary receivers utilize the same spectrum band. Another important scenario that could be seen in such systems is the one where the primary receivers use orthogonal spectrums. This situation could be seen in long term evolution (LTE) networks where the orthogonal frequency division multiple access (OFDMA) technique is used in the downlink transmission. Another application is in IEEE 802.22 wireless regional area networks (WRANs) where the OFDMA is a candidate access method for these networks. Furthermore, it is noticed that the mostly used scheduling scheme in multiuser CRNs is the opportunistic scheduling. A drawback of this scheme is the heavy load of channel estimations it requires in selecting the best secondary user among the available users. An efficient candidate which can reduce the channel estimation load between the secondary relay and users and the system complexity is the multiuser switched diversity (MUSwiD) selection scheme [12]. In this scheme, each user triggers a feedback only when its channel quality is greater than a certain threshold. Therefore, only the users with good channel quality are worth being considered to be scheduled [13].

According to authors knowledge, the scenario of cognitive AF relay networks with multiuser switched diversity selection and multiple primary receivers using orthogonal spectrums has not been presented yet. The contributions of this paper are as follows. *i*) We propose the new scenario of CRNs with multiple primary receivers using orthogonal spectrum bands. *ii*) Also, we introduce the MUSwiD user selection scheme to select among secondary users in the proposed scenario. *iii*) We provide a comprehensive analysis for evaluating the performance of the proposed scenario where some analytical approximations are derived for the outage probability and average symbol error probability (ASEP) for the independent non-identically distributed (i.n.i.d.) generic case of users channels. Furthermore, we study the behavior at the high SNR regime where the diversity order and coding gain are derived.

The rest of this paper is organized as follows. Section 2 presents the system and channel models. The performance evaluation is conducted in Section 3. Section 4 provides the asymptotic performance analysis. Some simulation and numerical results are discussed in Section 5. Finally, Section 6 concludes the paper.

2   System and Channel Models

Consider a dual-hop cognitive AF relay network consisting of one secondary source S, one AF secondary relay R, K secondary destinations or users D_k ($k = 1, \ldots, K$), and M primary receivers P_m ($m = 1, \ldots, M$) using orthogonal frequency bands. All nodes are assumed to be equipped with single antenna and the communication is assumed to operate in a half-duplex mode. Secondary users need to share the spectrum with the primary receiver whose channel satisfies the interference constraint and results in a best performance for the secondary system[1]. The communications take place in two phases. In the first phase, the secondary source sends its message x to relay under a transmit power constraint which guarantees that the interference with the primary receivers does not exceed a threshold $\mathcal{I}_p$. As a result, the source S must transmit at a power given by $P_s = \mathcal{I}_p / \min_m |g_{s,m}|^2$, $m = 1, \ldots, M$, where $g_{s,m}$ is the channel coefficient of the S $\to$ P_m link. In the second phase, R amplifies the received message from S with a variable gain G and forwards the amplified message to K users. The transmit power at R must also satisfy the interference constraint, it is defined as $P_R = \mathcal{I}_p / \min_m |g_{r,m}|^2$, $m = 1, \ldots, M$, where $g_{r,m}$ is the channel coefficient of the R $\to$ P_m link. Hence, the received message at D_k from R is given by $y_{r,k} = \sqrt{P_s} G h_{r,k} h_{s,r} x + G h_{r,k} n_{s,r} + n_{r,k}$, where $h_{s,r}$ and $h_{r,k}$ are the channel coefficients of the S $\to$ R and R $\to$ D_k links, respectively, $n_{s,r}$ and $n_{r,k}$ are the additive white Gaussian noise (AWGN) terms at R and D_k, respectively, with a power of N_0. We assume that the channel information of all links can

---

[1] Getting the best behavior of the secondary system in the sense of selecting the primary receiver which allows the secondary users to transmit at their max. power.

be perfectly estimated by the secondary users[2]. Also, it is assumed that the interference from the primary user is neglected[3]. As we are using a channel-state-information (CSI)-assisted AF relaying, the gain G can be expressed as

$$G^2 = 1 \bigg/ \left(\min_{m} |g_{r,m}|^2 \right) \left[\frac{|h_{s,r}|^2}{\left(\min_{m} |g_{s,m}|^2 \right)} + \frac{N_0}{\mathcal{I}_p} \right].$$ Thus, the end-to-end (e2e) SNR of

D_k can be written as [8]

$$\gamma_{S-R-D_k} = \frac{\frac{\frac{\mathcal{I}_p}{N_0} |h_{s,r}|^2}{\min_{m} |g_{s,m}|^2} \frac{\frac{\mathcal{I}_p}{N_0} |h_{r,k}|^2}{\min_{m} |g_{r,m}|^2}}{\frac{\frac{\mathcal{I}_p}{N_0} |h_{s,r}|^2}{\min_{m} |g_{s,m}|^2} + \frac{\frac{\mathcal{I}_p}{N_0} |h_{r,k}|^2}{\min_{m} |g_{r,m}|^2} + 1} \leq \gamma_k^{\text{up}} = \min \bigg(\underbrace{X/Y}_{\gamma_1}, \underbrace{X_k}_{\gamma_{2_k}} \bigg), \qquad (1)$$

where $X = \frac{\mathcal{I}_p}{N_0} |h_{s,r}|^2$, $Y = \min_{m} |g_{s,m}|^2$, and $X_k = \frac{\frac{\mathcal{I}_p}{N_0} |h_{r,k}|^2}{\min_{m} |g_{r,m}|^2}$. The MUSwiD user scheduling is achieved by selecting the user with the e2e SNR γ_k^{up} that satisfies a predetermined switching threshold. In the upcoming analysis, all channel coefficients are assumed to undergo i.n.i.d. Rayleigh fading and hence, the channel gains $|g_{s,m}|^2$, $|h_{s,r}|^2$, $|h_{r,k}|^2$, and $|g_{r,m}|^2$ follow exponential distribution with mean powers $\mu_{s,m}$, $\Omega_{s,r}$, $\Omega_{r,k}$, and $\mu_{r,m}$, respectively.

Referring to Figure 1, the MUSwiD selection scheme works as follows. In each scheduling period, the relay probes the secondary users in a sequential way so only a single user has an opportunity to send a feedback at one time. For each user to decide whether to send a feedback or not, a single feedback threshold is used for all users. This threshold could be assumed to be constant or it could be calculated to optimize a certain performance measure[4]. The order of the users is set by the relay and sent to all users each scheduling period. The second or even the kth user will not send any feedback signal to the relay unless it does not receive a flag from the previous user in the sequence within a certain time duration[5]. Suppose the users are arranged in a certain order, the first user compares its channel quality with the threshold. If it is higher than the threshold, the first user sends a feedback to the relay and a flag to other users signaling its presence. Otherwise, the first user keeps silent and the second (next) user compares its channel quality against the threshold. Again, if it exceeds the threshold, the second user sends a feedback to the relay and a flag to other users signaling its presence, otherwise the third user will get a chance. Once the relay detects a feedback from any user, it immediately selects

---

[2] Secondary users can know the channel information of the primary user by either a direct reception of pilot signals from a primary user [14].

[3] The interference is assumed to be represented by noise as in the case where the primary transmitters signal is generated by random Gaussian codebooks [15].

[4] In this paper, the switching threshold is numerically calculated to optimize the e2e outage probability. Also, a simple method is mentioned in Section 4 to obtain approximate but accurate values for the optimum switching threshold.

[5] The time duration of the feedback channel is not long and hence, the MUSwiD scheduling scheme does not cause additional delay to the scheduling process [16].

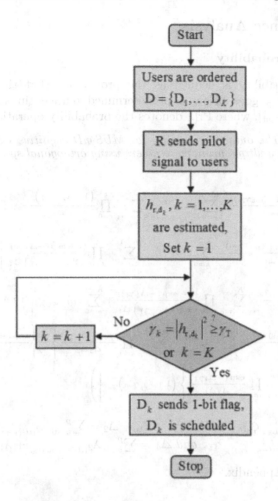

Fig. 1. Flowchart of the proposed MUSwiD user scheduling.

that user for the subsequent data reception and the whole user selection process ends. This process continues till a suitable user is found or all users are examined and found unacceptable. In this case, the MUSwiD scheme selects the last user for simplicity. To achieve fairness among users, the feedback sequence can be changed continuously. The feedback in MUSwiD systems is reduced significantly into only one feedback channel per resource unit instead of per-user feedback channels. Also, a user sends feedback only ahead of the resource units that it will be allocated instead of sending feedback for all resource units. This provides considerable savings in battery life of mobile terminals.

3   Performance Analysis

3.1   Outage Probability

The outage probability is defined as the probability that the SNR at the scheduled user γ_{up} goes below a predetermined outage threshold γ_{out}, i.e., $P_{out} = \Pr\left[\gamma_{up} \leq \gamma_{out}\right]$, where $\Pr[.]$ denotes the probability operation.

Lemma 1. *L.1 The outage probability for MUSwiD cognitive CSI-assisted AF relay system with multiple primary receivers using orthogonal spectrums is given by*

$$
\begin{aligned}
P_{out} \simeq & \sum_{i=0}^{K-1} \left\{ \frac{(1-\exp\left(-\zeta_{tot}\gamma_\mathsf{T}\right))}{\zeta_{tot}} + \sum_{k=0}^{K-1}(-1)^{k+1}\sum_{\substack{n_0<\ldots<n_k\\n(.)\neq i}}^{K-1}\prod_{t=0}^{k}\frac{\left(1+\lambda_{2n_t}\gamma_\mathsf{T}\right)^{-1}}{\Delta_1} - \left(1+\lambda_{2_i}\gamma_{out}\right)^{-1} \right. \\
& \times \left[\frac{(1-\exp\left(-\left(\lambda_1\gamma_{out}+\zeta_{tot}\right)\gamma_\mathsf{T}\right))}{\lambda_1\gamma_{out}+\zeta_{tot}} + \sum_{k=0}^{K-1}(-1)^{k+1}\sum_{\substack{n_0<\ldots<n_k\\n(.)\neq i}}^{K-1}\prod_{t=0}^{k}\frac{(1-\exp\left(-\Delta_1\gamma_\mathsf{T}\right))}{\left(1+\lambda_{2n_t}\gamma_\mathsf{T}\right)\Delta_1} \right] \left. \right\} (\pi_i\zeta_{tot}) \\
& + \sum_{l=0}^{K-1}\pi_l\zeta_{tot}\left(\sum_{q=0}^{K}\frac{(-1)^q}{q!}\sum_{m_1,\ldots,m_q}^{K}\prod_{z=1}^{q}\frac{(1-\exp\left(-\Delta_1\gamma_\mathsf{T}\right))}{\left(1+\lambda_{2m_z}\gamma_\mathsf{T}\right)\Delta_1} + \sum_{w=0}^{K-1}\pi_{((l-w))_K}\left[\left(1+\lambda_{2_l}\gamma_\mathsf{T}\right)^{-1} \right.\right. \\
& \times \left\{ \frac{\exp\left(-\left(\lambda_1\gamma_\mathsf{T}+\zeta_{tot}\right)\gamma_\mathsf{T}\right)}{\lambda_1\gamma_\mathsf{T}+\zeta_{tot}} + \sum_{p=0}^{w-1}(-1)^{p+1}\sum_{v_0<\ldots<v_p}^{w-1}\prod_{g=0}^{p}\frac{\exp\left(-\Delta_2\gamma_\mathsf{T}\right)}{\Delta_3\Delta_2} \right\} - \left\{ \frac{\exp\left(-\left(\lambda_1\gamma_{out}+\zeta_{tot}\right)\gamma_\mathsf{T}\right)}{\lambda_1\gamma_{out}+\zeta_{tot}} \right.\right. \\
& \left.\left.\left. + \sum_{p=0}^{w-1}(-1)^{p+1}\sum_{v_0<\ldots<v_p}^{w-1}\prod_{g=0}^{p}\frac{\exp\left(-\Delta_4\gamma_\mathsf{T}\right)}{\Delta_3\Delta_4} \right\} \left(1+\lambda_{2_l}\gamma_{out}\right)^{-1} \right] \right),
\end{aligned}
\tag{2}
$$

where $\zeta_{tot} = \sum_{m=1}^{M}\zeta_{s,m}$, $\Delta_1 = \lambda_1\gamma_\mathsf{T}+\zeta_{tot}$, $\Delta_2 = \sum_{u=0}^{p}\lambda_{2((l-w+v_u))_K}+\lambda_1\gamma_\mathsf{T}+\zeta_{tot}$, $\Delta_3 = 1 + \lambda_{2((l-w+v_g))_K}\gamma_\mathsf{T}$, and $\Delta_4 = \sum_{u=0}^{p}\lambda_{2((l-w+v_u))_K}+\lambda_1\gamma_{out}+\zeta_{tot}$.

Proof. lease see Appendix.

3.2   Average Symbol Error Probability

The ASEP can be expressed in terms of the cumulative distribution function (CDF) of γ_{up}, $F_{\gamma_{up}}(\gamma) = P_{out}(\gamma_{out} = \gamma)$ as

$$
\text{ASEP} \simeq \frac{a\sqrt{b}}{2\sqrt{\pi}}\int_0^\infty \frac{\exp\left(-b\gamma\right)}{\sqrt{\gamma}}F_{\gamma_{up}}(\gamma)d\gamma,
\tag{3}
$$

where a and b are modulation-specific constants.

Lemma 2. *L.2 The ASEP for MUSwiD cognitive CSI-assisted AF relay system with multiple primary receivers using orthogonal spectrums is given by*

$$
\text{ASEP} \simeq \frac{a\sqrt{b}}{2\sqrt{\pi}}\zeta_{\text{tot}} \Bigg\{ \sum_{i=0}^{K-1} \pi_i \Bigg(\frac{\sqrt{\pi}}{\sqrt{b}\zeta_{\text{tot}}} + \sum_{k=0}^{K-1}(-1)^{k+1} \sum_{\substack{n_0<\dots<n_k \\ n(.)\neq i}}^{K-1} \prod_{t=0}^{k} \frac{(1+\lambda_{2n_t}\gamma_{\text{T}})^{-1}}{\vartheta_1} - \Bigg[\Bigg(-\frac{\lambda_1}{\lambda_{2i}}+\zeta_{\text{tot}}\Bigg)^{-1} \Gamma(1/2)(\lambda_{2i})^{-1/2}
$$

$$
\times \Bigg\{ \exp\Bigg(\frac{b}{\lambda_{2i}}\Bigg)\Bigg(\Gamma(1/2,b/\lambda_{2i}) - \exp\Bigg(-\Bigg(\zeta_{\text{tot}}-\frac{\lambda_1}{\lambda_{2i}}\Bigg)\gamma_{\text{T}}\Bigg)\Gamma(1/2,(\lambda_1\gamma_{\text{T}}+b)/\lambda_{2i})\Bigg) + \Bigg(\frac{\lambda_{2i}\zeta_{\text{tot}}}{\lambda_1}\Bigg)^{-1/2}\exp\Bigg(\frac{b\zeta_{\text{tot}}}{\lambda_1}\Bigg)
$$

$$
\times \Bigg(\Gamma(1/2,(\lambda_1\gamma_{\text{T}}+b)\zeta_{\text{tot}}/\lambda_1)-\Gamma(1/2,b\zeta_{\text{tot}}/\lambda_1)\Bigg)\Bigg\} + \sum_{k=0}^{K-1}(-1)^{k+1}\sum_{\substack{n_0<\dots<n_k \\ n(.)\neq i}}^{K-1}\prod_{t=0}^{k}\frac{(1-\exp(-\vartheta_1\gamma_{\text{T}}))}{(1+\lambda_{2n_t}\gamma_{\text{T}})\vartheta_1}\Bigg]\Bigg) + \sum_{l=0}^{K-1}\pi_l
$$

$$
\times \Bigg(\sum_{q=0}^{K}\frac{(-1)^q}{q!}\sum_{m_1,\dots,m_q}^{K}\prod_{z=1}^{q}\frac{\exp(-\vartheta_1\gamma_{\text{T}})}{(1+\lambda_{2m_z}\gamma_{\text{T}})\vartheta_1} + \sum_{w=0}^{K-1}\pi_{((l-w))K}\Bigg[(1+\lambda_{2i}\gamma_{\text{T}})^{-1}\Bigg\{\frac{\exp(-\vartheta_1)}{\vartheta_1}+\sum_{p=0}^{w-1}(-1)^{p+1}\sum_{v_0,\dots,v_p}^{w-1}
$$

$$
\prod_{g=0}^{p}\frac{\exp(-\vartheta_3\gamma_{\text{T}})}{(1+\lambda_{2((l-w+v_g))K}\gamma_{\text{T}})\vartheta_3}\Bigg\} - \Bigg\{\Bigg(-\frac{\lambda_1}{\lambda_{2i}}+\zeta_{\text{tot}}\Bigg)^{-1}\Gamma(1/2)\frac{\exp(-\zeta_{\text{tot}}\gamma_{\text{T}})}{(\lambda_{12})^{1/2}}\Bigg(\exp\Bigg(\frac{\lambda_1\gamma_{\text{T}}+b}{\lambda_{2i}}\Bigg)\Gamma(1/2,(\lambda_1\gamma_{\text{T}}+b)/\lambda_{2i})
$$

$$
-\Bigg(\frac{\lambda_{2i}\zeta_{\text{tot}}}{\lambda_1}\Bigg)^{-1/2}\exp\Bigg(\frac{(\lambda_1\gamma_{\text{T}}+b)\zeta_{\text{tot}}}{\lambda_1}\Bigg)\Gamma(1/2,(\lambda_1\gamma_{\text{T}}+b)\zeta_{\text{tot}}/\lambda_1)\Bigg)\Bigg\} + \sum_{p=0}^{w-1}(-1)^{p+1}\prod_{g=0}^{p}\frac{\exp(-\vartheta_1\gamma_{\text{T}})}{1+\lambda_{2((l-w+v_g))K}\gamma_{\text{T}}}
$$

$$
\times \Bigg\{\Bigg(-\frac{\lambda_1}{\lambda_{2i}}+\zeta_{\text{tot}}\Bigg)^{-1}\Gamma(1/2)(\lambda_{2i})^{-1/2}\Bigg(\exp\Bigg(\frac{\lambda_1\gamma_{\text{T}}+b}{\lambda_{2i}}\Bigg)\Gamma(1/2,(\lambda_1\gamma_{\text{T}}+b)/\lambda_{2i}) - \exp\Bigg(\frac{(\lambda_1\gamma_{\text{T}}+b)\zeta_{\text{tot}}}{\lambda_1}\Bigg)
$$

$$
\times \Bigg(\frac{\lambda_{2i}\zeta_{\text{tot}}}{\lambda_1}\Bigg)^{-1/2}\Gamma(1/2,(\lambda_1\gamma_{\text{T}}+b)\zeta_{\text{tot}}/\lambda_1)\Bigg)\Bigg\}\Bigg]\Bigg)\Bigg\},
$$

$$(4)$$

where $\Gamma(.,.)$ is the incomplete Gamma function defined in [19, Eq.(8.350.2)], $\vartheta_1 = \lambda_1\gamma_{\text{T}}+\zeta_{\text{tot}}$ and $\vartheta_3 = 2\lambda_1\gamma_{\text{T}}+\zeta_{\text{tot}}$.

Proof. y replacing γ_{out} with γ in (2) and using the partial fraction expansion and the integration in (3) and with the help of [19, Eq.(3.361.2)] and [19, Eq.(3.383.10)], we get (4). ∎

4 Asymptotic Performance Analysis

To get more insights about the system performance and simplify the results, we study the behavior at high SNR values where the outage probability can be expressed as $P_{\text{out}}\approx (G_c\text{SNR})^{-G_d}$, where G_c and G_d denote the coding gain and diversity order of the system, respectively [17]. In the upcoming analysis, the $S \to P_m$ links, the $R \to D_k$ links, and the $R \to P_m$ links are assumed to be identical, that is $(\zeta_{s,1} = \dots = \zeta_{s,M} = \zeta_{s,p})$, $(\lambda_{r,1} = \dots = \lambda_{r,K} = \lambda_{r,d})$, and $(\zeta_{r,1} = \dots = \zeta_{r,M} = \zeta_{r,p})$, respectively. The conditional CDF of γ_{up} is given for the identical case of users channels as

$$
F_{\gamma_{\text{up}}}(\gamma|Y) = \begin{cases} [F_{\gamma^{\text{up}}}(\gamma_{\text{T}}|Y)]^{K-1}F_{\gamma^{\text{up}}}(\gamma|Y), & \gamma < \gamma_{\text{T}}; \\ \sum_{j=0}^{K-1}[F_{\gamma^{\text{up}}}(\gamma|Y) - F_{\gamma^{\text{up}}}(\gamma_{\text{T}}|Y)] \\ \times [F_{\gamma^{\text{up}}}(\gamma_{\text{T}}|Y)]^{j} + [F_{\gamma^{\text{up}}}(\gamma_{\text{T}}|Y)]^{K}, & \gamma \geq \gamma_{\text{T}}, \end{cases}
$$

$$(5)$$

where $F_{\gamma^{up}}(\gamma|Y)$ is the CDF of γ_k^{up} conditioned on $Y = \min_m |g_{s,m}|^2, m = 1, \ldots, M$ and it can be expressed for the case of identical first and second hop channels as

$$F_{\gamma_k^{up}}(\gamma|Y) = 1 - \frac{\exp(-\lambda_1 \gamma Y)}{(1 + \lambda_2 \gamma)}, \tag{6}$$

where λ_1 is as defined in the Appendix and $\lambda_2 = 1 \Big/ \left(\sum_{m=1}^{M} \zeta_{r,m} \Omega_{r,d} \frac{\mathcal{I}_p}{N_0}\right)$. As $\frac{\mathcal{I}_p}{N_0} \to \infty$, the CDF in (6) simplifies to $F_{\gamma_k^{up}}(\gamma|Y) \approx \lambda_1 Y \gamma$. Upon substituting this CDF in (5) and following the same procedure as in the Appendix, the outage probability at high SNR values can be evaluated with the help of [19, Eq. (3.351.1)] and [19, Eq.(3.351.2)] and recalling that $\lambda_1 = 1 \Big/ \left(\Omega_{s,r} \frac{\mathcal{I}_p}{N_0}\right)$ as

$$P_{out}^{\infty} = \left\{ \frac{\Omega_{s,r} M \zeta_{s,p}}{\Gamma(2, M\zeta_{s,p}\gamma_T)(\gamma_{out} - \gamma_T)} \frac{\mathcal{I}_p}{N_0} \right\}^{-1}. \tag{7}$$

Upon substituting the asymptotic outage probability in (3) and with the help of [19, Eq.(3.351.3)] and recalling that $\lambda_1 = 1 \Big/ \left(\Omega_{s,r} \frac{\mathcal{I}_p}{N_0}\right)$, the ASEP can be obtained at high SNR values as

$$\text{ASEP}^{\infty} = \left\{ \left(\Xi \left[\frac{\Gamma\left(\frac{3}{2}\right)}{b^{3/2}} - \frac{\Gamma\left(\frac{1}{2}\right)}{b^{1/2}}\gamma_T \right] \right)^{-1} \frac{\mathcal{I}_p}{N_0} \right\}^{-1}, \tag{8}$$

where $\Xi = \frac{a\sqrt{b}\Gamma(2, M\zeta_{s,p}\gamma_T)}{2\sqrt{\pi}M\zeta_{s,p}\Omega_{s,r}}$.

A simple but an accurate method to find approximate optimum switching threshold is by using $\min\left(\frac{\frac{\mathcal{I}_p}{N_0}\Omega_{s,r}}{(M\zeta_{s,p})^{-1}}, \frac{\frac{\mathcal{I}_p}{N_0}\Omega_{r,d}}{(M\zeta_{r,p})^{-1}}\right)$.

5 Simulation and Numerical Results

We can see from figure 2 that the asymptotic results perfectly converge to the analytical results as well as the Monte-Carlo simulations. It is obvious also that the used bound on the e2e SNR is indeed very tight; especially, at the high SNR region. Furthermore, we can see from this figure that the MUSwiD selection scheme has nearly the same performance as the opportunistic scheduling for very low SNR region; whereas, as we go further in increasing SNR, the opportunistic scheduling is clearly outperforming the MUSwiD scheme, as expected. In addition, we can see that for the MUSwiD scheme as K increases, the system performance becomes more enhanced; especially, at the range of SNR values that are comparable to the switching threshold γ_T. More importantly, for $K = 2, 3$, and 4, it is obvious that at both low and high SNR values, all curves asymptotically converge to the same behavior and no gain is achieved in the system performance with having more users. This is expected since when γ_T takes values much smaller or much larger than the average SNR, the system asymptotically

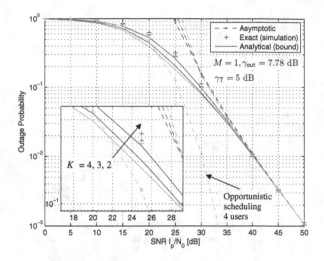

Fig. 2. P_{out} vs SNR for different values of K and $\mu_{s,p} = 30$, $\Omega_{s,r} = 0.8$, $\Omega_{r,p} = 0.1$, and $\Omega_{r,k} = 0.7$ for $k = 1, \ldots, 4$.

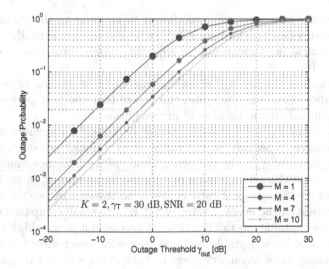

Fig. 3. P_{out} vs γ_{out} for different values of M and $\mu_{s,m} = 20$ for $m = 1, \ldots, M$, $\Omega_{s,r} = 0.8$, $\mu_{r,m} = 0.01$ for $m = 1, \ldots, M$, and $\Omega_{r,k} = 0.9$ for $k = 1, 2$.

converges to the case of two users and hence, having more users will have no effect on the system performance. Finally, the effectiveness of the MUSwiD scheme is in the reduction of CSI feedback load it offers compared to the opportunistic scheduling. In order to achieve this effectiveness with a slight reduction in the

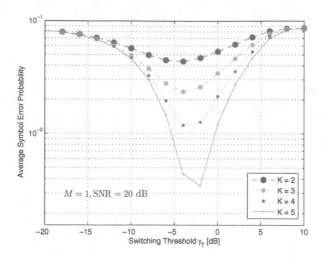

Fig. 4. ASEP vs γ_T for different values of K and $\mu_{s,p} = 20$, $\Omega_{s,r} = 0.1$, $\mu_{r,p} = 0.3$, and $\Omega_{r,k} = 0.2$ for $k = 1, \ldots, 5$.

multiuser diversity gain, γ_T should be chosen to be close to the average SNR. Due to their features and performance, MUSwiD systems are actually attractive options for practical implementation in emerging mobile broadband communication systems [16].

It is obvious from Figure 3 that as γ_{out} increases, worse the achieved performance. Furthermore, it is clear from this figure that the best performance is achieved with the maximum number of primary users M.

We can see from Figure 4 that increasing K leads to a significant gain in system performance; especially, in the range of γ_T values that are comparable to the average value of γ_k^{up}. On the other hand, as γ_T becomes much smaller or much larger than the average value of γ_k^{up}, the improvement in performance decreases, as all curves asymptotically converge to the case of two users. This is due to the fact that, if the average value of γ_k^{up} is very small compared to γ_T, all users will be unacceptable most of the time. Whereas, if it is very high compared to γ_T, all users will be acceptable and one user will be scheduled most of the time. Thus, having more secondary users in both cases will add no gain to the system performance.

The average number of channel estimations versus switching threshold γ_T is illustrated in Figure 5 for the case of 4 secondary users. The average number of channel estimations for the MUSwiD selection scheme is provided in [20]. We can see from this figure that as the channels of all users are required for its operation, the opportunistic scheduling is always of need for 4 channel estimations. On the other hand, we can see that the MUSwiD selection scheme needs to estimate at most 3 channels because when the first 3 users are found unacceptable, the

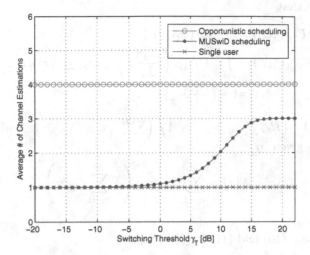

Fig. 5. Average number of channel estimations of the MUSwiD scheduling in comparison with the opportunistic scheduling with $K = 4$ and an average power/user path = 10 dB.

last checked user will be scheduled by the central unit regardless of its quality. Also, we can notice from this figure that as γ_T increases, the average number of channel estimations of users increases since it is more difficult to find a user with an acceptable quality.

6   Conclusion

The new scenario of MUSwiD cognitive AF relay network with multiple primary receivers using orthogonal spectrums was proposed in this paper. Analytical and asymptotic approximations for the outage and average symbol error probabilities were derived. Results illustrated that the diversity order of the proposed scenario is the same as its non-cognitive counterpart. Unlike the existing papers where the same spectrum band is shared by the primary receivers, increasing the number of primary receivers in the proposed scenario enhances the system behavior.

Appendix
Proof of Lemma 1

Herein, we first apply the conditional statistics on the fading channel from S to P. The CDF of γ_k^{up} conditioned on $Y = \min_m |g_{\mathsf{s},m}|^2, m = 1, \ldots, M$ can be written as

$$F_{\gamma_k^{\mathsf{up}}}(\gamma|Y) = 1 - (1 - F_{\gamma_1}(\gamma|Y))\left(1 - F_{\gamma_{2_k}}(\gamma|Y)\right). \tag{9}$$

It is easy to see that

$$F_{\gamma_1}(\gamma|Y) = 1 - \exp(-\lambda_1 \gamma Y), \tag{10}$$

$$F_{\gamma_{2_k}}(\gamma|Y) = \int_0^{\infty} F_{|h_{r,k}|^2}\left(\frac{N_0 \gamma}{\mathcal{I}_p} x\right) f_{\min_m |g_{r,m}|^2}(x)\, dx$$

$$= 1 - (1 + \lambda_{2_k} \gamma)^{-1}, \tag{11}$$

where $\lambda_1 = 1/\left(\Omega_{\mathsf{s},r} \frac{\mathcal{I}_p}{N_0}\right)$, $\lambda_{2_k} = 1/\left(\sum_{m=1}^{M} \zeta_{r,m} \Omega_{r,k} \frac{\mathcal{I}_p}{N_0}\right)$, and the PDF $f_{\min_m |g_{r,m}|^2}(x)$ is given by

$$f_{\min_m |g_{r,m}|^2}(x) = \sum_{m=1}^{M} \zeta_{r,m} \exp\left(-\sum_{m=1}^{M} \zeta_{r,m} x\right), \tag{12}$$

where $\zeta_{r,m} = 1/\mu_{r,m}$.

Upon substituting (10) and (11) in (9), we get

$$F_{\gamma_k^{\mathsf{up}}}(\gamma|Y) = 1 - \frac{\exp(-\lambda_1 \gamma Y)}{(1 + \lambda_{2_k} \gamma)}. \tag{13}$$

Upon substituting (13) in the conditional CDF of γ_{up} provided by [17], we get

$$F_{\gamma_{\mathsf{up}}}(\gamma|Y) = \begin{cases} \sum_{i=0}^{K-1} \pi_i \left(1 - \frac{\exp(-\lambda_1 \gamma Y)}{(1+\lambda_{2_i} \gamma)}\right) \underbrace{\prod_{\substack{k=0 \\ k \neq i}}^{K-1} \left(1 - \frac{\exp(-\lambda_1 \gamma_{\mathsf{T}} Y)}{(1+\lambda_{2_k} \gamma_{\mathsf{T}})}\right)}_{\mathcal{P}_1}, & \gamma < \gamma_{\mathsf{T}}; \\[4ex] \sum_{l=0}^{K-1} \pi_l \left\{ \underbrace{\prod_{q=1}^{K} \left(1 - \frac{\exp(-\lambda_1 \gamma_{\mathsf{T}} Y)}{(1+\lambda_{2_q} \gamma_{\mathsf{T}})}\right)}_{\mathcal{P}_2} + \sum_{w=0}^{K-1} \pi_{((l-w))_K} \right. \\[2ex] \left. \times \left[\left(1 - \frac{\exp(-\lambda_1 \gamma Y)}{(1+\lambda_{2_i} \gamma)}\right) - \left(1 - \frac{\exp(-\lambda_1 \gamma_{\mathsf{T}} Y)}{(1+\lambda_{2_i} \gamma_{\mathsf{T}})}\right)\right] \underbrace{\prod_{p=0}^{w-1} \left(1 - \frac{\exp(-\lambda_1 \gamma_{\mathsf{T}} Y)}{\left(1 + \lambda_{2_{((l-w+p))_K}} \gamma_{\mathsf{T}}\right)}\right)}_{\mathcal{P}_3} \right\}, & \gamma \geq \gamma_{\mathsf{T}}, \end{cases} \tag{14}$$

where K is the number of secondary destinations, γ_{T} is a predetermined switching threshold, $\pi_i, i = 0, \ldots, K-1$ is the probability that the ith destination or user is chosen as given by [17], and $((l-w))_K$ denotes $l - w$ modulo K.

With the help of the product identities in [18] and [8], the terms $\mathcal{P}_1$, $\mathcal{P}_2$, and $\mathcal{P}_3$ in (14) can be simplified as follows

$$\mathcal{P}_1 = 1 + \sum_{k=0}^{K-1} (-1)^{k+1} \sum_{\substack{n_0 < \ldots < n_k \\ n_{(.)} \neq i}} \prod_{t=0}^{k} \frac{\exp(-\lambda_1 \gamma_{\mathsf{T}} Y)}{(1 + \lambda_{2_{n_t}} \gamma_{\mathsf{T}})}, \tag{15}$$

where $\sum_{\substack{n_0 < \ldots < n_k \\ n_{(.)} \neq i}}^{K-1}$ is a short-hand notation for $\sum_{\substack{n_0=0 \\ n_0 \neq i}}^{K-k-1} \sum_{\substack{n_1=n_0+1 \\ n_1 \neq i}}^{K-k} \cdots \sum_{\substack{n_k=n_{k-1}+1 \\ n_k \neq i}}^{K-1}$.

$$\mathcal{P}_2 = \sum_{q=0}^{K} \frac{(-1)^q}{q!} \sum_{m_1,\ldots,m_q}^{K} \prod_{z=1}^{q} \frac{\exp\left(-\lambda_1 \gamma_{\mathsf{T}} Y\right)}{\left(1 + \lambda_{2_{m_z}} \gamma_{\mathsf{T}}\right)}, \tag{16}$$

where $\sum_{m_1,\ldots,m_q}^{K}$ is a short-hand notation for $\sum_{\substack{m_1=\ldots=m_q=1 \\ m_1 \neq \ldots \neq m_q}}^{K-k-1}$.

$$\mathcal{P}_3 = 1 + \sum_{p=0}^{w-1} (-1)^{p+1} \sum_{v_0 < \ldots < v_p}^{w-1} \prod_{g=0}^{p} \frac{\exp\left(-\lambda_1 \gamma_{\mathsf{T}} Y\right)}{\left(1 + \lambda_{2_{((l-w+v_g))_K}} \gamma_{\mathsf{T}}\right)}, \tag{17}$$

where $\sum_{v_0 < \ldots < v_p}^{w-1}$ is a short-hand notation for $\sum_{v_0=0}^{w-p-1} \sum_{v_1=v_0+1}^{w-p} \cdots \sum_{v_p=v_{p-1}+1}^{w-1}$.
Up to now, the outage probability can be expressed as

$$P_{\mathsf{out}} \simeq \int_0^\infty F_{\gamma_{\mathsf{up}}}(\gamma|Y) f_Y(y) dy, \tag{18}$$

where the PDF $f_Y(y)$ is given by

$$f_Y(y) = \sum_{m=1}^{M} \zeta_{\mathsf{s},m} \exp\left(-\sum_{m=1}^{M} \zeta_{\mathsf{s},m} y\right), \tag{19}$$

where $\zeta_{\mathsf{s},m} = 1/\mu_{\mathsf{s},m}$. Upon substituting (15), (16), and (17) in (14) and using (18), the outage probability can be evaluated as in (2).

Acknowledgments. This work is supported by King Fahd University of Petroleum & Minerals (KFUPM) through project number FT131009.

References

1. Haykin, S.: Cognitive radio: Brain-empowered wireless communications. IEEE J. Sel. Areas Commun. **23**, 201–220 (2005)
2. Goldsmith, A., Jafar, S., Maric, I., Srinivasa, S.: Breaking spectrum gridlock with cognitive radios: an information theoretic perspective. Proc. IEEE **97**(5), 894–914 (2009)
3. Laneman, J.N., Tse, D.N.C., Wornell, G.W.: Cooperative diversity in wireless networks: efficient protocals and outage behavior. IEEE Trans. Inf. Theory **50**(12), 3062–3080 (2004)
4. Chamkhia, H., Hasna, M.O., Hamila, R., Hussain, S.I.: Performance analysis of relay selection schemes in underlay cognitive networks with decode and forward relaying. In: IEEE Int'l Symp. on Personal, Indoor & Mobile Radio Commun. (PIMRC 2012), pp. 1552–1558, Australia (2012)
5. Xia, M., Aïssa, S.: Cooperative AF relaying in spectrum-sharing systems: performance analysis under average interference power constraints and Nakagami-m fading. IEEE Trans. Commun. **60**(6), 170–176 (2012)
6. Fredj, K.B., Aïssa, S.: Performance of amplify-and-forward systems with partial relay selection under spectrum-sharing constraints. IEEE Trans. Wireless Commun. **11**(2), 500–504 (2012)

7. Chen, J., Li, Z., Si, J., Huang, H., Gao, R.: Outage probability analysis of partial AF relay selection in spectrum-sharing scenario with multiple primary users. Elect. Lett. **48**(19), 1211–1212 (2012)
8. Bao, V.N.Q., Duong, T.Q., da Costa, D.B., Alexandropoulos, G.C., Nallanathan, A.: Cognitive amplify-and-forward relaying with best relay selection in non-identical Rayleigh fading. IEEE Commun. Lett. **17**(3), 475–478 (2013)
9. Guimarães, F.R.V., da Costa, D.B., Tsiftsis, T.A., Cavalcante, C.C., Karagiannidis, G.K.: Multiuser and multirelay cognitive radio networks under spectrum-sharing constraints. IEEE Trans. Veh. Tech. **63**(1), 433–439 (2014)
10. Zhai, C., Zhang, W.: Adaptive spectrum leasing with secondary user scheduling in cognitive radio networks. IEEE Trans. Wireless Commun. **12**(7), 3388–3398 (2013)
11. Fan, L., Lei, X., Duong, T.Q., Hu, R.Q., Elkashlan, M.: Multiuser cognitive relay networks: joint impact of direct and relay communications. IEEE Trans. Wireless Commun. **13**(9), 5043–5055 (2014)
12. Yang, L., Alouini, M.-S.: Performance analysis of multiuser selection diversity. IEEE Trans. Veh. Tech. **55**(6), 1848–1861 (2006)
13. Nam, H., Alouini, M.-S.: Multiuser switched diversity scheduling systems with per-user threshold. IEEE Trans. Commun. **58**(5), 1321–1326 (2010)
14. Ban, T., Choi, W., Jung, B., Sung, D.: Multi-user diversity in a spectrum sharing system. IEEE Trans. Wireless Commun. **8**(1), 102–106 (2009)
15. Etkin, R., Parekh, A., Tse, D.: Spectrum sharing for unlicensed bands. IEEE J. Sel. Areas Commun. **25**(3), 517–528 (2007)
16. Shaqfeh, M., Alnuweiri, H., Alouini, M.-S.: Multiuser switched diversity scheduling schemes. IEEE Trans. Commun. **60**(9), 2499–2510 (2012)
17. Simon, M.K., Alouini, M.-S.: Digital Communication over Fading Channels, 2nd ed. Wiley (2005)
18. Bithas, P.S., Karagiannidis, G.K., Sagias, N.C., Mathiopoulos, P.T., Kotsopoulos, S.A., Corazza, G.E.: Performance analysis of a class of GSC receivers over nonidentical Weibull fading channels. IEEE Trans. Veh. Tech. **54**(6), 1963–1970 (2005)
19. Gradshteyn, I.S., Ryzhik, I.M.: Tables of Integrals, Series and Products, 6th edn. Acadamic Press, San Diago (2000)
20. Yang, H.-C., Alouini, M.-S.: Order Statistics in Wireless Communications: Diversity, Adaptation, and Scheduling in MIMO and OFDM Systems, 1st ed. Cambridge University Press (2011)

An Efficient Secondary User Selection Scheme for Cognitive Networks with Imperfect Channel Estimation and Multiple Primary Users

Anas M. Salhab[✉]

Electrical Engineering Department, King Fahd University of Petroleum & Minerals,
Dhahran 31261, Saudi Arabia
salhab@kfupm.edu.sa

Abstract. In this paper, we study the performance of multiuser cognitive generalized order user scheduling networks with multiple primary users and imperfect channel estimation. The utilized generalized order user selection scheme is efficient in situations where a user other than the best user is erroneously selected by the scheduling unit for data reception as in imperfect channel estimation or outdated channel information conditions. In this scheme, the secondary user with the second or even the Nth best signal-to-noise ratio (SNR) is assigned the system resources in a downlink channel. In our paper, closed-form expressions are derived for the outage probability, average symbol error probability (ASEP), and ergodic channel capacity assuming Rayleigh fading channels. Also, to get more insights about the system performance, the behavior is studied at the high SNR regime where the diversity order and coding gain are derived and analyzed. The achieved results are verified by Monte-Carlo simulations. Main results illustrate that the number of primary users affects the secondary system performance through affecting only the coding gain. Also, findings illustrate that a zero diversity gain is achieved by the system and a noise floor appears in the results when the secondary user channels are imperfectly estimated. Finally, results show that the generalized order user scheduling in cognitive networks has exactly the same diversity order as when implemented in the non-cognitive counterparts.

Keywords: Multiuser cognitive networks · Generalized order user scheduling · Imperfect channel estimation · Rayleigh fading

1 Introduction

The multiuser diversity is achieved by taking advantage of the channel fading variations in wireless networks. More specifically, it was shown that selecting the user with the best instantaneous channel each transmitting or receiving time increases the chance of having the communication to occur over a good

A.M. Salhab— Member, IEEE.

© Institute for Computer Sciences, Social Informatics and Telecommunications Engineering 2015
M. Weichold et al. (Eds.): CROWNCOM 2015, LNICST 156, pp. 149–163, 2015.
DOI: 10.1007/978-3-319-24540-9_12

channel. As a way for improving the spectrum utilization efficiency in wireless networks, the cognitive radio has been proposed in [1]. In such networks, the secondary or cognitive users can share the spectrum with primary users via underlay, overlay, or interweave paradigms [2]. The underlay paradigm which is adopted in this paper allows secondary users to share the spectrum of primary users if the interference between them is below a certain threshold.

The opportunistic scheduling where the user with the best instantaneous channel is always selected by the scheduling unit for data communications was used in [3] to select among secondary users. The multiuser diversity gain and bit error rate for multiple access, broadcast, and parallel access channels in cognitive radio networks with opportunistic scheduling were derived in [4]. Recently, Wang *et al.* proposed in [5] a limited feedback based underlay spectrum sharing scheme where the opportunistic scheduling was used to select among secondary users in a downlink transmission. The opportunistic scheduling was also used in [6] to select among secondary users in an uplink transmission. Most recently, the secondary users were allowed to utilize the spectrum of primary users in an opportunistic way in [7]. The performance of multiuser cognitive radio networks with opportunistic scheduling and multiple primary receivers was studied in [8]. All the aforementioned papers assumed perfectly estimated channels.

Some scheduling fairness and power control schemes were presented in [9] for multiuser cognitive networks with opportunistic scheduling among the secondary users. In [10], Khan *et al.* derived the exact outage and error rate probabilities for multiuser cognitive networks with opportunistic scheduling and Nakagami-m fading channels. In general, two important issues need to be considered when designing any multiuser network: the sum-rate capacity and fairness among users. The maximum-rate schedulers such as the generalized order user scheduling maximizes the sum capacity at the expense of unfairness among users; whereas, the proportional fair user selection scheme satisfies fairness among users at the expense of system sum-rate [11]. Therefore, the selection of the scheduling scheme depends on the system requirements and nature of the system. Although the proportional fair scheduling could be helpful for users of weak channels, the loss happens in the throughput when this scheduling scheme is used can be large in situations where users are scattered across the cell [12].

From our reading to the literature on the area of multiuser cognitive networks, we noticed that the most commonly used secondary user selection scheme in these networks is the opportunistic scheduling. In this scheme, the user with the best instantaneous channel is selected every time for data transmission or reception. Also, we noticed that most of the papers on multiuser cognitive networks assumed perfectly estimated channels and ignored the effect of imperfect channel estimation on the system performance. There exists several situations in wireless networks where the opportunistic scheduling could fail, among which is the presence of imperfect channel state information where the scheduling unit could fail in error in selecting the best among the available users and in the presence of outdated channel information where the user which was the best at the selection time instant could not be the best at the transmission time instant. An efficient selection scheme which can deal with such situations is the generalized order user scheduling. In this scheme, the user with the second or even the Nth best channel is selected instead of the best user for transmitting or receiving data. This scheme was firstly proposed in literature to select among antennas [13], then, it was presented to select among relays in relay networks [14], and recently, it was used to select among users in multiuser relay networks [15]. Most of the previous papers consider the opportunistic scheduling and perfectly estimated channels.

In this paper, we study the performance of multiuser cognitive generalized order user selection network with multiple primary receivers in the presence of imperfect channel estimation. In the considered scheme, the secondary user with the first, the second, or even the Nth best signal-to-noise ratio (SNR) is allowed by the scheduling unit for data reception. Closed-form expressions are derived for the outage probability, average symbol error probability (ASEP), and ergodic channel capacity assuming independent non-identically distributed (i.n.i.d.) generic case of Rayleigh fading channels. Furthermore, the performance is studied at the high SNR regime where approximate expressions are derived for the outage probability and ASEP in addition to the derivation of the diversity order and coding gain of the system. The effect of number of primary users, number of secondary users, and channel estimation error on the system performance is illustrated via providing some simulation and numerical examples.

This paper is organized as follows. Section 2 presents the system and channel models. The exact performance evaluation is conducted in Section 3. Section 4 provides the asymptotic performance analysis. Some simulation and numerical results are discussed in Section 5. Finally, Section 6 concludes the paper.

2   System and Channel Models

The system under consideration consists of one secondary source S, K secondary destinations or users D_k ($k = 1, \ldots, K$), and M primary receivers P_m ($m = 1, \ldots, M$) using the same frequency band. All nodes are assumed to be equipped with single antenna. The secondary source sends its message x to K users under a transmit power constraint which guarantees that the interference with the primary users does not exceed a threshold $\mathcal{I}_{\mathsf{p}}$. To satisfy the primary interference constraint, the source S must transmit at a power given by $P_{\mathsf{s}} = \mathcal{I}_{\mathsf{p}}/\max_{m}|g_{\mathsf{s},m}|^2$, $m = 1, \ldots, M$, where $g_{\mathsf{s},m}$ is the channel coefficient of the S $\rightarrow$ P_m link. Therefore, the message at the kth destination D_k from the source S is given by $y_{\mathsf{s},k} = \sqrt{P_{\mathsf{s}}}h_{\mathsf{s},k}x + n_{\mathsf{s},k}$, where $h_{\mathsf{s},k}$ is the channel coefficient of the S $\rightarrow$ D_k link and $n_{\mathsf{s},k}$ represents the additive white Gaussian noise (AWGN) term at D_k with a power of N_0. We assume that perfect channel information including the interference channel is available at the secondary source[1].

---

[1] Secondary source can know the channel information of the primary users by either a direct reception of pilot signals from primary users [16], or by exchange of channel information between primary and secondary users through a band manager [17].

Also, we assume that no interference is introduced from the primary user on the secondary receivers[2]. All channel coefficients are assumed to be Rayleigh distributed, so the channel gains $|g_{s,m}|^2$ and $|h_{s,k}|^2$ follow exponential distribution with mean powers $\mu_{g_{s,m}}$ and $\Omega_{h_{s,k}}$, respectively.

The channel coefficient of the $S \rightarrow D_k$ channel can be written as [19]

$$h_{s,k} = \hat{h}_{s,k} + e_{h_{s,k}}, \tag{1}$$

where $\hat{h}_{s,k}$ is the estimate of the $S \rightarrow D_k$ link and $e_{h_{s,k}}$ is the channel estimation error, which is assumed to be complex Gaussian with zero mean and variance $\sigma^2_{e_{h_{s,k}}} = \Omega_{h_{s,k}} - \mathbb{E}[|\hat{h}_{s,k}|^2]$, with $\mathbb{E}[.]$ denoting the expectation operator. Also, $\hat{h}_{s,k}$ is complex Gaussian with zero mean and variance $\Omega_{\hat{h}_{s,k}} = \Omega_{h_{s,k}} + \sigma^2_{e_{h_{s,k}}}$. The above definition also applies to the $S \rightarrow P_m$ channel, i.e., $\hat{g}_{s,m} \sim \mathcal{CN}(0, \mu_{\hat{g}_{s,m}} = \mu_{g_{s,m}} + \sigma^2_{e_{g_{s,m}}})$.

Upon using the values $h_{s,k} = \hat{h}_{s,k} + e_{h_{s,k}}$ and $g_{s,m} = \hat{g}_{s,m} + e_{g_{s,m}}$, the signal at the kth user can be rewritten as

$$y_{s,k} = \sqrt{\frac{\mathcal{I}_p}{W + \sigma^2_{ew}}} \hat{h}_{s,k} x + \sqrt{\frac{\mathcal{I}_p}{W + \sigma^2_{ew}}} e_{h_{s,k}} x + n_{s,k}, \tag{2}$$

where $W = \max_{m} |\hat{g}_{s,m}|^2$, $m = 1, \ldots, M$ and σ^2_{ew} is the variance of the channel estimation error associated with the channel estimate $\max_{m} \hat{g}_{s,m}$.

From (2), the SNR of the $S \rightarrow D_k$ link can be easily obtained after simple manipulations as

$$\gamma_{S-D_k} = \frac{\bar{\gamma}|\hat{h}_{s,k}|^2}{W + \sigma^2_{ew} + \bar{\gamma}\sigma^2_{e_{h_{s,k}}}} = \gamma_k, \tag{3}$$

where $\bar{\gamma} = \mathcal{I}_p/N_0$. The generalized order user scheduling is performed by choosing the user which has the Nth best SNR γ_k. The estimation error variance can be made small by transmitting large number of pilots at medium to high SNRs [19][3].

3   Exact Performance Analysis

In this section, closed-form expressions are derived for the outage probability, ASEP, and ergodic channel capacity.

---

[2] This assumption is valid when the primary transmitter is in a location far from the secondary receiver [18].

[3] The variance of the channel estimation error can be also assumed to be inversely proportional to SNR as 1/SNR.

3.1   Outage Probability

In this section, we derive the outage probability of the considered system. The outage probability is defined as the probability that the SNR at the selected destination γ_{Sel} goes below a predetermined outage threshold γ_{out}, i.e., $P_{\mathsf{out}} = \Pr\left[\gamma_{\mathsf{Sel}} \leq \gamma_{\mathsf{out}}\right]$, where $\Pr[.]$ denotes the probability operation.

Theorem 1. *The outage probability for multiuser cognitive generalized order user selection network with multiple primary users and imperfect channel estimation is given by*

$$
P_{\mathsf{out}} = M\zeta_{\mathsf{s,p}} \sum_{i=1}^{M-1} \binom{M-1}{i} (-1)^i
$$

$$
\times \sum_{l=1}^{K} \lambda_{\mathsf{s},l} \sum_{\mathcal{P}} \left[\left\{ ((i+1)\zeta_{\mathsf{s,p}})^{-1} - \left(\Delta_1 \gamma_{\mathsf{out}} + (i+1)\zeta_{\mathsf{s,p}}\right)^{-1} \right\} (\Delta_1)^{-1} + \sum_{j=1}^{K-N} (-1)^j \right.
$$

$$
\left. \times \sum_{s_1 < ... < s_j} (\Delta_2)^{-1} \left\{ ((i+1)\zeta_{\mathsf{s,p}})^{-1} - \left(\Delta_2 \gamma_{\mathsf{out}} + (i+1)\zeta_{\mathsf{s,p}}\right)^{-1} \right\} \right]. \tag{4}
$$

Proof. To evaluate the outage probability, the cumulative distribution function (CDF) of γ_{Sel} is required to be obtained first. Herein, we first apply the conditional statistics on the fading channel from S to $\mathsf{P_{Sel}}$, where $\mathsf{P_{Sel}}$ is the primary receiver who has the best channel with S. The CDF of the SNR in (3) conditioned on $W = \max_{m} |\hat{g}_{\mathsf{s},m}|^2$, $m = 1, \ldots, M$ can be easily obtained as

$$
F_{\gamma_k}\left(\gamma|W\right) = 1 - \exp\left(-\lambda_{\mathsf{s},k}\gamma W\right), \tag{5}
$$

where $\lambda_{\mathsf{s},k} = \left(\sigma_{ew}^2 + \sigma_{e_{h_{\mathsf{s},k}}}^2 \bar{\gamma} + 1\right) \Big/ \left(\Omega_{\hat{h}_{\mathsf{s},k}} \bar{\gamma}\right)$.

The conditional probability density function (PDF) of the selected secondary user is given by [20]

$$
f_{\gamma_{\mathsf{Sel}}}(\gamma|W) = \sum_{l=1}^{K} f_{\gamma_l}(\gamma|W) \sum_{\mathcal{P}} \prod_{j=1}^{K-N} F_{\gamma_{i_j}}(\gamma|W) \prod_{w=K-N+1}^{K-1} \left(1 - F_{\gamma_{i_w}}(\gamma|W)\right), \tag{6}
$$

where $\sum_{\mathcal{P}}$ denotes the summation over all $n!$ permutations $(i_1, i_2, \ldots, i_K)$ of $(1, 2, \ldots, K)$ and N is the order of the selected user. Upon substituting (5) in (6), and using the binomial rule and applying the identity

$$
\prod_{j=1}^{K-N} (1 - t_j) = 1 + \sum_{j=1}^{K-N} (-1)^j \sum_{s_1 < ... < s_j} \prod_{n=1}^{j} t_{s_n}, \tag{7}
$$

with $\sum_{s_1 < ... < s_j}$ being a short hand-notation for $\sum_{s_1=1}^{K-N-j+1} \sum_{s_2=s_1+1}^{K-N-j+2} \cdots \sum_{s_j=s_{j-1}+1}^{K-N}$, (6) can be rewritten as

$$f_{\gamma_{\text{Sel}}}(\gamma|W)$$

$$= \sum_{l=1}^{K} \lambda_{\text{s},l} W \sum_{\mathcal{P}} \left[\exp\left(-\Delta_1 W\gamma\right) + \sum_{j=1}^{K-N} (-1)^j \sum_{s_1<...<s_j} \exp\left(-\Delta_2 W\gamma\right) \right], \quad (8)$$

where $\Delta_1 = \sum_{w=K-N+1}^{K-1} \lambda_{\text{s},i_w}$ and $\Delta_2 = \Delta_1 + \sum_{n=1}^{j} \lambda_{\text{s},s_n} + \lambda_{\text{s},l}$.
Assuming identical $\text{S} \to \text{P}_m$ channels, that is $\mu_{\hat{g}_{\text{s},1}} = \cdots = \mu_{\hat{g}_{\text{s},M}} = \mu_{\hat{g}_{\text{s},\text{p}}}$, the
CDF and PDF of W are respectively given by

$$F_W(w) = \left[F_{|g_{\text{s},\text{p}}|^2}(w)\right]^M = \left[1 - \exp\left(-\zeta_{\text{s},\text{p}}w\right)\right]^M,$$

$$f_W(w) = M f_{|g_{\text{s},\text{p}}|^2}(w) \left[F_{|g_{\text{s},\text{p}}|^2}(w)\right]^{M-1}$$

$$= M\zeta_{\text{s},\text{p}} \sum_{i=0}^{M-1} \binom{M-1}{i} (-1)^i \exp\left(-(i+1)\zeta_{\text{s},\text{p}}w\right), \quad (9)$$

where $\zeta_{\text{s},\text{p}} = 1/\mu_{\hat{g}_{\text{s},\text{p}}}$.

Up to now, the PDF of γ_{Sel} can be obtained using $\int_0^\infty f_{\gamma_{\text{Sel}}}(\gamma|W)f_W(w)dw$ as follows

$$f_{\gamma_{\text{Sel}}}(\gamma) = \sum_{i=0}^{M-1} \frac{\binom{M-1}{i}}{(-1)^{-i}} \sum_{l=1}^{K} \lambda_{\text{s},l} \sum_{\mathcal{P}} \left[\left(\Delta_1\gamma + (i+1)\zeta_{\text{s},\text{p}}\right)^{-2} \right.$$

$$\left. + \sum_{j=1}^{K-N} (-1)^j \sum_{s_1<...<s_j} \left(\Delta_2\gamma + (i+1)\zeta_{\text{s},\text{p}}\right)^{-2} \right] M\zeta_{\text{s},\text{p}}, \quad (10)$$

where [21, Eq.(3.381.4)] has been used in getting (10). The outage probability can be obtained by integrating (10) using $\int_0^{\gamma_{\text{out}}} f_{\gamma_{\text{Sel}}}(z)dz$ as given in (4).

3.2   Average Symbol Error Probability

In this section, we derive the ASEP of the considerd system. The ASEP can be written in terms of the CDF of γ_{Sel}, $F_{\gamma_{\text{Sel}}}(\gamma) = P_{\text{out}}(\gamma_{\text{out}} = \gamma)$ as

$$\text{ASEP} = \frac{a\sqrt{b}}{2\sqrt{\pi}} \int_0^\infty \frac{\exp\left(-b\gamma\right)}{\sqrt{\gamma}} F_{\gamma_{\text{Sel}}}(\gamma)d\gamma, \quad (11)$$

where a and b are modulation-specific parameters.

By replacing γ_{out} with γ in (4), and with the help of [21, Eq.(3.381.4)] and [21, Eq. (3.383.10)] and after some simple steps, the ASEP can be obtained as follows

$$\mathrm{ASEP} = \sum_{i=0}^{M-1} \binom{M-1}{i}(-1)^i$$

$$\times \sum_{l=1}^{K} \lambda_{\mathsf{s},l} \sum_{\mathcal{P}} \left[\frac{1}{\Delta_1} \left\{ \frac{((i+1)\zeta_{\mathsf{s},\mathsf{p}})^{-1}}{b^{1/2}} - \frac{\exp\left(\frac{b(i+1)\zeta_{\mathsf{s},\mathsf{p}}}{\Delta_1}\right) \Gamma\left(1/2, \frac{b(i+1)\zeta_{\mathsf{s},\mathsf{p}}}{\Delta_1}\right)}{\left(\Delta_1(i+1)\zeta_{\mathsf{s},\mathsf{p}}\right)^{1/2}} \right\}$$

$$+ \sum_{j=1}^{K-N} (-1)^j \sum_{s_1 < \ldots < s_j}$$

$$\times \frac{1}{\Delta_2} \left\{ \frac{((i+1)\zeta_{\mathsf{s},\mathsf{p}})^{-1}}{b^{1/2}} - \frac{\exp\left(\frac{b(i+1)\zeta_{\mathsf{s},\mathsf{p}}}{\Delta_2}\right) \Gamma\left(1/2, \frac{b(i+1)\zeta_{\mathsf{s},\mathsf{p}}}{\Delta_2}\right)}{\left(\Delta_2(i+1)\zeta_{\mathsf{s},\mathsf{p}}\right)^{1/2}} \right\} \right] \frac{a\sqrt{b}}{2} M\zeta_{\mathsf{s},\mathsf{p}},$$

$$(12)$$

where $\Gamma(.,.)$ is the incomplete Gamma function defined in [21, Eq.(8.350.2)].

3.3 Ergodic Channel Capacity

In this section, we derive the ergodic channel capacity of the considered system. The channel capacity can be written in terms of the PDF of γ_{Sel} as

$$C = \frac{1}{\ln(2)} \int_0^\infty \ln(1+\gamma) f_{\gamma_{\mathrm{Sel}}}(\gamma) d\gamma. \tag{13}$$

Upon substituting (10) in (13), and with the help of [21, Eq.(4.291.15)], the ergodic channel capacity can be obtained as

$$C = \sum_{i=0}^{M-1} \binom{M-1}{i}(-1)^i \sum_{l=1}^{K} \lambda_{\mathsf{s},l} \sum_{\mathcal{P}} \left[\frac{\ln\left(\frac{\Delta_1}{(i+1)\zeta_{\mathsf{s},\mathsf{p}}}\right)}{\Delta_1(\Delta_1 - (i+1)\zeta_{\mathsf{s},\mathsf{p}})} \right.$$

$$\left. + \sum_{j=1}^{K-N} (-1)^j \sum_{s_1 < \ldots < s_j} \frac{\ln\left(\frac{\Delta_2}{(i+1)\zeta_{\mathsf{s},\mathsf{p}}}\right)}{\Delta_2\left(\Delta_2 - (i+1)\zeta_{\mathsf{s},\mathsf{p}}\right)} \right] \frac{M\zeta_{\mathsf{s},\mathsf{p}}}{\ln(2)}. \tag{14}$$

4 Asymptotic Performance Analysis

To get more insights about the system behavior and to simplify the achieved expressions, we study in this section the performance at the high SNR regime where simple approximate expressions are derived for the outage probability and ASEP in addition to the derivation of the diversity order and coding gain of the system.

4.1 Outage Probability

The outage probability can be expressed at the high SNR regime as $P_{\text{out}} \approx (G_c \text{SNR})^{-G_d}$, where G_c and G_d denote the coding gain and diversity order of the system, respectively [22]. Obviously, G_c represents the horizontal shift in the outage probability performance relative to the benchmark curve $(\text{SNR})^{-G_d}$ and G_d refers to the increase in the slope of the outage probability versus SNR curve [22, Ch.14]. The parameters on which the diversity order depends will affect the slope of the outage probability curves and the parameters on which the coding gain depends will affect the position of the curves. In the upcoming analysis, the secondary users are assumed to have identical channels, that is $\lambda_{s,1} = \ldots = \lambda_{s,K} = \lambda_{s,d} = (\sigma_{e_W}^2 + \sigma_{e_{h_{s,d}}}^2 \bar{\gamma} + 1)/\bar{\gamma}\Omega_{\hat{h}_{s,d}}$, and the channels from the secondary source to primary users are also assumed to be identical $\zeta_{s,1} = \ldots = \zeta_{s,M} = \zeta_{s,p}$. The PDF of the selected N^{th} best user is given for identical users' channels as

$$f_{\gamma_{\text{Sel}}}(\gamma|W) \approx \binom{K-1}{N-1} K f_{\gamma_d}(\gamma|W) \left(F_{\gamma_d}(\gamma|W)\right)^{K-N} \left(1 - F_{\gamma_d}(\gamma|W)\right)^{N-1}. \quad (15)$$

As $\bar{\gamma} \to \infty$, the CDF in (5) simplifies to $F_{\gamma_d}(\gamma|W) \approx \lambda_{s,d} W \gamma$ and accordingly, the PDF simplifies to $f_{\gamma_d}(\gamma|W) \approx \lambda_{s,d} W$. Upon substituting the approximated CDF and PDF in (15) and following the same procedure as in Section 3.1, the outage probability at high SNR values can be evaluated with the help of [21, Eq.(3.351.3)] as

$$P_{\text{out}}^{\infty} = K\binom{K-1}{N-1} (\lambda_{s,d})^{K-N+1} M \sum_{i=0}^{M-1} \binom{M-1}{i}(-1)^i$$

$$\times \sum_{k=0}^{N-1} \binom{N-1}{k}(-1)^k (\lambda_{s,d})^k (k+K-N)!$$

$$\times ((i+1)\zeta_{s,p})^{-(k+K-N+1)}(\gamma_{\text{out}})^{k+K-N+1}. \quad (16)$$

The result in (16) is still dominant for the first term of the summation $k = 0$. With $\lambda_{s,d} = (\sigma_{e_W}^2 + \sigma_{e_{h_{s,d}}}^2 \bar{\gamma} + 1)/\bar{\gamma}\Omega_{\hat{h}_{s,d}}$, we may end up with three main cases. These cases are determined by the status of the estimation process of primary and secondary users' channels:

Case 1: $\sigma_{e_W}^2 = \sigma_{e_{h_{s,d}}}^2 = 0$ (perfect channel estimation)

For this case, $\lambda_{s,d}$ simplifies to $1/\bar{\gamma}\Omega_{\hat{h}_{s,d}}$ and the outage probability can be simplified as

$$P_{\text{out}}^{\infty} = \left\{ \left(\chi \sum_{i=0}^{M-1} \binom{M-1}{i}(-1)^i((i+1)\zeta_{s,p}\Omega_{\hat{h}_{s,d}})^{-(K-N+1)} \right. \right.$$

$$\left. \left. \times (\gamma_{\text{out}})^{K-N+1} \right)^{\frac{-1}{(K-N+1)}} \bar{\gamma} \right\}^{-K-N+1}, \quad (17)$$

where $\chi = K(K-N)!\binom{K-1}{N-1}M$.

Case 2: $\sigma^2_{e_{h_{s,d}}} \neq 0$ (imperfect channel estimation of secondary users' channels)
Here, the numerator of $\lambda_{s,d}$ can be approximated by $(\sigma^2_{e_{h_{s,d}}} \bar{\gamma})$ and hence, $\lambda_{s,d}$
simplifies to $\sigma^2_{e_{h_{s,d}}}/\Omega_{\hat{h}_{s,d}}$. As a result, the outage probability can be simplified as

$$P_{\text{out}}^{\infty} = \chi \sum_{i=0}^{M-1} \binom{M-1}{i}(-1)^i \left(\frac{(i+1)\zeta_{s,p}\Omega_{\hat{h}_{s,d}}}{\sigma^2_{e_{h_{s,d}}}\gamma_{\text{out}}} \right)^{-(K-N+1)}. \tag{18}$$

Case 3: $\sigma^2_{ew} = \sigma^2_{e_{h_{s,d}}} = 1/\text{SNR} = 1/\bar{\gamma}$ (imperfect channel estimation)
For this case, $\lambda_{s,d}$ simplifies to $1/\bar{\gamma}\Omega_{\hat{h}_{s,d}}$. As a result, the outage probability can
be simplified as obtained in (17) with the same coding gain and diversity order.

4.2 Average Symbol Error Probability

The asymptotic ASEP for the studied system can be obtained by replacing γ_{out}
by γ in (16) and then substituting the result in (11). Upon doing that, and with
the help of [21, Eq.(3.381.4)], we can easily get the following three cases:

Case 1: $\sigma^2_{ew} = \sigma^2_{e_{h_{s,d}}} = 0$ (perfect channel estimation)
For this case, the asymptotic ASEP can be obtained as

$$\text{ASEP}^{\infty} = \left\{ \left(\chi \sum_{i=0}^{M-1} \binom{M-1}{i}(-1)^i \left((i+1)\zeta_{s,p}\Omega_{\hat{h}_{s,d}} \right)^{-(K-N+1)} \right. \right.$$
$$\left. \left. \frac{\Gamma(K-N+3/2)}{(b)^{K-N+3/2}} \right)^{\frac{-1}{(K-N+1)}} \bar{\gamma} \right\}^{-(K-N+1)}. \tag{19}$$

Case 2: $\sigma^2_{e_{h_{s,d}}} \neq 0$ (imperfect channel estimation of secondary users' channels)
Here, the asymptotic ASEP can be obtained as

$$\text{ASEP}^{\infty} = \chi \sum_{i=0}^{M-1} \binom{M-1}{i}(-1)^i \left(\frac{(i+1)\zeta_{s,p}\Omega_{\hat{h}_{s,d}}}{\sigma^2_{e_{h_{s,d}}}} \right)^{-(K-N+1)} \frac{\Gamma(K-N+3/2)}{(b)^{K-N+3/2}}. \tag{20}$$

Case 3: $\sigma^2_{ew} = \sigma^2_{e_{h_{s,d}}} = 1/\text{SNR} = 1/\bar{\gamma}$ (imperfect channel estimation)
For this case, the asymptotic ASEP can be obtained to be similar to that found
in (19).

It clear from (17), (19) that the multiuser cognitive generalized order user
selection network with multiple primary users using the same spectrum band
and imperfect channel estimation has a coding gain that is affected by several
parameters such as K, N, $\Omega_{\hat{h}_{s,d}}$, M, $\zeta_{s,p}$, and γ_{out}; while the diversity order
is constant at $K - N + 1$. This is valid for the case where the channels are

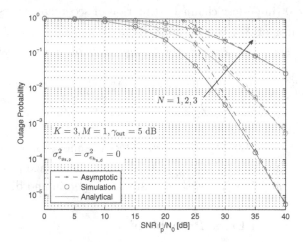

Fig. 1. P_{out} vs. SNR for different values of N and $\mu_{\hat{g}_{\text{s,p}}} = 15$ dB, $\Omega_{\hat{h}_{\text{s,k}}} = 0$ dB, $k = 1, \ldots, 3$.

perfectly estimated. Also, this applies when the estimation errors are inversely proportional to SNR as shown in Case 3. On the other hand, when the channels are imperfectly estimated with constant estimation errors, it is obvious from (18), (20) that the system has zero diversity order and a coding gain that is affected by the same previous parameters but now with the effect of the channel estimation error $\sigma^2_{e_{h_{\text{s,d}}}}$.

5   Simulation and Numerical Results

In this section, the achieved expressions are validated by Monte-Carlo simulations and some numerical examples are provided to illustrate the impact of several parameters on the system performance.

The effect of order of selected user N on the outage performance is illustrated in Figure 1. We can see from this figure that the asymptotic and analytical results perfectly match with Monte-Carlo simulations. Also, we can notice that as N increases, the diversity order of the system decreases and the system performance is more degraded. On the other hand, as N decreases, the diversity order increases and hence, better the achieved performance. These results on the diversity order of the generalized order selection scheme were achieved also when this scheme was implemented in non-cognitive systems.

Figure 2 studies the effect of number of primary users M on the system performance. Clearly, as M increases, worse the achieved performance. This is expected as having more primary users increases the probability of finding primary users of stronger channels and hence, having secondary users of smaller

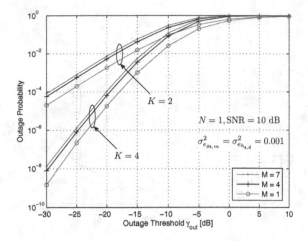

Fig. 2. P_{out} vs. outage threshold for different values of M, K and $\mu_{\hat{g}_{s,m}} = 15$ dB, $m = 1, \ldots, M$, $\Omega_{\hat{h}_{s,k}} = 0$ dB, $k = 1, \ldots, 4$.

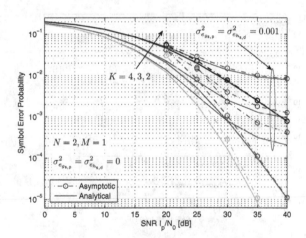

Fig. 3. ASEP vs. SNR for different values of K and $\mu_{\hat{g}_{s,p}} = 15$ dB, $\Omega_{\hat{h}_{s,k}} = 0$ dB, $k = 1, \ldots, 4$.

transmit power which degrades the system performance. As expected, more secondary users ($K = 4$) gives better performance compared to the case where $K = 2$.

The error probability performance of the studied system is shown in Figure 3 for different numbers of secondary users K. The figure is plotted for two cases: perfect channel estimation and imperfect channel estimation with constant estimation error variance. Again, it is clear that the asymptotic and analytical

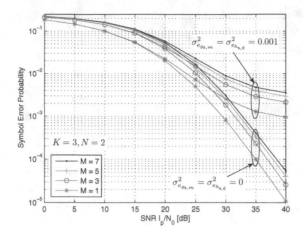

Fig. 4. ASEP vs. SNR for different values of M and $\mu_{\hat{g}_{s,m}} = 15$ dB, $m = 1, \ldots, M$, $\Omega_{\hat{h}_{s,k}} = 0$ dB, $k = 1, \ldots, 3$.

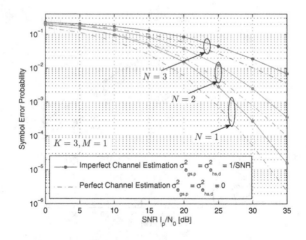

Fig. 5. ASEP vs. SNR for different values of N and $\mu_{\hat{g}_{s,p}} = 15$ dB, $\Omega_{\hat{h}_{s,k}} = 0$ dB, $k = 1, \ldots, 3$.

results perfectly match with Monte-Carlo simulations. Also, we can see from this figure that for the case of perfect channel estimation ($\sigma^2_{e_{h_{s,p}}} = \sigma^2_{e_{h_{s,d}}} = 0$), as K increases, the diversity order of the system increases and the system performance is more enhanced. Also, it is clear that as K decreases, the diversity order decreases and hence, worse the achieved performance. On the other hand, in the presence of channel estimation error ($\sigma^2_{e_{h_{s,p}}} = \sigma^2_{e_{h_{s,d}}} = 0.001$), zero diversity gain

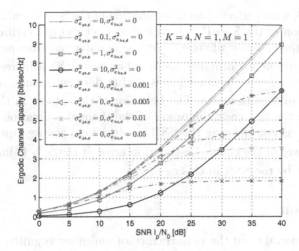

Fig. 6. Capacity vs. SNR for different values of $\sigma^2_{e_{g_{s,p}}}$, $\sigma^2_{e_{h_{s,d}}}$ and $\mu_{\hat{g}_{s,p}} = 15$ dB, $\Omega_{\hat{h}_{s,k}} = 0$ dB, $k = 1, \ldots, 4$.

is achieved by the system and a noise floor appears in the results due to the imperfect channel estimation effect on the system behavior. This can be easily concluded from the asymptotic results where the diversity order of the system becomes zero when $\sigma^2_{e_{h_{s,d}}} \neq 0$. In such case, any further increase in the SNR will add no enhancement to the system behavior.

Figure 4 studies the effect of number of primary users M on the error rate performance for the cases of perfect and imperfect channel estimations. Again, as M increases, worse the coding gain and hence, worse the achieved behavior. Also, as mentioned regarding Figure 3, it is clear in this figure that with imperfect channel estimation, the diversity gain of the system reaches zero and a noise floor appears in the results. This was illustrated in Case 2 of the asymptotic analysis section. Clearly, the diversity order of the system is not affected by the parameter M.

The effect of order of selected secondary user on the error probability performance is studied in Figure 5. The figure includes two cases: perfect channel estimation and imperfect channel estimation with an estimation error variance that is inversely proportional to SNR. The effect of channel estimation error on the system performance is obvious in this figure where worse behavior is achieved compared to the case where the channels are perfectly estimated. More importantly, for the case of imperfect channel estimation and as the variance of channel estimation error is assumed to be inversely proportional to SNR, the system can still achieve full diversity order when N decreases. This is also the case when the channels are perfectly estimated.

Figure 6 shows the ergodic channel capacity of the system for different values of $\sigma^2_{e_{g_{s,p}}}$, $\sigma^2_{e_{h_{s,d}}}$. Two cases are shown in this figure: imperfect channel estimation of the $S \rightarrow P$ link with perfect channel estimation of the $S \rightarrow D$ link; and perfect channel estimation of the $S \rightarrow P$ link with imperfect channel estimation of the $S \rightarrow D$ link. For the first case where $\sigma^2_{e_{g_{s,p}}}$ is taking different values and $\sigma^2_{e_{h_{s,d}}} = 0$, the system capacity or performance keeps enhancing as SNR increases. On the other hand, when $\sigma^2_{e_{g_{s,p}}} = 0$ and $\sigma^2_{e_{h_{s,d}}}$ is taking different values, a noise floor appears in all results of this case. The behavior of the system in the two cases is expected as in the first case, the power of the channel estimation error of the $S \rightarrow P$ link $\sigma^2_{e_{g_{s,p}}}$ is not affecting the SNR as clear from the asymptotic results; whereas, the power of the channel estimation error of the $S \rightarrow D$ link $\sigma^2_{e_{h_{s,d}}}$ is a multiplied factor by the SNR in this case.

6   Conclusion

In this paper, we evaluated the performance of multiuser cognitive generalized order user selection network with multiple primary receivers and imperfect channel estimation. Closed-form expressions were derived for the outage probability, average symbol error probability, and ergodic channel capacity assuming Rayleigh fading channels. Furthermore, the system performance was evaluated at the high SNR values. Main results showed that the number of primary receivers affects the system performance through affecting only the coding gain. Also, findings illustrated that zero diversity gain is achieved by the system and a noise floor appears in the results when the channels of secondary users are imperfectly estimated. Finally, results showed that the imperfect estimation of primary receivers' channels affects the system performance via affecting only the coding gain.

Acknowledgments. The author would like to acknowledge the support provided by the Deanship of Scientific Research (DSR) at King Fahd University of Petroleum & Minerals (KFUPM) for funding this work through project No. JF141008.

References

1. Haykin, S.: Cognitive radio: Brain-empowered wireless communications. IEEE J. Sel. Areas Commun. **23**, 201–220 (2005)
2. Goldsmith, A., Jafar, S., Maric, I., Srinivasa, S.: Breaking spectrum gridlock with cognitive radios: an information theoretic perspective. Proc. IEEE **97**(5), 894–914 (2009)
3. Foukalas, F., Khattab, T., Poor, H.V.: Adaptive modulation in multi-user cognitive radio networks over fading channels. In: Proc. 8th Int'l Conf. on Cognitive Radio Oriented Wireless Networks (CROWNCOM 2013), 8–10 July 2013, Washington DC, USA, pp. 226–230
4. Ratnarajah, T., Masouros, C., Khan, F., Sellathurai, M.: Analytical derivation of multiuser diversity gains with opportunistic spectrum charing in CR systems. IEEE Trans. Commun. **61**(7), 2664–2677 (2013)
5. Wang, Z., Zhang, W.: Exploiting multiuser diversity with 1-bit feedback for spectrum sharing. IEEE Trans. Commun. **62**(1), 29–40 (2014)

6. Ban, T.-W., Jung, B.C.: On the multi-user diversity with fixed power transmission in cognitive radio networks. IEEE Wireless Commun. Lett. **3**(1), 74–77 (2014)
7. Mosleh, S., Abouei, J., Aghabozorgi, M.R.: Distributed opportunistic interference alignment using threshold-based beamforming in MIMO overlay cognitive radio. IEEE Trans. Veh. Tech. (Accepted for publication)
8. Khan, F.A., Tourki, K., Alouini, M.-S., Qaraqe, K.A.: Performance analysis of an opportunistic multi-user cognitive network with multiple primary users. Wirel. Commun. Mob. Comput. (2014). doi:10.1002/wcm.2478
9. Song, Y., Alouini, M.-S.: On the secondary users opportunistic scheduling with reduced feedback. In: Proc. 3rd Int'l Conf. on Commun. and Net., 29 March-01 April 2012, Tunisia, pp. 1–6
10. Khan, F.A., Tourki, K., Alouini, M.-S., Qaraqe, K.A.: Outage and SER performance of an opportunistic multi-user underlay cognitive network. In: Proc. IEEE Int'l Symp. on Dynamic Spectrum Access Networks (DYSPAN 2012), 16–19 Oct. 2012, Washington DC, USA, p. 288
11. Yang, L., Kang, M., Alouini, M.-S.: On the capacity-fairness tradeoff in multiuser diversity systems. IEEE Trans. Veh. Tech. **56**(4), 1901–1907 (2007)
12. Yang, L., Alouini, M.-S.: Performance analysis of multiuser selection diversity. In: Proc. IEEE Int'l Conf. on Commun. (ICC 2004), pp. 3066–3070, Paris, France, June 2004
13. Choi, S., Ko, Y.C.: Performance of selection MIMO systems with generalized selection criterion over Nakagami-m fading channels. IEICE Trans. Commun. **E89–B**(12), 3467–3470 (2006)
14. Salhab, A.M., Al-Qahtani, F., Zummo, S.A., Alnuweiri, H.: Outage analysis of Nth-best DF relay systems in the presence of CCI over Rayleigh fading channels. IEEE Commun. Lett. **17**(4), 697–700 (2013)
15. Fan, L., Lei, X., Fan, P., Hu, R.Q.: Outage probability analysis and power allocation for two-way relay networks with user selection and outdated channel state information. IEEE Commun. Lett. **16**(5), 638–641 (2012)
16. Ban, T., Choi, W., Jung, B., Sung, D.: Multi-user diversity in a spectrum sharing system. IEEE Trans. Wireless Commun. **8**(1), 102–106 (2009)
17. Ghasemi, A., Sousa, E.: Fundamental limits of spectrum-sharing in fading environments. IEEE Trans. Wireless Commun. **6**(2), 649–658 (2007)
18. Kang, X., Liang, Y., Nallanathan, A.: Optimal power allocation for fading channels in cognitive radio networks: delay-limited capacity and outage capacity. In: Proc. IEEE Veh. Tech. Conf., pp. 1544–1548. Singapore, May 2008
19. Wang, C., Liu, T.C.-K., Dong, X.: Impact of channel estimation Error on the performance of amplify-and-forward two-way relaying. IEEE Trans. Veh. Tech. **61**(3), 1197–1207 (2012)
20. Vaughan, R.J., Venables, W.N.: Permanent expressions for order statistics densities. J. Roy. Statist. Soc. Ser. B **34**(2), 308–310 (1972)
21. Gradshteyn, I.S., Ryzhik, I.M.: Tables of Integrals, Series and Products, 6th edn. Acadamic Press, San Diago (2000)
22. Simon, M.K., Alouini, M.-S.: Digital Communication over Fading Channels, 2nd ed. Wiley (2005)

Implementing a MATLAB-Based Self-configurable Software Defined Radio Transceiver

Benjamin Drozdenko$^{(\boxtimes)}$, Ramanathan Subramanian,
Kaushik Chowdhury, and Miriam Leeser

Department of Electrical and Computer Engineering, Northeastern University,
360 Huntington Ave, Boston, MA 02115, USA
{bdrozdenko,rsubramanian,mel}@coe.neu.edu, krc@ece.neu.edu

Abstract. Software defined radio (SDR) transitions the communication signal processing chain from a rigid hardware platform to a user-controlled paradigm, allowing unprecedented levels of flexibility in parameter settings. However, programming and operating such SDRs have typically required deep knowledge of the operating environment and intricate tuning of existing code, which adds delay and overhead to the network design. In this work, we describe a bi-directional transceiver implemented in MATLAB that runs on the USRP platform and allows automated, optimal selection of the parameters of the various processing blocks associated with a DBPSK physical layer. Further, we provide detailed information on how to create a real-time multi-threaded design wherein the same SDR switches between transmitter and receiver functions, using standard tools like the MATLAB Coder and MEX to speed up the processing steps. Our results reveal that link latency and packet reception accuracy are greatly improved through our approach, making it a viable first step towards protocol design within an easily accessible MATLAB environment.

Keywords: Software defined radio · DBPSK · MATLAB · MATLAB coder · MEX · Reconfigurable computing

1 Introduction

Software Defined Radio (SDR) is a means of making radio programmable and multi-modal. It's a fundamental building block of dynamic spectrum access, in which the radio can sense unused spectrum and dynamically alter its transmission parameters to leverage this spectrum [1]. Apart from tunability in frequency, an SDR may also alter its transmission power, modulation, specific algorithms for channel estimation and packet decoding, among others, to best adapt to the changing environment, thereby giving it a "cognitive" ability [6].

In addition, timing is an important concern that needs to be addressed. To properly facilitate communications among nodes, a wireless system must be able to perform operations in a specific amount of time, a multiple of some small time unit. For this reason, we rely upon a construct that can send and receive a packet in a fixed slot time.

© Institute for Computer Sciences, Social Informatics and Telecommunications Engineering 2015
M. Weichold et al. (Eds.): CROWNCOM 2015, LNICST 156, pp. 164–175, 2015.
DOI: 10.1007/978-3-319-24540-9_13

In this paper, we propose a design approach that allows a user to solve the following problems associated with SDRs.

- *Complete knowledge of the processing chain*: Instead of demanding a deep user-knowledge in all aspects of signal processing (frequency compensation, automatic gain control) and communication (modulation/demodulation, bit scrambling, error detection), we allow the user to only insert a subset of parameters in MATLAB based on need and comfort level. We set up an optimization program that is executed in the initialization state, allowing an exhaustive search and detection of the optimal settings for the remaining parameters.
- *SDR processing latency*: A general problem in SDRs is that software processing is typically slow, as compared to hardware-executed instructions. Thus, not only must a pair of data exchanging SDRs exhibit minimal packet errors (or be able to recover from them), but also be able to complete the processing steps in real time. This constraint introduces complex design tradeoffs where each block of the transceiver needs to be optimized for minimum computational time at both ends. Our design incorporates time optimizations enabled by MATLAB Coder and MEX file generation, which considerably lowers processing time.
- *Bi-directional communication challenges*: Bi-directional data communication, which is our goal, requires precise time synchronization in a SDR environment, such that the transmitter is ready to receive incoming acknowledgements immediately on completing its one-way data transfer. While the link layer accounts for complete data frames, our design prefers a smaller USRP frame length to process smaller chunks at a time. Thus, the link layer countdown timers must be carefully set to allow for the additional lag in lower layer processing of the SDR.

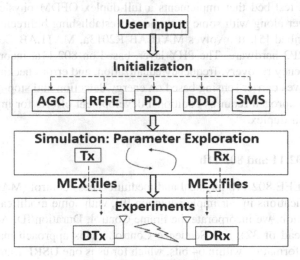

Fig. 1. System Architecture

1.1   Design Overview

Our system architecture and operational steps are shown in Fig. 1. In the initialization step, the system allows the user to set a set number of parameters for the entire transceiver chain. We next begin a parameter exploration stage in a simulation-only environment. The transmitter and receiver codes are executed with the user-supplied parameters as constants, and all other possible variations (both in terms of the settings of processing blocks as well as entire algorithms themselves) are considered. From this a feasible set of parameter options are presented that give 99% accuracy in the packet reception rate at the receiver. Note that this is a 'best case' scenario, as the actual wireless channel will introduce further channel outages. Once the user selects one of the possible feasible configurations returned by the search, the code is transferred to the actual USRP radios for over-the-air experiments.

Our approach involves first designing a number of (i) state diagrams to reflect the logical and time-dependent operational steps of our system and (ii) block diagrams to reflect the sequential order of operations. Furthermore, we structure the MATLAB code in a way that enables slot time-synchronized operations. For the eventual implementation, we use MATLAB Coder to generate C code. Finally, we compiled the C code into MEX executables that could be called directly from MATLAB on an Ubuntu 64-bit platform that serves as the host computer for the USRPs.

2   Background and Related Work

2.1   Prior and Existing SDR Programming Tools

An SDR-based test-bed that implements a full-duplex OFDM physical layer and a CSMA link layer along with some strategies for establishing bidirectional communications is described [5]. It involves MATLAB R2013a, MATLAB Coder on USRP-N210 and USRP2 hardware. The PHY layer, based on 802.11a, incorporates timing recovery, frequency recovery, frequency equalization, and error checking. The CSMA link layer involves carrier sensing based on energy detection and stop-and-wait ARQ. However, this approach requires additional development efforts for improving speed and enabling full-duplex.

2.2   IEEE 802.11 and 802.11b

We adopt the IEEE 802.11b physical and medium access control (MAC) layer frame structure specifications in our implementation [9], with some modifications. In MAC header information, we incorporate the Frame Control, Duration/ID, Address 1 and 2 (at 16 bits instead of 32), and Sequence Control. This approach maintains all the MAC header information within 64 bits, which for us is one USRP frame.

2.3   Differential Binary Phase Shift Keying (DBPSK):

We use DBPSK as the differential component enables us to recover a binary sequence from the phase angles of the received signal at any phase offset, without compensating for phase. In addition, DBPSK requires only coarse frequency offset compensation, without any close-loop techniques.   If residual frequency offset is less than DBPSK symbol rate, then the bit error rate (BER) approaches theoretical values.

3   Detailed System Design

To clearly identify the transmitting and receiving node for a given SDR pair, we use the terms *designated transmitter* (DTx) and *designated receiver* (DRx). This avoids ambiguity in describing a bi-directional communication link, where the *transmitter* must complete its packet transfer and then switch to a *receiver* role to get the acknowledgement (ACK). Thus, in the subsequent discussion, the DTx transitions between transmitter and receiver functions alternatively, and vice versa happens for the DRx.

3.1   State Diagrams

In implementing the CSMA/CA-based protocol at the intersection of the link and physical layers, we identify 4 main states (Fig. 2) at the DTx.

1. *Energy Detection State:* At START, a new packet arrives, and gets stored in a transmit buffer. The DTx begins sensing energy in the channel. The DTx decides to move either to a backoff state or to a transmit state depending on whether the channel is busy or not. A random amount of time is chosen uniformly from a progressively increasing time interval. DTx continually senses the channel and only when the channel is free, it decrements the backoff time, or freezes it otherwise. When the backoff time counts down to zero, the DTx attempts to transmit.
2. *Transmit (Tx) state*: In the transmit state, two possibilities exist. The transmission is successful (with the reception of an ACK), or transmission is not successful due to collision with transmissions (with no reception of ACK).
3. *Receive (Rx) state*: As soon as the transmission is completed, the DTx moves to Rx state, searching and decoding the PLCP header in the received ACK. The DTx then progresses to transmit a new frame and this continues till the last frame is successfully transmitted. On the other hand, if no ACK is received, the DTx enters the backoff state with an increased backoff time and re-attempts transmission.
4. *End of transmission state*: When transmission is successful, the DTx reaches the end of transmission (EOT) state. Now, the DTx might remain idle or progress to transmit another packet. In the latter case, the DTx re-sets its backoff time and moves into the backoff state for that duration.

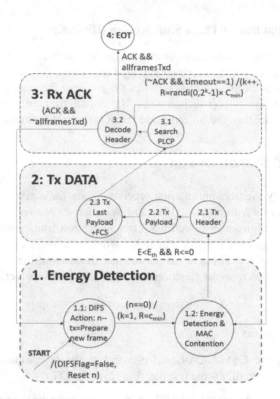

Fig. 2. State Chart for the Designated Transmitter (DTx)

Likewise, for the DRx we identified 3 main states. Unlike the DTx, the DRx does not perform energy detection.

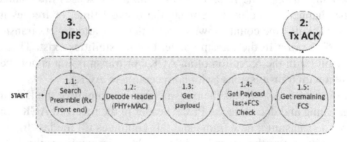

Fig. 3. State Chart for the Designated Receiver (DRx)

1. *Receive (Rx) state:* When the DRx succeeds in detecting the preamble, it decodes the PHY and MAC header and then progresses to extract the payload. When extracting the last set of payload bits, FCS is obtained and checked.
2. *Tx ACK:* The DRx sends out an ACK to the DRx when all the payload bits have been successfully received.
3. *DIFS:* The DRx waits for a fixed interval of time before moving to the reception of a new frame.

3.2   System Blocks

Within each of the substates in the state diagrams (Figs. 2 and 3), there are sequential operations that need to be performed. In order to simplify the logic of which operations must be performed in each state, we define a number of "blocks" to comprise the most common operations:

Table 1. Common Combinations of Operations for a Substate

RFFE	Radio Frequency Front End: Automatic Gain Control (AGC), frequency offset estimation and compensation, and raised cosine receive filter (RCRF)
PD	Preamble Detection: Find SYNC in received USRP frames
DDD	Despreading, Demodulation, and Descrambling
SMSRC	Scrambling, Modulation, Spreading, and Raised Cosine Transmit Filter (RCTF)

In each substate of DTx state 2 (Tx) and DRx state 2 (Tx ACK), SMSRC is performed prior to each transceiver (send and receive operation). In DTx substate 3.1 and DRx substate 1.1, RFFE and PD are performed after each transceive. In DTx substate 3.2 and DRx substates 1.2 to 1.5, RFFE and DDD are performed after each transceive.

4   Algorithms for System Blocks

4.1   RFFE System Block Algorithms

The components of this block recover a signal prior to preamble detection. These include the automatic gain control (AGC), frequency offset estimation and compensation, and raised cosine filtering. The ordering of these components is an important consideration, and through exhaustive simulations, we found the preceding order to be ideal. The AGC algorithm counters attenuation by raising the envelope of the received signal to the desired level. We chose to use a logarithmic loop method, as described in equations 1, 2, and 3: [4]

$$y(n) = e^{g(n)}x(n) \ . \tag{1}$$

$$e(n) = \ln(A) - \ln(z(m)) \ . \tag{2}$$

$$g(n+1) = g(n) + K\,e(n) \ . \tag{3}$$

where x is the input, y is the AGC output, z is the detector output, and K is the AGC step size. We use a rectifier detector method, as described in equation 4: [4]

$$z(m) = (1/N) \sum_{n=mN} |y(n)| \tag{4}$$

where N is the AGC update rate.

To accurately estimate the frequency offset between the receiver and the transmitter, we chose to use an FFT-based method that finds the frequency that maximizes the FFT of the squared signal:

$$f_{offset} = \operatorname{argmax}_f \mathcal{F}\{x^2\} \tag{5}$$

where $\mathcal{F}$ denotes the Fast Fourier Transform (FFT).

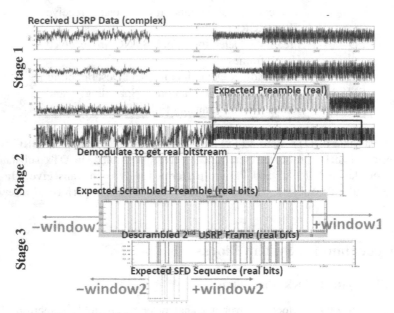

Fig. 4. The Three Stages of Preamble Detection: Coarse, Demodulated & SFD

4.2   PD System Block Algorithms

Preamble detection (PD) is performed in three stages, and we introduce a novel method that results in high accuracy. In the first stage, we perform a cross-correlation of the received complex data after raised cosine filtering to get an estimate of where the preamble starts, to give the so called synchronization delay. In the second stage, we compare the expected scrambled preamble to the demodulated bit stream. If they are not equal, we correlate a window of demodulated bit stream samples to the left and right of the maximum correlation index to fine-tune the synchronization delay. Finally, in stage three, we look for the Start Frame Delimiter (SFD) immediately after the preamble in the descrambled bit stream. If it is not in the expected place, we correlate a window of descrambled frame samples to the left and right to further fine-tune the synchronization delay (Fig. 4). Having multiple correlation stages ensures that we are able to find the preamble, and hence the start of the PLCP header information, as accurately as possible. However, this accuracy involves a tradeoff in the computational time.

4.3   Parameter Choices

There are a number of design parameters that must be carefully chosen (see Table 2), which are obtained through the initialization step described in section 1.

Table 2. Parameter Choices

Param	Block	Description	Value/ Range	Fixed/ Tunable
R_i, R_d	USRP	USRP Interpolation / Decimation Factor	500	Fixed
L_f	USRP	USRP Frame Length	64 bits	Fixed
L_p	Frame	#Octets per 802.11b Frame Payload	2012 octets	Fixed
K	RFFE	AGC Step Size	0.1 – 10	Tunable
N	RFFE	AGC Update Period	128 – 1408	Tunable
Δf	RFFE	Frequency Resolution	1 – 16 Hz	Tunable

4.3.1 Constant Parameters for USRP & 802.11b Frame
We recognize several parameters as being fixed because they cannot change during the course of a transception. The USRP N210 analog-to-digital converter (ADC) operates at a fixed rate of 100 MHz. The USRP interpolation-decimation rates control the factor by which we would like to slow down the rate of transmitting and receiving frames. For example, setting R_i and R_d to 500 ensures that a sample is processed every $500/100 \times 10^6 = 5$ μs. The USRP frame length should be minimized to make quick decisions with a small number of samples or bits. The number of octets per 802.11b frame payload should be maximized to decrease the header overhead.

4.3.2 Tunable Parameters for RFFE Block
Tunable parameters can be changed during the course of a transception. One example is the AGC step size, given by K in equation (3), which should be set to higher values for higher levels of attenuation or set low for lower attenuations. Another example is the AGC update period, which controls how quickly a received signal's envelope is able to converge to the desired level. Finally, the frequency offset estimation component's Frequency Resolution setting is an important design consideration. Since it is inversely proportional to the FFT length, a lower frequency resolution gives more accurate offset estimates, but also takes longer to compute.

4.4   Code Structure

Any 802.11-style wireless transceiver implementation must have the availability to perform operations based on some slot-based timing. We define this capability as *time slot-synchronized operations*. For example, before sending a data frame, a station must be able to wait for a backoff (BO) period. Interpreted MATLAB alone lacks the ability to perform time-sensitive operations in this manner, even with actively waiting. For this reason, we rely solely on the USRP for our timing. Our *transceive* function performs two actions: it gets a frame from, and puts a frame into the USRP buffers at fixed time intervals. Using the value for USRP interpolation/decimation defined in Section 4.2, we can calculate the slot time. Then, we can write our main program *while* loop so that it calls the *transceive* function once per loop, running helper functions to prepare data to transmit or process received data based on the active state, as shown in the following program code:

```
while ~endOfTransmission
  if (state==Tx)
    data2Tx = processData2Tx();
  end
  dataRxd = transceive(data2Tx);
  if (state==Rx)
    processRxdData(dataRxd);
  end
end
```

A slot time is defined as the smallest possible unit of time in which our SDR can make a decision. Our system sends or receives a data frame every slot time. The functions we define for processing the received data frame or preparing a new data frame to transmit must complete in less than a slot time to ensure timing accuracy.

5   Experiments

We use the USRP N210 platform as it allows us to define the parameters listed in Section 4.2, connect to a PC host using a gigabit Ethernet cable, and to use MATLAB [2]. We use the Ubuntu OS set with maximum send and receive buffer sizes for queues. This action ensures that there is enough kernel memory set aside for the network Rx/Tx buffers. We also set the maximum real-time priority for the *usrp* group to give high thread scheduling priority. The overall setup is given in Fig. 5.

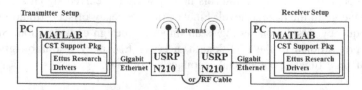

Fig. 5. Transceiver Hardware Setup

5.1   Communications System Toolbox USRP Support Package

We chose to use Communications System Toolbox System objects for the large part of our design [4]. The comm.AGC *System object* provides two Detector methods and two Loop methods whose functionality can be contrasted for received signals with varying attenuations. In addition, the PSK coarse frequency offset estimator allows us to shift between FFT-based options. These *System objects* facilitate easy generation of C code using MATLAB Coder. Here, the comm.SDRuTransmitter *System object* puts a frame on the USRP transmit buffer, and comm.SDRuReceiver gets a frame from the USRP receive buffer. However, this approach has some disadvantages; e.g., the frame length must now become fixed. Another issue is that running the step methods for these *System objects* is that they're single-threaded, whereas the USRP N210 is multi-threaded. On a single clock cycle, this allows to get a frame from the receive buffer or put a frame on the transmit buffer, but not both. Therefore, attempting to write MATLAB code that runs a *put* and *get* sequentially will result in an exponentially increasing delay, and eventually result in an overflow of the USRP buffer. To avoid this delay, we plan to explore parallelism and make the *transceive* function described in section 4.3 operate in a multi-threaded manner. We first generate C code from the MATLAB function using MATLAB Coder.

5.2   MATLAB Coder

MATLAB Coder is used for generating C code. In order to make the MATLAB code acceptable for C code generation, a number of actions must be taken beforehand. All variables are given a static size and type (including real or complex) that does not change in the course of the program. Since *System objects* cannot be passed into MEX functions, all *System objects* are declared as persistent variables. The first call to each function, tests whether the persistent variable is empty, and initializes each *System object* if true. The function code for the transceive, RFFE, DDD, and SMSRC blocks are all prepared in this same manner. We then compile the C code for each major block into a MATLAB executable (MEX) file, which can be called directly from MATLAB.

6   Results

The *transceive* function is at the core of our system design, since its ability to simultaneously receive and transmit a USRP frame at a near-constant time interval is key to our goal of slot time-synchronized operations. To compare its accuracy, we ran 2,000 time trials to see how long the *transceive* function takes from start to finish, and how this time difference changes over the course of a longer data bitstream. The timing using a *transceive* function in interpreted MATLAB and using C code compiled into a MEX are compared in Fig. 6. The timing exhibits some deviation: The function initially overshoots the expected time per USRP frame; on every subsequent iteration it then undershoots to make up for the time difference. Note that less undershooting is needed to compensate for initial overshoots, because the overshoot amounts have reduced significantly. The reason for this is that the MATLAB executable has more control over its timing.

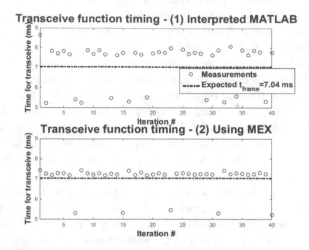

Fig. 6. Transceive function timing for interpreted MATLAB vs. MEX

The timing of the RFFE block for various values of the frequency resolution parameter in interpreted MATLAB and C code compiled into a MEX is shown in Fig. 7. We see that there is a general decrease in the average execution time for the RFFE block with increase in frequency resolution. For low frequency resolution values, the average execution time using MEX is longer than using interpreted MATLAB because it needs to use very large FFT lengths. However, in all cases, the standard deviation is always significantly less. Thus, MEX is a better option for the purpose of enforcing consistent RFFE execution times, which is required for slot time-synchronized operations.

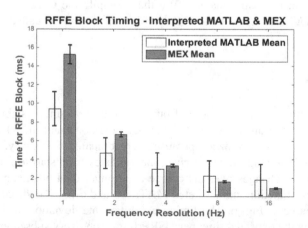

Fig. 7. RFFE block timing using interpreted MATLAB and MEX

Whereas the change to the frequency resolution parameter affects timing directly, the AGC parameters control how well a signal can be recovered under various attenuation levels. By performing a parameter sweep with different values for these parameters, we determined that a step size of 1 and an update period of 1408 minimizes frame misdetection.

7   Conclusion

We conclude that building our design around the concept of slot time-synchronized operations results in a system that adheres to our desired frame time and is able to reconfigure parameter values as needed. Using MEX is essential for realizing timing with little deviation from this frame time. In addition, using MEX is beneficial for improving the speed consistency of our system blocks, most notably RFFE, which can vary its frequency resolution parameter. As part of future work, we will continue towards the complete design of the MAC functions as well as implement our transceiver system design on the Xilinx Zynq-7000 All-Programmable System-on-Chip (APSoC).

Acknowledgments. This work is supported by MathWorks under the Development-Collaboration Research Grant A#: 1-945815398. We would like to thank Mike McLernon and Ethem Sozer for their continued support on this project.

References

1. Akyildiz, I.F., Mohanty, S., Vuran, M.C., Won-Yeol, V.: NeXt generation/dynamic spectrum access/cognitive radio wireless networks: A survey. Computer Networks 500(13), September 2006
2. Ettus Research, Inc.: USRP N200/N210 Networked Series
3. IEEE Std 802.11-2009: Part 11: Wireless LAN Medium Access Control (MAC) and Physical Layer (PHY) Specifications
4. MathWorks Documentation: Communications System Toolbox Documentation. USRP Support Package from Communications System Toolbox
5. Collins, T.: Multi-Node Software Defined Radio TestBed, NEWSDR 2014
6. Mitola III, J., Maguire Jr., G.Q.: Cognitive radio: making software radios more personal. IEEE Personal Communications Magazine 6(4), 13–18 (1999)
7. Luise, M., Reggiannini, R.: Carrier frequency recovery in all-digital modems for burst-mode transmissions. IEEE Trans. Commun. 43(3), 1169–1178 (1995)

Investigation of TCP Protocols in Dynamically Varying Bandwidth Conditions

Fan Zhou[1(✉)], Abdulla Al Ali[2], and Kaushik Chowdhury[1]

[1] Northeastern University, Boston, MA 02110, USA
zhou.fan1@husky.neu.edu, krc@ece.neu.edu
[2] Qatar University, Doha, Qatar
abdulla.alali@qu.edu.qa

Abstract. Cognitive radio (CR) networks experience fluctuating spectrum availability that impacts the end to end bandwidth of a connection. In this paper, we conduct an extensive simulation study of three different window-based TCP flavors- NewReno, Westwood+, and Compound, each of which has unique methods to determine the available bandwidth and scale the congestion window appropriately. These protocols also differ in their respective sensitivities to the metrics of round trip time, loss rate, residual buffer space, among others. These metrics exhibit divergent behavior in CR networks, as compared to classical wireless networks, owing to the frequent channel switching and spectrum sensing functions, and this influences the choice of the TCP protocol. Our ns-3 based simulation study reveals which specific rate control mechanism in these various TCP protocols are best suited for quickly adapting to varying spectrum and bandwidth conditions, and ensuring the maximum possible throughput for the connection.

Keywords: TCP evaluation · Dynamic bandwidth · Cognitive radio network

1 Introduction

Cognitive radio (CR) networks can potentially overcome the limitations of pre-assigned and static channelization by identifying unused spectrum, possibly in licensed bands, as well as switch the operation into these channels on a need basis [1]. While the research community has invested heavily in the design of the physical and link layers of the protocol stack, there is need for a systematic study of the end to end operation in the upper layers of the protocol stack, such as the transport layer. This is especially important as a desirable end-user experience, reliability of the data delivery between source and destination, and complete spectrum utilization within the short-time availability of the licensed spectrum can only be achieved through optimizations at the transport layer. In this paper, we report results from a methodical simulation study at the transport layer, focusing on three different variants of TCP that are window-based, i.e., the effective window containing the packets that may be transmitted by

© Institute for Computer Sciences, Social Informatics and Telecommunications Engineering 2015
M. Weichold et al. (Eds.): CROWNCOM 2015, LNICST 156, pp. 176–186, 2015.
DOI: 10.1007/978-3-319-24540-9_14

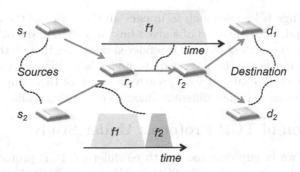

Fig. 1. End to end connection and varying spectrum in intermediate links

the source at any given time is altered depending upon observed delays, losses, bandwidth estimation and buffer space availability.

TCP has been extensively studied and analyzed over the past decades. Given its widespread adoption, identifying which of its specific rate control mechanisms are best suited in CR networks will yield useful insights in tuning them for these challenging environments. This is the first step towards determining whether minor changes to existing protocol flavors are sufficient to adapt TCP, or if an entirely new class of window-based approaches are needed to achieve satisfactory operation. We conduct the first comprehensive study of three different TCP flavors with this aim, focusing on high-bandwidth varying situations. Our previous works have explored enhancing TCP NewReno by leveraging extensive cross-layer and intermediate-node information [2] as well as extending equation-based protocol called TFRC [3]. However, both these approaches have limitations. First, the transport layer is envisaged to operate end to end from a classical networking viewpoint, and thus, enforcing dependence on the choices made at the lower layers, or mandating feedback from other intermediate nodes, brings about a radical change in the commonly accepted end-to-end paradigm, and results in a loss of generality. Similarly, TFRC is shown to be TCP-friendly in its classical incarnation, while our proposed triggers for the rate control equation in the modified TFRC-CR cause it to lose this important characteristic (though it gives higher network throughput and protection to licensed users). Thus, TFRC-CR can no longer fairly exist with other TCP flows, which limits its deployment opportunities. As a result, in this study, we examine other well known and unmodified TCP flavors that are compliant with the fairness criterion, as well as the end-to-end paradigm of the transport layer, and we determine which of these are best suited for CR networks 'as is'.

The simulation scenario that we will refer to in the subsequent discussion is shown in Fig. 1. Multiple sources (s_1, s_2) inject traffic into a chain network that may have several overlapping intermediate forwarding nodes (r_1, r_2). The connections end at respective destinations (d_1, d_2), which send back the acknowledgements (ACKs). The various links can be susceptible to high priority licensed user activity. All nodes perform synchronized spectrum sensing, and the licensed user coverage range is large enough to impact all the nodes of the chain. Thus, over time, the nodes at either end of a given link may switch to a different channel $(f_1, \ldots, f_n)$ if the current channel is rendered unusable, though the particular choice of the channels for the intermediate nodes is not known to the source. We assume that these channels vary greatly in terms of their upper and lower frequency bounds, as well have different amounts of existing traffic within them.

2   Discussion of TCP Protocols Under Study

In this section, we briefly describe the three different TCP protocols that we investigate in CR scenarios, i.e., (i) TCP NewReno, (ii) TCP Westwood+, and (iii) TCP Compound, mainly stressing on the features that will be useful for the dynamic spectrum environments.

2.1   TCP NewReno

This is the classical TCP version, and we use this to benchmark performance of the other protocols. The congestion window (*cwnd*) doubles in the *slow start* phase, and then increases linearly in the *congestion avoidance* phase. Timeouts caused by missing ACKs reduce the *cwnd* to 1, while triple duplicate ACKs cut the *cwnd* to half of its original value at the time of detecting the congestion. NewReno continuously updates its average round trip time (RTT) by maintaining a moving window over the last few sample RTT values, and thereby adjusts its estimate smoothly over time. We point the reader to [4] for further discussion on this protocol.

2.2   TCP Westwood+

Westwood+ deviates from NewReno in the sense that triple duplicate ACKs don't force the *cwnd* to half, but to $max(\frac{cwnd}{2}, EBW * RTT_{min})$ that allows the source to saturate bottleneck link, while draining the bottleneck queue buffer simultaneously [5]. EBW is the estimated bandwidth and RTT_{min} is the minimum RTT measured. We note that RTT_{min} refers to the time when there is no congestion in the network. Thus, the product of bottleneck link bandwidth and minimum RTT reflects the total capacity of the link. By reducing *cwnd* to $EBW * RTT_{min}$, packets in the buffer will be drained while the link capacity is kept full. Two features of this protocol make it attractive for use in DSA networks: First, for occasional packet losses that are channel-state induced, the estimated bandwidth would not change much. On the other hand, when spectrum changes happen leading to a massive increase or decrease in the available bandwidth, the EBW will change, which in turn allows for a more accurate setting for *cwnd*. Clearly, as the computation of EBW is a critical factor, the protocol uses information about the amount of data received during a certain period of time by passing the samples through a low pass filter using the so called *Tustin* approximation. This gives the protocol robustness against delayed ACKs. This feature is especially useful when licensed or priority users (PUs) of the spectrum cause interruptions in the connection.

2.3 TCP Compound (CTCP)

This protocol diverges from NewReno in the way the $cwnd$ changes in the congestion avoidance stage [6]. If $cwnd \geq 38$, a new parameter called as the delay window $dwnd$ is added to $cwnd$, i.e., $cwnd = cwnd + dwnd$. The $dwnd$ is itself defined in terms of a variable called as $diff$. Here, $diff$ is the difference between estimated current capacity and the theoretical capacity (when there is no congestion). It is computed by estimating all the buffered packets in the network, similar to a different protocol called TCP Vegas. When the latter is less than a threshold r, the $dwnd$ increases exponentially to quickly utilize the network resources. In network with high bandwidth-delay product, this increases the efficiency of bandwidth utilization. When $diff$ is larger than r, it suggests increasing accumulation of packets in the buffer. Hence, the $dwnd$ is decreased linearly until it reaches zero, and the performance of CTCP is degraded to that of TCP NewReno. This feature is critical in CR networks, as periodic spectrum sensing disconnects the connection temporarily, leading to a build up of packets from the source to the node immediately before the sensing node. Finally, if a packet loss is detected by triple duplicate ACK, $dwnd$ is set to a value to make sure the window size is decreased by a factor of $(1 - \beta)$. Clearly, the performance of CTCP is closely related with the choice of parameter values. For example, a balance must be struck to prevent the changes from being too aggressive or too conservative in $dwnd$. A set of recommended values of parameters is obtained in [6] through empirical studies.

3 Network Setup for the Evaluation Study

In this section, we describe the topology and simulation setup for CR network. All simulations are conducted through a packet level simulation in the open source ns3 simulator.

- *Traffic:* We assume a saturated scenario, where senders always have data to send. We inject new packets in the sending buffer as long as it is not completely full.
- *Topology:* We use the network topology as shown in Fig. 1. There are two senders $\{s_1, s_2\}$, two destinations $\{d_1, d_2\}$, and two intermediate routing nodes $\{r_1, r_2\}$, which we refer to as 'routers' to simplify the discussion. The link between first and second router is the bottleneck link with longer delay and lower bandwidth.
- *Bandwidth:* The bandwidth in the CR network is influenced by two factors: spectrum sensing and the activity of PUs. The transmission in either case is paused for a short period of time. Following this pause, the node might choose a new channel with different bandwidth. This may cause a sudden transition of the nodes into either channel with the same bandwidth (we call this as the *fixed* scenario where all channels are similar) or widely different bandwidth (we call this as the *varying* scenario where channels can be of unequal width that we select uniformly from a pre-decided range).

Note that we assume only the bottleneck router performs spectrum sensing and switching functions to clearly demonstrate the behavior of the TCP protocols.

- *Spectrum sensing:* We assume the bottleneck router senses the spectrum for 0.1s, in intervals of 5s. During spectrum sensing, the transmission is temporarily interrupted, and the router cannot receive nor send any packet. After spectrum sensing, the node might stay on original channel or use a new channel with different bandwidth, as we have discussed above.
- *Influence of PUs:* In a CR network, the transmission might be interrupted by sudden occurrence of the PU. We model the arrival of PUs as a Poisson process, and hence, the time interval between successive arrivals follows an Exponential distribution. We vary the mean time to model different extents of the interference and consequently, the network interruption.

4   Simulation Results and Observations

4.1   Impact of Fixed and Varying Bandwidth

Fixed Bandwidth. In this simulation, we study the performance of TCP in fixed bandwidth scenarios. We begin by considering a classical wireless network without CR functionality (classical) and plot real-time throughput of flow 1 (s_1, r_1, r_2, d_1) using TCP NewReno in Fig. 2. We see that the interruption caused due to periodic sensing in the same connection (CR with sensing) severely reduces the network throughput. The maximum throughput here is about 1.8Mbps, which is less than half of the minimum throughput classical case. We see a similar trend in Westwood+ and CTCP. The implication here is that even if the interruption time is 2% of the transmission time, the frequent timeout retransmit events continue to cause severe degradation to the performance to TCP.

Fig. 3 presents the throughput of three protocols for different values of the bottleneck channel bandwidth (though all channels have the same bandwidth) . We notice that the throughput of NewReno does not increase much with higher bandwidth availability. This is because NewReno's performance is limited by the slow rate of increase of the *cwnd* in the congestion avoidance stage. From Fig. 4, we see the three traces of *cwnd* are similar in trend, regardless of different bottleneck bandwidth selections.

Interestingly, Westwood+ outperforms NewReno and CTCP when bandwidth is less than 10Mbps. This is mainly because Westwood+ adjusts its slow start *cwnd* according to the product of estimated bandwidth and RTT_{min}. This is very effective in a CR network, as the estimated bandwidth is not immediately impacted by an occasional packet loss, and the *cwnd* is not directly halved. Moreover, after a channel switch, especially, if the new channel has lesser bandwidth than the previous one, the *cwnd* of TCP Westwood+ quickly sets itself to the new bandwidth. Unlike NewReno, it does not have to ramp up back from $\frac{cwnd}{2}$.

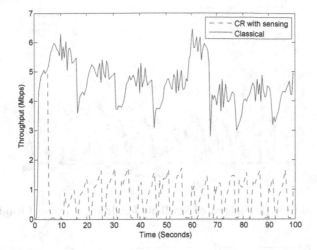

Fig. 2. Real-time throughput of NewReno in cognitive radio network and none-cognitive radio network

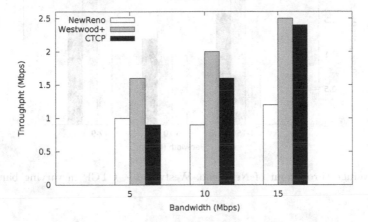

Fig. 3. Average throughput of NewReno, Westwood+, CTCP in fixed bandwidth model

We observe that CTCP has comparable throughput with Westwood+ when the channel bandwidth availability increases to 15Mbps. This is because the improvement of throughput in CTCP mainly comes from the addition of the *dwnd* in congestion avoidance stage. This mechanism is less effective in small bandwidth connections, as *cwnd* needs to be greater than the threshold of 38 to trigger inclusion of *dwnd*. Therefore, when spectrum switching allows the connection to enter into channels with much higher bandwidth, it is better to use CTCP as this results in faster rise of the end to end throughput. However, in low bandwidth conditions, CTCP performs similar to NewReno.

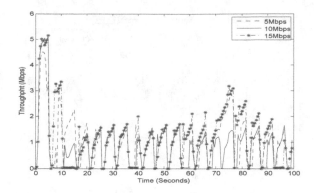

Fig. 4. Real-time throughput of NewReno for different channel bandwidths

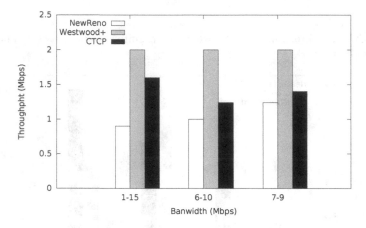

Fig. 5. Average throughput of NewReno, Westwood+, CTCP in varying bandwidth channels

Varying Bandwidth. In this scenario, we study the ability of TCP protocols to get adapt to different channels that have varying bandwidth. We assume that after each spectrum sensing stage, the node randomly picks a new channel. The bandwidth of this new channel is uniformly distributed between three ranges: [1Mbps,15Mbps], [6Mbps, 10Mbps] and [7Mbps, 9Mbps]. While the average bandwidth in these three ranges is the same, the wider the range, more is the diversity within the available bandwidth set.

Figure 5 shows the first flow's, i.e., (s_1, r_1, r_2, D_1), average throughput using the three TCP protocols in a multichannel environment where channels are picked from different varying ranges of bandwidth. We observe that the performance of three protocols is surprisingly stable. The throughput does not change much even if the available bandwidth fluctuates drastically. However, this stability is a direct outcome of their low efficiency in utilizing bandwidth in a CR

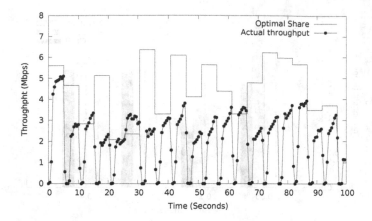

Fig. 6. Comparison of real-time throughput of Westwood+ and its optimal share (optimal share is Total Bandwidth/2)

network. It can be intuitively seen in Fig. 6 for the best performing protocol (Westwood+).

Note that the recovery of *cwnd* after a timeout event in Westwood+ is very fast (i.e., much faster than NewReno and Compound). However there still exists an upper limit on the throughput that Westwood+ can achieve. This large gap between the optimal throughput and actual throughput shows that there is a large proportion of bandwidth wasted. This implies that for a CR network, the bottleneck in the performance may lie not in the bandwidth being unavailable for the connection in the chosen channel, but in the less-than-optimal rate of increase of *cwnd* in the period between two transmission interruption events.

4.2 Spectrum Sensing Pattern

In this simulation, we investigate the impact of different spectrum sensing durations. The transmission (On) period time are set as 5s, 10s, 15s and spectrum sensing time (Off) durations are 0.05s, 0.1s, 0.2s and 0.4s. We perform an exhaustive set of simulation trials matching for every pair of On and Off time instances drawn from the above set.

Figure 7 shows the performance of each protocol with different Off time and fixed On time. Intuitively, increasing the sensing time will prolong the disconnection period of connection, and thus, we expect that the throughput will dramatically fall. However from Fig. 7, we see that this is not always the case. While the throughput drops sharply as sensing time increases from 0.05s to 0.1s, this trend is reversed when sensing time increases from 0.1s to 0.4s. To understand this phenomenon, we trace the behavior of *cwnd* in Westwood+, as shown in Fig. 8.

We find that transmission is not affected by the disconnection in the network when sensing time is small, say around 0.05s. As a result, there is no duplicate

184   F. Zhou et al.

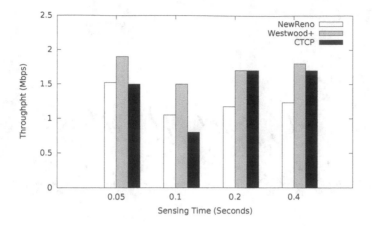

Fig. 7. Average throughput of NewReno, Westwood+, CTCP with different sensing times

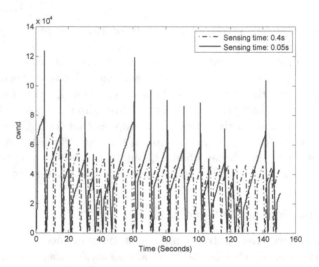

Fig. 8. *cwnd* of Westwood+ with different sensing times

ACKs or timeout event caused by packet loss during 45s to 60s in the simulation time. In other words, the interruption of the transmission at the affected link is completed hidden from the source node. Here, the bottleneck router alternatively inserts several packets from sender 1 and sender 2 into its own sending buffer queue. So correspondingly the senders will experience alternate bursty incoming ACKs and then this phase is followed by a phase of no ACKs. When there is no ACK coming in, the sender does not send any packet. When the periodic spectrum sensing coincides within the 'no ACK' periods, the sender remains oblivious to the reason why the ACKs are sparse.

When the sensing time crosses a certain threshold, say 0.1s, then timeouts happen after the spectrum sensing. In this situation, the throughput depends upon how fast *cwnd* can recover from the impact of timeout. Next, we study the performance of protocols with different On time between two spectrum sensing events. We see that there is a jump in the throughput for all three protocols when the On time increases from 5s to 10s. However, the throughput does not change to a similar extent when On time increases from 10s to 15s. We analyze the reason of this behavior in the next scenario.

4.3   Influence of PU

Here, we simulate the influence of random interference of the PU on the performance of TCP. After the PU appears, the affected router must do a spectrum switch. For simplicity, we assume that the time required to do this is fixed (0.1s). Also, to demonstrate the results with clarity, we use the fixed bandwidth model. At a high level, the influence of the PU is similar to that of spectrum sensing: the only difference is that the On time between two transmission interruptions is now an exponential random variable rather than a fixed constant.

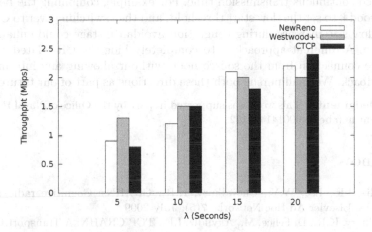

Fig. 9. Average throughput of NewReno, Westwood+, CTCP for various λ

Fig. 9 shows average throughput of flow 1. Overall, all three TCP protocols benefit from larger mean inter-arrival time λ. In addition, here we observe the similar phenomenon as in experiment B. That is for NewReno and Westwood, there is a sharp increase in throughput when the λ increase from 5s to 15s. However the throughput does not change much when λ increases from 15s to 20s. Note that for TCP Compound, the throughput has a stable increase with larger values of λ.

The increase of throughput between 5s to 15s correlates with fewer timeout events as these are PU-free durations. However for NewReno and Westwood+, as long as the transmission time is long enough for them to get out of slow start stage, further increase in throughput is throttled by the slow rate of change of *cwnd* (i.e., linear change) in the congestion avoidance stage. On the other hand, CTCP can benefit more from the longer uninterrupted transmission durations as it's *cwnd* grows faster in the congestion avoidance stage than NewReno and Westwood.

5   Discussion and Conclusion

While Westwood+ gives quick recovery of the *cwnd* after a timeout, its spectrum utilization is severely impacted by the slow increase of the *cwnd* in the congestion avoidance stage. In contrast, CTCP can utilize bandwidth better through the additive effect of the variable *dwnd* in the congestion avoidance stage. Also if the *cwnd* is not high enough, which may occur in low bandwidth channels or high PU activity channels, CTCP will degrade to NewReno, as it cannot reach the threshold to activate the addition of *dwnd*.

From the simulation study, we identify two ways to boost the performance of TCP protocol in a CR network. The first is to increase *cwnd* aggressively during the limited continuous transmission time. For example, combining the behavior of Westwood+ to set the slow start threshold, and then switching over to CTCP's fast window size increase during congestion avoidance stage could enhance the utilization. A different approach is to completely hide the CR-related activity inside the connection from the source node without allowing any intermediate node feedback. We shall pursue both these directions as part of our future work.

Acknowledgments. This work was supported in part by the Office of Naval Research under grant number N000141410192.

References

1. Akyildiz, I.F., Lee, W.Y., Chowdhury, K.R.: CRAHNs: Cognitive radio ad hoc networks. Elsevier Ad Hoc Networks **7**(5), July 2009
2. Chowdhury, K.R., Di Felice, M., Akyildiz, I.F.: TCP CRAHN: A Transport Control Protocol for Cognitive Radio Ad Hoc Networks. IEEE Trans. on Mobile Computing **12**(4), 790–803 (2011)
3. Al-Ali, A., Chowdhury, K.R.: An Equation-based Transport Protocol for Cognitive Radio Networks. Elsevier Ad Hoc Networks **11**(6), 1836–1847 (2013)
4. Floyd, S., Henderson, T.: The NewReno Modification to TCP's Fast Recovery Algorithm. Internet Engineering Task Force, Request for Comments (Experimental 2582), April 1999
5. Mascolo, S., Casetti, C., Gerla, M., Sanadidi, M.Y., Wang, R.: TCP westwood: bandwidth estimation for enhanced transport over wireless links. In: Proc. of the ACM Conf. on Mobile Computing and Networking, pp. 287–297 (2001)
6. Tan, K., Song, J., Zhang, Q., Sridharan, M.: A compound TCP approach for high-speed and long distance networks. In: MSR-TR-2005-86 (techreport) Microsoft Research, July 2005

Opportunistic Energy Harvesting and Energy-Based Opportunistic Spectrum Access in Cognitive Radio Networks

Yuanyuan Yao$^{(\boxtimes)}$, Xiaoshi Song, Changchuan Yin, and Sai Huang

Beijing Key Laboratory of Network System Architecture and Convergence,
Beijing University of Posts and Telecommunications, Beijing, China
{yyyao,songxiaoshi,ccyin,huangsai}@bupt.edu.cn

Abstract. The performance of large-scale cognitive radio (CR) networks with secondary users self-sustained by opportunistically harvesting radio-frequency (RF) energy from nearby primary transmissions is investigated. Using an advanced RF energy harvester, a secondary user is assumed to be able to collect ambient primary RF energy as long as it lies inside the harvesting zone of an active primary transmitter (PT). A variable power (VP) transmission mode is proposed, and a simple energy-based opportunistic spectrum access (OSA) strategy is considered, under which a secondary transmitter (ST) is allowed to transmit if its harvested energy is larger than a predefined transmission threshold and it is outside the guard zones of all active PTs. The transmission probability of the STs is derived. The coverage probabilities and the throughputs of the primary and the secondary networks, respectively, are characterized. The throughput can be increased by as much as 29%. Simulation results are provided to validate our analysis.

Keywords: Cognitive radio · Energy-based opportunistic spectrum access · Energy harvesting · Stochastic geometry · Transmit threshold

1 Introduction

Radio frequency (RF) energy harvesting holds promise for generating a small amount of electrical power to drive the circuits in wireless devices. Communication devices often have omni-directional antennas that propagate RF energy in all directions, and some of this power can be harvested to augment/replenish battery power in networks constituted of low-power devices such as wireless sensors [1].

Stochastic geometry theory [2] has been widely applied in the study of large-scale cognitive radio (CR) networks with energy harvesting. In [3], Dhillon *et al.*

This work was supported in part by the NSFC under Grants 61271257 and 61328102, the National Research Foundation for the Doctoral Program of Higher Education of China under Grant 20120005110007.

© Institute for Computer Sciences, Social Informatics and Telecommunications Engineering 2015
M. Weichold et al. (Eds.): CROWNCOM 2015, LNICST 156, pp. 187–198, 2015.
DOI: 10.1007/978-3-319-24540-9_15

developed a tractable model for K-tier heterogeneous cellular networks, where each base station is powered solely by a self-contained energy harvesting module. In [4], equipped with an advanced RF energy harvester, a secondary transmitter (ST) is assumed to be able to collect ambient RF energy from its nearest active primary transmitter (PT). However, it is assumed that the batteries of STs must be fully charged before their transmission, i.e., all the STs transmit with the same power, which theoretically limits the network capacity and in practise would result in a low level of convenience.

In general, since the energy arrivals are random and the energy storage capacities are finite, variable power (VP) transmission mode is more realistic. In this paper, we investigate the performance of a large-scale CR network with secondary users self-sustained by opportunistically harvesting RF energy from the primary transmissions. An energy-based OSA strategy is considered, under which STs use VP for transmission. Time is assumed to be slotted. In each time slot, a ST is considered to collect ambient primary RF energy if it lies inside the harvesting zone of an active PT, or start to transmit if its harvested energy is larger than a predefined transmission threshold and it is outside the guard zones of all active PTs, or be idle otherwise. By applying tools from stochastic geometry, the transmission probability of the STs is derived. Based on the results, we then characterize the coverage probabilities and throughputs of the primary and the secondary networks, respectively. Note that compared with [4], our proposed energy-based OSA protocol does not require the candidate STs to be fully charged before their transmissions, i.e., STs use VP for transmission, which considerably improves the reliability and stability of the CR network.

The remainder of this paper is organized as follows. The system model and performance metrics are introduced in Section 2. Section 3 investigates the transmit opportunity for the STs. The coverage performance of the primary and the secondary networks are studied in Section 4. Section 5 analyzes the primary and the secondary network throughputs, respectively. Simulation results are presented in Section 6. Finally, we conclude the paper in Section 7.

2   System Model

2.1   Network Model

We consider a large-scale CR network which consists of two mobile ad hoc networks, i.e., the primary network and the secondary network, on $\mathbb{R}^2$. The locations of the PTs and STs are assumed to follow two independent homogeneous Poisson point processes (HPPPs) with density μ'_p and μ_s, respectively, where we assume $\mu'_p \ll \mu_s$. For each PT, its associated primary receiver (PR) is located at a distance of d_p away in a random direction. Similarly, for each ST, its associated secondary receiver (SR) is located at a distance of d_s away in a random direction. We further assume that PTs use the same power P_p for data transmissions and STs use VP for data transmissions. The maximum transmit power of STs is P_s, which occurs when the batteries of STs are fully charged. In addition, $P_p \gg P_s$.

Time is partitioned into slots with unit duration. In each time slot, the PTs employ an Aloha type of medium access control (MAC) protocol and make independent decisions to access the spectrum with probability p_p. Then, according to the coloring theorem [2], the locations of the active PTs follow a HPPP with density $\mu_p = p_p \mu'_p$ [5]. We further denote $\Phi_p = \{X\}$ and $\Phi_s = \{Y\}$ as the point processes formed by the active PTs and STs, respectively, where X and Y denotes the coordinates of the PTs and STs, respectively.

Each PT is assumed to be associated with a guard zone to protect its intended receiver from STs' interference, and at the same time delivers RF energy to STs located in its harvesting zone. For the secondary network, a VP transmission mode is proposed and an energy-based OSA strategy is considered, under which a ST is allowed to transmit if its harvested energy is larger than a predefined transmission threshold, which is given by βP_s, and it is outside the guard zones of all active PTs, where β, $0 < \beta \leq 1$, is the transmission threshold coefficient. Note that the special case with $\beta = 1$ was considered in [4]. For simplicity, we refer to "active PTs" as PTs in the sequel.

The propagation channel is modeled as the combination of small-scale Rayleigh fading and large-scale path-loss given by $g(r) = hr^{-\alpha}$, where h denotes the exponentially distributed power coefficient with unit mean, r denotes the propagation distance, and $\alpha > 2$ is the path-loss exponent.

2.2 Energy Harvesting Model

The RF harvester in the ST is equipped with a power conversion circuit, which can transform the received electromagnetic wave from the PTs into direct-current (DC) power; as such the secondary network can utilize the harvested energy from RF signals to augment/replenish their power sources. The input power needs to be larger than a predesigned threshold, which is given by $P_p r_h^{-\alpha}$, for the circuit to harvest RF energy efficiently. r_h is defined as the radius of a disk which is called the harvesting zone and is centered at each PT, that is to say, a ST could harvest RF energy from its nearest PT provided it is inside the harvesting zone. Otherwise, the power received by a ST outside any harvesting zone is too small to activate the energy harvesting circuit, and thus is assumed negligible. We denote the probability that ST lies in a harvesting zone as p_h. Similar as in [4], we assume that the harvesting zones of different PTs do not overlap at most time. Thus we have

$$p_h = 1 - e^{-\pi r_h^2 \mu_p}. \tag{1}$$

Let η $(0 < \eta < 1)$ denote the harvesting efficiency, the distance between a ST and its nearest PT be given by D and $D \leq r_h$. Then, the average energy harvested by a ST in one slot can be obtained as $\eta P_p D^{-\alpha}$. Note that the harvested power has been averaged over the channel short-term fading within a slot.

2.3   ST Transmit Model

The STs access the spectrum of the primary network and cause interference to PRs. To protect the primary transmissions, the STs are prevented from transmitting when they lie in any of the guard zones, modeled as disks with a fixed radius r_g ($r_g \gg r_h$) centered at each PT. With the energy-based OSA strategy, the STs using the VP transmission mode are allowed to transmit under the following condition. The STs should be located outside any of the guard zones (the probability is denoted as p_g) and the power of STs should be larger than the transmission threshold βP_s (the probability is p_c). When the battery is charged larger than the transmission threshold and if it is outside all the guard zones, the ST will transmit all the stored energy in the next slot. Note that in our model the battery power level of every active ST is in the range $[\beta P_s, P_s]$, which is different from the model in [4]. Moreover, the point processes formed by the PTs change independently over different slots. Therefore, the events that a ST has been charged to the transmission threshold in one slot, and that it is outside all the guard zones in the next slot are independent. Accordingly, the transmit probability of the STs denoted by p_t is obtained by

$$p_t = p_c p_g. \tag{2}$$

The calculation of p_c will be discussed in Section 3, and p_g can be given similarly to p_h, as

$$p_g = e^{-\pi r_g^2 \mu_p}. \tag{3}$$

2.4   Performance Metric

In addition to transmission probability, two more performance metrics are studied in this paper, coverage probability and network throughput, which are specified as follows.

Coverage Probability. The coverage probability means the transmission non-outage probability, which is defined as the probability that a PR/SR decodes the received data packets successfully from its corresponding PT/ST. Specifically, given the signal-to-interference ratio (SIR), and a corresponding SIR target, denoted by θ, the coverage probability in the network is defined as $\tau = \Pr\{\text{SIR} \geq \theta\}$.

Network Throughput. The throughput of the primary network or the secondary network is the maximum rate the system can achieve with successful primary/secondary transmissions. Assume that the active PTs/STs follow a HPPP with average density μ. Consequently, the network throughput is given by $C = \mu\tau \log(1 + \theta)$.

3   Transmission Probability in Secondary Network

From (2), it can be observed that p_t depends on p_c and p_g. In this section, we first derive p_c, and then we characterize the transmission probability of STs.

The minimum power harvested by a ST in one slot is $\eta P_p r_h^{-\alpha}$, which occurs when the ST is at the edge of a harvesting zone. Therefore, the battery of an energy-harvesting ST can be charged to the transmission threshold within one slot time if $0 < \beta P_s \leq \eta P_p r_h^{-\alpha}$, thus this case is referred to as *single-slot charging*. Similarly, if $\eta P_p r_h^{-\alpha} < \beta P_s \leq 2\eta P_p r_h^{-\alpha}$, a ST needs at most two slots of harvesting to reach the transmission threshold, which is called *double-slot charging*. In either case, the battery power level can be exactly modeled by a finite-state Markov chain (MC), and the transmission probability p_t can be obtained accordingly. Otherwise, if $\beta P_s > 2\eta P_p r_h^{-\alpha}$, a ST needs at most N ($N > 2$) slots of harvesting to reach the transmission threshold, i.e., *multi-slot charging*. In this case, we can only obtain upper and lower bounds on p_t by using MC theory. However, since small value of P_s is of our interest, we will analyze the transmission probability in two different conditions as follows.

3.1 Single-Slot Charging

If $0 < \beta P_s \leq \eta P_p r_h^{-\alpha}$, i.e., $0 < \beta \leq \frac{\eta P_p r_h^{-\alpha}}{P_s}$, the battery is charged to the transmission threshold within one slot. Thus the power level can be characterized as two states $\{0, 1\}$, which are mapped to the power level 0 and the range $[\beta P_s, P_s]$, respectively. Accordingly, the state transition probability matrix denoted as $\mathbf{P}_1$ is obtained as

$$\mathbf{P}_1 = \begin{bmatrix} 1 - p_h & p_h \\ p_g & 1 - p_g \end{bmatrix}. \tag{4}$$

Therefore, we have the following proposition.

Proposition 1. *If $0 < \beta \leq \frac{\eta P_p r_h^{-\alpha}}{P_s}$, the transmission probability of a typical ST is obtained as*

$$p_t = p_c p_g = \frac{p_h}{p_h + p_g} p_g. \tag{5}$$

Proof. The probability p_c can be obtained by solving $\boldsymbol{\pi}_1 = \boldsymbol{\pi}_1 \mathbf{P}_1$, where $\boldsymbol{\pi}_1$ is the steady-state probability vector given by $\boldsymbol{\pi}_1 = \begin{bmatrix} \pi_1^0, \pi_1^1 \end{bmatrix}$. In this case, $p_c = \pi_1^1$. This completes the proof of Proposition 1.

From (5), it is observed that the transmit probability of a ST has no dependence on β. This is because once a ST lies in the harvesting zone, it is guaranteed to be charged to the threshold within one slot as the proposed condition $0 < \beta P_s \leq \eta P_p r_h^{-\alpha}$.

3.2 Double-Slot Charging

If $\eta P_p r_h^{-\alpha} < \beta P_s \leq 2\eta P_p r_h^{-\alpha}$, i.e., $\frac{1}{P_s}\eta P_p r_h^{-\alpha} < \beta \leq \frac{2}{P_s}\eta P_p r_h^{-\alpha}$, the battery of the ST needs at most two slots to reach the transmission threshold. We divide the harvesting zone into two parts as shown in Fig. 1, where T_p denotes a typical PT, a disk centered at T_p with radius h_1 denotes $\mathcal{H}_1$, and an annulus centered at T_p with radii $0 < h_1 < r_h$ denotes $\mathcal{H}_2$, and h_1 and r_h are the inner and outer diameter of the annulus, respectively. We can derive h_1 as $h_1 = \left(\frac{\beta P_s}{\eta P_p}\right)^{-\frac{1}{\alpha}}$. We consider T_s as a typical ST, and the average energy harvested by the T_s from T_p in one slot is $\eta P_p D^{-\alpha}$. If T_s is located inside the region $\mathcal{H}_1$, it will be charge to the range $[\beta P_s, P_s]$, else if T_s is located in the region $\mathcal{H}_2$, the power harvested is in the range $[\frac{1}{2}\beta P_s, \beta P_s)$.

Let us consider a three state MC with state space $\{0, 1, 2\}$, since $\eta P_p r_h^{-\alpha} \geq \frac{1}{2}\beta P_s$. In this case, the battery power level can be 0, in the range $[\frac{1}{2}\beta P_s, \beta P_s)$, or in the range $[\beta P_s, P_s]$, which are mapped to the states 0, 1, and 2, respectively. From Fig. 1, the state transition probability matrix denoted by $\mathbf{P}_2$ can be obtained as

$$\mathbf{P}_2 = \begin{bmatrix} 1 - p_h & p_h - p_1 & p_1 \\ 0 & 1 - p_h & p_h \\ p_g & 0 & 1 - p_g \end{bmatrix}. \tag{6}$$

Similarly to (1), the probability of $p_1 = \Pr\{T_s \in \mathcal{H}_1\}$ is obtained as $p_1 = 1 - e^{-\pi h_1^2 \mu_p}$. Based on the above analysis, we have the following proposition.

Proposition 2. *If* $\frac{1}{P_s}\eta P_p r_h^{-\alpha} < \beta \leq \frac{2}{P_s}\eta P_p r_h^{-\alpha}$, *the transmission probability of a typical ST is obtained as*

$$p_t = p_c p_g = \frac{p_h}{p_h + p_g\left(\frac{2p_h - p_1}{p_h}\right)} p_g. \tag{7}$$

Proof. The result in (7) can be obtained similarly as Proposition 1, i.e., by solving $\boldsymbol{\pi}_2 = \boldsymbol{\pi}_2 \mathbf{P}_2$, where $\boldsymbol{\pi}_2$ is the steady-state probability vector given by $\boldsymbol{\pi}_2 = [\pi_2^0, \pi_2^1, \pi_2^2]$, and we obtain $p_c = \pi_2^2$. This completes the proof of Proposition 2.

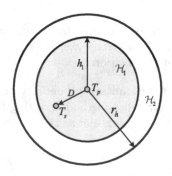

Fig. 1. The partitioned harvesting zone of double-slot charging.

Note that from (7), we can easily obtain that p_t is a decreasing function of β. An intuitive explanation of the above observation is that, if β grows, the time required for battery charging will get longer, thus leading to a lower p_t.

4 Coverage Probability

In this section, the coverage probabilities of the primary network and the secondary network are investigated. Note that due to the energy-based OSA, the point process developed by active STs does not follow a HPPP, and is difficult to characterize accurately. To simplify our analysis, following the assumptions in [5], we assume that the point process of active STs follows a HPPP, which will be verified by our simulation results. Let Φ_a denote the point process of active STs, I_p and I_s denote the aggregate interference at the origin from all PTs and active STs, respectively, which are modeled by *shot-noise processes* [6], given by

$$I_p = \sum_{X \in \Phi_p} h_X P_p |X|^{-\alpha}, \tag{8}$$

$$I_s = \sum_{Y \in \Phi_a} h_Y \overline{P}_s |Y|^{-\alpha}, \tag{9}$$

where $|X|, |Y|$ denote the distances from node X, Y to the origin, respectively, and $\{h_X\}$ and $\{h_Y\}$ are independent and identically distributed (i.i.d.) exponential random variables with unit mean. $\overline{P}_s$ denotes the average power of all the transmitting STs. Since the transmit power of the active STs are kept in a range $[\beta P_s, P_s]$, we have $\beta P_s < \overline{P}_s < P_s$. Intuitively, $\overline{P}_s$ is increasing with β. Similar to [4], we make the following approximations on the conditional distribution of the active STs, which will be verified by simulations in Section 6.

Assumption 1. *The point process formed by the active STs Φ_a follows a HPPP with density $p_t \mu_s$.*

4.1 Primary Network

To analyze the coverage probability of the primary network, we concentrate on a typical PR at the origin denoted by R_p with its intended PT denoted as T_p at a distance of d_p away. By using Slivnyak's theorem [7], in this case, the locations of the rest of the active PRs and PTs are both HPPPs with density μ_p. Therefore, the coverage probability of the primary network τ_p is given by

$$\tau_p = \Pr\{SIR_p \geq \theta_p\} = \Pr\left\{\frac{h_p P_p d_p^{-\alpha}}{I_p + I_s} \geq \theta_p\right\}, \tag{10}$$

where h_p is the channel power between R_p and its intended T_p. Then we have the following theorem.

Theorem 1. *Under Assumption 1, the average coverage probability of the primary network $\overline{\tau}_p$ is given by*

$$\overline{\tau}_p = \exp\left(-\left(\theta_p^{\frac{2}{\alpha}} d_p^2 \varphi\left(p_t \mu_s \left(\frac{\overline{P}_s}{P_p}\right)^{\frac{2}{\alpha}} + \mu_p\right)\right)\right), \tag{11}$$

where $\varphi = \pi \Gamma\left(1 + \frac{2}{\alpha}\right) \Gamma\left(1 - \frac{2}{\alpha}\right)$, and $\alpha > 2$ with $\Gamma(x) = \int_0^\infty t^{x-1} e^{-t} dt$ indicating the Gamma function.

Proof. The proof is omitted due to the space limitation. Please refer to [4].

Corollary 1. *Under Assumption 1 and the analysis above. the coverage probability of the primary network is upper-bounded and lower-bounded, respectively, by,*

$$\tau_p < \exp\left(-\left(\theta_p^{\frac{2}{\alpha}} d_p^2 \varphi\left(p_t \mu_s \left(\frac{\beta P_s}{P_p}\right)^{\frac{2}{\alpha}} + \mu_p\right)\right)\right), \tag{12}$$

$$\tau_p > \exp\left(-\left(\theta_p^{\frac{2}{\alpha}} d_p^2 \varphi\left(p_t \mu_s \left(\frac{P_s}{P_p}\right)^{\frac{2}{\alpha}} + \mu_p\right)\right)\right). \tag{13}$$

Proof. From (11), it can be observed that τ_p is a function of p_t and $\overline{P}_s$. An intuitive explanation of the above observation is that $\beta P_s < \overline{P}_s < P_s$. However, from Section 3, p_t is a constant for a given transmission threshold βP_s. Thus we can obtain (12) and (13) by substituting $\beta P_s < \overline{P}_s < P_s$ into (11).

4.2 Secondary Network

Under Assumption 1, to analyze the coverage probability of the secondary network, we concentrate on a typical SR at the origin denoted by R_s with its intended ST denoted as T_s at a distance of d_s away. By using Slivnyak's theorem, in this case, the locations of the rest of the active SRs and STs both follow HPPPs with density $p_t \mu_s$.

Since the STs cannot transmit if they are inside any guard zone of the PTs, to approximate τ_s, we consider the coverage probability conditioned on T_s being outside all the guard zones which means that there is no PT inside the disk centered at T_s with radius r_g, denoted as $\mathcal{G}_{T_s}^{r_g}$. Let the condition discussed above be denoted by $\zeta = \{\Phi_p \cap \mathcal{G}_{T_s}^{r_g} = \emptyset\}$. Then the coverage probability of the secondary network is given by

$$\tau_s = \Pr\left\{\frac{h_s P_{T_s} d_s^{-\alpha}}{I_p + I_s} \geq \theta_s \,|\, \zeta\right\}, \tag{14}$$

where h_s is the channel power between R_s and its intended T_s, P_{T_s} is the transmit power of the intended T_s, and $\beta P_s \leq P_{T_s} \leq P_s$. The active STs follow a HPPP with density $p_t \mu_s$ which means that none of the active STs are inside a guard zone, that is, $\Pr\{\zeta\} = 1$. Moreover, under the assumption $P_p \gg P_s$, it is a reasonable assumption that the interference from every PT inside $\mathcal{G}_{T_s}^{r_g}$ will cause an outage to the typical R_s at the origin. Thus, we have $\Pr\left\{h_s \geq \frac{\theta_s d_s^{\alpha}}{P_{T_s}} (I_p + I_s) \,|\, \bar{\zeta}\right\} \approx 0$. Then we have the following theorem.

Theorem 2. *Under Assumption 1, the average coverage probability of the secondary network $\overline{\tau}_s$ is obtained as*

$$\overline{\tau}_s = \exp\left(-\left(p_t \mu_s + \mu_p \left(\frac{\overline{P}_s}{P_p}\right)^{-\frac{2}{\alpha}}\right) \theta_s^{\frac{2}{\alpha}} d_s^2 \varphi\right). \tag{15}$$

Proof. The proof is omitted due to the space limitation.

Corollary 2. *Under Assumption 1 and the analysis above. The coverage probability of the secondary network is upper-bounded and lower-bounded, respectively, by,*

$$\tau_s < \exp\left(-\left(p_t\mu_s + \mu_p\left(\frac{P_s}{P_p}\right)^{-\frac{2}{\alpha}}\right)\theta_s^{\frac{2}{\alpha}}d_s^2\varphi\right), \tag{16}$$

$$\tau_s > \exp\left(-\left(p_t\mu_s + \mu_p\left(\frac{\beta P_s}{P_p}\right)^{-\frac{2}{\alpha}}\right)\theta_s^{\frac{2}{\alpha}}d_s^2\varphi\right). \tag{17}$$

Proof. This can be proved by applying a similar approach as used for the proof of Corollary 1.

5 Network Throughput

5.1 Primary Network

We characterize the throughput of the primary network as $C_p = \mu_p\tau_p\log(1+\theta_p)$. Note that the primary network throughput C_p mainly reflects the coverage probability τ_p. With (12) and (13), the throughput of the primary network is upper-bounded and lower-bounded, respectively, by,

$$C_p < \mu_p\log(1+\theta_p) \times \exp\left(-\left(\theta_p^{\frac{2}{\alpha}}d_p^2\varphi\left(p_t\mu_s\left(\frac{\beta P_s}{P_p}\right)^{\frac{2}{\alpha}} + \mu_p\right)\right)\right), \tag{18}$$

$$C_p > \mu_p\log(1+\theta_p) \times \exp\left(-\left(\theta_p^{\frac{2}{\alpha}}d_p^2\varphi\left(p_t\mu_s\left(\frac{P_s}{P_p}\right)^{\frac{2}{\alpha}} + \mu_p\right)\right)\right). \tag{19}$$

5.2 Secondary Network

We characterize the throughput of the secondary network as $C_s = \mu_s p_t\tau_s\log(1+\theta_s)$. The throughput of the secondary network C_s is a function of both p_t and τ_s. However, from Section 3, p_t is a constant for a given transmission threshold βP_s. Therefore, C_s is only dependent on τ_s. From (16) and (17), the throughput of the secondary network is upper-bounded and lower-bounded, respectively, by

$$C_s < \mu_s\log(1+\theta_s)\,p_t \times \exp\left(-\left(p_t\mu_s + \mu_p\left(\frac{P_s}{P_p}\right)^{-\frac{2}{\alpha}}\right)\theta_s^{\frac{2}{\alpha}}d_s^2\varphi\right), \tag{20}$$

$$C_s > \mu_s\log(1+\theta_s)\,p_t \times \exp\left(-\left(p_t\mu_s + \mu_p\left(\frac{\beta P_s}{P_p}\right)^{-\frac{2}{\alpha}}\right)\theta_s^{\frac{2}{\alpha}}d_s^2\varphi\right). \tag{21}$$

Remark 1: The maximum throughput of the secondary network is obtained with the transmission threshold coefficient $\beta^* = \frac{1}{P_s}\eta P_p r_h^{-\alpha}$, which will be verified by simulation in Section 6 by Fig. 3(b). Note that $\beta^* = \frac{1}{P_s}\eta P_p r_h^{-\alpha}$ can be write as $\beta^* P_s = \eta P_p r_h^{-\alpha}$, where $\beta^* P_s$ is exactly the transmission threshold. As mentioned in Section 3, the minimum power harvested by a ST in one slot is $\eta P_p r_h^{-\alpha}$, which means that, the secondary network throughputs are maximized over the energy-based OSA strategy if each candidate ST's harvested energy within one slot is larger than the transmission threshold.

6   Numerical Result

In this section, based on our theoretical analysis, we provide some numerical results and give some interpretations. Unless otherwise specified, we set the harvesting efficiency as $\eta = 0.1$ and the path-loss exponent as $\alpha = 4$.

Fig. 2(a) and Fig. 2(b) show the ST transmission probability p_t versus the ST transmission threshold coefficient β and the ST maximum transmit power P_s, respectively. From Fig. 2(a), it is observed that p_t is consistent with our analysis in Section 3. Furthermore, both Fig. 2(a) and Fig. 2(b) show that the transmit probability with $0 < \beta < 1$ outperforms the transmit probability with $\beta = 1$, which means that the performance of the energy-based OSA strategy outperforms that of the scheme in [4], since $\beta = 1$ means that the batteries of STs are fully charged.

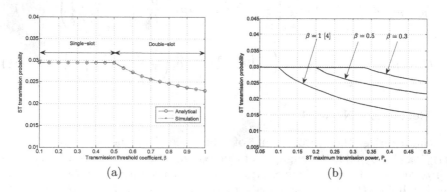

(a)   (b)

Fig. 2. (a) The ST transmission probability p_t versus the ST transmission threshold coefficient β; (b) The ST transmission probability p_t versus the ST maximum transmission power P_s.

In Fig. 3(a), we compare the analytical and simulated results on the coverage probability using the energy-based OSA scheme. We have following observations. First, the simulation results fall between the upper bounds and the lower bounds as expected, thus Assumption 1 is validated. Second, the coverage probability of the primary network τ_p is insensitive to β, this can be explained from (10), since

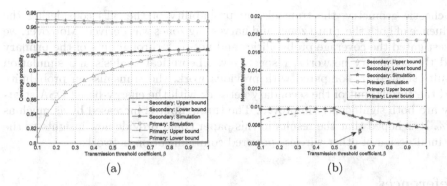

Fig. 3. (a) The network coverage probability versus the ST transmission threshold coefficient β; (b) The network throughput versus the ST transmission threshold coefficient β.

larger β increases the interference level from active STs (resulting in smaller τ_p) but at the same time reduces the ST transmission probability p_t and thus resulting in larger τ_p. Third, the coverage probability of the secondary network τ_s grows slightly with β, which can be explained theoretically from our result in (15), the numerator and the aggregate interference I_s in the denominator are both increasing with β, but the increment of I_s is negligible compared to I_p according to the condition $P_s \ll P_p$ and thus can be ignored. That is to say, compared with the scheme in [4], the coverage probabilities using energy-based OSA strategy are not changed significantly.

In Fig. 3(b), we compare the analytical and simulated results for the network throughput under the energy-based OSA scheme. Several observations follow. First, it is also observed that the simulated throughputs fall between the upper bounds and the lower bounds as expected. Second, the throughput of the primary network C_p is insensitive to β. This is because C_p mainly depends on τ_p and we have mentioned that τ_p is insensitive to β. Third, we show the maximum throughput of the secondary network is obtained with the transmission threshold coefficient $\beta^* = \frac{1}{P_s}\eta P_p r_h^{-\alpha}$. This can be explained as follows. On one hand, if $0 < \beta \leq \beta^*$, it can be observed that the transmit probability p_t is a constant, thus C_s depends only on τ_s. From the analysis above, τ_s grows slightly with β, therefore, C_s is increasing with β slightly. On the other hand, if $\beta > \beta^*$, C_s is dependent on both p_t and τ_s. However, p_t affects C_s more significantly than τ_s, and p_t is decreasing with β, which indicates that C_s is decreasing with β. Fourth, the network throughput with $0 < \beta < 1$ outperforms the throughput with $\beta = 1$, as much as 29% at $\beta^* = \frac{1}{P_s}\eta P_p r_h^{-\alpha}$, which confirms that the performance of the energy-based OSA strategy outperforms that of the scheme in [4].

7   Conclusion

In this paper, we proposed a VP transmission mode and an energy-based OSA strategy for opportunistic energy harvesting in CR networks. Using tools from stochastic geometry, the transmission probability of the STs considering the influence of both the guard zones and harvesting zones was derived. Moreover, we investigated the coverage probabilities and network throughputs of the primary and the secondary networks, respectively. Theoretical analysis and simulation results show that, compared with previous work, the transmission probability and the throughput of the secondary network with the energy-based OSA strategy are both significantly improved. The throughput is increased by as much as 29%. It is hoped that the results in this paper could provide new insights to the optimal design of other wireless powered communication networks.

References

1. Visser, H.J., Vullers, R.J.M.: RF Energy Harvesting and Transport for Wireless Sensor Network Applications: Principles and Requirements. Proc. IEEE **101**, 1410–1423 (2013)
2. Kingman, J.F.C.: Poisson Processes. Oxford University Press (1993)
3. Dhillon, H., Li, Y., Nuggehalli, P., Pi, Z., Andrews, J.: Fundamentals of Heterogeneous Cellular Networks with Energy Harvesting. IEEE Trans. Wireless Commun. **13**, 2782–2797 (2014)
4. Lee, S., Zhang, R., Huang, K.: Opportunistic Wireless Energy Harvesting in Cognitive Radio Networks. IEEE Trans. Wireless Commun. **12**, 4788–4799 (2013)
5. Song, X., Yin, C., Liu, D., Zhang, R.: Spatial Throughput Characterization in Cognitive Radio Networks with Threshold-Based Opportunistic Spectrum Access. IEEE J. Sel. Areas Commun. **32**, 2190–2204 (2014)
6. Haenggi, M., Ganti, R.K.: Interference in Large Wireless Networks. Found. Trends in Netw., 127–248 (2008)
7. Haenggi, M., Andrews, J., Baccelli, F., Dousse, O., Franceschetti, M.: Stochastic Geometry and Random Graphs for the Analysis and Design of Wireless Networks. IEEE J. Sel. Areas Commun. **27**, 1029–1046 (2009)

Channel Transition Monitoring Based Spectrum Sensing in Mobile Cognitive Radio Networks

Meimei Duan[✉], Zhimin Zeng, Caili Guo, and Fangfang Liu

Beijing Key Laboratory of Network System Architecture and Convergence School
of Information and Communication Engineering,
Beijing University of Posts and Telecommunications, Beijing 100876, China
mmduan.1276@163.com, zengzm@bupt.edu.cn, guocaili@gmail.com

Abstract. Spectrum sensing is a key technique for providing an opportunistic spectrum band in cognitive radio networks. The opportunistic spectrum is determined by the channel state. Mobility makes the problem of the traditional sensing mechanism more severe than in static scenarios. In this paper the channel transition monitoring based spectrum sensing mechanism is proposed. The proposed scheme not only reduces the influence of mobility on the current sensing mechanism, but also ensures reliability of the sensing and improves the spectrum efficiency. Our simulation results show that the proposed mechanism outperforms the traditional mechanism. Our method supplements the traditional sensing mechanism and enhances the efficiency of cognitive radio networks.

Keywords: Channel transition monitoring · Spectrum sensing · Mobile cognitive radio · Opportunistic spectrum

1 Introduction

Cognitive radio (CR) is one of the most promising ways to solve the spectrum scarcity problem; it opportunistically accesses a temporarily available licensed spectrum band [1–3]. Some research measurements for wireless radio spectrum show that the spectrum band is idle in range of 15% to 85% in both the time and spatial domain. For CR networks, the idle spectrum is called the opportunistic spectrum [4,5]. Spectrum sensing is a critical technology for finding the idle spectrum [6]. The task of the sensing mechanism is to sense the current spectrum and allocate the idle spectrum to the appropriate users. Additionally, it aims to reduce the overhead and interference to primary user (PU) as much as possible.

Currently, most of researches on traditional sensing mechanism focuses on periodic sensing mechanism, and some focuses on proactive sensing mechanism [7]. Research on periodic sensing employs the adaptive period to improve the spectrum efficiency [8]. This can reduce the sensing time according to the channel environment and greatly improve the throughput. However, the intrinsic problems–such as high overhead, wasting of the spectrum, and interference to the PU–still exist. Additionally, the above research has an implied condition that

© Institute for Computer Sciences, Social Informatics and Telecommunications Engineering 2015
M. Weichold et al. (Eds.): CROWNCOM 2015, LNICST 156, pp. 199–210, 2015.
DOI: 10.1007/978-3-319-24540-9_16

the CR user is always inside the primary protected region (PPR). This assumption is reasonable in large-scale coverage, such as digital television (DTV) base station, the coverage of which is about 50-60km [9]. The CR user's coverage is smaller than the PU's, and the sensing is always reliable. However, in small-scale-coverage PUs, such as wireless microphones (MWs), LTE networks and other networks, the sensing reliability of CR users is not always high because the users may leave the PPR. In these networks, user mobility has a significant effect on the spectrum sensing. The sensing capacity can be increased significantly in the presence of PU mobility [10]. The sensing accuracy exhibits threshold behavior that is a function of the sensing time when the users are mobile [11]. The sensor mobility information is exploited in the process of sensor localization with two range measurement models, namely, the time-of-arrival (TOA) model and the received signal strength (RSS) model [12]. A cognitive MAC protocol with mobility support (CM-MAC) is proposed in [13], addressing the decentralized control and local observation for spectrum management, where the CR mobile nodes move into the primary exclusive region. The study does not consider the impact of the mobility on the spectrum sensing mechanism.

In fact, mobility further increases the uncertainty of the spatial location. It makes the intrinsic problems of periodic sensing more serious. In mobile CR networks, in order to guarantee sensing accuracy, the periodic sensing should be more frequent than in static scenarios. Even though periodic sensing provides sensing reliability, it requires a short period to ensure that interference is minimized. The short sensing period increases the overhead and reduces the spectrum efficiency. Different opportunistic spectrums need different sensing techniques; especially in the spatial opportunistic spectrum, time sensing is not necessary.

We attempt to find a sensing mechanism that maximizes the opportunistic spectrum and reduces interference to the PU.

2   System Model and Problem Statement

In order to demonstrate the specific scenario, we provide a simple system model. Regardless of PU or CR user, the opportunistic spectrum is greatly affected by the user's mobility.

2.1   System Model

The CR network consists of PU and one or several CR users, as shown in Fig.1. t_1 and t_2 denote different times. The PPR is covered by R. In static scenario, for example, at time t_2, the CR user is always in the PPR. In mobile scenario, CR user is inside or outside the PPR. During the sensing operation, the distance between the PU and CR user changes over time, such as from the time t_1 to time t_2. If the CR user is inside the PPR, the distance is less than R. The opportunistic spectrum is the time opportunistic spectrum. If and only if the PUs is inactive, the CR user can occupy the spectrum band. Otherwise, the distance is larger than R, the opportunistic spectrum is the spatial opportunistic spectrum.

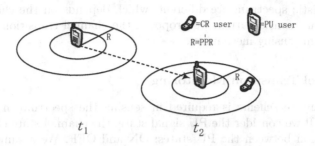

Fig. 1. The mobility system model of PU

Regardless of the PU's state, the CR user can always occupy the spectrum band. Therefore, at time t_1, the CR user accesses the spectrum directly without sensing. At time t_2, since the PU may be active or inactive, the CR user must sense.

The system model is very simple. But it can be extended to a complex system that is composed of multiple CR users and multiple PUs.

2.2 Problem Statement

In mobile scenario, as shown in the above system model, the intrinsic problems of the current sensing mechanism becomes more severe. Furthermore, the sensing reliability is very difficult to be achieved.

Firstly, owing the mobility, the channel is time-varying. In order to achieve accurate sensing, the sensing operation must be more frequent than in static scenarios. Consequently, the overhead is greater than in static scenarios. The wasting of the spectrum band and the interference to the PU are still present.

Secondly, the mobility makes the user change its location. When the user steps out of the PPR, it still senses the spectrum with traditional sensing method, which gives an incorrect result. If an individual user with an incorrect sensing result cooperates with other users, the cooperative performance is degraded.

Thirdly, the sensing technique is extended to be suitable for the spatial opportunistic spectrum. Most algorithms for spatial sensing depend on the localization and tracking of mobile users. The algorithms are very complex. Additionally, they add the complexity to the sensing .

The channel state is the most influential factor. As long as the channel state is determined, the user can easily access to it. In this paper, we investigate a new sensing mechanism.

3 Channel Transition Monitoring Based Spectrum Sensing

The objectives of spectrum sensing are to maximize the opportunistic spectrum, minimize the interference to the PU and reduce the overhead. The types of the opportunistic spectrum are different, which depends on the channel state. Taking this into consideration, we proposed the channel transition monitoring based spectrum sensing mechanism.

3.1 Channel Transition Monitoring

A mobility-unware scheme is required in sensing the spectrum in the mobile environment. If we consider the PU signal state, the channel state changes only at the transition between the PU states: ON and OFF. We assumed that the CR user and PU do not transmit simultaneously and the CR user can identify its own signal.

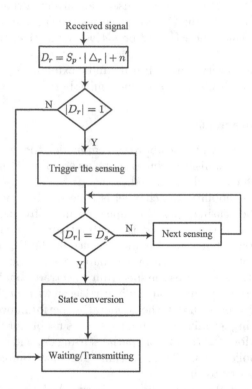

Fig. 2. The channel transition monitoring unit and its processing sequence

The channel transition monitoring should not affect the spectrum band. The monitoring unit is an internal processor in the receiver. The monitoring processing is parallel with the receiving processing and adopts the differential processing to differentiate the current channel state from the previous one, as shown in Fig.2. If the the monitoring result $|D_r|$ does not equal one, which means that the channel state has not changed, the CR users continue waiting or transmitting.

If the result $|D_r|$ equals one, which means that the channel state has changed, the CR user stops the current operation immediately and begins sensing to further determine the channel state. If the two results–i.e., the monitoring result D_r and the sensing result D_s–indicate that the channel state has changed, the next operation of the CR user is different from the previous operation. If the results indicate the different channel states, the CR user continues sensing until the two results indicate the channel state are the same. This also prevents sharp change in the channel from being affected by sudden change of background noise.

The received signal of the monitoring unit is $r(n)$ as follows:

$$r(n) = \eta h_p s_p(n) + (1 - \eta) h_s s_s(n) + n(n) \tag{1}$$

where $\eta \in \{0, 1\}$ denotes the absence or presence of the signal. h_p represents the channel coefficient between the PU and the CR user. s_p is the primary signal. h_s is the channel coefficient between the two CR users. s_s denotes the CR user's signal.

The received signal is $r(n-1)$. At the monitoring unit, the differential result $D_r(n)$ is as follows:

$$\begin{aligned} D_r(n) &= r(n) - r(n-1) \\ &= \eta h_p s_n + (1 - \eta) h_s s_s(n) - \eta' h_p s_{n-1} - (1 - \eta') h_s s_s + n' \end{aligned} \tag{2}$$

where n' denotes the differential noise. Given $s(n) = s(n-1)$, if the channel coefficient is constant, the above equation becomes:

$$D_r = h_p s_p(\eta - \eta') + h_s s_s(\eta' - \eta) + n' \tag{3}$$

$h_s s_s$ is known since s_s is given. Therefore, the above equation becomes:

$$D_r = S_p \cdot |\Delta_r| + n' \tag{4}$$

where $S_p = h_p s_p$ and $\Delta_r = \eta - \eta'$. When $\Delta_r = 0$, the channel state does not change. When $|\Delta_r| = 1$, the channel state has changed. $\Delta_r = 1$ represents the arrival of the PU's signal at the present time or the CR user steps into the PPR. The channel state changes from IDLE to BUSY. $\Delta_r = -1$ represents the departure of the PU's signal at the present time or the CR user steps out of the PPR. The channel state changes from BUSY to IDLE.

An appropriate threshold is selected to decide two states. When $|\Delta_r| = 1$, $|D_r| = 1$. The monitoring unit makes the CR user stop the current operation and immediately trigger the sensing process. With the CR sensing result, the next operation can be determined. When $\Delta_r = 0$, $|D_r| = 0$. No channel transition is occurring. The CR user continues the previous operation.

3.2 Spectrum Sensing Based on Channel Transition Monitoring

The result of the channel transition monitoring presents only at the transition of the channel. If and only if the result equals to 1, the sensing process is triggered to further determine the channel state, as shown in Fig.3 (b).

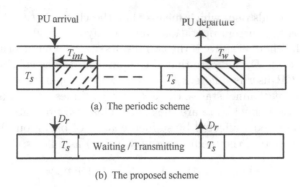

(a) The periodic scheme

(b) The proposed scheme

Fig. 3. The two sensing schemes: (a) the periodic sensing (b) the channel transition monitoring based sensing mechanism

Case 1: The uncooperative CR users

In the networks, the uncooperative user performs the channel transition monitoring when the PU's signal is received. The result $D_r = 0$ indicates that the channel state does not change. The CR user continues the previous operation: waiting or transmitting. In this process, the CR user's operation is not interrupted. The result $|D_r| = 1$ indicates that the channel state changes from IDLE to BUSY or vice versa. The initial result should be further verified by the sensing. Once the sensing operation is triggered, the CR user immediately stops the current operation. After the sensing result is determined, the CR user begins the next operation. The channel transition monitoring is only an internal operation that does not influence the spectrum band.

Case 2: The cooperative CR users

In current CR networks, multiple users cooperate with one another to overcome multiple pathes fading, shadowing fading, and receiver uncertainty. The channel transition monitoring based cooperative spectrum sensing adopts the proactive approach with "sound signal" in broadcasting way [7]. If the monitoring result shows that a channel transition has occurred, the CR user that monitors the spectrum sends the sound signal in broadcasting way to other users and requests them to perform the sensing. When the other users receive the sound signal, they immediately stop their current operation. Then they begin to sense the spectrum and send the local sensing result to the requesting user. The monitoring user combines the local sensing results and provides the global spectrum sensing result. When the monitoring result and the sensing result indicate same channel state, the CR user starts the next operation. If the current user does not need the spectrum, the spectrum information is stored in the spectrum database. Other users that need the spectrum query the database and access it.

It is pointed out that the user of the sensing schemes can be extended from the time opportunistic spectrum to the spatial opportunistic spectrum, or other potential opportunistic spectrum. The sensing schemes must provide reliable and efficient spectrum sensing. In the channel transition monitoring based spectrum sensing scheme, the sensing operation only occurs at the channel state transition. At other time, the CR users need to access the channel by searching the spectrum database. The approach does not waste the spectrum except the spectrum for sensing at the channel transition. The overhead, the wasting of the spectrum and interference are greatly reduced. The interference to PUs is related to the processing delay of the monitoring unit. When the channel is not in transition, the users that access the spectrum do not need to sense the spectrum. In order to obtain the potential spectrum, they only query the spectrum database. The delay involved in accessing the channel decreases.

It must be emphasized that the proposed scheme supplements the traditional spectrum sensing and improves the sensing performance. The only cost is the internal processor that monitors the channel transition.

4 Performance Analysis

Our channel transition monitoring based spectrum sensing mechanism provides the sensing operation at the end of the channel transition. The presence of a channel transition depends on the two factors. One is whether the PU changes its state. The other is whether the CR user leaves the PPR or vice versa. If the former leads to a channel transition, the detection performance is only related to the PU's OFF state. However, the latter will lead to a change in the opportunistic spectrum types.

4.1 Mobility-Enabled Sensing Capacity

In mobile scenarios, the CR user may be inside the PPR or outside the PPR. The mobility-aware detection capacity is as follows [10]:

$$C^{mob} = \zeta \rho W [(1 - P(I) + P_{off}P(I))] \tag{5}$$

where ζ, ρ, W and P_{off} represent the sensing efficiency, the spectral efficiency of the band, the bandwidth and the OFF state probability of the PU, respectively. From equation (5), $P(I)$ is the probability of that the CR user is inside the PPR. When $P(I) = 1$, the CR users are always inside the PPR. Therefore $C^{mob} = C^{statis}$. The probability of the PU's OFF state affects the capacity. The time sensing is adequate because there is only the time opportunistic spectrum. If $P(I) < 1$, the CR user locates inside or outside the PPR. Therefore, $C^{mob} > C^{statis}$. Both time sensing and spatial sensing are needed. In mobile scenarios, the value of capacity improvement is mainly from the value of the spatial opportunistic spectrum. The faster the users move in the certain coverage of the PPR, the shorter the sojourn time is. The smaller the probability within the PPR is, the larger the spatial opportunistic spectrum is. The detection capacity further improves.

4.2   Comparison Between the Two Scheme

Mobility results in the CR user being located in different positions. This leads to $P(I)$ and $1 - P(I)$. If $P(I) = 1$, the CR user is inside the PPR, and sensing reliability is provided. In general, the number of the PU's presence and departure are fewer than that of the sensing. From PU's arrival to PU's departure, the CR user has no opportunity to access the spectrum. However, from PU's departure to PU's arrival, the CR user can access the spectrum. In the periodic sensing, as shown in Fig.3(a), the CR transmission is frequently interrupted. T_{int} and T_w are inevitably greater than zeros. This leads to overhead, wasting of the spectrum and interference to the PUs. Whatever methods are adopted , the intrinsic problems can not be avoided. The proposed scheme, as shown in Fig.3(b), shows that the sensing operates only at the end of channel transition. It reduces the wasted spectrum band T_w, the interfering band T_{int}, and the number of the sensing. In a short, the proposed sensing scheme greatly decreases the sensing overhead.

If $P(I) < 1$, sometimes the CR user is outside the PPR. The users can access the spectrum band directly without sensing. The channel state is unchanging regardless of the PU's state. If the CR user is located inside the PPR, the user need to sense. The sensing result depends on the channel states: BUSY or IDLE. A channel state transition is caused by the arrival or departure of the PU's signal. The transition can also occur when the CR user's location changes from inside to outside the PPR or vice versa. Therefore, the channel transition monitoring based spectrum sensing is suitable for the time spectrum sensing as well as the spatial spectrum sensing. It not only greatly reduces sensing overhead but also utilizes the spectrum as much as possible.

5   Numerical Result

In this paper, we mainly consider the mobile CR networks, which have time and spatial opportunistic spectrums at different times. Simulation sets are based on the probabilities $P(Q)$, P_{off}, and different sizes of R etc..

The proposed scheme is based on the channel state transition, and the number of the PU's arrival or departure is an important factor. Fig.4 shows the impact of the number of the PU's arrival N_1 in two schemes on the throughput. The number of the PU's departure is $N_2 = N_1 - 1$. If the probability of the PU's OFF state is fixed, the throughput of the proposed scheme decreases as the number of the PU's departure/arrival increase; while the throughput of the periodic sensing is not greatly affected by the channel state transition. When the number of the departure/arrival are greater than a specific number, for example, $N = 7$, with $P_{off} = 0.5$, the throughput of the periodic sensing is greater than that of the proposed scheme. Because of the number of the PU's departure/arrival increasing, the interference increases. For the proposed scheme, if the number of the PU's departure/arrival increases, the time used to sense the spectrum increases, and the throughput decreases. However, the PU's state transition interval is longer

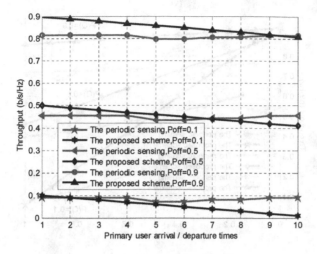

Fig. 4. The throughput vs. the number of the PU arrival/departure

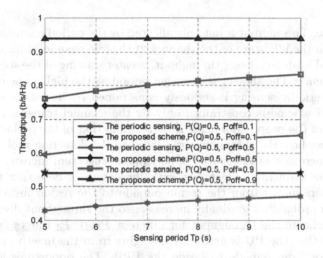

Fig. 5. The throughput relative to the sensing period

than the periodic sensing interval. In general, the throughput of the proposed scheme outperforms that of the periodic scheme.

Fig.5 shows the relationship between the throughput of the CR and the sensing period when $P(Q) = 0.5$. The active probabilities of PUs are varied. The number of the PUs arrival and departure are $N_1 = N_2 = 2$. The proposed scheme is robust in relation to the sensing period. It is dependent only on the state-changing number of the PUs. Therefore, if the number of PU arrival or departure is fixed, the throughput is determined only except influence of sensing time T_s. The periodic sensing mechanism adopts periodic sensing regardless of

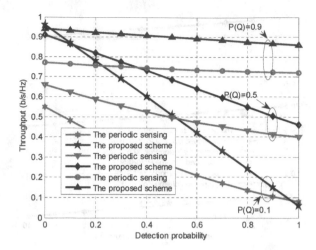

Fig. 6. The throughput v.s. detection probability

the PU's state. Throughput is not only affected by the periodic sensing time, but also related to the idle state of the channel. If the idle channel is fixed, a shorter sensing period leads to a lower throughput, greater wasting of the spectrum, and more interference. The shorter the sensing period is, the higher the overhead is. In a short, periodic sensing is seriously constrained by its intrinsic problems. The proposed scheme is constrained only by the channel transition state. The performance of the proposed scheme is better than that of the periodic sensing.

Fig.6 shows how the throughput is related to the detection probability. The throughput increases as the spatial opportunistic spectrum increases, i.e. $P(Q)$ increases. The throughput decreases with increasing of detection probability. If the CR users moves from the region outside to the region inside the PPR, the detection probability gradually increases, so the throughput decreases. The three lines indicate the throughput for different $P(Q)$. P_d ranges from 0 to 1, which means that the PU is gradually changing from the inactive state to the active state or from outside to inside the PPR. The opportunistic spectrum changes from the spatial domain to the time domain. For each sensing mechanism, the higher the idle spectrum probability, the higher is the throughput. For the specific probability of the idle spectrum, the throughput of the proposed sensing mechanism is higher than that of the periodic sensing mechanism. The far left points of the lines indicate the throughput at $P_{off} = 1$, when the maximum opportunistic spectrum exists. The improvement of the proposed scheme's throughput is greater than that of the periodic sensing from $P_d = 1$ to $P_d = 0$. Regardless of the behaviors of the users and the sensing techniques, the proposed scheme outperforms the periodic sensing.

Fig.7 shows how the throughput increases as the velocity of the users increases. The root cause is that mobility makes the user location changeable. The spatial opportunistic spectrum without sensing has greatly improved the

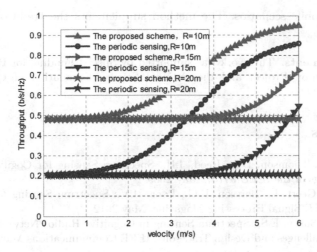

Fig. 7. The throughput related to the velocity of the users

throughput. The velocity affects the time inside or outside the PPR regardless of the mobility models. If R is fixed, the sojourn time decreases, and the time of outside the PPR increases as the velocity of the users increases. The more spatial opportunistic spectrum is obtained. For the same velocity with changing of R, the probability of being inside the PPR increases as R increases. The larger the coverage R is, the longer the sojourn time is, the shorter the time of outside the PPR is. The smaller spatial opportunistic spectrum is. Therefore, in small-scale-coverage PUs, such as MWs, the spatial opportunistic spectrum should be considered. In a short, in complex networks, the different opportunistic spectrums are the potential spectrums to enhance the throughput of CR networks.

In summary, in the proposed sensing mechanism, because the internal monitoring unit of the channel transition is used, the throughput is significantly improved and the interference to the PU is reduced. The wasting of the spectrum is shorter than that of the periodic mechanism. Therefore, the proposed scheme outperforms the traditional sensing mechanism.

6   Conclusion

In this paper, we proposed a spectrum sensing mechanism based on channel transition monitoring to improve the reliability of the sensing and the utilization efficiency of the opportunistic spectrum. Owing to user mobility, the users location varies over time. This leads to unreliability of the sensing. The proposed sensing mechanism based on channel transition monitoring not only reduces the influence of the intrinsic problems of the periodic sensing, but also overcomes the unreliability of the sensing. More importantly, it improves the utilization efficiency of the spectrum. The proposed sensing mechanism is suitable for both static and mobile scenarios. The method supplements the traditional sensing mechanism.

Acknowledgments. This work was supported by Special Funding for Beijing Common Construction Project and the national Science Foundation of China Grant No.61271177.

References

1. Simon, H., Thomson, D.J., Reed, J.H.: Spectrum Sensing for Cognitive Radios. Proceedings of the IEEE **97**(5), May 2009
2. Axell, E., Geert, L., Larsson, E.G., Poor, H.V.: Spectrum Sensing for Cognitive Radio. IEEE Signal Processing Magazine, May 2012
3. Amir, G., Sousa, E.S.: Spectrum Sensing in Cognitive Radios Networks: Requirements, Challenges and Design Trade-offs. IEEE Communications Magazine **46**(4), April 2008
4. Commission, F.C.: Spectrum Policy Task Force. Rep. ET. Docke. NO. 12-135, November 2002
5. Wang, B., Ray Liu, K.J.: Advances in Cognitive Radio Networks: A Survey. IEEE Journal of Selected Topics in Signal Processing **5**(1), February 2011
6. Tevfik, Y., Huseyin, A.: A survey of spectrum sensing algorithms for cognitive radio applications. IEEE Communications Surveys and Tutorials **11**(1), First Quarter (2009)
7. Mohammadkarimi, M.: Cooperative proactive spectrum sensing for cognitive radio networks. In: 6th International Conference on WiCOM (2010)
8. Liu, Q., Wang, X., Cui, Y.: Robust and Adaptive Scheduling of Sequential Periodic Sensing for Cognitive Radios. IEEE Journal On Selected Areas In Communications **2**(3), March 2014
9. Mishra, S.M.: Maximizing Avaible Spectrum for Cognitive Radios. Ph.D. Dissertation, UC Berkeley (2010)
10. Cacciapuoti, A.S., Akyildiz, I.F., Paura, L.: Primary-user mobility impact on spectrum sensing in cognitive radio networks. In: IEEE 22nd International Symposium on Personal, Indoor and Mobile Radio Communications (2011)
11. Cacciapuoti, A.S., Akyildiz, I.F., Paura, L.: Optimal Primary-User Mobility Aware Spectrum Sensing Design for Cognitive Radio Networks. IEEE Journal on Selected Areas in Communications **31**(11), November 2013
12. Salari, S., Shahbazpanahi, S., Ozdemir, K.: Mobility-Aided Wireless Sensor Network Localization via Semidefinite Programming. IEEE Transactions on Wireless Communications **12**(12), 5966–5978 (2013)
13. Hu, P., Ibnkahla, M.: A Cognitive MAC Protocol with Mobility Support in Cognitive Radio Ad Hoc Networks: Protocol Design and Analysis. Ad Hoc Networks **17**, June 2014

Power Minimization Through Packet Retention in Cognitive Radio Sensor Networks Under Interference and Delay Constraints: An Optimal Stopping Approach

Amr Y. Elnakeeb, Hany M. Elsayed, and Mohamed M. Khairy[✉]

Electronics and Electrical Communications Department, Faculty of Engineering,
Cairo University, Giza, Egypt
{amr.y.elnakeeb,helsayed,mkhairy}@ieee.org

Abstract. We consider the problem of power minimization in Cognitive Radio Sensor Networks (CRSN). The aim of this paper is twofold: First, we study the problem of packets retention in a queue with the aim of minimizing transmission power in delay-tolerant applications. The problem is classified as an optimal stopping problem. The optimal stopping rule has been derived as well. Optimal number of released packets is determined in each round through an Integer Linear Programming (ILP) optimization problem. This transmission paradigm is tested via simulations in an interference-free environment leading to a significant reduction in transmission power (at least 55%). Second, we address the problem of applying the scheme of packets retention through the Optimal Stopping Policy (OSP) to underlay CRSN where strict interference threshold does exist. Also, this problem is subjected to a delay constraint. Optimal number of released packets at this case is determined through another knapsack optimization problem that takes interference to Primary User (PU) into account. Extensive simulations that encompass dropped packet rate, Average Power per Transmitted Packet (APTP) and average consequent delay have been proposed. Simulations proved that our scheme outperforms traditional transmission method as far as dropped packet rate and APTP are substantially concerned, where end to end delay could be tolerated.

Keywords: Cognitive radio · Sensor network · Optimal stopping rule

1 Introduction

Due to the wide proliferation of mobile communication, there is an inevitable rapid growth in mobile traffic leading to the problem of spectrum scarcity. However, many spectrum studies showed that huge part of the spectrum is underutilized. This, consequently, led to the concept of Cognitive Radio (CR) [1] which received a great attention to alleviate the spectrum scarcity problem. It enables unlicensed users to communicate over the licensed bands assigned

© Institute for Computer Sciences, Social Informatics and Telecommunications Engineering 2015
M. Weichold et al. (Eds.): CROWNCOM 2015, LNICST 156, pp. 211–222, 2015.
DOI: 10.1007/978-3-319-24540-9_17

for the licensed users through one of two modes. First: Spectrum Sharing (SS), or underlay, where CR Secondary Users (SU) can operate on same bands licensed for PU provided that sufficient interference thresholds to PU are strictly maintained. Second: Opportunistic Spectrum Access (OSA), or overlay, where SU can dynamically exploit spectrum holes when PU are inactive [2]. Wireless Sensor Networks (WSN) have gained a great attention as a research area [3]. Due to the hardness of rechargeability of these networks, they have limited energy budget and hence a limited lifetime. Therefore, a considerable amount of research work has been exerted to mitigate the problem of energy limitation in WSN. Accordingly, energy minimization and lifetime maximization of WSN have been investigated in many research papers [4–7]. CRSN is a research trend that enables WSN to work in cognitive way. The essence of CRSN, its basic design principles, different architectures, applications, advantages and shortcomings have been well introduced in [8].

In this paper, we are concerned about energy minimization through formulation of an optimal stopping problem and finding out its stopping rule. Optimal stopping is concerned with the problem of taking a specific action at specific time based on sequentially observed previous states so as for maximizing the payoff, or minimizing the cost, or both. With the optimal stopping problem, there always exists an optimal stopping rule, where the decision is taken based on it. This type of problems usually arises in areas of statistics, where the action is taken with the aim of testing an hypothesis or estimating a parameter. Considering seminal and recent work, optimal stopping theory has been applied to opportunistic scheduling [9] and spectrum sensing [10],[11], but not to power minimization; which is the problem we will consider in this context. In[9], the authors studied optimal transmission scheduling policies in cognitive radio networks. They proposed a cooperative scheme that improves the primary network performance and allows secondary nodes to access the licensed spectrum in order to cooperate. In [10], the authors studied joint channel sensing and probing scheme and they proved that this scheme can achieve significant throughput gains over the conventional mechanism that uses sensing alone. However, in [11], authors studied the problem of optimizing the channel sensing parameters in the presence of sensing errors. They proposed suboptimal solutions that significantly reduce the complexity and maintain a near-optimal throughput. In this paper we apply the optimal stopping policy to the problem of power minimization in underlay cognitive radio sensor networks under interference and delay constraints.

The rest of this paper is organized as follows. Power minimization through packets retention via the optimal stopping approach is studied in Sect. 2, where the problem is formulated and the stopping policy is derived as well. In Sect. 3, the power minimization problem derived in Sect. 2 is extended to CRSN where interference threshold to PU does exist. Evaluating our performance is conducted through a simulation study which proves that our scheme through packets retention via OSP performs better than traditional transmission method as far as

APTP and successful packet reception are concerned. Finally, Sect. 4 concludes the paper.

2 Power Optimization Through Optimal Stopping Policy

2.1 Problem Formulation and Stopping Rule Derivation

In this paper, we focus on the problem of minimizing transmission power of nodes of any network through packet retention in a queue. Each node observes its power status round by round. Based on the observation sequence, it decides whether it sends its packet(s) instantaneously or further keeps it/them in the queue. To minimize the transmission power, each node makes the decision based on the result of comparing the instantaneous cost and the expected cost of future observations. The instantaneous cost is represented by the instantaneous power consumed if the packet is transmitted instantly. It depends directly on the instantaneous channel quality at this round for the observed nodes. On the other hand, the expected cost of future observations is the expected power the node will consume if it keeps the packet for more rounds taking into consideration how many packets are already existing in the queue. Consequently, this issue can be formulated as a sequential decision problem and can be investigated by applying the optimal stopping theory.

We are following the communication model presented in [4], the total power consumption consists of two components: power consumption of the amplifiers P_{PA} which depends on transmission power P_t with the relation

$$P_{PA} = (1 + \alpha)P_t. \tag{1}$$

where $\alpha = \dfrac{\xi}{\eta} - 1$ with η the drain efficiency of the power amplifier and ξ the peak-to-average power ratio(PAR), which depends on the modulation scheme and the associated constellation size. Transmission power P_t is given by the link-budget relationship, when the channel experiences a square-law path loss

$$P_t = \overline{E_b} \times \frac{R_b(4\pi d)^2}{G_t G_r \lambda^2} M_l N_f. \tag{2}$$

where $\overline{E_b}$ is the required energy per bit for a given BER requirement, R_b is the bit rate of the RF system, d is the transmission distance. G_t and G_r are the antenna gain of the transmitter and the receiver respectively, λ is the carrier wavelength, M_l is the link margin compensating the hardware process variations and other additive background noise or interference, N_f is the receiver noise figure defined as $N_f = \dfrac{N_r}{N_0}$ with N_0 the single-sided thermal noise power spectral density (PSD) at room temperature and N_r is the PSD of the total effective noise at the receiver input.

The other term in the total power consumption is the circuit power P_c. Finally, this gives the total energy consumption per bit as

$$E_{bt} = \frac{P_{PA} + P_c}{R_b}. \tag{3}$$

The instantaneous required BER per-packet, assuming BPSK modulation scheme is used, is also given by

$$\overline{P_b} = Q(\sqrt{2\gamma_b}) = e^{-\gamma_b} = e^{-\dfrac{\overline{E_b}|H|^2}{N_0}} \tag{4}$$

where $|H|^2$ is the instantaneous squared magnitude of the channel. Substituting from (4) into (2), and rearranging, we get the transmission power per node as follows

$$P_t = -ln(2\overline{P_b}) \times \frac{N_0}{|H|^2} \times \frac{R_b(4\pi d)^2}{G_t G_r \lambda^2} M_l N_f \tag{5}$$

Hence, and without loss of generality, we will consider only transmission power in our analysis as circuit power is, more or less, a constant that depends on the circuitry.

From now on, Q is defined as the queue size (Maximum number of packets could be kept) for each node in the network, and k is defined as the number of packets in the queue at round i. We intend to solve the stopping problem discussed above to minimize the cost represented by power by deriving an optimal rule that decides when to stop waiting for next rounds and transmit the packet(s) in the current round. Denote by $X_i^{(k)}$ the minimum cost the node can achieve at round i when k packets are in the queue.

$$X_i^{(k)} = \min\{P_{t_i}^k, E\{\min(P_{t_{i+1}}^{k+1}, P_{t_{i+2}}^{k+2}, \ldots, P_{t_{i+Q-k}}^Q)\}\} \tag{6}$$

where $P_{t_i}^k$ represents the instantaneous cost in round i (after the ith observation) when k packets are already existing in the queue. Also, $E\{\min(P_{t_{i+1}}^{k+1}, P_{t_{i+2}}^{k+2}, \ldots, P_{t_{i+Q-k}}^Q)\}$ represents the expected cost resulted by proceeding to keep the packet for next rounds till the queue is full. Inside the expectation operator, the minimum power scenario for keeping packets has to be chosen. To calculate $E\{\min(P_{t_{i+1}}^{k+1}, P_{t_{i+2}}^{k+2}, \ldots, P_{t_{i+Q-k}}^Q)\}$, we make the following mathematical analysis:

$$E\{\min(P_{t_{i+1}}^{k+1}, P_{t_{i+2}}^{k+2}, \ldots, P_{t_{i+Q-k}}^Q)\} = E\{\frac{1}{\max(\frac{1}{P_{t_{i+1}}^{k+1}}, \frac{1}{P_{t_{i+2}}^{k+2}}, \ldots, \frac{1}{P_{t_{i+Q-k}}^Q})}\} \tag{7}$$

Furthermore, since the function inside the expectation operator of (7) is a convex one ($f(x) = \dfrac{1}{x}$ is a convex function) [12], Jensen's inequality can be applied:

$$E\{\frac{1}{\max(\frac{1}{P_{t_{i+1}}^{k+1}}, \frac{1}{P_{t_{i+2}}^{k+2}}, ..., \frac{1}{P_{t_{i+Q-k}}^{Q}})}\} \geq \frac{1}{E\{\max(\frac{1}{P_{t_{i+1}}^{k+1}}, \frac{1}{P_{t_{i+2}}^{k+2}}, ..., \frac{1}{P_{t_{i+Q-k}}^{Q}})\}}$$

(8)

We will consider the lower bound of (8). Consequently, the aim now is to get $E\{\max(\frac{1}{P_{t_{i+1}}^{k+1}}, \frac{1}{P_{t_{i+2}}^{k+2}}, ..., \frac{1}{P_{t_{i+Q-k}}^{Q}})\}$. For simplicity, denote it by V_i^k.

According to (5), and extending for k to-be-transmitted packets, $P_{t_i}^k = \frac{C \times k}{|H_i|^2}$, where C is a constant equals to $-ln(2\overline{P_b}) \times N_0 \times \frac{R_b(4\pi d)^2}{G_t G_r \lambda^2} M_l N_f$.

Similarly, $P_{t_{i+1}}^{k+1} = \frac{C \times (k+1)}{|H_{i+1}|^2}$, and so on for all i and any k. Then,

$$V_i^k = E\{\max(\frac{1}{P_{t_{i+1}}^{k+1}}, ..., \frac{1}{P_{t_{i+Q-k}}^{Q}})\} = \frac{1}{C} \times E\{\max(\frac{|H_{i+1}|^2}{k+1}, ..., \frac{|H_{i+Q-k}|^2}{Q})\}$$

(9)

Since all transmission channels $|H_i|^2$, for all i, are assumed to be Rayleih-fading channels, any $|H|^2$ is exponentially distributed. Rewritting (9):

$$V_i^k = E\{\max(\frac{1}{P_{t_{i+1}}^{k+1}}, ..., \frac{1}{P_{t_{i+Q-k}}^{Q}})\} = \frac{1}{C} \times E\{\max(X_{i+1}^{k+1}, ..., X_{i+Q-k}^{Q})\}$$ (10)

Where X's are set of exponentially random variables. Let $F_X(x)$ be the Cumulative Density Function (CDF) of the variables X_i^k.

$$F_X(x) = 1 - e^{-\lambda x}$$

(11)

Let $F_q(v_i^k)$ be the Cumulative Density Function (CDF) of V_i^k. For Independent and Identically Distributed (iid) X's, $F_q(v_i^k)$ is simply given by:

$$F_q(v_i^k) = P[(x_{i+1}^{k+1} < v) \cap (x_{i+2}^{k+2} < v)..... \cap (x_{i+Q-k}^Q < v)] = \prod_{n=k(i)+1}^{Q} (1 - e^{-\lambda_n x})$$

(12)

Where $\lambda_n = n \times \lambda$ with λ the rate of the exponential distribution (assumed to be 1) and $k(i)$ is number of packets kept in the queue at round i.

Then,

$$V_i^k = \int_0^\infty [1 - \prod_{n=k(i)+1}^{Q} (1 - e^{-\lambda_n x})]dx = \frac{Q + 1 - k(i)}{Q + 1}$$

(13)

Going backward from equations (8) to (6) with the known value of V_i^k, (6) can be rewritten as

$$X_i^{(k)} = \min\{P_{t_i}^k, \frac{1}{C} \times \frac{Q + 1 - k(i)}{Q + 1}\}$$

(14)

Clearly from (14), the optimal stopping rule results in a threshold-comparison problem that compares the instantaneous transmission power with the minimum expected transmission power if the packet is kept in the queue. Also, it is clear that the stopping rule takes into consideration how many packets are already existing in the queue, denoted by k, besides the packet that has to be transmitted at this round. For instance, Table 1 shows the values of the thresholds for queue sizes of 3.

Table 1. THRESHOLDS FOR Q=3

Q=3	$k(i) = 0$	$k(i) = 1$	$k(i) = 2$	$k(i) = 3$
Threshold × C	$\frac{4}{4} = 1$	$\frac{4}{3} = 1.33$	$\frac{4}{2} = 2$	$\frac{4}{1} = 4$

Accordingly, for a queue of size Q, thresholds according to kept packets $k = 1, 2, ..., Q$ are given by $\frac{1}{C} \times \{\frac{Q+1}{Q+1}, \frac{Q+1}{Q}, \frac{Q+1}{Q-1}, \frac{Q+1}{Q-2}, \frac{Q+1}{Q-3}, ..., Q+1\}$

2.2 Queue Releasing

In this subsection we intend to optimize the releasing paradigm of the queue. That is, how many packets should be released in any round so as to minimize transmission power. If the number of packets already existing in the queue is k and one packet comes at this round, the node has the option of releasing all of the $k + 1$ packets, or releasing k packets and keeping one, or releasing $k - 1$ packets and keeping two, and so on till reaching the scenario of releasing no packets and keeping all of the $k + 1$ packets. Hence the available scenarios for transmission at round i can be mathematically written as follows:

$(k + 1) \times P_{t_i}$

$k \times P_{t_i} + E\{\min(P_{t_{i,1}}, P_{t_{i,2}}, ..., P_{t_{i,Q}})\}$

$(k - 1) \times P_{t_i} + E\{\min(P_{t_{i,2}}, P_{t_{i,3}}, ..., P_{t_{i,Q}})\}$

.....

.....

$0 \times P_{t_i} + E\{P_{t_{i,Q}}\}$

Where the second portion in any term of the above represents the expected transmission power if the packet(s) is/are kept till the queue is full. Actually, the second portion can be easily obtained from second portion of (14) for any value of Q and $k(i)$. Accordingly, we formulate an ILP optimization problem to choose the minimum transmission power scenario of all scenarios discussed above.

$$\underset{x}{\text{Minimize}} \quad \sum_{j=0}^{k+1} x_j[(k + 1 - j)P_{t_i} + E\{\min(P_{t_{i,j}}, P_{t_{i,j+1}}, ..., P_{t_{i,Q}})\}]$$

$$\text{subject to} \quad \sum_{j=0}^{k+1} x_j = 1,$$

$$x_j = \{0, 1\}, \text{ for all } j.$$

Where x represents on-off states that enables only one scenario from the available ones. ILP is NP-complete problem. Hence, there is no known polynomial algorithm which can solve the problem optimally. Optimal solution is still eluding researchers and a huge research effort has been exerted to find optimal solution for such problems either by heuristic algorithms [13,14] or by relaxation of the last constraint ($x_j = \{0,1\}$, for all$j$) [15]. We will follow [15], where the authors proposed a Linear Programming with Sequential Fixing (LPSF) algorithm that relaxes the last constraint. In this case, the formulation becomes a Linear Programming (LP) problem that is solvable in polynomial time. The algorithm is as follows:

i) Relaxing $x_j = \{0,1\}$, for all$j$ to take any continuous value between 0 and 1, and the problem is solved as a LP one. The solution to this LP problem is an upper bound on the optimal solution to our problem.

ii) Among all x_j, for all$j$, the largest one is picked up and denoted, for ease of identification, by x_k. x_k is set to 1. As a result, all x_h for $h \neq k$ is set to 0.

iii) A feasibility check is conducted on the resulting LP problem. An empty feasible region means that the first fixing in this iteration isn't correct. So, x_k is reset to 0 in a new LP and other x_h for $h \neq k$ become variables again.

iv) At this point, either LP problem constructed with $x_k = 1$ or with $x_k = 0$ has a feasible solution.

v) A new iteration starts following the same process above. The process is repeated until all x_j are set to either 0 or 1. We evaluate the performance of our proposed optimal stopping scheme for power minimization by conducting extensive simulation study. Simulations are conducted for a network of 100 nodes, and averaged over 100000 times. We assume channels are constant for one packet transmission. We assume Channel State Information (CSI) of all channels and distances from any node to the destination Base Station (BS) are well known at the BS where the decision is taken. We consider network parameters given in Table 2, in accordance with [4].

Table 2. SIMULATION PARAMETERS

Parameter	Value
$G_t G_r$	10 dB
η	0.35
f_c	15 MHz
$\overline{p_b}$	10^{-3}
M_l	10 dB
N_f	10 dB

Fig. 1 shows the amount of saved power through the policy of packet retention. The amount of saving starts with 55% for $Q = 1$ and increases monotonically for larger queue sizes. It is clear that power profile is constant for traditional transmission method (No queue), however through packet retention scheme power decreases as queue size increases. That's because as queue size increases, there are more chances for the nodes to keep the packets in the queue expecting better channel conditions in next rounds.

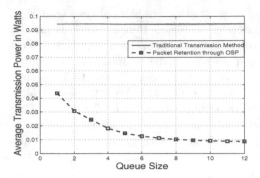

Fig. 1. Power Saving through OSP.

3 Applying OSP to CRSN Under Interference and Delay Constraints

3.1 Problem Formulation and Simulations

In the previous section, we discussed the problem of power minimization in any network through OSP in an interference-free environment. However, inducing the work of OSP to CRSN will have some effects considering transmission power and dropped packets rate. We assume underlay CRSN where SU are transmitting on the same bands licensed for PU as long as strict interference thresholds are well maintained. We formulate a knapsack optimization that chooses the minimum transmission power scenario for the CRSN and takes interference induced to PU into account. As mentioned earlier, we are dealing with delay-tolerant applications; though, we added to this formulation a delay constraint to show its effect. Denoting D_m^k as the delay which packet m undergoes when k packets are in the queue. D_m^k is updated within each round i based on how many rounds packet m has been kept in the queue till releasing. We assume for simplicity one SU interferes with one PU. The interfering channel from the SU to the PU is denoted by h_i^k (The interfering channel at round i when k packet are already in the queue), and it is assumed to be Rayleih-fading channel as well as the transmission channel H mentioned previously. The new problem is formulated as follows:

$$\underset{x}{\text{Minimize}} \quad \sum_{j=0}^{k+1} x_j[(k+1-j)P_{t_i} + E\{\min(P_{t_{i,j}}, P_{t_{i,j+1}}, \ldots, P_{t_{i,Q}})\}]$$

$$\text{subject to} \quad \sum_{j=0}^{k+1} x_j(k+1-j)P_{t_i} \times |h_i^k|^2 < I$$

$$D_m^k \le D_{max}, \quad \text{for each round } i,$$

$$\sum_{j=0}^{k+1} x_j = 1,$$

$$x_j = \{0,1\}, \quad \text{for all } j.$$

Where I is a strict interference threshold that must not be exceeded by the SU transmission, and D_{max} is the maximum delay that could be tolerated for each packet m. This problem is a knapsack optimization problem. Knapsack problem is a decision problem that is well known in combinatorial optimization. The knapsack problem is known, as ILP, to be NP-complete. We will use the same algorithm discussed in Sect 2 to solve it. We measure the performance of our scheme in terms of dropped packet rate and power saving. For convenience, a packet is considered dropped if its resulted interference exceeds the interference threshold I. Fig. 2 shows dropped packet rate percentage in case of traditional transmission method as well as transmission through OSP versus various interference threshold. We chose, without loss of generality, $Q = 8$. It is obvious that there is a significant decrease in the dropped packet rate through using OSP than using traditional method. In traditional transmission method, a packet is considered dropped if the resulted interference form its transmission exceeds the interference threshold instantaneously. However through the OSP, it can be kept in the queue expecting better interfering-channel conditions.

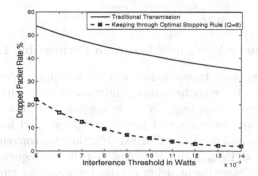

Fig. 2. Dropped Packet Rate in Traditional Transmission Method versus transmission through OSP with Q=8.

There is also an interesting point considering the comparison between traditional transmission method and transmission through OSP. In traditional transmission method, there are more dropped packets, and hence less power is consumed. However, in transmission through OSP, there are less dropped packets and more consumed power. To discriminate one scheme from the other, we consider the term Average Power per Transmitted Packet (APTP) which is defined as average consumed power divided by successfully received packets. APTP is the factor that makes one scheme outperforms the other. As shown in Fig. 3, in both transmission schemes, APTP increases as I increases because less packets are dropped and more power is consumed. However, transmission through OSP outperforms traditional method as it has less consumed power for any I. Traditional transmission method isn't affected by the queue size (Constant curves for the same Q). On the other hand, as Q increases, APTP decreases in transmission through OSP. Improvement in APTP swings from 4% (small Q size (Q=1)) to 23% (large Q size (Q=10)). Improvement for other queue sizes are in-between.

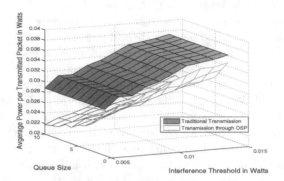

Fig. 3. APTP through traditional transmission method and OSP versus Queue Size and Interference Threshold.

3.2   Effect of Queue Size and Maximum Permissible Delay

As mentioned earlier, our transmission scheme is suitable for delay-tolerant applications such as mine reconnaissance, undersea explorations, environmental monitoring, and ocean sampling. In such applications power saving is an important issue as well as successful packet reception, and end to end delay isn't much important and could be afforded [16]. But, for convenience, we study the effect of both Q and D_{max} simultaneously on the consequent average delay. Fig. 4 shows average consequent delay per packet resulting from transmission through OSP when either one of the parameters Q or D_{max} is fixed, while the

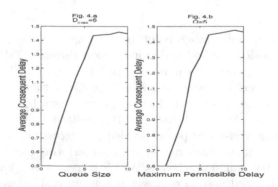

Fig. 4. Effect of Queue Size and Maximum Permissible Delay Parameters on Average Consequent Delay.

other is changing. It is clear that the average consequent delay is dominated by the fixed parameter of both. For instance, in Fig. 4.a, as Q increases, average consequent delay increases till $Q = 5$ which is the value of D_{max}, and then it starts to saturate. The same occurs in Fig. 4.b with fixing Q, and changing D_{max}. This phenomenon happens due to assuming, in our model, that packets are released according to their arrival in a First In First Out (FIFO) fashion. Hence, the smaller parameter dominates the effect on average consequent delay. Consequently, it is better to choose $Q = D_{max}$ to avoid the dominance of one parameter over the other on the average consequent delay.

4   Conclusion

We studied the power minimization problem through OSP fro delay-tolerant applications. Applying optimal stopping theory to packet retention and deriving the optimal stopping rule was the core of the work. We deduced that this transmission scheme outperforms traditional transmission method as far as power minimization is concerned. Also, it was shown that the improvement is overly significant; it reaches 55% for small queue sizes and increases monotonically as queue size increases.

We also extended the work of packet retention through OSP to CRSN where interference threshold to PU must not be exceeded by SU transmissions. Moreover, we studied the effect of queue size as well as the maximum permissible delay for a packet on the average consequent delay. Simulations were conducted in terms of dropped packet rate, APTP, and consequent delay.

Acknowledgments. This publication was made possible by NPRP grant # [5-250-2-087] from the Qatar National Research Fund (a member of Qatar Foundation). The statements made herein are solely the responsibility of the authors.

References

1. Mitola, J., Maguire Jr., G.Q.: Cognitive radio: making software radios more personal. IEEE Personal Communications **6**(4), 13–18 (1999)
2. Zhang, R., Liang, Y.-C.: Investigation on multiuser diversity in spectrum sharing based cognitive radio networks. IEEE Communications Letters **14**(2), 133–135 (2010)
3. Akyildiz, I., Su, W., Sankarasubramaniam, Y., Cayirci, E.: A survey on sensor networks. IEEE Communications Magazine **40**(8), 102–114 (2002)
4. Cui, S., Goldsmith, A., Bahai, A.: Energy-efficiency of MIMO and cooperative MIMO techniques in sensor networks. IEEE Journal on Selected Areas in Communications **22**(6), 1089–1098 (2004)
5. Farjow, W., Chehri, A., Mouftah, H., Fernando, X.: An energy-efficient routing protocol for wireless sensor networks through nonlinear optimization. In: 2012 International Conference on Wireless Communications in Unusual and Confined Areas (ICWCUCA), pp. 1–4, August 2012
6. Chen, Y., Zhao, Q.: On the lifetime of wireless sensor networks. IEEE Communications Letters **9**, 976–978 (2005)

7. Chen, Y., Zhao, Q.: Maximizing the lifetime of sensor network using local information on channel state and residual energy. In: Fourth International Symposium on Information Processing in Sensor Networks, IPSN 2005, March 2005
8. Akan, O., Karli, O., Ergul, O.: Cognitive radio sensor networks. IEEE Network **23**(4), 34–40 (2009)
9. Zheng, D., Ge, W., Zhang, J.: Distributed opportunistic scheduling for ad hoc networks with random access: An optimal stopping approach. IEEE Transactions on Information Theory **55**, 205–222 (2009)
10. Shu, T., Krunz, M.: Throughput-efficient sequential channel sensing and probing in cognitive radio networks under sensing errors, pp. 37–48, September 2009
11. Ewaisha, A., Sultan, A., ElBatt, T.: Optimization of channel sensing time and order for cognitive radios. In: 2011 IEEE Wireless Communications and Networking Conference (WCNC), pp. 1414–1419, March 2011
12. Shashi, M., Shouyang, W., Keung, L.K.: Generalized convexity and vector optimization, vol. 90 (2009)
13. Xiao-hua, X., An-bao, W., Ai-bing, N.: Notice of retraction competitive decision algorithm for 0–1 multiple knapsack problem. In: 2010 Second International Workshop on Education Technology and Computer Science (ETCS), vol. 1, pp. 252–255, March 2010
14. Islam, M., Akbar, M.: Heuristic algorithm of the multiple-choice multidimensional knapsack problem (MMKP) for cluster computing. In: 12th International Conference on Computers and Information Technology, ICCIT 2009, pp. 157–161, December 2009
15. Shu, T., Krunz, M.: Exploiting microscopic spectrum opportunities in cognitive radio networks via coordinated channel access. IEEE Transactions on Mobile Computing **9**, 1522–1534 (2010)
16. Keshtgary, M., Mohammadi, R., Mahmoudi, M., Mansouri, M.R.: Article: Energy consumption estimation in cluster based underwater wireless sensor networks using m/m/1 queuing model. International Journal of Computer Applications **43**, 6–10 (2012)

Modeling and Theory

Cooperative Spectrum Sensing using Improved p-norm Detector in Generalized κ-μ Fading Channel

Monika Jain$^{(\boxtimes)}$, Vaibhav Kumar, Ranjan Gangopadhyay, and Soumitra Debnath

Department of ECE, The LNM Institute of Information Technology, Jaipur, India
monikajain20690@gmail.com, vaibhav@lnmiit.ac.in,
{ranjan_iitkgp,soumitra_deb}@yahoo.com

Abstract. The classical energy detection (CED) system is a well-known technique for spectrum sensing in cognitive radio. Generalized p-norm detector for spectrum sensing in additive white Gaussian noise (AWGN) has been shown to provide improved performance over CED under certain conditions. Further, improved algorithm exists which works better than the classical energy detection algorithm. The present paper takes into account the combined benefit of the p-norm energy detector and the improved algorithm for spectrum sensing for individual cognitive user in a cooperative spectrum sensing system to achieve a significant performance gain in both AWGN and generalized κ-μ fading channels over the cooperative/ non-cooperative CED scheme.

Keywords: p-norm energy detector · Energy detection · Cooperative spectrum sensing · Cognitive radio · κ-μ fading channel

1 Introduction

Cognitive radio (CR) is considered as a promising solution to the radio spectrum under-utilization. Spectrum sensing is the key technology that enables the secondary users (SUs) to access the licensed frequency bands without affecting the quality-of-service (QoS) of the primary users (PUs). Various spectrum sensing techniques have been suggested [1,2], which include Energy detector, Matched filtering, Cyclostationary detection etc. Among all these techniques, the classical energy detector (CED) is the most popular because of its low implementation cost and less complexity. However, the performance of the energy detector is limited by high susceptibility of the detection threshold to noise uncertainty and interference level. An improved energy detector (IED) has been proposed [3], which outperforms the CED in AWGN channel with almost same algorithmic complexity without the need for a-priori information about the PU's signal format.

Another interesting improvement strategy for energy detection based on p-norm detector was first proposed by Chen [4], in which the classical energy

© Institute for Computer Sciences, Social Informatics and Telecommunications Engineering 2015
M. Weichold et al. (Eds.): CROWNCOM 2015, LNICST 156, pp. 225–234, 2015.
DOI: 10.1007/978-3-319-24540-9_18

detector was modified by replacing the squaring operation of the signal amplitude by arbitrary positive power p. The optimal p value depends on system
parameter settings viz. the probability of false alarm, the average signal-to noise
ratio, and the sample size in order to achieve a higher probability of detection.
The application of p-norm detector for spectrum sensing in fading channel and
diversity reception has been well investigated recently [5]. The performance of
p-norm detector for cooperative spectrum sensing has been carried out in [6],
where an optimized value of p and sensing threshold of each CR is obtained by
minimizing the total probability of error.

In the present work, we endeavor to evaluate the maximum achievable performance gain in a cooperative sensing system where each individual secondary
user utilizes the combined benefit of both the optimized p-norm detector and the
IED algorithm for spectrum sensing in generalized κ-μ fading channel. It is difficult to obtain analytically the optimized p-value for a given target performance
criterion and therefore a numerical evaluation is adopted.

The organization of the paper is as follows: Section 2 provides the mathematical details of the classical energy detector, improved energy detector, p-norm
energy detector and the improved p-norm energy detector with the derivation of
the performance parameters. The performance of the improved p-norm detector
in a generalized κ-μ fading channel is presented in section 3. Section 4 deals with
the mathematical details of the cooperative spectrum sensing. Section 5 provides
the detailed theoretical results of the improved p-norm energy detector as well
as the practical design guidelines. Finally the conclusion is drawn in section 6.

2   Spectrum Sensing

The spectrum sensing may be modeled as a binary hypothesis testing problem
as:

$$H_0 : y[n] = w[n]$$
$$H_1 : y[n] = h[n].s[n] + w[n]$$
$$(1)$$

where $y[n]$ is the signal sample detected by the secondary user, $s[n]$ is the signal
transmitted by the PU, $h[n]$ represents the channel fading coefficient, and $w[n]$
is a zero-mean additive white Gaussian noise (AWGN) with variance σ_w^2.

The hypotheses H_0 and H_1 correspond to the binary space, representing
the absence and the presence of the PU respectively. In order to analyze the
performance of the sensing scheme, the probability of false alarm, P_{fa}, and the
probability of detection, P_d need to be evaluated. The parameters are defined as
follows:

$$P_{fa} = P\left(H_1|H_0\right)$$
$$P_d = P\left(H_1|H_1\right)$$
$$(2)$$

where $P(\cdot|\cdot)$ denotes the conditional probability. The expression for these prob-
abilities are obtained in the next section.

2.1 Classical Energy Detector (CED)

In CED, if the received signal energy during a sensing event exceeds the predetermined threshold, the channel is considered as busy (H_1 is true), otherwise, the channel is idle (H_0 is true). The decision variable $T_i(y_i)$ at the ith sensing event can be represented as:

$$T_i(y_i) = \frac{1}{N} \sum_{n=1}^{N} \left| \frac{y_i(n)}{\sigma_w} \right|^2 \tag{3}$$

where N is the number of samples per sensing event, $y_i(n)$ is the nth received faded sample at the ith sensing event and σ_w is the standard deviation of the additive white Gaussian noise. The decision rule can be adopted as:

$$\begin{aligned} H_0 : \quad & T_i(y_i) < \lambda \\ H_1 : \quad & T_i(y_i) \geq \lambda \end{aligned} \tag{4}$$

where λ is the decision threshold. For the number of samples $N \gg 1$, the decision variable can be well approximated as a Gaussian distribution [3], i.e.,

$$T_i(y_i) = \begin{cases} \mathcal{N}\left(1, \frac{2}{N}\right) & : H_0 \\ \mathcal{N}\left((1+\gamma), \frac{2}{N}(1+\gamma)^2\right) & : H_1 \end{cases} \tag{5}$$

where $\gamma = \frac{\sigma_s^2}{\sigma_w^2}$ is the signal-to-noise ratio (SNR) of the received signal, σ_s^2 being the signal power. For the AWGN channel, P_{fa}^{CED} and P_d^{CED} can be expressed as [3]:

$$P_{fa}^{CED} = Q\left(\frac{\lambda - 1}{\sqrt{2/N}}\right) \tag{6}$$

$$P_d^{CED} = Q\left(\frac{\lambda - (1+\gamma)}{\sqrt{(2/N)(1+\gamma)^2}}\right) \tag{7}$$

where, $Q(x) = \int_x^{\infty} e^{-t^2} dt$ represents the Gaussian tail probability. From (6), the expression for λ directly follows:

$$\lambda = \sqrt{2/N} Q^{-1}\left(P_{fa}^{CED}\right) + 1 \tag{8}$$

2.2 Improved Energy Detector (IED)

The improved energy detector (IED), proposed in [3], is a modified version of CED, that provides better detection results without much additional complexity.

In IED, the decision for the presence of the primary user is done based on the average of last L test statistics T_i^{avg} at the ith interval, which is defined as:

$$T_i^{avg}(T_i) = \frac{1}{L} \sum_{l=1}^{L} T_{i-L+l}(y_{i-L+l}) \tag{9}$$

Out of these last L sensing events, $M \in [0, L]$ is the total number of events in which the primary signal was actually present. In IED algorithm, two additional checks are imposed to improve the detection probability as well as the probability of false alarm.

If $T_i(y_i) < \lambda$, the first additional check for $T_i^{avg}(T_i) > \lambda$ would improve the detection probability and the second additional check for $T_{i-1}(y_{i-1}) > \lambda$ would prevent the consequential false alarm degradation. Since $T_i^{avg}(T_i)$ is the average of independent and identically distributed Gaussian random variables, it is also normally distributed as:

$$T_i^{avg}(T_i) \sim \mathcal{N}(\mu_{avg}, \sigma_{avg}^2) \tag{10}$$

where, μ_{avg} and σ_{avg}^2 are obtained as [3]:

$$\mu_{avg} = \frac{M}{L}(1 + \gamma) + \frac{L - M}{L}$$
$$\sigma_{avg}^2 = \frac{M}{L^2}\left(\frac{2}{N}(1 + \gamma)^2\right) + \frac{L - M}{L^2}\left(\frac{2}{N}\right) \tag{11}$$

Based on the above assumption, the probability of false alarm, P_{fa}^{IED} and the probability of detection, P_d^{IED} can easily be derived as:

$$P_{fa}^{IED} = P_{fa}^{CED} + P_{fa}^{CED}(1 - P_{fa}^{CED})Q\left(\frac{\lambda_{IED} - \mu_{avg}}{\sigma_{avg}}\right)$$
$$P_d^{IED} = P_d^{CED} + P_d^{CED}(1 - P_d^{CED})Q\left(\frac{\lambda_{IED} - \mu_{avg}}{\sigma_{avg}}\right) \tag{12}$$

where λ_{IED} is the detection threshold in case of IED algorithm, that depends on the probability of false alarm, M, L and γ.

2.3   p-norm Energy Detector

The decision variable for the p-norm detector, $T_i^p(y_i)$ is obtained by modifying (3) as:

$$T_i^p(y_i) = \frac{1}{N} \sum_{n=1}^{N} \left|\frac{y_i(n)}{\sigma_w}\right|^p \tag{13}$$

It may be noted that $p = 2$ in (13) leads to the CED case. The decision statistics may again be well approximated by Gaussian distribution for $N \gg 1$ as follows:

$$T_i^p(y_i) = \begin{cases} \mathcal{N}\left(\mu_{0,p}, \sigma_{0,p}^2\right) & : H_0 \\ \mathcal{N}\left(\mu_{1,p}, \sigma_{1,p}^2\right) & : H_1 \end{cases} \tag{14}$$

where $\mu_{0,p}$ and $\mu_{1,p}$ are the means and $\sigma_{0,p}^2$ and $\sigma_{1,p}^2$ are the variances of the decision variable under the hypotheses H_0 and H_1 respectively. The above parameters are defined as follows [4]:

$$\mu_{0,p} = \frac{2^{p/2}}{\sqrt{\pi}} \Gamma\left(\frac{p+1}{2}\right) \tag{15}$$

$$\mu_{1,p} = \frac{2^{p/2}}{\sqrt{\pi}} \Gamma\left(\frac{p+1}{2}\right) \left(\sqrt{1+\gamma}\right)^p \tag{16}$$

$$\sigma_{0,p}^2 = \frac{2^p \Gamma\left(\frac{2p+1}{2}\right)}{N\sqrt{\pi}} - \frac{2^p}{N\pi} \left\{\Gamma\left(\frac{p+1}{2}\right)\right\}^2 \tag{17}$$

$$\sigma_{1,p}^2 = \left[\frac{2^p \Gamma\left(\frac{2p+1}{2}\right)}{N\sqrt{\pi}} - \frac{2^p}{N\pi} \left\{\Gamma\left(\frac{p+1}{2}\right)\right\}^2\right] (1+\gamma)^p \tag{18}$$

The probability of false alarm, P_{fa}^p and the probability of detection, P_d^p can be calculated as:

$$P_{fa}^p = Q\left(\frac{\lambda_p - \mu_{0,p}}{\sigma_{0,p}}\right)$$

$$P_d^p = Q\left(\frac{\lambda_p - \mu_{1,p}}{\sigma_{1,p}}\right) \tag{19}$$

where λ_p is the detection threshold in case of p-norm detector that depends on the probability of false alarm, $\mu_{0,p}$ and $\sigma_{0,p}$.

2.4 Improved p-norm Energy Detector

By replacing the squaring operation of the signal amplitude in IED by an arbitrary positive power p, $T_i^{avg}(T_i^p)$ may be well approximated by Gaussian distribution as:

$$T_i^{avg}(T_i^p) = \mathcal{N}\left(\mu_{avg,p}, \sigma_{avg,p}^2\right) \tag{20}$$

where $\mu_{avg,p}$ and $\sigma_{avg,p}^2$ being the mean and the variance of the decision variable $T_i^{avg}(T_i^p)$ defined as follows [3]:

$$\mu_{avg,p} = \frac{M}{L}\mu_{1,p} + \frac{L-M}{L}\mu_{0,p}$$

$$\sigma_{avg,p}^2 = \frac{M}{L^2}\sigma_{1,p}^2 + \frac{L-M}{L^2}\sigma_{0,p}^2 \tag{21}$$

Modifying (12) to the present case, one obtains the probability of false alarm, $P_{fa}^{IED,p}$ and the probability of detection, $P_d^{IED,p}$ in the following form:

$$
\begin{aligned}
P_{fa}^{IED,p} &= P_{fa}^p + P_{fa}^p \left(1 - P_{fa}^p\right) Q \left(\frac{\lambda_{IED,p} - \mu_{avg,p}}{\sigma_{avg,p}}\right) \\
P_d^{IED,p} &= P_d^p + P_d^p \left(1 - P_d^p\right) Q \left(\frac{\lambda_{IED,p} - \mu_{avg,p}}{\sigma_{avg,p}}\right)
\end{aligned}
\tag{22}
$$

where $\lambda_{IED,p}$ is the detection threshold for improved p-norm energy detector, which depends on the probability of false alarm

3   Spectrum Sensing over Generalized κ-μ Fading Channel

In case of fading channels, where the channel coefficient $h[n]$ varies, the probability of detection $P_d^{IED,p}$ in (22) gives a conditional probability for a given instantaneous signal-to-noise ratio, γ. To find the detection probability, this conditional probability should be averaged over the probability density function (pdf) of SNR i.e., $f(\gamma)$ [7]:

$$
P_{d_f}^{IED,p} = \int_0^\infty P_d^{IED,p}(\gamma) f(\gamma) d\gamma
\tag{23}
$$

Here, $P_{d_f}^{IED,p}$ represents the detection probability over the fading channel using the improved p-norm energy detection scheme. The integral in (23) is computed using MATLAB. In the following, the κ-μ generalized fading model [8], is described for computational purpose while evaluating (23).

For the κ-μ fading channel, the pdf of SNR is given as [8]:

$$
\begin{aligned}
f_{\kappa-\mu}(\gamma) = &\frac{\mu(1+\kappa)^{\frac{\mu+1}{2}}}{\kappa^{\frac{\mu-1}{2}} \exp[\kappa\mu]\sqrt{\gamma\bar{\gamma}}} \left(\frac{\gamma}{\bar{\gamma}}\right)^{\frac{\mu}{2}} \times \\
&\exp\left[-\mu(1+\kappa)\frac{\gamma}{\bar{\gamma}}\right] I_{\mu-1}\left[2\mu\sqrt{\kappa(1+\kappa)\frac{\gamma}{\bar{\gamma}}}\right]
\end{aligned}
\tag{24}
$$

where, $I_v(.)$ is the modified Bessel function of the first kind of order v. In this distribution, $\kappa(>0)$ represents the ratio between the total power in the dominant component and the total power in the scatter waves; $\mu(>0)$ is related to the multipath clustering and $\bar{\gamma}$ is the average SNR. Table 1 provides the values of κ and μ, for which the κ-μ distribution converges to some well-known wireless channel distributions.

In Table 1, m is the Nakagami shape parameter and K is the Rician-K parameter.

Table 1. Values of κ and μ for different known distributions

Type of distribution	κ	μ
Nakagami-m	$\to 0$	m
Rayleigh	$\to 0$	1
Rician	K	1
One sided Gaussian	$\to 0$	0.5

4   Cooperative Spectrum Sensing

In cooperative spectrum sensing there are multiple secondary users (SU), each SU sends its autonomous decision to a fusion center (FC) and the final decision about the presence of primary user (PU) is done at FC. In this work we have assumed that all the SUs behave identically (regarding SNR and threshold). Furthermore, we focus on the use of hard-decision based fusion rules e.g. OR, MAJORITY, and AND rules in the analysis. Since the binary decisions (H_0 or H_1) of all SUs are independent, the probability of detection in a cooperative scenario can be represented by [1]:

$$P_d^{coop} = \sum_{j=n}^{U} \binom{U}{j} \left(P_{d,i}^x\right)^j \left(1 - P_{d,i}^x\right)^{U-j} \tag{25}$$

where $P_{d,i}^x$ is the probability of detection for i^{th} individual node and U is the total number of SUs. In case of AWGN channel, $P_{d,i}^x = P_{d,i}^{IED,p}$ and for fading channel $P_{d,i}^x = P_{d_f,i}^{IED,p}$. Considering the optimal fusion rule i.e., OR [1], the probability of detection in a co-operative scenario can be evaluated by putting $n-1$. The expression for P_d^{IED} under OR fusion rule, therefore follows in a straightforward manner as:

$$P_{d,i,OR}^x = 1 - \left(1 - P_{d,i}^x\right)^U \tag{26}$$

The performance results of both, conventional and co-operative spectrum sensing in AWGN and fading channels are presented in the following section.

5   Results and Discussion

In this section, the results for the combined benefit of the improved algorithm as well as the p-norm detector are highlighted for cooperative spectrum sensing in κ-μ fading channel. For a given target false alarm probability, the threshold, λ is chosen for individual sensor node and for a given SNR, γ the optimal value of p is determined which yields the highest value of the probability of detection. The hard decision from the individual sensor node is sent to the FC, which combines the individual decisions using the OR rule.

To provide practical design guidelines for a spectrum sensing system with improved p-norm energy detector, Fig. 1 provides the surface plots for the probability of detection with the variation of p and the probability of false alarm for AWGN and κ-μ fading channel scenarios respectively for cooperative spectrum sensing, each for $N = 100$ and $\bar{\gamma} = -5$ dB. It is quite evident that p has a definitive role in order to achieve a higher detection probability in both AWGN and fading channel.

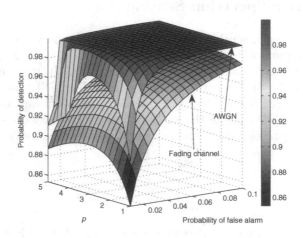

Fig. 1. Surface plot for the probability of detection as a function of p and probability of false alarm for IED over AWGN and κ-μ ($\kappa \to 0$, $\mu = 1$) fading for cooperative spectrum sensing with $N = 100$, $\bar{\gamma} = -5$ dB, $U = 4$.

In Fig. 2, a comparison of the receiver operating characteristics (ROCs) for CED and improved p-norm algorithm i.e., IED with optimal p value has been depicted for cooperative as well as non-cooperative spectrum sensing in the κ-μ channel ($\kappa \to 0$, $\mu = 1$). The optimal p-value is determined from the surface plot in Fig. 1, such that for a given $P_{fa,target}$ and $\bar{\gamma}$, the probability of detection becomes maximum. It is clearly evident that the IED with optimal p, outperforms the CED, in fading scenarios for both cooperative and non-cooperative spectrum sensing. As the probability of false alarm increases, the difference in the performance gain of IED with optimal p decreases for cooperative spectrum sensing, but the algorithm still retains its superiority in performance over CED.

In Fig. 3, the variation of the probability of detection against SNR at a fixed target false alarm probability of 10^{-3} has been shown for cooperative scenario as well as single user case. The optimal value of p has been determined in the same manner as in Fig. 2.

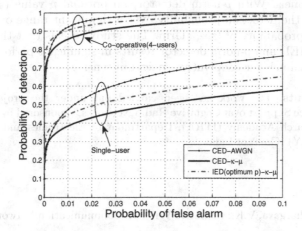

Fig. 2. ROCs for classical and improved p-norm energy detector over AWGN and κ-μ ($\kappa \rightarrow 0$, $\mu = 1$) fading channels in single user and cooperative scenarios with $N = 100$, $\overline{\gamma} = -5$ dB and $U = 4$.

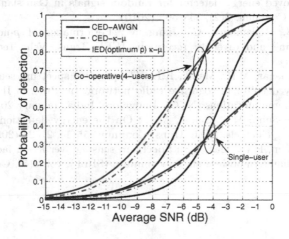

Fig. 3. Probability of detection as a variation of SNR for improved p-norm energy detector over κ-μ ($\kappa \rightarrow 0$, $\mu = 1$) fading in single user and cooperative scenario with $N = 100$, $P_{fa,target} = 10^{-3}$ and $U = 4$.

6 Conclusion

We have analyzed the sensing performance of an improved energy detector with the optimal p-norm value in a generalized κ-μ fading channel for cooperative spectrum sensing. The performance gain depends upon the various system design parameters e.g., SNR, the probability of false alarm and the number of samples per sensing event N. The IED algorithm outperforms CED in both AWGN and fading channels. With p-norm detector, an optimal p value ($\neq 2$) exists, that maximizes the detection probability over a significant range of SNRs with lower values of probability of false alarm. The study reveals that the combined benefit of both IED and p-norm detector results in significant performance gain in κ-μ fading channels for cooperative spectrum sensing.

Acknowledgments. The work has been carried out under the project, "Mobile Broadband Service Support over Cognitive Radio Networks," sponsored by Information Technology Research Academy (ITRA), Department of Electronics and Information Technology (DeitY), Govt. of India.

References

1. Hossain, E., Bhargava, V.K.: Cognitive Wireless Communication Networks. Springer (2007)
2. Wyglinski, A.M., Nekovee, M., Hou, T.: Cognitive Radio Communication and Networks. Academic Press (2010)
3. Lpez-Bentez, M., Casadevall, F.: Improved energy detection spectrum sensing for cognitive radio. IET Commun. **6**(8), 785–796 (2012)
4. Chen, Y.: Improved energy detector for random signals in Gaussian noise. IEEE Trans. Wireless Commun. **9**(2), 558–563 (2010)
5. Banjade, V.R.S., Tellambura, C., Jiang, V.: Performance of p-norm detector in AWGN, fading and diversity reception. IEEE Trans. Veh. Technol. **63**(7), 3209–3222 (2014)
6. Singh, A., Bhatnagar, M.R., Mallik, R.K.: Cooperative spectrum sensing with an improved energy detector in cognitive radio network. In: Proc. of IEEE National Conference on Communications (NCC), Bangalore, India, January 2011
7. Digham, F.F., Alouini, M.S., Simon, M.K.: On the energy detection of unknown signals over fading channels. IEEE Trans. Commun. **55**(1), 21–24 (2007)
8. Sanders, D., Glehn, F.V., Dias, U.S.: Spectrum sensing over $\kappa - \mu$ fading channel. In: XXIX Brazilian Symposium on Telecommun. (SBrT), vol. 2(5), October 2011

Kalman Filter Enhanced Parametric Classifiers for Spectrum Sensing Under Flat Fading Channels

Olusegun P. Awe$^{(\boxtimes)}$, Syed M. Naqvi, and Sangarapillai Lambotharan

Advanced Signal Processing Group, Loughborough University, Loughborough, UK
{o.p.awe,s.m.r.naqvi,s.lambotharan}@lboro.ac.uk

Abstract. In this paper we propose and investigate a novel technique to enhance the performance of parametric classifiers for cognitive radio spectrum sensing application under slowly fading Rayleigh channel conditions. While trained conventional parametric classifiers such as the one based on K-means are capable of generating excellent decision boundary for data classification, their performance could degrade severely when deployed under time varying channel conditions due to mobility of secondary users in the presence of scatterers. To address this problem we consider the use of Kalman filter based channel estimation technique for tracking the temporally correlated slow fading channel and aiding the classifiers to update the decision boundary in real time. The performance of the enhanced classifiers is quantified in terms of average probabilities of detection and false alarm. Under this operating condition and with the use of a few collaborating secondary devices, the proposed scheme is found to exhibit significant performance improvement with minimal cooperation overhead.

Keywords: Cognitive radio · spectrum sensing · Kalman filter · machine learning · fading channels

1 Introduction

Cognitive radio (CR) is an enabling technology for dynamic spectrum access that will form an integral part of future wireless devices [1]. A core requirement for the successful implementation of CR system is spectrum sensing. It enables CR devices to detect the presence or absence of primary user's (PU) signal in the licensed frequency bands so that secondary users (unlicensed users) can opportunistically utilize these frequency bands in a manner that no disruptive interference is caused to the PU's transmissions [1], [2]. Over the last one decade, several techniques have been proposed for performing spectrum sensing in CR systems, the most widely used of which are the energy detection, matched filtering and cyclostationary detector schemes [3].

In the energy detection method, during sensing interval the secondary user (SU) computes the accumulated energy of the signal received at its terminal

© Institute for Computer Sciences, Social Informatics and Telecommunications Engineering 2015
M. Weichold et al. (Eds.): CROWNCOM 2015, LNICST 156, pp. 235–247, 2015.
DOI: 10.1007/978-3-319-24540-9_19

within the band of interest and compares the result to the measurement obtained for the ambient noise. Although the technique has great potential for ubiquitous applications due to its relative simplicity and capability for blind signal detection, its performance often degrade when there is uncertainty about the actual ambient noise power [2], [3]. Matched filtering technique is implemented by correlating the received signal and known PU signal and the outcome is used as a basis for deciding whether the PU signal is present or not [4]. The scheme is capable of yielding excellent detection performance when the waveform of the PU signal is known apriori and there is perfect knowledge of the PU-SU channel. Evidently, this technique can not be successfully used in an alien radio frequency environment where a priori knowledge of the PU signal is lacking. Another constraint on the use of the method is that the PU and SU must be perfectly synchronized during sensing time which would be difficult to achieve especially when the signal-to-noise ratio (SNR) of the PU-SU channel is low. Cyclostationary detectors are built to take advantage of known, unique features that are usually present in transmitted PU signals (e.g. cyclic prefix, spreading codes, modulated carriers, etc.) which are repetitive in nature by using them as signatures for detecting the PU signal's presence or absence. Although the technique offers robust performance in the presence of noise uncertainty and low SNR, its use is limited to applications where the signal characteristics of the PU is known apriori which makes it impractical for use in scenarios involving frequency re-use. It also requires long sensing time and is characterized by high computational complexity [2].

In fairly recent times, machine learning (ML) techniques have gained attention for solving the spectrum sensing problem and the performance evaluation of this approach is found to be better than most of the traditional sensing methods. For example in [5], semi-supervised parametric classifiers based on the K-means clustering and expectation-maximization (EM) algorithms are proposed. The supervised support vector machine (SVM) binary classifier and K-nearest neighbour (kNN) techniques are also proposed in [4] and [5] while the unsupervised variational Bayesian (VB) learning based method is presented in [6]. The general idea behind all these ML schemes is to train the SUs by using features that are derived from the traffic pattern in the licensed band of interest so that if the SUs are operated in an environment similar to the one captured by the training features, the SUs can use the knowledge of the traffic pattern that has been acquired to distinguish between when the channel is being occupied by the PU and otherwise [7]. It should be noted, however, that in practical scenarios for the deployment of CRs, the SU or PU may be physically mobile and as such the channel conditions characterizing the training and operating environments may differ, thereby the CR might fail to reliably and efficiently detect the true status of the PU's activities under monitoring.

To the best of our knowledge, the deployment of parametric classifiers by mobile SUs for spectrum sensing purpose under time varying channel conditions has not been considered in the literature. In this paper, we investigate the performance of SUs that depend on these classifiers for spectrum sensing under flat fading channels and propose a novel Kalman filter channel estimation based technique for enhancing their performance under these practical operating conditions.

The rest of the paper is as organized as follows. The problem statement is presented in section 2. In section 3, we describe the system model, assumptions and proposed algorithms. The simulation results obtained are discussed in section 4 followed by conclusion in section 5.

2   Problem Statement

A spectrum sensing network consisting of a fixed PU transmitter (PU-TX), a collaborating sensor node (CSN) co-locating with the PU, N PU receivers (PU-RX), a secondary base station (SBS) which plays the role of a data clustering center as well as the SUs' coordinator and M SUs as illustrated in Fig. 1 is considered. It is assumed that the PU's activity is such that it switches alternately between active and inactive states allowing the SUs to be able to opportunistically use its dedicated frequency band and operate within the PU's coverage area. During the training phase, all SUs sense the energy of the PU-SU channel at their respective locations during both states and report it to the SBS where clustering is performed and appropriate decision boundary is generated. It is assumed that the training data from individual SU is independent but identically distributed.

Let us suppose that based on the decision boundary that is generated from the training data, the PU has been declared to be inactive while all the SUs are stationary. Consider also that SU-c3 that is initially at point 'A' is using the PU's band while having to transit to point 'B' as shown. We assume that the channel condition characterizing the SU's trajectory is flat fading (e.g. traveling through a heavily built-up urban environment). This description equally applies where multiple mobile SUs share the PU's band and are able to cooperate. Since the training process of a learning technique normally takes a long time, under this scenario it is impractical for the mobile SU(s) to undergo re-training while in motion owing to the dynamic nature of the channel gain and if sensing information is exchanged among SUs, it could be received incorrectly due to the channel fading and noise resulting in performance loss [8],[9]. In addition, significant amount of energy and other resources are required to communicate sensing results periodically to other users and in a bid to conserve resources, SUs may prefer not to share their results [10]. To be able to detect the status of the PU activities correctly and efficiently, the onus is therefore on the individual mobile SU as it travels to cater to making well informed decision by dynamically adjusting its decision boundary at the SBS in a manner that the changes in channel conditions are taken into consideration, doing so with minimal cooperation overhead.

To address this challenge, in this paper we propose a framework whereby each SU incorporates a channel tracking sub-system that is based on the Kalman filtering algorithm which enables the SU to obtain an online, unbiased estimate of the true channel gain as it travels. The estimated channel gain can then be used to generate energy features for updating its decision boundary in real time. To investigate the capability of the proposed scheme, without loss of generality, we adopt the energy vectors based K-means clustering platform earlier proposed in [5] due to its simplicity.

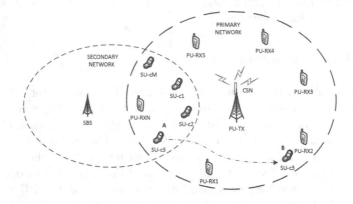

Fig. 1. A spectrum sensing system of a primary user and mobile secondary users networks.

3   System Model, Assumptions and Algorithms

Consider that the PU transmitter is located at a coordinate $\mathbf{x}_{pu}$ as shown in Fig. 1 and the mobile SU of interest SU-c3, is located initially at $\mathbf{x}_{su}^m$. During the training period, all SUs carry out sensing of the PU's channel at their respective locations and collectively report the estimated energy to the SBS where K-means clustering is performed and the cluster centroids are computed. The jointly reported sensing data can be used to obtain a high-dimensional decision plane at the SBS and can enable immobile SUs to be able to take advantage of space diversity and contain hidden node problem. Prior to SU-c3 being in motion, let $\phi(\mathbf{x}_{su}^m, n)$ represent the channel gain between the PU-TX and SU-c3 at a time instant n. Given that the PU signals are statistically independent, an estimate of the discrete-time signal received at the SU-c3 terminal can be written as

$$x_m(n) = \begin{cases} s(n)\phi(\mathbf{x}_{su}^m, n) + \eta_m(n), & H_1 : PU\ present \\ \eta_m(n), & H_0 : PU\ absent \end{cases} \tag{1}$$

where the channel coefficient $\phi(\mathbf{x}_{su}^m, n)$ is assumed to be zero-mean, unit-variance complex Gaussian random variable whose magnitude squared is the power attenuation $P_{\mathbf{x}_{pu}\to\mathbf{x}_{su}^m}^{att}$, between PU-TX and SU-c3 which can be described by

$$P_{\mathbf{x}_{pu}\to\mathbf{x}_{su}^m}^{att} = |\phi(\mathbf{x}_{su}^m, n)|^2$$
$$= L_p(\|\mathbf{x}_{pu} - \mathbf{x}_{su}^m\|_2) \cdot \delta_{\mathbf{x}_{pu}\to\mathbf{x}_{su}^m} \cdot \gamma_{\mathbf{x}_{pu}\to\mathbf{x}_{su}^m}, \tag{2}$$

where $\| \cdot \|_2$ implies Euclidean norm, $L_p(\rho) = \rho^{-d}$ is the path loss component over distance ρ, d is the path loss exponent, $\delta_{\mathbf{x}_{pu} \to \mathbf{x}_{su}^m}$ is the shadow fading component and $\gamma_{\mathbf{x}_{pu} \to \mathbf{x}_{su}^m}$ represents the small scale fading factors. The remaining parameters in (1) are $s(n)$ which is the instantaneous PU signal assumed to be circularly symmetric complex Gaussian with mean zero and variance $\mathbb{E}|s(n)|^2 = \sigma_s^2$ and $\eta_m(n)$, which is assumed to be an independent and identically distributed complex zero-mean Gaussian noise with variance $\mathbb{E}|\eta_m(n)|^2 = \sigma_\eta^2$. Throughout this consideration, the shadow fading effect is assumed to be quasi-static and the channel gain, $\phi(\mathbf{x}_{su}^m, n)$ is assumed to be time-invariant while SU-c3 is stationary at point 'A' during training and becomes a fading process as it transits from point 'A' at coordinate $\mathbf{x}_{su}^m$ to point 'B' at coordinate $\mathbf{x}_{su}^j$. We further assume that in order to reduce cooperation overhead, although the traveling SU is to be aided by the SBS and other collaborating device within the network, it is primarily responsible for the continuous monitoring of the PU's activities while using the PU's band and would vacate the band immediately when the PU becomes active.

3.1   Energy Vectors Realization for Secondary Users Training

During the training interval, given that the PU operates at a carrier frequency f_c and bandwidth ω, if the transmitted PU signal is sampled at the rate of f_s by each SU, the energy samples sent to the SBS for training purpose can be estimated as [11]

$$\psi_i = \frac{1}{N_s} \sum_{n=1}^{N_s} |x_m(n)|^2 \tag{3}$$

where $n = 1, 2, ..., N_s$ and $N_s = \tau f_s$ is the number of samples of the received PU signal used for computing the training energy sample at the SU while τ is the duration of sensing time for each energy sample realization. When the PU is idle, the probability density function (PDF) of ψ_i follows Chi-square distribution with $2N_s$ degrees of freedom and when N_s is large enough (say, $N_s \simeq 250$) [12], this PDF can be approximated as Gaussian through the central limit theorem with mean, $\mu_0 = \sigma_\eta^2$ and variance, $\sigma_0^2 = \frac{1}{N_s}[\mathbb{E}|\eta(n)|^4 - \sigma_\eta^4]$. However, for an additive white Gaussian noise, $\mathbb{E}|\eta(n)|^4 = 2\sigma_\eta^4$ so that we have $\sigma_0^2 = \frac{1}{N_s}\sigma_\eta^4$. Similarly, when the PU is active, the distribution of ψ_i can be approximated as Gaussian with mean, $\mu_1 = |\phi(\mathbf{x}_{su}^m, n)|^2\sigma_s^2 + \sigma_\eta^2$ and variance, $\sigma_1^2 = \frac{1}{N_s}[|\phi(\mathbf{x}_{su}^m, n)|^4\mathbb{E}|s(n)|^4 + \mathbb{E}|\eta(n)|^4 - (|\phi(\mathbf{x}_{su}^m, n)|^2\sigma_s^2 - \sigma_\eta^2)^2]$.

Let $\boldsymbol{\Psi} = \{\psi_1, ..., \psi_{\mathcal{L}}\}$ be the set of training energy vectors obtained at the SBS during the training period where $\psi_l \in R^q$, and q is the dimension of each training energy vector which corresponds to the number of collaborating SUs and antenna per SU. If $\boldsymbol{\Psi} \in \{H_0, H_1\}$ is fed into the parametric classifier, the output of the classifier is the cluster centroids (means) that can be used to generate the decision boundary which optimally separates the two clusters, H_0, H_1. This decision boundary can then be used for the classification of new data points when the classifier is deployed in an environment similar to where it has been trained given any desired false alarm probability. A simple K-means clustering algorithm for computing the cluster centroids at the SBS is shown in Algorithm 1. However, in the realistic deployment scenario under consideration involving a mobile SU which travels through a fading channel environment where frequent re-training is impractical, relying on the hitherto, optimal decision threshold obtained at the initial point of training would result in detection error. Therefore, in order to achieve high probability of detection and low false alarm, the cluster centroids computed at the SBS have to be continuously updated and the decision boundary adjusted correspondingly.

Algorithm 1. K-means Clustering Algorithm for CR Spectrum Sensing

---

1. $\forall\, m = 1, ..., M$, initialize cluster centroids
   $C_1, ..., C_K, \forall\, k = 1, ..., K$ given $\boldsymbol{\Psi}$, K.
2. **do repeat**
3.   **for** $k \leftarrow 1$ to K
4.   **do** $D_k \leftarrow \{\ \}$
5.   **for** $l \leftarrow 1$ to $\mathcal{L}$
6.   **do** $i \leftarrow argmin_i \|C_i - \psi_l\|^2$
7. $D_k \leftarrow D_k \cup \{\psi_l\}$
8.   **do** $C_k \leftarrow |D_k|^{-1} \sum_{\psi_l \in D_k} \psi_l, \forall\, k$
9. **until convergence**
10. $C_{H0} \leftarrow min\{|C_1|, ..., |C_K|\}$

3.2 Tracking Decision Boundary Using Kalman Filter Based Channel Estimation

In order to be able to track the changes in the cluster centroids under slow fading channel condition due to the mobility of the SU, we introduce the Kalman filtering technique to enable the mobile SU to obtain an online, unbiased estimate of the temporally correlated fading channel gain. Since the PU is assumed to be alternating between the active and inactive states, a collaborating sensor node (CSN) that is co-locating with the PU is activated during the SU's travel period. The sensor node's duty is to broadcast a signal known to the SUs (e.g. pilot signal) periodically during the PU's idle interval for the benefit of the mobile SUs to enable centroid update and avoid causing harmful interference to the PU's service. The role of the CSN in the proximity of the PU is similar to that of the *helper node* used for authenticating the PU's signal in [13] and the rationale behind incorporating a sensor node co-locating with the PU is to ensure that the channel between the PU and the mobile SU is captured by the CSN-to-mobile SU channel. It should be noted that our model is equally applicable in the case where there are multiple and/or mobile PUs and can accommodate any other collaborating sensor node selection method. The mobile SU on the other hand makes a prediction of the dynamic channel gain based on its speed of travel and combines this prediction with the noisy observation from the collaborating node via the Kalman filtering algorithm to obtain an unbiased estimate of the true channel gain.

Let the discrete-time observation at the mobile SU terminal due to the transmitted signal by the CSN be described by

$$z(t) = s(t)\phi(t) + \varrho(t) \tag{4}$$

where $s(t)$ is a known pilot signal, $\varrho(t)$ is a zero mean complex additive white Gaussian noise at the receiver with variance, σ_ϱ^2 and $\phi(t)$ is a zero mean circularly complex Gaussian channel gain with variance σ_ϕ^2, t is the symbol time index. If we let T_s be the symbol period of the pilot signal, the normalized Doppler frequency of the fading channel is $f_d T_s$ where f_d is the maximum Doppler frequency in Hertz defined by $f_d = \frac{v}{\lambda}$, v is the speed of the mobile and λ is the wavelength of the received signal. The magnitude of the instantaneous channel gain, $|\phi|$ is a random variable whose PDF is described by

$$p_\phi(\phi) = \frac{2\phi}{\nu} \exp(\frac{-\phi^2}{\nu}), \phi \geq 0 \tag{5}$$

where ϕ is the fading amplitude and $\nu = \overline{\phi^2}$ is its mean square value. Furthermore, the phase of $\phi(t)$ is assumed to be uniformly distributed between 0 and 2π. It should be noted, though, that by virtue of the location of CSN in the network, it is assumed that $\phi(t)$ also captures the channel gain between the PU-TX and SU-c3 during every observation interval. For the flat fading Rayleigh channel, the following Jake's Doppler spectrum is often assumed

$$S_\phi(f) = \begin{cases} \frac{1}{\pi f_d \sqrt{1-(f/f_d)^2}}, & |f| \leq f_d \\ 0, & |f| > f_d \end{cases} \tag{6}$$

where f is the frequency shift relative to the carrier frequency. The corresponding autocorrelation coefficient of the observation signal, $z(t)$ under this channel condition is given by [14]

$$R_\phi(\epsilon) = \mathbb{E}[\phi(\kappa) \cdot \phi^*(\kappa - \epsilon)]$$
$$= \sigma_\phi^2 J_0(2\pi f_d \epsilon) \tag{7}$$

for lag ϵ where $J_0(\cdot)$ is the zeroth order Bessel function of the first kind. It should be noted that in the actual deployment for cognitive radio, the idle time of the PU is long enough so that it is possible to periodically obtain the noisy observation (measurement) of the channel gain, $z(t)$ during the PU's idle time [15]. The mobile SU can apply the Kalman filter algorithm described in section 3.3 to obtain an unbiased estimate $\hat{\phi}$ of the true fading channel gain ϕ which can then be used to update the cluster centroids at the SBS and also for tracking the temporally dynamic optimal decision boundary. Since our target is to use the Kalman filtering to realize the best estimate $\tilde{\phi}$ of ϕ, a prediction of the dynamic evolution of the channel gain is required in addition to the noisy observation $z(t)$. For simplicity, we propose to use a first order autoregressive model $(AR-1)$ which has been shown to be sufficient to capture most of the channel tap dynamics in Kalman filter based channel tracking related problems [14]. It should be noted too that the $AR - 1$ model is widely acceptable as an approximation to the Rayleigh fading channel with Jake's Doppler spectrum [16], [17]. The $AR - 1$ model for approximating the magnitude of time varying complex channel gain can be expressed as

$$\phi_t^{AR-1} = \alpha \cdot \phi_{t-1}^{AR-1} + \zeta(t) \tag{8}$$

where t is the symbol index, $0 < \alpha < 1$ and $\zeta(t)$ is complex additive white Gaussian noise with variance $\sigma_\zeta^2 = (1 - \alpha^2)\sigma_\phi^2$. When $\alpha = 1$, the $AR - 1$ model for the dynamic evolution of ϕ in (8) becomes a random walk model [14]. One way of obtaining the coefficient of the $AR - 1$ model, α expressed as

$$\alpha = \frac{R_\phi^{AR-1}[1]}{R_\phi^{AR-1}[0]} \tag{9}$$

is by using correlation matching criterion whereby the autocorrelation function of the temporally correlated fading channel is matched with the autocorrelation function of the approximating AR model for lags 0 and 1 such that $R_\phi^{AR-1}[0] = R_\phi[0]$ and $R_\phi^{AR-1}[1] = R_\phi[1]$. However, if the evolution of the dynamic channel gain is modeled by a higher order AR process, the required coefficients can be obtained by solving the Yule-Walker set of equations [17].

Remarks: The optimal estimate of the channel gain that is obtained via the Kalman filter is sufficient to enable the mobile SU avoid frequent and total dependence on the CSN or other SUs for information regarding the status of PU-TX and the associated overhead.

3.3   Kalman Filtering Channel Estimation Process

At this point having obtained α, we combine the observation equation (4) and state evolution equation (8) to form a Kalman filter set of equations as [18]

$$\hat{\phi}_{t|t-1} = \alpha\hat{\phi}_{t-1|t-1} \tag{10}$$

$$M_{t|t-1} = \alpha^2 M_{t-1|t-1} + \sigma_\zeta^2 \tag{11}$$

$$K_t = \frac{M_{t|t-1}}{M_{t|t-1} + \sigma_\varrho^2} \tag{12}$$

$$\hat{\phi}_{t|t} = \hat{\phi}_{t|t-1} + K_t(z(t) - \alpha\hat{\phi}_{t|t-1}) \tag{13}$$

$$M_{t|t} = (1 - K_t)M_{t|t-1} \tag{14}$$

where K_t is the Kalman gain , $M_{t|t}$ is the variance of the prediction error and $\hat{\phi}_{t|t}$ is the desired optimal estimate of ϕ_t. It is pertinent to mention here that in the rare event that the PU is active for an unexpectedly prolonged period of time so that it becomes impossible to obtain an observation, the situation can be treated as missing observation. Suppose this occurs at a time t, the Kalman filtering prediction step described by (10) and (11) remains the same while the correction step in (13) and (14) will become

$$\hat{\phi}_{t|t} = \hat{\phi}_{t|t-1} \tag{15}$$

$$M_{t|t} = M_{t|t-1} \tag{16}$$

and if the period of missing observation is extremely prolonged, the significance on the detection of PU status is that the mobile SU loses its ability to track the fading channel for that period so that the only effect taken into consideration is the path loss. Consequently, it could be seen that even under this situation the proposed scheme does not perform worse than the alternative where the channel tracking is not considered (path loss only model). A simple algorithm for implementing the proposed enhanced classifier is presented in Algorithm 2.

Algorithm 2. Kalman Filter Enhanced Parametric Classifier based Spectrum Sensing Algorithm

---
1. Generate cluster centroids, $C_k \; \forall \; k = 1, ..., K$ at the SBS using Algorithm 1.
2. Initialize parameters α, $M_{t-1|t-1}$ and σ_ζ^2 at the SUs.
3. **if** SU begins motion, $t \leftarrow 1$
4. **repeat**
5. SU obtains $z(t)$ in (4) during PU's idle interval and computes $\hat{\phi}_{t|t}$ and $M_{t|t}$ using (10) - (14).
6. Compute new energy samples at SU using $\hat{\phi}_{t|t}$ in step 5 and update cluster centroids at the SBS.
7. Use updated centroids from step 6 to decide the PU status, H_0 or H_1.
8. $t \leftarrow t + 1$
9. **until** SU ends motion
10. **end if**

---

4   Simulation Results and Discussion

For simulation purpose, the average power of the fading process is normalized to unity and the mobile SU under consideration (SU-c3) is assumed to be equipped with an omnidirectional antenna while traveling at a constant velocity of 6 km/hr. We considered a single PU which operates alternately in the active and inactive modes, so that the number of clusters, K is 2. The symbol frequency of the PU is 10 ksymbol/s transmitted at the central carrier frequency of 1.8 GHz. As the SU travels, to model the effects of the scatterers, we assume that a total of 128 equal strength rays at uniformly distributed angles of arrival impinge on the receiving antenna, so we have a normalized Doppler frequency of 1e-3. During training the path loss exponent, d is assumed equals to 3 while the shadow

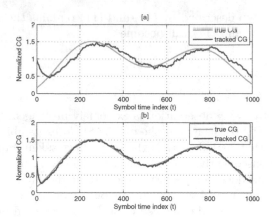

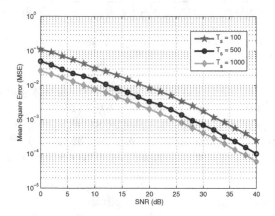

Fig. 2. Time varying channel gain (CG) tracked at [a] $SNR = 5\ dB$ and [b] $SNR = 20\ dB$.

Fig. 3. Mean square error performance of the AR-1 based Kalman filter at normalized Doppler frequency $= 1e\text{-}3$, tracking duration, $T_s = 100$, 500 and 1000 symbols.

fading component $\delta_{\mathbf{x}_{pu} \to \mathbf{x}_{su}^m}$ and the small scale fading factor, $\gamma_{\mathbf{x}_{pu} \to \mathbf{x}_{su}^m}$ are both assumed equal to 1, the PU signal is BPSK and transmit power is 1 Watt. The training energy samples at the SUs are computed using $N_s = 1000$. When SU is in motion, the waveform of the temporally correlated Rayleigh fading process to be tracked is generated using the modified Jake's model described in [19]. To test the enhanced classifier, we assume the mobile SU-c3's trajectory is at an approximately constant average distance to PU-TX throughout the duration of travel and energy samples for updating the centroids are computed using $N_s = 1000$.

In Fig. 2, we show the ability of the Kalman filter in tracking the true channel gain when the pilot signals are received from the CSN at SNR of 5 dB and 20

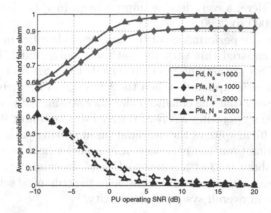

Fig. 4. Average probabilities of detection and false alarm vs SNR, tracking $SNR = 5$ dB, number of samples, $N_s = 1000$ and 2000, tracking duration = 1000 symbols.

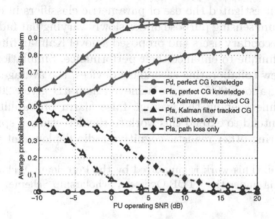

Fig. 5. Average probabilities of detection and false alarm vs SNR, tracking $SNR = 5$ dB, number of samples, $N_s = 2000$, tracking duration = 1000 symbols.

dB respectively over an observation window of 1000 symbol duration. It could be seen that as the pilot's SNR is increased, the performance of the tracker also improves. The mean square error performance of the AR-1 based Kalman filter is shown in Fig. 3 at normalized Doppler frequency of 1e-3 where at the same SNR the tracking error reduces for different duration of tracked pilot symbols (tracking duration). This shows that the longer the tracking duration the better the overall performance of the tracker. It is also seen that the average error reduces from 5e-2 to 16e-5 with increase in tracking SNR from 0 dB to 40 dB when the tracking duration, $T_s = 1000$. The effect of the number of PU's signal samples, N_s used for computing the energy features for training, tracking and testing on the average probabilities of detection (Pd_{Av}) and false alarm (Pfa_{Av})

is shown in Fig. 4. Here, a considerable improvement in Pd_{Av} is observed as N_s is increased from 1000 to 2000.

In Fig. 5, we show the performance of the enhanced classifier in terms of Pd_{Av} and Pfa_{Av} and compared this with the path loss only model. Here, the pilot symbols from the CSN are assumed to be received at the SNR of 5 dB each time the decision boundary is updated. When the PU's signal is received at SNR of 20 dB, it could be seen that the enhanced classifier attains Pd_{Av} of unity at zero Pfa_{Av} while at PU's operating SNR of 0 dB, Pd_{Av} of about 0.91 is achieved at Pfa_{Av} equals 0.07 in spite of the degradation in sensing path. This is in contrast to what obtains from the path loss only model where at the SNR of 20 dB, Pd_{Av} is only about 0.83 at a non-zero Pfa_{Av}. In summary, a performance improvement of about 20 percent is observable in the enhanced scheme with only very slight increase in overall system's complexity.

5   Conclusion

In this paper, we investigated the use of parametric classifiers in cognitive radio network for spectrum sensing purpose under slowly varying flat fading conditions involving mobile secondary users and proposed a novel Kalman filter based channel estimation technique to enhance their performance. Simulation results show that by utilizing few collaborating secondary devices, the average correct detection probability of about 90% can be achieved at 0 dB given 2000 samples of the PU signal, while keeping the average false alarm probability below 10%. In the future, we intend to extend this work to the detection of spatial spectrum hole while considering other realistic cognitive radio deployment scenarios.

Acknowledgments. This work is supported by the Petroleum Technology Development Fund of Nigeria under the PTDF Overseas Scholarship Scheme.

References

1. Haykin, S.: Cognitive Radio: Brain-empowered Wireless Communications. IEEE J. Sel. Areas Commun. **23**(2), 201–220 (2005)
2. Akyildiz, I.F., Lo, B.F., Balakrishnan, R.: Cooperative Spectrum Sensing in Cognitive Radio Networks: A Survey. Phys. Commun. J. **4**(1), 40–62 (2011)
3. Yucek, T., Huseyin, A.: A Survey of Spectrum Sensing Algorithms for Cognitive Radio Applications. IEEE Commun. Surv. Tut. **11**(1), 116–130 (2009)
4. Awe, O.P., Zhu, Z., Lambotharan, S.: Eigenvalue and support vector machine techniques for spectrum sensing in cognitive radio networks. In: Conf. Technol. Appl. Artif. Intell., Taipei, Taiwan, pp. 223–227 (2013)
5. Thilina, K., Saquib, N., Hossain, E.: Machine Learning Techniques for Cooperative Spectrum Sensing in Cognitive Radio Networks. IEEE J. Sel. Areas Commun. **31**(11), 2209–2221 (2013)
6. Awe, O.P., Naqvi, S.M., Lambotharan, S.: Variational bayesian learning technique for spectrum sensing in cognitive radio networks. In: 2nd IEEE Glob. Conf. Signal Inf. Process., Atlanta, GA, USA, pp. 1353–1357 (2014)

7. Bkassiny, M., Li, Y., Jayaweera, S.K.: A Survey on Machine Learning Techniques in Cognitive Radios. IEEE Commun. Surv. Tut. **15**(3), 1136–1159 (2013)
8. Biao, C., Ruixiang, J., Kasetkasem, T., Varshney, P.K.: Fusion of decisions transmitted over fading channels in wireless sensor networks. In: Conf. Rec. Thirty-Sixth Asilomar Conf. Signals, Syst. Comput., Pacific Grove, CA, USA, pp. 1184–1188 (2002)
9. Wang, T., Song, L., Han, Z., Saad, W.: Overlapping coalitional games for collaborative sensing in cognitive radio networks. In: IEEE Wirel. Commun. Netw. Conf., Shanghai, China, pp. 4118–4123 (2013)
10. Kondareddy, Y., Agrawal, P.: Enforcing cooperative spectrum sensing in cognitive radio networks. In: IEEE Glob. Tel. Conf., Houston, USA, pp. 1–6 (2011)
11. Liang, Y., Zeng, Y.: Sensing-Throughput Tradeoff for Cognitive Radio Networks. IEEE Trans. Wirel. Commun. **7**(4), 1326–1337 (2008)
12. Urkowitz, H.: Energy Detection of Unknown Deterministic Signals. Proc. IEEE **55**(4), 523–531 (1967)
13. Liu, Y., Ning, P., Dai, H.: Authenticating primary users' signals in cognitive radio networks via integrated cryptographic and wireless link signatures. In: IEEE Symp. Secur. Priv., Oakland, CA, USA, pp. 286–301 (2010)
14. Gerzaguet, R., Ros, L.: Self-adaptive stochastic rayleigh flat fading channel estimation. In: 18th Int. Conf. Digit. Signal Process., Fira, Greece, pp. 1–6 (2013)
15. Kim, S.J., DallAnese, E., Giannakis, G.B.: Cooperative Spectrum Sensing for Cognitive Radios Using Kriged Kalman Filtering. IEEE J. Sel. Top. Signal Process. **5**(1), 24–36 (2011)
16. Ros, L., Simon, E.P., Shu, H.: Third-order Complex Amplitudes Tracking Loop for Slow Flat Fading Channel Online Estimation. IET Commun. J. **8**, 360–371 (2014)
17. Baddour, K.E., Beaulieu, N.C.: Autoregressive Modeling for Fading Channel Simulation. IEEE Trans. Wirel. Commun. **4**(4), 1650–1662 (2005)
18. Kay, S.M.: Fundamentals of Statistical Signal Processing, Vol. II: Detection Theory.Signal Process, Up. Saddle River, NJ Prentice (1998)
19. Dent, P., Bottomley, G.E., Croft, T.: Jakes Fading Model Revisited. Electron. Lett. **29**(13), 1162–1163 (1993)

Differential Entropy Driven Goodness-of-Fit Test for Spectrum Sensing

Sanjeev Gurugopinath[1], Rangarao Muralishankar[2]($\boxtimes$), and H.N. Shankar[1]

[1] Department of EEE, CMR Institute of Technology, Bangalore 560037, India
{sanjeev.g,muralishankar,hnshankar}@cmrit.ac.in
http://www.cmr.ac.in
[2] Department of ECE, CMR Institute of Technology, Bangalore 560037, India

Abstract. We present a novel Goodness-of-Fit Test driven by differential entropy for spectrum sensing in cognitive radios. When the noise-only observations are Gaussian, it exploits the fact that the differential entropy of the Gaussian attains its maximum. We obtain in closed form the distribution of the test statistic under the null hypothesis and the detection threshold that satisfies a constraint on the probability of false-alarm using the Neyman-Pearson approach. Later, we discuss the use of this technique to the case of the noise process modeled as a mixture Gaussians. Through Monte Carlo simulations, we demonstrate that our detection strategy outperforms the existing technique in the literature which employs an order statistics based detector for a large class of practically relevant fading channel models and primary signal models, especially in the low Signal-to-Noise Ratio regime.

Keywords: Spectrum sensing · Goodness-of-fit · Differential entropy · MaxEnt principle · non-Gaussian noise

1 Introduction

Goodness-of-Fit Tests (GoFT) for Spectrum Sensing (SS) has received considerable attention in the recent past [1–4]. This approach may be gainfully employed in Cognitive Radio (CR) when the knowledge of the primary signal and the fading models is meagre. In its general form, the GoFT for SS compares a decision statistic to a threshold and rejects the null-hypothesis when the statistic exceeds the threshold. The detection threshold is chosen satisfying a constraint on the probability of false-alarm.

The authors in [1] present a GoFT based on the Anderson-Darling statistic (which we term here as the Anderson-Darling statistic based Detector (ADD)). This is shown to outperform the well-known radiometer or Energy Detector (ED) under low SNR regime with Rayleigh fading and constant primary signal. Later, it is shown that a combination of the Student's-t Test and the ADD, called the Blind Detector (BD) [2] is robust to noise variance uncertainty. The major disadvantages of these works are as follows:

© Institute for Computer Sciences, Social Informatics and Telecommunications Engineering 2015
M. Weichold et al. (Eds.): CROWNCOM 2015, LNICST 156, pp. 248–259, 2015.
DOI: 10.1007/978-3-319-24540-9_20

1. The underlying Anderson-Darling statistic is known to perform well only against another Gaussian with a shift in mean.
2. ADD does not perform well in many other relevant SS contexts, as for example, when the primary signal follows other signal models [5];
3. ADD is useful only where the observations under $\mathcal{H}_0$ are i.i.d. and
4. ADD is effective only with small number of observations.

Thus the utility of ADD and BD in SS is diminished.

In [3], the authors propose an Order Statistic based Detector (OSD) and show that it improves upon ADD under conditions discussed in the foregoing. Here, the performance of OSD detector is studied only for a constant primary model. Further, the threshold is set empirically. A Higher-Order statistics based Detector (HOD) proposed in [6] is shown to provide good performance under low SNR. Recently, a zero-crossings based GoFT in [4] demonstrates its robustness to uncertainties of the noise model and the parameters; its computational complexity equals that of the GoFT based on ED.

In this work, we propose a novel GoFT based estimate of the differential entropy in the received observations. We bring out the many advantages of this technique such as relative ease in computing the detection threshold, relaxation of the restriction of a constant primary signal and enhanced performance relative to OSD in several practically relevant scenarios. Additionally, the performance of the detector is studied for a bimodal, two parameter and mixed Gaussian noise model, one of practical relevance. In fact, this model is used, inter alia, to model a combination of Gaussian and Middleton's class A noise components [4] and co-channel interference (CCI) [7]. Further, a closed-form expression for the near-optimal detection threshold is derived.

The system model is described in § 2. The differential entropy estimate based detection is introduced and analyzed in § 3. In particular, the cases where (i) the noise process is purely Gaussian and (ii) follows a bimodal Gaussian are studied in § 3.1 and § 3.2 respectively. Simulation results are presented and discussed in § 4. Concluding remarks comprise § 5.

2   System Model

Consider a Cognitive Radio (CR) node collecting M observations from a primary transmitter operating in a particular frequency band. Based thereon, it decides whether the spectrum is occupied or vacant. The GoFT based SS problem is essentially a detection problem which rejects the hypothesis

$$Y_i \sim f_{\textsc{n}}, \quad i \in \mathcal{M} \triangleq \{1, \cdots, M\},$$

with the probability of false-alarm given by

$$p_f \triangleq \mathcal{P}\{\text{reject } \mathcal{H}_0 | \mathcal{H}_0\} \leq \alpha_f,$$

where $\alpha_f \in (0,1)$ is a fixed constant. In general, the noise distribution $f_{\textsc{n}}$ in the SS setup can be modeled by various distributions [4]. In this paper, we consider the following noise models. First, for the sake of simplicity and to study the baseline, we choose $f_N \sim \mathcal{N}(0, \sigma_n^2)$, where $\mathcal{N}(\mu, \sigma^2)$ represents a Gaussian distribution with mean μ and variance σ^2. This noise model is widely considered in most spectrum sensing applications. Later, we consider the bimodal, mixture Gaussian distribution to model the noise, which is seen to be useful in some applications in the communication domain [7]. To begin with, we assume that the noise variance is known perfectly. The statistics of the primary signal model and the fading channel between the primary transmitter and CR node, on the other hand, can be arbitrary. In the next section, we review the Ordered Statistic-based Detector (OSD) [3], which is known to be the best GoFT detector for testing f_N against a mean-change model.

3   Differential Entropy Estimate-Based GoFT

In this section, we present the main contribution of this paper, i.e., a simple detection strategy based on an estimate of the differential entropy in the observations. Given a continuous random variable, X, over the support $(-\infty, \infty)$, the differential entropy, $h(X)$, of X is defined as [8]

$$h(X) \triangleq -\int_{-\infty}^{\infty} f_X(x) \log(f_X(x)) dx$$

where $f_X(\cdot)$ is the probability density function of X.

3.1   Detection Under Gaussian Noise

The detection strategy proposed in this work exploits the fact that among all continuous distributions with finite mean and finite variance and on the support $(-\infty, \infty)$, the Gaussian noise yields maximum differential entropy. For this detector, the entropy when $Y_i \sim f_N, i \in \mathcal{M}$ (i.e., for observations under $\mathcal{H}_0$) will be low, as compared to the entropy if the primary signal is present i.e., $Y_i \nsim f_N$. It is known that under $\mathcal{H}_0$, i.e., when $Y_i \sim \mathcal{N}(0, \sigma_n^2)$ [8],

$$h(Y|\mathcal{H}_0) = \frac{1}{2} \log(2\pi e \sigma_n^2).$$

Now, the Differential Entropy estimate-based Detector (DED) is constructed as follows. Let

$$\widehat{Y_i} \triangleq \frac{1}{M} \sum_{i=1}^{M} Y_i \text{ and } \frac{1}{M-1} \sum_{i=1}^{M} (Y_i - \widehat{Y_i})^2$$

denote the sample mean and variance respectively in the observations. Then,

$$\widehat{h}(Y) \triangleq \frac{1}{2} \log \left\{ \frac{2\pi e}{M-1} \sum_{i=1}^{M} (Y_i - \widehat{Y_i})^2 \right\}$$

represents the maximum likelihood estimate of differential entropy in the observations. The test is of the form

$$\widehat{h}(Y) \underset{\sim\mathcal{H}_0}{\overset{\varkappa\mathcal{H}_0}{\gtrless}} \tau_{\mathrm{G}},$$

where τ_{G} is set such that a constraint on the probability of false-alarm is satisfied. See Appendix (A) for a procedure to find τ_{G}, given α_f.

3.2 Detection Under Mixed Gaussian Model

The mixed Gaussian noise model is considered in a variety of signal processing applications for communications. For instance, it is used to model a combination of thermal noise and man-made clutter noise [4], and the Co-Channel Interference (CCI) [7]. Some non-Gaussian noise processes can also be modeled as mixtures of Gaussian distributions [9]. The PDF of the mixture-Gaussian noise is [10]

$$f_{\mathrm{N}}(x) = \frac{1}{\sqrt{2\pi\sigma_n^2}} e^{-(x^2+\mu^2)/2\sigma_n^2} \cosh\left(\frac{\mu x}{\sigma_n^2}\right). \tag{1}$$

In general, the differential entropy of this two-component mixture-Gaussian model is expressible only as an integrable form, and is given by

$$h(Y|\mathcal{H}_0) = \frac{1}{2}\log(2\pi e\sigma_n^2) + \left(\frac{\mu}{\sigma_n}\right)^2 - \mathcal{I}.$$

The values of $\mathcal{I}$ for different μ and σ_n are available [10]. Moreover, closed-form expressions for tight upper and lower bounds on the entropy are reported [10]. Under $\mathcal{H}_0$,

$$h(Y|\mathcal{H}_0) \leq \frac{1}{2}\log(2\pi e\sigma_n) + \left(\frac{\mu}{\sigma_n}\right)^2 \left\{1 - \mathrm{erf}\left(\frac{\mu}{\sqrt{2\sigma_n^2}}\right)\right\} - \sqrt{\frac{2\mu^2}{\pi\sigma_n^2}} e^{-\mu^2/2\sigma_n^2} + \log 2,$$

$$h(Y|\mathcal{H}_0) \geq \frac{1}{2}\log(2\pi e\sigma_n) + \left(\frac{\mu}{\sigma_n}\right)^2 \left\{1 - \mathrm{erf}\left(\frac{\mu}{\sqrt{2\sigma_n^2}}\right)\right\} - \sqrt{\frac{2\mu^2}{\pi\sigma_n^2}} e^{-\mu^2/2\sigma_n^2}.$$

We choose the upper bound in the above equation as a test statistic for SS against the PDF of (1). Therefore, the test is of the form

$$\widehat{h}_{UB}(Y) \underset{\sim\mathcal{H}_0}{\overset{\varkappa\mathcal{H}_0}{\gtrless}} \tau_{\mathrm{MG}}.$$

Note that the above test is pessimistic, i.e., follows the worst-case design. Obtaining the exact PDF of the test statistic in this case is difficult. Therefore, we estimate the PDF of the test statistic and set the threshold through Monte Carlo simulations. In fact, the asymptotically optimal threshold in this setting has been derived, vide Appendix (B).

4   Simulation Results

We present the performance of DED vis-á-vis that of the OSD in the context of SS through extensive simulations under various primary signal models, fading models and noise models. We set the false-alarm, $\alpha_f = 0.05$. For performance comparison, we consider the low SNR regime (~ -10dB), as it is practically relevant. Fading models used are Nakagami-m, Weibull and Rayleigh. The Nakagami-m (and as a special case, Rayleigh) fading is a favored model for several indoor wireless communication contexts without line of sight [11]. For some applications in communication where the bandwidth is in excess of 900MHz, Weibull fading is found to be a good fit [11].

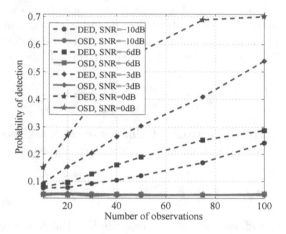

Fig. 1. Comparison of the proposed detector with OSD vs. M; different SNRs; Nakagami-m fading; shape parameter = 1; scale parameter = 0.5; Gaussian primary.

First, we consider the performance study under the Gaussian noise. Fig. 1 shows the performance comparison of the proposed detector DED with OSD vs. the number of observations M, for different values of SNR under Nakagami-m fading, with shape and scale parameters 1 and 0.5 respectively. The fading parameters were chosen arbitrarily. The primary signal is assumed to be Gaussian [4]; this is practically relevant in CR context owing to the errors due to synchronization and timing offsets. Clearly, DED outperforms OSD. The performance of OSD is non-trivial, i.e., it operates on the chance line in the receiver operating characteristics. Fig. 2 presents the results under the same setup as used in Fig. 1, except that the primary signal is constant. It is evident that the OSD is better than DED. It is significant to note that under a constant primary, such performance benefits of the OSD have been observed earlier too [3]. The deteriorated performance of DED is because of the scaling property of the entropy [8]. However, the constant primary model is highly constrained [12].

In Fig. 3, we present the performance comparison of DED and OSD as functions of the number of observations M under the Weibull fading, with shape and

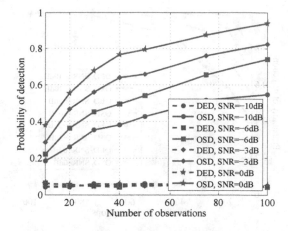

Fig. 2. Comparison of the proposed detector with OSD vs. M; different SNRs; Nakagami-m fading; shape parameter = 1; scale parameter = 0.5; constant primary.

scale parameters 1 and 2 respectively. The fading parameters are set arbitrarily. Again, the primary signal is Gaussian. Here, DED outperforms OSD across all M and SNR values. Fig. 4 shows the performance variation with constant primary signal. Clearly, OSD beats DED. As remarked, the constant primary signal assumption is removed from reality. Similar conclusions can be drawn from Fig. 5, which shows the performance comparison of DED and OSD vs. SNR under Rayleigh fading with parameter 1 and Gaussian primary signal.

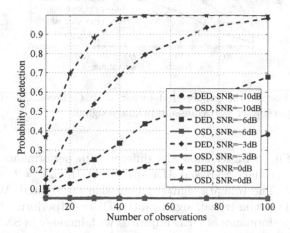

Fig. 3. Comparison of the proposed detector with OSD vs. M; different SNRs; Weibull fading; shape parameter = 1; scale parameter = 2; Gaussian primary signal.

Now, we present the results with bimodal Gaussian model for noise. We restrict our attention to Rayleigh fading and study the performance of DED

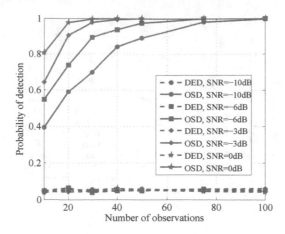

Fig. 4. Comparison of the proposed detector with OSD vs. M; different SNRs; Weibull fading; shape parameter $= 1$; scale parameter $= 2$; constant primary signal.

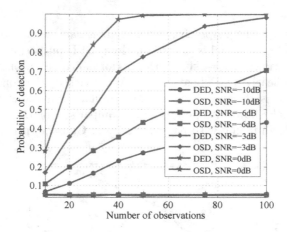

Fig. 5. Comparison of the proposed detector with OSD vs. M; different SNRs; Rayleigh fading; parameter $= 1$; Gaussian primary signal.

vis-á-vis OSD. Fig. 6 Fig. 7 show the difference in performance of DED and OSD vs. average primary SNR and M respectively. Here, $\mu = 2$ and the mixing parameter is 0.5. The primary signal is Gaussian distributed. While the performance of OSD is non-trivial, significantly, DED outperforms OSD. Though expectedly, the performance of DED improves with increase in SNR and M, this serious drawback of OSD lends credence to the proposition that its usefulness is restricted to the case of Gaussian noise and constant primary signal.

In this work, we relax the constraint on the choice of the noise distribution from a unimodal Gaussian to a bimodal Gaussian [7]. To test the utility of this choice, we compare the performance of DED under both Gaussian and bimodal Gaussian noise models, with a Gaussian distributed primary and Rayleigh fading

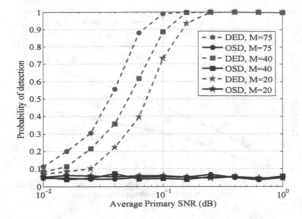

Fig. 6. Probability of detection vs. average primary SNR; different M; Rayleigh fading; mixture Gaussian model; Gaussian primary signal.

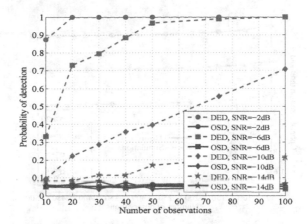

Fig. 7. Probability of detection vs. M; different average primary SNRs; Rayleigh fading; mixture Gaussian model; Gaussian primary signal.

(see Fig. 8) for different SNRs. It is seen that DED under the bimodal Gaussian noise performs better compared to its unimodal Gaussian counterpart. In particular, the performance of DED under bimodal Gaussian noise for -10dB SNR close to that under the unimodal Gaussian for -6dB SNR. Therefore, for a given p_d, the bimodal Gaussian model accommodates an additional 4dB SNR.

Fig. 9 shows the behavior of the optimal detection threshold of (4) taken over the number of observations M, seen as a function of σ_n^2. Clearly, the simulation results are in excellent agreement with the analytically derived results. Further, the detection threshold is independent of the average primary SNR, as we employ the Neyman-Pearson approach. Finally, the results shown in Fig. 10 validates the accuracy of our analysis, vide Appendix (B). That the analysis holds for

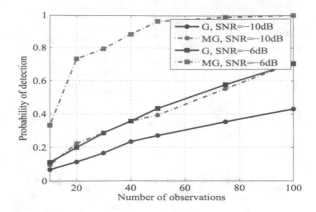

Fig. 8. Comparison of the variation of probability of detection vs. M; different primary SNRs; Rayleigh fading; Gaussian and mixture Gaussian noises; Gaussian primary.

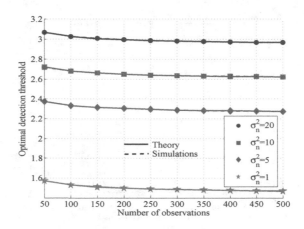

Fig. 9. Variation of the optimal threshold τ_{G} calculated through theory (derived in (4)) and simulations, with M, for various values of σ_n^2.

large M and $\mu(\geq 3)$ is borne out by the fact that the disparity between between the simulations and theory shrinks progressively.

5   Concluding Remarks

We proposed a novel spectrum sensing based on differential entropy estimate under the goodness-of-fit formulation. The distribution of the test statistic under the null hypothesis, and the detection threshold that satisfies a constraint on the probability of false-alarm were obtained in closed form. Through Monte Carlo simulations, it was shown that the proposed detector outperforms the ordered statistics based detector, significantly in the low SNR regime, under various

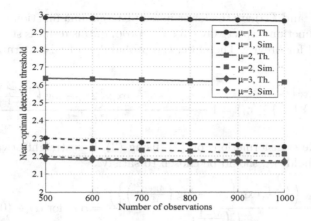

Fig. 10. Variation of the near-optimal threshold τ_G^{MG} calculated through theory and simulations, with M, for various values of μ.

fading and primary signal models. The results were compared with noise process being unimodal Gaussian vis-à-vis bimodal Gaussian. For a given probability of detection, this mixture model was shown to provide an additional leeway to the tune of 4dB in SNR over the corresponding unimodal Gaussian.

Appendix

A   Calculation of τ_G

We adopt one of the many ways to arrive at the result here. Under $\mathcal{H}_0$, since $Y_i \sim \mathcal{N}(0, \sigma_n^2)$, it follows from Cochran's Theorem that the unbiased estimate, $\mathcal{V}$, of the variance of Y_i follows a scaled, central χ^2 distribution with $M - 1$ degrees-of-freedom thus:

$$\mathcal{V} \triangleq \frac{1}{M-1} \sum_{i=1}^{M} (Y_i - \widehat{Y}_i)^2 \sim \frac{\sigma_n^2}{M-1} \chi_{M-1}^2,$$

which implies that the statistic $\widehat{h}(Y)$ can be written as

$$\widehat{h}(Y|\mathcal{H}_0) = \frac{1}{2} \log(2\pi e) + \frac{1}{2} \log \mathcal{V}. \tag{2}$$

Under $\mathcal{H}_0$, the statistic $\log \mathcal{V}$ follows a log-scaled, central χ^2 distribution with $M - 1$ degrees-of-freedom, represented by $\log \chi_{M-1}^2$. It can be shown that the Cumulative Distribution Function (CDF) of a random variable $X \sim \log \chi_n^2$, denoted by $F_X(\cdot)$ is

$$F_X(a) \triangleq \int_{-\infty}^{a} f_X(x) dx = \frac{\gamma_{\text{inc}}\left(\frac{n}{2}, e^{(a-\log 2)}\right)}{\Gamma\left(\frac{n}{2}\right)},$$

where $\gamma_{\text{inc}}(\cdot, \cdot)$, and $\Gamma(\cdot)$ are the lower incomplete gamma function and the standard gamma function respectively [13]. The proof of this result is straightforward and is omitted for brevity. Therefore, the probability of false-alarm, p_f is given by

$$p_f = \mathcal{P}\{\widehat{h}(Y|\mathcal{H}_0) \geq \tau_{\text{G}}\} = 1 - \frac{\gamma_{\text{inc}}\left(\frac{M-1}{2}, \exp\left\{2\tau_{\text{G}} - \log\left(\frac{4\pi e\sigma_n^2}{M-1}\right)\right\}\right)}{\Gamma\left(\frac{M-1}{2}\right)}. \tag{3}$$

Now, by simple transformations on (2), using (3), it is straightforward to show that the threshold, τ_{G}, should be chosen to satisfy

$$1 - \frac{\gamma_{\text{inc}}\left(\frac{M-1}{2}, \exp\left\{2\tau_{\text{G}} - \log\left(\frac{4\pi e\sigma_n^2}{M-1}\right)\right\}\right)}{\Gamma\left(\frac{M-1}{2}\right)} = \alpha_f, \text{ for } \alpha_f \in (0,1). \tag{4}$$

B Computing the Near-Optimal $\tau_{\text{G}}^{\text{MG}}$

It is known that if $\{Y_i, i \in \mathcal{M}\}$ represent a set of i.i.d. random variables from any distribution (possibly multimodal) with finite variance σ^2, then the random variable defined by

$$Y_s^2 \triangleq \frac{1}{M-1}\sum_{i=1}^{M}\left(Y_i - \widehat{Y}\right)^2,$$

has mean and variance in an asymptotic sense (as $M \to \infty$) given as follows [14]:

$$\mathbb{E}Y_s^2 = \sigma^2, \quad \text{var}(Y_s^2) = \sigma^4\left[\frac{2}{M-1} + \frac{\kappa}{M}\right], \tag{5}$$

where κ is the excess kurtosis and μ_4 is the fourth moment around the mean of the parent distribution. Therefore, in the case of the bimodal Gaussian distribution,

$$\mathbb{E}Y_s^2 = \sigma_n^2 + \mu^2, \quad \text{var}(Y_s^2) = (\sigma_n^2 + \mu^2)^2\left[\frac{2}{M-1} + \frac{\kappa}{M}\right]. \tag{6}$$

A closed form expression for the distribution of Y_s^2 for the bimodal Gaussian distribution is hard to obtain. However, it can be well approximated in the asymptotic sense by a Gaussian distribution with moments in (5) and (6).

For large values of μ (≥ 3), $h(Y|\mathcal{H}_0)$ can be approximated as [10]:

$$h(Y|\mathcal{H}_0) \approx \frac{1}{2}\log(2\pi e\sigma_n^2) + \log 2.$$

Hence, an estimate of the above entropy is given by

$$\widehat{h}(Y|\mathcal{H}_0) = \frac{1}{2}\log\left(\frac{2\pi e}{M-1}\sum_{i=1}^{M}(Y_i - \widehat{Y}_i)^2\right) + \log 2 = \frac{1}{2}\log(4\pi e Y_s^2).$$

Therefore, the probability of false-alarm, p_f, becomes

$$p_f = \mathcal{P}\left\{\widehat{h}(Y) \geq \tau_{\mathrm{G}}^{\mathrm{MG}} \,|\mathcal{H}_0\right\} \stackrel{(a)}{=} \mathcal{P}\left\{Y_s^2 \geq \frac{\exp(2\tau_{\mathrm{G}}^{\mathrm{MG}} - 1)}{4\pi}\right\}$$

$$= \mathcal{Q}\left[\frac{\frac{\exp(2\tau_{\mathrm{G}}^{\mathrm{MG}}-1)}{4\pi} - \mathbb{E}Y_s^2}{\sqrt{\mathrm{var}(Y_s^2)}}\right],$$

where $\stackrel{(a)}{=}$ denotes that the equality holds due to the fact that $\log(\cdot)$ is monotone and $\mathcal{Q}(\cdot)$ denotes the Q-function. Now, it is straightforward to show that, given $\alpha_f \in (0,1)$, the near-optimal threshold, $\tau_{\mathrm{G}}^{\mathrm{MG}}$, is

$$\tau_{\mathrm{G}}^{\mathrm{MG}} = 0.5 \log\left(4\pi e\left\{(\sigma_n^2 + \mu^2)\left[\mathcal{Q}^{-1}(\alpha_f)\sqrt{\left(\frac{2}{M-1} + \frac{\kappa}{M}\right)} + 1\right]\right\}\right). \qquad (7)$$

References

1. Wang, H., Yang, E.-H., Zhao, Z., Zhang, W.: Spectrum Sensing in Cognitive Radio Using Goodness of Fit Testing. IEEE Trans. Wireless Commun. **8**(11), 5427–5430 (2009)
2. Shen, L., Wang, H., Zhang, W., Zhao, Z.: Blind Spectrum Sensing for Cognitive Radio Channels with Noise Uncertainty. IEEE Trans. Wireless Commun. **10**(6), 1721–1724 (2011)
3. Rostami, S., Arshad, K., Moessner, K.: Order-Statistic Based Spectrum Sensing for Cognitive Radio. IEEE Commun. Lett. **16**(5), 592–595 (2012)
4. Gurugopinath, S., Murthy, C.R., Seelamantula, C.S.: Zero crossings based spectrum sensing under noise uncertainties. In: Proc. NCC (2014)
5. Nguyen-Thanh, N., Kieu-Xuan, T., Koo, I.: Comments and Corrections on Spectrum Sensing in Cognitive Radio using Goodness-of-Fit Testing. IEEE Trans. Wireless Commun. **11**(10), 3409–3411 (2012)
6. Denkovski, D., Atanasovski, V., Gavrilovska, L.: HOS Based Goodness-of-Fit Testing Signal Detection. IEEE Commun. Lett. **16**(3), 310–313 (2012)
7. Rohde, G., Nichols, J., Bucholtz, F., Michalowicz, J.: Signal estimation based on mutual information maximization. In: Proc. ACSSC, pp. 597–600 (2007)
8. Cover, T.M., Thomas, J.M.: Elements of Information Theory, 2nd edn. John Wiley and Sons, Inc. (2005)
9. Wang, Y., Wu, L.: Nonlinear Signal Detection From an Array of Threshold Devices for Non-Gaussian Noise. Digit. Signal Process. **17**(1), 76–89 (2007)
10. Michalowicz, J.V., Nichols, J.M., Bucholtz, F.: Calculation of Differential Entropy for a Mixed Gaussian Distribution. Entropy **10**(3), 200–206 (2008)
11. Hashemi, H.: The Indoor Radio Propagation Channel. Proceedings of the IEEE **81**(7), 943–968 (1993)
12. Axell, E., Leus, G., Larsson, E., Poor, H.V.: Spectrum Sensing for Cognitive Radio: State-of-the-Art and Recent Advances. IEEE Signal Process. Mag. **29**(3), 101–116 (2012)
13. Gradshteyn, I., Ryzhik, I.: Tables of integrals, series and products, 7th edn. Academic Press (2007)
14. Mood, A.M., Graybill, F.A., Boes, D.C.: Introduction to the Theory of Statistics, 3rd edn. McGraw-Hill (1974)

Experimental Results for Generalized Spatial Modulation Scheme with Variable Active Transmit Antennas

Khaled M. Humadi$^{(\boxtimes)}$, Ahmed Iyanda Sulyman, and Abdulhameed Alsanie

Department of Electrical Engineering, King Saud University, Riyadh, Saudi Arabia
{khumadi,asulyman,sanie}@ksu.edu.sa

Abstract. This paper presents experimental results on the performance of generalized spatial modulation scheme with variable active transmit antennas (VA-GSM). In the VA-GSM scheme, one or more transmit antennas are activated simultaneously to transmit the same complex symbol. The indices of the set of activated transmit antennas at any time instant are then used to convey extra information in addition to the transmitted complex symbol. The average bit error rate performance of the proposed scheme is evaluated experimentally, and the results agree closely with simulation. We also compare the VA-GSM with existing SM and GSM schemes.

Keywords: Spatial Modulation (SM) · Generalized Spatial Modulation (GSM) · Variable active transmit antennas

1 Introduction

Spatial modulation (SM) is a new promising transmission technique that uses antenna indices, in a multiple antenna system, as avenues for data transmissions in addition to the transmitted modulated symbols. The principle of wireless transmissions in which the information is carried by both the indices of the active antennas and the symbol transmitted through these active antennas was first investigated by Mesleh et al. in [1, 2] where only one antenna was activated at each time instant thereby avoiding inter-carrier-interference (ICI) between the transmit antennas. Following that work, many variants of this idea were introduced such as the space shift keying (SSK) modulation that uses only the spatial modulation concept without transmitting any symbol. That scheme reduces the system complexity by cancelling the amplitude/phase modulation required at the transmitter and receiver sides, but at the expense of some degradation in the system's spectral efficiency [3]. The idea of Trellis coded modulation (TCM) was applied to the spatial points available in SM in [4, 6]. That scheme, named Trellis Coded SM (SM–TCM), achieves both coding and diversity gains, and has a significant performance enhancements over SM especially in the presence of realistic channel conditions such as Rician fading and spatial correlation. A combination of both SM and space-time block coding (STBC) that takes advantage of the benefits of both schemes was proposed in [6].

A.I. Sulyman—Senior Member, IEEE.

© Institute for Computer Sciences, Social Informatics and Telecommunications Engineering 2015
M. Weichold et al. (Eds.): CROWNCOM 2015, LNICST 156, pp. 260–270, 2015.
DOI: 10.1007/978-3-319-24540-9_21

In [7], SM system was generalized to multiple active transmitting antennas and low complexity detection MIMO system. In [8] receiver-side SM scheme for single user, called the Generalized Pre-Coding Aided SM (GPASM), was introduced. Generalized form of SSK (GSSK) was proposed in [9] where a set of transmit antennas were activated at a time, and all the active antennas transmit ones (i.e no symbols were transmitted). The scheme increases the spectrum efficiency compared to SSK. Generalized SM (GSM) in which a fixed set of transmit antennas were activated simultaneously, with all active antennas transmitting the same constellation symbol, was proposed in [10, 11]. Transmitting the same symbol from all active antennas keep the key advantages of SM which are: avoiding ICI, reducing complexity, and using only one radio frequency (RF) chain for hardware implementation. In [12], SSK with variable number of active transmit antennas (VA-SSK) was introduced.

In this paper, we propose GSM scheme in which a variable set of active transmit antennas are used. Unlike the GSM where a fixed number of transmit antennas are used at any time instant to transmit the same constellation symbol, the proposed scheme will vary the number of activated transmit antennas. Also unlike VA-SSK where no symbols are transmitted, the proposed scheme transmits symbols which increases the system's spectral efficiency.

2   System Model

2.1   VA-GSM Modulator

Consider a MIMO system equipped with N_t transmit and N_r receive antennas, where we assume $N_t \geq N_r$. In this system, one or more transmit antennas are activated simultaneously at any time instant and all the activated antennas transmit the same constellation symbol. If the maximum number of activated transmit antennas at a time is N_a (where $N_a \leq N_t$), then one or two, or three, up to N_a transmit antennas can be activated simultaneously. Then, the total number of possible active transmit antenna combinations in this scheme is $\sum_{n=1}^{N_a} \binom{N_t}{n}$, where $(\cdot)$ denotes the binomial operation. These antenna combinations will be used to convey extra information. Because the number of active transmit antenna combinations that can be used for data transmission must be a power of two, only $g = 2^{k_l}$ active antenna combinations will be considered, where k_l is the number of bits transmitted per spatial symbol which is given by

$$k_l = \left\lfloor log_2 \left(\sum_{n=1}^{N_a} \binom{N_t}{n} \right) \right\rfloor, \tag{1}$$

where $\lfloor a \rfloor$ denotes the greatest integer less than or equal to a. For the case when $N_a = N_t$, Eq. (1) becomes

$$k_l = \left\lfloor log_2\left(\sum_{n=1}^{N_t}\binom{N_t}{n}\right)\right\rfloor$$

$$= \lfloor log_2(2^{N_t}-1)\rfloor = N_t - 1 \tag{2}$$

Equations (1) and (2) show that unlike SM in which the number of bits transmitted per spatial symbol increase logarithmically with increasing the number of transmit antennas N_t, the number of bits transmitted per spatial symbol in the proposed VA-GSM increases with increasing N_t and N_a. The total number of bits that can be transmitted using VA-GSM is given by

$$k_t = k_l + k_s, \tag{3}$$

where $k_s = log_2(M)$ and M is the constellation size.

Table 1. Transmitted vector $x_{l,m}$ for different input bits in VA-GSM (assuming BPSK modulation).

Bits mapping	l	m	$x_{l,m}$	Bits mapping	l	m	$x_{l,m}$
000 0	1	1	$[s_1\ 0\ 0\ 0]^T$	100 0	5	1	$[s_1\ s_1\ 0\ 0]^T$
000 1	1	2	$[s_2\ 0\ 0\ 0]^T$	100 1	5	2	$[s_2\ s_2\ 0\ 0]^T$
001 0	2	1	$[0\ s_1\ 0\ 0]^T$	101 0	6	1	$[s_1\ 0\ s_1\ 0]^T$
001 1	2	2	$[0\ s_2\ 0\ 0]^T$	101 1	6	2	$[s_2\ 0\ s_2\ 0]^T$
010 0	3	1	$[0\ 0\ s_1\ 0]^T$	110 0	7	1	$[s_1\ 0\ 0\ s_1]^T$
010 1	3	2	$[0\ 0\ s_2\ 0]^T$	110 1	7	2	$[s_2\ 0\ 0\ s_2]^T$
011 0	4	1	$[0\ 0\ 0\ s_1]^T$	111 0	8	1	$[0\ s_1\ s_1\ 0]^T$
$\underbrace{011}_{k_l}\ \underbrace{1}_{k_s}$	4	2	$[0\ 0\ 0\ s_2]^T$	$\underbrace{111}_{k_l}\ \underbrace{1}_{k_s}$	8	2	$[0\ s_2\ s_2\ 0]^T$

The transmitter divides the incoming data into blocks of k_t bits. The first k_s bits are mapped to a symbol $s_m \in \{s_1, s_2 \ldots, s_M\}$ in the constellation while the next k_l bits are used to select a set of active antenna combinations which is denoted here by the set $\Gamma_l, l \in \{1,2,\cdots g\}$. The set of active transmitting antennas Γ_l transmit the same complex symbol s_m while the other antennas transmit zeros. Then the received vector $y \in \mathbb{C}^{N_r \times 1}$ is given by

$$y = Hx_{l,m} + w, \tag{4}$$

Table 2. Comparing VA-GSM, GSM, and SM in term of number of available antenna combinations (assuming $N_r = 4$).

N_t	Scheme	$N_a = 1$	$N_a = 2$	$N_a = 3$	$N_a = 4$
4	SM	$g = 2^2$			
	GSM	$g = 2^2$	$g = 2^2$	$g = 2^2$	
	VA-GSM	$g = 2^2$	$g = 2^3$	$g = 2^3$	$g = 2^3$
8	SM	$g = 2^3$			
	GSM	$g = 2^3$	$g = 2^4$	$g = 2^5$	$g = 2^6$
	VA-GSM	$g = 2^3$	$g = 2^5$	$g = 2^6$	$g = 2^7$
16	SM	$g = 2^4$			
	GSM	$g = 2^4$	$g = 2^6$	$g = 2^9$	$g = 2^{10}$
	VA-GSM	$g = 2^4$	$g = 2^7$	$g = 2^9$	$g = 2^{11}$
64	SM	$g = 2^6$			
	GSM	$g = 2^6$	$g = 2^{10}$	$g = 2^{15}$	$g = 2^{19}$
	VA-GSM	$g = 2^6$	$g = 2^{11}$	$g = 2^{15}$	$g = 2^{19}$

where $x_{l,m} \in \mathbb{C}^{N_t \times 1}$ denotes the transmitted vector, $H \in \mathbb{C}^{N_r \times N_t}$ is the channel response between the transmitter and receiver, and w is the zero-mean Gaussian noise vector with variance $\sigma^2 = N_0/2$. Table 1 shows the mapping process at the transmitter in the proposed scheme for the case of four transmit antennas ($N_t = 4$), one or two active antennas at a time ($N_a = 2$), and Binary Phase shift keying (BPSK) modulation ($M = 2$). As shown in the last highlighted row of the table, the first bit ($k_s=1$) is mapped to a symbol in the BPSK constellation $s_m \in \{s_1, s_2\}$, while the next three bits (since in this case, $k_l = \lfloor log_2(\sum_{n=1}^{2} \binom{4}{n}) \rfloor = 3$ which gives $g = 2^3 = 8$) are used to select active antenna combinations $\Gamma_l, l \in \{1, 2, \cdots 8\}$. Therefore, the antenna selection in VA-GSM is based on the second part, k_l, of the incoming bit stream. For example, if we choose [0111] from Table 1, the first bit 1 is mapped to a symbol s_2 ($m = 2$) in the BPSK constellation while the next three bits 011 select the fourth transmit antenna combination ($l = 4$) which means the fourth antenna is activated to transmit the symbol s_2 while the other antennas transmit zeros as described by the transmitted vector $x_{l,m} = x_{4,2} = [0\ 0\ 0\ s_2]^T$. If we choose 1111, the transmitted vector is $x_{l,m} = x_{8,2} = [0\ s_2\ s_2\ 0]^T$ which means the second symbol is transmitted through the eighth transmit antenna combination where the first and fourth antennas are selected to transmit the same copy of the BPSK symbol s_2. The received signal at each receiving antenna can be described as

$$y_j = h_{j,l} s_m + w_j, j = 1, 2, \cdots, N_r, \tag{5}$$

where $h_{l,j} = \sum_{i \in \Gamma_l} h_{j,i}$ is the summation of channel responses from all active transmit antennas Γ_l used at the current transmission instant to the j-th receiving antenna, and w_j is the Gaussian noise at the j-th receiving antenna.

Table 2 compares VA-GSM method with the conventional SM and the GSM approach proposed in [10]. It is clear from the table that the proposed VA-GSM provides more active transmit antenna combinations except for $N_t = 64$ which is the same as GSM because we limit the number of activated transmit antennas N_a in this illustration to $N_a \leq N_r = 4$. If we increase N_a, VA-GSM will have more active antenna combinations than the GSM for this case as will.

2.2 Maximum Likelihood (ML) Detection

At the receiver, ML optimum detector computes the Frobenius distances between the received signal and the set of possible transmitted vectors. Then, the ML detection operation can be expressed as

$$[\hat{l}, \hat{m}] = \arg \min_{\substack{m \in \{1,2,...,M\} \\ l \in \{1,2,...,g\}}} \left\{ \|\mathbf{y} - \mathbf{H}\mathbf{x}_{l,m}\|_F^2 \right\}$$

$$= \arg \min_{\substack{m \in \{1,2,...,M\} \\ l \in \{1,2,...,g\}}} \left\{ \sum_{j=1}^{N_r} |y_j - h_{l,j}s_m|^2 \right\} \tag{6}$$

where $\|.\|_F$ denotes the Frobenius norm while $\hat{l}, \hat{m}$ denote the indices of the estimated set of activated transmit antennas and the transmitted complex symbol respectively. If the complex symbol s_m is transmitted over the antenna combination Γ_l, then the correct decision is obtained when $\hat{l} = l$ and $\hat{m} = m$.

3 System Performance Analysis

3.1 Upper Bound on the Average Bit Error Rate

The average bit error rate (ABER) performance can be analytically estimated using the well-known union bounding technique [13]. The ABER in the proposed method can be upper bounded as

$$\text{ABER} \leq \frac{1}{2^{k_t}} \sum_{\hat{l}} \sum_{\hat{m}} \sum_{l} \sum_{m} \frac{d_H(x_{l,m}, x_{\hat{l},\hat{m}}) E\{\Pr(x_{l,m} \to x_{\hat{l},\hat{m}})\}}{k_t}, \tag{7}$$

where $d_H(x_{l,m}, x_{\hat{l},\hat{m}})$ is the Hamming distance between the transmitted symbol $x_{l,m}$ and the estimated symbol $x_{\hat{l},\hat{m}}$ and $\Pr(x_{l,m} \to x_{\hat{l},\hat{m}})$ denotes pairwise error probability when estimating $x_{\hat{l},\hat{m}}$ while $x_{l,m}$ is transmitted which is given by

$$\Pr(x_{l,m} \to x_{\hat{l},\hat{m}}) = \Pr\left[\|\mathbf{y} - \mathbf{H}\mathbf{x}_{l,m}\|_F^2 > \|\mathbf{y} - \mathbf{H}\mathbf{x}_{\hat{l},\hat{m}}\|_F^2 \right]$$

$$= Q\left(\sqrt{\frac{\gamma\|\mathbf{H}\boldsymbol{\delta}\|^2}{2}}\right) \tag{8}$$

where $\boldsymbol{\delta} = (\mathbf{x}_{\hat{l},\hat{m}} - \mathbf{x}_{l,m})$ and $\gamma = E_s/N_0 = 1/\sigma^2$ is the SNR between the transmitter and the receiver, and $E_s = E\left\{\|\mathbf{Hx}_{l,m}\|^2\right\} = 1$. Using the alternative integral expression for the Q-function we get

$$\Pr(x_{l,m} \rightarrow x_{\hat{l},\hat{m}}) = \frac{1}{\pi}\int_0^{\pi/2} \exp\left(-\frac{\|\mathbf{H}\boldsymbol{\delta}\|^2}{4\sigma^2\sin^2\theta}\right) d\theta. \tag{9}$$

3.2   Complexity Analysis

The receiver complexity is estimated in terms of the number of real multiplicative operations required by the ML detector. Note that computing the Euclidian distance $|y_j - h_{l,j}s_m|^2$ requires 2 complex multiplications, where each complex multiplication requires 4 real multiplications. Thus, from Eq. (6), the computational complexity of VA-GSM receiver is given by

$$C = 8N_r g 2^m, \tag{10}$$

which is the same as SM and GSM if we assume that the same number of bits are transmitted in these schemes at a time.

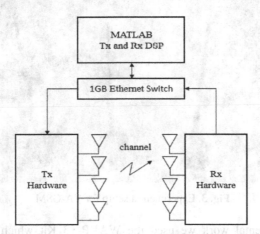

Fig. 1. Block diagram of test-bed setup for the experiment.

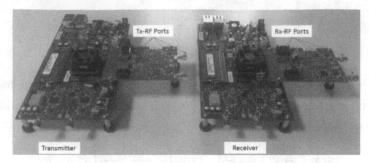

Fig. 2. WARP v3 Kits used in the hardware experiment.

4   Experimental Setup

Fig. 1 shows the block diagram of the software and hardware test-bed setup used in the experiment. As shown in this figure, the transmitted and received data are processed using MATLAB installed on a PC. The PC and the two hardware radio devices used are connected through a 1GB Ethernet Switch.

Fig. 3. Experimental setup for VA-GSM.

In our experimental work we used the WARP v3 Kit which is shown in Fig. 2.WARP v3 Kit is a software defined radio (SDR) platform developed by Rice University and Mango Communications. It is built on a Xilinx Virtex-6 LX240T FPGA with four programmable RF interfaces operating at 2.4 and 5 GHz with a 40 MHz bandwidth. The WARP v3 Kit was selected to be used with the platform due to its accessibility and ease of interface with MATLAB. The transmitted waveforms are sent directly to the transmit buffers and the received waveforms are extracted from the receive buffer using MATLAB. Synchronization, Modulation, coding, channel estimation, and equalization are performed in MATLAB. The physical layout of the experimental test-bed is shown in Fig. 3.

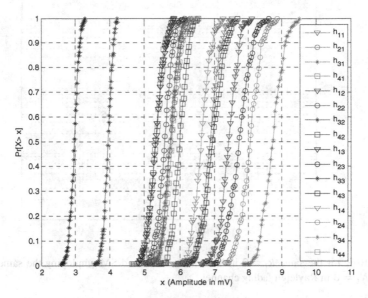

Fig. 4. CDFs for each of the fast-fading coefficients h_{ji} of all channel paths in the experiment.

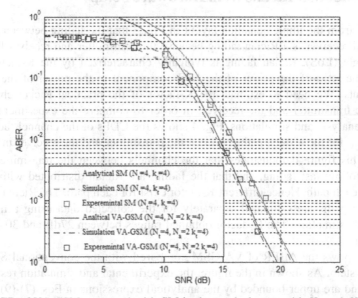

Fig. 5. ABER of VA-GSM compared with SM in the test-bed setup with $N_t = 4, N_a = 2$ and $N_r = 4$. BPSK modulation is used for VA-GSM while 4QAM modulation is used for SM. The simulation and analytical results are obtained with K=28dB to match the experimental Rician channel.

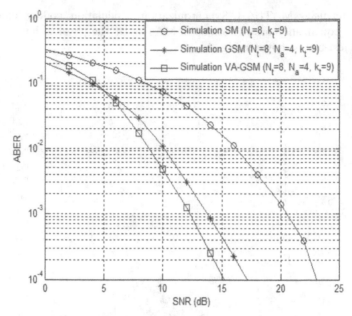

Fig. 6. ABER for the same $k_t = 9$ in VA-GSM, GSM and SM employing the same transmit antennas $N_t = 8$ in Rayleigh fading channel.

5   Validation Results from Hardware Setup

It is clear from the implementation layout in Fig. 3 that the fading between the transmitter and receiver is a Rician fading because the transmitter and receiver have clear line-of-sight (LoS). Rician fading is typically characterized by the K-factor, which denotes the ratio of the power of the LoS component to the power of the scattered components. We need to estimate the K-factor of the practical Rician channel between the transmitter and the receiver, in order to compare the experimental results with the analysis and simulations. Fig. 4 shows the CDFs of the channel fading coefficients h_{ji} for the sixteen paths between the transmitter and the receiver in the experiment. This figure demonstrates that the channels between the transmitter and the receiver follow fast fading, and that the fading are Rician distributed with different mean for each path. Hence different K-factor exist for each path. The Rician K-factors for all the paths are estimated separately for the collected data using a maximum-likelihood estimation. The K- factor obtained ranges between 27dB and 30 dB in this experiment.

Fig. 5 shows the ABER of VA-GSM compared with the conventional SM in 4x4 MIMO system. As shown in the figure, the experimental and simulation results agree closely, and are upper-bounded by the analytical expressions in Eqs. (7)-(9). Because we use 4 transmit antennas, the number of available antenna combinations for VA-GSM is $g = 2^{k_l} = 8$, and for SM is $g = 4$ as shown in Table 2. Therefore, the number of bits per spatial symbol $k_l = 3$ for VA-GSM while $k_l = 2$ for SM. To get the same transmitted bits at a time for both method, we used Binary Phase shift keying (BPSK) modulation for the proposed GSM ($k_s = 1$) and QPSK for SM ($k_s = 2$). It is clear from the figure that for the same transmitted bits $k_t = k_l + k_s = 4$, VA-GSM performs about 1.7 dB better than SM at ABER = 10^{-4}.

Fig. 6 compares the ABER of VA-GSM with GSM and SM in Rayleigh channels by simulation when using the same number of transmit antennas $N_t = 8$, and the same transmitted bits $k_t = 9$. To achieve $k_t = 9$ for each of these schemes, it was necessary to use QPSK for VA-GSM, 8PSK for GSM, and 64QAM for SM. The number of active antennas $N_a = 4$ which is fixed in GSM and variable (between 1 and 4) for the case of VA-GSM. As shown in the figure, for the same number of transmit antennas and the same transmitted bits at a time, VA-GSM has a better BER performance over both GSM and SM which can be attributed to the difference in the constellation size required to reach the same number of bits transmitted.

6   Conclusion

This work proposes generalized spatial modulation scheme using variable active transmit antennas (VA-GSM). The proposed scheme uses the variations in the active transmit antennas to increase the number of available spatial symbols and hence the number of transmitted spatial bits. The experimental results on the ABER performance of the system confirm the analytical bound as well as the computer simulations, for both the VA-GSM and Spatial modulation (SM) schemes. The simulation results shows that VA-GSM has better average bit error rate (ABER) performance over both the GSM and SM, when using the same number of transmit antennas and the same number of transmitted bits for the three schemes.

Acknowledgment. This work was supported by NSTIP strategic technologies programs (no 11-ELE1854-02) in the Kingdom of Saudi Arabia.

References

1. Mesleh, R., Haas, H., Ahn, C.W., Yun, S.: Spatial modulation-a new low complexity spectral efficiency enhancing technique. In: Proc. ChinaCom 2006, pp. 1–5 (2006)
2. Mesleh, R.Y., Haas, H., Sinanovic, S., Ahn, C.W., Yun, S.: Spatial modulation. IEEE Trans. Veh. Technol. **57**(4), 2228–2241 (2008)
3. Jeganathan, J., Ghrayeb, A., Szczecinski, L., Ceron, A.: Space shift keying modulation for MIMO channels. IEEE Trans. Wireless Commun. **8**(7), 3692–3703 (2009)
4. Mesleh, R., Di Renzo, M., Haas, H., Grant, P.M.: Trellis coded spatial modulation. IEEE Trans. Wireless Commun. **9**(4), 2349–2361 (2010)
5. Başar, E., Aygölü, Ü., Panayırcı, E., Poor, H.V.: Super-orthogonal trellis-coded spatial modulation. IET Commun. **6**(17), 2922–2932 (2012)
6. Basar, E., Aygolu, U., Panayirci, E., Poor, H.V.: Space-time block coded spatial modulation. IEEE Trans. Commun. **59**(3), 823–832 (2011)

7. Wang, J., Jia, S., Song, J.: Generalised spatial modulation system with multiple active transmit antennas and low complexity detection scheme. IEEE Trans. on Wireless Commun. **11**(4), 1605–1615 (2012)
8. Zhang, R., Yang, L.-L., Hanzo, L.: Generalised pre-coding aided spatial modulation. IEEE Trans. on Wireless Commun. **12**(11), 5434–5443 (2013)
9. Jeganathan, J., Ghrayeb, A., Szczecinski, L.: Generalized space shift keying modulation for MIMO channels. In: Proc. IEEE 19th International Symp. Personal, Indoor Mobile Radio Commun., Cannes, France, pp. 1–5 (2008)
10. Younis, A., Serafimovski, N., Mesleh, R., Haas, H.: Generalised spatial modulation. In: IEEE Conf. on Signals, Systems and Computers (2010)
11. Younis, A., Basnayaka, D.A., Haas, H.: Performance analysis for generalised spatial modulation. In: 20th European Wireless Conf., pp. 1–6 (2014)
12. Osman, O.: Variable antenna-space shift keying for high spectral efficiency in exponentially correlated channels. Int. Journal of Commun. Sys. (2014)
13. Proakis, J.G.: Digital Communications, 4th edn. McGraw-Hill, New York (2000)

Low Complexity Multi-mode Signal Detection for DTMB System

Xue Liu[✉], Guido H. Bruck, and Peter Jung

Department of Communication Technologies,
University of Duisburg-Essen, Duisburg, Germany
{xue.liu,guido.bruck,peter.jung}@kommunikationstechnik.org

Abstract. The Digital Terrestrial Multimedia Broadcast (DTMB) system utilizes three modes of Pseudo-noise (PN) sequences as Frame Header (FH) that can be exploited for spectrum sensing on TV White Space (TVWS). As a consequence, multi-mode signal detection is required at the receiver. In this paper, Neyman-Person (NP) lemma based, Multi-mode Local Sequence (MLS) detectors are proposed for multi-mode signal detection. The theoretical and quantitative performance analysis of MLS detectors are mainly concentrated and fully studied by utilizing the addition and auto-correlation properties of PN sequence. In addition, the single-mode detector for the DTMB system is presented as the optimal detector and taken as a benchmark for MLS detectors. Both theoretical analyses and simulation results show that MLS detectors are capable of multi-mode signal detection and satisfy the predefined sensing requirements.

Keywords: Digital terrestrial multimedia broadcast · TV white space · Spectrum sensing · Pseudo-noise sequence

1 Introduction

TV White Space (TVWS) refers to the frequency band where the radio spectrum is not used by primary users [4]. It can be potentially accessed by secondary users via Cognitive Radio (CR) [7], which results in increased overall spectrum efficiency and innovative new services. The requirement for TVWS spectrum sensing allows the coexistence of secondary users without harmful interference to primary users. For instance, the sensing requirements for the Advanced Television Systems Committee (ATSC) system [8] defined by IEEE 802.22 [2] are as low as −21 dB within 20 ms sensing time when the false-alarm probability (P_{FA}) is below 10% and the detection probability (P_D) is above 90%.

Digital Terrestrial Multimedia Broadcast (DTMB) is a Chinese digital terrestrial television standard [1,9]. Applying spectrum sensing for the DTMB system can extend the application of TVWS. Although specific requirements for the DTMB spectrum sensing are not available so far, the same requirements of the ATSC system are also adequate for the DTMB system.

© Institute for Computer Sciences, Social Informatics and Telecommunications Engineering 2015
M. Weichold et al. (Eds.): CROWNCOM 2015, LNICST 156, pp. 271–281, 2015.
DOI: 10.1007/978-3-319-24540-9_22

The DTMB system specifies three major modes of Pseudo-noise (PN) sequences padding with PN420/595/945 as Frame Header (FH) which is shown in Figure 1. They can be utilized for signal synchronization and channel estimation at the receiver. Several detectors have been derived, which exploited properties of PN sequence, for example the auto-correlation detector is proposed in [3], and in [10] the cross-correlation property is employed. The detector proposed in [6] is an extension of an auto-correlation detector. It takes advantage of max-mean-ratio as test statistic that is a true constant false-alarm rate (CFAR) detector owing to its "self-normalizing" feature, which removes the dependence on the noise power. All the aforementioned algorithms are designated for single-mode signal. They are not capable of detecting multi-mode signals. Therefore a low time complexity and implementation-friendly detector for multi-mode signals is necessary for the DTMB system, which is the main focus of this manuscript.

In this paper, we focus on the spectrum sensing algorithm for DTMB multi-mode signals. By exploiting addition and auto-correlation properties of PN sequence, two Multi-mode Local Sequence (MLS) combination schemes are proposed and analyzed quantitatively in detail. Theoretical analysis and simulation results show that MLS detectors can achieve multi-mode signal detection with a tolerable loss of sensing performance.

The rest of this paper is organized as what follows. The single-mode signal detector for DTMB system is introduced in Section 2. MLS based multi-mode signal detection methodologies are analyzed in detail in Section 3. The simulation results are demonstrated in Section 4. Finally Section 5 concludes the manuscript.

2   Single-Mode Signal Detector

In what follows, the optimal detector of single-mode signal is derived from first principles. Regarding a known deterministic signal for Additive White Gaussian Noise (AWGN) scenario, the optimal detector is Neyman-Person (NP) test, known as matched filter [5].

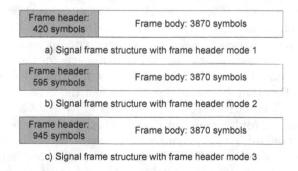

a) Signal frame structure with frame header mode 1

b) Signal frame structure with frame header mode 2

c) Signal frame structure with frame header mode 3

Fig. 1. Frame structure of DTMB system

Let $\mathcal{H}_0$ be the null hypothesis, i.e., a primary user is absent, and $\mathcal{H}_1$ be the alternate hypothesis, i.e., DTMB signals presence. Therefore, the hypothesis testing problem can be written as

$$\mathcal{H}_1 : \underline{r}_i = \underline{c}_i + \underline{n}_i, \qquad i = 0, 1, \ldots, N-1, \tag{1}$$

$$\mathcal{H}_0 : \underline{r}_i = \underline{n}_i, \qquad i = 0, 1, \ldots, N-1, \tag{2}$$

where $\underline{c}$ denotes one of original three modes PN sequences. N is the number of received signal frames. $\underline{r}_i$ denotes the received signal vector. $\underline{n}_i$ denotes the AWGN vector and it is a complex circularly symmetric Gaussian random variable with distribution $\mathcal{CN}(\mathbf{0}, \sigma_n^2 \mathbf{I})$.

The NP detector is in favors of $\mathcal{H}_1$ if the Likelihood Ratio Test (LRT) exceeds a threshold, like

$$L(\underline{r}) = \frac{p(\underline{r}; \mathcal{H}_1)}{p(\underline{r}; \mathcal{H}_0)} > \gamma. \tag{3}$$

Since

$$p(\underline{r}; \mathcal{H}_1) = \frac{1}{(\pi \sigma_n^2)^{NM}} \exp\left[-\frac{1}{\sigma_n^2} \sum_{i=0}^{N-1} \| \underline{r}_i - \underline{c}_i \|^2 \right], \tag{4}$$

and

$$p(\underline{r}; \mathcal{H}_0) = \frac{1}{(\pi \sigma_n^2)^{NM}} \exp\left[-\frac{1}{\sigma_n^2} \sum_{i=0}^{N-1} \| \underline{r}_i \|^2 \right], \tag{5}$$

we have

$$T(\underline{r}) = \frac{1}{\sigma_n^2} \sum_{i=0}^{N-1} \Re\left(\underline{c}_i^H \underline{r}_i \right) > \ln \gamma + \frac{N\varepsilon}{2\sigma_n^2}, \tag{6}$$

$$\Rightarrow T(\underline{r}) = \sum_{i=0}^{N-1} \Re\left(\underline{c}_i^H \underline{r}_i \right) \geq \gamma'. \tag{7}$$

where ε is the energy of $\underline{c}$. $T(\underline{r})$ is the test statistic of optimal detector of single-mode signal.

Thus, under both hypotheses, the distribution of $T(\underline{r})$ is given by

$$T(\underline{r}) \sim \begin{cases} \mathcal{N}\left(0, N\sigma_n^2 \varepsilon/2\right) & \text{under } \mathcal{H}_0 \\ \mathcal{N}\left(N\varepsilon, N\sigma_n^2 \varepsilon/2\right) & \text{under } \mathcal{H}_1. \end{cases} \tag{8}$$

Therefore, P_{FA} and P_D can be derived from (8), and shown below,

$$P_{\mathrm{FA}} = \Pr\{T > \gamma'; \mathcal{H}_0\} = Q\left(\frac{\gamma'}{\sqrt{N\sigma_n^2 \varepsilon/2}} \right) \tag{9}$$

$$P_{\mathrm{D}} = \Pr\{T > \gamma'; \mathcal{H}_1\} = Q\left(\frac{\gamma' - N\varepsilon}{\sqrt{N\sigma_n^2\varepsilon/2}}\right),$$ (10)

where Q is the right-tail probability function of the standard normal distribution. The new threshold is derived by

$$\gamma' = \sqrt{\frac{N\sigma_n^2\varepsilon}{2}} Q^{-1}\left(P_{\mathrm{FA}}\right).$$ (11)

Substituting (11) to (10), we have

$$P_{\mathrm{D}} = Q\left(Q^{-1}\left(P_{\mathrm{FA}}\right) - \sqrt{\frac{2N\varepsilon}{\sigma_n^2}}\right).$$ (12)

It is known that $N\varepsilon/\sigma_n^2$ is Energy Noise Ratio (ENR) [5]. It presents the ratio between the whole energy from received signal and the noise power spectrum density.

3   Multi-mode Signal Detector

In this section, multi-mode detectors are derived from the NP test for AWGN scenario. The key observations for deducing the multi-mode detector are the addition and auto-correlation properties of PN sequence. In the DTMB system, PN255 and PN511 sequences are cyclically extended to sequences of length 420 and 945 for Mode 1 and Mode 3 frames, respectively. Due to reduction of the analysis and time complexity, only PN255, PN511 and PN595 are utilized for MLS design.

3.1   Multi-mode Local Sequence Design

PN sequences are quasi-orthogonal, which means

$$\frac{\underline{c}_i(p)^{\mathrm{H}}\underline{c}_j(q)}{M} \approx \begin{cases} 1 & p = q \text{ and } i = j \\ 0 & \text{others}, \end{cases}$$ (13)

where M is the length of $\underline{c}$, $\underline{c}_i(p)$, $\underline{c}_j(q)$ are different PN sequences from different modes and with different phases. Based on this observation, three different multi-mode local sequences with serial, parallel, and mixed structures are introduced for the DTMB system [11] below.

The structure of Serial Multi-Mode Local Sequence (SMLS) is shown in Figure 2 and its vector form is represented as

$$\underline{c}_{\mathrm{S}} = \begin{bmatrix} \underline{c}_1 \\ \underline{c}_2 \\ \underline{c}_3 \end{bmatrix}.$$ (14)

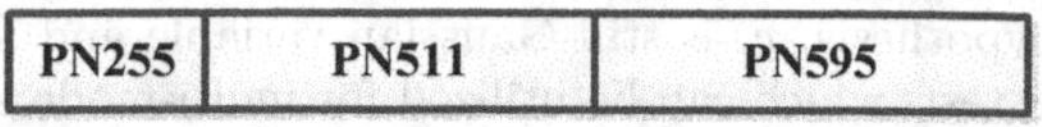

PN255
PN511
PN595

Fig. 2. Structure of serial multi-mode local Sequence

$\underline{c}_1$
$\underline{c}_{3,1}$
$\underline{c}_{3,2}$
PN255
PN511
PN595
0..0
$\underline{c}_{2,1}$
$\underline{c}_{2,2}$

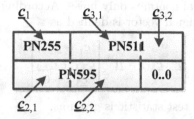

Fig. 3. Structure of mix multi-mode local sequence

For ease of notation and calculation, the MLS is divided into different parts. The structure of Mix Multi-mode Local Sequence (MMLS) is shown in Figure 3, and its vector form is represented as

$$\underline{c}_{\mathrm{M}} = \begin{bmatrix} \underline{c}_1 + \underline{c}_{2,1} \\ \underline{c}_{3,1} + \underline{c}_{2,2} \\ \underline{c}_{3,2} + \mathbf{0} \end{bmatrix}. \tag{15}$$

Similarly, the Parallel Multi-mode Local Sequence (PMLS) is also divided into parts and shown in Figure 4 and its vector form is represented as

$$\underline{c}_{\mathrm{P}} = \begin{bmatrix} \underline{c}_1 + \underline{c}_{2,1} + \underline{c}_{3,1} \\ \underline{c}_{2,2} + \underline{c}_{3,2} + \mathbf{0} \\ \underline{c}_{2,3} + \mathbf{0} \end{bmatrix}, \tag{16}$$

where $\underline{c}_1$ and $\underline{c}_2$ must be padded with $\mathbf{0}$ until the same length as $\underline{c}_3$.

Since the mount of calculation is proportional to the length of sequence, the SMLS has the same time complexity as parallel single-mode detectors, which is not satisfied with the requirement of time complexity and therefore analysis and simulation of the SMLS detector is left out.

$\underline{c}_1$
$\underline{c}_{3,1}$
$\underline{c}_{2,1}$
PN255
0..0
PN511
0
PN595
$\underline{c}_{3,2}$
$\underline{c}_{2,2}$
$\underline{c}_{2,3}$

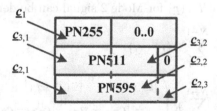

Fig. 4. Structure of parallel multi-mode local sequence

Under either hypothesis, $\underline{r}_i$ is still Gaussian variable and $T(\underline{r})$ satisfies the requirement of NP test, which can be utilized for multi-mode signal detection.

3.2   Mix Multi-mode Local Sequence Detector

Under $\mathcal{H}_0$, received signal contains only noise. According to (7), inner product of MMLS and received signal vector is defined as

$$\langle \underline{r}^*, \underline{c}_M \rangle = \underline{n}^H \begin{bmatrix} \underline{c}_1 + \underline{c}_{2,1} \\ \underline{c}_{3,1} + \underline{c}_{2,2} \\ \underline{c}_{3,2} + \mathbf{0} \end{bmatrix}. \tag{17}$$

Substituting (17) to (7), test statistic is given as,

$$T_M(\underline{n}) = \sum_{i=0}^{N-1} \Re\left(\underline{n}_i^H \underline{c}_M\right)$$

$$= \sum_{i=0}^{N-1} \Re\begin{pmatrix} \underline{n}_{i,1}^H \underline{c}_1 + \underline{n}_{i,1}^H \underline{c}_{2,1} \\ + \underline{n}_{i,2}^H \underline{c}_{3,1} + \underline{n}_{i,2}^H \underline{c}_{2,2} \\ + \underline{n}_{i,3}^H \underline{c}_{3,2} + \underline{n}_{i,3}^H \mathbf{0} \end{pmatrix}. \tag{18}$$

Under $\mathcal{H}_0$, we have

$$\mathrm{E}\left(T_M(\underline{n}); \mathcal{H}_0\right) = 0 \tag{19}$$

$$\mathrm{Var}\left(T_M(\underline{n}); \mathcal{H}_0\right) = 1491 N \sigma_n^2. \tag{20}$$

The $T_M(\underline{n})$ consists of both auto-correlation of single mode PN sequence and its cross-correlation of other overlapping sequences. In addition, for the given MMLS, and referencing to (11), the threshold is fixed for different modes of signal.

Under $\mathcal{H}_1$, taking Mode 2 signals as the primary signal, each received signal consists of noise, a Mode 2 PN sequence and some data from frame body. The structure of MMLS and received Mode 2 signal is shown in Figure 5. Similarly, inner product of MMLS and received signal is defined as

$$\langle \underline{r}^*, \underline{c}_M \rangle = \begin{bmatrix} \underline{c}_{2,1} + \underline{n}_1 \\ \underline{c}_{2,2} + \underline{n}_2 \\ \underline{d} + \underline{n}_3 \end{bmatrix}^H \begin{bmatrix} \underline{c}_1 + \underline{c}_{2,1} \\ \underline{c}_{3,1} + \underline{c}_{2,2} \\ \underline{c}_{3,2} + \mathbf{0} \end{bmatrix}. \tag{21}$$

From (21), test statistic $T_M(\underline{r})$ for Mode 2 signal can be derived and defined as

$$T_M(\underline{r}) = \sum_{i=0}^{N-1} \Re\left(\underline{r}_i^H \underline{c}_M\right)$$

$$= \sum_{i=0}^{N-1} \Re\begin{pmatrix} \left[\underline{c}_{2,1} + \underline{n}_1\right]^H \left[\underline{c}_1 + \underline{c}_{2,1}\right] \\ + \left[\underline{c}_{2,2} + \underline{n}_2\right]^H \left[\underline{c}_{3,1}^H + \underline{c}_{2,2}\right] \\ + \left[\underline{d} + \underline{n}_2\right]^H \left[\underline{c}_{3,2} + \mathbf{0}\right] \end{pmatrix}. \tag{22}$$

Therefore, we have

$$E\left(T_M\left(\underline{r}\right);\mathcal{H}_1\right) = 660N. \tag{23}$$

The mean value of $T_M\left(\underline{r}\right)$ consists of the real part of auto-correlation of PN595 and its correlation value with PN255 and part of PN511. By the similar deviation, we have

$$\mathrm{Var}\left(T_M\left(\underline{r}\right);\mathcal{H}_0\right) = 1491N\sigma_n^2 + 353550N. \tag{24}$$

Note that multi-mode detectors work in a very low Signal Noise Ratio (SNR) scenario, where energy of signal is much lower than σ_n^2, which means that the variance of test statistic under $\mathcal{H}_1$ can be estimate by its variance under $\mathcal{H}_0$.

Finally, distributions of test statistic $T_M\left(\underline{r}\right)$ for two hypotheses are shown below.

$$T_M\left(\underline{r}\right) \sim \begin{cases} \mathcal{N}\left(0, 1491N\sigma_n^2\varepsilon\right) & \text{under } \mathcal{H}_0 \\ \mathcal{N}\left(660N, 1491N\sigma_n^2\varepsilon\right) & \text{under } \mathcal{H}_1. \end{cases} \tag{25}$$

3.3   Parallel Multi-mode Local Sequence Detector

Under $\mathcal{H}_0$, each received signal contains only noise. According to (7), inner product of PMLS and received signal vector is defined as

$$\langle \underline{r}^*, \underline{c}_P \rangle = \underline{n}^H \left(\begin{bmatrix} \underline{c}_1 \\ 0 \end{bmatrix} + \begin{bmatrix} \underline{c}_{3,1} \\ \underline{c}_{3,2} \end{bmatrix} + \begin{bmatrix} \underline{c}_{2,1} \\ \underline{c}_{2,2} \\ \underline{c}_{2,3} \end{bmatrix} \right). \tag{26}$$

By the similar deviation, we have

$$E\left(T_P\left(\underline{n}\right);\mathcal{H}_0\right) = 0 \tag{27}$$

$$\mathrm{Var}\left(T_P\left(\underline{n}\right);\mathcal{H}_0\right) = 1440N\sigma_n^2. \tag{28}$$

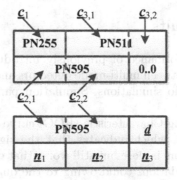

Fig. 5. Mix multi-mode local sequence and received DTMB Mode 2 signal, where each received signal consists of three parts, PN595, data frame and noise, which is divide into three vectors.

Likewise, the variance of $T_P(\underline{n})$ consists of both auto-correlation of PN595 and its cross-correlation of overlapping sequences from PN255 and PN511. In addition, the threshold for PMLS is also unique, which leads to a fix threshold for different modes of signal.

Under hypothesis $\mathcal{H}_1$, taking Mode 2 signal as primary signal. Similarly, inner product of PMLS and received signal is defined as

$$\langle \underline{r}^*, \underline{c}_P \rangle = \begin{bmatrix} \underline{c}_{2,1} + \underline{n}_1 \\ \underline{c}_{2,2} + \underline{n}_2 \\ \underline{c}_{2,3} + \underline{n}_3 \end{bmatrix}^H \left(\begin{bmatrix} \underline{c}_1 \\ 0 \end{bmatrix} + \begin{bmatrix} \underline{c}_{3,1} \\ \underline{c}_{3,2} \\ 0 \end{bmatrix} + \begin{bmatrix} \underline{c}_{2,1} \\ \underline{c}_{2,2} \\ \underline{c}_{2,3} \end{bmatrix} \right). \tag{29}$$

Substituting (29) to (7), test statistic $T_P(\underline{r})$ for Mode 2 signal is shown as

$$T_P(\underline{r}) = \sum_{i=0}^{N-1} \Re\left(\underline{r}_i^H \underline{c}_P \right)$$

$$= \sum_{i=0}^{N-1} \Re \left(\begin{array}{c} \left[\underline{c}_{2,1} + \underline{n}_1 \right]^H \left[\underline{c}_1 + \underline{c}_{2,1} + \underline{c}_{3,1} \right] \\ + \left[\underline{c}_{2,2} + \underline{n}_2 \right]^H \left[\underline{c}_{3,2} + \underline{c}_{2,2} \right] \\ + \left[\underline{c}_{2,3} + \underline{n}_2 \right]^H \underline{c}_{2,3} \end{array} \right). \tag{30}$$

By the similar derivation, we have

$$\mathrm{E}\left(T_P(\underline{r}); \mathcal{H}_1\right) = 630N \tag{31}$$

$$\mathrm{Var}\left(T_P(\underline{r}); \mathcal{H}_1\right) = 1440N\sigma_n^2 + 350150N. \tag{32}$$

Finally, with the same estimation of variance above under $\mathcal{H}_1$, the distributions of test statistic $T_P(\underline{r})$ for two hypotheses are shown below.

$$T_P(\underline{r}) \sim \begin{cases} \mathcal{N}\left(0, 1440N\sigma_n^2\varepsilon\right) & \text{under } \mathcal{H}_0 \\ \mathcal{N}\left(630N, 1440N\sigma_n^2\varepsilon\right) & \text{under } \mathcal{H}_1. \end{cases} \tag{33}$$

4   Simulation Results

To show the overall performance of previously introduced algorithms, signal sensitivity comparisons between multi-mode detectors and single-mode detectors are given via Monte Carlo simulations. Simulation parameters are showed in Table 1.

The Receiver Operating Characteristic (ROC) curve for DTMB Mode 1 signal is shown in Figure 6, which indicates that the signal sensitivity of both MMLS and PMLS detectors is close but still worse that of signal-mode detector. The reason of such degradations could owing to the quasi-orthogonal property of the PN sequence with finite length. It will results in the decrease of ENR.

The signal sensitivity in terms of probability of detection over SNR is depicted in Figure 7 and Figure 8. For DTMB Mode 2 signals, taking 20 signal frames,

Table 1. Simulation parameters

Parameter	Value
Signal mode	PN255, PN511, PN595
Number of received frames	20
Target probability of false-alarm	10%
Propagation channel	AWGN

the single-mode detector can achieve the signal sensitivity in term of SNR as low as -35.43 dB and $P_{MD} = 10\%$. In the same situation, MMLS and PMLS detectors are capable of -32.67 dB and -31.81 dB, respectively. For DTMB Mode 1 signals, taking the same number of received frames, a sensitivity of -32.09 dB, -22.09 dB and -19.09 dB can be reached by single-mode detector, MMLS and PMLS detectors, respectively when $P_{MD} = 10\%$. The sensing time of 20 frames takes approximately 11 ms, which is much less than the predefined 20 ms. Therefore the sensing performance of both multi-mode detectors can satisfy the predefined sensitivity requirements of the DTMB system.

All the figures show that the MMLS detector has slight better sensing performance than PMLS detector. The reason is that the MMLS has larger ENR under $\mathcal{H}_1$. According to the discussion in Section 2, the ENR is proportional to the ratio of mean value and variance of test statistic. For Mode 2 signal, from distributions given in (25) and (33), we can find that the ratio of MMLS is 0.442 that is larger than 0.4375 of PMLS.

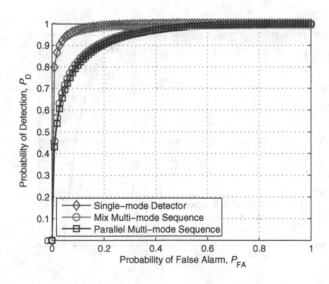

Fig. 6. ROC curves of single-mode detector, MMLS detector and PMLS detector for DTMB Mode 1 signal when SNR $= -34$ dB.

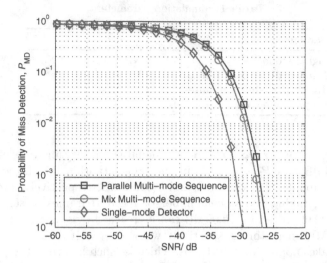

Fig. 7. Sensing performance of DTMB Mode 2 Signal when $N = 20$, $P_{\mathrm{FA}} = 10\%$.

Comparing with other modes of primary signals, MLS detectors suffer more performance degradation from receiving Mode 1 signal. The reason is that with the same number of received frames, the ENR of PN255 is lower than that of PN595 and PN511. Therefore Mode 1 signal is easier susceptible to the quasi-orthogonality of PN sequence.

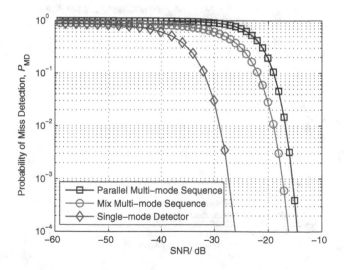

Fig. 8. Sensing performance of DTMB Mode 1 Signal when $N = 20$, $P_{\mathrm{FA}} = 10\%$.

5 Conclusion

In this paper, we have proposed two MLS detectors and their theoretical models for the DTMB multi-mode signal detection. The single-mode detector, which is derived from NP test, for the DTMB system was presented and thoroughly studied. The sensing performance of single-mode detector is the upper bound, which can be the benchmark for other detectors. The MLS multi-mode signal detectors, which are also based on Neyman-Pearson test, were fully investigated by quantitative analyses. The sensing performance was obtained through Monte Carlo simulations. Both simulation and theoretical analyses showed that MMLS and PMLS detectors can achieve the multi-mode signal detection. Due to the quasi-orthogonality of PN sequence, a tolerable performance loss of multi-mode detectors is not unavoidable.

References

1. Framing Structure, Channel Coding and Modulation for Digital Television Terrestrial Broadcasting System (2006)
2. Standard for Cognitive Wireless Regional Area Networks (RAN) for Operation in TV Band, July 2011
3. Chen, H.-S., Gao, W., Daut, D.G.: Spectrum sensing for DMB-T systems using PN frame headers. In: Proc. IEEE Int. Conf. Communications ICC 2008, pp. 4889–4893 (2008)
4. Federal Communication Commission, et al.: Second Report and Order and Memorandum Opinion and Order in the Matter of Unlicensed Operation in the TV Broadcast Bands Additional Spectrum for Unlicensed Devices Below 900 MHz and in the 3 GHz Band. Document, 260 (2008)
5. Kay, S.M.: Fundamentals of Statistical Signal Processing: Detection Theory, vol. ii, p. 7, Upper Saddle River, New Jersey (1998)
6. Lei, C., Jing, Q., Viessmann, A., Kocks, C., Bruck, G.H., Jung, P., Hu, R.Q.: A spectrum sensing prototype for tv white space in china. In: Proc. IEEE Global Telecommunications Conf., GLOBECOM 2011, pp. 1–6 (2011)
7. Mitola, J., et al.: Cognitive radio: An Integrated Agent Architecture for Software Defined Radio. Doctor of Technology, Royal Inst. Technol. (KTH), Stockholm, Sweden, pp. 271–350 (2000)
8. Shellhammer, S., Chouinard, G.: Spectrum Sensing Requirements Summary. IEEE P802 **22**, 802–822 (2006)
9. Song, J., Yang, Z., Yang, L., Gong, K., Pan, C., Wang, J., Wu, Y.: Technical Review on Chinese Digital Terrestrial Television Broadcasting Standard and Measurements on Some Working Modes **53**(1), 1–7 (2007)
10. Xu, A., Shi, Q., Yang, Z., Peng, K., Song, J.: Spectrum sensing for DTMB system based on PN cross-correlation. In: IEEE International Conference on Communications, pp. 1–5 (2010)
11. Yang, F., Peng, K., Song, J., Pan, C., Yang, Z.: Guard-interval mode detection method for the chinese DTTB system. In: Proc. Int. Conf. Communications, Circuits and Systems ICCCAS 2008, pp. 216–219 (2008)

Best Relay Selection for DF Underlay Cognitive Networks with Different Modulation Levels

Ahmed M. ElShaarany[1], Mohamed M. Abdallah[1,2]([✉]),
Salama Ikki[3], Mohamed M. Khairy[1], and Khalid Qaraqe[2]

[1] Electronics and Electrical Communications Department,
Faculty of Engineering, Cairo University, Giza, Egypt
{ahmed.m.elshaarany,mkhairy}@ieee.org
[2] Texas A&M University at Qatar, Doha, Qatar
{mohamed.abdallah,khalid.qaraqe}@qatar.tamu.edu
[3] Department of Electrical Engineering, Lakehead University,
Thunder Bay, ON, Canada
sikki@lakeheadu.ca

Abstract. In an underlay setting, a secondary user shares the spectrum with a primary user under the condition that the interference at this primary user is lower than a certain threshold. The said condition limits the transmission power and therefore, limits the coverage area. Hence, to reach remote destinations, relaying the signal between the source and destination can be an adequate solution to enhance the secondary network's performance. Selective relaying in underlay cognitive networks has been studied in many previous literatures. The source and relay nodes in most of this literature use the same modulation level. The use of multiple modulation levels by the transmitting terminals has not been explored comprehensively from the physical layer point of view. In this paper, the error performance of a secondary cognitive network with a source and multiple decode and forward (DF) relays using different modulation levels sharing the spectrum with a nearby primary user has been investigated. In particular, a closed form expression for the error probability for two scenarios have been obtained. In the first scenario, where the relays have fixed transmission power, we additionally present an approximate error probability expression that is exact at high signal-to-noise ratio. In the second scenario, where the relays adjust their transmission power such that the interference at the primary user is below a certain threshold with a defined tolerable error, it is referred to as the interference outage scenario.

Keywords: Underlay cognitive radio · Relay selection · Performance analysis · Different modulation levels

1 Introduction

The continuous pursuit of higher data rates rises day by day due to the increase of wireless applications, wireless multimedia and interactive wireless services.

© Institute for Computer Sciences, Social Informatics and Telecommunications Engineering 2015
M. Weichold et al. (Eds.): CROWNCOM 2015, LNICST 156, pp. 282–294, 2015.
DOI: 10.1007/978-3-319-24540-9_23

This lead to the emergence of more and more wireless technologies every day. These technologies are inefficiently utilizing the usable spectrum. Based upon reports published by the Federal Communications Commission (FCC) [1], the spectrum utilization efficiency reaches percentiles as low as 15%. Such low utilization compelled researchers to find and exploit new techniques to make use of the unused spectrum in a cognitive fashion[2,3].

In short, in a cognitive network, the secondary (unlicensed) user can make use of the unused spectrum portions by the primary user. These unused spectrum portions are known as spectrum holes [3]. If the primary user is to acquire its proprietary spectrum back, the secondary user searches for a new spectrum hole or stops transmitting. This manner enjoins that the secondary user applies spectrum sensing techniques. In-band operation of both the primary and secondary user is possible but demands complicated interference cancellation methods. This method is known as the overlay operation method.

The more simple in-band operation is the underlay operation method. In the latter method, the primary and secondary user operate in the same band on condition that the interference on the primary user is below a certain threshold [2]. This method of operation limits the coverage area of the secondary network due to the constrained power of the secondary transmitter. Consequently, relaying the signal is suggested as an adequate solution to solve the limited coverage area problem.

Two of the most famous relaying operating modes is the amplify and forward (AF) mode and the decode and forward (DF) mode[4]. In AF mode, the signal is received by the relay, amplified by a factor and afterwards forwarded to the destination. In DF mode, the relay decodes the received signal, reproduces it and then forwards the regenerated signal to the destination.

It is worthy to note that although DF mode may suffer from computational delay, it gives a slightly better performance than AF mode[4]. In order to efficiently utilize the spectrum, selective relaying was recently suggested in which a single best relay is selected to relay the signal from the source to the destination[5]. This best relay is selected based on the signal to noise (SNR) it can provide at the destination.

However, in most of the previous literature, the selection was either from a set of relays that were all AF relays or from a set of DF relays that all use the same modulation level. For the relay selection algorithm in [6], the authors based the relay selection criteria on the quotient of the SNR to the interference induced by the relay to the primary user. The best relay selected in this criteria is the relay with the maximum quotient among the relays operating in AF mode in the secondary underlay network. The authors derived closed form expressions for the outage probability and bit error probability.

Another relay selection algorithm proposed in [7] where the best selected relay, operating in DF mode, satisfies an outage probability constraint at the primary network. The authors in [8] propose a selection scheme that in which the relays operate also in DF mode and takes into consideration that an interference constraint at the primary user is not violated. The authors also derived the outage probability. The secondary nodes in both [7] and [8] have the ability to adjust their transmission powers to avoid violating the interference constraint. The authors in [9] suggest a cooperative network in which the source and relay nodes employ different modulation levels. The relays have fixed power and operate in DF mode.

Most of the aforementioned selection techniques are SNR-based and assume that all nodes are using the same modulation level. However, for the case where the modulation level is different, selection based on BER can be more appropiate than SNR-based. The reason for that is that when we have different modulation levels, simply selecting the signal with the highest SNR is not optimal. SNR-based selection does not take into consideration the error flexibility of each received signal. Therefore, in case of different modulation levels, the system performance where BER-based selection is employed is the optimal choice. For a non-cognitive setting, the authors derived a closed form expression for the BER and they only considered the scenario where the relays have fixed transmission power.

In this paper, we propose an extension to the work done in [9]. We suggest an underlay cognitive network in which the source and DF relays use different modulation levels. We derive the corresponding closed form BER expression for two scenarios. In the first scenario, we assume that the relays have a fixed transmission power. In the second scenario, we assume that the relays can adjust their transmission power to satisfy a certain average interference at the primary user.

2   System Model

The system consists of a secondary source S that broadcasts its signal to a secondary destination D. The transmitted signal is passed on to the destination through K DF relays R_k, $k = 1, 2, ..., K$ as illustrated by Fig. 1. These secondary nodes are sharing the spectrum with a primary user P. Each node is assumed to have a single antenna. Communication is achieved over two time slots. In the first time slot, the source S transmits an N-bit packet with power P_s using $M_s - QAM$ modulation scheme to the K relays and the destination with channel gains h_{1k}, h_0, respectively. The channel gain from relay k to the destination is h_{2k}. Each relay is assumed to transmit with a maximum power of $P_{R_k D}$. Each hop suffers from additive white Gaussian noise (AWGN) with zero mean and variance N_0. The channel gains are modeled as a Rayleigh distribution. Hence, the instantaneous SNRs in the hops $S - D$, $S - R_k$, and $R_k - D$ are independent exponential random variables (rv) and are given by $\gamma_{SD} = \frac{P_s |h_0|^2}{N_0}$, $\gamma_{SR_k} = \frac{P_s |h_{1k}|^2}{N_0}$, and $\gamma_{R_k D} = \frac{P_{R_k D} |h_{2k}|^2}{N_0}$, respectively. The average SNRs in the hops $S - D$, $S - R_k$, and $R_k - D$ are denoted by $\bar{\gamma}_{SD}$, $\bar{\gamma}_{SR_k}$, and $\bar{\gamma}_{R_k D}$ respectively. The relays receive the packet, decode it and check its correctness through cyclic redundancy check (CRC). A decoding set $\mathcal{DS}$ of candidate relays is formed which contains relays that have received the packet correctly. The relays are also assumed to use $M_{R_k} - QAM$ modulation. Therefore, the BER as

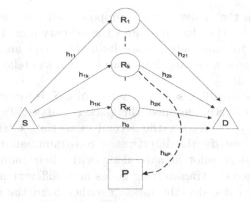

Fig. 1. System Model: A secondary underlay cognitive network close to a primary user

a function of the end-to-end SNR for the square Gray-coded $M - QAM$ is given by [10] as $\text{BER}_{M_i}(\gamma_{iD}) \approx c_{M_i} Q\left(\sqrt{2d^2_{M_i}\gamma_{iD}}\right)$ where

$$(c_{M_i}, d_{M_i}) = \begin{cases} (1,1), & M_i = 2, \\ \left(\dfrac{2-2/\sqrt{M_i}}{\log_2 \sqrt{M_i}}, \sqrt{\dfrac{3}{2(M_i-1)}}\right), & M_i \geq 4. \end{cases}$$

Since the secondary network is operating in an underlay setting, it is important to operate under strict interference limits so that the secondary network does not affect the primary user. Therefore, it is important to propose relay selection algorithms in an according manner. The cognitive selection algorithms, along with their corresponding performance analysis, are explained in the next subsections.

2.1   Fixed Power Underlay Relay Selection

As we previously mentioned, an underlay cognitive network dictates to operate under strict interference limits so that the primary user is not affected. As a result, we set an interference threshold λ. An interference at the primary receiver above this threshold is unacceptable.

In this selection algorithm, each relay is assumed to know whether the interference it generates at the primary receiver satisfies the interference constraint or not. The interference generated by a relay on the primary user is given by $I_{R_k P} = P_{R_k D}|h_{kP}|^2$. Therefore, the interference from the kth relay to the primary user follows an exponential distribution and its probability density function (pdf) is given by

$$p_{I_{R_k P}}(x) = \frac{1}{\mu_{R_k P}} e^{-\frac{x}{\mu_{R_k P}}}, \tag{1}$$

where $\mu_{R_k P}$ is the average value of the interference of the kth relay on the primary user. We assume that the interfering channels are generated with $\mu_{R_k P} = \alpha \bar{\gamma}_{R_k D} = \alpha \frac{P_{R_k D}}{N_0}$, where α is a constant>0.

For simplicity in the analysis of the proposed scheme, we assume that the source is non-cognitive (i.e. does not affect the primary user) and the relays are cognitive. In order to avoid interference from the relays on the primary user higher than λ, we must take only into consideration the relays that satisfy the interference constraint.

Therefore, a new decoding set $\mathcal{DS}^*$, subset of $\mathcal{DS}$, is formed which contains relays that have correctly decoded the packet and satisfy the interference constraint (i.e. $I_{R_k P} \leq \lambda$). In the second time slot, after ruling out the relays that do not satisfy the interference constraint, all the relays in the $\mathcal{DS}^*$ send independent pilot signals along with their modulation levels to the destination. Since the transmitting nodes have different modulation levels, then the destination decodes the message either from the source or one of the relays based upon the biased SNR (i.e. BER-based selection where the destination selects to decode the message from one node only).

Thus, the destination calculates the approximate SNRs from the relays and the sources. According to the received SNRs and modulation levels of the relays in the $\mathcal{DS}^*$, $\{M_{R_k} | k \in \mathcal{DS}^*\}$, and the modulation level of the source, M_S, the destination chooses to decode from one of the candidate relays in $\mathcal{DS}^*$ or directly from the source by comparing the received weighted SNR's and selecting the SNR that minimizes the BER.

As a result, the instantaneous BER, according to the BER-based selection, at the secondary destination is given by

$$
\mathrm{BER}_{comp,\,inst} \approx
$$

$$
\begin{cases}
c_{M_S} Q\left(\sqrt{2d_{M_S}^2 \gamma_{SD}}\right), & \gamma_{SD} \geq \rho_i \gamma_{R_i D}, i \in \mathcal{DS}^* \\[3mm]
c_{M_{R_i}} Q\left(\sqrt{2d_{M_{R_i}}^2 \gamma_{R_i D}}\right), & \begin{array}{l} \gamma_{SD} < \rho_i \gamma_{R_i D}, \text{and} \\ \gamma_{R_j D} < \beta_{ij} \gamma_{R_i D}, \\ j \neq i, j \in \mathcal{DS}^*, \end{array}
\end{cases}
\tag{2}
$$

where $\rho_i = d_{M_{R_i}}^2 / d_{M_S}^2$ is a biasing factor between the relays and the source and $\beta_{ij} = d_{M_{R_i}}^2 / d_{M_{R_j}}^2, i, j = 1, 2, ..., K$ is the biasing factor between the relays.

It is obvious that if all nodes have the same modulation level, then, BER-based selection algorithm becomes SNR-based selection algorithm, i.e., $\rho_i = \beta_{ij} = 1$. Hence, the average BER of this selection scheme can be written as

$$
\mathrm{BER} = \left(\prod_{k=1}^{K} \mathrm{PER}_{SR_k}\right) \mathrm{BER}_{SD} + \sum_{r=1}^{K} \sum_{m=1}^{|P_r(\mathcal{S}_{all})|} \left[\left(\prod_{e_i \in P_{r,m}(\mathcal{S}_{all})} (1 - \mathrm{PER}_{SR_{e_i}})\right)\right.
$$

$$
\times \left(\prod_{e_o \notin P_{r,m}(\mathcal{S}_{all})} \mathrm{PER}_{SR_{e_o}}\right) \left(\prod_{e_i \in P_{r,m}(\mathcal{S}_{all})} P_{\lambda_{R_{e_i}} P} \mathrm{BER}_{comp_{P_{r,m}(\mathcal{S}_{all})}}\right)
$$

$$
+ \prod_{e_i \in P_{r,m}(\mathcal{S}_{all})} (1 - P_{\lambda_{R_{e_i}} P}) \mathrm{BER}_{SD} + \sum_{\substack{l_1, l_2, ..., l_K \in \{0,1\} \\ l_1 l_2 ... l_K \neq 1 \\ (1 - l_1)(1 - l_2)...(1 - l_K) \neq 1}} \Theta_{l_1, l_2, ..., l_K} \mathrm{BER}_{comp\{i, \forall l_i = 1\}} \left.\right)\Bigg],
$$

$$
\tag{3}
$$

where

- $\mathcal{S}_{all}$ is the set of all relays indices, i.e., $\mathcal{S}_{all} = 1, 2, ..., K$,
- $P_r(\mathcal{S}_{all})$ is the r-element power set of $\mathcal{S}_{all}$,
- $P_{r,m}(\mathcal{S}_{all})$ is the m-th element of $P_r(\mathcal{S}_{all})$ as defined in [9],
- $|P_r(\mathcal{S}_{all})|$ represents the cardinality of $P_r(\mathcal{S}_{all})$,
- PER_{SR_k} is the average packet error rate in $S - R_k$ link,
- BER_{SD} is the average BER in $S - D$ link,
- $\text{BER}_{comp\mathcal{DS}}$ is average BER conditioned on the $\mathcal{DS}$ at the destination.
- $P_{\lambda_{R_k P}}$ is the probability that $I_{R_k P}$ is less than λ and is given by

$$P_{\lambda_{R_k P}} = \Pr(I_{R_k P} < \lambda) = 1 - e^{-\frac{\lambda}{\mu_{R_k P}}}. \qquad (4)$$

- $\Theta_{l_1, l_2, ..., l_K}$ is defined as

$$\Theta_{l_1, l_2, ..., l_K} = \left[\left(P_{\lambda_{R_1 P}} l_1 + (1 - P_{\lambda_{R_1 P}})(1 - l_1) \right) \left(P_{\lambda_{R_2 P}} l_2 + (1 - P_{\lambda_{R_2 P}})(1 - l_2) \right) \right.$$
$$\left. ... \times \left(P_{\lambda_{R_K P}} l_K + (1 - P_{\lambda_{R_K P}})(1 - l_K) \right) \right]. $$
$$(5)$$

The average BER between nodes i and j for M-QAM in case of a Rayleigh fading channel can be estimated as

$$\text{BER}_{ij} \approx \int_0^\infty c_{M_i} Q\left(\sqrt{2 d_{M_i}^2 \gamma_{iD}} \right) \frac{1}{\bar{\gamma}_{ij}} e^{\frac{\gamma_{ij}}{\bar{\gamma}_{ij}}} d\gamma_{ij} = \frac{1}{2} c_{M_i} \left(1 - \sqrt{\frac{d_{M_i}^2 \bar{\gamma}_{ij}}{1 + d_{M_i}^2 \bar{\gamma}_{ij}}} \right). \qquad (6)$$

Assuming symbol errors occur independently in the N-bit packet, PER is then given by

$$\text{PER}_{SR_i} = 1 - (1 - \text{SER}_{SR_i})^{\frac{N}{\log_2 M_S}}$$
$$\approx 1 - \left(1 - \frac{1}{2} c_{M_S} \log_2(M_S) \left(1 - \sqrt{\frac{d_{M_S}^2 \bar{\gamma}_{SR_i}}{1 + d_{M_S}^2 \bar{\gamma}_{SR_i}}} \right) \right)^{\frac{N}{\log_2 M_S}}, \qquad (7)$$

where for Gray-coded constellations $\text{SER} \approx (\log_2 M_S)\text{BER}$[10]. BER_{comp} for a certain set of relays is given by [9, Eq. 19]. Therefore, by substituting (4), (5), (6), (7), and [9, Eqs. 19] in (3), we get a closed form expression for the average BER in case of fixed power underlay relay selection given by (8) in the next page, where $HM\{.\}$ is the harmonic mean; the set is defined as $\mathcal{S} = \{\rho_i \bar{\gamma}_{R_i D}\}$, $\mathcal{S}_x = \{\bar{\gamma}_{SD} \rho_i^{-1}, \bar{\gamma}_{R_j D} \beta_{ij}^{-1}\}$, $j \neq i$, $i, j = 1, 2, ..., K$, $P_{k,y}(\mathcal{S})$ is the y-th element of the k-element power set of $\mathcal{S}$, and $P_{k,y}(\mathcal{S}_x)$ is the y-th element of the k-element power set of $\mathcal{S}_x$. The following function was used $I(a, b, c) = \int_0^\infty a Q\left(\sqrt{2bt} \right) \frac{1}{c} e^{-\frac{t}{c}} dt = \frac{a}{2} \left(1 - \sqrt{\frac{bc}{1+bc}} \right)$.

$$
\text{BER} = \prod_{k=1}^{K} \left[1 - \left(1 - \frac{1}{2} c_{M_S} \log_2(M_S) \left(1 - \sqrt{\frac{d_{M_S}^2 \bar{\gamma}_{SR_k}}{1 + d_{M_S}^2 \bar{\gamma}_{SR_k}}} \right) \right)^{\frac{N}{\log_2 M_S}} \right] \text{BER}_{SD}
$$

$$
+ \sum_{r=1}^{K} \sum_{m=1}^{|P_r(S_{all})|} \left[\prod_{e_i \in P_{r,m}(S_{all})} \left(\left(1 - \frac{1}{2} c_{M_S} \log_2(M_S) \left(1 - \sqrt{\frac{d_{M_S}^2 \bar{\gamma}_{SR_{e_i}}}{1 + d_{M_S}^2 \bar{\gamma}_{SR_{e_i}}}} \right) \right)^{\frac{N}{\log_2 M_S}} \right) \right.
$$

$$
\times \prod_{e_o \notin P_{r,m}(S_{all})} \left(1 - \left(1 - \frac{1}{2} c_{M_S} \log_2(M_S) \left(1 - \sqrt{\frac{d_{M_S}^2 \bar{\gamma}_{SR_{e_o}}}{1 + d_{M_S}^2 \bar{\gamma}_{SR_{e_o}}}} \right) \right)^{\frac{N}{\log_2 M_S}} \right)
$$

$$
\times \left(\prod_{e_i \in P_{r,m}(S_{all})} \left(1 - e^{-\frac{\lambda}{\mu R_{e_i} P}} \right) \right) \left[I \left(c_{M_S}, d_{M_S}^2, \bar{\gamma}_{SD} \right) + \sum_{k=1}^{K} \sum_{y=1}^{\binom{K}{k}} (-1)^k \right.
$$

$$
\times I \left(\frac{c_{M_S}}{\bar{\gamma}_{SD}}, d_{M_S^2}, \frac{k+1}{HM\{\bar{\gamma}_{SD}, P_{k,y}(S)\}} \right) \left(\frac{k+1}{HM\{\bar{\gamma}_{SD}, P_{k,y}(S)\}} \right)
$$

$$
+ \sum_{i=1}^{K} \left[I \left(c_{M_{R_i}}, d_{M_{R_i}}^2, \bar{\gamma}_{R_i D} \right) + \sum_{k=1}^{K} \sum_{y=1}^{\binom{K}{k}} (-1)^k I \left(\frac{c_{M_{R_i}}}{\bar{\gamma}_{R_i D}}, d_{M_{R_i}^2}, \frac{k+1}{HM\{\bar{\gamma}_{R_i D}, P_{k,y}(S_x)\}} \right) \right.
$$

$$
\left. \left. \left(\frac{k+1}{HM\{\bar{\gamma}_{R_i D}, P_{k,y}(S_x)\}} \right) \right] \right] + \prod_{e_i \in P_{r,m}(S_{all})} e^{-\frac{\lambda}{\mu R_{e_i} P}} \left(\frac{1}{2} c_{M_S} \left(1 - \sqrt{\frac{d_{M_S}^2 \bar{\gamma}_{SD}}{1 + d_{M_S}^2 \bar{\gamma}_{SD}}} \right) \right)
$$

$$
+ \sum_{\substack{l_1, l_2, \ldots, l_K \in \{0,1\} \\ l_1 l_2 \ldots l_K \neq 1 \\ (1-l_1)(1-l_2)\ldots(1-l_K) \neq 1}} \left[\left(P_{\lambda_{R_1} P} l_1 + (1 - P_{\lambda_{R_1} P})(1 - l_1) \right) \left(P_{\lambda_{R_2} P} l_2 + (1 - P_{\lambda_{R_2} P})(1 - l_2) \right) \right.
$$

$$
\times \ldots \left(P_{\lambda_{R_K} P} l_K + (1 - P_{\lambda_{R_K} P})(1 - l_K) \right) \Big] \text{BER}_{comp\{i, \forall l_i = 1\}} \Big) \Big].
$$

$$
\tag{8}
$$

As an example, the average BER in the case of two relays for the fixed power underlay selection algorithm is given by,

$$
\text{BER} = \text{PER}_{SR_1} \text{PER}_{SR_2} \text{BER}_{SD} + (1 - \text{PER}_{SR_1}) \text{PER}_{SR_2}
$$

$$
\times \left(P_{\lambda_{R_1} P} \text{BER}_{comp\{1\}} + (1 - P_{\lambda_{R_1} P}) \text{BER}_{SD} \right) + (1 - \text{PER}_{SR_2}) \text{PER}_{SR_1}
$$

$$
\times \left(P_{\lambda_{R_2} P} \text{BER}_{comp\{2\}} + (1 - P_{\lambda_{R_2} P}) \text{BER}_{SD} \right) + (1 - \text{PER}_{SR_1})(1 - \text{PER}_{SR_2})
$$

$$
\times \left(P_{\lambda_{R_1} P} P_{\lambda_{R_2} P} \text{BER}_{comp\{1,2\}} + (1 - P_{\lambda_{R_1} P})(1 - P_{\lambda_{R_2} P}) \text{BER}_{SD} \right.
$$

$$
\left. + P_{\lambda_{R_1} P}(1 - P_{\lambda_{R_2} P}) \text{BER}_{comp\{1\}} + P_{\lambda_{R_2} P}(1 - P_{\lambda_{R_1} P}) \text{BER}_{comp\{2\}} \right).
$$

$$
\tag{9}
$$

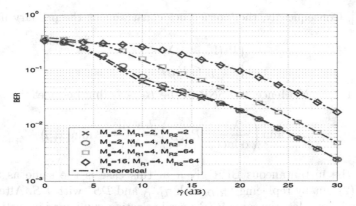

Fig. 2. BER performance of fixed power underlay relay selection algorithm for two relay setting, where $\bar{\gamma}_{SR_1} = \bar{\gamma} + 10$, $\bar{\gamma}_{SR_2} = \bar{\gamma} + 10$, $\bar{\gamma}_{SD} = \bar{\gamma} - 10$, $\bar{\gamma}_{R_1 D} = \bar{\gamma}$, $\bar{\gamma}_{R_2 D} = \bar{\gamma}$, $\lambda = 10$, $\alpha = 0.7$ and $N = 264$ bits.

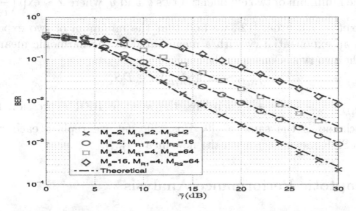

Fig. 3. BER performance of Interference Outage-based selection algorithm for two relay setting, where $\bar{\gamma}_{SR_1} = \bar{\gamma} + 10$, $\bar{\gamma}_{SR_2} = \bar{\gamma} + 10$, $\bar{\gamma}_{SD} = \bar{\gamma} - 10$, $\bar{\gamma}_{R_1 D} = \bar{\gamma}$, $\bar{\gamma}_{R_2 D} = \bar{\gamma}$, $\lambda = 10$, $\alpha = 0.7$, $\varepsilon = 0.05$ and $N = 264$ bits.

2.2   Interference Outage Based Selection

In the fixed power underlay relay selection explained in section II.A, in order to uphold an acceptable interference at the primary user P, the knowledge of the interference channels is needed at the relays. The interference channel knowledge helps in determining which relays satisfy the interference constraint to include them in $\mathcal{DS}^*$. However, this may sometimes be difficult to achieve and cost a lot of feedback. Therefore, to avoid the need for having the interference channel knowledge at the relays and the feedback it requires, we suggest the interference outage based selection scheme. In this scheme, we adjust the transmission power each relay in $\mathcal{DS}$ to be $P^*_{R_k D}$ so that, on the average, the interference generated by the relays on the primary user P is below a certain threshold with a tolerated error ε [11]. Consequently, the interference constraint at the primary user P is given by

$$\begin{cases} \Pr(I_{R_kP} > \lambda) \leq \varepsilon \\ P^*_{R_kD} \leq P_{R_kD} \end{cases} \tag{10}$$

where $k = 1, 2, ..., K$. Hence, the SNR of the $R_k - D$ link becomes

$$\gamma^*_{R_kD} = \min\left(\frac{\lambda N_0}{\alpha \ln(\frac{1}{\varepsilon})}, P_{R_kD}\right)\frac{|h_{R_kD}|^2}{N_0} = \min(x, y) \tag{11}$$

Therefore, the instantaneous BER given in this case is the same as the one defined in (2) but by replacing γ_{R_iD} with $\gamma^*_{R_iD}$ and $\mathcal{DS}^*$ with $\mathcal{DS}$. Afterwards, in the second time slot, the same BER-based selection, explained in section II.A, is applied to the decoding set $\mathcal{DS}$ after adjusting the transmission power of the relays. From (11), it is obvious that the SNR of each relay to the destination becomes the minimum of two exponential rv's x and y, where $x \sim \exp\left(\frac{\lambda}{\alpha \ln(\frac{1}{\varepsilon})}\right)$ and $y \sim \exp(\bar{\gamma}_{R_kD})$. From [12], we find that the minimum of two exponential rv's is also an exponential rv with a mean equal to the harmonic mean of the means of the two rv's. Consequently,

$$\gamma^*_{R_iD} \sim \exp(\bar{\gamma}^*_{R_iD}), i \in \mathcal{DS}, \tag{12}$$

where $\bar{\gamma}^*_{R_iD} = \frac{1}{\frac{1}{\bar{\gamma}_{R_iD}} + \frac{\alpha \ln(\frac{1}{\varepsilon})}{\lambda}}$. Therefore, the average BER expression for this selection scheme is the same as [9, Eqs. 20] but we replace each relay to destination SNR mean $\bar{\gamma}_{R_iD}$ with the new mean $\bar{\gamma}^*_{R_iD}$.

3   Asymptotic Performance Analysis

In this section, we derive an asymptotic BER expression for the fixed power underlay relay selection scheme that is accurate at high SNR values (i.e. as the SNR goes to infinity) for the sake of having more information about the system's performance.

In order to simplify the derived expression, we assume that all relays decode the received packet correctly. Consequently, the BER in (3) is modified to be

$$\mathrm{BER} = \prod_{k=1}^{K} P_{\lambda_{R_kP}}\mathrm{BER}_{comp\mathcal{DS}} + \prod_{k=1}^{K}(1 - P_{\lambda_{R_kP}})\mathrm{BER}_{SD}$$

$$+ \sum_{\substack{l_1,l_2,...,l_K \in \{0,1\} \\ l_1 l_2...l_K \neq 1 \\ (1-l_1)(1-l_2)...(1-l_K) \neq 1}} \Theta_{l_1,l_2,...,l_K}\mathrm{BER}_{comp\{i, \forall l_i = 1\}}, \tag{13}$$

where the asymptotic approximations for BER_{SD} and $\mathrm{BER}_{comp\mathcal{DS}}$ are found in [13] and [9], respectively when the average SNRs are expressed as $\bar{\gamma}_{SD} = \sigma^2_{SD}\mathrm{SNR}$ and $\bar{\gamma}_{RD} = \sigma^2_{RD}\mathrm{SNR}$:

$$\mathrm{BER}_{SD} \overset{\mathrm{SNR}\to\infty}{\approx} \frac{c_{M_S}}{4d^2_{M_S}\sigma^2_{SD}\mathrm{SNR}} \tag{14}$$

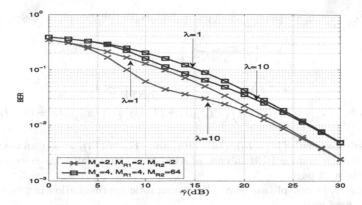

Fig. 4. BER performance of fixed power underlay relay selection algorithm for two relay setting with different interference thresholds, where $\bar{\gamma}_{SR_1} = \bar{\gamma} + 10$, $\bar{\gamma}_{SR_2} = \bar{\gamma} + 10$, $\bar{\gamma}_{SD} = \bar{\gamma} - 10$, $\bar{\gamma}_{R_1 D} = \bar{\gamma}$, $\bar{\gamma}_{R_2 D} = \bar{\gamma}$, $\alpha = 0.7$ and $N = 264$ bits.

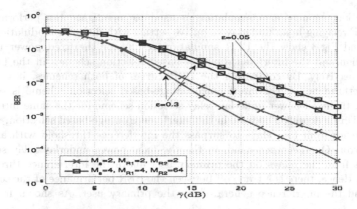

Fig. 5. BER performance of Interference Outage-based selection algorithm for two relay setting with different tolerable errors, where $\bar{\gamma}_{SR_1} = \bar{\gamma} + 10$, $\bar{\gamma}_{SR_2} = \bar{\gamma} + 10$, $\bar{\gamma}_{SD} = \bar{\gamma} - 10$, $\bar{\gamma}_{R_1 D} = \bar{\gamma}$, $\bar{\gamma}_{R_2 D} = \bar{\gamma}$, $\lambda = 10$, $\alpha = 0.7$ and $N = 264$ bits.

$$\text{BER}_{compDS} \overset{\text{SNR}\to\infty}{=} \left[\left[\prod_{i=1}^{K} \frac{\rho_i^{-1}}{\sigma_{R_i D}^2} \right] \frac{c_{M_S} \Gamma(K+1.5)}{2\sqrt{\pi}\sigma_{SD}^2 (1+K) \left(d_{M_S}^2\right)^{K+1}} \right.$$

$$\left. + \sum_{i=1}^{K} \frac{\rho_i c_{M_i} \Gamma(K+1.5)}{2\sqrt{\pi}\sigma_{SD}^2 \sigma_{R_i D}^2 (1+K) \left(d_{M_i}^2\right)^{K+1}} \times \left[\prod_{\substack{j=1 \\ j \neq i}}^{K} \frac{\beta_{ij}}{\sigma_{R_j D}^2} \right] \right] \frac{1}{\text{SNR}^{K+1}}. \qquad (15)$$

As for the expressions in the approximated BER equation, such as $P_{\lambda_{R_k P}}$, $(1 - P_{\lambda_{R_k P}})$, and $\Theta_{l_1, l_2, \dots, l_K}$, they can approximated to be

$$P_{\lambda_{R_k P}} \overset{\text{SNR}\to\infty}{=} \Pr(\text{I}_{R_k P} < \lambda) = 1 - e^{-\frac{\lambda}{\sigma_{R_k P}}} \overset{\text{SNR}\to\infty}{=} 1 - e^{-\frac{\lambda}{\alpha\overline{\text{SNR}}}}$$

$$\overset{\text{SNR}\to\infty}{=} 1 - \left(1 - \frac{\lambda}{\alpha\overline{\text{SNR}}}\right) = \frac{\lambda}{\alpha\overline{\text{SNR}}}$$

$$1 - P_{\lambda_{R_k P}} \overset{\text{SNR}\to\infty}{=} 1 - \frac{\lambda}{\alpha\overline{\text{SNR}}} \approx 1 \tag{16}$$

$$\Theta_{l_1,l_2,\ldots,l_K} = \left[\left(P_{\lambda_{R_1 P}} l_1 + (1 - l_1)\right)\left(P_{\lambda_{R_2 P}} l_2 + (1 - l_2)\right)\ldots\left(P_{\lambda_{R_K P}} l_K + (1 - l_K)\right)\right]. \tag{17}$$

by substituting with (14), (15), (16), (16), and (17) in (13), we get an asymptotic expression for the average BER in case of fixed power underlay relay selection given by (19) in the next page.

For instance, in case of two relays, the asymptotic approximation is given by (20) given in the next page.

4   Simulation Results

In this section, Monte-Carlo simulation is used to investigate the performance of selective DF relaying in an underlay cognitive setting with different modulation levels.

Fig.2 and Fig.3 show the BER simulation results for the fixed power underlay relay selection and the interference outage-based selection schemes in the two relay setting, respectively. By comparing the BER curves of both schemes, it is obvious that the performance of the interference outage-based selection scheme is relatively better than the fixed power underlay relay selection scheme for the same interference threshold. This is expected because in the interference outage-based selection, we allow the interference from the relays to surpass the interference threshold with a defined tolerable error. On the other hand, in the the fixed power underlay relay selection, interference from the relays on the primary user above the interference threshold is intolerable. Hence, there is a trade off between the BER performance of the secondary network and the interference generated on the primary user. As shown from both

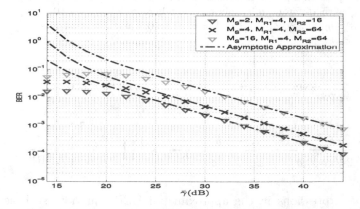

Fig. 6. Asymptotic BER performance of fixed power underlay relay selection algorithm for two relay setting, where $\bar{\gamma}_{SR_1} = \bar{\gamma} + 10$, $\bar{\gamma}_{SR_2} = \bar{\gamma} + 10$, $\bar{\gamma}_{SD} = \bar{\gamma} - 10$, $\bar{\gamma}_{R_1 D} = \bar{\gamma}$, $\bar{\gamma}_{R_2 D} = \bar{\gamma}$, $\lambda = 10$, $\alpha = 0.7$ and $N = 264$ bits.

$$\text{BER} \stackrel{\text{SNR}\to\infty}{=} \left(\frac{\lambda}{\alpha\text{SNR}}\right)^K \stackrel{\text{SNR}\to\infty}{\text{BER}_{comp_{DS}}} + \frac{c_{M_S}}{4d_{M_S}^2 \sigma_{SD}^2 \text{SNR}}$$

$$+ \sum_{\substack{l_1,l_2,\dots,l_K \in\{0,1\} \\ l_1 l_2 \dots l_K \neq 1 \\ (1-l_1)(1-l_2)\dots(1-l_K)\neq 1}} \left[\left(P_{\lambda_{R_1 P}} l_1 + (1-l_1)\right) \times \left(P_{\lambda_{R_2 P}} l_2 + (1-l_2)\right)\dots \qquad (19)$$

$$\times \left(P_{\lambda_{R_K P}} l_K + (1-l_K)\right)\right] \stackrel{\text{SNR}\to\infty}{\text{BER}_{comp\{i,\forall l_i=1\}}}.$$

$$\text{BER} \stackrel{\text{SNR}\to\infty}{=} \frac{1}{16\sigma_{R_1 D}^2 \sigma_{R_2 D}^2 \sigma_{SD}^2 d_{MR_1}^2 d_{MR_2}^2 d_{M_S}^2} \left[\frac{40\lambda^2 \left(c_{M_S} + c_{M_1} + c_{M_2}\right)}{3\alpha^2 \text{SNR}^3}\right.$$

$$\left.+3\left(c_{M_S} + c_{M_1}\right)\sigma_{R_2 D}^2 d_{MR_2}^2 + 3\left(c_{M_S} + c_{M_2}\right)\sigma_{R_1 D}^2 d_{MR_1}^2\right]\frac{1}{\text{SNR}^2} + \frac{c_{M_S}}{4\sigma_{SD}^2 d_{M_S}^2 \text{SNR}}$$

$$(20)$$

figures, the derived theoretical results are in complete agreement with the simulation results.

It is worthy to mention that at low and medium SNR's, the interference at the primary receiver is relatively low. This means that more relays satisfy the interference constraint enabling selection between multiple signals and providing improved BER. Whereas at high SNR's, high interference is generated at the primary receiver causing the number of relays satisfying the interference constraint to decrease and therefore, the signal is received directly from the source giving a higher BER. This is clear from the bottom two curves in Fig.2 and the bottom most curve in Fig.4 where the curve drops at low and medium SNR's and goes back up again at high SNR's.

In Fig.4, we demonstrate the BER performance of the fixed power-based algorithm under different interference thresholds (i.e. different values of λ). As it is obvious, the higher the interference threshold, the better the BER performance. This can be explained by noting that when the interference threshold is high, the probability of finding a relay that satisfies the interference constraint is high allowing selection between different signals and hence, the better the BER. On the contrary, for a low interference threshold, the probability of finding relays that satisfy the interference constraint becomes lower to the limit that no relays satisfy the interference constraint and hence, the signal is received from the source only.

In Fig.5, we plot the performance of the interference outage based selection algorithm under different tolerable error values (i.e. different values of ε). As it is evident from Fig.5, the higher the value of the tolerable error, the more interference we allow from the relays on the primary user and consequently, the better the BER performance of the secondary network.

In Fig.6, the asymptotic BER expression is shown for two relays. We verify that the derived asymptotic BER equation is accurate at high SNRs.

5 Conclusion

This paper proposes two relay selection schemes in an underlay cognitive setting. In both selection algorithms, the destination chooses the best link from a set of candidate links. The links are assumed to have different modulation levels. The first selection scheme depends on the interference channel knowledge at the relays to rule out the relays that violate the interference constraint at the primary user. In the second selection algorithm, the relays do not need the knowledge of the interference channel provided that they adjust their transmission power to keep the interference at the primary user less than a certain threshold with a certain tolerable error. Closed form expressions of the average BER is derived for both selection schemes. The Monte-Carlo based simulations validate the derived theoretical expressions.

Acknowledgments. This publication was made possible by NPRP grant # [5-250-2-087] from the Qatar National Research Fund (a member of Qatar Foundation). The statements made herein are solely the responsibility of the authors.

References

1. FCC: ET Docket No 03–222 Notice of proposed rule making and order, December 2003
2. Akyildiz, I.F., Lee, W.Y., Vuran, M.C., Mohanty, S.: Next generation/dynamic spectrum access/cognitive radio wireless networks: A survey. Computer Networks **50**, 2127–2159 (2006)
3. Haykin, S.: Cognitive radio: brain-empowered wireless communications. IEEE Journal on Selected Areas in Communications **23**, 201–220 (2005)
4. Yu, M., Li, J.: Is amplify-and-forward practically better than decode-and-forward or vice versa? In: Proceedings of IEEE International Conference on Acoustics, Speech, and Signal Processing (ICASSP 2005), vol. 3, pp. iii/365–iii/368 (2005)
5. Bletsas, A., Khisti, A., Reed, D., Lippman, A.: A simple cooperative diversity method based on network path selection. IEEE Journal on Selected Areas in Communications **24**, 659–672 (2006)
6. Hussain, S., Abdallah, M., Alouini, M., Hasna, M., Qaraqe, K.: Best relay selection using snr and interference quotient for underlay cognitive networks. In: 2012 IEEE International Conference on Communications (ICC), pp. 4176–4180 (2012)
7. Zou, Y., Zhu, J., Zheng, B., Yao, Y.D.: An adaptive cooperation diversity scheme with best-relay selection in cognitive radio networks. IEEE Transactions on Signal Processing **58**, 5438–5445 (2010)
8. Lee, J., Wang, H., Andrews, J., Hong, D.: Outage probability of cognitive relay networks with interference constraints. IEEE Transactions on Wireless Communications **10**, 390–395 (2011)
9. Sokun, H., Sediq, A.B., Ikki, S., Yanikomeroglu, H.: Selective DF relaying in multi-relay networks with different modulation levels. In: 2014 IEEE International Conference on Communications (ICC), pp. 5035–5041 (2014)
10. Sklar, B.: Digital Communications, 2nd edn. Prentice Hall (2001)
11. Tourki, K., Qaraqe, K., Alouini, M.S.: Outage analysis for underlay cognitive networks using incremental regenerative relaying. IEEE Transactions on Vehicular Technology **62**, 721–734 (2013)
12. Bertsekas, D.P., Tsitsiklis, J.N.: Introduction to Probability. Athena Scientific
13. Sediq, A.B., Yanikomeroglu, H.: Performance analysis of selection combining of signals with different modulation levels in cooperative communications. IEEE Transactions on Vehicular Technology **60**, 1880–1887 (2011)

Spectrum-Sculpting-Aided PU-Claiming in OFDMA Cognitive Radio Networks

Yi Ren[✉], Chao Wang, Dong Liu, Fuqiang Liu, and Erwu Liu

School of Electronic and Information Engineering, Tongji Uniniversity,
Shanghai, People's Republic of China
{092755,chaowang}@tongji.edu.cn,
liu-dong@live.cn, liufuqiang@tongji.edu.cn, erwu.liu@ieee.org

Abstract. We consider a cognitive radio network where the primary user (PU) supports applications with multiple quality of service (QoS) requirements via orthogonal frequency division multiple access (OFDMA). To let the secondary users (SUs) be aware of PU's QoS level, and hence provide PU with sufficient protection, we propose a spectrum-sculpting-aided PU-claiming scheme that does not demand strict PU-SU synchronization. Specifically, the PU deliberately inserts one zero-subcarrier into its subcarriers, the position of which represents the QoS requirement of the PU. Through a two-step sensing procedure, each SU then can estimate the QoS requirement and adjust its sensing or accessing strategies accordingly. Simulation results exhibit the advantages of the proposed scheme, in terms of both weighted received interference (WRI) and SU's throughput[1].

Keywords: Spectrum sculpting · Cognitive radio · Multi-QoS · OFDM · PU-claiming

1 Introduction

The explosive increase of wireless devices and services in recent years will potentially lead to a scarcity in spectrum resource. On the other hand, as reported by FCC, the conventional fixed spectrum allocation policy results in a heavy spectrum underutilization, which exacerbates the shortage of spectrum resource. In order to handle this issue, the concept of cognitive radio (CR), which gives secondary users (SUs) the ability to probe and access specific spectrum bands when they are not occupied by primary users (PUs), has been proposed and studied intensively [1].

Previously, most works on spectrum sensing and accessing mainly assume that PU only has one quality-of-service (QoS) level, hence the SU usually has a constant configuration in terms of detection probability or power constraints.

This work was supported in part by the Key Program of National Natural Science Foundation of China under Grant No. 61331009 and the Fundamental Research Funds for the Central Universities No. 0800219236.

© Institute for Computer Sciences, Social Informatics and Telecommunications Engineering 2015
M. Weichold et al. (Eds.): CROWNCOM 2015, LNICST 156, pp. 295–307, 2015.
DOI: 10.1007/978-3-319-24540-9_24

However, multi-QoS system is also an important scenario, especially when some modern wireless communication systems (e.g. LTE or WiMAX [2]) which have large amount subscribed users are considered as PU. In this paper we consider a more complex primary network which will have applications with different QoS requirement levels, and these distinct applications will be dynamically served in PU's allocated bands. Under such condition, the constant constraints for SU may be not satisfactory. For instance, for some delay-sensitive or real-time applications, even small interference caused by SU brings enormous economic cost, while many data-service applications such as email and downloading can tolerate more interference. Thus in multi-QoS systems, the sensing strategies of SU should be adaptive according to PU's QoS requirements to provide a better protection for PU, and potentially, increase its own throughput. Enabling information sharing between PU and SU (we call this *PU-claiming*) may be an effective way to realize such an adaptation. However, as primary network and secondary network are usually heterogeneous, and also because PU's signal is usually weak at SUs, such PU-claiming may be hard to realize. Conventionally, some researches assume that PU can periodically broadcast beacon-frames to SUs to deliver some important information to SUs (as mentioned in [3], [4] and [5]). However, such beacon-based PU-claiming scheme faces three practical roadblocks: the conundrum of strict synchronization between PU and SU; sometimes, PU's weak signal power at SU and the considerable resource squander of PU.

To address these problems, in this paper we propose a spectrum-sculpting-aided PU-claiming scheme for orthogonal frequency division multiple access (OFDMA) systems. Specifically, inspired by spectrum notching [6], which has been intensively discussed in NC-OFDM (non-continuous OFDM) systems for interference controlling, we require the PU to deliberately insert one zero-subcarrier (named as *spectrum valley*) into its subcarriers of each service. The position of the spectrum valley is used to represent the QoS requirement. When each SU conducts a two-step spectrum sensing and attains such information, it can adjust its sensing parameters accordingly to provide corresponding protection to the PU. Since the QoS requirements of PU's applications are inherently attached in the spectrum structure and can be changed dynamically, a strict synchronization between PU and SUs is not necessary (i.e. SU can attain such information no matter when it intends to conduct spectrum sensing). Through mathematical analysis and numerical simulations, we show that the proposed scheme brings a significant performance improvement in terms of *weighted received interference* (WRI) over the conventional scheme proposed in[7]. And from SU's perspective, SUs with PU-claiming can also have potential to gain a higher throughput even through PU may possibly have some extremely interference-sensitive applications.

2   System Description

2.1   System Model

We consider a CR-OFDMA system with one centralized primary network (e.g., LTE systems). The PU's base station (BS) uses N_B frequency channels tosupport diverse services, such as voice, data traffic and real-time applications, to its sub-scribed users. Fig.1-A illustrates a typical frequency-domain structure of LTE signal. The entire band of N_f adjacent subcarriers is divided into N_B channels, each containing L'_d subcarriers. Several unused subcarriers, termed *virtual carrier (VC)*, locate at the edges of the band to avoid energy leakage. The PU's transmitting signal is expressed as

$$s_i(t) = \frac{1}{N_f} \sum_{n=0}^{N_f-1} S_i(n)G(n)e^{\frac{j2\pi nt}{N_f}}, \quad t \in \{0, \cdots, N_f - 1\}, \tag{1}$$

where N_f is the symbol size, and $S_i(n)$ is the frequency-domain transmitted signal at the nth subcarrier in the ith symbol. $G(n)$, called *transmitted kernel function*, represents the spectrum structure of OFDM signal, is set to be 1 if the nth subcarrier is occupied by a PU application, and be 0 otherwise. To simplify analysis, the processes of adding cyclic prefix and windowing are not considered, because they have no influence on the frequency-domain structure of PU's signal.

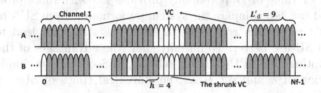

Fig. 1. Spectrum sculpting of PU's transmitted OFDM signal

An SU locating in the vicinity of PU is allowed to conduct spectrum sensing and opportunistically access the channels that are not occupied by PU. Constrained by sampling rate, each SU can perform sensing on only one channel at a time instant. Due to SU's imperfect spectrum sensing, PU will inevitably be interfered if any SU mistakenly treats an occupied channel as unused. Clearly, applications with higher QoS requirements are more sensitive to such interference, i.e., they have higher cost compared to applications with lower QoS requirements when suffered by interference. Similar to the priority table defined in [2], we can use an integer variable $k \in \{1, 2, \cdots, K\}$ to classify those applications and represent the QoS requirement level of current application ($k = K$ denotes the most sensitive application).

2.2   Performance Metrics

The performance of the CR system can be considered from two perspectives. From PU's viewpoint, we use WRI ($\mathcal{W}$), i.e. the weighted received interference, to evaluate the impact of SUs' interference on the PU network. The WRI is defined as follows and can be used to indicate the performance cost or economical cost sacrificed by the PU for permitting SUs' access:

$$\mathcal{W} = \sum_{k=1}^{K} \omega_k \cdot I_k \cdot M_k, \tag{2}$$

where I_k denotes the interference intensity experienced by the level-k applications and ω_k is the associated weighting factor. M_k can be considered as the level-k application's cumulative operating time in the observation duration (or the probability that a level-k application may be served). Note that the value of ω_k mainly depends on the customers' requirement, which can be attained by the telecom operators via long term observing, so as the M_k. Then the SU can buy these message from the telecom operators or just learn from its history observations. Depending on the objective, there are different ways to define I_k. For instance, one can use SU's miss detection probability or the average interference power seen by the PU to serve as I_k. Generally speaking, as those applications with higher QoS requirement are usually more sensitive to SU's interference, hence the corresponding ω_k will be larger comparing with that of low level applications. Then $\mathcal{W}$ can be used to assess the performance of secondary access strategies in CR networks: the approaches applied by SU that lead to a smaller value of $\mathcal{W}$ potentially provides PU with more protection. On the other hand, we can compare the system performance from SU's perspective: satisfying PU's protection requirements level, the strategies lead to a higher throughput of SU are better. To initially show the advantages of the proposed spectrum-sculpting-aided PU-claiming scheme, in Section 5 we simply set ω_k as a linear function of k and set I_k as the statistical times of SUs' disturbance toward the PU.

3   Spectrum-Sculpting-Aided PU-claiming

As stricter protection is required by those higher-level applications while more access opportunities embed in the spectrum which serving lower-level applications, a simple way to enhance the system performance is to feed the information regarding k to SU. Armed with such knowledge, the SU can adapt its sensing and accessing strategies accordingly.

3.1   PU's Spectrum Sculpting

To establish PU-claiming more efficiently, we consider moving a VC into each channel. In other words, in the PU's signal structure, each channel now contains $L_d = L_d' + 1$ data subcarriers and one of them is a zero-subcarrier (termed *spectrum valley*). The potential locations of the spectrum valley are selected in such a way that each position can be mapped to a QoS requirement level. Without loss of generality, we assume that the first K subcarriers (assume $K \leq L_d$) in each channel are used to place the spectrum valley. Let $\bar{h} \in \{1, \cdots, K\}$ denote the position index of the spectrum valley in the signal of current application. The frequency-domain signal structure when $\bar{h} = 4$ is illustrated in Fig. 1-B. It should be noted that moving VC into the channels results in a slight reduction

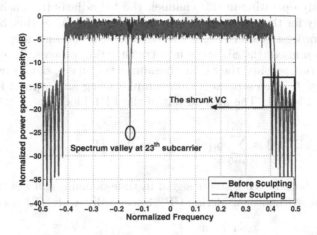

Fig. 2. Spectrum sculpting on one channel of OFDMA signal.

of the width of VC. Nevertheless, we observe that in many OFDMA systems the VC's width is not fixed but selected within a certain range (i.e. 159-183 in [8] when $N_f = 1024$). Fig.2 illustrates how PU sculpt its spectrum to make a spectrum valley. We can see the width of VC in the right side is shrunk by spectrum sculpting. Nevertheless, due to the sharp out-band decrease of one subcarrier's power spectral density, the power at the edge of this band (after sculpting) is still less than -18dB, which is almost the same comparing with the energy leakage before spectrum sculpting. In addition, some time-domain windowing techniques can effectively restrain the leakage of side-lobe [9]. And in practice, there should not always be signals transmitting in its adjacent channel. Hence the resulting energy leakage brought by the smaller VC should not be noticeable.

3.2   SU's Two-step Spectrum Sensing

Regarding SU, akin to [4], we also formulate SU's behavior by two kinds of frames: transmitting frame with the duration T_t and sleeping frame with the duration T_s. SU will conduct a *two-step spectrum sensing* to detect target channel's condition. And attain the value of k which is embedded to PU's signal structure. Particularly, the SU first applies an energy detection algorithm to determine whether this channel is occupied by the PU. Note that SU will store the data samples used for first step sensing in a samples queue (can contain totally N_Q transmitting frames) as long as the result of first-step detection is yes. Knowing the presence of PU, the SU will then detect the spectrum valley's position by searching for the subcarrier with the smallest energy among the first K subcarriers of the channel. Generally speaking, as the switches of PU's application level is far slower than the switches of PU's presence/absence status [10], the samples queue can easily store enough data samples for the second-step sensing. Since the result of this step indicates the QoS requirement level of the service currently operating in this channel, the SU adjusts its sensing parameters accordingly for the next sensing to reduce the WRI. After this, SU will step into sleeping process. Otherwise, if the detection in the first step shows that the channel is unused, the SU steps into transmitting process. To conduct the two-step spectrum sensing, the SU firstly sample the analog signals using a sampling frequency f_s, then separates data samples using an L_d-point fast Fourier transform (FFT) block. The mth output of the FFT block is:

$$Y_m(i) = \sum_{n=0}^{L_d-1} y_i[n] \cdot e^{-j\frac{2\pi mn}{L_d}} \quad m \in \{1, \cdots, L_d\}, \tag{3}$$

where $y_i[n]$ is the sampled discrete signal in time-domain, and i is the index of OFDM symbol (without considering CP). Specifically, when PU is active:

$$Y_m(i) = \sum_{t=0}^{L_d-1} (\sum_{l=0}^{L_d-1} (h(l)s_i(t-l)) + v(t))e^{-j\frac{2\pi mt}{L_d}}. \tag{4}$$

Hence we can divide the mth output of FFT block as follows:

$$\begin{aligned} Y_m(i) &= V(n) & \text{CASE} - \text{I} \\ Y_m(i) &= H(m)S_i(m) + V(n) & \text{CASE} - \text{II}, \end{aligned} \tag{5}$$

where $H(m)$, $S_i(m)$ and $V(n)$ are frequency-domain form of channel response, signal and noise. We assume a frequency-flat channel, i.e. $H(m) = H$ to make a initial analysis. The influence of fading channel will be left for future. In equation (5), the CASE-I represents the scenario that $\mathcal{H} = 0$ (PU not active in practice), or $\mathcal{H} = 1$ and $G(m) = 0$ (PU is active and the mth subcarrier is spectrum valley or VC). And the CASE-II represents that $\mathcal{H} = 1$ and $G(m) = 1$ (PU is active and the mth subcarrier is data carrier). Let τ be the sensing time and N_r be the number of data samples used for the first-step sensing in one frame ($N_r = \tau f_s$), the mth subcarrier's energy can be expressed as:

$$T(m) = \frac{L_d}{N_r} \sum_{i=1}^{\lfloor \frac{N_r}{L_d} \rfloor} |Y_m(i)|^2. \tag{6}$$

Assume $E[|V(t)|^2] = \sigma_v^2$, $E[|S_i(m)|^2] = \sigma_s^2$ and let $\rho = \frac{\sigma_s^2}{\sigma_v^2}$ represent the transmitting signal-to-noise ratio. Using the central limit theorem (CLT), the energy $T(m)$ in CASE-I and CASE-II can be approximated by Gaussian variables. Specifically, in CASE-I, $T(m) \sim \mathcal{N}(\mu_0, \sigma_0^2)$, where $\mu_0 = \sigma_v^2$ and $\sigma_0^2 = \frac{L_d}{N_r}\sigma_v^4$ while in CASE-II, $T(m) \sim \mathcal{N}(\mu_1, \sigma_1^2)$ where $\mu_1 = (1 + H^2\rho)\sigma_v^2$ and $\sigma_1^2 = \frac{L_d}{N_r}(1 + H^2\rho)^2\sigma_v^4$.

Use variable $\hat{\mathcal{H}}$ to represent the SU's detection decision. The probability that the SU correctly detects the existence of the PU when it is active (i.e. the *detection probability*) is expressed as $P_d := P_r\{\hat{\mathcal{H}} = 1 | \mathcal{H} = 1\}$. Similarly, the *false alarm probability* is $P_f := P_r\{\hat{\mathcal{H}} = 1 | \mathcal{H} = 0\}$. Different from the conventional energy detection algorithm that calculates sum energy in the time domain, we apply an energy detector at each SU using energy accumulated from all the subcarriers of current channel, i.e.,

$$\lambda := \frac{1}{L_d} \sum_{m=1}^{L_d} T(m) - \epsilon_k \qquad (7)$$

where ϵ_k is the decision threshold knowing the level of current application is k. $\lambda > 0$ claims $\hat{\mathcal{H}} = 1$ and $\lambda < 0$ claims $\hat{\mathcal{H}} = 0$. Thanks to Parseval theorem, this frequency-domain detector leads to the same result as conventional time-domain energy detector, because this energy detector collects the energy of all the outputs (including VC and spectrum valley). Again, using CLT we can see that $\lambda_{|\mathcal{H}=1} \sim N(\bar{\mu}_1, \bar{\sigma}_1^2)$, where $\bar{\mu}_1 = \mu_1$ and $\bar{\sigma}_1^2 = \frac{\sigma_1^2}{L_d}$. Similarly, $\lambda_{|\mathcal{H}=0} \sim N(\bar{\mu}_0, \bar{\sigma}_0^2)$, where $\bar{\mu}_0 = \mu_0$ and $\bar{\sigma}_0^2 = \frac{\sigma_0^2}{L_d}$. Given P_d, after some mathematical manipulations, we can express P_f, which mainly influences SU's throughput, as:

$$P_f = Q\left(\frac{\bar{\sigma}_1}{\bar{\sigma}_0} Q^{-1}(\bar{P}_d) + \frac{\bar{\mu}_1 - \bar{\mu}_0}{\bar{\sigma}_0}\right). \qquad (8)$$

where Q-function $Q(\cdot)$ is complementary cumulative distribution function of standard Gaussian distribution $Q(x) = \frac{1}{\sqrt{2\pi}} \int_x^\infty e^{-\frac{t^2}{2}} dt$.

If the above process shows the channel is currently occupied, the SU starts detecting the position of the spectrum valley, $\bar{h}$, by comparing the energy values of different subcarriers and choosing the one with the smallest energy. Using P_s to represent the probability that the SU correctly locates the spectrum valley, given that it successfully detects the existence of PU in the first step, we can have:

$$\begin{aligned} P_s &= P_r\{\hat{h} = k | \bar{h} = k\} \\ &= P_r\{\tilde{T}(k) < \tilde{T}(m),\ m \in [1, k], m \neq k\}. \end{aligned} \qquad (9)$$

As the second-step sensing uses the samples stored in samples queue, which contains the samples of N_Q frames. Using CLT, we can obtain $\tilde{T}_I := \tilde{T}(m)_{|CASE-I} \sim N(\tilde{\mu}_0, \tilde{\sigma}_0^2)$ and $\tilde{T}_{II} := \tilde{T}(m)_{|CASE-II} \sim N(\tilde{\mu}_1, \tilde{\sigma}_1^2)$. The expectation and variance of these two cases are: $\tilde{\mu}_0 = \sigma_v^2$, $\tilde{\sigma}_0^2 = \frac{L_d}{N_Q N_r} \sigma_v^4$, $\tilde{\mu}_1 = (1 + H^2 \rho)\sigma_v^2$ and $\tilde{\sigma}_1^2 = \frac{L_d}{N_Q N_r}(1 + H^2 \rho)^2 \sigma_v^4$. Given the probability distribution function (PDF) of $\tilde{T}(m)$ under both CASE-I and CASE-II, the equation (9) can be calculated by multiple integral. Specifically, we use $\{x_1, x_2, \ldots, x_K\}$ to represent these K random variables. Without loosing generality, we assume x_1 and $\tilde{T}_I$ have identical distribution and x_m ($m = 2, 3, \ldots, K$) has identical distribution with $\tilde{T}_{II}$. The joint probability distribution function of these K random variables can be expressed by: $f_T(x_1, \ldots, x_K)$. Following the assumption that random variables $x_m,\ m = 2, 3, \ldots, K\}$ are identically independently distributed (i.i.d), applying the distribution of $\tilde{T}_I$ and $\tilde{T}_{II}$, we can have:

$$f_T(x_1, \ldots, x_K) = \frac{1}{(2\pi\tilde{\sigma}_0^2)^{\frac{1}{2}}} e^{-\frac{(x_1 - \tilde{\mu}_0)^2}{2\tilde{\sigma}_0^2}} \cdot \prod_{i=2}^{Q-1} \frac{1}{(2\pi\tilde{\sigma}_1^2)^{K-1}} e^{-\sum_{i=2}^{K-1} \frac{(x_i - \tilde{\mu}_1)^2}{2\tilde{\sigma}_1^2}}. \qquad (10)$$

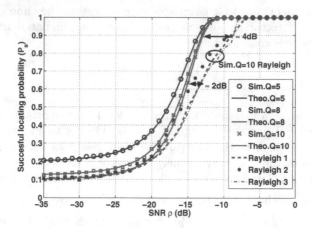

Fig. 3. Probability of successfully location the spectrum valley.

Then P_s can be expressed as:

$$P_s = \int_{\mathbb{R}^{\mathbb{T}}} f_T(x_1, \ldots, x_K) dx_1 dx_2 \ldots dx_K, \qquad (11)$$

where $\mathbb{R}^{\mathbb{T}}$ is the field of integration in which $\{x_1 < x_m, \ m = 2, 3, \ldots, K]\}$ will hold (i.e. $\mathbb{R}^{\mathbb{T}} = \bigcap_{m=2}^{K}\{x_1 < x_m\}$). Hence the equation (11) can be decomposed as:

$$P_s = \int_{-\infty}^{\infty} \int_{a}^{\infty} \cdots \int_{a}^{\infty} f_T(x_1, \ldots, x_K) dx_K \ldots dx_2 dx_1. \qquad (12)$$

Finally, substituting equation (10) and Q-function to equation (12), the P_s can be estimated as:

$$P_s = \int_{-\infty}^{\infty} \frac{1}{\sqrt{2\pi}\tilde{\sigma}_0} e^{-\frac{(a-\tilde{\mu}_0)^2}{2\tilde{\sigma}_0^2}} Q^{K-1}\left(\frac{a - \tilde{\mu}_1}{\tilde{\sigma}_1}\right) da. \qquad (13)$$

Additionally, the probability that the SU wrongly claims the spectrum valley's position as $\hat{h} = l$ (when in practice, $\bar{h} = k$) is represented by $P_e := P_r\{\hat{h} = l, l \neq k | \bar{h} = k\} = \frac{1-P_s}{K-1}$, because the energy of all the data subcarriers' are identically independently distributed (i.i.d). Fig.3 illustrates the simulation results and mathematical approach of analyzing the probability of successfully locating the spectrum valley (N_f=64 and 20000 data samples are used). From Fig.3, we can see the simulation results fit the estimation well, and the value of P_s converges to one when SNR is large while it converges to $\frac{1}{K}$ when SNR is low. In other words, P_s will be always no less than P_e, and the worst case is that SU randomly selects a position from the L_d potential subcarriers. Considering the influence of fading on our proposed scheme, we simulated the signal experienced three different Rayleigh fading channels. Specifically, the channel named Rayleigh 1 has 6 taps ([0 -8 -17 -27 -25 -31]dB), and the delay of these factor is [0 4 8 12 16 20]; Rayleigh 2 has same taps with Rayleigh 1, but the delay factor changes to [0 1 2 3 4 5]; Rayleigh 3 only has 4 taps ([0 -8 -17 -27]dB) and the delay vector is [0 4 8 12]. Comparing Rayleigh 1 and 2, we can see the larger the delay factor is, the worse the P_s is, which coincides with our intuition that larger time-domain spread will harm the detection performance more. Comparing Rayleigh 1 and 3, we find that the performance does not change so much. That is because the first 4 taps of Rayleigh 3 and 1 are the same. The gap between P_s under AWGN and Rayleigh 1 channel ranges from zero to approximately 4dB, and larger gap exists when SNR is small.

3.3 Weakness Discussion

Comparing with conventional beacon based PU-claiming schemes, the proposed method can skillfully solve some conundrums such as synchronization, high dynamic and low received power. However, this scheme inherently has its short comings. First, the specific squander and sacrifice of PU is hard to qualified. Because moving one VC will keep the spectrum efficiency of the PU in target channel constant, however, the shrunk VC will cause interference to adjacent channels. Second, this scheme need to be optimized to compete with more serious fading scenarios. Which will be left for our future work.

4 Performance Analysis

Knowing the position of the spectrum valley, the SU is able to attain the level of current application, and thus the knowledge of ω_k and $\bar{I}_k$ (here we use $\bar{I}_k$ to represent the requirement of PU's level-k application) from a pre-defined table in standards or protocols. Then, SU can adjust its sensing and transmission strategies accordingly to avoid causing a large W and improve its throughput under PU's constraints. Assume I_k is the interference probability observed at PU, then SU must set appropriate P_d^k to ensure $I_k \leq \bar{I}_k$. On the other hand, as the interference intensity is defined as *SU's miss detection probability* in this paper, the relationship between P_d^k and I_k can be expressed as follows:

$$1 - I_k = P_s P_d^k + P_e \sum_{l=1,l\neq k}^{K} P_d^l, \tag{14}$$

which means that the interference caused by the SU applying our scheme to the level-k application can be divided into K sub-conditions: one is the SU correctly detect the level-k application being served while another $K - 1$ are the scenario that SU wrongly claims a level-l application as level-k.

In this paper, three different kinds of SUs are considered: 1) *no PU-claiming (NC)*: SU does not know PU's QoS level and choose a constant P_d for all applications; 2) our *spectrum-sculpting-aided PU-claiming (SC)* and 3) *genie-aided PU-claiming (GC)*: SU perfectly knows PU's exact QoS level, which can be considered as the extreme scenario for SC that $P_s \rightarrow 1$. Note that NC-SU and GC-SU will just have $I_k = 1 - P_d^k$.

Table 1. Performance of 3 WRI with 50000 sec. observation.

	WRI				Comparison (%)	
	SC	GC	NC	Mth.	GC/NC	SC/GC
A	66.413	62.578	86.665	65.667	72.22	106.1
B	54.404	51.193	73.008	53.208	70.12	106.2
C	29.665	28.056	40.807	29.254	68.75	105.7

Firstly, we will observe the system performance from PU's perspective, i.e. the value of weighted received interference defined in equation (2). To calculate I_k, we observe one channel for a duration T_c (assumed to be sufficiently large), within which only one type of application is served. The probability that PU occupies the channel is denoted by $\mathcal{P}_1 = P_r\{\mathcal{H} = 1\}$. Use D_{01} and D_{11} to represent the numbers of times that the events $\{\hat{\mathcal{H}} = 0|\mathcal{H} = 1\}$ and $\{\hat{\mathcal{H}} = 1|\mathcal{H} = 1\}$ occur within the duration T_c, respectively. We have:

$$\frac{D_{11}}{D_{01}+D_{11}} = P_d^k$$
$$T_t D_{01} + T_s D_{11} = T_c \mathcal{P}_1. \tag{15}$$

When PU is active, SU's miss detection causes interference. Using the above relations, I_k can be expressed as:

$$I_k := \frac{D_{01}}{T_c} = \frac{(1 - P_d^k)\mathcal{P}_1}{T_t(1 - P_d^k) + T_s P_d^k}. \tag{16}$$

Secondly, from SU's perspective, we can assume PU is strict and requires all the level of its applications should be well protected, i.e., SU must ensure $I_k \le \bar{I}_k$, $\forall k$. Under such condition, NC-SU have to choose $P_d = 1 - \min(\bar{I}_k)$, $\forall k$, and SC-SU must apply equation (14) to PU's constraints to calculate the according P_d^k. Then, similar with the process of analyzing SU's interference, we use D_{00} and D_{10} to represent the numbers of times that the events $\{\hat{\mathcal{H}} = 0|\mathcal{H} = 0\}$ and $\{\hat{\mathcal{H}} = 1|\mathcal{H} = 0\}$ occur within the duration T_c, respectively:

$$\frac{D_{10}}{D_{00}+D_{10}} = P_f^k$$
$$T_t D_{00} + T_s D_{10} = T_c(1 - \mathcal{P}_1), \tag{17}$$

where P_f^k is calculated by applying P_d^k to equation (8). Then, using R_0 to represent SU's data rate and by solving the equation set (17), the SU's throughput can be expressed as:

$$\mathcal{C} = \sum_{k=1}^{K} M_k \left(\frac{D_{00}(T_t - \tau)R_0}{T_c T_t} \right) = \frac{(T_t - \tau)R_0}{T_t} \sum_{k=1}^{K} M_k \frac{(1 - \mathcal{P}_1)(1 - P_f^k)}{T_t(1 - P_f^k) + T_s P_f^k}. \tag{18}$$

5   Simulations and Comparisons

In this section, numerical simulations are conducted to evaluate the enhancement of our proposed scheme from both PU's perspective (compare WRI) and SU's

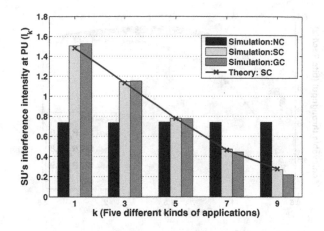

Fig. 4. Average I_k under 50000 seconds simulation, when k=1, 3, 5, 7, 9

perspective (compare throughput). We employ Monte-Carlo simulations, which observe one SU's access behavior in a centralized PU network for 50000 seconds. We set N_f=1024, K=10, L'_d=48, N_B=30, f_s=20MHz, τ=2ms, SNR=-12.5dB, T_t=200ms, $M_k|_{\forall k}=\frac{1}{K}$ and $\mathcal{P}_1$=0.5.

Firstly, we compare the value of $\mathcal{W}$ when all SUs throughput are equal. To calculate the WRI, we consider a simple example that the detection probability at SU is a linear function of k, i.e., $\bar{I}_k = \frac{1.99}{9} - \frac{0.19}{9}k$ which results in $\bar{I}_1$=0.2 and $\bar{I}_{10}$=0.01. In addition, simply let $\omega_k = k$.

Table 1 compares the WRI of the simulation results of the three schemes and SC-SU's mathematical estimation based on (16). Three different sleeping frame lengths are considered: A (T_s=10ms), B (T_s=20ms), and C (T_s=40ms). Intuitively, SU with larger T_s conducts spectrum sensing more infrequently, and thus has less chances to interfere the PU. From the first four columns, we can see that $\mathcal{W}$ reduces by increasing T_s in each scheme. The column "GC/NC" shows the improvement of $\mathcal{W}$ when PU-claiming introduced to the system, while "SC/GC" shows the gap between actual condition and ideal condition. Clearly, the estimation error of the spectrum valley's position does not significantly reduce the performance of the proposed PU-claiming scheme. Fig.4 depicts the I_k of five different types of applications with k=1, 3, 5, 7, 9 when T_s=40ms. We assume that in the NC scheme the SU conducts spectrum sensing with a fixed detection probability since it does not know the PU's QoS requirement. Hence in different applications, it introduces interference to the PU with a stable frequency. From the figure we can see that using PU-claiming, the frequency that SU introduces interference to PU depends on the QoS requirement level. By this means, the protection provided to PU and the opportunities provided for SU can be much better balanced. Again, since the performances of the GC and SC schemes are close, estimation error at the second detection step does not have obvious impact to the system performance. Finally, simulation results coincide with mathematical analysis. The advantages of the proposed scheme are clearly exhibited.

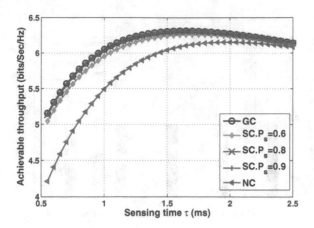

Fig. 5. Comparison of achievable throughput of different kinds of SUs.

Then, we consider a scenario that PU is strict, in which SUs must ensure all PU's applications are satisfied. Regarding SC-SU, for a given P_s, it must apply equation (14) to PU's constraints ($I_k \leq \bar{I}_k$), then use the calculated P_d^k to sense and access, and use the corresponding P_f^k to estimate SU's throughput. Similar with previous simulation, we set $R_0 = 10$ and observe the influence of τ. Assume $K = 6$ and M_k has a Gaussian distribution. To make an initial analysis, we compare the scenario that P_s does not vary with sensing time τ. Note that the value of P_s can be ensured by extending the length of the samples queue (N_Q). In Fig.5, the throughput of these three kinds of SUs are compared. NC-SU must ensure the most sensitive application, hence it has the worst performance. Knowing the knowledge of current application, we see the throughput will have a significant improvement. In this figure, there is a tradeoff between SU's sensing time τ and its achievable throughput: The throughput firstly increases when sensing time τ becomes larger, that is because longer τ implies more data samples used for energy detection. However, too long sensing time will shrink the time of SU's data delivery, which in turn lead the achievable throughput decreases. Another interest thing illustrated in this figure is that, we see when P_s is larger than 0.8, the performance of SC-SU is already very close to that of GC-SU, when P_s equal 0.9, their curves almost coincide. That is to say under some conditions, P_s has no need to be so closed to one (say 0.999).

6   Conclusion

Considering the fact that different PU applications may have different QoS requirements in modern wireless communication systems, we established PU-claiming to share the QoS level between PU and SU. With the help of such knowledge, SUs should adjust their transmission and sensing strategies to provide sufficient protection to the PU. To provide the SUs with the QoS requirements, we use a spectrum sculpting techniques to create a specific spectrum valley in each PU application's data frame. Via a two-step sensing procedure, the SU estimates the spectrum valley's position if it determines that the interested channel is occupied by PU. Using some initial simple examples, we provided a clear exhibition of the advantages of the proposed scheme. Note that such a scheme can enable information sharing under a relatively low SNR (say less than -8dB), using the specific *signal structure* in frequency domain. As the implementation of the scheme is easy at PU, PU's some other important status information can also be shared in this way without squander its time-domain resource. Hence the scheme will potentially be meaningful directions for further research.

References

1. Liang, Y.-C., Chen, K.-C., et al.: Cognitive radio networking and communications: An overview. IEEE Trans. Veh. Technol. **60**(7), 3386–3407 (2011)
2. Alasti, M., Neekzad, B., Hui, J., Vannithamby, R.: Quality of service in WiMAX and LTE networks. IEEE Commun. Mag. **48**(5), 104–111 (2010)
3. Hoang, A.T., Liang, Y.-C., Islam, M.H.: Maximizing throughput of cognitive radio networks with limited primary users' cooperation. In: Proc. IEEE ICC (2007)
4. Zhang, W., Yeo, C.K., Li, Y.: A MAC sensing protocol design for data transmission with more protection to primary users. IEEE Trans. Mobile Comput. **12**(4), 621–632 (2013)
5. Hoang, A.T., Liang, Y.-C., Islam, M.H.: Power control and channel allocation in cognitive radio networks with primary users' cooperation. IEEE Trans. Mobile Comput. **9**(3), 348–360 (2010)
6. Mahmoud, H.A., Yücek, T., Arslan, H.: OFDM for cognitive radio: merits and challenges. IEEE Wireless Commun. Mag. **16**(2), 6–15 (2009)
7. Liang, Y.-C., Zeng, Y., et al.: Sensing-throughput tradeoff for cognitive radio networks. IEEE Trans. Wireless Commun. **7**(4), 1326–1337 (2008)
8. IEEE 802.16 Working Group. IEEE Standard for Local and Metropolitan Area Networks, Part 16: Air Interface for Fixed Broadband Wireless Access Systems Amendment3: Advanced Air Interface, IEEE Std, vol. 802 (2011)
9. IEEE 802.11-2007: Wireless LAN medium access control (MAC) and physical layer (PHY) specifications, IEEE 802.11 LAN Standards (2007)
10. López-Benítez, M., Casadevall, F.: Time-dimension models of spectrum usage for the analysis, design and simulation of cognitive radio networks. IEEE Trans. Veh. Technol. **62**(5), 2091–2104 (2013)

Sensing-Throughput Tradeoff for Cognitive Radio Systems with Unknown Received Power

Ankit Kaushik[1]([✉]), Shree Krishna Sharma[2], Symeon Chatzinotas[2], Björn Ottersten[2], and Friedrich Jondral[1]

[1] Communications Engineering Lab, Karlsruhe Institute of Technology (KIT), Karlsruhe, Germany
{ankit.kaushik,friedrich.jondral}@kit.edu
[2] SnT - Securityandtrust.lu, University of Luxembourg, Walferdange, Luxembourg
{shree.sharma,symeon.chatzinotas,bjorn.ottersten}@uni.lu

Abstract. Understanding the performance of the cognitive radio systems is of great interest. Different paradigms have been extensively analyzed in the literature to perform secondary access to the licensed spectrum. Of these, Interweave System (IS) has been widely investigated for performance analysis. According to IS, sensing is employed at the Secondary Transmitter (ST) that protects the Primary Receiver (PR) from the interference induced. Thus, in order to control the interference at the PR, it is required to sustain a certain level of probability of detection. In this regard, the ST requires the knowledge of the received power. However, in practice, this knowledge is not available at the ST. Thereby performing analysis considering the prior knowledge of the received power is too idealistic, thus, do not depict the actual performance of the IS. Motivated by this fact, an estimation model that includes received power estimation is proposed. Considering a sensing-throughput tradeoff, we apply this model to characterize the performance of the IS. Most importantly, the proposed model captures the estimation error to determine the distortion in the system performance. Based on analysis, it is illustrated that the ideal model overestimates the performance of the IS. Finally, it is shown that with an appropriate choice of the estimation time, the severity in distortion can be effectively regulated.

1 Introduction

For future wireless technologies, cognitive radio communication is emerging as a possible solution to the problem of spectrum scarcity. The available cognitive radio paradigms in the literature can be categorized into interweave, underlay and overlay [1]. In Interweave Systems (IS), the Secondary Users (SUs) utilize the licensed spectrum opportunistically by exploiting spectral holes in different domains such as time, frequency, space and polarization, whereas in Underlay Systems (US), SUs are allowed to use the primary spectrum as long as they respect the interference constraints of the Primary Receivers (PRs). On the other hand, Overlay Systems (OS) allow the spectral coexistence of two or more wireless networks by employing advanced transmission and coding strategies.

© Institute for Computer Sciences, Social Informatics and Telecommunications Engineering 2015
M. Weichold et al. (Eds.): CROWNCOM 2015, LNICST 156, pp. 308–320, 2015.
DOI: 10.1007/978-3-319-24540-9_25

Due to its ease in deployment, IS is mostly preferred for performing analysis among these paradigms. In this context, this paper focuses on the performance analysis of the ISs considering a hardware deployment where sensing is employed at the Secondary Transmitter (ST).

1.1 Motivation

Sensing is an integral part of the IS. At the ST, sensing is necessary for detecting the presence and absence of a primary signal, thereby protecting the PRs against harmful interference. Sensing at the ST is accomplished by listening to the power received from the PT. For detecting a primary signal, several techniques such as Energy Detection (ED), matched filtering, cyclostationary and feature-based detection exist [2,3]. Because of its versatility towards unknown primary signals, ED has been extensively investigated in the literature [4–8]. According to ED, the decision is accomplished by comparing the power received at the ST to a threshold. In reality, the ST encounters a variation in the received power due to the thermal noise at the receiver and fading in the channel. This leads to sensing errors described as misdetection or false alarm. The characterization of sensing errors as probability of detection and probability of false alarm has been studied in [9]. These sensing errors limit the performance of the IS.

In particular, probability of detection is critical for the primary system because it precludes the PR from the interference induced by the ST. On the other side, probability of false alarm accounts for the throughput attained by the secondary system at the Secondary Receiver (SR). In this regard, ST has to sustain a desired probability of detection and optimize its throughput. This phenomenon is characterized as a sensing-throughput tradeoff by Liang et al. [10]. According to it, the ST is able to determine a suitable sensing time that achieves an optimum throughput for a given received power. Several contributions have considered the performance of IS based on sensing errors [10–12]. However, the analysis described in the literature is too idealistic and not feasible for deployment, as it considers the perfect knowledge of the received power at the ST.

With the presence of channel and noise in the system, the received power is never known accurately, thus, needs to be estimated at the ST. Considering a hardware deployment, it is important to determine the performance of the IS based on received power estimation. A similar analysis is performed in [13], where the authors employ the received power estimation to control transmit power at the ST deployed as an underlay system. However, in this paper, we intend to capture the effect of estimation on the performance of an IS. Now, to realize received power estimation at the ST, it is necessary to allocate a certain time interval for the estimation within the frame duration. With the introduction of this estimation time, the system performance differs from its ideal behaviour. Additionally, the employed estimation process itself induces a certain level of error in the system. Hence, in order to understand the performance of the IS, it is necessary to consider the aforementioned aspects.

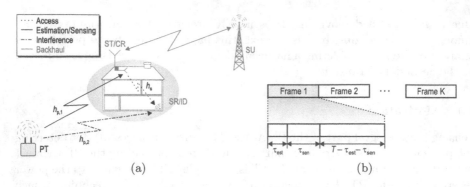

Fig. 1. (a) A scenario demonstrating the interweave paradigm. (b) Frame structure of interweave system with received power estimation.

1.2 Contributions

To realize the received power estimation, we consider a new frame structure. According to it, in a single frame, the ST performs (i) received power estimation, (ii) sensing, and (iii) data transmission. To perform analysis based on this new frame structure, we propose a novel estimation model. Most importantly, we evaluate the system performance with the inclusion of the estimation time and the errors occurred due to estimation. Based on the analytical expressions, we analyze the sensing-throughput tradeoff for the proposed estimation model. Finally, we determine the confidence intervals for the estimated received power. Particularly, based on these intervals, we capture the distortion in the performance based on the upper and lower bounds. This distortion, however, depends on the choice of design parameters depicted as probability of confidence and estimation duration.

1.3 Organization

The rest of the paper is organized as follows: Section 2 describes the system model that includes the interweave scenario and the signal model. Section 3 investigates the sensing-throughput tradeoff for the estimation model and derives upper and lower bounds for the performance parameters. Section 4 analyzes the numerical results based on the obtained expressions. Finally, Section 5 concludes the paper.

2 System Model

2.1 Interweave Scenario

Cognitive Relay (CR) [14] characterizes a small cell deployment that fulfills the spectral requirements for Indoor Devices (IDs). Fig. 1a illustrates a snapshot of a CR scenario to depict the interaction between the CR with PT and ID, where CR and ID represents the ST and SR respectively. In [14], the challenges involved while deploying the CR as an IS were presented. For simplification, a constant false alarm rate was considered in the system model. Now, we extend the analysis to employ a constant detection rate.

The medium access for the IS is slotted, where the time axis is segmented into frames of length T. The frame structure is analog to the one illustrated in [10]. However, unlike [10], the proposed frame structure uses τ_{est} to estimate and τ_{sen} to sense the received power, where τ_{est}, τ_{sen} correspond to time intervals and $\tau_{est} + \tau_{sen} < T$, cf. Fig. 1b. To incorporate the effect of fading in the model, we assume that the channel remains constant for T. Hence, characterized by the fading process, each frame witnesses a different received power. Therefore, to sustain a desired probability of detection, it is important to perform estimation τ_{est} followed by sensing τ_{sen} for each frame. The remaining time $T - (\tau_{est} + \tau_{sen})$ is utilized for data transmission.

2.2   Signal Model

In the estimation and sensing phase, the received signal at the ST is sampled with a sampling frequency of f_s for given hypotheses, that depicts the presence ($\mathcal{H}_1$) and absence ($\mathcal{H}_0$) of the primary signal, is given by

$$y_{rcvd}[n] = \begin{cases} \sqrt{h_{p,1}} \cdot x_p[n] + w[n] & : \mathcal{H}_1 \\ w[n] & : \mathcal{H}_0, \end{cases} \tag{1}$$

where $x_p[n]$ corresponds to a discrete sample at the PT, $h_{p,1}$ represents the power gain for the channel and $w[n]$ is circularly symmetric complex Additive White Gaussian Noise (AWGN) at the ST. $x_p[n]$ is an i.i.d. (independent and identically distributed) random process. As the channel $h_{p,1}$ is independent to x_p and $w[n]$ is an i.i.d. Gaussian random process with zero mean and variance $\mathbb{E}\left[|w[n]|^2\right] = \sigma^2$, the y_{rcvd} is also an i.i.d. random process. The true received power is defined as

$$\bar{P}_{rcvd} = \mathbb{E}\left[|\sqrt{h_{p,1}} \cdot x_p[n]|^2\right]. \tag{2}$$

Based on (2), the received SNR at the ST is $\gamma_{rcvd} = \frac{\bar{P}_{rcvd}}{\sigma^2} - 1$.

Now, the data transmission at the ST is conditioned over the probability of detection (P_d). In this context, the received signal at the SR is given by

$$y_s[n] = \begin{cases} \sqrt{h_s} \cdot x_s[n] + \sqrt{h_{p,2}} \cdot x_p[n] + w[n] & : 1 - P_d \\ \sqrt{h_s} \cdot x_s[n] + w[n] & : P_d, \end{cases} \tag{3}$$

where $x_s[n]$ is an i.i.d. random process and corresponds to discrete signal transmitted by the ST. Further, h_s and $h_{p,2}$ represent the power gains for channel, cf. Fig. 1a. The received SNRs over the links ST-SR and PT-ST are $\gamma_s = \frac{\mathbb{E}\left[|\sqrt{h_s} \cdot x_s[n]|^2\right]}{\sigma^2}$ and $\gamma_p = \frac{\mathbb{E}\left[|\sqrt{h_{p,2}} \cdot x_p[n]|^2\right]}{\sigma^2}$.

In the estimation phase, the estimated power received at the ST is given as [4]

$$P_{\text{rcvd}} = \frac{1}{\tau_{\text{est}} f_{\text{s}}} \sum_{n}^{\tau_{\text{est}} f_{\text{s}}} |y_{\text{rcvd}}[n]|^2. \tag{4}$$

P_{rcvd} determined in (4) using $\tau_{\text{est}} f_{\text{s}}$ samples follows a non central chi-squared distribution [15]. Considering large number of samples, thereby following similar approach as in [9], we apply the central limit theorem to approximate the distribution for P_{rcvd} as Gaussian distribution

$$P_{\text{rcvd}} \sim \mathcal{N}\left(\bar{P}_{\text{rcvd}}, \frac{2}{\tau_{\text{est}} f_{\text{s}}} \bar{P}_{\text{rcvd}}^2\right). \tag{5}$$

Following the estimation of the received power, the ST performs sensing for a duration of τ_{sen}, cf. Fig. 1b. The test statistics $T(\mathbf{y})$ at the ST is evaluated as

$$T(\mathbf{y}) = \frac{1}{\tau_{\text{sen}} f_{\text{s}}} \sum_{n}^{\tau_{\text{sen}} f_{\text{s}}} |y_{\text{rcvd}}[n]|^2 \underset{\mathcal{H}_0}{\overset{\mathcal{H}_1}{\gtrless}} \epsilon, \tag{6}$$

where ϵ is the threshold and $\mathbf{y}$ is a vector with $\tau_{\text{sen}} f_{\text{s}}$ samples. The probability of detection P_{d} and the probability of false alarm P_{fa} corresponding to (6) is determined as [9]

$$P_{\text{d}}(\epsilon, \tau_{\text{sen}}, \bar{P}_{\text{rcvd}}) = \mathcal{Q}\left(\frac{\epsilon - \bar{P}_{\text{rcvd}}}{\sqrt{\frac{2}{\tau_{\text{sen}} f_{\text{s}}}} \bar{P}_{\text{rcvd}}}\right), \tag{7}$$

$$P_{\text{fa}}(\epsilon, \tau_{\text{sen}}) = \mathcal{Q}\left(\frac{\epsilon - \sigma^2}{\sqrt{\frac{2}{\tau_{\text{sen}} f_{\text{s}}}} \sigma^2}\right), \tag{8}$$

where $\mathcal{Q}(\cdot)$ represents the Q-function [16]. Subsequently, by sustaining the P_{d} above a certain desired level $\bar{P}_{\text{d}}$

$$P_{\text{d}}(\epsilon, \tau_{\text{sen}}, \bar{P}_{\text{rcvd}}) \geq \bar{P}_{\text{d}}, \tag{9}$$

the ST precludes the interference to the primary system. Consequently, an optimum performance is achieved when the ST operates at the desired level, i.e., $P_{\text{d}} = \bar{P}_{\text{d}}$. Hence, using (7) and (9), the threshold is evaluated as

$$\epsilon(\bar{P}_{\text{d}}, \tau_{\text{sen}}, \bar{P}_{\text{rcvd}}) = \left(\mathcal{Q}^{-1}(\bar{P}_{\text{d}})\sqrt{\frac{2}{\tau_{\text{sen}} f_{\text{s}}}} + 1\right) \bar{P}_{\text{rcvd}}. \tag{10}$$

2.3   Assumptions

As a preliminary step, for the proposed model, we consider only the estimation of $\bar{P}_{\text{rcvd}}$ at the ST. Hence, in this paper, it is assumed that the ST acquires the perfect knowledge about γ_p and γ_s from the SR over a feedback channel. The inclusion of the imperfect knowledge of γ_p and γ_s in the proposed model poses an interesting research direction. Moreover, we consider that all transmitted signals are subjected to distance dependent path loss and the small scale fading gains. The coherence time for the channel gain is greater than the frame duration. However, we may still encounter scenarios where the coherence time exceeds the frame duration, in such cases our characterization depicts a lower performance bound. With no loss of generality, we consider that the channel gain ($h_{p,1}$, $h_{p,2}$ and h_s) includes the distance dependent path loss and the small scale gain. Finally, we target short term performance, according to which the performance parameters are optimized for each frame.

3   Sensing-Throughput Analysis

3.1   Ideal Model (IM)

According to Liang *et al.* [10], the secondary throughput subject to a desired probability of detection $\bar{P}_d$ is given by

$$\tilde{R}_s(\tilde{\tau}_{sen}) = \max_{\tau_{sen}} R_s(\tau_{sen}) = \frac{T - \tau_{sen}}{T}\Big[C_0(1 - P_{fa}(\epsilon, \tau_{sen}))P(\mathcal{H}_0)$$

$$+ C_1(1 - P_d(\epsilon, \tau_{sen}, \bar{P}_{rcvd}))P(\mathcal{H}_1)\Big], \tag{11}$$

$$\text{s.t. } P_d(\epsilon, \tau_{sen}, \bar{P}_{rcvd}) \geq \bar{P}_d,$$

$$\text{where } C_0 = \log_2(1 + \gamma_s) \text{ and } C_1 = \log_2\left(1 + \frac{\gamma_s}{\gamma_p + 1}\right).$$

$P(\mathcal{H}_0)$ and $P(\mathcal{H}_1)$ are the probabilities of occurrence for the respective hypothesis. Based on (11), the ST is able to determine the suitable sensing time $\tau_{sen} = \tilde{\tau}_{sen}$ such that an optimum throughput $\tilde{R}_s(\tilde{\tau}_{sen})$ is achieved. According to (11), the performance parameters for the IM are defined as $\tilde{R}_s$, P_d and P_{fa}.

3.2   Estimation Model (EM)

The system described in [10] is good for performing analysis, however, to determine $\tilde{\tau}_{sen}$ at the ST requires the knowledge of the received power $\bar{P}_{rcvd}$. Considering a hardware deployment, this information is not available at the ST. Unless estimated, it is not possible to determine $\tilde{\tau}_{sen}$. According to the EM, the ST estimates the $\bar{P}_{rcvd}$ for a duration of τ_{est} as P_{rcvd} and based on its value, the ST determines $\tilde{\tau}_{sen}$ for the given frame. The samples needed for estimation can be utilized for sensing as well. However, for analytical tractability, in the proposed model the estimation and sensing are considered to be disjoint in time. Now, with the introduction of τ_{est}, the actual performance of the IS deviates from its ideal performance. Moreover, the estimation itself causes distortion in the actual performance of the IS. As a part of the proposed model, these aspects are dealt in the following subsections.

3.3 Actual Performance

With the introduction of the EM, we first characterize the performance of the IS. To realize this, it is considered that the ST perfectly estimates the P_{rcvd}, that is $P_{\mathrm{rcvd}} = \bar{P}_{\mathrm{rcvd}}$. In accordance with the proposed model, the sensing-throughput tradeoff with perfect estimation is determined as

$$\tilde{R}_{\mathrm{s}}^{\mathrm{P}}(\tilde{\tau}_{\mathrm{sen}}^{\mathrm{P}}) = \max_{\tau_{\mathrm{sen}}} R_{\mathrm{s}}^{\mathrm{P}}(\tau_{\mathrm{sen}}) = \frac{T - (\tau_{\mathrm{est}} + \tau_{\mathrm{sen}})}{T}\left[C_0(1 - \mathrm{P}_{\mathrm{fa}}(\epsilon, \tau_{\mathrm{sen}})) \right.$$

$$\left. \mathrm{P}(\mathcal{H}_0) + C_1(1 - \mathrm{P}_{\mathrm{d}}(\epsilon, \tau_{\mathrm{sen}}, \bar{P}_{\mathrm{rcvd}}))\mathrm{P}(\mathcal{H}_1) \right], \tag{12}$$

$$\text{s.t. } \mathrm{P}_{\mathrm{d}}(\epsilon, \tau_{\mathrm{sen}}, \bar{P}_{\mathrm{rcvd}}) \geq \bar{\mathrm{P}}_{\mathrm{d}}.$$

According to the (12), for a given $\bar{\mathrm{P}}_{\mathrm{d}}$ and estimation of $\bar{P}_{\mathrm{rcvd}}$ in the interval τ_{est}, ST is able to determine the threshold as $\epsilon(\bar{\mathrm{P}}_{\mathrm{d}}, \tau_{\mathrm{sen}}, \bar{P}_{\mathrm{rcvd}})$. Finally, based on the new sensing-throughput tradeoff (12), ST evaluates the suitable sensing time as $\tau_{\mathrm{sen}} = \tilde{\tau}_{\mathrm{sen}}^{\mathrm{P}}$ that achieves the optimum throughput $\tilde{R}_{\mathrm{s}}^{\mathrm{P}}$. However, $\tilde{\tau}_{\mathrm{sen}} \neq \tilde{\tau}_{\mathrm{sen}}^{\mathrm{P}}$ due to the inclusion of the estimation time in the considered sensing-throughput tradeoff. According to the EM, the performance parameters that characterize the performance of the IS are $\tilde{R}_{\mathrm{s}}^{\mathrm{P}}$, $\mathrm{P}_{\mathrm{d}}^{\mathrm{P}}$ and $\mathrm{P}_{\mathrm{fa}}^{\mathrm{P}}$. Now, with the perfect estimation of P_{rcvd}, the constraint in (12) is sustained, hence, $\mathrm{P}_{\mathrm{d}} = \bar{\mathrm{P}}_{\mathrm{d}}$. However, with $\tilde{\tau}_{\mathrm{sen}} \neq \tilde{\tau}_{\mathrm{sen}}^{\mathrm{P}}$, $\mathrm{P}_{\mathrm{fa}}^{\mathrm{P}}$ and $\tilde{R}_{\mathrm{s}}^{\mathrm{P}}$ witness a deviation from their ideal behaviour.

3.4 Distortion

Previously, we determined the effect of estimation time on the performance. In this section, we extend the analysis by considering the influence of estimation error on the system performance. In this context, based on (5), we characterize a confidence interval $[P_{\mathrm{rcvd}}^{\mathrm{L}}, P_{\mathrm{rcvd}}^{\mathrm{U}}]$ for a certain choice of probability of confidence P_{c} and τ_{est} as

$$P_{\mathrm{rcvd}} = \begin{cases} P_{\mathrm{rcvd}}^{\mathrm{L}} = \left(\mathcal{Q}^{-1}\left(\frac{\mathrm{P}_{\mathrm{c}}+1}{2}\right)\sqrt{\frac{2}{\tau_{\mathrm{est}}f_{\mathrm{s}}}} + 1\right)\bar{P}_{\mathrm{rcvd}} \\ P_{\mathrm{rcvd}}^{\mathrm{U}} = \left(\mathcal{Q}^{-1}\left(1 - \frac{\mathrm{P}_{\mathrm{c}}+1}{2}\right)\sqrt{\frac{2}{\tau_{\mathrm{est}}f_{\mathrm{s}}}} + 1\right)\bar{P}_{\mathrm{rcvd}}, \end{cases} \tag{13}$$

where $\mathcal{Q}^{-1}(\cdot)$ is the inverse-Q function [16]. Hence, we utilize this confidence interval to depict the maximum estimation error in the estimated received power, thereby characterizing maximum distortion in the performance parameters $\mathrm{P}_{\mathrm{d}}^{\mathrm{P}}$, $\mathrm{P}_{\mathrm{fa}}^{\mathrm{P}}$ and $\tilde{R}_{\mathrm{s}}^{\mathrm{P}}$. For $\mathrm{P}_{\mathrm{c}} = 0.95$, the $P_{\mathrm{rcvd}}^{\mathrm{L}}$ and $P_{\mathrm{rcvd}}^{\mathrm{U}}$ are equivalent to the lower and upper bounds of the $\bar{P}_{\mathrm{rcvd}}$.

This confidence interval further depicts the distortion in the system parameters ϵ and $\tilde{\tau}_{\mathrm{sen}}^{\mathrm{P}}$. Hence, as an intermediate step, we first characterize the distortion in terms of these system parameters. Subject to the received powers as $P_{\mathrm{rcvd}}^{\mathrm{L}}$ and $P_{\mathrm{rcvd}}^{\mathrm{U}}$, the expressions for the threshold are evaluated as

$$\epsilon^{\mathrm{L}} = \epsilon(\bar{\mathrm{P}}_{\mathrm{d}}, \tau_{\mathrm{sen}}, P_{\mathrm{rcvd}}^{\mathrm{L}}) \text{ and } \epsilon^{\mathrm{U}} = \epsilon(\bar{\mathrm{P}}_{\mathrm{d}}, \tau_{\mathrm{sen}}, P_{\mathrm{rcvd}}^{\mathrm{U}}). \tag{14}$$

Clearly, due to difference in received power estimated at the ST, the expressions in (14) differ from the one illustrated in (10). By inserting the thresholds ϵ^L and ϵ^U in (12), the suitable sensing times computed at the ST are represented as

$$\tilde{\tau}_{\text{sen}}^L \text{ and } \tilde{\tau}_{\text{sen}}^U. \tag{15}$$

As a result, (14) and (15) clearly illustrates the distortion in the system parameters. Now, as a final step, we characterize the distortion in the performance parameters in terms of the distorted system parameters and the true received power. Consequently, we represent the distortion in $\tilde{R}_s^P$, P_d^P and P_{fa}^P as a function of $(\tilde{\tau}_{\text{sen}}^L, \epsilon^L)$ and $(\tilde{\tau}_{\text{sen}}^U, \epsilon^U)$ subject to true received power, i.e., $\bar{P}_{\text{rcvd}}$.

Following the above discussion, the distortion in the P_d^P, in terms of upper and lower bounds, due to the inclusion of estimation error in the received power is determined as

$$P_d^P = \begin{cases} P_d^L = P_d(\epsilon^L, \tilde{\tau}_{\text{sen}}^L, \bar{P}_{\text{rcvd}}) \\ P_d^U = P_d(\epsilon^U, \tilde{\tau}_{\text{sen}}^U, \bar{P}_{\text{rcvd}}). \end{cases} \tag{16}$$

It is evident that the distortion in the P_d^P results in the violation of the regulatory constraint, c.f. (9). If this constraint is not sustained, it may result in harmful interference at the PR. Hence, using (16), we are able to characterize the situations where the IS may degrade the performance of the primary system.

On similar basis, the distortion in P_{fa}^P in terms of upper and lower bound is depicted as

$$P_{\text{fa}}^P = \begin{cases} P_{\text{fa}}^L = P_{\text{fa}}(\epsilon^L, \tilde{\tau}_{\text{sen}}^L) \\ P_{\text{fa}}^U = P_{\text{fa}}(\epsilon^U, \tilde{\tau}_{\text{sen}}^U). \end{cases} \tag{17}$$

Finally, including the distortion in the probabilities P_d^P and P_{fa}^P from (16) and (17) and the system parameters $(\epsilon, \tilde{\tau}_{\text{sen}}^P)$, the distortion in the optimum throughput $\tilde{R}_s^P$ in terms of upper and lower bound is determined as

$$\tilde{R}_s^P = \begin{cases} \tilde{R}_s^L = \frac{T-(\tau_{\text{est}}+\tilde{\tau}_{\text{sen}}^L)}{T}\left[C_0(1 - P_{\text{fa}}(\epsilon^L, \tau_{\text{sen}}^L))P(\mathcal{H}_0) \right. \\ \left. +C_1(1 - P_d(\epsilon^L, \tilde{\tau}_{\text{sen}}^L, \bar{P}_{\text{rcvd}})P(\mathcal{H}_1)\right] \\ \tilde{R}_s^U = \frac{T-(\tau_{\text{est}}+\tilde{\tau}_{\text{sen}}^U)}{T}\left[C_0(1 - P_{\text{fa}}(\epsilon^U, \tau_{\text{sen}}^U))P(\mathcal{H}_0) \right. \\ \left. +C_1(1 - P_d(\epsilon^U, \tilde{\tau}_{\text{sen}}^U, \bar{P}_{\text{rcvd}})P(\mathcal{H}_1)\right]. \end{cases} \tag{18}$$

Based on the expressions (16), (17) and (18) characterized by the EM, it is possible to depict the distortion for the IS from its actual performance. Moreover, the severity in distortion can be controlled through P_c and τ_{est}. In particular, it is important to select τ_{est} appropriately such that the distortion in the performance doesn't exceed beyond a certain level. This aspect is investigated more deeply in the next section.

4   Numerical Analysis

In this section, the performance of the IS for the EM is analyzed. In this regard, we perform simulations to: (i) validate the expressions obtained in Section 3, (ii) provide a mathematical justification to the Gaussian approximation considered in Section 2. Although, the expressions derived using our sensing-throughput analysis are general and applicable to all cognitive radio systems, however, the parameters are selected in such a way that they closely relate to the deployment scenario described in Fig. 1a. Unless stated explicitly, the following choice of the parameters is considered for the analysis, $f_s = 1\,\text{MHz}$, $h_{p,1} = h_{p,2} = -100\,\text{dB}$, $h_s = -80\,\text{dB}$, $T = 100\,\text{ms}$, $P_c = 0.95$, $\bar{P}_d = 0.9$, $\sigma^2 = -100\,\text{dBm}$, $\gamma_{\text{rcvd}} = -10\,\text{dB}$, $\gamma_p = -10\,\text{dB}$, $\gamma_s = 10\,\text{dB}$ and $P(\mathcal{H}_1) = 1 - P(\mathcal{H}_0) = 0.2$, $\tau_{\text{est}} = 5\,\text{ms}$.

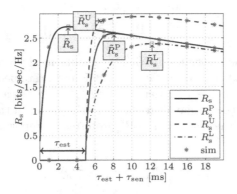

Fig. 2. Sensing-throughput tradeoff for the ideal and estimation models with $\gamma_{\text{rcvd}} = -10\,\text{dB}$ and $\tau_{\text{est}} = 5\,\text{ms}$.

Firstly, the analysis in terms of sensing-throughput tradeoff based on (11) and (12) for the IM and the EM is performed. The curves R_s and R_s^P in Fig. 2 depict the throughput based on the IM and the EM (actual performance) at the ST. Due to the inclusion of received power estimation in the frame structure, the ST produces no throughput at the SR for the interval τ_{est}. The sensing times $\tilde{\tau}_{\text{sen}} = 3.11\,\text{ms}$ and $\tilde{\tau}_{\text{sen}}^P = 3.06\,\text{ms}$ are evaluated, which yield the optimum throughputs as $\tilde{R}_s = 2.73\,\text{bits/sec/Hz}$ and $\tilde{R}_s^P = 2.59\,\text{bits/sec/Hz}$, cf. Fig. 2. This variation is due the inclusion of τ_{est} in the sensing-throughput analysis. Hence, for the given choice of τ_{est} at ST, the ideal model overestimates the optimum throughput by $\approx 5\%$.

Next, sensing-throughput analysis is performed, considering that the $\bar{P}_{\text{rcvd}}$ is estimated as P_{rcvd}^L or P_{rcvd}^U at ST. $\gamma_{\text{rcvd}} = -10\,\text{dBm}$ corresponds to $\bar{P}_{\text{rcvd}} = 1.10 \cdot 10^{-10}\,\text{mW}$. With $P_c = 0.95$ and $\tau_{\text{est}} = 5\,\text{ms}$, the confidence intervals are determined as $P_{\text{rcvd}}^L = 1.05 \cdot 10^{-10}\,\text{mW}$ and $P_{\text{rcvd}}^U = 1.14 \cdot 10^{-10}\,\text{mW}$, cf. (13). Fig. 2 demonstrates the throughput corresponding to distorted system parameter ϵ^L and ϵ^U, cf. (14). The suitable sensing times are evaluated as $\tilde{\tau}_{\text{sen}}^L = 7.21\,\text{ms}$ and

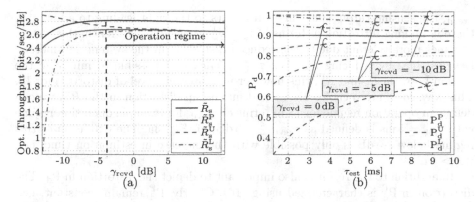

Fig. 3. (a) Distortion in optimum throughput versus the γ_{rcvd} with $P_c = 0.95$ and $\tau_{\text{est}} = 5\,\text{ms}$. (b) Distortion in probability of detection versus the estimation time for different $\gamma_{\text{rcvd}} = \{-10, -5, 0\}\text{dB}$.

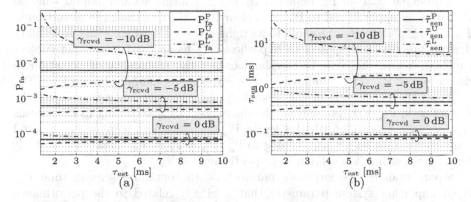

Fig. 4. (a) Distortion in probability of false versus the estimation time for different $\gamma_{\text{rcvd}} \in \{-10, -5, 0\}\text{dB}$. (b) Distortion in optimum sensing time versus the estimation time for different $\gamma_{\text{rcvd}} \in \{-10, -5, 0\}\text{dB}$.

$\tilde{\tau}_{\text{sen}}^{\text{U}} = 1.76\,\text{ms}$. Finally, the distortion in the $\tilde{R}_s^{\text{P}}$ corresponding to the distortion in the system parameters is demonstrated. The lower bound and upper bound on the optimum throughput $\tilde{R}_s^{\text{P}}$ are determined as $\tilde{R}_s^{\text{L}} = 2.38\,\text{bits/sec/Hz}$ and $\tilde{R}_s^{\text{U}} = 2.84\,\text{bits/sec/Hz}$. It corresponds to 12.82% underestimation and 4.07% overestimation of the optimum throughput. Therefore, based on the analytical expressions determined under EM, it is possible to determine the severity of distortion in the system performance.

Hereafter, we consider the theoretical expressions for the analysis. Next, we determine the variation of the optimum throughput against the $\gamma_{\text{rcvd}} \in [-13, 10]\text{dB}$ at ST with $\tau_{\text{est}} = 5\,\text{ms}$. It is evident from Fig. 3a that the distortion in the $\tilde{R}_s^{\text{P}}$ decreases with increase in γ_{rcvd}. For $\gamma_{\text{rcvd}} > -5\,\text{dB}$ and $\tau_{\text{est}} = 5\,\text{ms}$, the level of distortion is negligible. This is due to the fact that with increase in P_{rcvd}, P_{fa} shifts to a very low value such that $1 - P_{\text{fa}} \approx 1$. Moreover, for large

γ_{rcvd}, the sensing-throughput curvature shifts to the right leading to a low $\tilde{\tau}_{\text{sen}}^{\text{P}}$, thereby making distortion in terms of $\tilde{\tau}_{\text{sen}}^{\text{L}}$, $\tilde{\tau}_{\text{sen}}^{\text{U}}$ insignificant. Hence, the system becomes more tolerant due to the reduced distortion in the system parameters. Besides that, by reducing τ_{est}, i.e., $\tau_{\text{est}} < 5\,\text{ms}$, it is possible to minimize the margin between $\tilde{R}_{\text{s}}$ and $\tilde{R}_{\text{s}}^{\text{P}}$, this however increases the level of distortion for IS. This way, for a given choice of τ_{est} and maximum distortion in the $\tilde{R}_{\text{s}}^{\text{P}}$, we can define an operation regime for IS in terms of γ_{rcvd}, for example operation regime with $\tau_{\text{est}} = 5\,\text{ms}$ is defined as $\gamma_{\text{rcvd}} \geq -5\,\text{dB}$, cf. Fig. 3a. The extension of this regime below $-5\,\text{dB}$ is only possible with the increase in estimation time, i.e., $\tau_{\text{est}} > 5\,\text{ms}$.

In addition to the $\tilde{R}_{\text{s}}^{\text{P}}$, it is also important to depict the distortion in P_{d}. The distortion in $\text{P}_{\text{d}}^{\text{P}}$ is characterized using (16). Clearly, $\text{P}_{\text{d}}^{\text{P}}$ remains constant, i.e., $\text{P}_{\text{d}}^{\text{P}} = \bar{\text{P}}_{\text{d}}$ with τ_{est} and for $\gamma_{\text{rcvd}} \in \{-10, -5, 0\}\text{dB}$. Fig. 3b reveals the distortion in $\text{P}_{\text{d}}^{\text{P}}$ in terms of $\text{P}_{\text{d}}^{\text{L}}$ and $\text{P}_{\text{d}}^{\text{U}}$. It is evident that the distortion is small for a large value of γ_{rcvd} and it decreases with increase in τ_{est}. Most importantly, it is observed that an $\text{P}_{\text{d}}^{\text{U}}$ depicted from $P_{\text{rcvd}}^{\text{U}}$ forms a lower bound whereas $\text{P}_{\text{d}}^{\text{L}}$ forms an upper bound to $\text{P}_{\text{d}}^{\text{P}}$. It is clear from the fact that, the distortion in $\bar{P}_{\text{rcvd}} \leq P_{\text{rcvd}}^{\text{U}}$ shifts the threshold to its right side $\epsilon \geq \epsilon^{\text{U}}$, whereas the probability density function corresponds to hypothesis $\mathcal{H}_1$ has a fixed expression subject to $\bar{P}_{\text{rcvd}}$, hence, this shift in threshold causes the probability of detection to decrease, i.e., $\text{P}_{\text{d}}^{\text{U}} \leq \text{P}_{\text{d}}^{\text{P}}$. Similarly, $\epsilon^{\text{L}} \leq \epsilon$ corresponds to upper bound $\text{P}_{\text{d}}^{\text{P}} \leq \text{P}_{\text{d}}^{\text{L}}$.

From the perspective of the secondary user, it is interesting to depict the distortion in $\text{P}_{\text{fa}}^{\text{P}}$. Fig. 4a analyzes the distortion in $\text{P}_{\text{fa}}^{\text{P}}$ according to (17) versus τ_{est} for $\gamma_{\text{rcvd}} = \{-10, -5, 0\}\text{dB}$. Clearly, $\text{P}_{\text{fa}}^{\text{P}}$ decreases with increase in γ_{rcvd} and remains constant with γ_{rcvd}. Analog to $\text{P}_{\text{d}}^{\text{P}}$, $\text{P}_{\text{fa}}^{\text{U}} \leq \text{P}_{\text{fa}}^{\text{P}} \leq \text{P}_{\text{fa}}^{\text{L}}$, hence, $\text{P}_{\text{fa}}^{\text{L}}$ and $\text{P}_{\text{fa}}^{\text{U}}$ form an upper and lower bounds, respectively.

Apart from the performance parameters, the optimum sensing time $\tilde{\tau}_{\text{sen}}^{\text{P}}$ is an important system parameter that is closely related to the performance parameters. Hence, Fig. 4b reveals the distortion in the $\tilde{\tau}_{\text{sen}}^{\text{P}}$ versus the τ_{est} for $\gamma_{\text{rcvd}} \in \{-10, -5, 0\}\text{dB}$, cf. (15). Similar to $\text{P}_{\text{d}}^{\text{P}}$ and $\text{P}_{\text{fa}}^{\text{P}}$, $\tilde{\tau}_{\text{sen}}^{\text{L}}$ and $\tilde{\tau}_{\text{sen}}^{\text{U}}$ represents the upper and lower bound to $\tilde{\tau}_{\text{sen}}^{\text{P}}$. It is obvious from the fact that larger value of estimated received power $P_{\text{rcvd}}^{\text{U}}$ shifts the curvature in the sensing-throughput, that depicts the optimum sensing time, to a lower value, therefore, $\tau_{\text{sen}}^{\text{U}} \leq \tau_{\text{sen}}^{\text{P}} \leq \tau_{\text{sen}}^{\text{L}}$, cf. Fig. 4b.

5   Conclusion

In this paper, we considered the deployment of a cognitive radio as an interweave system. For sustaining a minimum probability of detection, it requires the knowledge of the received power at the ST. To acquire this knowledge, an estimation has been included within the frame duration. In this regard, we proposed an estimation model that characterizes the actual performance of the IS. More specifically, the distortion in terms of bounds on the performance parameters has been captured based on the analytical expression. Moreover, it has been indicated that the severity in distortion can be confined by regulating the estimation time. Through theoretical and numerical analysis, it has been demonstrated that for a given choice of estimation time, the distortion in the performance parameters limits the operation regime, defined in terms of the received signal to noise ratio, at the ST.

In future, we plan to depict the exact expression of distortion instead of performance bounds. To pursue this, an outage constraint will be applied on the probability of detection in place of received power, that is $P(P_d \leq \bar{P}_d)$. In this way, we shall determine the estimation time subject to the new outage constraint.

References

1. Goldsmith, A., Jafar, S., Maric, I., Srinivasa, S.: Breaking Spectrum Gridlock With Cognitive Radios: An Information Theoretic Perspective. Proceedings of the IEEE **97**(5), 894–914 (2009)
2. Axell, E., Leus, G., Larsson, E., Poor, H.: Spectrum sensing for cognitive radio: State-of-the-art and recent advances. IEEE Signal Processing Magazine **29**(3), 101–116 (2012)
3. Sharma, S., Chatzinotas, S., Ottersten, B.: Exploiting polarization for spectrum sensing in cognitive satcoms. In: CROWNCOM, pp. 36–41, June 2012
4. Urkowitz, H.: Energy detection of unknown deterministic signals. Proceedings of the IEEE **55**(4), 523–531 (1967)
5. Kostylev, V.: Energy detection of a signal with random amplitude. In: ICC, vol. 3, pp. 1606–1610 (2002)
6. Digham, F., Alouini, M.-S., Simon, M.K.: On the energy detection of unknown signals over fading channels. In: ICC, vol. 5, pp. 3575–3579, May 2003
7. Herath, S., Rajatheva, N., Tellambura, C.: Unified approach for energy detection of unknown deterministic signal in cognitive radio over fading channels. In: ICC Workshops, pp. 1–5, June 2009
8. Mariani, A., Giorgetti, A., Chiani, M.: Energy detector design for cognitive radio applications. In: 2010 International Waveform Diversity and Design Conference (WDD), pp. 000053–000057, August 2010
9. Tandra, R., Sahai, A.: SNR Walls for Signal Detection. IEEE Journal of Selected Topics in Signal Processing **2**(1), 4–17 (2008)
10. Liang, Y.-C., Zeng, Y., Peh, E., Hoang, A.T.: Sensing-Throughput Tradeoff for Cognitive Radio Networks. IEEE Transactions on Wireless Communications **7**(4), 1326–1337 (2008)
11. Cardenas-Juarez, M., Ghogho, M.: Spectrum Sensing and Throughput Tradeoff in Cognitive Radio under Outage Constraints over Nakagami Fading. IEEE Communications Letters **15**(10), 1110–1113 (2011)
12. Sharkasi, Y., Ghogho, M., McLernon, D.: Sensing-throughput tradeoff for OFDM-based cognitive radio under outage constraints. In: ISWCS, pp. 66–70, August 2012

13. Kaushik, A., Sharma, S.K., Chatzinotas, S., Ottersten, B., Jondral, F.K.: Estimation-Throughput tradeoff for underlay cognitive radio systems. In: IEEE Int. Conf. on Communications (ICC) - Cognitive Radio and Networks Symposium, June 2015 (to appear)
14. Kaushik, A., Mueller, M., Jondral, F.K.: Cognitive Relay: Detecting Spectrum Holes in a Dynamic Scenario. In: ISWCS, pp. 1–2, April 2013
15. Kay, S.: Fundamentals of Statistical Signal Processing: Detection theory. Prentice Hall Signal Processing Series. Prentice-Hall PTR (1998)
16. Gradshteyn, I.S., Ryzhik, I.M.: Table of Integrals, Series, and Products, 6th edn. Academic Press, San Diego (2000)

Cooperative Spectrum Sensing for Heterogeneous Sensor Networks Using Multiple Decision Statistics

Shree Krishna Sharma$^{(\boxtimes)}$, Symeon Chatzinotas, and Björn Ottersten

SnT - securityandtrust.lu, University of Luxembourg, 2721 Kirchberg, Luxembourg
{shree.sharma,symeon.chatzinotas,bjorn.ottersten}@uni.lu

Abstract. The detection of active Primary Users (PUs) in practical wireless channels with a single Cognitive Radio (CR) sensor is challenging due to several issues such as the hidden node problem, path loss, shadowing, multipath fading, and receiver noise/interference uncertainty. In this context, Cooperative Spectrum Sensing (CSS) is considered a promising technique in order to enhance the overall sensing efficiency. Existing CSS methods mostly focus on homogeneous cooperating nodes considering identical node capabilities, equal number of antennas, equal sampling rate and identical Signal to Noise Ratio (SNR). However, in practice, nodes with different capabilities can be deployed at different stages and are very much likely to be heterogeneous in terms of the aforementioned features. In this context, we propose a novel decision statistics-based centralized CSS technique using the joint Probability Distribution Function (PDF) of the multiple decision statistics resulting from different processing capabilities at the sensor nodes and compare its performance with various existing cooperative schemes. Further, we provide a design guideline for the network operators to facilitate decision making while upgrading a sensor network.

Keywords: Cooperative Spectrum Sensing · Cognitive Radio · Joint PDF · Heterogeneous sensor networks

1 Introduction

Cognitive Radio (CR) communications is considered a promising solution in order to address the spectrum scarcity problem caused by the high demand of data rates and current frequency allocation policies. The most commonly used spectrum sharing paradigms in the literature are interweave, underlay, and overlay [1,2]. Out of several spectrum awareness techniques for enabling these paradigms, Spectrum Sensing (SS) is an important mechanism in order to exploit the spectral gaps in the underutilized primary spectrum so that they can be used by Secondary Users (SUs) in order to enhance the overall spectral efficiency of the system. Several SS techniques such as Matched filter, Energy Detection (ED), Cyclostationary Detection (CD), Autocorrelation-based detection (AD),

© Institute for Computer Sciences, Social Informatics and Telecommunications Engineering 2015
M. Weichold et al. (Eds.): CROWNCOM 2015, LNICST 156, pp. 321–333, 2015.
DOI: 10.1007/978-3-319-24540-9_26

Eigenvalue-based Detection, etc. have been proposed in the literature for CR systems [3]. These techniques have different operational requirements, advantages and disadvantages from the practical perspectives and can be broadly categorized into: (i) knowledge-aware, (ii) semi-blind, (ii) blind SS techniques [4,5].

In practical wireless fading channels, the SS efficiency of the aforementioned techniques may be degraded due to the hidden node problem, path loss, shadowing, multipath fading and receiver noise/interference uncertainty issues. In this context, Cooperative Spectrum Sensing (CSS) has been considered as a promising approach [6–8]. The main concept behind CSS is to enhance the sensing performance by exploiting the observations captured by spatially located CR users. The CSS gain is achieved by sharing the information gathered by the cooperating users, thus making the combined decision more reliable than the individual decisions. Despite the its several advantages [6], CSS requires a control channel for each cooperating node to report its sensed information to the Fusion Center (FC) and this channel is usually bandwidth limited. Thus, cooperation burden can be a critical issue from a practical perspective. In this context, we are interested in studying the decision statistics-based centralized CSS which can reduce the signalling burden compared to the one of sample-based CSS and at the same time achieves the desired level of sensing performance.

Most of the existing CSS literature considers a CR network with homogeneous sensor nodes and assumes identical capabilities, equal number of antennas, equal sampling rate and identical received Signal to Noise Ratio (SNR) for all the cooperating nodes. However, in practice, the nodes with different capabilities can be deployed at different stages and are very much likely to be heterogeneous in terms of the aforementioned features. In this context, it's an important challenge to investigate suitable CSS techniques which can provide better sensing performance and low signalling overhead in heterogeneous environments. Further, most of the existing decision and data fusion techniques in the CSS context use a single type of detector (ED in many cases) as local and CSS mechanisms. However, in heterogeneous environments, different nodes can employ separate decision statistics since they may have different capabilities. The issue of data fusion considering different decision statistics for CSS has not been addressed in the literature.

To address the aforementioned issues, this paper investigates the combination of ED and eigenvalue-based decision statistics in order to achieve reliable sensing in heterogeneous environments. More specifically, we propose a novel CSS technique based on the joint Probability Distribution Function (PDF) of multiple decision statistics. We consider that these multiple decision statistics arise from the different processing capabilities of the heterogeneous cooperating nodes. We evaluate and compare the detection performance of the proposed approach with the existing cooperative approaches. Moreover, we provide a design guideline for the network operators to facilitate decision making while upgrading a sensor network.

The remainder of this paper is structured as follows. Section 2 presents the system and signal models, and further describes the considered local and cooperative detection techniques. Section 3 proposes a novel CSS approach for the considered heterogeneous environment. Section 4 evaluates and compares the performance of the proposed approach with several existing approaches with the help of numerical results. Finally, Section 5 concludes the paper.

2   System and Signal Model

We consider a large scale CR network consisting of N_c number of heterogeneous cooperating nodes which communicate with a FC as depicted in Fig. 1. The considered heterogeneous environment is motivated from the real world practical scenarios. If we consider the deployment of nodes in a sensor network, different nodes may be deployed at different stages as the available technology evolves. It is in general both impractical and wasteful to replace all the existing nodes with the new nodes while implementing a new technology. The heterogeneity of the nodes can be in the form of capability of nodes, number of samples acquired during the sensing time (sampling rate and sampling time can be different for different nodes), number of antennas equipped in the nodes and the received PU SNR. For simplicity of analysis in this paper, we assume the same received PU SNR at each cooperating node, and fixed number of samples. Further, we consider the following two categories of nodes on the basis of their capabilities: (i) existing nodes which are capable of performing a simple sensing algorithm such as ED (We call these nodes the first generation nodes.), and (ii) new sensor nodes which are capable of performing advanced sensing techniques such as eigenvalue-based algorithms (We call these nodes the second generation nodes).

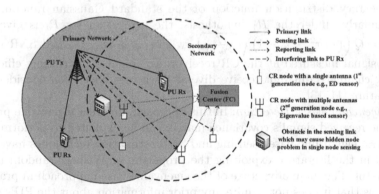

Fig. 1. Schematic of the considered CSS scenario with heterogeneous nodes

Let $N = \lceil \tau f_s \rceil$ be the number of observations collected by each node in the time duration of τ, f_s being the sampling frequency. We denote the hypothesis of the PU absence and the PU presence by H_0 and H_1 respectively. When each cooperating node performs local sensing independently, the SS problem can be written as

$$y_k(n) = z_k(n), \qquad\qquad H_0 \tag{1}$$
$$y_k(n) = h_k s(n) + z_k(n), \quad H_1$$

where $s(n)$ is the nth sample of the transmitted PU signal, h_k denotes the complex gain of the channel between the PU and the kth node, and $z_k(n)$ is the Additive White Gaussian Noise (AWGN) at the receiver of the kth cooperating node.

We assume that the sensing channel remains constant during the period of sensing and the transmitted PU symbols are independent and identically distributed (i.i.d.) complex circularly symmetric (c.c.s.) symbols. Further, we assume that the reporting channels are ideal as considered in various literature [4, 9]. The decision statistic used for testing the above hypothesis problem depends on the employed detection technique. For a detection technique based on a single decision statistic, let us denote by T, the probability of detection (P_d) and the probability of false alarm (P_f) can be calculated as: $P_d = Pr(T > \lambda|H_1)$, and $P_f = Pr(T > \lambda|H_0)$, where $Pr(\cdot)$ denotes the probability.

1. Sensing Techniques: In the CSS, the local nodes may employ any one of the sensing techniques such as ED, CD, AD, eigenvalue detector, etc. in order to capture the information about the presence or absence of the PU signal. In this paper, we consider the following two detection techniques in the considered heterogeneous environment.

i. Energy Detection: For the ED detector, the decision statistic for the kth cooperating node is given by $T_k = \frac{1}{N}\sum_{n=1}^{N}|y_k(n)|^2$. For the Circularly Symmetric Complex Gaussian (CSCG) noise with variance σ_z^2, the expression for P_f can be written as [10]: $P_f(\lambda, \tau) = Q\left(\left(\frac{\lambda}{\sigma_z^2} - 1\right)\sqrt{\tau f_s}\right)$, where $Q(.)$ is the complementary distribution function of the standard Gaussian random variable. Similarly, under the H_1 hypothesis, the expression for P_d is given by; $P_d(\lambda, \tau) = Q\left(\left(\lambda/\sigma_z^2 - \gamma_p - 1\right)\sqrt{\frac{\tau f_s}{2\gamma_p + 1}}\right)$, where γ_p is the received SNR of the primary signal measured at the CR receiver. To enhance the sensing efficiency in fading channels, different receive diversity schemes have been considered in the literature [11–13].

ii. Eigenvalue-based Detection: In this approach, different eigenvalue properties of the received signal's covariance matrix can be exploited to perform sensing. Several eigenvalue-based sensing and SNR estimation techniques have been proposed in the literature exploiting the properties of Wishart random matrices [4, 14–16]. The main advantage of the eigenvalue-based approach in practical scenarios is that it does not require any prior information about the PU's signal and the channel.

After collecting N samples using M receive dimensions, we form the $M \times N$ received signal matrix $\mathbf{Y}$ and define sample covariance matrices of the received signal and the noise as: $\mathbf{R_Y}(N) = \frac{1}{N}\mathbf{Y}\mathbf{Y}^H$ and $\mathbf{R_Z}(N) = \frac{1}{N}\mathbf{Z}\mathbf{Z}^H$. Under the H_0 hypothesis, $\mathbf{R_Y}(N) = \mathbf{R_Z}(N)$. By using different eigenvalue properties of $\mathbf{R_Y}(N)$ such as Maximum Eigenvalue (ME) [17], Signal Condition Number (SCN) [4, 15], Scaled Largest Eigenvalue (SLE) [14], etc., the presence or absence of the PU signal can be decided.

2. Data/Decision Fusion Schemes: Based on the employed cooperative decision mechanism, the nodes can forward one of the following parameters (i) hard decision (single bit), (ii) decision statistics, (iii) quantized decision fusion (multiple bits), and (iv) all the samples (measurements) collected over the sensing duration. Based on these parameters, the existing fusion mechanisms in the FC can be categorized into the following.

a. Hard Decision Fusion: In this scheme, each node forwards a single bit decision to the FC i.e., 1 for the PU signal presence case and 0 for the PU signal absence case. The requirement of the limited bandwidth is the main advantage of this approach. The FC can employ any one of the "AND", "OR", or majority decision rules. The expressions for P_d and P_f with the "OR" decision rule are given by [9]; $P_d = 1 - \prod_{m=1}^{N_c}(1 - P_{d,m})$, and $P_f = 1 - \prod_{m=1}^{N_c}(1 - P_{f,m})$, where $P_{d,m}$ and $P_{f,m}$ being the P_d and the P_f of the mth sensing node while using the local sensing technique. Further, the expressions for P_d and P_f using the "AND" decision rule can be written as: $P_d = \prod_{m=1}^{N_c} P_{d,m}$, and $P_f = \prod_{m=1}^{N_c} P_{f,m}$.

b. Decision Statistics-based Soft Data Fusion: In this approach, the nodes forward the decision statistics i.e., energy in the ED context, without performing any decision and the FC makes the decision by combining them using different combining methods such as Equal Gain Combining (EGC), Maximum Ratio Combining (MRC), and Selection Combining (SC) [18]. This method provides better performance than the hard combination schemes but requires a larger bandwidth for the reporting channels [8].

c. Quantized (soft hardened) Decision Fusion: In this scheme, the nodes send the local decision in the form of multiple bits instead of a single bit in the hard decision fusion scheme and the FC provides different weights to these decisions while making the final decision. This method provides better detection performance than the hard decision scheme at the expense of the signalling overhead [9].

d. Samples-based Soft Data Fusion: In this approach, the nodes forward all the samples captured during the period of sensing. The main disadvantage of this technique is that it requires high bandwidth of the reporting links. Although decision statistics-based data fusion and sample-based data fusion seem to provide similar performance in the ED context (as illustrated in Section 4), they may provide different performance for other decision statistics.

While comparing the existing SS techniques, it can be noted that the ED technique is simple to implement but is susceptible to noise variance uncertainty [19]. This drawback can be addressed by using blind eigenvalue-based techniques such as SCN, SLE, John's Detection (JD) method, Spherical Test (ST) detector, etc [5]. However, these techniques are complex in comparison to the ED technique and require an Eigenvalue Decomposition (EVD) operation in order to calculate the decision statistics. Assuming that newly deployed nodes are capable of performing EVD operation and the existing nodes can only perform the ED detection, the research problem is how to make reliable sensing decision by exploiting different decision statistics originating in heterogeneous nodes of a CR network. In this context, we apply the marginal approximation-based approach to calculate decision thresholds based on the joint PDF of multiple decision statistics. This process is carried out at the beginning of the system operation and fixed thresholds (e.g., based on the look-up tables) can be used while combining the instantaneous decision statistics forwarded by the cooperating nodes.

3   Proposed Cooperative Sensing Method

1. *Multivariate Preliminaries*: The probability density function of a single normal random variable x is given by

$$f(x) = \frac{1}{\sqrt{2\pi}\sigma} \exp\left(-\frac{1}{2\sigma^2}(x-\mu)^2\right), \tag{2}$$

where μ and σ^2 are mean and the variance of the distribution respectively. The density functions of two random variables x_1 and x_2 which are correlated by a correlation coefficient ρ is given by

$$f(x_1, x_2) = \frac{1}{2\pi\sigma_1\sigma_2\sqrt{1-\rho^2}} \exp\left(-\frac{(x_1-\mu_1)^2}{2\sigma_1^2(1-\rho^2)} + \frac{(x_2-\mu_2)^2}{\sigma_2^2} - \frac{2\rho(x_1-\mu_1)(x_2-\mu_2)}{\sigma_1\sigma_2}\right). \tag{3}$$

The set of points for which the values of x_1 and x_2 give the same value for the density function $f(x_1, x_2)$ can be defined as an isodensity contour and is given by [20]

$$\frac{(x_1-\mu_1)^2}{\sigma_1^2} + \frac{(x_2-\mu_2)^2}{\sigma_2^2} - 2\rho\frac{(x_1-\mu_1)(x_2-\mu_2)}{\sigma_1\sigma_2} = P. \tag{4}$$

The above equation defines an ellipse with the centroid (μ_1, μ_2), which is the locus of points representing the combinations of the values of x_1 and x_2 with the same probability, defined by the constant P. For various values of P, we can obtain a family of concentric ellipses having different cross sections of the density surface with planes at various elevations. The angle joining the axis center with the centroid of the distribution, let us denote by θ, is independent of the value of P and depends on the values of σ_1, σ_2 and ρ. The steepness of this line depends on the correlation i.e., the higher the correlation, the steeper is the line.

$$f(\mathbf{x}) = (2\pi)^{-1}|\mathbf{\Sigma}|^{-\frac{1}{2}} e^{-\frac{1}{2}\mathbf{X}'\mathbf{\Sigma}^{-1}\mathbf{X}}, \tag{5}$$

where $\mathbf{X} = \mathbf{x} - \boldsymbol{\mu}$ and $\mathbf{\Sigma} = \begin{pmatrix} \sigma_1^2 & \rho\sigma_1\sigma_2 \\ \rho\sigma_1\sigma_2 & \sigma_2^2 \end{pmatrix}$. The bivariate distribution function (5) can be generalized as n-variate distribution in the following way [20].

$$f(\mathbf{x}) = (2\pi)^{-n/2}|\mathbf{\Sigma}|^{-\frac{1}{2}} e^{-\frac{1}{2}\mathbf{X}'\mathbf{\Sigma}^{-1}\mathbf{X}}. \tag{6}$$

In (6), an ellipsoid is formed for a fixed value of density $f(\mathbf{x})$. It should be noted that $\mathbf{X}'\mathbf{\Sigma}^{-1}\mathbf{X}$ is a chi-square (χ^2) variate and the inequality $\mathbf{X}'\mathbf{\Sigma}^{-1}\mathbf{X} \leq \chi^2$ defines any point within the ellipsoid.

2. *Signal Detection using Multiple Decision Statistics*: Despite the important application of joint PDF for PU signal detection, it has received limited attention in the CR literature. The contribution in [21] has proposed a method of

constructing a joint PDF under the H_1 hypothesis assuming that the joint PDF under the H_0 hypothesis is known. Recently, authors in [22] have exploited the moments of joint and marginal distributions of extreme eigenvalues in order to find out the decision threshold of a SCN-based detector. In [22], the joint PDF of the maximum and the minimum eigenvalues is approximated by a dependent Gaussian distribution function. The signal detection using multiple decision statistics can be considered as a generalization of the univariate representation [23]. Instead of being represented as a point on a line in the univariate case, a multivariate observation becomes a point in multi-dimensional space. The signal detection using multiple decision statistics can be performed either using the multivariate PDF or multivariate Cumulative Distribution Function (CDF) evaluated under the H_0 hypothesis. For example, for the signal detection based on bivariate CDF, the instantaneous test pair (T_1, T_2) can be checked whether it lies within the contour plot of the bivariate PDF corresponding to a predetermined P_f or not in order to decide on the presence or absence of the PU signal. Similarly, for the signal detection based on the bivariate PDF, the decision can be taken by testing the probability of instantaneous pair (T_1, T_2) lying within an ellipsoid corresponding to a target P_f as mentioned before. In Section 4, we present results evaluated based on the bivariate PDF-based approach using energy and eigenvalue-based decision statistics.

Let $f(T_1, T_2)$ and $F(T_1, T_2)$ denote the joint PDF and joint CDF of two decision statistics T_1 and T_2. Then the expression for P_f can be written as

$$P_f = 1 - F(\lambda_1, \lambda_2) = 1 - \int_{-\infty}^{\lambda_1} \int_{-\infty}^{\lambda_2} f(T_1, T_2) dT_2 dT_1, \tag{7}$$

where λ_1 and λ_2 correspond to thresholds obtained from the joint distribution of T_1 and T_2, respectively. From (7), it can be observed that we need to find two decision thresholds unlike a single decision threshold in the univariate signal detection problem. Therefore, the main issue in the bivariate detection problem is the calculation of decision threshold pair (λ_1, λ_2) in order to satisfy the desired probability of false alarm constraint. While applying the univariate decision rule, we should be able to calculate the threshold pair using the following expression

$$(\lambda_1, \lambda_2) = F^{-1}(1 - P_f), \tag{8}$$

where F^{-1} denote the inverse of the joint CDF of the decision statistics T_1 and T_2. If we can find the unique values of λ_1 and λ_2 corresponding to the inverse of the joint CDF, we can find the optimum pair of threshold to take the decision. However, there exist multiple pairs of (λ_1, λ_2) which yield the same P_f. Therefore, the bivariate detection problem becomes complex in comparison to the univariate detection problem. To address this issue, different approximation methods can be exploited in order to obtain the approximated threshold pair. In this paper, we focus on marginal approximation method[1] in order to derive the decision thresholds for two decision statistics. It should be noted that in our

---

[1] Investigating other suitable approximation methods based on the joint PDF of multiple decision statistics is our ongoing work.

case, the decision statistics are independent random variables and the joint PDF of two independent random variables is equal to the product of their marginal distributions.

Using marginal approximation, the values of decision thresholds λ_{1m} and λ_{2m} can be obtained using $\lambda_{1m} = F_{T_1}^{-1}(1 - P_f)$ and $\lambda_{2m} = F_{T_2}^{-1}(1 - P_f)$, where F_{T_1} and F_{T_2} are marginal CDFs of T_1 and T_2, respectively and can be obtained from their joint PDF $f(T_1, T_2)^2$ in the following way

$$F_{T_1}(\lambda_1) = \lim_{\lambda_2 \to \infty} \int_{-\infty}^{\lambda_1} \int_{-\infty}^{\lambda_2} f(T_1, T_2) dT_2 dT_1. \tag{9}$$

$$F_{T_2}(\lambda_2) = \lim_{\lambda_1 \to \infty} \int_{-\infty}^{\lambda_1} \int_{-\infty}^{\lambda_2} f(T_1, T_2) dT_2 dT_1. \tag{10}$$

After obtaining marginal thresholds λ_{1m} and λ_{2m}, we apply the following two different decision rules in order to take the decision about the presence or the absence of the PU signal.

i. Logical OR Rule: In this method, the expressions for P_d and P_f can be written as

$$P_d = Pr(T_1 > \lambda_{1m}|H_1 \text{ OR } T_2 > \lambda_{2m}|H_1),$$
$$P_f = Pr(T_1 > \lambda_{1m}|H_0 \text{ OR } T_2 > \lambda_{2m}|H_0). \tag{11}$$

ii. Logical AND rule: In this scheme, the expressions for P_d and P_f are given by

$$P_d = Pr(T_1 > \lambda_{1m}|H_1 \text{ AND } T_2 > \lambda_{2m}|H_1),$$
$$P_f = Pr(T_1 > \lambda_{1m}|H_0 \text{ AND } T_2 > \lambda_{2m}|H_0). \tag{12}$$

4   Numerical Results

4.1   Comparison of Existing CSS Schemes

In this subsection, we compare the sensing performance of different hard decision and soft data fusion techniques in terms of their Receiver Operating Characteristics (ROC) i.e., P_d versus P_f plot. Figure 2(a) presents the ROC comparison of different techniques with parameters (Number of antennas in each node $(K)= 1$, $N = 50$, $N_c = 10$, $SNR = -10$ dB). Further, we present theoretical results for hard fusion and local sensing in order to validate our approach. In this experiment, we use an AWGN channel and the ED as the local sensing technique. Further, we use the EGC-based soft data fusion technique, and "OR" and "AND" rules-based decision fusion techniques for the comparison purpose. It should be noted that in all the simulated results, the distribution of decision statistics under the H_0 hypothesis was computed by accumulating decision statistics over 10^4 noise only realizations and then the decision threshold was calculated using the constructed distribution for a certain P_f value. From the figure

---

[2] In the considered heterogeneous environment, we assume that only joint PDF of decision statistics is available at the FC.

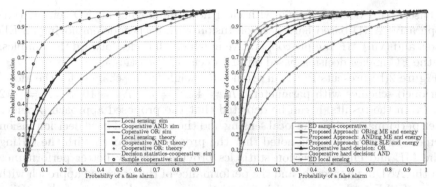

Fig. 2. (a) Performance comparison of different cooperative schemes in an AWGN channel ($K = 1$, $N = 50$, $N_c = 10$, $SNR = -10$ dB), (b) Performance comparison of the proposed approach with other cooperative schemes considering different processing capabilities and equal number of antennas in a Rayleigh fading channel ($K = 4$, $N = 50$, $N_c = 10$, $SNR = -12$ dB).

(Fig. 2(a)), it can be noted that the CSS with "AND" and "OR" rules-based decision fusion schemes perform better than the local ED sensing and the performance of "AND"-based decision fusion technique is better at the lower P_f values than the "OR"-based combining method as depicted in the literature [9]. Additionally, we can note that the CSS with EGC i.e., decision statistics-based soft data fusion performs better the hard decision fusion techniques. Another important observation from Fig. 2(a) is that the ED decision statistics-cooperative provides similar performance as that of the ED sample-cooperative scheme.

4.2   Performance Evaluation of the Proposed Approach

In this subsection, we compare the performance of the proposed approach with the existing CSS schemes considering heterogeneity firstly in terms of processing capabilities, and then in terms of processing capabilities and the number of antennas as described below.

   1. Heterogeneity in terms of Processing Capabilities: In order to compare the performance of the proposed approach with different schemes, we consider the following cases: (i) ED local sensing, (ii) Proposed joint PDF-based approach with ORing of the ME and energy statistics, obtained from the marginal approximation, in which half of the nodes forward energy and other half forward ME decision statistics, (iii) Proposed joint PDF-based approach with ANDing of the ME and energy, obtained from the marginal approximation, and the decision statistics forwarding as in case (ii), (iv) Proposed joint PDF-based approach with ORing of the SLE and energy, obtained from the marginal approximation, in which half of the nodes forward energy and other half forward SLE decision statistics, (v) ED sample-cooperative scheme in which all nodes forward the samples, and then the FC calculates the energy decision statistics and takes decision based on the threshold calculated under the H_0 hypothesis, (vi) Cooperative

OR-based hard decision fusion with each node employing an ED sensor, (vii) Cooperative AND-based hard decision fusion with each node employing an ED sensor, and (viii) ME sample-cooperative in which all nodes forward the samples, and the FC makes decision based on ME decision statistics.

Figure 2(b) presents ROC curves in a Rayleigh fading channel for the aforementioned schemes with parameters ($K = 4$, $N = 50$, $N_c = 10$, $SNR = -12$ dB). From the results, it can be depicted that the proposed approach provides better performance than the local sensing and the considered hard decision fusion schemes. Further, it can be noted that the proposed approach with the ORing of the ME and energy decision statistics provides slightly better performance than the ANDing approach, and the performance of the proposed approach with ORing of SLE and energy decision statistics is worse than that of the ORing and Anding schemes with energy and the ME decision statistics. During simulation, it has been noted that the ED sample-cooperative and the ED decision statistics-cooperative provide similar sensing performance as noted in Fig. 2(a). In addition, From Fig. 2(b), it is noted that the performance of the proposed approach with the ORing of the ME and energy decision statistics is slightly worse than the ED sample-cooperative scheme. However, the proposed scheme has low signalling burden compared to the sample-cooperative ED scheme as well as the decision statistics-based ED scheme since the transmission of the eigenvalue-based decision statistics e.g., the ME, requires less bandwidth than that of forwarding all the samples and the total energy.

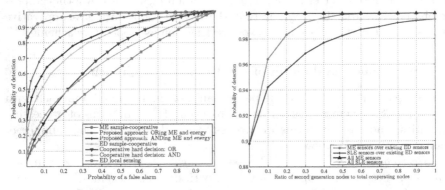

Fig. 3. (a) Performance comparison of different cooperative schemes in an AWGN channel ($K = 1$, $N = 50$, $N_c = 10$, $SNR = -10$ dB), (b) P_d versus the ratio of the second generation nodes to the total number of nodes ($K = 1$ for ED sensors and $K = 4$ for ME/SLE sensors, $N = 100$, $N_c = 10$, $SNR = -10$ dB, $P_f = 0.1$).

2. Heterogeneity in terms of Processing Capabilities and Number of Antennas: The results in Fig. 2(b) consider the heterogeneity of nodes in terms of their processing capabilities but assume the same number of antennas for all the sensors. To better illustrate the performance in the considered heterogeneous scenario, we present ROC curves for the aforementioned schemes in Fig. 3(a) with parameters ($K = 1$ for ED sensor and $K = 4$ for ME sensor, $N = 50$, $N_c = 10$, $SNR = -12$ dB). It can be noted that the proposed approach with the ORing and the ANDing of the ME and the energy decision statistics perform better than the sample-cooperative ED scheme in the considered scenario. Further, from Fig. 3(a), it can be noted that the ME sample-cooperative provides the best performance but it requires a large signalling overhead for forwarding all the samples to the FC.

3. Discussion on Cooperative Signalling Burden: As mentioned earlier, in a sample-based CSS scheme, cooperating nodes have to forward all the samples collected during the period of sensing to the FC. Subsequently, the FC may employ any of the SS schemes to take the decision about the presence or absence of the active PU. However, in the considered decision statistics-based approach, the cooperating nodes need to forward only the decision statistics i.e., energy value for the ED. It can be noted that the number of quantized bits required to deliver all the samples to the FC is significantly higher than the number of bits required to send only the decision statistics. In the proposed approach using multiple decision statistics, some of the nodes employ the eigenvalue based decision statistics and the rest employ the energy. Although hard decision fusion scheme has low signalling burden compared to the proposed one, its sensing performance is far worse than that of the proposed one as noted in Figs. 2(b) and 3(a). From the results presented in Fig. 3(a), it can be concluded that in one hand, the proposed approach saves a lot of signalling resources since it requires the transmission of only decision statistics instead of all samples, and on the other hand, it provides better sensing performance than the one of ED sample/decision statistics-based CSS scheme considered in most of the literature.

4. Design Guideline for Upgrading Sensor Networks The advantage of the proposed approach in the considered heterogeneous scenario has been further illustrated in Fig. 3(b). In this result, we plot the P_d versus the ratio of the second generation nodes to the total number of nodes with parameters ($N = 100$, $N_c = 10$, $SNR = -10$ dB, $P_f = 0.1$). Further, we consider a single antenna for the first generation sensors and multiple antennas for the second generation sensors. It can be noted from the result that as the number of the second generation nodes in the network increases, the detection performance increases and becomes more or less constant beyond a certain value of this ratio. For example, for the combination of sensors with the ME and energy decision statistics, the performance becomes constant at the ratio of 0.6. This means that 60 % new nodes in the network will be sufficient to have reliable sensing performance and it's not necessary to replace all the existing nodes of the network. In other words, the network designer can choose this ratio based on the desired performance criteria and network parameters such as the number of antennas, sampling rate etc.

5 Conclusions and Future Works

In this paper, we have considered an interesting problem of CSS with heterogeneous nodes having different capabilities. A novel technique based on the joint

PDF has been proposed in order to combine multiple decision statistics forwarded by heterogeneous cooperating nodes at the FC. The performance of the proposed technique has been compared with several existing data/decision fusion techniques. It has been noted that there exists a trade-off between the detection performance and the bandwidth overhead in the reporting channels while using various cooperative schemes and the proposed scheme provides less overhead than the overhead required by the cooperative schemes based on the sample forwarding. Further, a design guideline for the network operators has been suggested. In our future work, we target to extend this work for the CSS with more than two decision statistics and to explore suitable cooperative techniques for the scenarios with cooperating nodes having several heterogeneous features.

Acknowledgments. This work was supported by the National Research Fund, Luxembourg under the CORE project "SEMIGOD".

References

1. Goldsmith, A., et al.: Breaking spectrum gridlock with cognitive radios: An information theoretic perspective. Proc. IEEE **97**(5), 894–914 (2009)
2. Sharma, S.K., Chatzinotas, S., Ottersten, B.: Satellite cognitive communications: interference modeling and techniques selection. In: 6th ASMS and 12th SPSC, pp. 111–118, September 2012
3. Axell, E., et al.: Spectrum sensing for cognitive radio: State-of-the-art and recent advances. IEEE Signal Process. Magazine **29**(3), 101–116 (2012)
4. Zeng, Y., Liang, Y.C.: Eigenvalue-based spectrum sensing algorithms for cognitive radio. IEEE Trans. Commun. **57**(6), 1784–1793 (2009)
5. Chatzinotas, S., Sharma, S.K., Ottersten, B.: Asymptotic analysis of eigenvalue-based blind spectrum sensing techniques. In: IEEE ICASSP, pp. 4464–4468, May 2013
6. Akyildiz, I., Lo, B., Balakrishnan, R.: Coperative spectrum sensing in cognitive radio networks: A survey. Physical Commun. **4**, 40–62 (2011)
7. Chen, X., Chen, H.H., Meng, W.: Cooperative communications for cognitive radio networks-from theory to applications. IEEE Comm. Surveys & Tutorials (99), 1–13 (2014)
8. Quan, Z., et al.: Collaborative wideband sensing for cognitive radios. IEEE Signal Process. Mag. **25**(6), 60–73 (2008)
9. Teguig, D., et al.: Data fusion schemes for cooperative spectrum sensing in cognitive radio networks. In: Commun. and Info. Systems Conf., 2012 Military, pp. 1–7, October 2012
10. Liang, Y.C., et al.: Sensing-throughput tradeoff for cognitive radio networks. IEEE Trans. Wireless Commun. **7**(4), 1326–1337 (2008)
11. Digham, F.F., Alouini, M.S., Simon, M.K.: On the energy detection of unknown signals over fading channels. IEEE Trans. Commun. **55**(1), 21–24 (2007)
12. Sharma, S.K., Chatzinotas, S., Ottersten, B.: Exploiting polarization for spectrum sensing in cognitive SatComs. In: Porc. CROWNCOM, pp. 36–41, September 2012
13. Sharma, S.K., Chatzinotas, S., Ottersten, B.: Spectrum sensing in dual polarized fading channels for cognitive satcoms. In: 2012 IEEE GLOBECOM, pp. 3419–3424, December 2012

14. Wang, P., et al.: Multiantenna-assisted spectrum sensing for cognitive radio. IEEE Trans. Veh. Technol. **59**(4), 1791–1800 (2010)
15. Sharma, S.K., Chatzinotas, S., Ottersten, B.: Eigenvalue based sensing and SNR estimation for cognitive radio in presence of noise correlation. IEEE Trans. Veh. Technol. **62**(8), 1–14 (2013)
16. Sharma, S.K., Chatzinotas, S., Ottersten, B.: SNR estimation for multi-dimensional cognitive receiver under correlated channel/noise. IEEE Trans. Wireless Commun. **12**(12), 6392–6405 (2013)
17. Sharma, S.K., Chatzinotas, S., Ottersten, B.: Maximum eigenvalue detection for spectrum sensing under correlated noise. In: Proc. IEEE ICASSP, pp. 4464–4468, May 2014
18. Nallagonda, S., et al.: Performance of cooperative spectrum sensing with soft data fusion schemes in fading channels. In: IEEE Annual India Conf. (INDICON), 2013 Annual IEEE, pp. 1–6, December 2013
19. Tandra, R., Sahai, A.: SNR walls for signal detection. IEEE J. Sel. Topics Signal Porcess. **2**(1), 4–17 (2008)
20. Gatignon, H.: Statistical Analysis of Management Data, chap. 2. Springer (2010)
21. Kay, S., Ding, Q., Emge, D.: Joint pdf construction for sensor fusion and distributed detection. In: 13th Conf. Info. Fusion (FUSION), pp. 1–6, July 2010
22. Shakir, M.Z., Rao, A., Alouini, M.S.: On the decision threshold of eigenvalue ratio detector based on moments of joint and marginal distributions of extreme cigenvalues. IEEE Trans. Wireless Commun. **12**(3), 974–983 (2013)
23. Wickens, T.D.: Elementary Signal Detection Theory, chap. 10. Oxford University Press (2001)

A Cognitive Subcarriers Sharing Scheme for OFDM Based Decode and Forward Relaying System

Naveen Gupta[✉] and Vivek Ashok Bohara

WiroComm Research Lab, Indraprastha Institute of Information Technology
(IIIT-Delhi), New Delhi, India
{naveeng,vivek.b}@iiitd.ac.in

Abstract. This paper analyzes the performance of a proposed subcarriers sharing scheme. According to the scheme, secondary system helps the primary system via two phase Decode and Forward orthogonal frequency division multiplexing based relaying. If primary (licensed user) is unable to achieve its target rate then secondary transmitter (which is located within a critical distance from primary transmitter) will provide few subcarriers to primary receiver, to fulfill the requirement of the primary system and remaining subcarriers can be used by secondary (cognitive) system for its own data transmission. If secondary transmitter is located at or beyond the critical distance from primary transmitter then no spectrum sharing is possible. The analytic expression of outage probability of the primary and secondary system has been computed. Through theoretical and simulation results it has been shown that the primary outage probability with cooperation (while secondary transmitter acts a partial relay) is less than the outage probability for direct transmission. Therefore opportunistic spectrum sharing can be achieved by secondary system.

Keywords: OFDM · Opportunistic spectrum sharing · Cooperative relaying · Outage probability

1 Introduction

Due to advent of new wireless communication techniques, the demand for additional bandwidth is increasing every other day. Researchers and technologists are seeking the solutions to cope up with the problem of bandwidth scarcity. Cognitive radio technology introduced by [1] has provided an alternative to solve the problem of spectrum scarcity and under-utilization. Cognitive radio networks support opportunistic spectrum sharing (OSS) [2] via granting secondary system (low priority) to share the spectrum of the primary user (higher priority) to efficiently utilize the radio spectrum. Practical implementation of OSS in Long Term Evolution-Advanced (LTE-A) has been proposed in [3]. According to [3], OSS has played a significant role with carrier aggregation technique to improve the performance of the LTE-A system.

© Institute for Computer Sciences, Social Informatics and Telecommunications Engineering 2015
M. Weichold et al. (Eds.): CROWNCOM 2015, LNICST 156, pp. 334–345, 2015.
DOI: 10.1007/978-3-319-24540-9_27

Recently cooperative relaying has been incorporated to facilitate spectrum sharing schemes in cognitive radio system [4]. In cooperative relaying one or more relays are used to improve the performance of a system via space and time diversity. Relay based on Amplify and Forward (AF), Decode and Forward (DF) or Compressed and Forward (CF) protocol [5] acts as a virtual antenna for the primary system.

Cooperative spectrum sharing for single carrier system has been proposed in [6]. In this work, the regenerated primary signal at secondary transmitter is combined with the secondary signal by providing fraction of secondary power to the primary signal and remaining power to the secondary signal. Cooperative relaying for two phase cognitive system has been given in literature [7]. Some existing work on cognitive radio is based on interference limited systems [8] in which primary system has capacity to handle additional interference from other systems without affecting its quality of service. Combination of multi-carrier modulation such as OFDM and DF relaying has been proposed in literature [9] for selective subcarrier pairing and power allocation. In [9] more secondary system power will be provided to the better channel. The optimization of subcarrier power to maximize the cognitive system throughput without providing excessive interference to the primary system has been given in [10]. Similarly, a protocol for OFDM AF relaying has been given in [11]. In this work, a joint optimization problem is formed for subcarrier pairing and power allocation, where secondary system uses fraction of its subcarriers to boost the performance of primary system. Results are shown for outage probability w.r.t secondary system power. However, in [11] authors have not determined the exact numbers of subcarriers relayed via secondary transmitter to primary receiver, to fulfill the target rate requirement of primary system. Another work based on OFDM-AF relaying [12] illustrates that the secondary system can do opportunistic communication by providing half subcarriers to primary receiver to achieve the target rate of primary system and remaining half subcarriers can be used for secondary communications.

In this paper, we have proposed an opportunistic subcarriers sharing scheme based on OFDM-DF relaying. We have computed the outage probability as a performance metrics for primary and secondary systems for different numbers of subcarriers. In this scheme, it is assumed that both primary and secondary system uses OFDM based modulation. According to the scheme, if primary system is not able to achieve the target rate of transmission due to poor link quality then advanced secondary system supports the primary system via converting into a two phase cooperative relaying based on DF protocol. Here secondary transmitter which behaves as a relay for primary system will receive the signal broadcast by primary transmitter over N subcarriers in phase I, decode it, and forward few (D) subcarriers to the primary receiver in phase II in order to satisfy the target rate requirement of primary system. The remaining (N-D) subcarriers can be used by secondary transmitter to transmits its own signal to secondary receiver. Hence secondary system while working as a relay for primary system will get opportunistic spectrum access in exchange of fulfilling the target rate requirement of primary system. The number of subcarriers (D) allocated to the primary system depend on the required target rate or quality of services of the primary system. If primary target rate changes, D also changes.

Primary transmitter (PT) and primary receiver (PR) are the components of the primary system while secondary system comprises of secondary transmitter (ST) and secondary receiver (SR) as shown in figure 1. Here we assume that the secondary system follows the same radio protocol as primary system (eg. source coding, channel coding, synchronization). Results interpret the requirement of subcarriers relayed via ST to PR to fulfill the target rate of primary system. As numbers of relayed subcarriers increases, primary outage probability decreases while secondary outage probability increases. Outage probability is also a factor of distance between primary system and ST. As discussed later, for a given co-linear model (where PT, ST, PR and SR are co-linear), as distance between secondary transmitter and primary receiver decreases, primary outage probability decreases and less number of subcarriers would be required to forward the primary signal. But at the same time, distance between primary and secondary transmitter increases and decoding of primary signal at ST becomes the limiting factor for outage probability and no more performance improvement will be achieved by increasing the number of subcarriers after a threshold. There will be a critical distance, defined as a distance between PT and ST above which no opportunistic spectrum access is possible. Theoretical and simulation results are provided for outage probability for both primary and secondary system. It has been shown through the results that, as long as ST lies within the critical distance, outage probability of proposed scheme will be less as compare to the direct transmission. Excellent agreement between the theoretical and simulation results validate the analytical results of the proposed scheme.

The remainder of this paper is organized as follows. Section 2 discusses the system model. Section 3 demonstrates the rate and outage probability analysis for direct and cooperative communication for both primary and secondary system. Simulation results and discussion are provided in section 4. Section 5 concludes the paper.

2   System Model

In our system model we consider secondary system as an advance relay with cooperative relaying functionality. OFDM is used as a modulation scheme to modulate primary & secondary signal over N subcarriers. All transmission and reception nodes *i.e.* PT, PR, ST, SR comprise single antenna. Primary system has authority to operate in some license band while secondary system can do only opportunistic spectrum access in the primary license band, when primary is unable to achieve its target rate of communication due to various channel impairments.

Here Rayleigh frequency flat fading model has been considered for primary and secondary system. The channel coefficient corresponds to PT→PR is $\Psi_{1,k}$ over subcarrier $k(1 \leq K \leq N), \Psi_i \sim \mathcal{CN}(0, \zeta_i^{-l})$ where ζ is the normalized distance between transmitter and receiver. Normalization is done w.r.t. distance between PT→PR, which is set to $\zeta = 1$ and path loss component is denoted by l. Similarly channel coefficients for primary to secondary transmitter, secondary transmitter to primary receiver and secondary transmitter to secondary receiver is denoted as $\Psi_{2,k}, \Psi_{3,k}, \Psi_{4,k}$ respectively. The channel instantaneous gain for each subcarrier is defined as $\gamma_{1,k} = |\Psi_{1,k}|^2, \gamma_{2,k} = |\Psi_{2,k}|^2, \gamma_{3,k} = |\Psi_{3,k}|^2$ and $\gamma_{4,k} = |\Psi_{4,k}|^2$. The additive white Guassian noise at each receiver is denoted as $n_{1,k}, n_{2,k}, n_{4,k} \sim \mathcal{CN}(0, \sigma^2)$. The primary and secondary signal is denoted as $s_{p,k}$ and $s_{s,k}$ respectively with zero mean and $E\{s_{p,k}^* s_{p,k}\} = E\{s_{s,k}^* s_{s,k}\} = 1$.

Here total transmission has been divided into two time phases. In phase I, PT will broadcast signal to PR, ST & SR. In phase II, PT remains silent and ST will decode and forward few subcarriers to PR, keeping in mind to fulfill the requirement of primary target rate on top priority. With remaining number of subcarriers ST can do opportunistic spectrum access to transmit its own signal to SR. ST will transmit orthogonal subcarriers to PR & SR to avoid interference between them.

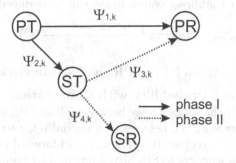

Fig. 1. System model

3 Rate and Outage Analysis for Direct and Cooperation

3.1 Primary Outage Probability for Direct Transmission

In phase 1, signal $s_{p,k}$ is broadcast by PT, received by PR and overheard by SR & ST. Received signal at PR over subcarrier k is denoted as $\phi_k^{pr,1}$ which is equal to,

$$\phi_k^{pr,1} = (p_{p,k})^{\frac{1}{2}} \Psi_{1,k} s_{p,k} + n_{1,k} \tag{1}$$

where $n_{1,k}$ denotes AWGN noise over subcarrier k and $p_{p,k}$ is the power of each subcarrier. The total power available at PT is sum of all subcarrier power i.e. $\sum_{k=1}^{N} p_{p,k}$. Let's the available bandwidth at the primary system is divided into N orthogonal subcarriers and primary signals are transmitted over N subcarriers. The instantaneous rate for all N subcarriers is given as,

$$R_N = \sum_{k=1}^{N} \log_2 \left(1 + \frac{p_{p,k}\gamma_{1,k}}{\sigma_1^2}\right). \tag{2}$$

The target rate of primary system is R_T and outage occur if $R_N < R_T$

$$P_{out} = Pr(R_N < R_T) \tag{3}$$

Without any loss of generality, let's assume that all subcarriers comprises same power $p_{p,1} = p_{p,2} = p_{p,N} = p_p$ and channel coefficients of all subcarriers are same for one OFDM symbol $\gamma_{1,1} = \gamma_{1,2} = \gamma_{1,N} = \gamma_1$.
So (2) can be deduce to,

$$R_N = N\log_2\left(1 + \frac{p_p\gamma_1}{\sigma_1^2}\right). \tag{4}$$

The outage probability,

$$P_{out} = Pr\left(\gamma_1 < \frac{(2^{R_T/N} - 1)\sigma_1^2}{p_p}\right). \tag{5}$$

where $\gamma_1 \sim \exp\left(\zeta_1^l\right)$ is exponentially distributed i.i.d. random variable. So the outage probability for the direct transmission can be deduced to

$$P_{out} = 1 - \exp\left(-\frac{\sigma_1^2(2^{R_T/N} - 1)\zeta_1^l}{p_p}\right) \tag{6}$$

where ζ_1 is distance between PT and PR and l is path loss component.

3.2 Primary Outage Probability with Cooperation

With cooperation ST behaves as a relay between PT and PR to provide diversity gain to the primary system. ST behaves as a partial relay to decode & forward primary data to primary receiver. It will decode and forward D subcarriers to PR and remaining N-D subcarriers to SR. Signal received by Secondary Transmitter in phase I is

$$\phi_k^{st,1} = (p_{p,k})^{\frac{1}{2}}\Psi_{2,k}s_{p,k} + n_{2,k} \tag{7}$$

where $n_{2,k}$ is the AWGN noise over subcarrier k. The instantaneous rate of signal received at ST with N subcarriers is,

$$R_N^{Pt-St} = \sum_{k=1}^{N}\log_2\left(1 + \frac{p_{p,k}\gamma_{2,k}}{\sigma_2^2}\right). \tag{8}$$

Taking same assumption as (2), eq. (8) will be deduce to,

$$R_N^{Pt-St} = \frac{N}{2}\log_2\left(1 + \frac{p_p\gamma_2}{\sigma_2^2}\right) \tag{9}$$

where $\frac{1}{2}$ is due to transmission in two phases. Out of total N subcarriers received from PT, ST will decode & forward only D subcarriers to PR while remaining N-D subcarriers will be transmitted to SR. The instantaneous rate at PR after Maximum ratio combining (MRC) of two phases transmission with a condition of successful decoding of primary signal $s_{p,k}$ at ST is,

$$R_{PR}^{MRC} = \frac{1}{2}\sum_{k=1}^{D}\log_2\left(1 + \frac{p_{p,k}\gamma_{1,k}}{\sigma_1^2} + \frac{p_{s,k}\gamma_{3,k}}{\sigma_1^2}\right) + \frac{1}{2}\sum_{k=1}^{N-D}\log_2\left(1 + \frac{p_{p,k}\gamma_{1,k}}{\sigma_1^2}\right) \tag{10}$$

where factor $\frac{1}{2}$ is due to two phase transmission and $p_{s,k}$ is the power of each subcarrier belonging to secondary system. Let all subcarriers carries same power and channel coefficients of a path are same for all subcarriers $i.e.$ $\gamma_{1,1} = \gamma_{1,2} = \gamma_{1,N} = \gamma_1, \gamma_{2,1} = \gamma_{2,2} = \gamma_{2,N} = \gamma_2, \gamma_{3,1} = \gamma_{3,2} = \gamma_{3,N} = \gamma_3$. So (10) can be deduce to,

$$R_{PR}^{MRC} = \frac{D}{2}\log_2\left(1 + \frac{p_p\gamma_1}{\sigma_1^2} + \frac{p_s\gamma_3}{\sigma_1^2}\right) + \frac{N-D}{2}\log_2\left(1 + \frac{p_p\gamma_1}{\sigma_1^2}\right). \tag{11}$$

On the other hand when ST would unable to decode the primary signal received in phase I of transmission then there will be no transmission from ST to PR in phase II. But PR would still be able to receive the primary signal from direct link. Thus the outage probability is,

$$P_{out}^p = Pr(R_N^{Pt-St} > R_T)Pr(R_{PR}^{MRC} < R_T) + Pr(R_N^{Pt-St} < R_T)Pr(\frac{1}{2}R_N < R_T) \tag{12}$$

where R_N can be found from (2).

$$Pr\left(\frac{1}{2}R_N < R_T\right) = Pr\left(\gamma_1 < \frac{\rho_1\sigma^2}{p_p}\right) = 1 - e^{-\frac{\zeta_1^l\sigma^2}{p_p}\rho_1} \tag{13}$$

$$Pr(R_N^{Pt-St} < R_T) = Pr\left(\gamma_2 < \frac{\rho_1\sigma^2}{p_p}\right) = 1 - e^{-\frac{\zeta_2^l\sigma^2}{p_p}\rho_1}. \tag{14}$$

Similarly,

$$Pr(R_N^{Pt-St} > R_T) = Pr\left(\gamma_2 > \frac{\rho_1\sigma^2}{p_p}\right) = e^{-\frac{\zeta_2^l\sigma^2}{p_p}\rho_1} \tag{15}$$

where $\gamma_1 \sim \exp\left(\zeta_1^l\right)$, $\gamma_2 \sim \exp\left(\zeta_2^l\right)$, and $\rho_1 = 2^{\frac{2R_T}{N}} - 1$

$$Pr(R_{PR}^{MRC} < R_T)$$

$$= Pr\left(\left(1 + \frac{p_p\gamma_1}{\sigma_1^2} + \frac{p_s\gamma_3}{\sigma_1^2}\right)^D \left(1 + \frac{p_p\gamma_1}{\sigma_1^2}\right)^{N-D} < 2^{2R_T}\right). \tag{16}$$

Let $\sigma_1^2 = \sigma_2^2 = \sigma_3^2 = \sigma^2$,

$$= Pr\left(\left(1 + \frac{p_p\gamma_1}{\sigma^2} + \frac{p_s\gamma_3}{\sigma^2}\right)^D \left(1 + \frac{p_p\gamma_1}{\sigma^2}\right)^{N-D} < 2^{2R_T}\right). \tag{17}$$

Let $\frac{p_p\gamma_1}{\sigma^2} + \frac{p_s\gamma_3}{\sigma^2} >> \sigma^2$, Now (17) deduce to

$$= Pr\left((p_p\gamma_1 + p_s\gamma_3)^D (p_p\gamma_1)^{N-D} < 2^{2R_T}\sigma^{2N}\right) \tag{18}$$

$$= Pr\left((p_p\gamma_1 + p_s\gamma_3) < \frac{\left(2^{2R_T}\sigma^{2N}\right)^{\frac{1}{D}}}{(p_p\gamma_1)^{\frac{N-D}{D}}}\right) \tag{19}$$

$$= Pr\left(\gamma_3 < \frac{\Lambda^{\frac{1}{D}}}{p_s(p_p\gamma_1)^{\frac{N-D}{D}}} - \frac{p_p\gamma_1}{p_s}\right) \tag{20}$$

where $\Lambda = 2^{2R_T}\sigma^{2N}$

Let's

$$\frac{\Lambda^{\frac{1}{D}}}{p_s\,(p_p\gamma_1)^{\frac{N-D}{D}}} - \frac{p_p\gamma_1}{p_s} = \beta\,(\gamma_1) \tag{21}$$

where $\gamma_3 \sim \exp\left(\zeta_3^l\right)$ and for exponential random variable,

$$\beta\,(\gamma_1) > 0 \tag{22}$$

$$\frac{\Lambda^{\frac{1}{D}}}{p_s\,(p_p\gamma_1)^{\frac{N-D}{D}}} - \frac{p_p\gamma_1}{p_s} > 0. \tag{23}$$

After simplifying (23) we will get,

$$\gamma_1 < \frac{\Lambda^{\frac{1}{N}}}{p_p} = \alpha(\text{let}). \tag{24}$$

As γ_1 and γ_3 are independent exponential random variable, it's joint probability density function can be represented as $\zeta_1^l e^{-\zeta_1^l\gamma_1}\zeta_3^l e^{-\zeta_3^l\gamma_3}$

$$Pr(\gamma_3 < \beta\,(\gamma_1)) = \int_{\gamma_1=0}^{\alpha}\int_{\gamma_3=0}^{\beta(\gamma_1)} \zeta_1^l e^{-\zeta_1^l\gamma_1}\zeta_3^l e^{-\zeta_3^l\gamma_3}\,d\gamma_1 d\gamma_3 \tag{25}$$

$$= \int_{\gamma_1=0}^{\alpha} \zeta_1^l e^{-\zeta_1^l\gamma_1}\left(1 - e^{-\zeta_3^l\beta(\gamma_1)}\right) \tag{26}$$

$$= 1 - e^{-\zeta_1^l\alpha} - \zeta_1^l\Upsilon_1 \tag{27}$$

where,

$$\Upsilon_1 = \int_{\gamma_1=0}^{\alpha} e^{\left(\delta_2\gamma_1 - \frac{\delta_1}{\gamma_1^{\frac{N}{D}-1}}\right)}\,d\gamma_1 \tag{28}$$

$$\delta_1 = \frac{\zeta_3^l\Lambda^{\frac{1}{D}}}{p_s p_p^{\frac{N}{D}-1}}, \quad \delta_2 = \frac{\zeta_3^l p_p}{p_s} - \zeta_1^l \tag{29}$$

Equation (28) is intractable, however, if we substitute, $\delta_2 = 0$ i.e. $\frac{\zeta_1^l}{\zeta_3^l} = \frac{p_p}{p_s}$, so eq. (28) can be reduced to,

$$\Upsilon_1 = \int_{\gamma_1=0}^{\alpha} e^{-\delta_1\gamma_1^{-\left(\frac{N}{D}-1\right)}}\,d\gamma_1. \tag{30}$$

Let's substitute $\gamma_1^{-\frac{N-D}{D}} = t$. So eq. (30) reduces to,

$$\Upsilon_1 = \frac{D}{N-D}\int_{\alpha^{-\frac{N-D}{D}}}^{\infty} t^{-\frac{N}{N-D}} e^{-\delta_1 t}\,dt. \tag{31}$$

From [13], eq. (31) can be solved as, [1]

$$\Upsilon_1 = (-1)^{n+1}\delta_1^n\frac{Ei(-\delta_1 u)}{n!} + \frac{e^{-\delta_1 u}}{u^n}\sum_{k=0}^{n-1}\frac{(-1)^k\delta_1^k u^k}{n(n-1)...(n-k)} \tag{32}$$

---

[1] However, in this paper we have solved eq. (28) numerically to obtain the theoratical plots.

where, $n = \frac{D}{N-D}$ and $u = \alpha^{-\frac{N-D}{D}}$

Hence from (13),(14),(15),(27) the primary outage probability,

$$P_{out}^p = e^{-\frac{\zeta_2^l \sigma^2}{p_p}\rho_1}(1 - e^{-\zeta_1^l \alpha} - \zeta_1^l \Upsilon_1) + \left(1 - e^{-\frac{\zeta_2^l \sigma^2}{p_p}\rho_1}\right)\left(1 - e^{-\frac{\zeta_1^l \sigma^2}{p_p}\rho_1}\right). \quad (33)$$

3.3 Critical Distance Analysis

If we consider only primary system (i.e. no secondary system exists) then outage probability of the direct link with target rate R_T is given by (6). With proposed scheme, the outage probability of the primary system (with few subcarriers D) should be always less than or equal to the outage probability of direct link. From eq. (6) and (33) we have,

$$P_{out} > P_{out}^p \quad (34)$$

For a given target rate, there will be always a critical distance ζ_2^* (distance between PT and ST) above which no spectrum sharing is possible. In other words, if ST is located at or beyond critical distance from PT then primary outage probability with cooperation will always be greater than or equal to direct outage probability even ST would work as a pure relay.

$$1 - e^{\left(-\frac{\sigma_1^2(2^{R_T/N}-1)\zeta_1^l}{p_p}\right)} < e^{-\frac{\zeta_2^l \sigma^2}{p_p}\rho_1}(1 - e^{-\zeta_1^l \alpha} - \zeta_1^l \Upsilon_1)$$

$$+ \left(1 - e^{-\frac{\zeta_2^l \sigma^2}{p_p}\rho_1}\right)\left(1 - e^{-\frac{\zeta_1^l \sigma^2}{p_p}\rho_1}\right) \quad (35)$$

From eq. (35)

$$\zeta_2^* = \left[\frac{p_p}{\rho_1 \sigma^2}ln\left(\frac{\Phi_1 - \Phi_2}{\Phi_3 - \Phi_2}\right)\right]^{\frac{1}{l}} \quad (36)$$

where $\Phi_1 = 1 - e^{-\zeta_1^l \alpha} - \zeta_1^l \Upsilon_1, \Phi_2 = \left(1 - e^{-\frac{\zeta_2^l \sigma^2}{p_p}\rho_1}\right)$ and $\Phi_3 = 1 - e^{\left(-\frac{\sigma_1^2(2^{R_T/N}-1)\zeta_1^l}{p_p}\right)}$.

From eq. (36), the critical distance ζ_2^* for different target rates can be calculated. The critical distances for some of the predefined target rates with $p_p = 20$ dB, $p_s = 30$ dB and best possible value of D i.e. (D=N=32) has been given in table 1.

Table 1.

R_T	8	16	32	64
ζ_2^*	3.4	2.68	1.92	1.12

3.4 Secondary Outage Probability

In phase II of transmission ST will transmit N-D subcarriers to SR via $\Psi_{4,k}$ link. The instantaneous rate is given by,

$$R_S = \frac{1}{2}\sum_{k=1}^{N-D} \log_2\left(1 + \frac{p_{s,k}\gamma_{4,k}}{\sigma_4^2}\right). \quad (37)$$

Let's assume that all subcarriers comprise same power $p_{s,1} = p_{s,2} = p_{s,N} = p_p$ and channel coefficients of all subcarriers are same for one OFDM symbol $\gamma_{4,1} = \gamma_{4,2} = \gamma_{4,N} = \gamma_4$. Equation (37) can be deduced as,

$$R_S = \frac{1}{2}(N - D)\log_2\left(1 + \frac{p_s\gamma_4}{\sigma_4^2}\right). \tag{38}$$

Let the target rate for secondary system is defined as R_{T_s}. Outage occur if $Pr(R_S < R_{T_s})$

$$Pr(R_S < R_{T_s}) = Pr\left(\gamma_4 < \frac{\rho_2\sigma_4^2}{p_s}\right) = 1 - e^{-\frac{\zeta_4^l\sigma_4^2}{p_s}\rho_2} \tag{39}$$

where $\gamma_4 \sim \exp\left(\zeta_4^l\right)$ and $\rho_2 = 2^{\frac{2R_{T_s}}{N-D}} - 1$

4   Simulation Results and Discussion

We have done simulation for the outage probability with respect to number of subcarriers under specified settings. For the ease of analysis, PT, PR, ST & SR are considered to be collinear. PT is located at distance (0,0) in two dimensional plane while PR is located at (1,0) *i.e.* $\zeta_1 = 1$. SR lies in between PT and PR at coordinates (0.75,0) and ST moves along the X axis. Theoretical and simulation results of the outage probability for $\zeta_2 = 0.5, \zeta_2 = 1.2$ & $\zeta_2 = 1.92$ with respect to number of relayed subcarriers D are given. we have chosen target rate $R_T = N = 32$ and subcarrier power $p_p = 10dB, p_s = 30dB$. The path loss exponent has set to be $l = 4$.

Figure 2 shows the theoretical and simulation result of the outage probability of the primary system vs number of subcarrier required for cooperation. Theoretical results are strongly following the simulated one. From figure 2 we

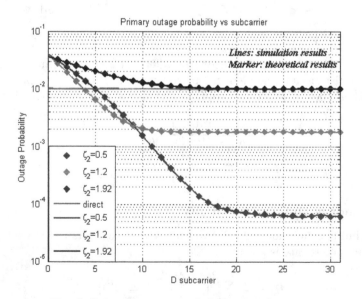

Fig. 2. Primary outage probability vs subcarriers

can see that as number of forwarded subcarriers D from ST to PR increases, the outage probability decreases. For $\zeta_2 = 0.5$, and D > 5 the outage probability with cooperation is less than the direct transmission. Hence with this given power and target rate profile, if ST will forward only 5 subcarriers to PR then outage probability with cooperation will be less than the outage probability with direct tranmission. So secondary system has opportunity to do spectrum access of licensed primary band to transmit its own data to secondary receiver. With this protocol out of total given N=32 subcarriers, the remaining N − D subcarriers can be used by ST to do opportunistic spectrum access. For higher values of D it is quite obvious that primary outage probability will reduce as providing more subcarriers for the relaying of primary signal & less subcarriers used for the secondary system transmission. However for D > 20, the outage probability gets stagnant. This is due to the fact that when D approaches it's maximum value, the outage probability with MRC $Pr\left(R_{PR}^{MRC} < R_T\right)$ attains very small value and the decoding of primary signal at ST becomes the only limiting factor for outage probability i.e $Pr(R_N^{Pt-St} < R_T)Pr(\frac{1}{2}R_N < R_T)$. Thus further increasing D will not impact primary outage probability.

For $\zeta_2 = 1.2$, distance between PT→ST increases but distance between ST→PR decreases i.e. $\zeta_3 = 0.2$. Therefore for small value of D, the outage probability is less than previous case, here only 4 subcarriers are required to achieve same outage probability as direct transmission. But for D > 10, the outage probabilty is constant. This happens because of high distance between PT and ST, no further successful decoding of primary signal occur at ST. For $\zeta = 1.92$ (critical distance), no opportunistic spectrum access is possible, as outage probability with cooperation is always higher than direct transmission. Figure 3 shows the relationship between D and ζ_2 for $R_T = 32$. From figure 3 we can find the exact number of subcarriers D forwarded via ST to PR, when outage probability of

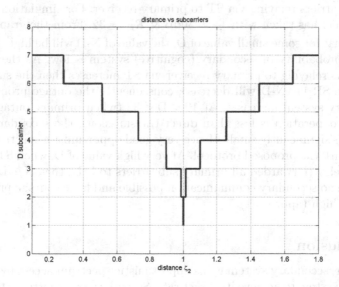

Fig. 3. Subcarriers vs ζ_2

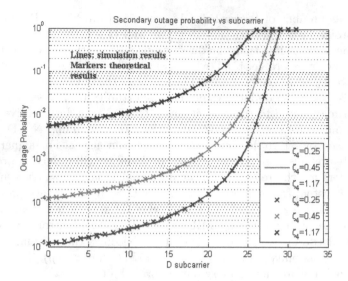

Fig. 4. Secondary outage probability vs subcarriers

proposed cooperation scheme would be less than direct transmission. At $\zeta_2 = 0.1$ distance between PT-ST is 0.1 consequently distance between ST-PR will be 0.9, therefore ST would required to forward minimum 7 subcarriers to PR to achieve the performance better than direct transmission. As the value of ζ_2 increases from $0.1 < \zeta_2 < 1$, distance between ST-PR will decreases, so required number of subcarriers i.e. D will also decreases. However for $\zeta_2 > 1$, ST is moving away from PR, therefore value of D increases with ζ_2. Figure 4 shows the 2-D graph of secondary system outage probability for $\zeta_4 = 0.25$, $\zeta_4 = 0.45$ & $\zeta_4 = 1.17$ w.r.t D subcarriers relaying via ST to primary receiver. For simulation collinear distance model has taken with $\frac{P_s}{\sigma_4^2} = 30$dB & $R_{T_s} = 32$. From the graph we can clearly see that for some small value of D, the value of N-D will be high, therefore the outage probability of secondary (cognitive) system is low. As the numbers of subcarriers relaying to primary receiver via ST increases then the subcarriers forwarded to SR i.e. N-D will decrease, consequently the outage probability of the secondary system increases. At $10 < D < 32$ (when primary outage probability with cooperation is less than direct transmission), the secondary outage probability is significantly small. Hence very good opportunistic spectrum access is possible with the proposed protocol. At very high value of D, when ST behaves like a pure relay (forwarded all primary subcarriers to PR), then $N - D = 0$ and there will be no secondary communication possible and hence outage probability will be very high (approx=1).

5   Conclusion

In this paper, secondary system gains opportunistic spectrum access by assisting the primary system to achieve its target rate. Secondary transmitter acts as a DF relay to help the primary system by relaying few subcarriers to primary receiver to fulfill the required target rate of primary system while remaining subcarriers can be used for secondary transmission. There is a critical distance between PT and ST, if ST is located at or beyond critical distance from PT then there will be no opportunistic spectrum sharing. We have shown that depending on the distance between PT and ST, it is possible to find exact number of subcarriers that should be forwarded by ST to PR for outage probability with cooperation to be less than direct transmission.

References

1. Mitola, J., Maguire Jr, G.Q.: Cognitive radio: making software radios more personal. IEEE Personal Communications 6(4), 13–18 (1999)
2. Zhang, R., Liang, Y.-C.: Exploiting multi-antennas for opportunistic spectrum sharing in cognitive radio networks. IEEE Journal of Selected Topics in Signal Processing 2(1), 88–102 (2008)
3. Osa, V., Herranz, C., Monserrat, J., Gelabert, X.: Implementing opportunistic spectrum access in lte-advanced. EURASIP Journal on Wireless Communications and Networking 2012(1), 99 (2012). http://jwcn.eurasipjournals.com/content/2012/1/99
4. Jia, J., Zhang, J., Zhang, Q.: Cooperative relay for cognitive radio networks. In: INFOCOM 2009, pp. 2304–2312. IEEE, April 2009
5. Gunduz, D., Yener, A., Goldsmith, A., Poor, H.: The multi-way relay channel. In: IEEE International Symposium on Information Theory, ISIT 2009, pp. 339–343, June 2009
6. Han, Y., Pandharipande, A., Ting, S.H.: Cooperative decode-and-forward relaying for secondary spectrum access. IEEE Transactions on Wireless Communications 8(10), 4945–4950 (2009)
7. Bohara, V., Ting, S.H., Han, Y., Pandharipande, A.: Interference-free overlay cognitive radio network based on cooperative space time coding. In: 2010 Proceedings of the Fifth International Conference on Cognitive Radio Oriented Wireless Networks Communications, CROWNCOM, pp. 1–5, June 2010
8. Bohara, V., Ting, S.H.: Measurement results for cognitive spectrum sharing based on cooperative relaying. IEEE Transactions on Wireless Communications 10(7), 2052–2057 (2011)
9. Boostanimehr, H., Bhargava, V.: Selective subcarrier pairing and power allocation for df ofdm relay systems with perfect and partial csi. IEEE Transactions on Wireless Communications 10(12), 4057–4067 (2011)
10. Shaat, M., Bader, F.: Joint subcarrier pairing and power allocation for df-relayed ofdm cognitive systems. In: 2011 IEEE Global Telecommunications Conference, GLOBECOM 2011, pp. 1–6, December 2011
11. Lu, W.D., Gong, Y., Ting, S.H., Wu, X.L., Zhang, N.-T.: Cooperative ofdm relaying for opportunistic spectrum sharing: Protocol design and resource allocation. IEEE Transactions on Wireless Communications 11(6), 2126–2135 (2012)
12. Gupta, N., Bohara, V.A.: Outage analysis of cooperative ofdm relaying system with opportunistic spectrum sharing. In: 2014 International Conference on Advances in Computing, Communications and Informatics, ICACCI, pp. 2803–2807, September 2014
13. Gradshteyn, I.S., Ryzhik, I.M.: Table of integrals, series, and products, 7th edn. Elsevier/Academic Press, Amsterdam (2007)

Efficient Performance Evaluation for EGC, MRC and SC Receivers over Weibull Multipath Fading Channel

Faissal El Bouanani[1](✉) and Hussain Ben-Azza[2]

[1] ENSIAS, Mohammed V University, Rabat, Morocco
elbouanani@ensias.ma
[2] ENSAM, Moulay Ismail University, Meknes, Morocco

Abstract. The probability density function (PDF) of the output SNR (Signal to Noise Ratio) at a receiver operating under Weibull fading multipath channel is unknown in closed form and exists only as a complicated multiple integrals or approximated by a series of functions, or recently, by a single function whose evaluation time is not negligible. Our main result is a new simple approximate closed-form of the SNR PDF at the output of three types of receivers over Weibull multipath fading channels. The advantage of this new expression is that its evaluation time is less compared to all previous results. Based on this expression, approximate analytical expressions of the outage probability (OP), the average bit error rate (BER) for several M-ary modulation techniques, and the average channel capacity (CC) are derived in terms of only one particular hypergeometric function, known as Fox-H function. Numerical results have been validated by simulation and compared with recent results.

Keywords: Fox H-function · Meijer G-function · Maximal ratio combining · Equal gain combining · Weibull fading · Shannon capacity · Bit error rate

1 Introduction

In wireless digital communication system, the diversity techniques are used to combine the original signal copies arrived often from different paths, especially in an urban environment, at the receiver. The choice of a combining method has a great influence on system performance. Several diversity techniques, such as maximal-ratio combining (MRC), selection combining (SC), and equal-gain combining (EGC) are used in many wireless communication system. In MRC method, all received signals at the input of the receiver are multiplied by their channel gains conjugates. This receiver is known to be optimal for all multipath fading environment. In coherent EGC reception, each received signal on its branches is weighted with the equal gain. Although the complexity of EGC is

© Institute for Computer Sciences, Social Informatics and Telecommunications Engineering 2015
M. Weichold et al. (Eds.): CROWNCOM 2015, LNICST 156, pp. 346–357, 2015.
DOI: 10.1007/978-3-319-24540-9_28

acceptable, its performances are usually less efficient than the MRC ones. In SC technique, as long as the receiver select only one branch that having the greatest SNR, its performances are always lower than those of EGC.

The Weibull distribution is often used for describing the fading amplitude in both indoor and outdoor communication environments. As the analytical expression of the SNR PDF at the output of either coherent EGC or MRC receiver is unknown and difficult or even impossible to find, an another way is to derive a closed-form approximation. Recently, many papers have been written on the performance of both MRC, EGC, SC, and GSC (Generalized SC) receivers over Weibull fading channels. In [1], a closed-form of the moment generating function (MGF) of the SNR at the output of both MRC and SC receivers was derived in terms of power and finite series of Meijer G-functions [2] respectively. In [3], average symbol error rate and MGF of the output SNR at EGC and MRC combiners are derived in terms of product and infinite series of Meijer G-functions. The expression of the SNR at EGC output was investigated in [4]. In [5], joint PDF, CDF, and MGF of SNR are expressed for SC receiver. In [6], various statistical characteristics of the output SNR at EGC receiver, and average bit error rate (BER) for 3-branch MRC are derived as infinite series of Meijer G-function. In [7], CC for MRC over several types of fading channels comprising Weibull case was derived as a series of Meijer G-function. In [8]-[10], some other contributions dealing with SNR PDF at the output of the receiver and other system performance indicators have been presented in close approximation by only one Fox H-function for MRC, EGC, and GSC respectively. However, to ensure the convergence of this function and to be close to the exact expression of the performance indicator, the value of the parameter λ, appearing in the denominator of the first argument of all derived Fox H-functions, should be too big. Although the evaluation of this function is easy using some mathematical software such as Mathematica and Maple, it becomes extremely slow when its first argument is very small (close to 0), and sometimes lead to numerical instabilities and erroneous results.

Hence, it is highly desirable to find another approximate closed-form that resolves the problem of evaluation time complexity of the said function when λ approaches infinity. Thus, in this paper, we present a new, stable and low complexity, approximate closed-form of SNR PDF at the output of MRC, EGC, and SC receivers over Weibull multipath fading channels, and so other performance criteria, in terms of only one simple Fox H-function.

The rest of this paper is structured as follows. In section 2, the description of the studied receiver operating in Weibull multipath fading environment and some statistical characteristics are presented. Section 3 present a novel formula, with low evaluation time, for the PDF and the CDF of the SNR at the output of the receiver. In Sections 4-5, BER for various M-ary modulation techniques and CC of the studied equivalent channel are derived in terms of only one Fox H-function and Meijer G-function with small computational complexity. In section 6, the results are illustrated and verified by comparison with the recent results. Our main conclusions are summarized in the final section.

2 Receiver Description and Preliminary Statistics

We consider a digital wireless communication system with L-branch MRC, EGC, or SC receiver operating in a Weibull multipath fading environment. The SNR of the combined signal at the output of the studied receiver is given by

$$\gamma = \delta \left(\sum_{i=1}^{L} \gamma_i^{\alpha} \right)^{\frac{1}{\alpha}} \tag{1}$$

where

- $1/\alpha$ is a positive integer,
- $\gamma_i = \frac{E_s}{N_0} R_i^2$ is the instantaneous SNR per symbol on the ith branch,
- E_s denotes the average energy per symbol and N_0 denotes the power spectral density of thermal noise,
- L denotes the number of combined diversity branches,
- R_i denotes the fading amplitude corresponding to the ith received signal at the combiner, assumed to be Weibull distributed with the shape parameter β_i and the scale parameter ω_i:

$$R_i = \left(N_{1i}^2 + N_{2i}^2 \right)^{\frac{1}{\beta_i}} \tag{2}$$

where N_{1i} and N_{2i} are two normally distributed variates with mean and variance 0 and $\frac{\omega_i^{\beta_i}}{2}$, respectively. The values of parameters δ and α vs. receiver are summarized in Table 1.

Table 1. Values of δ and α for some known receivers

Receiver	MRC	EGC	SC
δ	1	$1/L$	1
α	1	$1/2$	$+\infty$

In the following, the Weibull distribution is denoted $W(\omega_i, \beta_i)$. Its PDF is

$$f_{R_i}(r) = \frac{\beta_i}{\omega_i} \left(\frac{r}{\omega_i} \right)^{\beta_i - 1} \exp\left[-\left(\frac{r}{\omega_i} \right)^{\beta_i} \right], r \geq 0 \tag{3}$$

so, the nth moment of R_i is given by

$$\mu_n^{(R_i)} = \omega_i^n d_n(\beta_i) \tag{4}$$

where $d_k(\beta_i) = \Gamma(1 + k/\beta_i)$, and $\Gamma(\cdot)$ is the gamma function. Accordingly, the scale parameter of R_i is

$$\omega_i = \sqrt{\frac{\mu_2^{(R_i)}}{d_2(\beta_i)}} \tag{5}$$

and the average of the ith SNR is then

$$\overline{\gamma}_i = \frac{E_s}{N_0}\omega_i^2 d_1\left(\frac{\beta_i}{2}\right) \tag{6}$$

Since the square of a Weibull RV is a Weibull RV, and according to (4) and (6), γ_i is a Weibull distribution $W\left(\frac{E_s}{N_0}\omega_i^2, \frac{\beta_i}{2}\right)$.

3   On the Sum of Weibull RVs

In this section, based on the recently derived expressions for the sum of Weibull distributed RVs [9]-[10], a new closed-form approximation of the output SNR PDF having low evaluation time is presented.

Lemma 1. *If* X *is a Weibull distribution* $W(\omega, \beta)$, *then its PDF can be expressed as a Fox H-function*

$$f_X(x) = \frac{1}{\omega}H_{0,1}^{1,0}\left(\frac{x}{\omega}\left|\begin{matrix}\cdot\\(1-\frac{1}{\beta},\frac{1}{\beta})\end{matrix}\right.\right), \quad x \geq 0 \tag{7}$$

Proof. See Appendix A.   □

Lemma 2. *Let* $Z_{\alpha,i} = \gamma_i^\alpha$. *Then* $Z_{\alpha,i}$ *is a Weibull distribution* $W\left(\left[\frac{\overline{\gamma}_i}{d_2(\beta_i)}\right]^\alpha, \frac{\beta_i}{2\alpha}\right)$

Proof. See Appendix A.   □

3.1   PDF of the Output SNR γ

As the exact expression of the SNR PDF at the output of both MRC and coherent EGC receivers operating under Weibull fading channel is unknown or even impossible to find explicitly, the main idea is to derive, as much as possible, a closed-form approximation with low evaluation time.

Theorem 1. *Let* $Z^{(\alpha)} = \sum_{i=1}^{L} Z_{\alpha,i}$. *The SNR PDF at the output of the studied receiver can be approximated by a Fox H-function*

$$f_\gamma(\gamma) \approx \frac{\alpha\left(\Psi_{\alpha,L}\left(\frac{\gamma}{\delta}\right)^\alpha\right)^{\Phi_{\alpha,L}}}{\gamma\Gamma(\Phi_{\alpha,L})}\exp\left(-\Psi_{\alpha,L}\left(\frac{\gamma}{\delta}\right)^\alpha\right) \tag{8}$$

where $\Phi_{\alpha,L} = \dfrac{\left(\mu_1^{\left(Z^{(\alpha)}\right)}\right)^2}{\mu_2^{\left(Z^{(\alpha)}\right)}-\left(\mu_1^{\left(Z^{(\alpha)}\right)}\right)^2}$ *and* $\Psi_{\alpha,L} = \dfrac{\mu_1^{\left(Z^{(\alpha)}\right)}}{\mu_2^{\left(Z^{(\alpha)}\right)}-\left(\mu_1^{\left(Z^{(\alpha)}\right)}\right)^2}$.

Proof. A very tight closed-form approximation of the PDF of $Z^{(\alpha)}$ was derived in terms of Meijer G-function as [8]-[10]

$$f_{Z^{(\alpha)}}(z) \approx \frac{\Gamma(a+1)}{\lambda\Gamma(b+1)} G_{1,1}^{1,0}\left(\frac{z}{\lambda}\middle|\begin{matrix} a \\ b \end{matrix}\right), \quad z \geq 0 \tag{9}$$

where

$$a = \left(\lambda - \frac{\mu_2^{\left(Z^{(\alpha)}\right)}}{\mu_1^{\left(Z^{(\alpha)}\right)}}\right)\Psi_{\alpha,L} - 1, b = \Phi_{\alpha,L} - \frac{\mu_2^{\left(Z^{(\alpha)}\right)}\Psi_{\alpha,L}}{\lambda} - 1, \tag{10}$$

and λ is a real number.

Now, using the Jacobian of the transformation, the PDF of $\gamma = \delta\left[Z^{(\alpha)}\right]^{\frac{1}{\alpha}}$ is

$$f_\gamma(\gamma) = \frac{\alpha\gamma^{\alpha-1}}{\delta^\alpha} f_{Z^{(\alpha)}}\left[\left(\frac{\gamma}{\delta}\right)^\alpha\right] \tag{11}$$

Substituting (9) into (11), we get

$$f_\gamma(\gamma) \approx \frac{\Gamma(a+1)}{\gamma\Gamma(b+1)} \frac{1}{2\pi j} \int_{\mathcal{C}} \frac{\Gamma\left(b+1+\frac{s}{\alpha}\right)}{\Gamma\left(a+1+\frac{s}{\alpha}\right)} \left(\frac{\gamma}{\lambda^{\frac{1}{\alpha}}\delta}\right)^{-s} ds \tag{12}$$

where $\mathcal{C}$ denotes an infinite complex contour of integration, and j is an imaginary number such that $j^2 = -1$.

The PDF of the output SNR given in (9) can be expressed either as a sum of LHP (Left Half-Plane) or RHP (Left Right-Plane) residues if $\lambda > \mu_2^{(Z)}/\mu_1^{(Z)}$ [9]. In addition, this expression becomes closer to the exact one for large values of λ. However, the evaluation time of the Meijer G-function takes a lot of time if its first argument is close to zero. Hence, to get a close approximation with low evaluation time complexity of this function, we will eliminate the λ-term in this expression by finding the limit of this PDF as λ approaches $+\infty$.

We have from (10)

$$\frac{\Gamma\left(b+1+\frac{s}{\alpha}\right)}{\Gamma(b+1)} \sim \frac{\Gamma(\Phi_{\alpha,L}+\frac{s}{\alpha})}{\Gamma(\Phi_{\alpha,L})} \quad \text{as } \lambda \to +\infty \tag{13}$$

On the other hand, using the stirling's formula [15, 6.1.39], we get from (10)

$$\frac{\Gamma(a+1)\lambda^{\frac{s}{\alpha}}}{\Gamma(a+1+\frac{s}{\alpha})} \sim \Psi_{\alpha,L}^{-\frac{s}{\alpha}} \quad \text{as } \lambda \to +\infty \tag{14}$$

Substituting (13) and (14) into (12), yields

$$f_\gamma(\gamma) \approx \frac{1}{\gamma\Gamma(\Phi_{\alpha,L})} \frac{1}{2\pi j} \int_{\mathcal{C}} \Gamma(\Phi_{\alpha,L}+\frac{s}{\alpha}) \left(\Psi_{\alpha,L}^{1/\alpha}\frac{\gamma}{\delta}\right)^{-s} ds \tag{15}$$

$$= \frac{1}{\gamma\Gamma(\Phi_{\alpha,L})} H_{0,1}^{1,0}\left(\frac{\Psi_{\alpha,L}^{1/\alpha}\gamma}{\delta}\middle|\begin{matrix} \cdot \\ (\Phi_{\alpha,L},\frac{1}{\alpha}) \end{matrix}\right)$$

Now, using [11, eq.(07.34.03.0228.01)] and the change of variable $s' = \Phi_{\alpha,L} + \frac{s}{\alpha}$, we get (8) which concludes the proof of the theorem. $\qquad\square$

Remark 1. The terms $\Phi_{\alpha,L}$ and $\Psi_{\alpha,L}$ are explicitly expressed for i.i.d Weibull fading channel as

$$\Phi_{\alpha,L} = \frac{\left(\sum_{i=1}^{L} \overline{\gamma}_i^{\alpha}\right)^2}{\sum_{i=1}^{L} \overline{\gamma}_i^{2\alpha}} \frac{d_1^2\left(\frac{\beta}{2\alpha}\right)}{d_2\left(\frac{\beta}{2\alpha}\right) - d_1^2\left(\frac{\beta}{2\alpha}\right)}, \Psi_{\alpha,L} = \frac{\sum_{i=1}^{L} \overline{\gamma}_i^{\alpha}}{\sum_{i=1}^{L} \overline{\gamma}_i^{2\alpha}} \frac{d_1\left(\frac{\beta}{2\alpha}\right) d_2^{\alpha}(\beta)}{d_2\left(\frac{\beta}{2\alpha}\right) - d_1^2\left(\frac{\beta}{2\alpha}\right)}$$

and for the same average SNR of the signal at each input branch ($\overline{\gamma}_i = \overline{\gamma}_1$ for all i)

$$\Phi_{\alpha,L} = \frac{L d_1^2\left(\frac{\beta}{2\alpha}\right)}{d_2\left(\frac{\beta}{2\alpha}\right) - d_1^2\left(\frac{\beta}{2\alpha}\right)}, \Psi_{\alpha,L} = \frac{d_1\left(\frac{\beta}{2\alpha}\right) d_2^{\alpha}(\beta) \overline{\gamma}_1^{-\alpha}}{d_2\left(\frac{\beta}{2\alpha}\right) - d_1^2\left(\frac{\beta}{2\alpha}\right)}$$

Remark 2. In uncorrelated fading channel, the exact value of the average SNR at the output of the receiver is expressed using the multinomial theorem, lemma 2, and (4) as

$$\overline{\gamma} \equiv \delta \mu_{1/\alpha}^{(Z^{(\alpha)})} = \delta \sum_{\sum_{j=1}^{L} i_j = 1/\alpha} \frac{(1/\alpha)!}{\prod_{j=1}^{L} i_j!} \prod_{j=1}^{L} \left(\frac{\overline{\gamma}_{i_j}}{d_2(\beta_{i_j})}\right)^{\alpha i_j} d_{2\alpha i_j}(\beta_{i_j}) \qquad (16)$$

On the other side, it can be approximated by placing (15) into [12, eq. (2.8)]

$$\overline{\gamma} \approx \frac{\delta \Gamma(\Phi_{\alpha,L} + \frac{1}{\alpha})}{\Psi_{\alpha,L}^{1/\alpha} \Gamma(\Phi_{\alpha,L})} \qquad (17)$$

3.2 Outage Probability

The outage probability is the key metric to characterize the performance limits of wireless communication systems. It's defined in terms of SNR CDF as [13]

$$P_{out} = F_{\gamma}(\gamma_{\min}) \qquad (18)$$

where $\gamma_{\min}$ is the minimum SNR threshold that ensure a reliable communication and equivalent channel is not in outage.

Proposition 1. *The Outage probability for the studied receiver is expressed as*

$$P_{out} \approx \frac{1}{\Gamma(\Phi_{\alpha,L})} H_{1,2}^{1,1}\left(\Psi_{\alpha,L}^{1/\alpha} \frac{\gamma_{\min}}{\delta} \,\middle|\, \begin{matrix}(1,1)\\(\Phi_{\alpha,L},\frac{1}{\alpha}),(0,1)\end{matrix}\right) \qquad (19)$$

Proof. Using (15) and [11, eq.(06.05.16.0002.01)], the CDF of γ is given by

$$F_{\gamma}(\gamma) \approx \frac{1}{2\pi j \Gamma(\Phi_{\alpha,L})} \int_C \Gamma\left(\Phi_{\alpha,L} + \frac{s}{\alpha}\right) \left(\frac{\Psi_{\alpha,L}^{1/\alpha}}{\delta}\right)^{-s} \int_0^{\gamma} t^{-s-1} dt\, ds \qquad (20)$$

$$= \frac{1}{2\pi j \Gamma(\Phi_{\alpha,L})} \int_C \frac{\Gamma(\Phi_{\alpha,L} + \frac{s}{\alpha})\Gamma(-s)}{\Gamma(1-s)} \left(\Psi_{\alpha,L}^{1/\alpha} \frac{\gamma}{\delta}\right)^{-s} ds$$

Which completes the proof of the proposition.   □

Remark 3. by applying the relation [15, q.(6.1.20)]

$$\Gamma(\Phi_{\alpha,L} + \frac{s}{\alpha}) = (2\pi)^{\frac{1-1/\alpha}{2}} \prod_{k=0}^{\frac{1}{\alpha}-1} \frac{\Gamma(\alpha(\Phi_{\alpha,L} + k) + s)}{\alpha^{\Phi_{\alpha,L} + \frac{s}{\alpha} - \frac{1}{2}}} \tag{21}$$

furthermore, OP can be rewritten in terms of Meijer G-function as

$$P_{out} \approx \frac{(2\pi)^{\frac{1-1/\alpha}{2}} G^{\frac{1}{\alpha},1}_{1,\frac{1}{\alpha}+1}\left((\alpha\Psi_{\alpha,L})^{1/\alpha}\frac{\gamma_{\min}}{\delta}\middle|\begin{array}{c} 1 \\ \Delta_{\alpha,L};0 \end{array}\right)}{\Gamma(\Phi_{\alpha,L})\alpha^{\Phi_{\alpha,L}-\frac{1}{2}}} \tag{22}$$

with $\Delta_{\alpha,L} = \{\alpha\Phi_{\alpha,L}, \alpha(\Phi_{\alpha,L}+1), ..., \alpha(\Phi_{\alpha,L}+\frac{1}{\alpha}-1)\}$

4   Average Bit Error Probability

Proposition 2. *The BER of several M-ary modulation techniques using the studied combiner in Weibull multipath fading environment*

$$\overline{P}_e \approx \frac{1}{2\Gamma(\varrho)\Gamma(\Phi_{\alpha,L})} H^{1,2}_{2,2}\left(\frac{\Psi^{1/\alpha}_{\alpha,L}}{\delta\theta}\middle|\begin{array}{c} (1,1), (1-\varrho,1) \\ (\Phi_{\alpha,L},\frac{1}{\alpha}); (0,1) \end{array}\right) \tag{23}$$

where ϱ and θ are parameters depending on modulation scheme [9].

Proof. The BER for several *M*-ary modulation scheme over fading channel is given by [14, eq.(13)]

$$\overline{P}_e = \frac{\theta^\varrho}{2\Gamma(\varrho)} \int_0^\infty \gamma^{\varrho-1} e^{-\theta\gamma} F_\gamma(\gamma)\,d\gamma \tag{24}$$

Substituting (20) into (24), yielding

$$\overline{P}_e \approx \frac{\theta^\varrho}{4\pi j\Gamma(\varrho)\Gamma(\Phi_{\alpha,L})} \int_C \frac{\Gamma(\Phi_{\alpha,L}+\frac{s}{\alpha})\Gamma(-s)}{\Gamma(1-s)} \left(\frac{\Psi^{1/\alpha}_{\alpha,L}}{\delta}\right)^s \int_0^\infty \gamma^{\varrho-s-1} e^{-\theta\gamma}\,d\gamma\,ds \tag{25}$$

$$= \frac{1}{4\pi j\Gamma(\varrho)\Gamma(\Phi_{\alpha,L})} \int_C \frac{\Gamma(\Phi_{\alpha,L}+\frac{s}{\alpha})\Gamma(-s)\Gamma(\varrho-s)}{\Gamma(1-s)} \left(\frac{\Psi^{1/\alpha}_{\alpha,L}}{\delta\theta}\right)^{-s}\,ds$$

That concludes the proof of the proposition.   □

Remark 4. Replacing (21) into (25), we get

$$\overline{P}_e \approx \frac{(2\pi)^{\frac{1-\frac{1}{\alpha}}{2}} \alpha^{\frac{1}{2}-\Phi_{\alpha,L}}}{2\Gamma(\varrho)\Gamma(\Phi_{\alpha,L})} G^{\frac{1}{\alpha},2}_{2,\frac{1}{\alpha}+1}\left(\frac{(\alpha\Psi_{\alpha,L})^{\frac{1}{\alpha}}}{\delta\theta}\middle|\begin{array}{c} 1, 1-\varrho \\ \Delta_{\alpha,L};0 \end{array}\right) \tag{26}$$

5   Average Shannon Capacity

Proposition 3. *Let B_w be the channel bandwidth. The CC for the studied receiver in the case of Weibull multipath fading channels is close to*

$$\overline{C} \approx \frac{B_w}{\Gamma(\Phi_L)\ln 2} H_{2,3}^{3,1}\left(\frac{\Psi_{\alpha,L}^{\frac{1}{\alpha}}}{\delta} \left| \begin{array}{l} (0,1)\,;(1,1) \\ (\Phi_{\alpha,L},\frac{1}{\alpha})\,,(0,1)\,,(0,1) \end{array} \right.\right) \qquad (27)$$

Proof. The average capacity of the studied channel is given by

$$\overline{C} = B_w \int_0^\infty \log_2\left(1+\gamma\right) f_\gamma(\gamma)\,d\gamma \qquad (28)$$

Using [11, eq.(07.34.03.0456.01)], it can be rewritten using a Mellin-Barnes contour integral

$$\overline{C} \approx \frac{B_w}{2\pi j\Gamma(\Phi_{\alpha,L})\ln 2} \int_{\mathcal{C}} \Gamma(\Phi_{\alpha,L}+\frac{s}{\alpha})\left(\frac{\Psi_{\alpha,L}^{1/\alpha}}{\delta}\right)^{-s}\left(\int_0^\infty \gamma^{-s-1}G_{2,2}^{1,2}\left[\gamma \left| \begin{array}{c} 1,1 \\ 1,0 \end{array}\right.\right] d\gamma\right) ds \qquad (29)$$

The Mellin transform in (29) can be evaluated using [11, eq. (07.34.21.0009.01)]

$$\int_0^\infty \gamma^{-s-1}G_{2,2}^{1,2}\left[\gamma \left| \begin{array}{c} 1,1 \\ 1,0 \end{array}\right.\right] d\gamma = \frac{\Gamma(1-s)\Gamma^2(s)}{\Gamma(1+s)} \qquad (30)$$

Substituting (30) into (29), concludes the proof of the proposition.   □

Remark 5. Replacing (21) into (29), the CC can also be expressed in terms of Meijer G-function

$$\overline{C} \approx \frac{B_w\,(2\pi)^{\frac{1-\frac{1}{\alpha}}{2}}\,\alpha^{\frac{1}{2}-\Phi_{\alpha,L}}}{\Gamma(\Phi_{\alpha,L})\ln 2}\,G_{2,\frac{1}{\alpha}+2}^{\frac{1}{\alpha}+2,1}\left(\frac{(\alpha\Psi_{\alpha,L})^{\frac{1}{\alpha}}}{\delta} \left| \begin{array}{l} 0;1 \\ 0,0,\Delta_{\alpha,L} \end{array}\right.\right) \qquad (31)$$

6   Performance Evaluation Results

In this section, the results containing the Meijer G-function were evaluated using Mathematica software. All the Monte Carlo simulations are established by generating $10^8 L$ Weibull distributed random numbers over the studied SNR range, subdivided into 10^4 subintervals of equal length. We have assumed an exponentially decaying power delay profile (PDP) $\overline{\gamma}_i/\overline{\gamma}_1 = \exp\left[-\varphi(i-1)\right]$ where φ is the average fading power decay factor [16], and $\overline{\gamma}_1 = 1$ except for figure 4 and 5.

In Fig. 1, the analytical expression (8) and the simulated PDF versus γ are plotted, for the L-branch EGC and dual-branch MRC. It can be seen that the MRC curves are closer to the simulated ones than those of EGC.

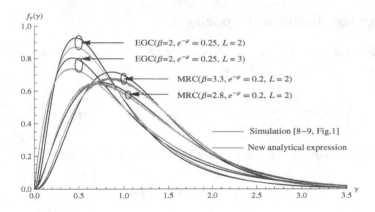

Fig. 1. PDF of the output SNR at both L-branch EGC and 2-branch MRC receiver

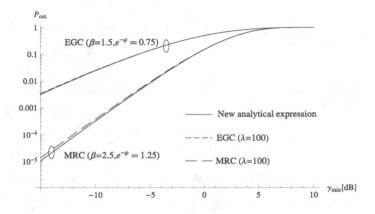

Fig. 2. Outage probability of dual-branch EGC/MRC receiver

Fig. 2 depicts the OP for dual-branch MRC/EGC receiver, given by (31) for i.i.d. Weibull fading channels. The plotted curves are compared for EGC and MRC with those plotted from [8, eq.(28)] and [9, eq.(13)], respectively. It can be observed that the new derived expression is very close to the previous one for great values of λ.

In Fig. 3, the normalized CC, given by (31), versus the diversity order L, is plotted and compared with the previous result [8, eq.(37)] for EGC receiver. The new analytical expression of capacity is very close to the recent one (plotted in dashed line). Besides, the evaluation time of this new expression is very low since the λ-term, sufficiently large, is eliminated.

Fig. 4 compares the new expression of BER for both BPSK and BFSK modulation with 4-branch MRC receiver, plotted from (16) and (26), with the previous one [9, eq.(20)]. It can be seen that the new approximate expression becomes

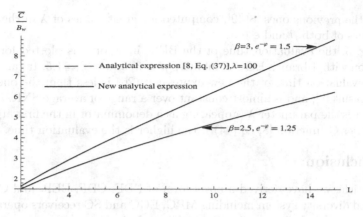

Fig. 3. Nomalized average Shannon capacity of L-branch EGC receiver

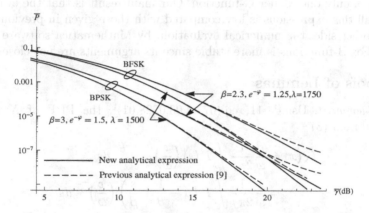

Fig. 4. Analytical expression of BER for BPSK/BFSK modulation with MRC receiver

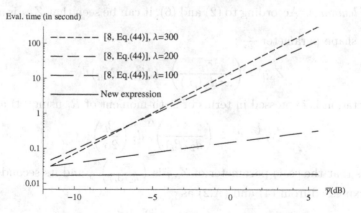

Fig. 5. Evaluation time of the BER for BPSK modulation with EGC receiver

closer to the previous ones [8]-[9], computed for great values of λ in the case of small values of both β and $e^{-\varphi}$.

In Fig. 5, the evaluation time of the BER, in second, is plotted for BPSK modulation with 4-branch EGC combiner, $\beta = 3$, and $e^{-\varphi} = 1.5$. It can be seen that the evaluation time of the new expression (26) is less than the one of the previous result [8], and is almost constant over a range of average SNR. Besides, the greater is the parameter λ (appearing as a denominator in the first argument of the Meijer G-function [8, eq.(44)]), the higher is the evaluation time.

7 Conclusion

In this article, we have derived a novel form of PDF, CDF, BER, and CC for a generalized diversity system including MRC, EGC and SC receivers operating in Weibull multipath fading environment. All analytical expressions were derived in terms of only one Meijer G-function. Our main result is that the evaluation time of all this expressions is low compared with those given in previous work. On the other side, the numerical evaluation, by Mathematica software, of the derived Fox H-functions is more stable since its arguments are not close to 0.

A Proofs of Lemmas

Proof of lemma 1. Using [11, eq.(07.34.03.0228.01)], the PDF of X can be expressed from (3)

$$f_X(x) = \frac{\beta}{\omega} \frac{1}{2\pi j} \int_C \Gamma(s) \left(\frac{x}{\omega}\right)^{\beta-1-\beta s} ds \qquad (A.1)$$

$$= \frac{1}{\omega} \frac{1}{2\pi j} \int_C \Gamma\left(1 - \frac{1}{\beta} + \frac{s}{\beta}\right) \left(\frac{x}{\omega}\right)^{-s} ds$$

which concludes the proof of lemma 1. □

Proof of lemma 2. According to (2) and (6), it can be seen that $Z_{\alpha,i}$ is a Weibull RV with shape parameter $\frac{\beta_i}{2\alpha}$:

$$Z_{\alpha,i} = \left(\frac{\overline{\gamma}_i}{\omega_i^2 d_2(\beta_i)}\right)^\alpha (N_{1i}^2 + N_{2i}^2)^{\frac{2\alpha}{\beta_i}} \qquad (A.2)$$

Its expectation is expressed in terms of 2αth-moment of R_i using (4) and (A.2)

$$\mu_1^{(Z_{\alpha,i})} = \left(\frac{\overline{\gamma}_i}{d_2(\beta_i)}\right)^\alpha d_1\left(\frac{\beta_i}{2\alpha}\right)$$

It follows that the scale parameter of $Z_{\alpha,i}$ is $\left(\frac{\overline{\gamma}_i}{d_2(\beta_i)}\right)^\alpha$, and its second moment is then expressed from (4) and (A.2) as

$$\mu_2^{(Z_{\alpha,i})} = \left(\frac{\overline{\gamma}_i}{d_2(\beta_i)}\right)^{2\alpha} d_2\left(\frac{\beta_i}{2\alpha}\right) \qquad (A.3)$$

that concludes the proof of the lemma 2. □

References

1. Cheng, J., Tellambura, C., Beaulieu, N.C.: Performance Analysis of Digital Modulations on Weibull Fading channels. IEEE Transactions on Wireless Communications **3**(4), 1124–1133 (2004)
2. Hai, N.T., Yakubovich, S.B.: The Double Mellin-Barnes Type Integrals and their Applications to Convolution Theory. World Scientific, Singapore (1992)
3. Karagiannidis, G.K., Zogas, D.A., Sagias, N.C., Kotsopoulos, S.A., Tombras, G.S.: Equal-Gain and Maximal-Ratio Combining Over Nonidentical Weibull Fading Channels. IEEE Trans. Comm. **4**(3), 841–846 (2005)
4. Zogas, D.A., Sagias, N.C., Tombras, G.S., Karagiannidis, G.K.: Average output SNR of equal-gain diversity receivers over correlative Weibull fading channels. European Transactions on Telecommunications **16**, 521–525 (2005)
5. Sagias, N.C., Karagiannidis, G.K., Bithas, P.S., Mathiopoulos, P.T.: On the Correlated Weibull Fading Model and Its Applications, pp. 2149–2153. IEEE (2005)
6. Papadimitiou, Z.G., Bithas, P.S., Mathiopoulos, P.T., Sagias, N.C., Merakos, L.: Triple-branch MRC diversity in Weibull fading channels. In: Signal Design and Its Appl. in Commun., IWSDA 2007, Chengdu, pp. 247–251, September 23–27, 2007
7. El Bouanani, F., Ben-Azza, H., Belkasmi, M.: New Results for the Shannon channel capacity over generalized multipath fading channels for MRC diversity. EURASIP Journal on Wireless Communications and Networking **2012**, 336 (2012)
8. El Bouanani, F., Ben-Azza, H.: Unified analysis of EGC diversity over Weibull fading channels. Wiley Internat. Journal of Commun. Systems (IJCS), December 2014
9. El Bouanani, F.: A new closed-form approximations for MRC receiver over nonidentical Weibull fading channels. In: Internat. Wireless Commun. & Mobile Computing Conference, IWCMC, Nicosia, Cyprus, pp. 600–605, August 4–8, 2014
10. El Bouanani, F., Ben-Azza, H., Belkasmi, M.: Novel results for the spectral efficiency and symbol error rate of GSC receiver over identical and uncorrelated Weibull fading channels. In: World Congress on Computer Applications and Information Systems, WCCAIS, Hammamet, Tunisia, January 17–19, 2014, pp. 1–5 (2014)
11. Wolfram Research, Inc. Mathematica, Edition: Version 10.0, Champaign, Illinois, Wolfram Research, Inc. (2014)
12. Mathai, A.M., Haubold, H.J., Saxena, R.K.: The H-function: Theory and Applications. Springer, New York (2010)
13. Simon, M.K., Alouini, M.S.: Digital Communication over Fading Channels. John Wiley & Sons, Hoboken (2005)
14. Ansari, I.S., Al-Ahmadi, S., Yilmaz, F., Alouini, M.-S., Yanikomeroglu, H.: A new formula for the BER of binary modulations with dual-branch selection over generalized-K composite fading channels. IEEE Transactions on Communications **59**(10), 2654–2658 (2011)
15. Abramowitz, M., Stegun, I.A.: Handbook of Mathematical Functions with formula, graphs, and mathematical tables. National Bureau of Standards, Applied Mathematical Series, vol. 55, June 1964
16. Kong, N., Milstein, L.B.: SNR of generalized diversity selection combining with nonidentical Rayleigh fading statistics. IEEE Tr. Com. **48**, 1266–1271 (2000)

Power Control in Cognitive Radio Networks Using Cooperative Modulation and Coding Classification

Anestis Tsakmalis$^{(\boxtimes)}$, Symeon Chatzinotas, and Björn Ottersten

SnT - securityandtrust.lu, University of Luxembourg, Luxembourg, Luxembourg
{anestis.tsakmalis,symeon.chatzinotas,bjorn.ottersten}@uni.lu

Abstract. In this paper, a centralized Power Control (PC) scheme aided by interference channel gain learning is proposed to allow a Cognitive Radio (CR) network to access the frequency band of a Primary User (PU) operating based on an Adaptive Coding and Modulation (ACM) protocol. The main idea is the CR network to constantly probe the band of the PU with intelligently designed aggregated interference and sense whether the Modulation and Coding scheme (MCS) of the PU changes in order to learn the interference channels. The coordinated probing is engineered by the Cognitive Base Station (CBS), which assigns appropriate CR power levels in a binary search way. Subsequently, each CR applies a Modulation and Coding Classification (MCC) technique and sends the sensing information through a control channel to the CBS, where all the MCC information is combined using a fusion rule to acquire an MCS estimate of higher accuracy and monitor the probing impact to the PU MCS. After learning the normalized interference channel gains towards the PU, the CBS selects the CR power levels to maximize total CR network throughput while preserving the PU MCS and thus its QoS. The effectiveness of the proposed technique is demonstrated through numerical simulations.

Keywords: Cognitive Radio · Centralized power control · Spectrum sensing · Cooperative Modulation and Coding Classification · Adaptive coding and modulation

1 Introduction

Radio Spectrum is well known to be a limited resource. Ever since its first commercial usage, regulations for limiting services to specific frequency bands have been enforced. This rulemaking process assumes that a static assignment of services to frequency bands not only facilitates the financial exploitation of the Radio Spectrum, but also limits interference and supports the construction of cheap and less complicated transceivers, a major technological restraint.

Nowadays though, the burst in service demand has led us to rethink the static nature of this architecture. Taking into account also the fact that some frequency

© Institute for Computer Sciences, Social Informatics and Telecommunications Engineering 2015
M. Weichold et al. (Eds.): CROWNCOM 2015, LNICST 156, pp. 358–369, 2015.
DOI: 10.1007/978-3-319-24540-9_29

bands are being underutilized and that others accommodate services resilient to interference, the research community proposed the idea of the Dynamic Spectrum Access (DSA) [1]. Some DSA techniques suggest the use of frequency bands by unlicensed users, also called Secondary Users (SUs), when the licensed ones (PUs) are absent or even their coexistence as long as the received interference by PUs is below a certain threshold. This flexible structure enables us to exploit the Radio Spectrum resource more efficiently. A candidate technology to reach this objective and enhance the operation of the SUs is the Cognitive Radio (CR) [2]. The main functionalities of the CR are the Spectrum Sensing (SS), which consists of methods detecting the existence or type of licensed primary signal, and the PC, the adaptive adjustment of transmit power. Unlicensed SUs equipped with these CR mechanisms can apply DSA techniques and help us resolve both Radio Spectrum underutilization and congestion.

One important SS mechanism is the identification of the PU signal type. An interesting approach to address this problem could be the classification of the modulation and coding scheme (MCC) [3,4]. As far as the modulation classification is concerned, features like the signal cumulants of 2nd, 3rd, 4th, 6th and 8th order which have distinctive theoretical values among different modulation schemes [5] are estimated and then fed into a powerful classification tool, the Support Vector Machine (SVM) [6]. For the coding identification part, the exploited statistical features are the log-likelihood ratios (LLRs) of the received symbol samples [7,8]. The detection technique in this case involves the comparison of the average LLRs of the error syndromes derived from the parity-check relations of each code.

The second CR enhancement mentioned before is the PC strategy based on which the SUs are accessing the frequency band of the PU. This vast topic has been thoroughly investigated from many aspects depending on the system model, the optimization variables, the objective functions, the constraints and other known or unknown parameters. An interesting approach to the PC problem tackled by the research community within the wireless network context has been the centralized one. Based on this, a central decision maker, the Base Station, gathers local information from the users through a control channel, elaborates an intelligent selection of their operational parameters, such as their transmit power, channel or time schedule, and communicates it to them. In this general context, the research community has formulated and tackled PC problems to achieve common or different signal to interference plus noise ratio ($SINR$) requirements, maximum total system throughput, maximum weighted throughput, maximum worst user throughput or minimum transmit power, subject to QoS constraints from individual users, like $SINR$, data rate or outage probability. In the CR regime, the centralized PC problem retains its basic form but with some small alterations. One critical modification is the knowledge of the interference channels from the CR transmitters to the PU receivers. Previous work has considered perfect CR-to-PU channel knowledge [9,10], limited-rate feedback from the PUs on CR-to-PU channel gains [11], imperfect CR-to-PU channel knowledge [12] and CR-to-PU channel uncertainty knowledge attained through SS or channel gain cartography [13].

An even more challenging PC problem in CR networks is the one without any prior knowledge of the interference channels and cooperation from the PU link. The additional burden in this case is learning the CR-to-PU channels using eavesdropped information from the PU feedback channel. A solution for one SU coexisting with one PU was given in [14] based on a probing and sensing model. Nevertheless, the most sophisticated methods suitable for learning the interference channel gains of multiple SUs through probing with the use of even binary feedback are derived from multiple input multiple output (MIMO) and beamforming research scenarios. Previous researchers have exploited a slow random exploration algorithm [15], the one-bit null space learning algorithm [16] and an analytic center cutting plane method (ACCPM) based learning algorithm [17].

In this paper, a centralized PC method aided by interference channel power gain learning is demonstrated which concerns multiple SUs and a PU and maximizes the total SU throughput subject to maintaining the PU QoS. This case study considers the PU link changing its MCS based on an ACM protocol and operating in its assigned band together with a CR network accessing this band and having knowledge of this ACM protocol. Our idea is to apply an algorithm in order to first estimate the interference channel power gains by exploiting SS feedback and finally maximize the total SU throughput. This CR-to-PU channel knowledge is acquired by having the coexisting cognitive SUs constantly probing in a binary search trial and error way and checking whether the CR network caused the PU MCS to change. The detection of the PU MCS is performed in a cooperative way at the CBS which gathers the MCC feedback from all the SUs through a control channel and combines them using a hard decision fusion rule. The proposed DSA application concerns only the SU system without adding any complexity in the infrastructure or a control channel between the PU system and the SU one in order to exchange information about the channel gains or the induced interference.

The remainder of this paper is structured as follows: Section 2 provides the system model and the problem formulation. Section 3 introduces the cooperative MCC. Section 4 analyzes the interference channel power gain learning. Section 5 shows the results obtained by the combination of the above. Finally, Section 6 gives the concluding remarks and future work in this topic.

2   System Model and Problem Formulation

Initially, the outputs of the MCC procedure and the way they are employed have to be described. All the SUs are equipped with a secondary omnidirectional antenna only for sensing the PU signal and an MCC module which enables them to identify the MCS of the PU. Specifically, each SU collects PU signal samples using a standard sensing period T_S, estimates the current MCS and transmits it through a control channel to the CBS. The MCS observation of the SU_i over the n_{th} sensing period is expressed as MCS_i^n and a detailed description about its estimation can be found in [3,4].

Furthermore, at system level a PU link and N SU links exist in the same frequency band as shown in Fig. 1. As far as the interference to the PU link is

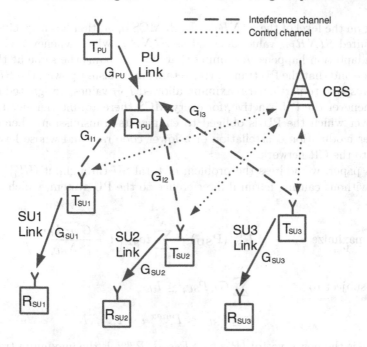

Fig. 1. The PU system and the CR network

concerned, this is caused by the transmitter part of each SU link to the receiver of the PU link. Considering strong interference links, this may have a severe effect on the MCS chosen by the PU link. In addition, a channel access method allows SU links not to interfere with each other. In this scenario, the unknown interference channel gains and the PU channel gain are static and no fading channel models are considered.

Here we focus on channel power gains G, which in general are defined as $G = \|g\|^2$, where g is the channel gain. From this point, we will refer to channel power gains as channel gains. The aggregated interference to the PU side is defined as:

$$I_{PU} = \sum_{i=1}^{N} G_{I_i} P_{SU_i} \tag{1}$$

where G_{I_i} is the SU_i-to-PU interference channel gain and P_{SU_i} is the SU_i transmit power. Additionally, the $SINR$ of the PU is defined in as:

$$SINR_{PU} = 10 \log \left(\frac{G_{PU} P_{PU}}{I_{PU} + N_{PU}} \right) \text{ dB} \tag{2}$$

where G_{PU} is the PU link channel gain, P_{PU} is the PU transmit power and N_{PU} is PU receiver noise power. From a PU system perspective, an ACM scheme is applied with a set of possible MCS's. The ACM protocol changes the MCS of the PU link to more or less robust modulation constellations and coding rates depending on the level of the $SINR_{PU}$. Each MCS operation has a specific minimum required $SINR_{PU}$ value, denoted as $SINR_{th}$, which whenever violated, an MCS adaptation happens. Assuming that N_{PU} remains the same at the PU receiver side and that the PU transmitter retains its transmit power, the $SINR_{th}$ values correspond to particular maximum allowed I_{PU} values, designated as I_{th}. Hence, whenever the PU is active, for every MCS there are interference thresholds I_{th} over which the PU is obliged to change its transmission scheme to a lower order modulation constellation or a lower code rate and whose levels are unknown to the CR network.

In this paper, we address the problem of total SU throughput (U_{SU}^{tot}) maximization without causing harmful interference to the PU system, which can be written as:

$$\underset{\mathbf{P_{SU}}}{\text{maximize}} \quad U_{SU}^{tot}(\mathbf{P_{SU}}) = \sum_{i=1}^{N} log_2 \left(1 + \frac{G_{SU_i} P_{SU_i}}{N_{SU_i}} \right) \tag{3a}$$

$$\text{subject to} \quad \sum_{i=1}^{N} G_{I_i} P_{SU_i} \leq I_{th} \tag{3b}$$

$$0 \leq P_{SU_i} \leq P_i^{max} \quad i = 1, \dots, N \tag{3c}$$

where $\mathbf{P_{SU}}$ is the power vector $[P_{SU_1}, \dots, P_{SU_N}]$, P_i^{max} is the maximum transmit power level of the SU_i transmitter, G_{SU_i} is the channel gain of the SU_i link and N_{SU_i} is the noise power level of the SU_i receiver. The channel gain parameters G_{SU_i} and the noise power levels N_{SU_i} are considered to be known to the CR network and not changing in time. An observation necessary for tackling this problem is that the G_{I_i} gains normalized to I_{th} are adequate for defining the interference constraint. Therefore, the new version of (3b), will be:

$$\sum_{i=1}^{N} G_{I_i}^{norm} P_{SU_i} \leq 1 \tag{4}$$

where $G_{I_i}^{norm} = \frac{G_{I_i}}{I_{th}}$.

This optimization problem is convex and using the Karush-Kuhn-Tucker (KKT) approach a capped multilevel waterfilling (CMP) solution is obtained [18] for each SU_i of the closed form:

$$P_{SU_i}^* = \begin{cases} P_i^{max} & \text{if } \frac{1}{\lambda G_{I_i}^{norm}} - \frac{N_{SU_i}}{G_{SU_i}} \geq P_i^{max} \\ 0 & \text{if } \frac{1}{\lambda G_{I_i}^{norm}} - \frac{N_{SU_i}}{G_{SU_i}} \leq 0 \\ \frac{1}{\lambda G_{I_i}^{norm}} - \frac{N_{SU_i}}{G_{SU_i}} & \text{otherwise} \end{cases} \tag{5}$$

where λ is the KKT multiplier of the interference constraint (4) and which can be determined as presented in [18]. Once, all the parameters of the optimization problem are established, the aforementioned analytical solution can be directly calculated. In the following sections, we deal with the learning of the unknown parameters described in the constraint (4).

3   Cooperative Modulation and Coding Classification

A general description of cooperative SS is that each SU performs a SS technique independently, forwards its observation to the CBS via a control channel and finally the CBS using a fusion rule combines this information to get to a decision. In this paper, a hard decision fusion of observations obtained by MCC is considered using a plurality voting system [19]. Based on this voting system, the CBS collects all the MCC feedback over the n_{th} sensing period and decides the MCS of the PU, denoted as MCS^n. Let $C = \{c_1, .., c_K\}$ denote the set of the MCS candidates of the ACM protocol, which are considered to be equiprobable, K the number of elements of this set and V_{c_j} the vote tally associated with the class c_j.

During the voting procedure, the CBS first gathers the votes of the n_{th} period, which in our case are the $MCS_1^n, ..., MCS_N^n$ and support elements of the class set C. All the votes are of same importance and no use of weight factors is made. With every vote MCS_i^n, the CBS increases by one the vote tally V_{c_j} of the c_j class supported by this vote. After casting every vote of the n_{th} period to the corresponding vote tally, the CBS identifies the MCS^n as:

$$MCS^n = \arg\max_{c_j \in C} V_{c_j}. \tag{6}$$

Even though plurality voting is a simple and not sophisticated method which elects the MCS value that appears more often than all of the others, it produces the correct voting output under the condition that some SUs have sensing channels of moderate quality. Its equivalent voting system for binary data fusion, the majority one, has been used by the research community to improve the detection and false alarm probabilities with satisfactory results.

4   Interference Hyperplane Learning

From here on, the equality extreme of the constraint (4) will be referred to as the interference hyperplane. In this section, a binary search probing method is described for estimating the interference hyperplane. First of all, a binary indicator is defined which shows whether the CR network is generating I_{PU} above or below the I_{th} based on the MCS^n fusion output of the MCS_i^n observations. Whenever the CBS detects a deterioration of the MCS from the $(n-1)_{th}$ to the n_{th} period, the indicator I^n changes state as shown below:

$$I^n = \begin{cases} 1 \text{ if } MCS^n \neq MCS^{n-1} \\ 0 \text{ if } MCS^n = MCS^{n-1} \end{cases}. \tag{7}$$

In addition, the feasible set of this problem is defined as $\Omega^N = \{\mathbf{P_{SU}} | 0 \leq P_{SU_i} \leq P_i^{max}, i = 1, \ldots, N\}$, an N-dimensional rectangle with 2^N corners. The objective of this section is to find a geometric method for determining this hyperplane which crosses Ω^N. The only exploitable feedback of this method is the I^n which specifies whether the SU power allocation $\mathbf{P_{SU}}$, chosen by the CBS, just before the beginning of the n_{th} period causes or not harmful interference. In other words, if the SU power allocation before the beginning of the n_{th} period is expressed as $\mathbf{P^n_{SU}} = [P^n_{SU_1}, ..., P^n_{SU_N}]$, I^n demonstrates whether this chosen point in Ω^N is above or below the interference hyperplane.

Thus, the main challenge of this method is how to intelligently select a series of points in Ω^N, for which we only know whether they are above or below the desired hyperplane, in order to estimate this hyperplane. Also, this series has to be limited within the N-dimensional rectangle, since the CBS cannot assign power levels beyond this region. Another challenge that has to be taken into account is the total number of these probing/testing points. The more trial points are used, the more MCS deteriorations the CR network likely causes, which is a considerable damage to the PU QoS. So, this geometric method must find the interference hyperplane with the lowest number of trial points possible.

The core idea for solving the problem is the limitation of the feedback I^n. A binary indicator would be ideal to determine a threshold in the 1-dimensional case by using binary search. Still, in the N-dimensional case a binary search-like method must have some kind of directivity to identify the hyperplane-threshold. Hence, the question becomes how can binary search be applied in this scenario. Basically, to detect an N-dimensional plane one has to find N linearly independent points upon it. Furthermore, if each point belongs to a 1-dimensional ordered set, like a line segment, the binary indicator I^n could be used for a binary search upon the set to find this point. Consequently, for this idea to work, N line segments which cross the hyperplane need to be found and with the lowest number of trial points possible.

To locate N line segments crossing the hyperplane, a number of end points need to be known with some of them below the N-dimensional plane and the rest above it. Considering that any combination of points from different sides creates line segments which cross the hyperplane, if points above and below the N-dimensional plane belong respectively to groups A and B and N_A and N_B are the number of points in these groups, then the number of possible line segments is $N_A N_B$. As mentioned before, the required number of line segments is N, but since the lowest number of trial points possible is demanded the problem is to find N_A and N_B points minimizing $N_A + N_B$ while $N_A N_B \geq N$.

Taking into account some facts from the nature of this problem, the aforementioned end point search can be simplified. Given that the interference hyperplane crosses Ω^N, there is always a known point below this N-dimensional plane, the $[0, ..., 0]$, and one above it, the $[P_1^{max}, ..., P_N^{max}]$. So, in the worst case scenario, $N - 1$ more points are needed to define N line segments crossing the hyperplane. To simplify the end point search, it is proposed to examine randomly the corners of the Ω^N. After these segments are found, binary searches are performed on each one of them so as to detect the N intersection points of the line segments and hence the interference hyperplane.

A detailed description of the binary search method on a line segment with arbitrary end points should also be given. Assuming 2 points, $\mathbf{p_1}$ and $\mathbf{p_2}$, in the N-dimensional space, every point $\mathbf{p}(\theta)$ lying on the line segment defined by them is expressed using the parametric equation $\mathbf{p}(\theta) = \theta\mathbf{p_1} + (1-\theta)\mathbf{p_2}$, where $\theta \in [0,1]$. So, basically the binary search is performed within the θ region $[0,1]$.

Once, the intersection points of the line segments and the interference hyperplane are estimated the G_{I_i} gains normalized to I_{th} can be found as the solution of an $N \times N$ system using the equality of the constraint (4):

$$
\begin{bmatrix} G_{I_1}^{norm} \\ G_{I_2}^{norm} \\ \vdots \\ G_{I_N}^{norm} \end{bmatrix} = \begin{bmatrix} \mathbf{P_1^{cross}} \\ \mathbf{P_2^{cross}} \\ \vdots \\ \mathbf{P_N^{cross}} \end{bmatrix}^{-1} \begin{bmatrix} 1 \\ 1 \\ \vdots \\ 1 \end{bmatrix} \tag{8}
$$

where $\mathbf{P_i^{cross}}$, $i = 1, \ldots, N$, are the intersection points as row vectors.

Algorithm 1. Interference hyperplane learning geometric algorithm

---

Sense MCS^0
$n = 1$
Transmit $\mathbf{P_{SU}^1} = [P_1^{max}, ..., P_N^{max}]$
Sense MCS^1
if $I^1 = 0$ **then**
    Let SUs transmit at maximum
else
    repeat
        $n = n + 1$
        Transmit at $\mathbf{P_{SU}^n}$, a random corner point of Ω^N
        Sense MCS^n and cast point $\mathbf{P_{SU}^n}$ to group A or B
    until $N_A N_B \geq N$
    Combine points in A and B to create line segments
    for $k = 1, \ldots, N$ **do**
        Select a line segment with endpoints $\mathbf{P_A^k} \in A$ and $\mathbf{P_B^k} \in B$
        repeat
            $n = n + 1$
            Transmit at $\mathbf{P_{SU}^n}$, the midpoint of $\mathbf{P_A^k}$ and $\mathbf{P_B^k}$
            Sense MCS^n
            if $I^n = 0$ **then**
                $\mathbf{P_B^k} = \mathbf{P_{SU}^n}$
            else
                $\mathbf{P_A^k} = \mathbf{P_{SU}^n}$
            end if
        until $\|\mathbf{P_A^k} - \mathbf{P_B^k}\| \leq \epsilon$
        Define $\mathbf{P_k^{cross}}$ as the midpoint of $\mathbf{P_A^k}$ and $\mathbf{P_B^k}$
    end for
    Calculate normalized $G_{I_i}^{norm}$ using (8)
end if

Additionally, it is necessary to determine the maximum probing/testing points needed to detect the intersection points and thus the interference hyperplane, since it was explained that a large number of probing/testing points could degrade the PU QoS. Supposing that each binary search is performed with accuracy ϵ, it is well known that the maximum attempts for each line segment of length d_k, where $k = 1, \ldots, N$, are $\lceil log_2(\frac{d_k}{\epsilon}) \rceil$. Even though the lengths d_k cannot be precisely estimated, because a random selection of corner points is performed and so the line segments do not have a standard length, an upper bound can be derived for the total binary search attempts of the procedure. The maximum length a line segment can have in Ω^N, d_{max}, is of the diagonal defined by the points $[0, \ldots, 0]$ and $[P_1^{max}, \ldots, P_N^{max}]$ and calculated as $\sqrt{\sum_{i=1}^{N}(P_i^{max})^2}$. Therefore, the following

$$\sum_{k=1}^{N}\lceil log_2(\frac{d_k}{\epsilon}) \rceil \leq N\lceil log_2(\frac{d_{max}}{\epsilon}) \rceil \tag{9}$$

holds and presents an upper boundary of $\mathcal{O}(Nlog_2(N))$ performing trials. This result proves the scalability of this geometric algorithm, presented in Algo. 1, which can be used even when the SUs of the CR network are large in number.

A simple example of how this geometric algorithm progresses in time for $N = 2$ SUs is given in Fig. 2. The binary searches were performed on the line segments OB and BC in order to find their intersection points with the interference line, E and D. Once, these points are obtained it is easy to define the interference line.

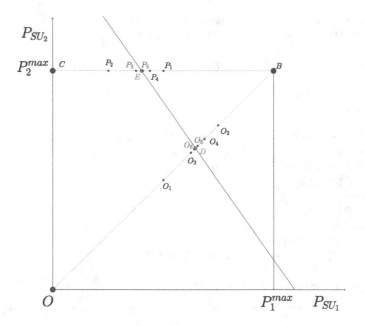

Fig. 2. A 2D graphical example of the geometric algorithm

5   Results

Following, the performance of the aforementioned geometric algorithm and the probing progress of each P_{SU_i} vs time are presented. For testing the performance, a case of CR network with $N = 3$ SUs was considered. The following diagrams in Fig. 3 represent geometrically the probing/testing point coordinates which gradually converge to the coordinates of the intersection points $\mathbf{P}_i^{cross}$.

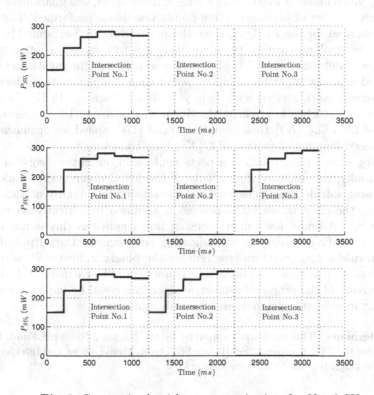

Fig. 3. Geometric algorithm progress in time for $N = 3$ SUs

As seen in Fig. 3, testing power allocation points are tried with a time step of 200ms and it is considered that $P^{max} = 300mW$ and $T_S = 100ms$ for every SU. Also, at each time step the output of the cooperative MCC process is assumed to be correctly estimated and thus the binary indicator I^n contributing in the geometric algorithm is always accurate. An important aspect of the algorithm is the convergence time which for the examined CR network is $3200ms$. After the interference hyperplane learning process finishes, the CBS is able to directly find the optimal power allocation based on (5).

6   Conclusions

In this paper, a centralized Power Control (PC) scheme aided by an interference hyperplane learning algorithm is proposed to allow a CR network and a PU coexist in a frequency band. The leading idea of this algorithm is to exploit the SS feedback from the cooperative MCC procedure and perform consecutive binary searches in the power allocation set to define points of the interference hyperplane and thus the hyperplane itself. The algorithm is of $\mathcal{O}(Nlog_2(N))$ time complexity, which makes it ideal even for large CR networks, and guarantees that the minimum number of probing/testing points possible is performed. The last remark is essential for the PU QoS, since the more trials are performed by the CBS for the hyperplane learning, the more likely it is to surpass the interference hyperplane and deteriorate the PU MCS. Moreover, it must be mentioned that the proposed learning algorithm achieves better time complexity than the ones used in previous work related with binary feedback learning. In [15], a slow convergence rate stochastic gradient algorithm was utilized, [16] suggested an algorithm of $\mathcal{O}(N^2log_2(N))$ time complexity and [17] applied an optimization technique based learning algorithm of $\mathcal{O}(N^2)$ time complexity.

Improving some of the problem aspects could lead our future work in this subject. Initially, an enhanced fusion rule of the MCC observations could be suggested using soft decision rules based on the sensing link quality of each SU. Furthermore, the cooperative MCC process is assumed to perfectly recognize the PU MCS, but under low quality sensing link conditions this is not true. An introduction of a reliability factor indicating how accurate the output of the MCS fusion rule is and therefore how reliable the binary indicator I^n is could be useful so that binary searches using uncertainty could be carried out. Finally, an online version of the proposed geometric algorithm could be implemented to tackle fading interference channels and not only static ones.

Acknowledgments. This work was supported by the National Research Fund, Luxembourg under the CORE project "SeMIGod: SpEctrum Management and Interference mitiGation in cOgnitive raDio satellite networks".

References

1. Zhao, Q., Sadler, B.: A Survey of Dynamic Spectrum Access. IEEE Signal Processing Magazine, 79–89 (2007)
2. Mitola, J.: Cognitive radio an integrated agent architecture for software defined radio, Ph.D. dissertation, KTH Royal Institute of Technology Stockholm, Stockholm, Sweden (2000)
3. Tsakmalis, A., Chatzinotas, S., Ottersten, B.: Modulation and coding classification for adaptive power control in 5g cognitive communications. In: IEEE International Workshop on Signal Processing Advances in Wireless Communications (SPAWC) (2014)

4. Tsakmalis, A., Chatzinotas, S., Ottersten, B.: Automatic modulation classification for adaptive power control in cognitive satellite communications. In: 7th Advanced Satellite Multimedia Systems Conference (ASMS) and 13th Signal Processing for Space Communications Workshop (SPSC) (2014)
5. Ramkumar, B.: Automatic Modulation Classification for Cognitive Radios Using Cyclic Feature Detection. IEEE Circuits and Systems Magazine, 27–45 (2009)
6. Vapnik, V.N.: The Nature of Statistical Learning Theory. Springer (1999)
7. Xia, T., Wu, H.: Novel Blind Identification of LDPC Codes Using Average LLR of Syndrome a Posteriori Probability. IEEE Transactions on Signal Processing, 632–640 (2014)
8. Moosavi, R., Larsson, E.: A Fast scheme for blind identification of channel codes. In: IEEE Global Telecommunications Conference (GLOBECOM), pp. 1–5 (2011)
9. Mitliagkas, I., Sidiropoulos, N., Swami, A.: Convex approximation-based joint power and admission control for cognitive underlay networks. In: International Wireless Communications and Mobile Computing Conference (IWCMC), pp. 28–32 (2008)
10. Zhang, R., Liang, Y.C.: Exploiting Multi-Antennas for Opportunistic Spectrum Sharing in Cognitive Radio Networks. IEEE Journal of Selected Topics in Signal Processing, 88–102 (2008)
11. Marques, A., Wang, X., Giannakis, G.: Dynamic Resource Management for Cognitive Radios Using Limited-Rate Feedback. IEEE Transactions on Signal Processing, 3651–3666 (2009)
12. Mitliagkas, I., Sidiropoulos, N., Swami, A.: Joint Power and Admission Control for Ad-Hoc and Cognitive Underlay Networks: Convex Approximation and Distributed Implementation. IEEE Transactions on Wireless Communications, 4110–4121 (2011)
13. Dall'Anese, E., Kim, S., Giannakis, G., Pupolin, S.: Power Control for Cognitive Radio Networks Under Channel Uncertainty. IEEE Transactions on Wireless Communications, 3541–3551 (2011)
14. Bajaj, I., Gong, Y.: Cross-channel estimation using supervised probing and sensing in cognitive radio networks. In: IEEE International Conference on Communications (ICC), pp. 1–5 (2011)
15. Banister, B.C., Zeidler, J.R.: A Simple Gradient Sign Algorithm for Transmit Antenna Weight Adaptation With Feedback. IEEE Transactions on Signal Processing, 1156–1171 (2003)
16. Noam, Y., Goldsmith, A.J.: The One-Bit Null Space Learning Algorithm and Its Convergence. IEEE Transactions on Signal Processing, 6135–6149 (2013)
17. Xu, J., Zhang, R.: Energy Beamforming With One-Bit Feedback. IEEE Transactions on Signal Processing, 5370–5381 (2014)
18. Zhang, L., Liang, Y.C., Xin, Y.: Joint Beamforming and Power Allocation for Multiple Access Channels in Cognitive Radio Networks. IEEE Journal on Selected Areas in Communications, 617–629 (1994)
19. Parhami, B.: Voting Algorithms. IEEE Transactions on Reliability, 617–629 (1994)

Symbol Based Precoding in the Downlink of Cognitive MISO Channel

Maha Alodeh$^{(\boxtimes)}$, Symeon Chatzinotas, and Björn Ottersten

Interdisciplinary Centre for Security, Reliability and Trust,
University of Luxembourg, 4, Alphonse Weicker, 2721 Luxembourg, Luxembourg
{maha.alodeh,symeon.chatzinotas,bjorn.ottersten}@uni.lu

Abstract. This paper proposes symbol level precoding in the downlink of a MISO cognitive system. The new scheme tries to jointly utilize the data and channel information to design a precoding that minimizes the transmit power at a cognitive base station (CBS); without violating the interference temperature constraint imposed by the primary system. In this framework, the data information is handled at symbol level which enables the characterization the intra-user interference among the cognitive users as an additional source of useful energy that should be exploited. A relation between the constructive multiuser transmissions and physical-layer multicast system is established. Extensive simulations are performed to validate the proposed technique and compare it with conventional techniques.

Keywords: Constructive interference · Underlay cognitive radio · MISO system

1 Introduction

The combination of the spectrum scarcity and congestion has motivated researchers to propose more innovative techniques to tackle these challenges. Fixed spectrum allocation techniques assign certain bands to certain applications, which may no longer efficiently used [1]. Solving the problem would require changing the regulations which is a complicated and lengthy procedure. With that in mind, the paradigm of cognitive radios has been proposed as a promising agile technology that can revolutionize the future of telecommunication by "breaking the gridlock of the wireless spectrum" [2]. The key idea of their implementation is to allow opportunistic transmissions to share the wireless medium. Thus, two initial hierarchical levels have been defined: primary level and cognitive level (the users within each level are called primary users (PU) and cognitive users (CU) respectively). The interaction between these two levels is determined

M. Alodeh—This work is supported by Fond National de la Recherche Luxembourg (FNR) projects, project Smart Resource Allocation for Satellite Cognitive Radio (SRAT-SCR) ID:4919957 and Spectrum Management and Interference Mitigation in Cognitive Radio Satellite Networks SeMiGod.

© Institute for Computer Sciences, Social Informatics and Telecommunications Engineering 2015
M. Weichold et al. (Eds.): CROWNCOM 2015, LNICST 156, pp. 370–380, 2015.
DOI: 10.1007/978-3-319-24540-9_30

by the agility of the cognitive level and the predefined constraints imposed by the primary level [3]. Overlay, underlay and interweave are three general implementations which regulate the coexistence terms of both systems. The first two implementations allow simultaneous transmissions, which leads to better spectrum utilization in comparison to the last one, which allocates the spectrum to the cognitive system by detecting the absence of the primary one [4].

The form of integration in this work is defined by cooperation between the two levels in the cognitive interference channel. The cooperation can aid the primary network to satisfy the quality of service (QoS) or enhance the rate of its own users by backhauling its data through the cognitive system[5]-[8]; CBSs can operate as relays for primary messages and as regular base stations to serve their cognitive users. The cognitive system benefits by providing a service to its users. This kind of cognitive implementation fits with practical overlay cognitive definition, as the PU is being served from both the PS and the CBSs by performing relaying between them to make primary data accessible by the CBS. Sometimes the primary symbols are not available to the cognitive system, as a result the cognitive system needs to take the sufficient precautions to protect the primary system from the interference created by its own transmissions. It should be noted that we assume that the CBSs are equipped with multiple antennas to handle multi-user transmissions, and to enable interference mitigation.

The conventional look at interference can be shifted from a degradation factor into a favorable one if we handle the transmitted data frame at symbol level. At this level, the interference can be classified into: constructive and destructive ones. This classification is initially proposed in [12]; instead of fully inverting the channel to grant zero interference among the spatial streams, the proposed precoding suggests keeping the constructive interference while removing the destructive part by partial channel inversion. This technique is proven to outperform the traditional zero forcing precoding. A more advanced technique is proposed in [13], where an interference rotation is examined to make the interference constructive for all users. Moreover, a modified maximum ratio transmissions technique that performs unitary rotations to create constructive interference among the interfering multiuser streams is proposed [15]. Furthermore, a connection between symbol based constructive interference precoding and PHY multicast is established in [15]-[16].

In this work, we utilize the symbol level precoding in underlay MISO cognitive radio scenarios. We shape the interference between the cognitive users to provide constructive characteristics without violating the interference temperature constraints on the primary receivers.

2 System and Signals Model

We consider a cognitive radio network which shares the spectrum resource with a primary network in the underlay mode as fig. (1). The primary network consists of a primary base station (PBS), equipped with N_p antennas, serving a single primary user. The cognitive network has a single CBS, equipped with M antennas, serving K CU. Each CU is equipped with a single antenna. Throughout this

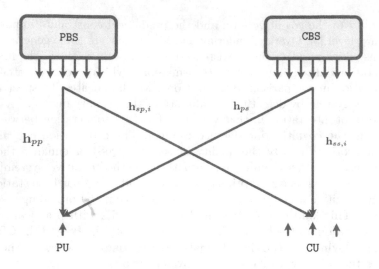

Fig. 1. System model

paper, we consider that $K \leq M - 1$ and that the primary user is equipped with a single antenna. Due to the sharing of the same frequency band, the received signal at the primary user is interfered by the signals transmitted from CBS. Similarly, the received signals at the CUs are interfered by the signal transmitted from the PBS.

Assume that in one time slot, a block of information symbols $\mathbf{d} = [d_1, d_2, ..., d_K]^T$ are sent from the CBS in which d_k, $k = 1, ..., K$ is the desired signal for user k. We assume that $\mathbf{d}$ contains uncorrelated unit-power M-PSK entries. With a proper beamforming (which will be specified later), the transmit signal is given by

$$\mathbf{x} = \mathbf{W}\mathbf{d} \tag{1}$$

where $\mathbf{W} = [\mathbf{w}_1, \mathbf{w}_2, ..., \mathbf{w}_K]$ denotes the transmit precoding matrix for the cognitive system while $\mathbf{w}_k \in \mathbb{C}^{M \times 1}$ denotes the beamforming vector for k^{th} CU. The received signal at the kth user, denoted by $y_{s,k}$, is given by

$$y_{s,k} = \mathbf{h}_{ss,k}\mathbf{w}_k d_k + \sum_{j \in K, j \neq k} \mathbf{h}_{ss,k}\mathbf{w}_j d_j + \mathbf{h}_{sp,k}\mathbf{g}^p d_p + n_k, \tag{2}$$

and the received signal at PU's receiver is given by

$$y_p = \mathbf{h}_{pp}\mathbf{g}d_p + \sum_{j \in K} \mathbf{h}_{ps}\mathbf{w}_j d_j + n \tag{3}$$

where $\mathbf{h}_{ss,k} \in \mathbb{C}^{1 \times M}$ and $\mathbf{h}_{sp,k} \in \mathbb{C}^{1 \times N_p}$ are the channels between the CBS and the PBS respectively and the kth CU. While $\mathbf{h}_{pp}$ and $\mathbf{h}_{ps}$ denote the channel between the PBS and PU, CBS and PU respectively. The transmitted power of the primary user is denoted by p_p, $\mathbf{g} \in \mathbb{C}^{N_p \times 1}$ represents the precoding vector used by the PBS, and d_p represents the transmitted symbol from the PBS and it is not available at CBS. Finally, $n_k \sim \mathcal{CN}(0, \sigma^2)$ and $n \sim \mathcal{CN}(0, \sigma^2)$ are additive i.i.d. complex Gaussian noise with zero mean and variance σ_k^2 at the kth CU and PU respectively. The channel state information (CSI) $\mathbf{h}_{ps}$ and $\mathbf{h}_{ss,j}$ are available at the CBS.

3   Constructive Interference Definition

The interference is a random deviation which can move the desired constellation point in any direction. To address this problem, the power of the interference has been used in the past to regulate its effect on the desired signal point. The interference among the multiuser spatial streams leads to deviation of the received symbols outside of their detection region. However, in symbol level precoding (e.g. M-PSK) this interference pushes the received symbols further into the correct detection region and, as a consequence it enhances the system performance. Therefore, the interference can be classified into constructive or destructive based on whether it facilitates or deteriorates the correct detection of the received symbol. For BPSK and QPSK scenarios, a detailed classification of interference is discussed thoroughly in [12]. In this section, we describe the required conditions to have constructive interference for any M-PSK modulation.

3.1   Constructive Interference Definition

Assuming both the data symbols and CSI are available at the CBS, the unit-power created interference from the kth data stream on j^{th} user can be formulated as:

$$\rho_{jk} = \frac{\mathbf{h}_{ss,j}\mathbf{w}_k}{\|\mathbf{h}_{ss,j}\|\|\mathbf{w}_k\|}. \tag{4}$$

Since the adopted modulations are M-PSK ones, a definition for constructive interference can be stated as

Lemma 1 *[16]. For any M-PSK modulated symbol d_k, it is said to receive constructive interference from another simultaneously transmitted symbol d_j which is associated with $\mathbf{w}_j$ if and only if the following inequalities hold*

$$\angle d_j - \frac{\pi}{M} \le \arctan\left(\frac{\mathcal{I}\{\rho_{jk}d_k\}}{\mathcal{R}\{\rho_{jk}d_k\}}\right) \le \angle d_j + \frac{\pi}{M}, \tag{5}$$

$$\mathcal{R}\{d_k\}.\mathcal{R}\{\rho_{jk}d_j\} > 0, \mathcal{I}\{d_k\}.\mathcal{I}\{\rho_{jk}d_j\} > 0. \tag{6}$$

Corollary 1 *[16]. The constructive interference is mutual. If the symbol d_j constructively interferes with d_k, then the interference from transmitting the symbol d_k is constructive to d_j.*

4 Constructive Interference Exploitation

4.1 Relaxed Interference Constraint

The precoding aims at exploiting the constructive interference among the cognitive users without violating the interference temperature constraint imposed by the primary system $\mathcal{I}_{th}$. The optimization can be formulated as

$$\mathbf{w}_1, ..., \mathbf{w}_K = \underset{\mathbf{w}_1,...,\mathbf{w}_K}{\arg\min} \quad \| \sum_{k=1}^{K} \mathbf{w}_k d_k \|^2 \tag{7}$$

$$s.t.\ \mathcal{C}_1 : \angle(\mathbf{h}_{ss,j} \sum_{k=1}^{K} \mathbf{w}_k d_k) = \angle(d_j), \forall j \in K$$

$$\mathcal{C}_2 : \frac{\|\mathbf{h}_{ss,j} \sum_{k=1}^{K} \mathbf{w}_k d_k\|^2}{\sigma^2 + \|\mathbf{h}_{sp,j}\mathbf{g}\|^2} \geq \zeta_j \quad, \forall j \in K$$

$$\mathcal{C}_3 : \|\mathbf{h}_{ps} \sum_{k=1}^{K} \mathbf{w}_k d_k\|^2 \leq \mathcal{I}_{th}$$

The first two sets of constraints $\mathcal{C}_1$ and $\mathcal{C}_2$ grant the reception of the data symbols with certain SNR level ζ_j. The third constraint $\mathcal{C}_3$ is to protect the PU from the cognitive systems transmissions. In order to solve (7), we formulate it by using $\mathbf{x} = \sum_{k=1}^{K} \mathbf{w}_k d_k$ as the following

$$\mathbf{x} = \underset{\mathbf{x}}{\arg\min} \quad \|\mathbf{x}\|^2 \tag{8}$$

$$s.t. \quad \mathcal{C}_1 : \quad \frac{\mathbf{h}_{ss,j}\mathbf{x} + \mathbf{x}^H\mathbf{h}_{ss,j}^H}{2} = \sqrt{\psi_i \zeta_j} \mathcal{R}\{d_j\} \quad, \forall j \in K$$

$$\mathcal{C}_2 : \quad \frac{\mathbf{h}_{ss,j}\mathbf{x} - \mathbf{x}^H\mathbf{h}_{ss,j}^H}{2i} = \sqrt{\psi_i \zeta_j} \mathcal{I}\{d_j\} \quad, \forall j \in K$$

$$\mathcal{C}_3 : \quad \|\mathbf{h}_{ps}\mathbf{x}\|^2 \leq \mathcal{I}_{th}.$$

where $\psi_j = \sigma^2 + \|\mathbf{h}_{sp,j}\mathbf{g}\|^2$. To solve the problem, the corresponding Lagrange function can be expressed as

$$\mathcal{L}(\mathbf{x}) = \|\mathbf{x}\|^2 \tag{9}$$

$$+ \sum_j \mu_j \left(-0.5i(\mathbf{h}_{ss,j}\mathbf{x} - \mathbf{x}^H\mathbf{h}_{ss,j}^H) - \sqrt{\psi_j \zeta_j} \mathcal{I}\{d_j\} \right)$$

$$+ \sum_j \alpha_j \left(0.5(\mathbf{h}_{ss,j}\mathbf{x} + \mathbf{x}^H\mathbf{h}_{ss,j}^H) - \sqrt{\psi_j \zeta_j} \mathcal{R}\{d_j\} \right)$$

$$+ \lambda \left(\mathbf{x}^H\mathbf{h}_{ps}^H\mathbf{h}_{ps}\mathbf{x} - \mathcal{I}_{th} \right).$$

The KKT conditions can be derived as

$$\frac{d\mathcal{L}(\mathbf{x},\mu_j,\alpha_j,\lambda)}{d\mathbf{x}^*} = \mathbf{x} + \sum 0.5i\mu_j \mathbf{h}_{ss,j}^H + \sum_j 0.5\alpha_j \mathbf{h}_{ss,j}^H + \lambda \mathbf{h}_{ps}^H \mathbf{h}_{ps} \mathbf{x}$$

$$\frac{d\mathcal{L}(\mathbf{x},\mu_j,\alpha_j,\lambda)}{d\mu_j} = -0.5i(\mathbf{h}_{ss,j}\mathbf{x} - \mathbf{x}^H \mathbf{h}_{ss,j}^H) - \sqrt{\psi_j \zeta_j} \mathcal{I}\{d_j\}, \forall j \in K$$

$$\frac{d\mathcal{L}(\mathbf{x},\mu_j,\alpha_j,\lambda)}{d\alpha_j} = 0.5(\mathbf{h}_{ss,j}\mathbf{x} + \mathbf{x}^H \mathbf{h}_{ss,j}^H) - \sqrt{\psi_j \zeta_j} \mathcal{R}\{d_j\}, \forall j \in K$$

$$\frac{d\mathcal{L}(\mathbf{x},\mu_j,\alpha_j,\lambda)}{d\lambda} = \left(\mathbf{x}^H \mathbf{h}_{ps}^H \mathbf{h}_{ps} \mathbf{x} - \mathcal{I}_{th} \right) \tag{10}$$

By equating $\frac{d\mathcal{L}(\mathbf{x},\mu_j,\alpha_j,\lambda)}{d\mathbf{x}^*}$ to zero, we can formulate $\mathbf{x}$ as the following expression

$$\mathbf{x} = (\mathbf{I} + \lambda \mathbf{h}_{ps}^H \mathbf{h}_{ps})^{-1} \left(\sum_j 0.5i\mu_j \mathbf{h}_{ss,j}^H + \sum_j 0.5\alpha_j \mathbf{h}_{ss,j}^H \right). \tag{11}$$

By substituting (11) in the set of (10) to form the set of equations (12), we can find the solution of λ, α_j and μ_j that satisfies the constraints.

$$\left(\sum_j 0.5\mu_j^* \mathbf{h}_{ss,j} + \sum_j 0.5\alpha_j^* \mathbf{h}_{ss,j} \right)(\mathbf{I} + \lambda \mathbf{h}_{ps}^H \mathbf{h}_{ps})^{-1} \quad \mathbf{h}_{ps}^H \mathbf{h}_{ps} \quad (\mathbf{I} + \lambda \mathbf{h}_{ps}^H \mathbf{h}_{ps})^{-1} \left(\sum_j 0.5i\mu_j \mathbf{h}_{ss,j}^H + \sum_j 0.5\alpha_j \mathbf{h}_{ss,j}^H \right) \leq \mathcal{I}_{th}$$

$$-0.5i(\mathbf{h}_{ss,1}(\mathbf{I} + \lambda \mathbf{h}_{ps}^H \mathbf{h}_{ps}^H)^{-1} \left(\sum_j 0.5i\mu_j \mathbf{h}_{ss,j}^H + \sum_j 0.5\alpha_j \mathbf{h}_{ss,j}^H \right) + 0.5i(\sum_j \mu_j \mathbf{h}_{ss,j} + \sum_j 0.5\alpha_j \mathbf{h}_{ss,j})(\mathbf{I} + \lambda \mathbf{h}_{ps}^H \mathbf{h}_{ps})^{-1} \mathbf{h}_{ss,1} = \sqrt{\psi_1 \zeta_1} \mathcal{I}\{d_1\}$$

$$0.5(\mathbf{h}_{ss,j}(\mathbf{I} + \lambda \mathbf{h}_{ps}^H \mathbf{h}_{ps}^H)^{-1} \left(\sum_j 0.5i\mu_j \mathbf{h}_{ss,j}^H + \sum_j 0.5\alpha_j \mathbf{h}_{ss,j}^H \right) + 0.5(\sum_j \mu_j \mathbf{h}_{ss,j} + \sum_j 0.5\alpha_j \mathbf{h}_{ss,j})(\mathbf{I} + \lambda \mathbf{h}_{ps}^H \mathbf{h}_{ps})^{-1} \mathbf{h}_{ss,j} = \sqrt{\psi_1 \zeta_1} \mathcal{R}\{d_1\}$$

$$\vdots$$

$$-0.5i(\mathbf{h}_{ss,K}(\mathbf{I} + \lambda \mathbf{h}_{ps}^H \mathbf{h}_{ps}^H)^{-1} \left(\sum_j 0.5i\mu_j \mathbf{h}_{ss,j}^H + \sum_j 0.5\alpha_j \mathbf{h}_{ss,j}^H \right) + 0.5i(\sum_j \mu_j \mathbf{h}_j + \sum_j 0.5\alpha_j \mathbf{h}_{ss,j})(\mathbf{I} + \lambda \mathbf{h}_{ps}^H \mathbf{h}_{ps}^H)^{-1} \mathbf{h}_{ss,K} = \sqrt{\psi_K \zeta_K} \mathcal{I}\{d_K\}$$

$$0.5(\mathbf{h}_{ss,K}(\mathbf{I} + \lambda \mathbf{h}_{ps}^H \mathbf{h}_{ps}^H)^{-1} \left(\sum_j 0.5i\mu_j \mathbf{h}_{ss,j}^H + \sum_j 0.5\alpha_j \mathbf{h}_{ss,j}^H \right) + 0.5(\sum_j \mu_j \mathbf{h}_{ss,j} + \sum_j 0.5\alpha_j \mathbf{h}_{ss,j})(\mathbf{I} + \lambda \mathbf{h}_{ps}^H \mathbf{h}_{ps}^H)^{-1} \mathbf{h}_{ss,K} = \sqrt{\psi_K \zeta_1} \mathcal{R}\{d_K\} \tag{12}$$

4.2 Zero Interference Constraint

If the PU cannot handle any interference, the cognitive transmissions should be in the null space of the channel between CBS and PU. The null space can be defined as

$$\mathbf{\Pi}_{\perp \mathbf{h}_{ps}} = \mathbf{I} - \frac{\mathbf{h}_{ps}^H \mathbf{h}_{ps}}{\|\mathbf{h}_{ps}\|^2}. \tag{13}$$

We design the output vector $\mathbf{x}$ to span the null space of $\mathbf{h}_{ps}$ as the following

$$\mathbf{x} = \mathbf{\Pi}_{\perp \mathbf{h}_{ps}} \hat{\mathbf{x}}. \tag{14}$$

$$\arg \min \mathbf{w}_1, \mathbf{w}_2, ..., \mathbf{w}_K \quad \left\| \sum_{k=1}^{K} \mathbf{w}_k d_k \right\|^2$$

$$s.t. \; \mathcal{C}1 : \angle(\mathbf{h}_j \sum_{k=1}^{K} \mathbf{w}_k d_k) = \angle(d_j), \forall j \in K$$

$$\mathcal{C}2 : \left\| \mathbf{h}_j \sum_{k=1}^{K} \mathbf{w}_k d_k \right\|^2 \geq \sigma^2 \zeta_j \quad , \forall j \in K$$

$$\mathcal{C}3 : \left\| \mathbf{h}_{ps} \sum_{k=1}^{K} \mathbf{w}_k d_k \right\|^2 = 0.$$

The previous optimization can be written as The Lagrange function of this optimization problem

$$\mathcal{L}(\hat{\mathbf{x}}) = \|\hat{\mathbf{x}}\|^2 \tag{15}$$
$$+ \sum_j \hat{\mu}_j \left(-0.5i(\mathbf{h}_{ss,j}\hat{\mathbf{x}} - \hat{\mathbf{x}}^H \mathbf{h}_{ss,j}^H) - \sqrt{\psi_j \zeta_j} \mathcal{I}\{d_j\} \right)$$
$$+ \sum_j \hat{\alpha}_j \left(0.5(\mathbf{h}_{ss,j}\hat{\mathbf{x}} + \hat{\mathbf{x}}^H \mathbf{h}_{ss,j}^H) - \sqrt{\psi_j \zeta_j} \mathcal{R}\{d_j\} \right).$$

The KKT condition can be written as

$$\frac{d\mathcal{L}(\mathbf{x},\mu_j,\alpha_j,\lambda)}{d\hat{\mathbf{x}}^*} = \hat{\mathbf{x}} + \sum 0.5i\mu_j \mathbf{h}_{ss,j}^H + \sum_j 0.5\alpha_j \mathbf{h}_{ss,j}^H + \lambda \mathbf{h}_{ps}^H \mathbf{h}_{ps}\hat{\mathbf{x}}$$

$$\frac{d\mathcal{L}(\mathbf{x},\mu_j,\alpha_j,\lambda)}{d\mu_j} = -0.5i(\mathbf{h}_{ss,j}\hat{\mathbf{x}} - \hat{\mathbf{x}}^H \mathbf{h}_{ss,j}^H) - \sqrt{\psi_j \zeta_j} \mathcal{I}\{d_j\}, \forall j \in K$$

$$\frac{d\mathcal{L}(\mathbf{x},\mu_j,\alpha_j,\lambda)}{d\alpha_j} = 0.5(\mathbf{h}_{ss,j}\hat{\mathbf{x}} + \hat{\mathbf{x}}^H \mathbf{h}_{ss,j}^H) - \sqrt{\psi_j \zeta_j} \mathcal{R}\{d_j\}, \forall j \in K$$

$$\tag{16}$$

The solution for the previous optimization problem can be written as

$$\hat{\mathbf{x}} = \sum_j 0.5i\hat{\mu}_j \mathbf{h}_{ss,j}^H + \sum_j 0.5\hat{\alpha}_j \mathbf{h}_{ss,j}^H \tag{17}$$

where $\hat{\mu}_j, \hat{\alpha}_j$ can be found by solving the set of equation (19). Hence, the final formulation for the solution at zero interference temperature constraint

$$\mathbf{x} = \left(\mathbf{I} - \frac{\mathbf{h}_{ps}^H \mathbf{h}_{ps}}{\|\mathbf{h}_{ps}\|^2} \right) \left(\sum_j 0.5i\hat{\mu}_j \mathbf{h}_{ss,j}^H + \sum_j 0.5\hat{\alpha}_j \mathbf{h}_{ss,j}^H \right) \tag{18}$$

5   Theoretical Upper-Bound

The theoretical upper-bound can be formulated by dropping the phase constraint $\mathcal{C}_1$ of (7). The optimal input covariance $\mathbf{Q}$ can be found by solving the following optimization problem:

$$0.5K\|h_{s1}\|(\sum_k(-\mu_k+\alpha_k i)\|h_{ss,k}\|\rho_{1k} \quad - \quad \sum_k(-\mu_k+\alpha_k i)\|h_{ss,k}\|\rho^*_{1k}) = \sqrt{\psi_1\zeta_1}\mathcal{I}(d_1)$$
$$0.5K\|h_{ss,1}\|(\sum_k(-\mu_k i-\alpha_k)\|h_{ss,k}\|\rho_{1k} \quad + \quad \sum_k(-\mu_k i-\alpha_k)\|h_{ss,k}\|\rho^*_{1k}) = \sqrt{\psi_1\zeta_1}\mathcal{R}(d_1)$$

$$\vdots$$

$$0.5K\|h_{ss,K}\|(\sum_k(-\mu_k+\alpha_k i)\|h_{ss,k}\|\rho_{Kk} - \sum_k(-\mu_k+\alpha_k i)\|h_{ss,k}\|\rho^*_{Kk}) = \sqrt{\psi_K\zeta_K}\mathcal{I}(d_K)$$
$$0.5K\|h_{ss,K}\|(\sum_k(-\mu_k i-\alpha_k)\|h_{ss,k}\|\rho_{Kk} + \sum_k(-\mu_k i-\alpha_k)\|h_{ss,K}\|\rho^*_{Kk}) = \sqrt{\psi_K\zeta_K}\mathcal{R}(d_K) \quad (19)$$

$$\mathbf{Q} = \arg\min_{\mathbf{Q}} \quad tr(\mathbf{Q})$$
$$s.t. \ \mathbf{h}_{ss,j}\mathbf{Q}\mathbf{h}^H_{ss,j} = \psi_j\gamma_j \forall j \in K.$$
$$\mathbf{h}_{ps}\mathbf{Q}\mathbf{h}^H_{ps} \leq \mathcal{I}_{th}. \tag{20}$$

where $\mathbf{Q} = \mathbf{x}\mathbf{x}^H$. This problem resembles the multicast problem [10] with additional interference temperature constraint to suit the constraint imposed by the primary system.

6   Numerical Results

In order to assess the performance of the proposed transmissions schemes, Monte-Carlo simulations of the different algorithms have been conducted to study the performance of the proposed techniques and compare to the state of the art techniques. The adopted channel model is assumed to be as the following

- $\mathbf{h}_{pp} \sim \mathcal{CN}(0, \sigma^2_{pp}\mathbf{1}_{1\times M})$, where $\mathbf{1}_{1\times M}$ is vector of all ones and of size $1 \times M$.
- $\mathbf{h}_{ps} \sim \mathcal{CN}(0, \sigma^2_{ps}\mathbf{1}_{1\times M})$
- $\mathbf{h}_{ss,j} \sim \mathcal{CN}(0, \sigma^2_{ss}\mathbf{1}_{1\times M}), \forall j \in K$
- $\mathbf{h}_{sp,j} \sim \mathcal{CN}(0, \sigma^2_{sp,j}\mathbf{1}_{1\times M}), \forall j \in K$
- To study the performance of the system at the worst case scenario, when all users have a strong channel with respect to its direct and interfering base stations $\sigma^2_{pp} = \sigma^2_{ps} = \sigma^2_{sp,j} = \sigma^2_{ss} = \sigma^2$.

In the figures, we denote the proposed cognitive technique that exploits the constructive interference by (CCIPM), while S denotes the strict interference constraints $\mathcal{I}_{th} = 0$. CBS has 3 antennas and serves 2 cognitive users. We compare the performance of the proposed scheme (CCIPM) to the scheme in [13] tailored to cognitive scenario by solving the following optimization

$$\mathbf{W}_{CCIZF} = \min_{\mathbf{W}} \mathbb{E}\{\|\mathbf{R}_\phi - \mathbf{H}_{ss}\mathbf{W}\mathbf{d}\|^2\}$$
$$s.t. \ \|\mathbf{W}\|^2 \leq P$$
$$\mathbf{h}_{ps}\mathbf{W}\mathbf{W}^H\mathbf{h}^H_{ps} = 0 \tag{21}$$

where $\mathbf{R}_\phi$ is defined in [13]. We utilize the energy efficiency metric to assess the performance of the proposed technique as

$$\eta = \frac{\sum_{j=1}^K R_j}{\|\mathbf{x}\|^2}, \tag{22}$$

R_j denotes the rate of the jth user. In all figures, we depict the performance of constructive interference zero forcing (CCIZF). For the sake of comparison, the transmit power of the CCIZF solutions can be scaled until all users achieve the target rate.

In Fig. 2, the energy efficiency with respect to the average channel is depicted, the used modulation is QPSK and the strict interference constraint imposed by the primary system. We compare the performance of the CCIPM to the theoretical upper-bound and CCIZF. It can be noted that the CCIZF curve saturates at the low-mid SNR regime, while the curves of CCIPM and multicast have higher growth in terms of energy efficiency in the same regime. Moreover, it can be noted that CCIPM outperforms CCIZF at different channel strength values.

In Fig. 3, we depict the energy efficiency at different target rates. For the constructive interference schemes, we assign the target rate with its corresponding MPSK modulation. It can be noted that the theoretical upper bounds for the scenario of the strict and the relaxed interference constraints have the same power consumption at target rate equals to 1 bps/Hz. However, this result does not hold for constructive interference technique. Moreover, it can be noted that the gap between the theoretical bound and the CCIPM is fixed at the both scenario for all target rates.

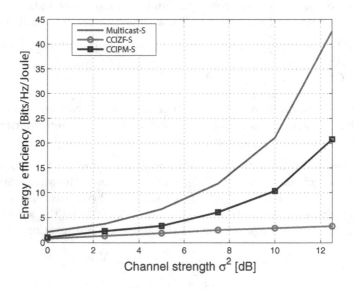

Fig. 2. Energy efficiency vs channel strength. The adopted modulation is QPSK for CCIZF and CCIPM, $\zeta = 4.7712dB$.

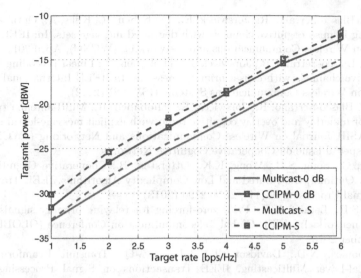

Fig. 3. Transmit power vs target rate. $\sigma^2 = 10dB$

7 Conclusions

In this paper, we propose symbol-level precoding techniques for the downlink of cognitive underlay system. These techniques exploits the availability of the CSI and the data symbols to constructively correlate the transmission for the cognitive users without violating the interference temperature at the primary users. This enables interference exploitation among the cognitive multiuser transmissions assuming M-PSK modulation. The designed precoder aims at minimizing the transmitted power at CBS while granting a certain received SNR at each cognitive users. From the numerical results section, it can be concluded that the CBS consumes less power at the relaxed interference constraints. Finally, a comparison with the theoretical upperbound and the state-of-the art techniques is illustrated.

References

1. Federal Communication Commission: Spectrum Policy Task Force. ET document no. 02–135, November 2002
2. Goldsmith, A., Jafar, S.A., Maric, I., Srinivasa, S.: Breaking spectrum gridlock with cognitive radios: An information theoretic perspective. IEEE **97**(5), 894–914 (2009)
3. Haykin, S.: Cognitive Radio: Brain-Empowered Wireless Communications. IEEE Journal on Selected Areas in Communications **23**, 201–222 (2005)
4. Srinivasa, S., Jafar, S.A.: Soft Sensing and Optimal Power Control for Cognitive Radio. IEEE Transactions on Wireless Communications **9**(12), 3638–3649 (2010)

5. Lv, J., Blasco-Serrano, R., Jorswieck, E., Thobaben, R., Kliks, A.: Optimal beamforming in miso cognitive channels with degraded message sets. In: IEEE Conference on Wireless Communications and Networking (WCNC), April 2012
6. Lv, J., Blasco-Serrano, R., Jorswieck, E., Thobaben, R.: Linear precoding in MISO cognitive channels with causal primary message. In: IEEE International Symposium on Wireless Communications Systems (ISWCS) (2012)
7. Lv, J., Blasco-Serrano, R., Jorswieck, E., Thobaben, R.: Multi-antenna transmission for underlay and overlay cognitive radio with explicit message-learning phase. EURASIP Journal on Wireless Communications and Networking (JWCN), July 2013. special issue on Cooperative Cognitive Networks
8. Zheng, G., Song, S.H., Wong, K.K., Ottersten, B.: Cooperative Cognitive Networks: Optimal, Distributed and Low-Complexity Algorithms. IEEE Transaction on Signal Processing 61(11), 2778–2790 (2013)
9. Song, S.H., Letaief, K.B.: Prior zero-forcing for relaying primary signals in cognitive network. In: IEEE Global Telecommunication Conference (GLOBECOM), December 2010
10. Sidropoulos, N.D., Davidson, T.N., Luo, Z.-Q.: Transmit Beamforming for Physical-Layer Multicasting. IEEE Transactions on Signal Processing 54(6), 2239–2251 (2006)
11. Jindal, N., Luo, Z.-Q.: Capacity limits of multiple antenna multicast. In: IEEE International Symposium on Information Theory (ISIT), pp. 1841–1845, June 2006
12. Masouros, C., Alsusa, E.: Dynamic Linear Precoding for the exploitation of Known Interference in MIMO Broadcast Systems. IEEE Transactions on Communications 8(3), 1396–1404 (2009)
13. Masouros, C.: Correlation Rotation Linear Precoding for MIMO Broadcast Communications. IEEE Transaction on Signal Processing 59(1), 252–262 (2011)
14. Alodeh, M., Chatzinotas, S., Ottersten, B.: Data aware user selection in the cognitive downlink MISO precoding systems. Invited Paper to IEEE International Symposium on Signal Processing and Information Technology (ISSPIT), December 2013
15. Alodeh, M., Chatzinotas, S., Ottersten, B.: A multicast approach for constructive interference precoding in MISO downlink channel. In: Proceedings of International Symposium in Information Theory (ISIT) (2014). arXiv:1401.6580v2 [cs.IT]
16. Alodeh, M., Chatzinotas, S., Ottersten, B.: Constructive Multiuser Interference in Symbol Level Precoding for the MISO Downlink Channel. IEEE Transactions on Signal Processing (2015). arXiv:1408.4700 [cs.IT]

A Discrete-Time Multi-server Model
for Opportunistic Spectrum Access Systems

Islam A. Abdul Maksoud$^{(\boxtimes)}$ and Sherif I. Rabia

Department of Engineering Mathematics and Physics, Faculty of Engineering,
Alexandria University, Alexandria, Egypt
{islammax,sherif.rabia}@alexu.edu.eg

Abstract. In opportunistic spectrum access communication systems, secondary users (SUs) exploit the spectrum holes not used by the primary users (PUs) and cease their transmissions whenever primary users reuse their spectrum bands. To study the mean time an SU spends in the system we propose a discrete-time multi-server access model. Since periodic sensing is commonly used to protect the PU, discrete-time models are more convenient to analyze the performance of the SU system. Additionally, a multi-server access model is assumed in order to give the SU the capability to access a channel that is not occupied by a PU or any other SUs. We derive the probability generating function of the number of connections in the system. Then we derive a formula for the mean response time of an SU. In the numerical results we show the relationship between the mean response time and the SU traffic intensity. In addition we show the effect of changing the number of channels in the system and the PU traffic intensity on the mean response time of an SU.

Keywords: Discrete-time queueing · Multi-server · Cognitive radio · Opportunistic spectrum access

1 Introduction

The conventional static spectrum access policy had caused a paradoxical spectrum scarcity problem. Although some bands are rarely accessed by their licensed holders, some other bands (e.g. ISM band) are overloaded with traffic [1]. In order to properly manage the problem, another newly dynamic policy is proposed. One of the most appealing models of the new dynamic policy is the opportunistic spectrum access (OSA) model [1]. In the OSA model, unlicensed users (called secondary users or SUs) access the frequency bands owned by the licensed users (called primary users or PUs) only when the PUs do not use their frequency bands and cease their connections when the PUs reuse their channels. In order to realize the OSA model with both acceptable PU protection and SU performance, the SU should be able to perform the following main spectrum management functionalities: spectrum sensing, spectrum decision, spectrum sharing, and spectrum mobility [1],[2].

In literature, there exists a rich body of research works analyzing the performance of OSA systems employing queueing. These works can be classified into continuous-time models and discrete-time models. In continuous-time models, e.g. [3]-[7], opportunities

© Institute for Computer Sciences, Social Informatics and Telecommunications Engineering 2015
M. Weichold et al. (Eds.): CROWNCOM 2015, LNICST 156, pp. 381–388, 2015.
DOI: 10.1007/978-3-319-24540-9_31

can be exploited whenever they appear in the time axis. PUs and SUs can start their connection at any time and depart the system at any time. However, in discrete-time models, e.g. [8]-[13], the time axis is divided into equal time slots. Primary or secondary users are allowed to access the system (i.e., start their connections or transmissions) only at the beginning of a time slot. Although continuous-time models are easier in analysis, discrete-time models are more realistic for OSA systems. This is because an SU usually has one transceiver to either detect the PU or transmit its data. As a result, the most common protocol is to perform periodic sensing to detect the PU, then transmit the data in between the sensing periods if a spectrum hole is available.

The models analyzing the performance of OSA systems can also be classified into single-server access and multi-server access models. In single-server-access models, e.g. [12],[14]-[17], SUs access a certain channel based on a probability profile or a deterministic profile. If there is no PU accessing the channel, then all SUs accessing the same channel either contend on the channel using conventional MAC protocols or queue until they have the right to start their transmissions. However, in multi-server-access models, SUs access only the channels that have no PUs or any other SUs accessing them at the moment (i.e., completely idle). If there is a buffer in multi-server access models it will be a single global buffer for all the channels instead of a local buffer at each channel as in single-server access models.

Since there is an abundance of single-server discrete-time queueing literature, most papers in OSA working with discrete-time models are restricted to single-server access, e.g. [9],[10],[12],[13]. To the best of the authors' knowledge, the only discrete-time model applying multi-server access is [11]. However, [11] is with no buffer. That meant higher forced termination probability for the ongoing SU connections. In this paper, we present a discrete-time model with multi-server access but with an infinite buffer. The infinite buffer vanishes the forced termination probability but with the cost of increasing the delay of some connections. One of the main objectives of the paper is to study the mean time in system of an SU connection and its relation with different parameters, namely, the overall SU traffic, the number of channels in the system and the PU traffic.

In [18],[19], Bruneel and Laevens investigated the discrete-time queueing analysis of an infinite buffer multi-server system with the number of available servers changing randomly over time. The queueing analysis was in terms of the probability generating function (PGF) of the system contents [18] and the PGF of the delay [19]. They assumed general i.i.d bulk arrivals for the number of arrivals during a single time slot. A single arrival is a packet, where a packet constitutes the amount of data transmitted in a single time slot. In this paper, we utilize the model already available in [18],[19], however, we extend the model to be applied to users whose connection is constituted of a geometric number of packets instead of the deterministic service time assumed in [18],[19].

2   System Model

2.1   Primary Network

We assume a primary network with multiple PUs and a PU base station (PU-BS). The PUs are working on a frequency band that is equally divided into M homogeneous channels. The network is time-slotted with all PUs synchronized to the same time slot structure with the help of the PU-BS. That is, any PU can only access a channel at the beginning of a time slot.

It is assumed that the occupation of the channels by PUs is independent and identically distributed (i.i.d). Among adjacent time slots, the PU activity is assumed to be independent, i.e., the state of the primary activity in a certain time slot is independent of the state of the primary activity on that channel in previous time slots. We assume that the probability to have channel i occupied by a PU at time slot n is a constant probability p.

2.2 Secondary Network

We assume a centralized network of SUs which have capability to exploit the time slots not used by PUs in the M-channel band. SUs must first synchronize to the time slot structure of the PU network. With the SU network thoroughly synchronized to the PU network, SUs can sense the PU activity at the channels only at the beginning of the time slot. In this paper, we assume that, at each time slot, all SUs see the same spectrum opportunities and that the sensing results are perfect. In addition, any sensing overhead or delay is neglected.

We assume that SUs connection requests arrive to the SU base station (SU-BS) as a Poisson process with an arrival rate λ. If there is a channel available to the incoming request, it immediately uses that channel. If no channel is available, the SU request is saved in an infinite buffer at the SU-BS. Whenever one channel or more become available, the SU-BS assigns available channels to the requests waiting in the buffer in a first come first serve (FCFS) discipline. The length of the SU connection is assumed to be a random variable with a geometric distribution for the number of packets. A packet is the amount of data transmitted in a single time slot. The average length of the SU connection is assumed to be $1/s$. We assume that each SU is equipped with one transceiver, thus, each SU connection is assigned to at most one channel at a time.

If an SU is using a channel at a certain time slot and a PU occupies the channel on the next time slot, then if the SU connection has still some remaining packets not transmitted, the SU immediately switches to another available channel. This is called the handoff process. If no other channel is available, this indicates the failure of the handoff process. However, instead of dropping the failed-handoff connection, we propose suspending the SU connection while keeping a request at the infinite buffer in the SU-BS to complete the connection. The position of that request in the buffer is based on the timing the SU sent its request to the SU-BS as a new arrival (i.e., when it first entered the system). Priority is always given to the request with the earliest arrival time as a new connection. By doing so, priority is given to interrupted connections over newly coming ones. This is often desirable in communication systems, as more delay in an ongoing connection is more annoying for the user than the delay encountered before the connection is setup.

In this paper, we also make the assumption that arrivals at a certain time slot cannot be transmitted before the next time slot.

3   System Analysis

In this paper, we use the same analysis in [18] to model an OSA network with respect to the performance of SUs. However, instead of assuming the user's connection as a single packet [18],[19], by a single and simple tweak we extend the model to be applied to users whose connection is constituted of a geometric number of packets.

The discrete-time system they investigated can be described in detail as follows. They assumed general i.i.d bulk arrivals with a PGF $A(z)$ for the number of packets arriving during a single time slot. In addition, they assumed a general i.i.d distribution for the number of available servers during a single time slot with a PGF $C(z)$, where transitions in the number of available servers can only occur at the boundaries between consecutive slots. Additionally, they put the assumption that packets arriving in a particular time slot cannot be transmitted during this same slot.

Given that the mean bulk size $\bar{a}$ is less than the mean number of available servers $\bar{c}$, the analysis yields the following expression of the steady state PGF of the system contents,

$$V(z) = \frac{(\bar{c}-\bar{a})A(z)(z-1)}{z^M - z^M C(1/z)A(z)} \prod_{i=1}^{M-1} \frac{z-z_i}{1-z_i} \,,$$

where M is the maximum number of available servers during a time slot, and z_i's are the zeros of the denominator excluding $z_0 = 1$, i.e., z_i's are all the roots (except $z_0 = 1$) of the complex equation

$$z^M - z^M C(1/z)A(z) = 0 \,.$$

Since we assume Poisson arrival bulks, we have

$$A(z) = e^{-\lambda_s(1-z)} \,,$$

where $\lambda_s = \lambda \Delta T$ is the mean number of secondary arrivals during a single slot and ΔT is the time slot duration. Additionally, since we assume i.i.d Bernoulli PU occupation of the channels with probability of occupation p in a time slot, the PGF of the number of available channels during a time slot, denoted $\acute{C}(z)$, can be expressed as

$$\acute{C}(z) = [(1 - q) + qz]^M \,,$$

where $q = 1 - p$.

Since the service time of a packet in [18],[19] was a single time slot, the PGF of the number of available servers during a single time slot $C(z)$ therein played the role of describing the number of actual departures at the end of each time slot. However, in this paper, we aim at extending the model to geometric service time. That means that the number of available servers only describes the number of potential departures in a time slot. In order to describe the number of actual departures during a time slot in the new model we must include the parameter s of the geometric distribution of number of packets in a connection (i.e., the service time). Thus, for the case of geometric number of packets of each connection, the PGF of the number of departures during a time slot can be described as

$$C(z) = [(1 - qs) + qsz]^M .$$

Accordingly, by applying the same analysis in [18], the PGF of the steady state number of users in the system can be described as

$$V(z) = \frac{(Mqs - \lambda_s)(z-1)}{z^M e^{\lambda_s(1-z)} - [(1-qs)z+qs]^M} \prod_{i=1}^{M-1} \frac{z - z_i}{1 - z_i},$$

provided that $Mqs > \lambda_s$, where z_i's are all the roots (except $z_0 = 1$) of the complex equation

$$z^M e^{\lambda_s(1-z)} - [(1 - qs)z + qs]^M = 0 . \tag{1}$$

4 Performance Analysis

In this section, we compute the mean response time of an SU connection in the system. The response time is defined as the total time an SU connection spends in the system. Because we assume that an arriving connection cannot depart in the same time slot it came in, the calculation of the response time is divided into two parts. Starting from the time instant the SU arrived to the system, the first part of the response time is the time until the next time slot. Then, from the beginning of that time slot until the departure of the SU is the second part. The mean value of the first part is $\Delta T/2$. For the second part, we calculate it by using Little's theorem as follows. The mean number of connections in the system can be computed from the PGF by calculating $V'(1)$. The expression for $V'(1)$ is found in [19] to be,

$$V'(1) = \sum_{i=1}^{M-1} \frac{1}{1-z_i} - M + A'(1) + \frac{A''(1) + C''(1) + 2C'(1)\left(1 - A'(1)\right)}{2\left(C'(1) - A'(1)\right)} .$$

For our model, we have $A'(1) = \lambda_s$, $A''(1) = \lambda_s^2$, $C'(1) = Mqs$, and $C''(1) = M(M - 1)(qs)^2$. Then, by using Little's result, the second part of the mean response time can be expressed as $[V'(1)/A'(1)]\,\Delta T$. Hence, the mean SU response time can be expressed as

$$\text{mean SU response time} = \left(\frac{1}{2} + \frac{V'(1)}{A'(1)}\right) \Delta T .$$

5 Numerical Results

To illustrate the above analysis, we present the following numerical example to study the mean SU response time vs. the effective SU traffic intensity $I = \bar{a}/\bar{c}$. We consider two scenarios. In the first scenario, the PU traffic is fixed while the maximum number of available channels is varied. In the second scenario, the maximum number of available channels is fixed while PU traffic is changed. We note that in order to find all the roots of (1), we replaced the right hand side with the 50th degree Taylor polynomial, then solved the resulting polynomial equation.

5.1   Scenario 1

Fig.1 shows the relation between the SU mean response time and the SU effective traffic intensity with a fixed PU probability of time-slot occupation $p = 0.5$. The mean number of packets in an SU connection is assumed to be 4 packets. The relation is drawn for different number of channels, namely, 4,8,12 and 16 channels. The figure obviously shows that the mean SU response time is nearly stable until SU traffic intensity ≈ 0.7. However, afterwards the mean response time increases rapidly as we approach the instability region (i.e., $\bar{a} \geq \bar{c}$). As expected, increasing the number of available channels decreases the mean SU response time. Additionally, increasing the number of available channels widens the interval for which the mean SU response time is stable and makes the curve more acute.

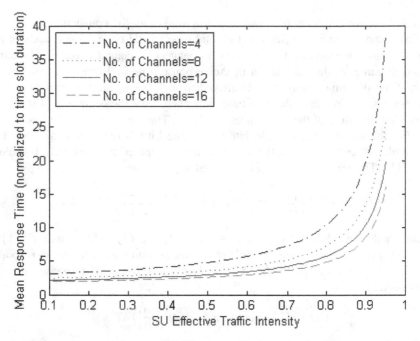

Fig. 1. SU mean response time vs. SU effective traffic intensity for different number of channels in the system.

5.2   Scenario 2

Fig.2 shows the relation between the SU mean response time and the SU effective traffic intensity with a fixed maximum number of available channels $M = 12$. As in scenario 1, the mean number of packets in an SU connection is assumed to be 4 packets. The relation is drawn for different PU traffic activity. Namely, the probability of PU time-slot occupation p is taking the values 0.2, 0.4, 0.6 and 0.8. Expectedly, the figure shows an increase of the mean SU response time as the primary traffic increases.

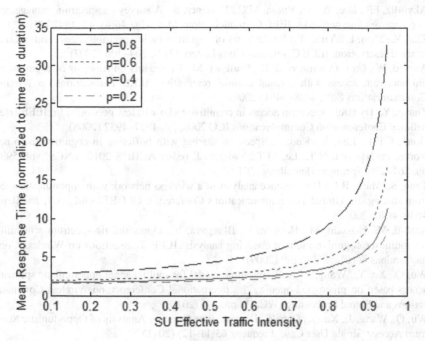

Fig. 2. SU mean response time vs. SU effective traffic intensity for different PU probability of time slot occupation.

6   Conclusion and Future Work

In this paper we have assumed a discrete-time multi-server model to analyze the performance of the SUs with geometric service time. We first derived the PGF of the number of SU connections in the system. Then, we derived a formula for the mean response time of an SU. In the numerical results we have presented the relation between the mean SU response time and the SU traffic intensity in different scenarios of either changing the number of channels in the system or changing the PU traffic intensity. It is shown that the mean SU response time keeps nearly stable for moderate SU traffic intensities, however, the mean response time increases rapidly for large values of SU traffic intensities as the instability region is approached. As expected, the mean SU response time increases when the number of channels in the system decreases or when the primary traffic increases. In the future work, we aim at deriving the PGF of the delay of an SU connection.

References

1. Akyildiz, I.F., Lee, W.-Y., Vuran, M.C., Mohanty, S.: NeXt generation dynamic spectrum access cognitive radio wireless networks: a survey. Computer Networks **50**(13), 2127–2159 (2006)

2. Akyildiz, I.F., Lee, W.-Y., Vuran, M.C., Mohanty, S.: A survey on spectrum management in cognitive radio networks. IEEE Communications Magazine **46**(4), 40–48 (2008)
3. Zhu, X., Shen, L., Yum, T.S.P.: Analysis of cognitive radio spectrum access with optimal channel reservation. IEEE Communications Letters **11**(4), 304–306 (2007)
4. Ahmed, W., Gao, J., Suraweera, H., Faulkner, M.: Comments on Analysis of cognitive radio spectrum access with optimal channel reservation. IEEE Transactions on Wireless Communications **8**(9), 4488–4491 (2009)
5. Zhang, Y.: Dynamic spectrum access in cognitive radio wireless networks. In: IEEE International Conference on Communications ICC 2008, pp. 4927–4932 (2008)
6. Hong, C.P.T., Lee, Y., Koo, I.: Spectrum sharing with buffering in cognitive radio networks. In: Nguyen, N.T., Le, M.T., Świątek, J. (eds.) ACIIDS 2010. LNCS, vol. 5990, pp. 261–270. Springer, Heidelberg (2010)
7. Tang, S., Mark, B.L.: Performance analysis of a wireless network with opportunistic spectrum sharing. In: Global Telecommunications Conference, GLOBECOM 2007, 26-30, pp. 4636–4640 (2007)
8. Rashid, M., Hossain, M., Hossain, E., Bhargava, V.: Opportunistic spectrum scheduling for multiuser cognitive radio: a queueing analysis. IEEE Transactions on Wireless Communications **8**(10), 5259–5269 (2009)
9. Wu, Q., Xu, Y., Wang, J.: A discrete-time model for multichannel opportunistic spectrum access based on preempted queuing. In: International Conference on Wireless Communications and Signal Processing (WCSP), pp. 1–6 (2010)
10. Wu, Q., Wang, J., Xu, Y., Gao, Z.: Discrete-time Queuing Analysis of Opportunistic Spectrum Access: Single User Case. Frequenz **65**(11–12) (2011)
11. Gelabert, X., Sallent, O., Pérez-Romero, J., Agustí, R.: Spectrum sharing in cognitive radio networks with imperfect sensing: A discrete-time Markov model. Computer Networks **54**(14), 2519–2536 (2010)
12. Suliman, I., Lehtomaki, J.: Queueing analysis of opportunistic access in cognitive radios. In: Second International Workshop on Cognitive Radio and Advanced Spectrum Management, CogART 2009, pp. 153–157 (2009)
13. Sheikholeslami, F., Nasiri-Kenari, M., Ashtiani, F.: Optimal Probabilistic Initial and Target Channel Selection for Spectrum Handoff in Cognitive Radio Networks. IEEE Transactions on Wireless Communications **14**(1), 570–584 (2015)
14. Shiang, H.P., van der Schaar, M.: Queuing-based dynamic channel selection for heterogeneous multimedia applications over cognitive radio networks. IEEE Transactions on Multimedia **10**(5), 896–909 (2008)
15. Wang, C.-W., Wang, L.-C., Feng, K.-T.: A queueing-theoretical framework for QoS-enhanced spectrum management in cognitive radio networks. IEEE Wireless Communications **18**(6), 18–26 (2011)
16. Wang, C.-W., Wang, L.-C., Chang, C.-J.: Modeling and analysis for spectrum handoffs in cognitive radio networks. IEEE Transactions on Mobile Computing **11**(9), 1499–1513 (2012)
17. Wang, C.-W., Wang, L.-C.: Analysis of Reactive Spectrum Handoff in cognitive radio networks. IEEE Journal on Selected Areas in Communications **30**(10), 2016–2028 (2012)
18. Bruneel, H.: A general model for the behaviour of infinite buffers with periodic service opportunities. European Journal of Operational Research **16**(1), 98–106 (1984)
19. Laevens, K., Bruneel, H.: Delay analysis for discrete-time queueing systems with multiple randomly interrupted servers. European Journal of Operational Research **85**(1), 161–177 (1995)

HW Architecture and Implementations

A Hardware Prototype of a Flexible Spectrum Sensing Node for Smart Sensing Networks

Ahmed Elsokary$^{(\boxtimes)}$, Peter Lohmiller,
Václav Valenta, and Hermann Schumacher

Institute of Electron Devices and Circuits, Ulm University,
Albert-Einstein-Allee 45, 89081 Ulm, Germany
ahmed.elsokary@uni-ulm.de

Abstract. In this paper we present a prototype for a spectrum sensing node for a cognitive radio sensing network. Our prototype consists of a custom down-conversion front-end with an RF input frequency range from 300 MHz to 3 GHz and a Power Spectral Density (PSD) estimation algorithm implemented on a Virtex-6 Field Programmable Gate Array (FPGA). The base-band processing part is capable of calculating the PSD for a bandwidth upto 245.76 MHz achieving a resolution of 60 kHz and an online variable averaging functionality with a maximum of 32767 averages. Real time performance and calculation of the PSD for real world signals in the GSM downlink, DECT and the UHF DVB-T bands is demonstrated.

Keywords: Spectrum sensing · Multiband · Cognitive radio · FPGA implementation · Hardware prototype · Periodogram · PSD · FFT

1 Introduction

Cognitive radio is an appealing concept for solving the spectrum resource scarcity problem caused by the current static allocation of frequency bands [1][2]. In this context, the Primary User (PU) is the licensed user with priority to use a frequency channel. A Secondary User (SU) is a cognitive user who can reuse frequency white spaces that are channels unused by PUs. Once a PU transmission is detected, the SU must leave the used frequency band immediately to avoid interference. A key requirement for the CR concept is a reliable spectrum sensing [3] with the aim of robust PU detection.

Multiband spectrum sensing deals with the detection of multiple white spaces simultaneously. The main advantage is the quick detection of PUs and offering multiple opportunities to the SU. An important decision metric for multiband detection is the PSD, that describes the average density of the power distribution within the detection bandwidth.

This paper deals with design, implementation and experimental deployment of a spectrum sensing node prototype. The future goal is the hardware implementation of a multiband distributed spectrum sensing network, where cooperative

© Institute for Computer Sciences, Social Informatics and Telecommunications Engineering 2015
M. Weichold et al. (Eds.): CROWNCOM 2015, LNICST 156, pp. 391–404, 2015.
DOI: 10.1007/978-3-319-24540-9_32

detection takes place between all nodes. Each node contains a flexible front-end and a bank of sensing algorithms for decision evaluation. The realized sensing node in this paper consists of a customized RF front-end and a PSD evaluation implemented on FPGA. The system diagram is shown in Fig. 1.

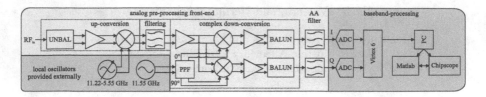

Fig. 1. Frontend connected to the base-band processing.

The main goal is the optimization of the sensing speed and the evaluation in practical scenarios. Our prototype shows a sensitivity of -107 dBm and a bandwidth of 245.76 MHz with a frequency resolution of 60 kHz. Economic use of target FPGA resources was achieved through optimization of arithmetic and memory operations. The RF front-end covers an RF input frequency range from 300 MHz up to 3 GHz. The flexibility in the hardware prototype is achieved through the reconfiguration of parameters such as averaging and windowing in real time, and a tunable external local oscillator to select the target frequency band during operation. The prototype is tested and verified in realistic scenarios.

2   PSD Evaluation

The PSD is evaluated by averaging over several periodograms. This reduces the estimation error as more averages are taken [4]. The periodogram is the squared magnitude of the Discrete Fourier Transform (DFT) of the received signal. The L-point DFT of a discrete time domain vector x of length L is calculated as follows:

$$X_k = \sum_{n=0}^{L-1} x(n)^{\frac{-j2\pi nk}{L}}, k = 0, 1, ..., L-1 \tag{1}$$

A window function is used to reduce spectral leakage, on the other hand it degrades the frequency resolution. Different window functions that offer different trades between resolution and leakage reduction are available [5]. In this paper, we use the Blackman window as it offers the highest sidelobe suppression. The modified periodogram calculation becomes

$$X_k = \sum_{n=0}^{L-1} w(n) \bullet x(n)^{\frac{-j2\pi nk}{L}}, k = 0, 1, ..., L-1, \tag{2}$$

where w is the window function of size L.

The PSD is estimated by

$$\hat{P} = \frac{1}{N_{avg}} \sum_{m=0}^{N_{avg}-1} |\hat{X}|_m^2, \tag{3}$$

where

$$|\hat{X}|_m^2 = Re\{\hat{X}_m\}^2 + Im\{\hat{X}_m\}^2, \tag{4}$$

$$\hat{X} = \{X_0, X_1, X_2, ...X_{L-1}\}.$$

This is a special case of the Welch estimator described in [4] where there is no overlap between the time domain signals used to calculate the averaged periodograms.

3 Implementation Methodology

This section discusses the implementation of the PSD evaluation on the FPGA and the performed optimization to operate at the target clock frequency while choosing the precision to achieve a high sensitivity. In this section, D refers to the complex data path between the consecutive arithmetic blocks. F denotes complex signals that get stored in a First In First Out (FIFO) memory. Complex in this context implies the concatenation of the real and imaginary parts of the signal. Each block is enabled by a 1-bit signal *en*, it issues a signal *valid* when it starts to stream its output into the following block. L refers to the Fast Fourier Transform (FFT) size and N_{avg} refers to the number of calculated spectral averages. A delay of m clock cycles is denoted by z^{-m}.

3.1 Main Building Blocks

A block diagram of the calculation is shown in Fig. 2. The DFT is practically realized using the FFT. A window function is applied to the FFT to reduce spectral leakage. The FFT is followed by magnitude evaluation and reordering that performs the bit reversed indexing of the FFT output to show the PSD in natural order of the frequency bins. This is followed by averaging of the calculated PSD for a user selected number of averages. The implementation is done in VHDL with generic parameters for each module for PSD calculation. This enables manual optimization and achieving an improved performance on the FPGA. Xilinx ISE 14.5 was used for synthesis and bit file generation for the FPGA.

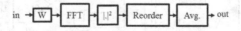

Fig. 2. Block diagram for PSD evaluation.

3.2 Fast Fourier Transform

An FFT was designed based on the radix-2^2 architecture. The radix-2^2 algorithm introduced in [6] offers a simple butterfly structure similar to the radix-2 algorithm and a low number of multipliers similar to the radix-4. This technique was used in our design to conserve area and reduce complexity. Single-path Delay Feedback (SDF) architecture is used to reduce the control complexity and memory requirements. The implementation methodology in [7] was used for the FFT implementation, where we use arithmetic optimization and extended pipelining to increase clock frequency. Fig. 3 shows the block diagram of the FFT.

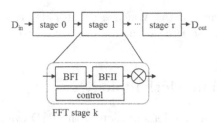

Fig. 3. Block diagram of the FFT unit.

It consists of consecutive stages where each stage comprises two butterfly calculation units (BFI, BFII) and a complex multiplier to carry out the twiddle factor multiplications. Twiddle factors are the complex factors multiplied by the input samples in the DFT calculation. The number of stages r is equal to $\lceil \log_4 L \rceil$. The last stage does not contain a multiplier and contains only one butterfly (BFI) when the FFT length is an odd power of 2. When the length is an even power of 2, it contains both BFI and BFII.

Butterfly Unit. The difference between the two butterfly units is the BFII multiplication by $-j$ which is carried by multiplexers to swap the real and imaginary parts. The structure of BFII is shown in Fig. 4. The selection lines to the multiplexers from the control unit choose the samples that are multiplied by $-j$.

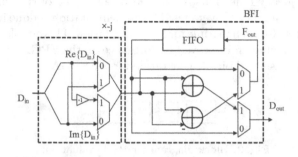

Fig. 4. Butterfly unit BFII.

Twiddle Factor Multiplication. The twiddle factors were pre-calculated and stored in ROMs on the FPGA. A MATLAB script was written to write the ROM initialization file for a generic FFT length and precision. The method for the pre-calculation was mentioned in [7]. Each stage k has a ROM of size $N/2^{2k}$. The implemented complex multiplier for that purpose is shown in Fig. 5. The complex multiplier was fully pipelined to achieve a minimum critical path delay. The real multipliers need three pipeline levels.

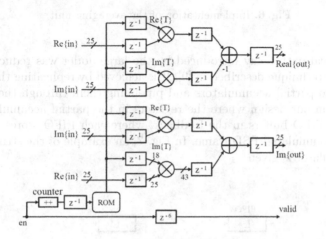

Fig. 5. Twiddle factor multiplier implementation.

3.3 Reordering Unit

The reordering unit was designed according to the methodology shown in [8]. The design uses consecutive stages each with a memory feedback for reordering. It achieves a lower clock cycle delay than direct reordering schemes that use simple storing in a RAM and manipulating the read address.

3.4 Magnitude Evaluation Unit

This unit was realized by two multipliers and an adder, for performing addition of the squared real and imaginary parts of the calculated FFT according to Eq. (4).

3.5 Averaging Unit

An array accumulator was designed to perform the spectral averaging while streaming the input PSD frames continuously. The control structure is shown in Fig. 6. The first counter of size $\log_2 L$ counts the output PSD points. The second counter of size $\log_2 N_{avg,max}$ counts the number of accumulations. Resetting the FIFO is accomplished by adding 0 to the incoming frame to replace the current stored accumulation value.

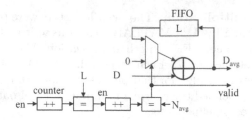

Fig. 6. Implementation of the averaging unit.

The critical path delay introduced by the large adder was reduced using a conventional technique described in [9]. It is achieved by replicating the feedback path into two partial accumulators and pipelining them through the carry bit. It was used in our design where the registers in the partial accumulators were extended to FIFO buffers in the feedback, where each FIFO stores half of the bits of the accumulated PSD frame. In Fig. 7, an example of the structure for a 4-bit accumulator is given.

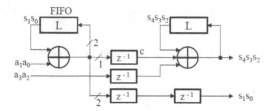

Fig. 7. Pipelined averaging to reduce the adder critical path delay.

3.6   Arithmetic Operations

The arithmetic operations were optimized by exploiting the upper bounds for the expected intermediate calculations for the economic use of the arithmetic resources. The fixed point format [10] was used for the representation of the signal values. The notation $Q(1, i, f)$ refers to a fixed point format where the number is a signed number and i bits are assigned to the integer part and f bits to the fractional part.

Fig.[8-10] show the format used for the window function, butterfly additions and twiddle factor multiplications, respectively. The operator $tr(n)$ refers to the truncation of n Least Significant Bits (LSBs). A division by 2 to avoid overflow was realized by a shift-right after the butterfly addition. The multipliers on the target FPGA operate on 25×18 bit operands [11]. Therefore, a bit size of 25 bits was chosen for each of the real and imaginary parts of D to use an optimum number of multipliers achieving the desired sensitivity.

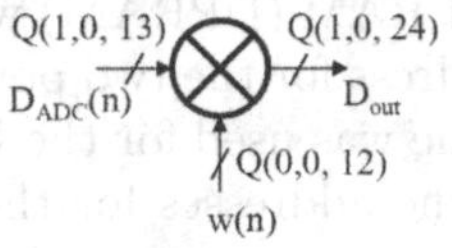

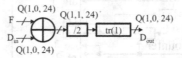

Fig. 8. Number format for the window function multiplication.

Fig. 9. Number format for the butterfly additions.

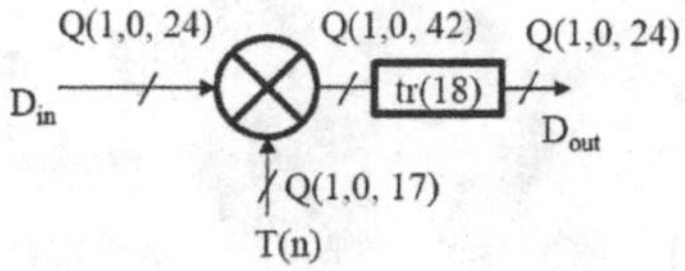

Fig. 10. Number format for the twiddle factor multiplication.

Fig. 11 shows the format used for the squaring and accumulation operations. $sxt(n)$ refers to sign extension by n bits. No truncation is performed to preserve the sensitivity. Therefore, the implemented accumulator needs a sign extension by $\log_2(N_{avg,max} + 1)$ bits to avoid overflow.

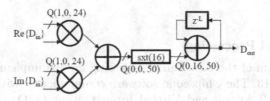

Fig. 11. Number format for magnitude evaluation and accumulation.

3.7 Memory Optimization

To realize large depth FIFO buffers in the data path, the buffer implementation was optimized to achieve a low routing delay on the FPGA. As using a large number of slice registers results in high routing path delay, RAM resources were used instead. The idea was mentioned in [12] as a RAM based shift register.

For each buffer, a Dual Port RAM (DPRAM) was used with custom manipulation of the read and write address for the two ports to realize a FIFO functionality. A read before write scheme was used for the RAM implementation. Fig. 12 shows the control scheme for the addresses for the two RAM ports, where d is the desired FIFO depth.

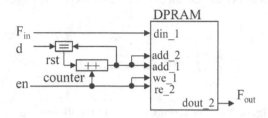

Fig. 12. Realization of a FIFO memory using RAM

4   Sensing Node

4.1   Analog Front-End

As shown in Fig. 1, the front-end used in this work is based on an up/down-heterodyne architecture. It is capable of converting an input RF frequency range from 300 MHz to 3 GHz into the base-band. The dedicated front-end relies on external local oscillator sources and provides a base-band bandwidth of 100 MHz per I and Q channel. A detailed description of the analog front-end is given in [13]. For all the real world tests, a discone omni-directional antenna (frequency range 300 MHz to 3 GHz) was connected to the front-end. The front-end is followed by a dual channel Analog to Digital Converter (ADC) on the FMC150 evaluation board that provides the digital input of I and Q channels for the Virtex-6 FPGA.

4.2   Base-Band Processing

The block diagram of the base-band processing for the implemented prototype is shown in Fig. 13. The Chipscope software communicates with the Integrated Logic Analyzer (ILA) [14] and Virtual Input Output (VIO) [15] Xilinx cores. A software interface was written to control Chipscope from MATLAB to update VIO with the user options for averaging and windowing, and to visualize the output in real time.

The communication speed between the FPGA and the PC was the bottleneck for the update rate (approx. 1 s) to the PC due to the JTAG interface used by Chipscope. However, this only affects the data display update rate and not the calculation speed, as some calculated data has to be discarded to cope with the limited communication buffer size. In a real system, the communication between the sensing nodes can be implemented with a higher data throughput than JTAG to assist decision making speed.

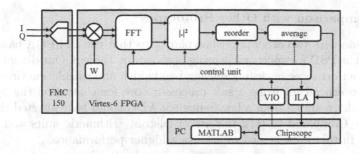

Fig. 13. Block diagram for base-band processing for the sensing node.

4.3 Overview of the Hardware Results

The detection performance is shown in Table 1. The target clock frequency is 245.76 MHz to operate in streaming mode from the ADC. For the FFT size of 4096, the Place And Route (PAR) process estimated a maximum clock frequency of 262 MHz.

Table 1. Detection performance

Parameter	Value
f_s	245.76 MHz
Resolution	60 kHz
FFT size	4096
Calculation time /clock cycle	$10204 + 4096 \times Navg$
$f_{clk,max}$ (PAR.)	262 MHz

The FPGA resource utilization is shown in Table 2. The low usage of the DSP48E1 slices is a result of the performed arithmetic optimization. The RAM resources were used to realize the ROM units used for storing the twiddle factors, and to realize large delays in the data path for optimum performance. The RAM resources used for buffering the output data to the PC are not shown.

Table 2. Design FPGA resource utilization (map results)

Resource	Number used	Utilization(%) XC6VLX240T-1
DSP48E1 slices	26	3.3
RAM36K	25	6
RAM18K	15	1
Logic Slices	1496	4

4.4 Comparison with Other Prototypes

Comparison with two other published platforms that use an FPGA based calculation of the PSD for spectrum sensing is shown in Table 3. Our design achieves superior performance regarding sensing bandwidth and calculation time. In [16], an extra overhead for using a soft processor core for control on the Spartan-6 FPGA and streaming from a low frequency ADC could have affected the detection time. Compared to [17], the use of custom arithmetic units and memory functions in this work helped to achieve a higher performance.

Table 3. Comparison with existing platforms

Platform	fclk (MHz)	bandwidth (MHz)	resolution (kHz)	calculation time(ms)
This work	245.76	245.76	60	$0.042 + 0.017 N_{avg}$
[16]	12.8	12.8	25	$0.54(N_{avg}=10)$
[17]	125	62.5	122	N/A

5 Real Time Testing

5.1 Detection of DVB-T Signals

The detector was tested in the UHF digital TV band for a bandwidth of 245.76 MHz around a central frequency of 580 MHz. The TV signals in the received bandwidth can clearly be observed as shown in Fig. 14.

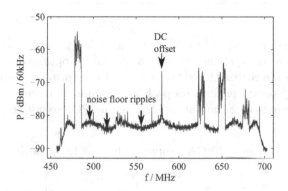

Fig. 14. PSD output from the FPGA based PSD calculation. A ripple in the noise floor is observed. The calculation is performed using 100 averages.

This measurement is done for $N_{avg} = 100$. A ripple in the noise floor can be noticed due to the frequency response of the front-end. This ripple was calibrated by disconnecting the antenna, recording the PSD for 32767 averages and dividing the calculated PSDs by the corresponding noise power for each bin.

The result of that calibration is shown in Fig. 15. The three 8 MHz wideband signals at carrier frequencies 482 MHz, 626 MHz and 650 MHz are the DVB-T channels 22, 40 and 43 respectively which are received in the city of Ulm [18]. Image signals due to IQ mismatch can be also seen at 18 dB below the corresponding signal powers. Digital calibration to compensate the IQ mistmatch is a remaining future task.

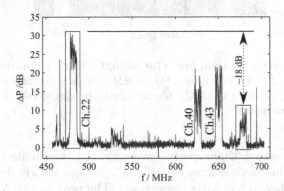

Fig. 15. Calibrated PSD output to achieve a flat noise floor. The calculation is performed using 100 averages. Three TV channels can be observed at 482 MHz, 626 MHz and 650 MHz with their corresponding image signals, with 18 dB image suppression.

5.2   Detection of Signals in the Band from 1788 to 2033 MHz

To assess the detection speed, the prototype was tested in the frequency range 1788 to 2033 MHz which contains two known signals which apply frequency hopping; Global System for Mobile communication (GSM) downlink and Digital Enhanced Cordless Telecommunications (DECT) signals.

GSM downlink signals apply frequency hopping with time slots equal to 576.9 μs according to [19]. DECT phone signals in the band from 1880 to 1900 MHz have 5 ms frames [20]. Fig. 16 shows the spectrogram that is displayed real time from the FPGA calculation.

The detection time here is 1.76 ms for $N_{avg} = 100$. The frequency hopping of DECT signals can be distinguished more clearly than GSM downlink signals which exhibit faster changes. Image signals from high power DECT signals can also be seen approx. 18 dB below the corresponding signal powers.

5.3   Sensitivity

To assess the sensitivity, a sinusoidal signal with adjustable power is connected to the RF front-end's input. The central frequency was set to 580 MHz. The signal is generated at 600 MHz. It mimics a narrowband low power carrier in

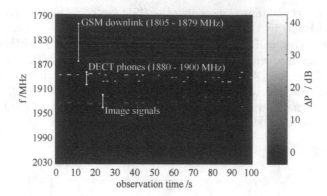

Fig. 16. Spectrogram for the band from 1788 to 2033 MHz for a duration 100 seconds. The update rate is 1 PSD measurement /s. GSM downlink signals can be observed as well as frequency hopping of DECT phone signals. Images of DECT phone signals appear due to IQ mismatch

the TV band. By taking 32000 spectral averages, the lower bound for detection was -107 dBm. The result is shown in Fig. 17. We can clearly distinguish two peaks, at 580 MHz and 600 MHz respectively. The peak at 580 MHz is due to the DC offset of the ADC.

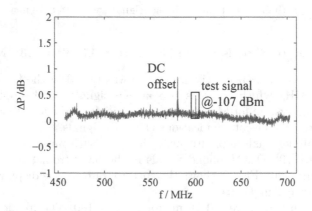

Fig. 17. Sensitivity test on a sinusoidal signal with input power -107 dBm and f = 600 MHz. N_{avg} is set to 32000.

The IEEE 802.22 standard for Wireless Regional Area Networks (WRAN) requires a sensitivity of -107 dBm for a signal within a 200 kHz bandwidth [21]. As the PSD is calculated at a resolution bandwidth of 60 kHz, the system needs to be sensitive to a -112 dBm sinusoidal signal to meet the standard for microphone signals. The sensitivity is limited so far by the fairly high noise figure of the front-end [13], which is due to an unexpected reduced gain in the up-conversion stage as well as due to the losses introduced through the realization into a single module. A future re-design, including a low loss input BALUN as well as a re-design of the high-IF filter is expected to further improve the sensitivity of the system by lowering the front-end noise figure.

6   Conclusion and Future Work

In this paper, the design, implementation and real time testing of a sensing node that calculates the PSD are shown. A MATLAB based interface controls the FPGA design and displays the calculated PSD in real-time. The flexibility of the node is demonstrated for spectrum sensing in different frequency bands with different averaging options. The RF front-end down-converts signals in the range of 300 MHz to 3 GHz to baseband. The achieved baseband detection bandwidth is 245.76 MHz, with a frequency resolution of 60 kHz. The detection is tested on UHF TV signals, GSM downlink signals and DECT phone signals. A sensitivity test is performed and the node was able to detect signals at -107 dBm. Future work includes digital IQ mismatch compensation that is needed to cancel the image signals that appear during testing. Moreover, the noise figure of the front-end needs to be lowered in order to achieve a higher sensitivity. More integration is targeted by incorporating the local oscillators in the front-end chip, and implementation of wideband frequency sweeping over the full range of the front-end RF bandwidth. The next step is the decision implementation on FPGA and the deployment of several sensing nodes for evaluation of detection algorithms in collaborative sensing scenarios.

Acknowledgement. This project is partially funded by the German Research Foundation (DFG) under grant no. VA941/1-1.

References

1. Shin, K.G., Kim, H., Min, A.W., Kumar, A.: Cognitive Radios for Dynamic Spectrum Access: From Concept to Reality. IEEE Wireless Communications Magazine **17** (2010)
2. Valenta, V., Maršálek, R., Baudoin, G., Villegas, M., Suarez, M., Robert, F.: Survey on spectrum utilization in europe: measurements, analyses and observations. In: Fifth International Conference on Cognitive Radio Oriented Wireless Networks and Communications, CROWNCOM (2010)
3. Yücek, T., Arslan, H.: A Survey of Spectrum Sensing Algorithms for Cognitive Radio Applications. IEEE Communications Surveys and Tutorials **11** (2009)
4. Manolakis, D.G., Ingle, V.K., Kogon, S.M.: Statistical and Adaptive Signal Processing: Spectral Estimation, Signal Modeling, Adaptive Filtering and Array Processing. McGraw Hill (2000)
5. Harris, F.J.: On the Use of Windows for Harmonic Analysis with Discrete Fourier Transform. Proceedings of the IEEE **66** (1978)

6. He, S., Torkelson, M.: A new approach to pipeline FFT processor. In: IEEE Parallel Processing Symposium (1996)
7. Cortés, A., Vélez, I., Zalbide, I., Irizar, A., Sevillano, J.F.: An FFT Core for DVB-T/DVB-H Receivers. VLSI Design. Hindawi Publishing Corporation (2008). doi:10.1155/2008/610420
8. Garrido, M., Grajal, J., Gustafsson, O.: Optimum Circuits for Bit Reversal. IEEE Transactions on Circuits and Systems II: Express Briefs (2011)
9. Chappell, M., McEwan, A.: A low power high speed accumulator for DDFS applications. In: IEEE International Symposium on Circuits and Systems ISCAS (2004)
10. Parhami, B.: Computer Arithmetic: Algorithms and Hardware Designs. Oxford University Press (2010)
11. Xilinx Inc: Virtex-6 FPGA DSP48E1 Slice User Guide (2011)
12. Alfke, P.: Creative Uses of Block RAM. White Paper: Virtex and Spartan FPGA Families, Xilinx (2008)
13. Lohmiller, P., Elsokary, A., Chartier, S., Schumacher, H.: Towards a broaband front-end for cooperative spectrum sensing networks. In: European Microwave Conference, EuMC (2013)
14. Xilinx Inc.: LogiCORE IP ChipScope Pro Integrated Logic Analyzer (ILA)(v1.04a) (2011)
15. Xilinx Inc.: LogiCORE IP ChipScope Pro Virtual Input/Output (VIO) (1.04a) (2011)
16. Riess, S., Brendel, J., Fischer, G.: Model-based implementation for the calculation of power spectral density in an FPGA system. In: 7th International Conference on Signal Processing and Communication Systems, ICSPCS (2013)
17. Povalač, K., Maršálek, R., Baudoin, G., Šrámek, P.: Real-time implementation of periodogram based spectrum sensing detector in TV bands. In: 20th International Conference Radioelektronika, RADIOELEKTRONIKA (2010)
18. Südwestrundfunk: Das Programmangebot in Baden-Württemberg (2009)
19. ETSI: Digital cellular telecommunications system (phase 2+) physical layer on the radio path: General description. Technical report (1998)
20. ETSI: Digital Enhanced Cordless Telecommunications (DECT) - TR103089 v1.1.1. Technical report (2013)
21. Stevenson, C.R., Cordeiro, C., Sofer, E., Chouinard, G.: Functional requirements for the 802.22 WRAN standard. IEEE 802.22-05/0007r46 (2005)

Development of TV White-Space LTE Devices Complying with Regulation in UK Digital Terrestrial TV Band

Takeshi Matsumura[1(✉)], Kazuo Ibuka[1], Kentaro Ishizu[1], Homare Murakami[1,2], Fumihide Kojima[1,2], Hiroyuki Yano[1], and Hiroshi Harada[2,3]

[1] NICT, Wireless Network Research Institute, Smart Wireless Laboratory,
3-4 Hikarino-Oka, Yokosuka, Kanagawa 239-0847, Japan
matsumura@nict.go.jp
[2] NICT, Social ICT Research Center, 4-2-1 Nukui-Kitamachi, Koganei, Tokyo 184-8795, Japan
[3] Department of Communications and Computer Engineering, Graduate School of Infomatics,
Kyoto University, Yoshida-Konoe-cho, Sakyo-ku, Kyoto 606-8501, Japan

Abstract. Recently, white-space communication system has been widely promoted as one of the spectrum sharing technologies for further efficient use of spectrum resource. In particular, white-spaces in the TV bands (TVWS, TV white-spaces) have attracted attentions worldwide and regulations have been established in the U.S.A, the U.K., Singapore, etc. In the U.K., pilot program has been started based on regulations established by the Ofcom and trial operation of TVWS communication systems in cooperation with authenticated white-space databases has been performed from July 2014. In this trial, to avoid interference with primary users such as TV broadcasters and wireless microphones, draft rules established by ETSI are employed as specifications of radio characteristics. In this study, to verify the feasibility of spectrum expansion of mobile communication systems by utilizing the TVWS, the authors have developed TVWS LTE devices complying with the Ofcom rules and performed field experiment in the U.K. Prototyped devices have been licensed by the Ofcom and TVWS LTE system has been successfully demonstrated in London urban area. As a result, it is confirmed that the existing LTE system can be operated in the U.K. TVWS without interference with the primary users. This is the first report on prototype of TVWS LTE devices conforming to the ETSI draft specification.

Keywords: LTE · Spectrum sharing · TV white-spaces · Frequency conversion · Field experiment

1 Introduction

Recently, wireless communication services such as video streaming and social networking have been widely used with the explosive spread of portable devices such as smart phones and tablet computers, and thus, the demand for high-speed and large-capacity communications are increasing day by day. In such a situation, data traffic in

© Institute for Computer Sciences, Social Informatics and Telecommunications Engineering 2015
M. Weichold et al. (Eds.): CROWNCOM 2015, LNICST 156, pp. 405–416, 2015.
DOI: 10.1007/978-3-319-24540-9_33

mobile communication will continuously increase and further expansion of the assigned frequency bands is strongly required. ITU (International Telecommunication Union) estimated that a bandwidth (BW) of 1,280–1,720 MHz including current allocated frequency bands is necessary for mobile communication system by 2020 [1]. However, it is difficult to squeeze out a wide space from current crowded frequency bands, especially under 6 GHz that is suitable for mobile communication system. Therefore, technologies which allow more efficient use of limited spectrum resource are entreated and the white-space access is expected as one of such technologies. In particular, white-spaces in TV bands (TVWS, TV White-spaces) have attracted attentions due to a wide allocated frequency band for TV broadcasters and superior radio wave propagation and penetration characteristics in the UHF band, and thus, TVWS communication systems have been globally researched and developed. Furthermore, operational rules for practical use of the TVWS have been established in the U.S.A, the U.K. and Singapore, etc [2]-[4].

Toward the next generation of mobile communication system, TVWS utilization is also one of significant options, and operational scenarios, use cases and simulation-level system evaluations about LTE (Long Term Evolution) communication systems have been studied [5]-[8]. Furthermore, prototype of TVWS LTE system including devices was reported in 2013 [9]. This system incorporates spectrum sensing technology for enabling TVWS communication, however, white-space database (WSDB) approach is mainly employed as a basic direction in established rules.

In the U.K., the Ofcom (Office of Communication) has conducted a pilot program for trial operation of TVWS communication systems in cooperation with the WSDB based on the established rules from July 2014, and a series of trials have been performed [10]. In this study, the authors have also prototyped TVWS LTE eNB (enhanced Node B) and UE (User Equipment) conforming to the Ofcom rules, based on the frequency conversion technology [11]. Prototyped TVWS devices are licensed by the Ofcom and developed TVWS LTE system including the devices is operated in cooperation with the WSDB authenticated by the Ofcom. Also, by participating in the Ofcom trial, field experiment was performed in the King's College London (KCL), Denmark Hill Campus. In this trial, prototyped TVWS LTE system has been successfully demonstrated in London urban area.

The rest of this paper is organized as follows. TVWS regulation in the U.K. is summarized in Section 2. Performance of prototyped TVWS eNB and UE are characterized in Section 3. Prototyped TVWS LTE system and results of field experiment in London are described in Section 4. Section 5 is conclusion.

2   TVWS Regulation in the U.K

2.1   Requirements for TVWS Devices

In the U.K., a frequency band of 470–790 MHz is allocated to the digital terrestrial TV (DTT) and its channel BW is 8 MHz. In the pilot program for the secondary use of TVWS, TVWS devices are required to comply with the Ofcom regulation, and TVWS communication system is also required to operate in cooperation with the

WSDB authenticated by the Ofcom. Requirements for TVWS devices are defined by the draft specification developed by ETSI (European Telecommunications Standards Institute) and TVWS devices are licensed by the Ofcom based on this ETSI draft specification [12].

For avoidance of interference with the primary users, limitation of adjacent channel leakage power (ACLP) is defined as one of required specifications for TVWS devices, but different in each country and region. For example, the limitation of ACLP is defined by the absolute value according to operational conditions in the U.S.A [2]. On the other hand, the limitation of ACLP is defined by the relative value to the output power, in the U.K. [12]. In the ETSI draft specification, the adjacent channel leakage power ratio (ACLR) is defined as the relative value of following equation.

$$P_{OOB} \text{ (dBm / (100 kHz))} \leq \max\{ P_{IB} \text{ (dBm / (8 MHz))} - \atop ACLR \text{ (dB)}, -84 \text{ (dBm / (100 kHz))} \}, \tag{1}$$

where P_{OOB} is leakage power in a 100 kHz BW outside using DTT channel and P_{IB} is output power in a 8 MHz BW of using DTT channel [12]. Here, the ACLR is defined by the suppression ratio of P_{OOB} to P_{IB}. Table 1 shows required ACLR specification. Depending on the ACLR performance, TVWS devices are classified into five "Device Emission Classes," i.e. device classes. In the pilot program in the U.K., before starting radio communication in the TVWS, TVWS devices inform the WSDB of device information including geo-location and the device class, and then, receive an available DTT channel list including information of permitted maximum power for each channel from the WSDB. At this time, received parameters vary according to the device class.

Table 1. ACLR for different device emission classes [12].

Where P_{OOB} falls within nth adjacent DTT channel	ACLR (dB)				
	Class 1	Class 2	Class 3	Class 4	Class 5
$n = \pm 1$	74	74	64	54	43
$n = \pm 2$	79	74	74	64	53
$n \geq +3, n \leq -3$	84	74	84	74	64

In the case that the signal BW is wider than 8 MHz and multiple DTT channels are used, limitation of P_{OOB} is calculated from (1) with the lowest P_{IB} in used channels. Fig. 1 shows channel allocation of TVWS LTE system to DTT channels with a signal BW of 5 MHz, 10 MHz and 20 MHz. Here, it is required to use 2 and 3 contiguous DTT channels for 10 MHz and 20 MHz LTE system, respectively. In the case of 20 MHz LTE system, center DTT channel is fully occupied, while a BW of only 6 MHz is occupied in two DTT channels at both ends, and thus, P_{IB} of DTT channels at both ends is theoretically 1.25 dB decreased in comparison with the center DTT channel. Therefore, limitation of P_{OOB} is calculated from P_{IB} in DTT channels at both ends.

In addition, transmitter unwanted emissions outside the TV band, transmitter intermodulation and receiver spurious emissions are defined as well as common radio standard, in the ETSI draft specification [12].

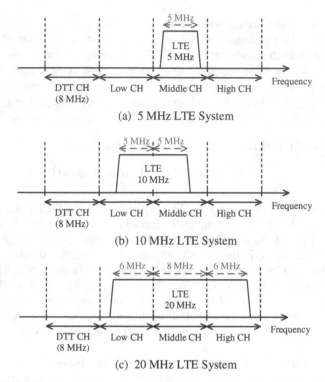

(a) 5 MHz LTE System

(b) 10 MHz LTE System

(c) 20 MHz LTE System

Fig. 1. TVWS LTE channel allocation to DTT channels (based on 8 MHz wide channels).

2.2   ACLR Requirement in 3GPP Standard

Prototyped TVWS eNB and UE employs frequency conversion technology which allows LTE communication in the TVWS by converting the frequency band of existing LTE system to the UHF band [11]. In our TVWS LTE system, 2.6 GHz LTE system is used as an existing LTE system. Therefore, RF characteristics of prototyped TVWS eNB and UE comply with the existing LTE standard regulated by 3GPP, and ACLR characteristics are also based on the existing LTE system. Table 2 shows ACLR requirement for UE in 3GPP standard in comparison with that in the ETSI draft specification for the TVWS. In the 3GPP standard, 29.2 dB is defined as the ACLR limit for 5 MHz, 10 MHz and 20 MHz LTE system [13]. However, measurement frequency, i.e. frequency offset (Δf), and measurement BW are different between the 3GPP standard and the ETSI draft specification. As easier comparison, by converting ACLR of the 3GPP standard to that with a 100 kHz measurement BW, ACLR(/100kHz) is calculated to 46.2 dB. This is about 3.2 dB lower than that in the ETSI draft specification. This means the possibility of TVWS utilization by existing LTE system, although Δf in the

3GPP standard is larger than that in the ETSI draft specification. In addition, the ACLR of eNBs in the 3GPP standard is more stringent than that of UEs, and thus, the ACLR of eNBs also complies with the ETSI draft specification.

Table 2. Comparison of ACLR requirement [12],[13].

Signal BW	3GPP standard			ETSI specification		
	Δf	ACLR	Measurement BW	Δf	ACLR	Measurement BW
MHz	MHz	dB	MHz	MHz	dB	MHz
5	5	29.2	4.5	4-12	43	0.1
10	10	29.2	9	8-16	43	0.1
20	20	29.2	18	12-20	43	0.1

3   RF Performance of Prototyped TVWS Devices

We prototyped TVWS eNB and UE as shown in Fig. 2 by utilizing the frequency conversion technology, as reported in the previous work [11]. To comply with the TVWS regulation in the U.K., some alterations have been made to previous design; e.g. channel allocation to the DTT bands as depicted in Fig. 1. In our system, therefore, the FDD (Frequency Division Duplex) LTE system with a signal BW of 20 MHz can be operated, when 2 blocks of 3 contiguous DTT channels involving a moderate duplex spacing are available.

In this section, performance of the prototyped TVWS eNB and UE in the TDD (Time Division Duplex) mode are characterized for each signal BW, and licensed by the Ofcom based on measured RF characteristics. Device class of each device is "Class 5." Table 3 summarizes required specification and typical measurement result of prototyped TVWS UE in the TDD mode. Detailed measurement processes and results are described in each sub-section. All characteristics are basically measured in the measurement processes elaborated in [12], except for transmission intermodulation.

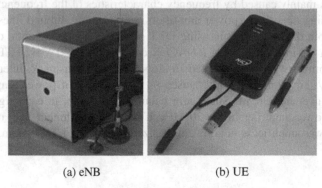

(a) eNB   (b) UE

Fig. 2. Prototyped TVWS LTE devices [11].

3.1   Output Power

Output power is measured by spectrum analyzer (SA) with the following settings: a resolution BW (RBW) of 10 kHz, a video BW (VBW) of 30 kHz, RMS (Root Mean Square) detector and max hold trace mode. Measured frequency range is set to 470 MHz + RBW/2 to 790 MHz − RBW/2. Measurement result is compensated by comparing the summation of measured power for all the sampling points to the power measured by a calibrated power meter for the entire TV band. Subsequently, P_{IB} is calculated as summation of measured powers for all the sampling points in each DTT channel with a BW of 8 MHz. P_{OOB} and P_0 that is power spectral density in a BW of 100 kHz are calculated by adding up 10 contiguous measured powers at an arbitrary frequency.

LTE system using multiple DTT channels such as 10 MHz and 20 MHz LTE are required to calculate P_{IB} and P_0 for all using DTT channels as shown in Table 3. In the case of 20 MHz LTE system, occupied BW of three DTT channels are different, and thus, P_{IB} of a center DTT channel is theoretically about 1.25 dB higher than that of DTT channels at both ends. In the measurement result, there is about 2–3 dB difference between center DTT channel and others, and slightly higher than the theoretical value. This is mainly caused by unflatten spectrum of prototyped LTE devices.

3.2   ACLR

As abovementioned, ACLR in Table 3 is calculated from the lowest P_{IB} in all using DTT channels. As a result, it is confirmed that ACLR of prototyped TVWS LTE UEs complies with the ETSI draft specification for class 5 TVWS devices in all of 5 MHz, 10 MHz and 20 MHz LTE systems. Therefore, it is also confirmed that the existing LTE system can be used in the U.K. TVWS without interference with primary users, since the prototyped devices just convert the frequency of existing LTE system to the UHF band.

Fig. 3 shows typical spectrum mask of prototyped TVWS UE for 5 MHz, 10 MHz and 20 MHz LTE system in the TDD mode. Dashed lines are limited lines of P_{OOB} that are calculated from measured P_{IB} and ACLR requirement for class 5 TVWS devices shown in Table 1. Inclination is observed in measured spectrum masks. This asymmetry is mainly caused by frequency characteristics of the frequency conversion circuit including mixers and power amplifiers. In this study, limited lines of P_{OOB} are calculated from the lowest P_{IB} in using DTT channels in consideration of this asymmetry (P_{IB} in the low CH is used for ACLR calculation in 20 MHz LTE system). In addition, asymmetry is also observed in slopes of spectrum mask, inter alia, for 20 MHz LTE system. This is mainly caused by nonlinearity of power amplifier in the frequency conversion circuit. Since gain balance is different in each signal BW and gain of the power amplifier in the 20 MHz LTE system is higher than others, nonlinearity of power amplifier is notably emphasized in 20 MHz LTE system.

Table 3. Required specification and typical RF characteristics of prototyped TVWS UE in TDD mode.

Item	Conditions	Unit	Spec.	5 MHz LTE			10 MHz LTE			20 MHz LTE		
Center frequency		MHz	–	482	626	762	494	630	758	498	634	738
Maximum output power		dBm	–	14.3	14.8	14.5	13.9	14.6	14.3	9.0	10.8	9.7
P_{IB}	Low CH	dBm/ 8 MHz	–				10.8	11.4	11.4	2.8	4.6	3.8
	Middle CH		–				10.9	11.7	11.2	5.6	7.5	6.4
	High CH		–				–	–	–	3.6	5.5	4.2
P_0	Low CH	dBm/ 100 kHz	–				-4.8	-4.5	-4.6	-13.3	-11.5	-12.2
	Middle CH		–	-1.8	-1.0	-1.3	-4.9	-3.8	-4.5	-12.5	-11.0	-11.8
	High CH		–				–	–	–	-12.7	-10.9	-12.0
ACLR	n=-1	dB	≥43	45.3	50.7	53.3	44.0	49.5	50.9	55.3	57.5	56.1
	n=+1	dB	≥43	45.4	49.9	53.8	44.9	51.6	52.9	50.0	51.4	51.3
	n=-2	dB	≥53	–	71.4	73.8	62.5	62.8	65.3	60.9	62.5	61.3
	n=+2	dB	≥53	62.8	69.6	72.9	62.1	60.4	63.5	56.0	56.7	57.5
	n≤-3	dB	≥64	–	81.7	79.7	–	71.7	73.9	–	67.1	65.8
	n≥+3	dB	≥64	74.4	81.8	80.2	68.8	69.7	72.2	65.3	67.1	66.5
Transmitter unwanted emissions outside TV bands	30–47MHz	dBm/ 100 kHz	≤–36	-59.8	-60.1	-56.1	-59.6	-59.6	-60.2	-57.5	-60.0	-59.6
	47–74MHz		≤–54	-58.8	-59.6	-56.4	-59.3	-59.4	-59.7	-57.1	-59.7	-59.0
	74–87.5MHz		≤–36	-59.8	-60.3	-56.2	-59.7	-60.2	-59.9	-54.4	-59.4	-60.1
	87.5–118MHz		≤–54	-58.9	-59.6	-56.5	-59.2	-59.5	-58.7	-55.8	-58.4	-59.6
	118–174MHz		≤–36	-59.2	-59.8	-56.0	-59.2	-58.4	-59.4	-59.4	-58.6	-59.3
	174–230MHz		≤–54	-57.5	-59.4	-54.9	-58.2	-59.0	-58.8	-58.6	-59.1	-59.1
	230–470MHz		≤–36	-38.8	-57.9	-55.1	-43.8	-58.4	-56.9	-38.8	-58.0	-58.2
	790–862MHz		≤–54	-58.9	-58.4	-54.6	-57.3	-58.2	-56.3	-58.5	-58.2	-56.4
	862–1,000MHz		≤–36	-48.4	-58.8	-54.8	-48.9	-58.3	-57.3	-49.2	-58.5	-58.1
	1–4GHz	dBm/ 1 MHz	≤–30	-48.7	-48.5	-43.2	-48.1	-45.6	-45.4	-37.4	-44.9	-55.4
3rd order reverse intermodulation attenuation (*RIM3*)	Lower	dB	≥45	48.4	56.2	57.0	53.0	59.3	62.3	55.6	58.9	56.0
	Upper	dB	≥45	46.8	51.3	55.0	52.6	57.8	59.8	50.0	50.5	49.6
Receiver spurious emissions	30MHz–1GHz	dBm/ 100 kHz	≤–57	-90.3	-89.8	-87.1	-93.2	-93.2	-92.3	-83.5	-89.2	-82.9
	1–4GHz	dBm/ 1 MHz	≤–47	-63.2	-69.2	-65.2	-70.2	-69.1	-56.4	-61.8	-65.8	-60.2

3.3   Unwanted Emissions and Intermodulation

Transmitter unwanted emissions outside the TV band and receiver spurious emissions comply with the ETSI draft specification as shown in Table 3.

Transmitter intermodulation is measured by using a typical measurement system as shown in Fig. 4, in lieu of the measurement system in [12]. In this measurement, TVWS UE, signal generator (SG) for unwanted signal generation and SA are connected to each other through a power divider. Transmission frequency of the TVWS UE is set to f_w and output power is set to maximum output power in Table 3. Unwanted signal with a frequency of f_{un} is input from the SG and its signal level of P_u is set to −20 dBm, even though signal level of the unwanted signal is defined to 40 dB lower than that of transmission signal in the ETSI draft specification. During the measurement with this setup, 3rd intermodulation with a signal level of *PIM3* is observed at a frequency of $2 \times f_w - f_{un}$ on the SA. Then, the third-order reverse intermodulation attenuation of *RIM3* is calculated by substitution of P_{IB}, P_u and *PIM3* to the following equation.

$$RIM3 = 2 \times P_{IB} + P_u - PIM3, \tag{2}$$

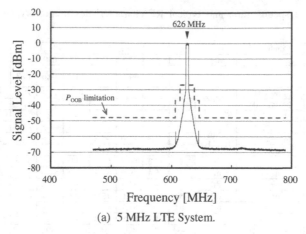

(a) 5 MHz LTE System.

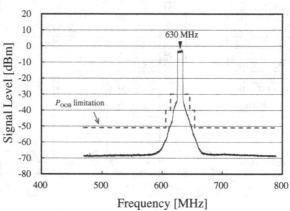

(b) 10 MHz LTE System.

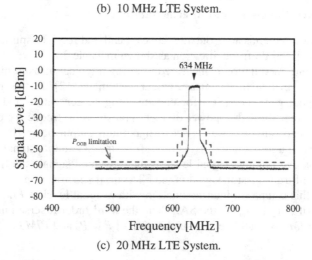

(c) 20 MHz LTE System.

Fig. 3. Spectrum mask of prototyped TVWS UE in the TDD mode.

In this measurement, unwanted signal is normally input to the adjacent DTT channel; i.e. 8 MHz offset. However, in TVWS systems that use multiple DTT channels, 3rd order intermodulation will be generated in occupied frequency band. Therefore, f_{un} is set to $f_w \pm 12$ MHz and $f_w \pm 16$ MHz for 10 MHz and 20 MHz LTE system, respectively. For this measurement methods and results, consensus was obtained from the Ofcom.

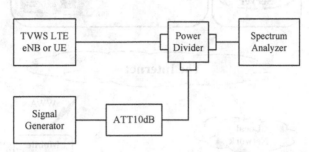

Fig. 4. Measurement system for transmitter intermodulation.

4   Field Experiment

Fig. 5 shows architecture of the developed TVWS LTE communication system demonstrated in the TVWS pilot program conducted by the Ofcom. As this platform is elaborated in more detail in [14], we briefly introduce the overview in this section. The TVWS eNB is operated as an LTE base station, and the TVWS UE is operated as an LTE terminal in this system. Therefore, the TVWS eNB is used as a fixed device and the TVWS UE is used as a non-fixed device. In conformance with the ETSI draft specification [12] and the Ofcom WSDB specification [15], TVWS devices are categorized into two types of "Master device" and "Slave device," and "Master device" should negotiate with the WSDB listing server and the WSDB before starting the TVWS communication. Procedures for the negotiation are generally as follows. First, "Master device" acquires a list of authenticated WSDBs from the WSDB listing server and select one WSDB for parameter exchange. Then, "Master device" informs the WSDB of "device parameters" including "device emission class," and receives "operational parameters" such as time validity and available maximal output power from the WSDB. Here, both devices are categorized as "Master device" in this system. In general, non-fixed devices are operated as "Slave devices," however, slave devices need to obtain "operational parameters" from master devices by using the TVWS communication system without directly access to the WSDB, and thus, slave devices cannot transmit any signal during the reception of "operational parameters." By categorizing both WS eNB and UE as "master devices," in this system, the existing LTE standard is used without any change while avoiding the difficulty to incorporate such parameter exchange in the current LTE system information.

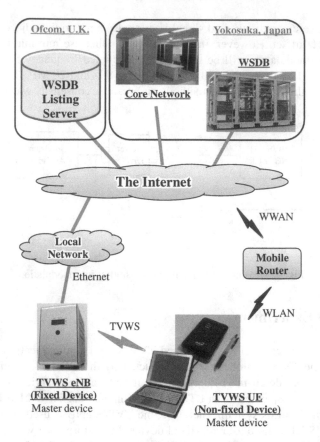

Fig. 5. Developed TVWS LTE communication system in the Ofcom TVWS pilot program.

Field experiment with TVWS eNB and UE was performed in the London urban area. Connectivity and maximum throughput were measured in the Denmark Hill Campus of the KCL where the TVWS eNB was placed on the rooftop of the seven-story building and the TVWS UE moved campus area around the building. Fig. 6 shows maximum downlink throughput in the TDD mode as a function of RSRP (Reference Signal Receiver Power) in the conditions of a signal BW of 20 MHz and a SISO (Single-Input Single-Output) communication link. Despite the high-interference and high-temperature environment resulting in the performance degradation, a maximum downlink throughput of 19.5 Mbit/s was successfully achieved in the field experiment. In the FDD mode, a maximum downlink throughput of about 45.4 Mbit/s was obtained in the same measurement conditions. In addition, a downlink throughput of more than 2 Mbit/s was achieved with a line-of-site distance of 250 m in the TDD mode with a signal BW of 5 MHz, while the TVWS UE was moving in the campus.

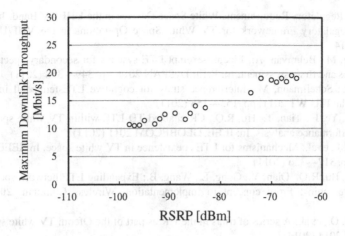

Fig. 6. Maximum downlink throughput as a function of RSRP.

5   Conclusion

In this study, TVWS eNB and UE complying with the ETSI draft specification were developed by utilizing frequency conversion technology, and licensed by the Ofcom. Even in the 10 MHz and 20 MHz LTE systems that use multiple DTT channels, it was confirmed that all the RF characteristics including ACLR complied with the ETSI draft specification.

Also, the TVWS LTE system, which can operate in the cooperation with the WSDB authenticated by the Ofcom, was developed and trial operation of this system was performed in London urban area by participating in the Ofcom pilot program. Although measured maximum throughput was degraded due to tough measurement environment, decent maximum throughputs of 19.5 Mbit/s and 45.4 Mbit/s were achieved in TDD and FDD modes, respectively. Consequently, it is also confirmed that the existing LTE system can be operated in the U.K. TVWS without any interference with the primary users and TVWS LTE system is expected as one of the technologies to extend spectrum access opportunities towards new generation of mobile communication system.

Acknowledgement. This research was conducted under a contract of R&D on "dynamic spectrum access in multiple frequency bands for efficient radio resource utilization" with the Ministry of Internal Affairs and Communications, Japan. The authors thank for Dr. Oliver Holland, Shuyu Ping, Dr. Reza Akhavan and other colleagues from the King's College London for supporting the TVWS LTE experiment in the TVWS pilot program.

References

1. ITU: Estimated spectrum bandwidth requirements for the future development of IMT-2000 and IMT-Advanced. ITU-R Report M.2078 (2006)
2. FCC: Unlicensed Operation in the TV Broadcast Bands, Third Memorandum Opinion and Order. FCC. 12-36, Apil 5, 2012

3. Ofcom: Regulatory Requirements White Space Devices in the UHF TV Band. July 2012
4. IDA: Regulatory Framework for TV White Space Operations in the VHF/UHF Bands. June 2014
5. Rahman, M., Behravan, A.: License-exempt LTE systems for secondary spectrum usage: scenarios and first assessment. In: IEEE DySPAN 2011, pp. 349—358 (2011)
6. Zhao, Z., Schellmann, M.: Interference study for cognitive LTE-femtocell in TV white spaces. In: ITU WT 2011, pp. 153—158 (2011)
7. Xiao, J., Ye, F., Tian, T., Hu, R.Q.: CR enabled TD-LTE within TV white space: system level performance analysis. In: IEEE, GLOBECOM 2011 (2011)
8. Beluri, M., et al.: Mechanisms for LTE coexistence in TV white space. In: IEEE, DySPAN 2012, pp. 317—326 (2011)
9. Xiao, J., Hu, R.Q., Qian, Y., Gong, L., Wang, B.: Expanding LTE network spectrum with cognitive radios: From concept to implementation. Wireless Commun. **20**(2), 12–19 (2013)
10. Holland, O., et al.: A series of trials in the UK as part of the Ofcom TV white spaces pilot. In: CCS 2014 (2014)
11. Matsumura, T., Ibuka, K., Ishizu, K., Murakami, H., Harada, H.: Prototype of FDD/TDD dual mode LTE base station and terminal adaptor utilizing TV white-spaces. In: CROWNCOM 2014, pp. 317—322 (2014)
12. ETSI: White Space Devices (WSD); Wireless Access Systems operating in the 470 MHz to 790 MHz TV broadcast band. ETSI EN 301 598, April 2014
13. 3GPP: Technical Specification Group Radio Access Network; User Equipment (UE) radio transmission and reception (TDD). 3GPP TS 25.102
14. Ibuka, K., et al.: Development and field experiment of white-spaces LTE communication system in UK digital terrestrial TV band. In: IEEE 81st VTC (2015)
15. Ofcom: TV white spaces: Pilot Database Provider Contract. February 4, 2014

Feasibility Assessment of License-Shared Access in 600~700 MHz and 2.3~2.4GHz Bands: A Case Study

Yao-Chia Chan[1], Ding-Bing Lin[1], and Chun-Ting Chou[2(✉)]

[1] Graduate Institute of Computer and Communication Engineering,
National Taipei University of Technology, Taipei, Taiwan
yaochia.chan@gmail.com, dblin@ntut.edu.tw
[2] Graduate Institute of Communication Engineering,
National Taiwan University, Taipei, Taiwan
chuntingchou@ntu.edu.tw

Abstract. License-Shared Access (LSA) has been regarded as a feasible solution for spectrum sharing. In LSA, regulators coordinate the spectrum use between incumbents — the users who hold the licenses and have the exclusive access to the spectrum — and secondary licensees that need authorization before their access. In the midst of 2014, Taiwanese government proposed a draft of Frequency Provision Plan, in which 600~700MHz and 2.3~2.4GHz bands are considered to be opened for more flexible usage, presumably including LSA. The benefit that secondary licensees obtain from LSA depends on the behaviors of the incumbents. This paper evaluates the feasibility of LSA in these bands in Taiwan. Our experiment is based on a 26-day spectrum measurement in Taipei, Taiwan. The behaviors of the incumbents are analyzed in both temporal and spectral domains. The results show that in 600~700MHz band, only narrowband incumbents were detected during small and sporadic time intervals. No incumbent activity was observed in 2.3~2.4GHz band. The experiment shows that low spectrum usage in these bands allows LSA licensees to provide services with predictable quality of service (QoS).

Keywords: Spectrum observation · License-Shared Access (LSA) · Spectrum sharing

1 Introduction

The demand for frequency spectrum has never been higher since broadband wireless communication becomes an integrated part of our daily life. Operators require more bandwidth to support higher data rates and service more clients. The regulation of the spectrum, unfortunately, falls way behind such a trend. Under the current spectrum regulation, the Ultra-High Frequency (UHF) band is sliced into fragments, where some of them are allocated to satellite, civil, and military uses. The spectrum frag-

C.-T. Chou—This work was supported in part by National Taiwan University under Grants G0685 and the Ministry of Science and Technology (Taiwan) under Grants MOST 103-2221-E-002-100.

M. Weichold et al. (Eds.): CROWNCOM 2015, LNICST 156, pp. 417–426, 2015.
DOI: 10.1007/978-3-319-24540-9_34

mentation makes contiguous bands scarce and even unavailable, thus preventing the adoption of new wireless technologies that need higher bandwidth.

The frequency bands allocated for licensed uses scatter in the UHF band. Most of them are used in restricted areas or within limited time spans of a day. The rest of the time in most of the areas can be regarded as idle, and could be reused without causing interference to incumbents. Spectrum reuses at fixed locations during predictable time periods are most suitable for License-Shared Access (LSA) [1]. Proposed and developed in Europe in the last few years, LSA improves the utilization of licensed bands by sharing the frequency bands with regulated secondary users. According to the architecture proposed in European Communication Committee (ECC) report [2], LSA involves three parties including regulator, incumbents and secondary licensees. The first coordinates the band usage between incumbents and secondary licensees based on predefined rules, which not only guarantee the availability of licensed bands for incumbents, but also offer a predictable level of QoS to secondary licensees.

In midst of 2014, Taiwan Ministry of Transportation and Communication announced a draft of Frequency Provision Plan, in which the UHF sub-bands may be released for commercial mobile communication [3]. In addition to these bands, the draft also includes some bands that have never been licensed for commercial communication, including 608~698MHz band and 2.3~2.4GHz band. These bands, if being used in an integrated way, could provide a considerable amount of spectrum for mobile communication without incurring a formidable expense in spectrum refarming/reclaiming, if the idea of LSA can be applied.

The feasibility of LSA relies heavily on the behaviors of incumbents. Such behaviors vary with the bands and the regions where the incumbents are located. In [4], spectrum usage from 30MHz to 3GHz in Chicago, US, had been studied from 2007 to 2010. The results of the three-year observation are used to not only assess the long-term trend in spectrum usage, but also capture significant events such as the release of digital dividends in 700MHz band after the analog-digital TV conversion. Their results confirm the presence of some TV white spaces in 30~1000MHz band. In a six-month spectrum observation in 300~4900MHz band in Bristol, UK, [5] the usage is divided into three categories including continuously occupied, permanently free, and temporally occupied, according to the temporal occupancy pattern. Their measurement shows that TV broadcast bands contribute most of the temporally-occupied usage. In [6] and [7], cellular radio and ISM bands had been observed at multiple locations in London, UK, for a week. The results suggest that GSM bands are less occupied in suburb areas or during the weekends. In [8], the measurements of 2.3~2.4GHz bands were carried out in Chicago, US, and in Turku, Finland, respectively, for two weeks. This band in Turku, according to the results, was occupied in an unpredictable manner due to wireless cameras, while in Chicago the occupancy pattern was relatively stable. The results suggest that even though only a smaller proportion of the band is available in Chicago, the more stable occupancy patterns make LSA a feasible solution in that area.

However, to the best of our knowledge, few reports regarding the usage of 600~700MHz or 2.3~2.4GHz bands were made available to the public in Asian countries. In this paper, we present the results of a 26-day spectrum measurement in

Taipei, Taiwan --- a typical modern Asian metropolitan area. The received power and the time duration of the spectrum usage in 600~700MHz and 2.3~2.4GHz bands were recorded. The behaviors are analyzed in time and frequency domains to assess the feasibility of LSA in these bands. The results show that in 600~700MHz band, only some narrowband incumbents were detected during small and sporadic time intervals. No incumbent activity was observed in 2.3~2.4GHz band. The experiment shows that low spectrum usage in these bands allows LSA licensees to provide services with predictable QoS in Taipei, let alone other areas in the rest of Taiwan.

The rest of the paper is organized as follows. The architecture of LSA proposed by European ECC is introduced in Section 2. The settings of our measurement are presented in Section 3 along with a snapshot of the spectrum usage in these two bands. The spectrum usage of both bands are analyzed in both time and frequency domains in Section 4. The feasibility of LSA is also discussed. Finally, conclusions are given in Section 5.

2   A Review on the Architecture of LSA

The concept of LSA is intended to accommodate users within a licensed spectrum in different time or locations, so that they may share the spectrum and operate with some QoS guarantee. Users are usually divided into two categories including incumbent users and secondary licensees. Their spectrum usage is coordinated by the regulator based on some predefined rules. Incumbent users are licensed to use the spectrum. From the perspective of traditional regulators, incumbent users are the only legitimate users in the designated bands. In the architecture of LSA, they still have the highest priority in using the spectrum. Secondary licensees do not have the license as the incumbent users. They can use the spectrum only when they are authorized by the regulator and when no incumbent user is present. Whenever an incumbent user reclaims the spectrum for its own use, all licensees are required to evacuate within a short time, thus keeping the spectrum clear.

The relationships between regulators, incumbents, and secondary licensees are depicted in Fig. 1. In addition to the three parties, two more components are needed in the architecture: LSA repository and LSA controller. The LSA repository is a database containing the information of incumbents and licensees, including their geolocations, radio-frequency (RF) characteristics, and intended operation frequencies and times. The LSA controller, on the other hand, is served as a gatekeeper between licensees and the repository. The controller receives queries from the licensees, and releases the information of the vacant bands and the spectrum-sharing authorization to the licensees. It also sends evacuation requests to the licensees whose authorizations are reclaimed on the (re)entry of incumbents. In addition to message delivery, the controller also protects the fidelity of the database content from being erroneously or even maliciously modified.

The licensees are authorized to use LSA-regulating bands based on the policy of regulators, the status of incumbent spectrum usage, and the locations and RF characteristics of the licensees. A licensee sends its location and RF characteristics (such as transmitting power, antenna type, etc.) along with its query for vacant bands to the repository through the controller. The regulator replies the licensee with candidates of bands. The licensee selects one or several bands from the candidates and obtains the authorization. If a nearby incumbent emerges and sends a request of an immediate use of the authorized bands, the regulator recalls the bands even if such a request is not in repository schedule. The authorization is reclaimed, and the licensee is required to evacuate within a short time. The status of LSA-regulating bands is updated in the repository whenever they are authorized or reclaimed.

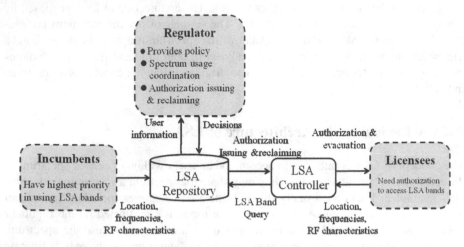

Fig. 1. The architecture of LSA

Although LSA licensees are free from contention when compared with licensed-exempt access, their transmission quality heavily depends on the behaviors of the incumbents. If the behaviors are stable or even predictable, the licensees may plan their usage of the authorized bands in advance without sudden evacuation. In contrast, if the incumbent usage is dynamic, the time interval in which the licensees can share the bands becomes unpredictable and the licensees will be subject to frequent evacuation and service disruption. Such uncertainty in using the authorized bands could void the advantages of LSA as it not only incurs radio resource management overhead but also reduces the utilization of the authorized bands.

As indicated in [9], incumbents can be divided into two types including government and commercial. These two types of users have distinct behaviors of spectrum usage. Government incumbents such as national defense, public safety, meteorology, and science do not have individual rights to access the spectrum. Instead, they share the spectrum to perform their duties. Their access to spectrum is coordinated and regularly reviewed by the regulators and usually is sporadic in both time and spatial domains. In contrast, spectrum usage for commercial incumbents is more contiguous. In general, the spectrum with government incumbents is favorable for LSA. In this paper, we evaluate two spectrum bands that are currently allocated for government incumbents in Taiwan. The goal is to not only characterize the usage pattern of these bands but also provide regulation insights to the regulator based on our quantitative measurement results.

3   Measurement Setup

Although the identities of the users in 600~700MHz and 2.3~2.4GHz bands in Taiwan are not fully revealed by the regulator, both bands are not licensed for commercial use. The activities of the incumbents affect the potential advantage of the licensees under LSA. To gain better knowledge regarding the incumbent activities, we conducted a 26-day spectrum measurement from 09/13/2014 to 10/08/2014. A spectrum observatory is set up on the rooftop of Barry Lam Hall, National Taiwan University, which is located in the center of Taipei metropolitan area.

Our observatory consists of a spectrum analyzer (Rohde & Schwarz FSV-4), a wideband antenna (Rohde & Schwarz HE500), and a computer for control and storage. In order to provide better frequency resolution, resolution bandwidth was set as 100Hz with a noise floor of approximately -133dBm. The root-mean-square detector of the analyzer was chosen to measure the average spectral power within signal acquisition. The control computer recorded spectral power measurement in the respective bands every 3 minutes. As a result, a total of 12,500 records were collected for each band.

Fig. 2 show a two-second snapshot of our spectrum measurement for both bands. Their neighboring bands were also measured and are plotted for comparison. For example, the neighboring 700~800MHz band is also given for comparison with the 600~700MHz band as shown in Fig. 2(a). While the identities of the incumbents in 600~700MHz band are unclear, their neighboring band mostly belongs to commercial use in Taiwan. For example, 703~748MHz and 758~803MHz are licensed to 4G Long-Term Evolution (LTE) operators, and the rest (i.e., 748~758MHz) is assigned to unlicensed wireless microphones [10]. Compared with the occupancy patterns in the 700~800MHz band, the 600~700MHz band is rather vacant with some scattering spikes. The spikes indicate the presence of narrow-band incumbents in the band. The sparse spikes confirm the existence of relatively clean bands that are potentially available for LSA.

Fig. 2(b) shows a two-second measurement snapshot of the 2.3~2.4GHz band and its neighboring 2.4~2.5GHz band. The latter is globally known as the Industrial, Scientific, and Medical (ISM) band, in which wireless local area networks (WLAN) and Bluetooth devices operate. As shown in the figure, this band is occupied with several wideband WLAN signals, and one Bluetooth signal somewhere near 2.4GHz. In contrast, no spectral activity except noise was detected in the 2.3~2.4GHz band. The results show that either the incumbents of this band remained silent during our measurement period, or there was no incumbent at all. In any case, the band is perfect for LSA to provide predictable QoS support.

Although Fig. 2 suggests the potentials of LSA in these bands, a more long-term spectral observation is needed as the behaviors of incumbents may vary at a larger time scale. In what follows, our complete (26 days) measurement results are presented and analyzed.

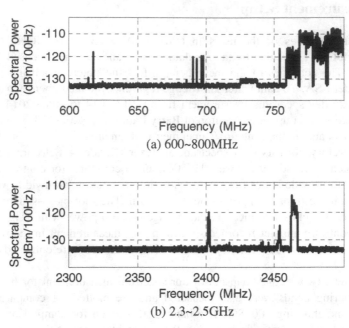

(a) 600~800MHz

(b) 2.3~2.5GHz

Fig. 2. Snapshots of the measured spectrum

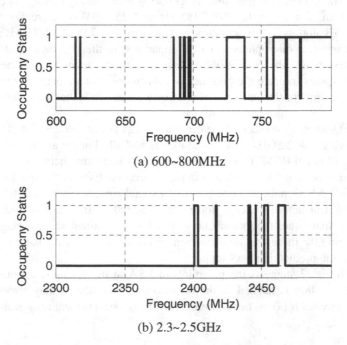

(a) 600~800MHz

(b) 2.3~2.5GHz

Fig. 3. The corresponding spectrum occupancy status of Fig 2 (1 for occupied and 0 for vacant)

4   Variation in Spectrum Occupancy

In order to quantify the incumbents' behaviors based on the measurement results, the occupancy state of a measured spectrum must be determined first. Since there is no prior knowledge about the RF characteristics of the incumbents in these bands, energy detection is applied to determine the occupation status. The measured spectral power at each frequency is compared with the noise floor. The noise floor is estimated with Forward Consecutive Mean Excision (FCME) [11]. Since the spectral power of an incumbent signal may vary within its transmission bandwidth, direct comparison between the spectral power and the noise floor leads to an underestimate in spectrum usage. Therefore, the occupancy status is determined with Localization Algorithm with Double-thresholding and Adjacent Channel Clustering (LAD-ACC) [12]. Take the spectrum snapshots in Fig. 2 as an example. The corresponding occupancy status determined with LAD-ACC is shown in Fig. 3.

The spectrum of the incumbents may vary in both time and frequency domains. To capture the statistics of spectrum occupancy in each domain, we define so-called temporal and spectral occupancy rates as follows. Let $O(f, t)$ be the indicator of occupation status at frequency f ($f = 1, 2, ..., N_F$) and time t ($t = 1, 2, ..., N_T$). The temporal occupancy rate is defined as

$$R_T(f) = \frac{1}{N_T} \sum_{t=1}^{N_T} O(f,t),   \tag{1}$$

which is the proportion of the measurement time occupied by the incumbents at frequency f. On the other hand, the spectral occupancy rate is defined as

$$R_F(t) = \frac{1}{N_F} \sum_{f=1}^{N_F} O(f,t),   \tag{2}$$

which is the proportion of the spectrum occupied by the incumbent at time t. It is often referred to as duty cycle in literature [13]. The occupancy statistics in different domains have different interpretations. A high temporal occupancy rate usually indicates frequent incumbent usage. In this case, the licensees may not benefit from LSA due to frequent evacuation. On the other hand, a high spectral occupancy rate indicates that a large number of incumbents operate simultaneously in these bands, or a few incumbents with large transmission bandwidth exist. Either of the cases makes the corresponding bands unsuitable for LSA with predictable QoS support.

The temporal occupancy rates of our 26-day measurement results are calculated for 600~700MHz band and shown in Fig. 4(a). The rate is about 70% at frequencies around 685MHz, roughly 20%~25% at some frequencies in the lower and upper parts of 600MHz band, and is almost zero for the rest of the band. The result shows that the majority of the incumbents in 600~700MHz band used spectrum in a concentrated manner; that is, few of them used other frequencies and the majority of 600~700MHz was left unused during the 26-day observation. The rates evaluated in 2.3~2.4GHz band are uniformly zero, as shown in Fig. 4(b). The uniformity implies either the incumbents remained dormant, or there were no incumbent near our observatory.

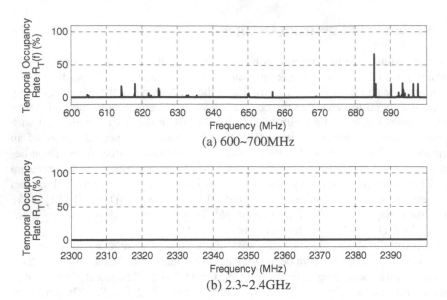

(a) 600~700MHz

(b) 2.3~2.4GHz

Fig. 4. Temporal occupancy rate

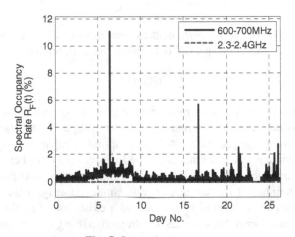

Fig. 5. Spectral occupancy rate

The spectral occupancy rates are also evaluated with all the 26-day records of spectrum measurement. Fig. 5 shows the result in a chronological order with two traces corresponding to the respective bands. The spectral occupancy rates in 600~700MHz fluctuate with a period of roughly a day. The rates are higher from Day 4 to Day 9, but they remain below 3% except for two days, which are less than 12%, on Day 7 and Day 17. The low occupancy in the frequency domain for most of the observation indicates there was no wideband incumbent user observed during our 26-day measurement period. On the other hand, the spectral occupancy rates are zero in 2.3~2.4GHz band. The zero rates in spectral occupancy, together with the zero rates in temporal occupancy in Fig. 4(b), confirm the absence of the incumbents in this band during our measurement.

As mentioned in Section II, the incumbent behaviors significantly affect the performance of LSA. Based on the temporal and spectral occupancy rates observed in 600~700MHz, it is found that the incumbents occupied less than 2% of this band for most of the time. The concentration of the incumbents in spectrum usage shows that the licensees could benefit from LSA in these white spaces without frequent evacuation. In addition, zero occupancy in 2.3~2.4GHz indicates no active incumbents found during the measurement period. The result shows that LSA licensees can provide predictable QoS or even guaranteed QoS in this band in Taipei and presumably the rest of Taiwan.

5   Conclusion

Both 600~700MHz and 2.3~2.4GHz bands, according to the draft of Taiwan Frequency Provision Plan, are considered to be opened for commercial mobile communication in the near future. The two bands are currently assigned or licensed to some incumbents. Given that the expense of spectrum refarming/reclaiming is very high, LSA seems to be a good alternative for efficient spectrum usage. The usage behaviors of the incumbents, however, determine the feasibility of LSA. To assess the feasibility of LSA in Taiwan, we conducted a 26-day spectrum observation in Taipei, Taiwan, in the midst of 2014. Our measurement results show that in 600~700MHz band, the temporal occupancy rates are considered high only at 685MHz (about 70%), moderate at a few others (about 20~25%), and almost zero for the rest of the band. The result indicates that the incumbents only use some fixed frequencies for their narrowband transmission. In 2.3~2.4GHz band, both temporal and spectral occupancy rates are zero throughout the 26-day observation period. The results of this band indicate that the entire band is purely white space. Such low spectrum usage, along with predictable patterns in temporal and frequency domains, makes LSA an economic and feasible solution in Taiwan.

References

1. Mustonen, M., et al.: Cellular architecture enhancement for supporting the European licensed shared access concept. IEEE Wireless Communications 21(3), 37–43 (2014)
2. ECC Report 205 License Shared Access (LSA) (2014)
3. Frequency Provision Plan (Draft) (in Chinese). http://www.motc.gov.tw/ch/home.jsp?id =15&mcustomize=multimessages_view.jsp&dataserno=201408220001
4. Taher, T.M., Bacchus, R.B., Zdunek, K.J., Roberson, D.A.: Long-term spectral occupancy findings in Chicago. In: 2011 IEEE Symposium on New Frontiers in Dynamic Spectrum Access Networks, pp. 100–107. IEEE (2011)
5. Harrold, T., Cepeda, R., Beach, M.: Long-term measurements of spectrum occupancy characteristics. In: 2011 IEEE Symposium on New Frontiers in Dynamic Spectrum Access Networks, pp. 83–89. IEEE (2011)

6. Palaios, A., Riihijarvi, J., Holland, O., Achtzehn, A., Mahonen, P.: Measurements of spectrum use in London: exploratory data analysis and study of temporal, spatial and frequency-domain dynamics. In: 2012 IEEE Symposium on New Frontiers in Dynamic Spectrum Access Networks, pp. 154–165. IEEE (2012)
7. Palaios, A., Riihijarvi, J., Holland, O., Mahonen, P.: A week in London: spectrum usage in metropolitan London. In: 2013 IEEE International Symposium on Personal Indoor and Mobile Radio Communications, pp. 2522–2527. IEEE (2013)
8. Hoyhtya, M., et al.: Measurements and analysis of spectrum occupancy in the 2.3–2.4 GHz band in Finland and Chicago. In: International Conference on Cognitive Radio Oriented Wireless Networks and Communications, pp. 95–101. IEEE (2014)
9. Mustonen, M., et al.: Considerations on the licensed shared access (LSA) architecture from the incumbent perspective. In: International Conference on Cognitive Radio Oriented Wireless Networks and Communications, pp. 150–155. IEEE (2014)
10. Amandement on the table of radio frequency allocations of the Republic of China (in Chinese). http://www.motc.gov.tw/ch/home.jsp?id=15&mcustomize= multimessages_viewjsp&dataserno=201411140001
11. Umebayashi, K., Takagi, R., Ioroi, N., Suzuki, Y., Lehtomaki, J.J.: Duty cycle and noise floor estimation with welch FFT for spectrum usage measurements. In: International Conference on Cognitive Radio Oriented Wireless Networks and Communications, pp. 73–78. IEEE (2014)
12. Lehtomaki, J.J., Vuohtoniemi, R., Umebayashi, K.: On the Measurement of Duty Cycle and Channel Occupancy Rate. IEEE Journal on Selected Areas of Communications 31(11), 2555–2565 (2013)
13. Recommendation ITU-R SM.2256 Spectrum occupancy measurements and evaluation (2012)

Dynamic Cognitive Radios
on the Xilinx Zynq Hybrid FPGA

Shanker Shreejith[1], Bhaskar Banarjee[1],
Kizheppatt Vipin[2], and Suhaib A. Fahmy[1]([✉])

[1] School of Computer Engineering,
Nanyang Technological University, Singapore, Singapore
{shreejit1,bhaskar.banarjee,sfahmy}@ntu.edu.sg
[2] Mahindra École Centrale, Hyderabad, India

Abstract. Cognitive radios require an intelligent MAC layer coupled
with a flexible PHY layer. Most implementations use software defined
radio platforms where the MAC and PHY are both implemented in software, but this can result in long processing latency, and makes advanced
baseband processing unattainable. While FPGA based SDR platforms
do exist, they are difficult to use, requiring significant engineering expertise, and adding dynamic behaviour is even more difficult. Modern hybrid
FPGAs tightly couple an FPGA fabric with a capable embedded processor, allowing the baseband to be implemented in hardware, and the
MAC in software. We demonstrate a platform that enables radio designers to build dynamic cognitive radios using the Xilinx Zynq with partial
reconfiguration, enabling truly dynamic, low-power, high-performance
cognitive radios with abstracted software control.

Keywords: Cognitive radio platforms · Field programmable gate arrays

1   Introduction

Cognitive radios can adapt to channel conditions to effectively utilise available
radio frequency spectrum. Their adoption is driven by increasing demand for pre-
cious frequency spectrum while statically allocated spectrum is often significantly
under-utilised by primary users. Designing cognitive radio systems requires consideration on multiple fronts. High performance baseband processing is necessary to support advanced wireless standards with high data throughput. Yet,
the baseband should be modifiable at runtime for a wide range of deployment
scenarios. Additionally, an easily programmable software stack is necessary to
provide higher level functions and programmability (cognitive logic) by application experts. A radio that combines these features can respond to environmental
changes to maximise radio performance, as shown in Fig. 1. This performance
and flexibility should be achieved within a low power budget to enable deployment in a range of scenarios.

The flexibility requirement has often meant general purpose processors are
chosen for cognitive radio implementations. However, processors are not ideally

© Institute for Computer Sciences, Social Informatics and Telecommunications Engineering 2015
M. Weichold et al. (Eds.): CROWNCOM 2015, LNICST 156, pp. 427–437, 2015.
DOI: 10.1007/978-3-319-24540-9_35

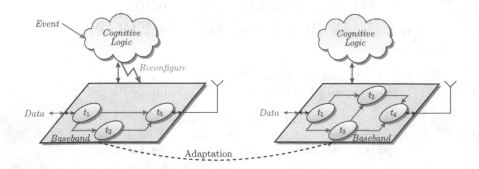

Fig. 1. Baseband adaptation under the control of software cognitive logic, in response to an external event.

suited to the high throughput signal processing required for baseband processing, and as a result, experimental radios are often implemented on fully-functional desktop computers that consume significant power, precluding deployment in scenarios with restricted power, space, and portability requirements. As a result much cognitive radio systems research has been restricted to investigations in labs.

Field programmable gate arrays (FPGAs) have been used for signal processing for over two decades. They enable highly advanced baseband systems to be implemented within a low computational power budget, by exploiting the fine-grained parallelism found in such algorithms. FPGAs are also volatile devices that can be reprogrammed with different functionality at runtime. However, designing FPGA based systems has remained difficult for non-experts.

Recently, new platforms have emerged that couple high performance embedded processors with a flexible FPGA fabric on a single die. In such systems, the processor can host a fully functioning software stack while the baseband can be implemented in the reconfigurable fabric, with high throughput connectivity between them. This represents a promising platform for cognitive radio systems, offering both high computational performance and flexibility that can be leveraged from higher software layers.

We have developed a prototyping system incorporating the Xilinx Zynq hybrid FPGA, allowing us to combine software programmability for upper layers of the radio with a high-performance flexible baseband implemented in hardware. The software portion of the radio has full control of baseband configuration, and we have abstracted control to enable radio experts to leverage advanced features like FPGA partial reconfiguration. In this paper, we present our platform and a case study, before characterising its dynamic properties. We show that with the proposed abstraction layer, it is possible to bring together the flexibility of software control with the performance of a hardware baseband.

2   Background and Related Work

Modern radio protocols rely on flexibility to make efficient use of limited communication spectrum, resulting in the need for highly adaptable baseband and RF processing systems. Research on cognitive radio platforms has focussed primarily on software platforms with component level architectures to afford flexibility as in the case of GNU Radio [1] and Iris [2]. The use of software enables dynamic configurability of the baseband, coupled with an easily programmable MAC layer. While these platforms have been useful for prototyping and academic research, the overhead of implementing advanced baseband processing in software running on general purpose processors means prototyping advanced systems is unrealistic.

Field programmable gate arrays (FPGAs) are silicon devices with a programmable architecture that is flexible enough to implement arbitrary custom datapaths [3]. To implement a datapath circuit, the designer describes it using a hardware description language like Verilog. This design is synthesised and converted, using vendor tools, to a set of configurations that describe how the basic components are to be set up. This "bitstream" is loaded into the FPGA configuration memory to implement the described circuit. As FPGAs are ideally suited to parallel algorithms with large amounts of regular computation, they have been widely used in software radio systems [4]. What makes modern FPGAs attractive for cognitive radios is that besides the high performance of a static datapath implementation, they offer flexibility too.

Multiple radio test beds have leveraged FPGA capabilities for acceleration, like the WARP project from Rice University [5] and the SDC Testbed from Drexel [6]. Iris [2] was also extended with FPGA baseband processing support [7], demonstrating the ability to minimise power consumption in the baseband as channel conditions change [8]. KAUR [9] closely couples a general purpose processing platform, a Xilinx Virtex-II Pro FPGA, and RF front-end in a compact form factor with software and hardware API functions for managing computation. CRUSH [10] integrates a Xilinx Virtex 4 platform with GNU Radio and the USRP front end, offering the performance benefits of a custom baseband datapath, but none of the programmability benefits of FPGAs. CRKIT [11] also aims at integrating hardware baseband processing with software radio management in a system-on chip on an FPGA. It hosts multiple hardware baseband processing chains with support for switching between them and adapting parameters at run-time. Fundamentally, these platforms do offer the performance benefits of hardware, but in many cases, these are limited by the latencies of software-hardware communication due to distinct subsystems being used for the two components, or outdated embedded processor integration on older FPGAs.

Beyond performance, FPGAs offer the advantage of flexibility as they can be reprogrammed with different hardware depending on requirements. Partial Reconfiguration (PR) is an advanced technique that allows parts of the hardware to be modified at runtime while other parts continue to run, enabling designers to swap modules at any given time. Though PR provides adaptability and power

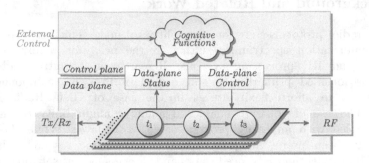

Fig. 2. Conceptual view of intelligent cognitive radio separated into control and data planes.

savings, it requires explicit management of reconfiguration and synchronisation, which is difficult in heterogeneous radio systems. Designing efficient PR systems is non-trivial and is an area explored only by FPGA experts. Iris has explored the use of partial reconfiguration, but the software portions of the radio are deployed on a PowerPC hard processor on the Virtex 5 FPGA, resulting in low performance [8]. Furthermore, the use of soft and hard processors in FPGAs remains difficult for anyone other than FPGA designers.

In this paper, we present a radio platform based on the hybrid Xilinx Zynq FPGA that offers a strongly integrated processing system and reconfigurable fabric. It enables low latency data movement and close integration between higher layers of the radio stack and the computational baseband processing with predicable performance. In [12], the authors showed that the tight coupling and predictable latency afforded by moving even MAC layer functions into hardware can improve radio performance. We believe hybrid FPGAs offer an ideal architecture for integrating the computational capabilities of hardware processing with high level management of dynamic radio behaviour. However, design complexity must be addressed if such platforms are to be adopted by the radio community. The team behind Iris demonstrated an initial attempt at using the Zynq processor to run a radio management system, but with no hardware support [13].

We present the first platform to demonstrate interacting software on the ARM processor and a reconfigurable hardware baseband in programmable logic on the Xilinx Zynq. We abstract the low-level baseband management operations, allowing the software radio designer to use high-level function calls to cause parametric and structural reconfiguration of the baseband, simplifying the management process. Our platform integrates a high speed partial reconfiguration controller to allow reconfiguration of the radio baseband with very low latency.

3 System Architecture

Intelligent cognitive radio designs can be conceptualised as two independent planes, as shown in Fig. 2: the *data plane* which performs the baseband modulation/demodulation and the *control plane* that performs radio control functions; the *cognitive* part. In essence, the control plane implements medium access schemes, enabling the same channel to be used by multiple devices with collision avoidance or detection, and higher layer protocols that ensure reliable transmission. With cognitive radios, the control plane takes on more complex tasks like monitoring channel conditions, triggering sensing, and modifying the configuration of the baseband in response to varying conditions, e.g., by switching the modulation scheme, modifying coding, or changing the baseband transmission standard entirely. The data plane responds to such requests by altering its functions, and should hence support all required types of signal processing to support the various possible modes the system may operate in.

The control plane may use complex intelligent algorithms. Ideally, this cognitive part of the radio should offer maximum flexibility, easy programmability, and being control-intensive, is suited to implementation in software running on a processor. The data plane, however, deals with heavy computational processing on streams of data samples, and so when implemented in software, suffers from long computational latency, and a reliance on powerful processors for advanced baseband schemes or radio standards.

We propose that the strength of FPGAs in data processing be leveraged as in the case of some of the platforms discussed in Section 2, and the data plane be implemented in custom hardware. The key novelty in our platform is to enable dynamic modification of the data plane from the control plane without the need for detailed FPGA knowledge. Previous attempts at building such platforms have used software running on a soft processor on the FPGA fabric with the data plane also in the FPGA as custom hardware blocks, but radio control and reconfiguration still required low-level FPGA knowledge. Essentially, the designer would need to prepare a set of valid hardware configurations for the data plane, store this configuration information in off-chip memory, then manage the loading of the required configurations at runtime through the low-level driver provided by the FPGA vendor. This meant that only FPGA designers could use these systems, and the software programmability was still at a very low-level.

Hybrid FPGAs like the Xilinx Zynq present a compact and efficient architecture for building such software/hardware systems. They tightly couple a highly capable dual-core ARM processing system with a reconfigurable fabric, providing computational capability and flexibility for both the control and data planes. These platforms offer the benefit of a fully functional software side with the ability to add hardware processing in the FPGA fabric. However, managing hardware adaptation on FPGAs at runtime remains difficult and is achieved by a complex sequence of operations, requiring low-level control and knowledge of the underlying hardware.

The unique feature of our proposed framework is that it abstracts such low-level details from the user while also integrating efficient high-speed dynamic

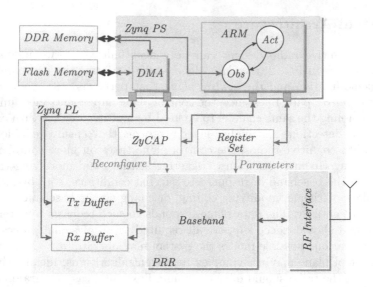

Fig. 3. Proposed cognitive radio architecture on the Zynq.

partial reconfiguration for hardware-level support of baseband adaptation. The simplified architecture of our platform is shown in Fig. 3. The Zynq allows a clear partition between the control plane and data plane with its hybrid architecture comprising the processing system (PS) and the programmable logic (PL). A high speed datapath enables data to be moved to/from the external interfaces (like DDR memory or Ethernet in the PS region) to the data plane in the PL region using dedicated direct memory access (DMA).

The logic in the PL implements the baseband, antenna interfaces, buffers and control/status registers for efficient interaction with the control software running on the ARM core. A key benefit of our platform is that the RF interface is directly connected to the baseband chain, avoiding the need for a round-trip in software as required by some other platforms. This ensures minimal latency and high throughput. The *Tx* and *Rx* buffers form the high-speed data interface between the PS and baseband processing system in the PL. DMA-based data movement allows high speed full-duplex data streaming to the Tx/Rx buffers from the software or other interfaces within the PS, like Ethernet. Baseband control and status monitoring is established via the *Register Set*, that provides configuration and status information for both *Tx* and *Rx* interfaces. The register set also sets the parameters of the RF interface, providing a unified view of all the parameters a radio designer may wish to modify at runtime.

A baseband radio chain with tunable parameters is loaded into the PL at system start-up. The control registers in the register set configure the parameters of the baseband, allowing them to be altered at run-time. This provides fast adaptation without requiring any changes in the physical design of the baseband block, and is called parametric reconfiguration. This works for small changes like modifying the carrier modulation scheme or selecting a new coding scheme.

The antenna interface provides the interface to off-the-shelf RF boards, and we initially support the Analog Devices AD-FMCOMMS4-EBZ (based on the AD9364) that interfaces with the FPGA using an FPGA Mezzanine Card (FMC) connector.

The control plane implements the *Observe (Obs)–Decide–Act (Act)* loop, which observes events, either through changes in values in the register set, or external events triggered from software. If modification of the baseband is required, it can trigger parametric and/or physical reconfiguration. Higher layer protocols may be integrated on top of this to provide a complete network stack.

While parametric reconfiguration allows us to modify some aspects of the baseband, any significant changes, e.g. changing from sensing mode to transmission, require more significant changes in hardware. This is achieved using partial reconfiguration, which is standard across the entire range of Zynq devices and other 7-series FPGAs from Xilinx. The baseband processing chain is implemented in a partially reconfigurable region (PRR) within the PL, enabling its physical implementation (and thus function) to be modified at runtime beyond just parametric changes. Effectively, the whole PRR can be replaced with new hardware blocks at runtime by writing new configuration bits into the FPGA's configuration memory.

The standard PR flow supported by Xilinx requires extensive understanding of FPGA architecture and programming such reconfiguration is complex, requiring low-level access to memory addresses, and understanding of configuration *bitstream* details, making it difficult for radio designers. Furthermore, the reconfiguration speed attainable with the supported flow is very slow, resulting in considerable latency when switching between different baseband modes.

To address this challenge, we use ZyCAP, a custom-designed reconfiguration manager that enables seamless management of partial reconfiguration from the ARM processor via its own software/libraries [14]. ZyCAP is added as a peripheral to the PS and connected to the internal configuration circuitry of the FPGA. At the software level, the ZyCAP driver manages the low level commands for reconfiguration, memory organisation for the different bitstreams, and performance enhancements like bitstream caching, all of which are abstracted from the system designer through high-level API functions.

The application designer is able to call different physical configurations of the baseband using function calls like `set_baseband(receive1)`. The driver handles all the steps required for physical reconfiguration. ZyCAP provides non-blocking operation, which returns from the reconfiguration call immediately, allowing processor load to be minimised, and hence supporting more complex cognitive algorithms. By using DMA-based bitstream reconfiguration, ZyCAP also minimises reconfiguration time by a factor of 20 times or more.

4   Case Study

To evaluate the capabilities of our platform, we have implemented a dynamically modifiable radio on a Xilinx ZC702 evaluation board. We have created a

Fig. 4. Test setup for DVB-S/C case study.

hypothetical scenario with two baseband standards based on the digital video broadcasting (DVB) cable (DVB-C) and satellite (DVB-S) standards. The two baseband processing chains are distinct and implemented as custom hardware designed in Verilog. Each configuration supports a number of parameters that modify the coding mechanism (like modes of convolutional/differential coding in DVB-S/C) or a variation in the code rate (2/3, 3/4, or 7/8). The parameter changes for both baseband configurations are present as multiplexed hardware and the active path is chosen by setting multiplexer control signals in the register set. This allows low latency parametric adaptation, representing system changes that may be required to adapt to instantaneous channel conditions using the same baseband scheme. Switching between baseband schemes requires partial reconfiguration of the FPGA.

The baseband output is interfaced to the Analog Devices AD-FMCOMMS4-EBZ FMC module with a tunable operating frequency, providing a highly flexible air interface. The transceiver is configured over an SPI interface from the PS providing complete software control over the data-plane (from baseband to RF). Fig. 4 shows the laboratory setup for evaluating the case study.

Our experiments aim at quantifying overall data-path latency and the delay incurred for data-plane adaptation (both parametric and full reconfiguration adaptation). For our experiments, we have simple software control in C that initialises the baseband modules and initialises transmission of data: no medium access control is implemented. Baseband adaptation is managed from software by modifying the transmit and receive status registers, that can trigger a parametric reconfiguration or a physical hardware reconfiguration. In a full cognitive radio

Table 1. Resource utilisation on ZC-7020.

Function	LUTs	FFs	BRAMs	DSPs
DVB-S	1487	2812	0	24
DVB-C	1109	2580	0	24
PRR	5400	8000	50	40
RF I/F	13389	21086	15	69
Reconfig	806	620	0	0
Total	15682	24518	15	93
(%)	29.5%	23%	10.6%	42.3%

implementation, more complex software can be used to decide on the correct configuration, and would use the same abstracted interface. We keep the software simple to provide us with meaningful latency numbers in our experiments.

Table 1 shows the resource utilisation of the different modules in the case study. The PRR is large enough to include all resources required for the DVB-S or DVB-C baseband scheme, with PR used to switch between them. It still consumes a minimal amount of resources considering the simplicity of the baseband in this case. The entire design does not consume more than 42% of the resources (DSPs) in the relatively small Zynq XC7020 device. More complex baseband schemes based on OFDM would consume more resources, but our initial experiments have shown that a flexible OFDM baseband consumes just over half the resources on this same device.

Table 2 shows the end-to-end latency of the data-plane for transmitting one complete frame of 188 Bytes. The path delay is composed of two components: the delay for loading data from external memory (DDR) into the internal buffers and the processing delay of the baseband logic. The packet from the external DDR memory is loaded into (or read from) the Tx (Rx) buffer through a dedicated DMA into the baseband, enabling high speed uninterrupted data flow. The path latency of the processing chain depends on the different baseband configurations and different parameter settings chosen by the control software at runtime. We can see that overall latency is dominated by the baseband logic and is largest in case of DVB-S with the 1/2 coding rate. These latencies are 400× less than what can be achieved by implementing the baseband in software (in C) running on the ARM processor in the Zynq, with the DVB-S baseband consuming 39.18 ms to encoding each frame at 1/2 code rate. It is also worth noting that data movement from external memory to the hardware baseband consumes only a fraction of the total time, and can be effectively hidden by overlapping data movement with baseband operation.

To determine the latency incurred during parametric and physical reconfiguration, the Tx/Rx status registers were used to trigger changes in the system from the control plane. Parametric adaptation incurred a delay of 150 ns from the time the register values were changed and hence detected by the software

Table 2. End-to-end latency of the Data Plane at 100 MHz.

DDR Latency	Baseband	Code Rate	Latency
		1/2	96.25 μs
		2/3	72.19 μs
3.70 μs	DVB-S	3/4	64.19 μs
		5/6	57.73 μs
		7/8	54.98 μs
	DVB-C	NA	24.06 μs

(periodic polling). The delay incurred accounts only for the write path delay from the processing system to the Tx/Rx control registers in the register set as these controls are directly wired to multiplexers controlling the different paths. A complete baseband adaptation using PR incurred a delay of 786 μs from status decode, primarily due to ZyCAP achieving a reconfiguration throughput of 380 MB/s — nearly 95 percent of the theoretical bandwidth. This is more than 3× faster than the normal blocking reconfiguration control possible in the Zynq.

5   Conclusion

Cognitive radio systems require highly flexible hardware support for implementing adaptive and computationally complex baseband functions, with further constraints on the power budget for mobile applications. Hybrid FPGAs like the Xilinx Zynq show promise for such platforms as they closely integrate adaptability at the hardware level with a computationally capable processing system. However, managing runtime adaptation through reconfiguration and exploiting the benefits of partial reconfiguration on the Zynq is generally too difficult for radio designers used to software, instead requiring experienced FPGA engineers. In this paper, we have presented a cognitive radio prototyping platform based on the Xilinx Zynq which uses a high level reconfiguration management system to abstract low level details of hardware management from the application designers.

We demonstrated a case study with a DVB baseband, showing that hardware level adaptation (including parametric and full baseband reconfiguration) can be achieved with minimal latency, while still being abstracted. This opens the door to radio designers with no FPGA experience to benefit from the capabilities of new hybrid architectures like the Zynq to build dynamic radios with minimal latency, high baseband performance and true hardware reconfigurability.

We are working on a public release of our framework and developing a library of baseband blocks for flexible OFDM cognitive radios in the hope that more radio designers will be able to benefit from this technology.

References

1. GNU Radio. http://www.gnuradio.org/ (accessed March 2015)
2. Sutton, P.D., Lotze, J., Lahlou, H., Fahmy, S.A., Nolan, K.E., Ozgul, B., Rondeau, T.W., Noguera, J., Doyle, L.E.: Iris: An architecture for cognitive radio networking testbeds. IEEE Communications Magazine 48(9), 114–122 (2010)
3. Kuon, I., Tessier, R., Rose, J.: FPGA architecture: Survey and challenges. Foundations and Trends in Electronic Design Automation 2(2), 135–253 (2008)
4. Cummings, M., Haruyama, S.: FPGA in the software radio. IEEE Communications Magazine 37(2), 108–112 (1999)
5. Amiri, K., Sun, Y., Murphy, P., Hunter, C., Cavallaro, J.R., Sabharwal, A.: WARP, a unified wireless network testbed for education and research. In: IEEE International Conference on Microelectronic Systems Education, pp. 53–54 (2007)
6. Shishkin, B., Pfeil, D., Nguyen, D., Wanuga, K., Chacko, J., Johnson, J., Kandasamy, N., Kurzweg, T.P., Dandekar, K.R.: SDC testbed: software defined communications testbed for wireless radio and optical networking. In: International Symposium on Modeling and Optimization in Mobile, Ad Hoc and Wireless Networks, pp. 300–306 (2011)
7. Lotze, J., Fahmy, S.A., Noguera, J., Ozgul, B., Doyle, L.E., Esser, R.: Development framework for implementing FPGA-based cognitive network nodes. In: Proceedings of the IEEE Global Telecommunications Conference (GLOBECOM) (2009)
8. Fahmy, S.A., Lotze, J., Noguera, J., Doyle, L.E., Esser, R.: Generic software framework for adaptive applications on FPGAs. In: IEEE Symposium on Field Programmable Custom Computing Machines, pp. 55–62 (2009)
9. Minden, G.J., Evans, J.B., Searl, L., DePardo, D., Petty, V.R., Rajbanshi, R., Newman, T., Chen, Q., Weidling, F., Guffey, J., Datla, D., Barker, B., Peck, M., Cordill, B., Wyglinski, A.M., Agah, A.: KUAR: a flexible software-defined radio development platform. In: IEEE International Symposium on New Frontiers in Dynamic Spectrum Access Networks (DySPAN), pp. 428–439 (2007)
10. Eichinger, G., Chowdhury, K., Leeser, M.: Crush: cognitive radio universal software hardware. In: International Conference on Field Programmable Logic and Applications (FPL), pp. 26–32 (2012)
11. Le, K., Maddala, P., Gutterman, C., Soska, K., Dutta, A., Saha, D., Wolniansky, P., Grunwald, D., Seskar, I.: Cognitive radio kit framework: experimental platform for dynamic spectrum research. ACM SIGMOBILE Mobile Computing and Communications Review 17(1), 30–39, January
12. Di Francesco, P., McGettrick, S., Anyanwu, U.K., O'Sullivan, J.C., MacKenzie, A.B., DaSilva, L.A.: A split MAC approach for SDR platforms. IEEE Transactions on Computers 64(4), 912–924 (2015)
13. van de Belt, J., Sutton, P.D., Doyle, L.E.: Accelerating software radio: iris on the synq SoC. In: IFIP/IEEE International Conference on Very Large Scale Integration (VLSI-SoC), pp. 294–295 (2013)
14. Vipin, K., Fahmy, S.A.: ZyCAP: Efficient partial reconfiguration management on the Xilinx Zynq. IEEE Embedded Systems Letters 6(3), 41–44 (2014)

Next Generation of Cognitive Networks

A Novel Algorithm for Blind Detection of the Number of Transmit Antenna

Mostafa Mohammadkarimi[✉], Ebrahim Karami, and Octavia A. Dobre

Memorial University of Newfoundland, St. John's, NL, Canada
{m.mohammadkarimi,ekarami,odobre}@mun.ca

Abstract. In this paper, a novel algorithm is proposed to blindly detect the number of transmit antennas by exploiting the time-diversity of the fading channel. It employs a second-order moment and a fourth-order statistic of the received signal when the transmission occurs over a time-varying multiple-input single-output channel. When compared with information theoretic algorithms, it does not require the number of received antennas be larger than the number of transmit antennas, and when compared with existing feature-based algorithms, it does not require *a priori* information about the transmitted signal, such as preambles or pilots. Simulation results show that the proposed algorithm exhibits a good performance over a wide range of signal-to-noise-ratios (SNRs), and the probability of correct detection approaches one at low SNR values for various numbers of transmit antennas. Furthermore, it is robust to the modulation format and carrier frequency offset, and exhibits a good performance in the presence of the noise power mismatch and spatially correlated fading.

Keywords: Number of antenna detection · Second-order moment · Fourth-order statistic · Time-diversity

1 Introduction

With the advent of multiple-input multiple-output (MIMO) systems, the problem of the number of transmit antennas detection has emerged in both military and commercial communications, such as spectrum surveillance, electronic warfare, and cognitive radio [1–6]. For the cognitive radio systems, the coexistence of the secondary users (SUs) and the primary users (PUs) equipped with multiple antennas ameliorates when the SUs have *a priori* information about the PUs number of transmit antennas, as the interference tolerated by the PUs from the SUs depends on that; hence, such knowledge allows the SUs to better adjust their transmit power to avoid destructive interference to the PUs [3]. Furthermore, the radio front end has a complexity, size and price that scales with the number of transmit antennas. Recently, the antenna selection technique was proposed to alleviate this cost and at the same time to capture many of the advantages of

© Institute for Computer Sciences, Social Informatics and Telecommunications Engineering 2015
M. Weichold et al. (Eds.): CROWNCOM 2015, LNICST 156, pp. 441–450, 2015.
DOI: 10.1007/978-3-319-24540-9_36

MIMO systems [7–9]; in this case, detecting and tracking the number of transmit antennas is of interest to eliminate the need for additional signaling, which introduces overhead and transmission latency [5].

There are two main approaches for the detection of the number of transmit antennas: information-theoretic [1,2] and feature-based [3–6]. The Akaike information criterion (AIC) and the minimum description length (MDL) algorithms are two well-known information-theoretic methods. With these algorithms, the problem of the number of transmit antennas detection is formulated as a model order selection problem, which relies on the rank estimation of the received signal correlation matrix. However, such algorithms usually suffer from high computational complexity, as they require the eigen-decomposition of the sample covariance matrix. Further, they fail to detect the number of transmit antennas when this is larger than the number of received antennas. On the other hand, the existing feature-based algorithms rely on a priori information about the transmitted signals, e.g., pilot patterns [3,4] or preamble sequences [5,6]. As such information is actually not available at the blind receiver, it represents the main drawback of these feature-based algorithms.

A novel feature-based algorithm for the blind detection of the number of transmit antennas is presented in this paper, where a single receive antenna is used. The proposed algorithm employs a second-order moment and a fourth-order statistic of the received signal, and exploits the time-diversity of the fading channel. In contrast with the information theoretic algorithms, it does not require the number of received antennas be larger than the number of transmit antennas, and when compared with the existing feature-based algorithms, it does not require a priori information about the transmitted signals.

The rest of the paper is organized as follows. The signal model is presented in Section 2, the proposed algorithm is introduced in Section 3, simulation results are provided in Section 4, and conclusions are drawn in Section 5.

Notation: Throughout the paper, bold-faced letters are used for vectors, $[.]^{\dagger}$ represents the transpose operator, $(.)^*$ denotes the complex conjugate, $n!$ is the factorial of n, $E_x[.]$ is the statistical expectation of the random variable x, and $\hat{x}$ is the estimate of x.

2   System Model

A multiple-input single-output (MISO) block fading channel with n_t transmit antennas is considered, where n_t is unknown at the receive-side [10]. We assume that the receiver observes N_b blocks, each with a length of N_c symbols. Each block is affected by independent and identically distributed (i.i.d.) fading characterized by an $(1 \times n_t)$ matrix $\mathbf{H}_b$, $b = 1, ..., N_b$, and corrupted by additive white Gaussian noise. With the assumption of the Clarke-Jakes Doppler spectrum, the block length is $N_c = \lfloor 0.2/f_d T_s \rfloor$, where f_d and T_s are the maximum Doppler frequency and symbol period, respectively [10]. Thus, the received complex-valued signal can be expressed as

$$r_{k,b} = \mathbf{H}_b \mathbf{s}_{k,b} + w_{k,b} \quad k = 1, ..., N_c, \quad b = 1, ..., N_b, \tag{1}$$

$$\mu_{21,b} = \sum_{m_1=1}^{n_t} \left|h_b^{(m_1)}\right|^2 E_s\left[|s_{k,b}^{(m_1)}|^2\right] + \sum_{m_1=1}^{n_t} \sum_{\substack{m_2=1 \\ m_2 \neq m_1}}^{n_t} h_b^{(m_1)} \left(h_b^{(m_2)}\right)^* E_s\left[s_{k,b}^{(m_1)} \left(s_{k,b}^{(m_2)}\right)^*\right]$$

$$+ \sum_{m_1=1}^{n_t} \left(h_b^{(m_1)} E_{s,w}\left[s_{k,b}^{(m_1)} (w_{k,b})^*\right] + \left(h_b^{(m_1)}\right)\tilde{E}_{s,w}\left[\left(s_{k,b}^{(m_1)}\right)^* w_{k,b}\right]\right) + E_w\left[|w_{k,b}|^2\right]$$

$$(2)$$

where $r_{k,b}$ is the kth received symbol in the bth observation block, $\mathbf{s}_{k,b} = [s_{k,b}^{(1)}, s_{k,b}^{(2)}, ..., s_{k,b}^{(n_t)}]^\dagger$ represents the transmitted symbols from the n_t transmit antennas, whose variance $E_s[|s_{k,b}^{(m)}|^2] = \sigma_s^2$, $m = 1, ..., n_t$ is unknown at the receive-side, $w_{k,b}$ is complex additive white Gaussian noise with zero-mean and variance σ_w^2 assumed to be known at the receive-side, and $\mathbf{H}_b = [h_b^{(1)}, h_b^{(2)}, ..., h_b^{(n_t)}]$ denotes the channel coefficients, with $h_b^{(j)}$, $j = 1, ..., n_t$ as the channel coefficient between the jth transmit antenna and the receive antenna for the bth observation block. It is assumed that the channel coefficients in each block are independent complex-valued Gaussian random variables with zero-mean and variance $E_{\mathbf{H}_b}[|h_b^{(j)}|^2] = \sigma_h^2$, where σ_h^2 is unknown at the receive-side.

3   Number of Transmit Antennas Detection

The proposed algorithm for the number of transmit antennas detection exploits a second-order moment and a fourth-order statistic of the received signal, along with the time-diversity of the fading channel, as subsequently presented.

Let us first consider the second-order moment and the fourth-order statistic of the received signal within an observation block. By using (1) and the linearity property of the statistical expectation, one can express the second-order moment/ one conjugate, $\mu_{21,b} \triangleq E_{s,w}[|r_{k,b}|^2]$, as in (2). With the assumptions that the additive noise, $w_{k,b}$, is independent of the transmitted symbols, $s_{k,b}^{(m)}$, $m = 1, 2, ..., n_t$, the symbols transmitted with different antennas are independent, i.e., $E_s[s_{k,b}^{(m_1)} s_{k,b}^{(m_2)}] = \sigma_s^2 \delta(m_1 - m_2)$, with $\delta(.)$ as the Dirac delta function, and by using that $E_s[s_{k,b}^{(m)}] = 0$ for the symmetric constellation points, $\mu_{21,b}$ is further expressed as

$$\mu_{21,b} = \sigma_s^2 \sum_{m=1}^{n_t} |h_b^{(m)}|^2 + \sigma_w^2. \tag{3}$$

Similarly, for the fourth-order/ two-conjugate statistic, $\omega_{42,b} \triangleq E_{s,w}[|r_{k,b}|^4] - 2(E_{s,w}[|r_{k,b}|^2)^2$, [1] one can easily obtain

---

[1] Note that $\omega_{42,b}$ is related to the fourth-order/ two-conjugate cumulant, with a difference of $\mu_{20,b}\mu_{22,b}$, where $\mu_{20,b}$ and $\mu_{22,b}$ are the second-order/ zero-and two-conjugates, respectively.

$$\omega_{42,b} = \omega_{42}^s \sigma_s^4 \sum_{m=1}^{n_t} |h_b^{(m)}|^4, \tag{4}$$

where ω_{42}^s denotes the fourth-order/ two conjugate statistic for unit variance constellations.

With the channel coefficients corresponding to different transmit antennas being independent[2] complex-valued zero- mean Gaussian random variables with variance σ_h^2, i.e., $E_{\mathbf{H}_b}[\mathbf{H}_b^T \mathbf{H}_b] = \sigma_h^2 \mathbf{I}$, and employing the following property of a complex Gaussian random variable $x \sim CN\left(0, \sigma_x^2\right)$ that [11]

$$E_x\left[|x|^{2n}\right] = n!\sigma_x^{2n}, \tag{5}$$

the expectations of the second-order moment and fourth-order statistic in (3) and (4) over channel distributions are

$$\mu_{21} \overset{\Delta}{=} E_{\mathbf{H}_b}[\mu_{21,b}] = \sigma_s^2 \sum_{m=1}^{n_t} E_{\mathbf{H}_b}\left[|h_b^{(m)}|^2\right] + \sigma_w^2$$
$$= n_t \sigma_h^2 \sigma_s^2 + \sigma_w^2 \tag{6}$$

and

$$\omega_{42} \overset{\Delta}{=} E_{\mathbf{H}_b}[\omega_{42,b}] = \omega_{42}^s \sigma_s^4 \sum_{m=1}^{n_t} E_{\mathbf{H}_b}\left[|h_b^{(m)}|^4\right]$$
$$= 2n_t \omega_{42}^s \sigma_s^4 \sigma_h^4. \tag{7}$$

Furthermore, with the modulation type and noise power[2] known at the receive-side, by employing (6) and (7), n_t can be straightforwardly expressed as

$$n_t = \frac{2\omega_{42}^s\left(\mu_{21} - \sigma_w^2\right)^2}{\omega_{42}}. \tag{8}$$

In practice, the statistical moments are estimated by time averages [12]. Furthermore, an unbiased estimator is of interest, as on average, the expected value of the parameter being estimated equals its actual value. For (8), the following unbiased estimators are employed to estimate the corresponding statistics, i.e., μ_{21}, $\zeta \overset{\Delta}{=} (\mu_{21})^2$ and ω_{42}, respectively.

$$\hat{\mu}_{21} = \frac{1}{N_b N_c} \sum_{b_1=1}^{N_b} \sum_{k_1=1}^{N_c} |r_{k_1,b_1}|^2, \tag{9}$$

$$\hat{\zeta} = \frac{1}{N_b\left(N_b - 1\right)N_c\left(N_c - 1\right)} \sum_{b_1=1}^{N_b} \sum_{\substack{b_2=1 \\ b_2 \neq b_1}}^{N_b} \sum_{k_1=1}^{N_c} \sum_{\substack{k_2=1 \\ k_2 \neq k_1}}^{N_c} |r_{k_1,b_1}|^2 |r_{k_2,b_2}|^2, \tag{10}$$

---

[2] Note that the deviation from this assumption is considered later in the paper, in Section 4.

$$\hat{\omega}_{42} = \frac{1}{N_b N_c} \sum_{b_1=1}^{N_b} \sum_{k_1=1}^{N_c} |r_{k,b}|^4$$
$$- \frac{2}{N_b N_c (N_c-1)} \sum_{b_1=1}^{N_b} \sum_{k_1=1}^{N_c} \sum_{\substack{k_2=1 \\ k_2 \neq k_1}}^{N_c} |r_{k_1,b_1}|^2 |r_{k_2,b_1}|^2. \tag{11}$$

It can be easily shown that $E[\hat{\mu}_{21}] = \mu_{21}$, $E[\hat{\zeta}] = (\mu_{21})^2$, and $E[\hat{\omega}_{42}] = \omega_{42}$, where $E[.] \triangleq E_{\mathbf{H}_b}[E_{s,w}[.]]$. It is worth noting that $(\hat{\mu}^{(1)})^2$ cannot be employed for the estimation of ζ, as it results in a biased estimator.

With (8), (9), (10), and (11), one obtains the following decision statistic for the number of transmit antennas,

$$\Psi = \frac{2\omega_{42}^s (\hat{\zeta} - 2\hat{\mu}_{21}\sigma_w^2 + \sigma_w^4)}{\hat{\omega}_{42}}. \tag{12}$$

It can be easily noticed that Ψ is a continuous random variable, whereas n_t takes discrete values; hence, regions of decision need to be set up to estimate the number of transmit antennas, along with their corresponding thresholds. Since

$$E[\Psi] \approx \frac{2\omega_{42}^s E[\hat{\zeta} - 2\hat{\mu}_{21}\sigma_w^2 + \sigma_w^4]}{E[\hat{\omega}_{42}]} = \frac{2\omega_{42}^s (\mu_{21} - \sigma_w^2)^2}{\omega_{42}} = n_t, \tag{13}$$

the decision is made according to the following criterion:

$$\Gamma_{n_t-1} < \Psi \leq \Gamma_{n_t} \to \hat{n}_t = n_t \qquad n_t = 1,2,3,\ldots \tag{14}$$

where $\Gamma_0, \Gamma_1, \Gamma_2, \ldots$ represent the decision thresholds, with $\Gamma_0 = -\infty$ and $n_t < \Gamma_{n_t} < n_t + 1$. A formal description of the proposed algorithm is presented below.

Algorithm 1

1. Acquire the measurement $r_{k,b}$, $k = 1, \ldots, N_c$, $b = 1, \ldots, N_b$
2. Compute the decision statistic Ψ according to (12)
3. Initialize $i = 1$
4. Set the threshold value Γ_i
   If $\Gamma_{i-1} < \Psi \leq \Gamma_i$
   $\qquad \hat{n}_t = i$
   else
5. Increment $i = i + 1$ and go to step 4
   end

---

4 Simulation Results

In this section, we examine the detection performance of the proposed algorithm through several simulation experiments.

4.1 Simulation Setup

We consider a system employing spatial multiplexing transmission scheme, with $N_c = 100$ (e.g., $f_d = 200$ Hz and $T_s = 10$ μs). Unless otherwise mentioned, $N_b = 100$ and the modulation is quadrature phase-shift-keying (QPSK). The channel coefficients are independent complex Gaussian random variables with zero-mean and variance σ_h^2. The additive white noise is modeled as a complex Gaussian random variable with zero-mean and variance σ_w^2. The average SNR per transmit antennas is defined as $\gamma \triangleq 10 \log\left(\frac{\sigma_h^2 \sigma_s^2}{\sigma_w^2}\right)$dB. Without loss of generality, we consider $\sigma_h^2 \sigma_s^2 = 1$. The thresholds to make a decision are set as $\Gamma_{n_t} = n_t + 1/2$, $n_t = 1, 2, \dots$. The overall detection performance is presented in terms of the probability of correct detection, $P(\hat{n}_t = m | n_t = m)$, $m = 1, 2, \dots, 4$, and the average probability of correct detection, $P_c = \frac{1}{3} \sum_{m=1}^{3} P(\hat{n}_t = m | n_t = m)$, obtained from 1000 Monte Carlo trials for each m.

4.2 Simulation Results

Fig. 1 shows $P(\hat{n}_t = m | n_t = m)$ versus SNR for different number of transmit antennas, n_t, $n_t = 1, \dots, 4$, and different N_b values. As can be seen, the proposed algorithm exhibits a good performance over a wide range of SNRs for $N_b = 100$ and 1000, and the probability of correct detection goes to one even at negative SNRs for $N_b = 1000$. The performance improves as either N_b or SNR increases, which can be easily explained, as each leads to a reduced estimation error of the statistics in (9), (10), and (11). Additionally, the probability of correct detection decreases as the number of transmit antenna increases; this is because the variance of the decision statistic Ψ in (12) increases with n_t, as confirmed by simulation experiments.

In Fig. 2, the effect of the noise power mismatch, i.e, $\hat{\sigma}_w^2 - \sigma_w^2$, on the probability of correct detection is illustrated at SNR=10 dB. As can be observed, the proposed algorithm is relatively robust to the noise power mismatch. This can be easily explained, as the effect of the noise power mismatch on the test statistic Ψ in (12) is not significant for a large enough observation interval.

Fig. 3 shows the effect of the frequency offset normalized to the data rate, Δf, on P_c. As can be seen, the proposed algorithm is completely robust to the carrier frequency offset. This is because such an effect is eliminated through the absolute value operator in the definition of the second-order moment and the fourth-order statistic.

Fig. 4 presents the effect of the modulation format on the average probability of correct detection, P_c. As can be seen, while the proposed algorithm is relatively robust to the modulation format at positive SNRs, a better performance is achieved for M-ary PSK when compared with M-ary quadrature amplitude modulation (QAM) at negative SNRs. This can be explained, as the effect of the modulation format, ω_{42}^s is not totally eliminated through $\hat{\omega}_{42}$ for M-ary QAM due to less accurate estimates, particularly at negative SNRs.

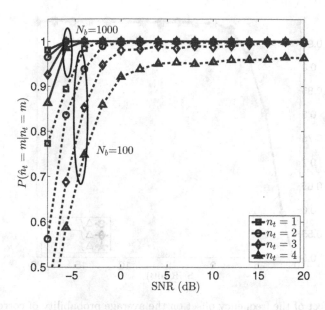

Fig. 1. The probability of correct detection, $P(\hat{n}_t = m|n_t = m)$ versus SNR for different n_t and N_b values.

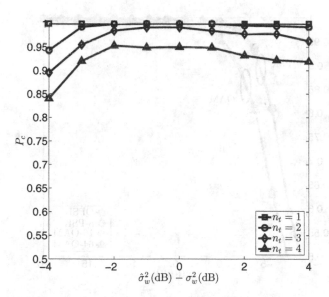

Fig. 2. The effect of the noise power mismatch on the probability of correct detection, $P(\hat{n}_t = m|n_t = m)$ at SNR=10 dB.

448   M. Mohammadkarimi et al.

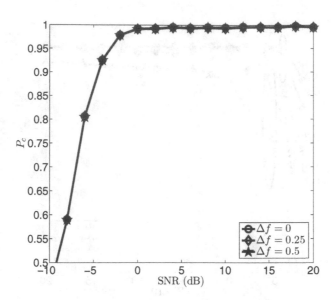

Fig. 3. The effect of the frequency offset on the average probability of correct identification, P_c.

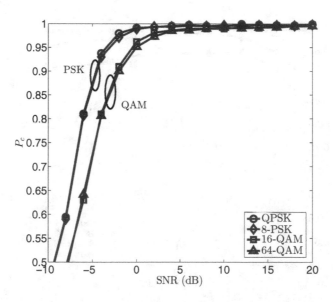

Fig. 4. The effect of the modulation format on the average probability of correct identification, P_c.

Fig. 5 shows the effect of the spatially correlated fading on P_c versus SNR for a correlation coefficient $\rho = 0, 0.4, 0.6,$ and 0.8. As can be observed, the performance of the proposed algorithm is robust to the spatial correlation for $\rho < 0.6$. This can be explained, as for low values of ρ, $E_{\mathbf{H}_b}\left[\sum_{m=1}^{n_t} |h_b^{(m)}|^{2l}\right], l = 1, 2,$ remains approximately equal to $l!2n_t\sigma_h^{2l}$ (see (6) and (7)), and (8) remains valid for the number of transmit antennas detection.

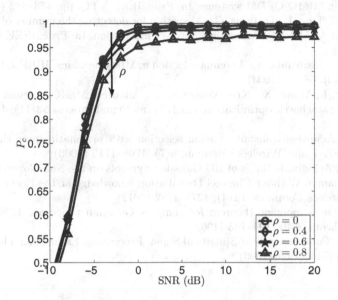

Fig. 5. The effect of the spatially correlated fading on the average probability of correct identification, P_c.

5 Conclusion

A novel feature-based algorithm was introduced for the detection of the number of transmit antennas. This relies on a second-order moment and a fourth-order statistic of the received signal, and exploits the time diversity of the fading channels, while employing a single receive antennas. The proposed algorithm attains a good performance at low SNRs, being robust to the carrier frequency offset and relatively robust to the modulation format. Additionally, it exhibits a good performance in the presence of noise power mismatch and spatially correlated fading.

References

1. Somekh, O., Simeone, O., Bar-Ness, Y., Su, W.: Detecting the number of transmit antennas with unauthorized or cognitive receivers in MIMO systems. In: Proc. IEEE MILCOM, pp. 1–5 (2007)

2. Shi, M., Bar-Ness, Y., Su, Wei.: Adaptive estimation of the number of transmit antennas. In: Proc. IEEE MILCOM, pp. 1–5 (2007)
3. Oularbi, M.-R., Gazor, S., Aissa-El-Bey, A., Houcke, S.: Enumeration of base station antennas in a cognitive receiver by exploiting pilot patterns. IEEE Commun. Lett. **17**(1), 8–11 (2013)
4. Oularbi, M-R., Gazor, S., Aissa-El-Bey, A., Houcke, S.: Exploiting the pilot pattern orthogonality of OFDMA signals for the estimation of base stations number of antennas. In Proc. IEEE WOSSPA, pp. 465–470 (2013)
5. Ohlmer, E., Ting, L., Fettweis, G.: Algorithm for detecting the number of transmit antennas in MIMO-OFDM systems. In: Proc. IEEE VTC, pp. 478–482 (2008)
6. Ohlmer, E., Ting, L., Fettweis, G.: Algorithm for detecting the number of transmit antennas in MIMO-OFDM systems: receiver integration. In: Proc. IEEE VTC, pp. 1–5 (2008)
7. Sanayei, S., Nosratinia, A.: Antenna selection in MIMO systems. IEEE J. Commun. Mag. **42**(10), 68–73 (2004)
8. Berenguer, I., Wang, X., Krishnamurthy, V.: Adaptive MIMO antenna selection via discrete stochastic optimization. IEEE Trans. Signal Process **53**(11), 4315–4329 (2005)
9. Yuan, J.: Adaptive transmit antenna selection with pragmatic space-time trellis codes. IEEE Trans. Wireless Commun. **5**(7), 1706–1715 (2006)
10. Rusek, F.: Achievable Rates of IID Gaussian Symbols on the Non-Coherent Block-Fading Channel Without Channel Distribution Knowledge at the Receiver. IEEE Trans. Wireless Commun. **11**(4), 1277–1282 (2012)
11. Reed, I.: On a moment theorem for complex Gaussian processes. IEEE Trans. Inform. Theory. **8**(3), 194–195 (1962)
12. Kay, S.M.: Fundamentals of Statistical Signal Processing: Estimation Theory, vol. I. Pearson Education (1993)

Localization of Primary Users by Exploiting Distance Separation Between Secondary Users

Audri Biswas[1]([✉]), Sam Reisenfeld[1], Mark Hedley[2],
Zhuo Chen[2], and Peng Cheng[2]

[1] Department of Engineering, Macquarie University, Sydney, NSW 2109, Australia
{Audri.biswas,Sam.Reisenfeld}@mq.edu.au
[2] Digital Productivity Flagship, CSIRO, Marsfield, NSW 2122, Australia
{Mark.Hedley,Zhuo.Chen,Peng.Cheng}@csiro.au

Abstract. Accurate localization of Primary Users (PUs) is an extremely useful procedure which can improve the performance of Cognitive Radio (CR) by more efficient dynamic allocation of channels and transmit powers for unlicensed users. In this paper, we analyze the performance of a Compressive Sensing (CS) method which simultaneously yields the PU transmitter locations and transmit powers for any channel in a Cognitive Radio Network (CRN). Additionally, we propose a novel approach of selectively eliminating Secondary User (SU) power observations from the set of SU receiving terminals such that pairs of the remaining SUs are separated by a minimum geographic distance. The modified algorithm demonstrates substantial performance improvements compared to random deployment of receiving terminals. Simulations were run for both the cases of uniform and Gaussian distributions for the SU random locations. The simulation results indicate that the new approach significantly reduced the number of received power measurements from SU terminals required to achieve a particular level of performance.

Keywords: Cognitive Radio · Radio Environment Map · Compressive sensing · Localization power · Measurements

1 Introduction

The spectrum scarcity along with inefficient spectrum usage has motivated the development of *Cognitive Radio* (CR). The increasing demand of high data rates due to large numbers of portable hand-held devices initiated significant research in the field of interference mitigation and effective spectral utilization. CR provides a promising solution to the existing problem by efficiently using the under-utilized spectrum to facilitate services by *Dynamic Spectrum Sharing* (DSS) for both licensed and unlicensed users. CR technology is based on the concept of learning the state of channel use of PUs, and subsequent efficient allocation of channels and transmit parameters to SUs. This allocation takes into account maximum acceptable interference levels to PUs and the throughput and performance requirements of SUs.

© Institute for Computer Sciences, Social Informatics and Telecommunications Engineering 2015
M. Weichold et al. (Eds.): CROWNCOM 2015, LNICST 156, pp. 451–462, 2015.
DOI: 10.1007/978-3-319-24540-9_37

In a Cognitive Radio Network, both PUs and SUs share the same channels. Since SUs have lower priority, the channel use is constrained by a maximum acceptable level of interference to PUs. Many efforts have been made in previous literature [1][2] to tackle the issue of interference mitigation but only a few research papers have been published on channel collision avoidance based on the utilization of a *Radio Environment Map* (REM). To generate a REM, the locations of the transmitters and their transmit power levels need to be accurately estimated. From this estimation, the received power level throughout a two dimensional area may be estimated. For the REM, the received power levels interpolated over a two dimensional geographic area are obtained through the use of analytic equations for signal propagation.

In CR, the REM is extremely useful in secondary user channel and transmit parameter selection. This selection must be made with the dual requirements of SU communication effectiveness and bounded interference to PUs. The bounded interference to PUs can only be maintained if the PU locations and received power levels from other PUs, are known by SUs. Therefore an accurate REM is crucial for effective CR operation.

In [3], a cooperative algorithm is formulated that takes the received signal strength at each SU to create a weighting function and uses it to compute the location of multiple PUs. Although it has relatively low computational complexity, it requires a high density of SUs, and the performance degrades with channel fading. The work in [4] and[5] is based on the concept of using sectorized antennas to detect Direction of Arrival (DOA) of a signal. The phase information of a received signal is exploited to estimate the position of PUs. However, this technique might not be feasible for a practical CRN implementation due to antenna requirements which may be impractical for portable devices.

In this paper we adopt a Compressive Sensing (CS) technique to retrieve the locations of multiple transmitting PUs in a CRN. The approach relies on a location fingerprinting approach, where a certain geographic area is discretized into equally spaced grid points. The PUs are assumed to be positioned at a subset of the grid points. The SUs are also assumed to be positioned at some known locations in the area of interest. Each SU measures Received Signal Strength (RSS) from target PUs. From this set of measurements, there is an attempt to recover the PU locations and transmit power levels. It is usually the case that the number of PUs is much smaller than the number of grid points. Consequently, the set of equations for power levels transmitted by PUs is underdetermined and there are many possible solutions. When the number of PUs is much smaller than the number of grid points, the sparsest solution for the set of equations yields accurate power levels at the correct grid points. Compressive sensing can be used to obtain the data required for the formulation of the REM. Similar techniques were used in [6], [7], [8] and [9].

In a physical system, some of the SUs will be closely geographically located. Having closely placed SUs introduces correlated observations which increases the observation coherence. This may have a negative impact upon the performance of CS algorithm. To improve the performance of the CS algorithm, we propose a novel approach. The measurements of closely spaced SUs are removed from the set such as to increase the minimum distance separation between adjacent SUs in the measurement set. Our method achieved superior detection of multiple PUs with significantly fewer SU measurements, compared to random deployment of SUs.

In this paper, the locations of SUs are specified by two dimensional vectors. Both the cases of uniform distribution and Gaussian distribution were considered for the random assignment of SU positions. Irrespective of distribution used, our novel approach of pre-selecting SU power measurements appears to reduce the number of measurements required to achieve reliable detection. Section 2 discusses the background of compressive sensing. Sections 3-5 describe the system model. Section 6 presents the simulation results which validate the effectiveness of our proposed method. The conclusion is given in Section 7.

2   Compressive Sensing

The CS technique is an approach for the solution of an under-determined set of equations for which the solution vector is known to be sparse. Some data vectors are sparse while others can be made more sparse by an appropriate basis transformation. A typical example would be the time frequency pair. A signal, which is a linear combination of several frequency components, can be easily retrieved by exploiting the sparsity in frequency domain. The complex Fourier Transform basis functions can be used to represent the time domain signal with few non-zero coefficients. In such case the CS algorithm can be used to obtain a sparsest solution vector to a set of underdetermined equations. The sparse vector, $x_{N \times 1}$ is the solution with the minimum number of non-zero elements. If $y_{M \times 1}$ is the raw observation vector obtained by the SU power measurments, there exist the following relationship,

$$y = \phi x, \tag{1}$$

where $\phi_{M \times N}$ is a measurement matrix, representing the power propagation losses from each grid point to each SU. In [7] it states that, a matrix ϕ satisfies *Restricted Isometry Property* (RIP) condition, when all subsets of S columns chosen from ϕ are nearly orthogonal. Once this is true, there is a high probability of completely recovering the sparse vector with at least $M = CK \times \log_e(N/K)$ measurements (where K is the number of PUs and C is a positive constant) using l_1 -minimization algorithm [10]. This can be can be expressed as,

$$min \, \|x\|_1 = min \sum_i |x_i|$$

subject to

$$y = \phi x. \tag{2}$$

This formulation is valid for a noiseless scenario but when external noise is considered the algorithm is modified to a *Second-Order Cone Program* for an optimized solution for a defined threshold [10]. This can be stated as,

$$min \, \|\boldsymbol{x}\|_1 = min \sum_i |x_i|$$

subject to

$$\|\boldsymbol{y} = \phi\boldsymbol{x}\|_2 \le \varepsilon, \qquad (3)$$

where $\|\cdot\|_p$ is the l_p-norm and ε is the relaxation constraint for measurement errors. The sparest solution for $\boldsymbol{x}$ is the solution with minimum $\|\boldsymbol{x}\|_0$. However, the CS algorithm is effective because the same solution vector usually has minimum l_0 norm and minimum l_1 norm.

3   System Model

Let us consider a square area discretized into equally spaced $P \times P$ grid where, K PUs are randomly positioned at unique grid points. For simplicity of illustration, we assume that each PU is assigned a single dedicated sub-channel to carry out duplex communication with the base station. Now to observe radio environment and detect the free spectrum, M SUs are deployed randomly in the area of interest. Unlike [6] and [8] the SUs are not placed on the grid points. We adapted a more realistic approach of allowing the SUs to be placed at some known locations in the area. They have the added flexibility of being positioned at non-discretized points on the map. The SUs are controlled and managed by a central node called the *Fusion Centre* (FC). There exist a common control channel between central node and SUs for effective communication of RSS observations and channel allocation information. The FC processes the signal level measurements and manages SU channel allocation. The most crucial assumption in the model is that, spatial coordinates of both the grid points and SUs are known a priori by the FC which receives sensing information from each individual SU. The received power at a SU is a function of distance between the PU and SU as well as shadowing loss. The wireless channels are corrupted by noise and are also considered to be affected by lognormal shadowing. The simplified path-loss model as a function of distance may be described as,

$$Pathloss_{dB}(d) = K_1 dB + 10\eta log_{10}(\frac{d}{d_0}) + \alpha, \qquad (4)$$

where,

d is transmission distance in meters,

d_0 is the reference distance of the antenna far field,

K_1 is a dimensionless constant,

η is the propagation loss exponent,

α is the shadowing loss in dB.

K_1 is a unit-less constant that relies on the antenna characteristics and average channel attenuation and $K_1 dB = 10log_{10}(K_1)$ [11]. α accounts for the random attenuation of signal strength due to shadowing where α in dB scale is a Gaussian random variable with zero mean and standard deviation $\sigma_{dB} = 5.5$dB [3]. This model was used in [3] for both multipath and shadowing characterization.

4 Localization Using Compressive Sensing

This section combines the location dependent RSS information at each SU to formulate a sparse matrix problem, which can then be solved using the CS method to obtain the exact location of PUs in a CRN. Our grid layout consists of N grid points, with grid resolution w in both x-axis and y-axis. The N grid points are located at $\{V_n, 1 \leq n \leq N\}$, where V_n is a two dimensional position vector. The M SUs are located at $\{U_m, 1 \leq m \leq M\}$, where U_m is also a two dimensional position vector. Earlier in Section 3 we mentioned K PUs are positioned only at K discrete grids where $K < N$. The FC is assumed to have prior knowledge of V_n and U_m. Using the distance information and signal propagation model described in (4) a measurement matrix Φ is constructed. The entries of the matrix are the channel gain and are expressed using the following equations,

$$d_{mn} = \| U_m - V_n \|_2, \tag{5}$$

$$\Phi_{mn} = 10^{\frac{-Pathloss_{dB}(d_{mn})}{10}}, \tag{6}$$

where d_{mn} is the distance between m^{th} SU and n^{th} grid point and Φ_{mn} is the pathloss between m^{th} SU and n^{th} grid point. Let Y be a $M \times 1$ column vector where the mth element, Y_m, represents the summation of received power from K PUs on m^{th} SU.

$$Y_m = \sum_{k=1}^{K} Q_{m,k}, \tag{7}$$

where,

$$Q_{m,k} = 10^{\frac{Q_{m,k,dB}}{10}}$$

and,

$$Q_{m,k,dB} = P_{k,t} - Pathloss_{dB}(d_{mk})$$

where, $Q_{m,k}$ is the power received at SU m which
was transmitted by PU k,
$P_{k,t}$ is the power transmitted by user k,
and, $d_{m,k}$ is the distance between SU m and PU k.

Equation (6) and (7) may be combined to formulate a CS problem similar to (2). It is assumed that the FC has complete knowledge of Φ. Therefore,

$$Y = \Phi X \tag{8}$$

with $X_{N \times 1}$ being a $N \times 1$ column vector that is to be recovered using CS approach described in Section 2. In a realistic scenario, the observations are corrupted with noise power vector P_n. The elements of P_n are statistically independent with variance σ_n^2, and are chi-square distributed with 1 degree of freedom. We can include the effect of additive noise by,

$$Y_n = \Phi X + P_n. \tag{9}$$

Since the model assumes having only few PUs on a large grid size N, the vector $X_{N \times 1}$ satisfies the sparsity requirement for accurate recovery using a CS algorithm. Due to its sparse condition, the vector will have only few nonzero elements representing the transmit powers while the indices corresponding to nonzero elements indicate the grid points on which transmitting PUs are located. Hence using a single compressed sensing problem we can jointly estimate both the locations and transmit powers of multiple PUs by solving (3) described in Section 2. From the estimation, FS can approximate the received power level throughout a two dimensional area, using the path loss model in (4).

5   Data Processing

Based on the problem formulation in Section 4, $Y_{M \times 1}$ is a power observation vector with each row representing sum of RSS received from K PUs on m^{th} SU, and $\Phi_{M \times N}$ is the measurement matrix with channel gain from each grid point. The small grid separation adds large coherence between the columns of the measurement matrix and this may violate the RIP condition[12]. A matrix transformation may be employed to increase the incoherence between the columns. We adopt a data processing technique described in [6] and [8] to decorrelate the rows which are the observation of signal strength from grid points on each SU. Let T be a processing operator,

$$T = QR^+ \tag{10}$$

where, $Q = orth(\Phi^T)^T$. The built in function of Matlab, $orth(B)$ returns an orthonormal basis of the range of B, and B^T returns the transpose of B. R^+ is the Moore-Penrose pseudoinverse of a matrix R, where $R = \Phi$. Applying the operator T on both sides of (9) yields,

$$QR^+(Y_n) \; = \; QR^+\Phi X + QR^+ P_n \; = \; Q\Phi^+\Phi X + QR^+ P_n \; = \; Ax + \omega$$

$$Y' = AX + \omega. \tag{11}$$

Let Y' be $QR^+(Y_n)$, the noisy processed observation vector. $A = Q\Phi^+\Phi$ be the processed measurement matrix and $\omega = QR^+ P_n$ is the processed measurement noise. The row vectors are being orthogonalised by Q while the columns are decorrelated by the influence of $\Phi^+\Phi$. Hence we can claim that matrix A satisfies the RIP condition. Note that [6] and [8] considered $\Phi^+\Phi = \mathbb{I}_N$, as a diagonal identity matrix. Although $\Phi^+\Phi$ acts like an identity on a portion of the space in the sense that it is symmetric. However it is not an identity matrix. After applying the processing operator, CS may be used to recover the sparse vector from processed observation Y', via l_1-minimization program [6].

6   Simulation and Results

The localization accuracy of the CS algorithm can be effected by certain external factors such as *Signal to Noise Ratio* (SNR), shadowing, density of SUs and distribution of SUs. This section analyses the dependency of these factors on the performance parameters of three l_1 constrained optimization algorithms (L1-Magic, OMP and CoSAMP) to produce an accurate result. L1 Magic, CoSAMP, and OMP are three numerical algorithms for constrained l_1 vector optimization [13], [14] and [15]. The performance parameters are categorized as,

$$Detection Ratio = \left[\frac{PU_{Det}}{PU_{Total}} \right]$$

$$Normalized\ Error\ Per\ Grid = \frac{1}{N} \left\| X_{org} - X_{est} \right\|$$

where PU_{Det} is the number of detected PUs; PU_{Total} is the sum of the PUs in the network; X_{org} is the original sparse vector; X_{est} the recovered vector using CS algorithms. The average absolute error between the vectors X_{org} and X_{est} is obtained by simulation. This is used to evaluate the accuracy of the algorithms to reconstruct a sparse vector with minimum non-zero coefficient. Furthermore to study the impact of each factor, the simulation is analyzed independently to demonstrate the robustness and reliability of the algorithms.

6.1   Simulation Setup

The simulation is carried out on a 43×43 (i.e. $N = 1849$) square grid with a grid separation of 80m. Among the 1849 grid points, 10 PUs are uniformly distributed on the grid points. The transmit power is random and uniformly distributed over the range of 1 to 5 Watts. The scenario consists of 160 SUs with a two dimensional, zero mean, Gaussian spatial distribution with standard deviation σ_{sd}. The shadowing factor is log normal distributed.

Simulation (I) - Impact of SNR. Signal to noise ratio is one the crucial factors effecting the performance of each algorithm. SNR is calculated at the receiver as the ratio of sum of received powers at a SU to σ_n^2. Where,

σ_n^2 is the variance of the additive, zero mean, Gaussian noise.

Then,

$$SNR(dB) = 10 \log_{10}(\frac{1}{M} \sum_{i=1}^{M} \frac{Y_i}{\sigma_n^2}).$$

Y_i is the received RSS from all transmitting PUs at i^{th} SU. As the received signal power is position dependent, SNR will vary with respect to the positioning of

SUs. Such scenario prompted us to take the average SNR over M elements of the observation vector. Fig 1(a) and (b) shows the plots for detection ratio of PUs and normalized error per grid versus average received SNR in dB. As shown in Fig. 1 (a) when SNR < 12dB, L1-Magic performs better than CoSAMP however when SNR > 15dB, CoSAMP outperforms L1-Magic and OMP. At a higher SNR = 25dB, both CoSAMP and L1-Magic achieved a detection ratio of 1 while OMP is at 0.6. Fig. 1(b) shows that, with gradual increase in SNR, CoSAMP generates fewer normalized errors per grid compared to L1-Magic and OMP. Even at a low SNR = 15dB, CoSAMP produces 50% and 54% less errors compared to L1-Magic and OMP.

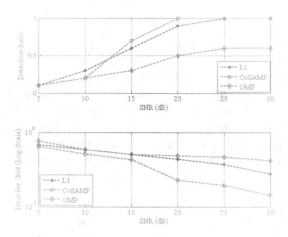

Fig. 1. (a) SNR vs detection ratio and (b) SNR vs normalized error per grid

Similation (II) - Sampling Ratio. Sampling ratio $\frac{M}{N}$ is another major factor that has a significant impact on the performance of these algorithms. In this simulation we start with 200 SUs to detect the position of 10 PUs, where at each iteration 20 SUs are randomly removed to observe the effect of reduced sampling points. The SNR is kept constant at 25dB. The plots in Fig.2 follows a similar trend as in Fig. 1. At very low sampling ratio of 0.05, almost all three algorithms fails to recover an accurate sparse solution as solving an undermined system with such small number of measurements is not feasible regardless of any methods used. However with increase in sampling ratio CoSAMP achieves detection ratio of 1 using 10% less SUs compared to L1-Magic. OMP seems to require higher number of SUs to meet the accuracy of CoSAMP and L1-Magic. Similar conclusion can be drawn from Fig. 2(b), where the graph of normalized error per grid for CoSAMP as a function of sampling ratio decreases much rapidly compared to the other two algorithms. Results from simulation (I) and (II) indicate that, CoSAMP is more robust and can perform with superior results compared to other two algorithms. The next set of simulations will be carried out using CoSAMP algorithm only.

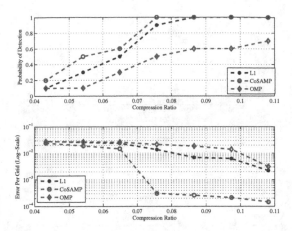

Fig. 2. (a) Sampling ratio vs detection ratio (b) sampling ratio vs normalized error per grid

Refinement of Secondary User Set. From the theory of CS we learn that, columns with higher incoherence increases the probability of accurate recovery in CS [12]. In a typical scenario, we randomly distribute the SUs in the area of interest. This might cause a few SUs to be placed very closely to each other. Though close sensor spacing might be useful for some localization algorithms such as weighted centroid [3], this condition produces similar observations at different SUs and does not yield good performance using CS. The effectiveness of our technique to refine the SU measurement set was verified by Matlab. To test our refinement technique, we extracted two sets of SU positions from both a uniform and Gaussian spatial distribution. Our Matlab script takes the 2D position matrix of SUs and $min-dist$ (minimum distance separation parameter) as an input and generates a refined set of SUs such that each SU is separated from an adjacent SU by $min-dist$. We assume in Section III that the spatial coordinates of the SUs are known a *priori* to FC. The known coordinates are then used to calculate the distance between pairs of SUs. The script identifies pairs of SUs with $min-dist$ separation and removes one of the SUs from each pair. The script iterates through a loop until all SUs have a $min-dist$ or greater separation between them. Fig. 3 shows that, as we sweep across minimum distance separation between SUs, the number of SU curve deceases even while maintaining a detection ratio of 1 for both sets of SUs. Our novel approach achieved reduction in the number of SU measurements by 21% and 30% for uniform distribution and normal distribution, respectively. The algorithm fails at a minimum distance separation of 400m for uniform and 600m for normal. This because at that point there are insufficient measurements to solve an undetermined linear system.

Simulation (III) - SU Distribution. In this section we observe the impact of the spread of a particular spatial distribution, used to obtain location of

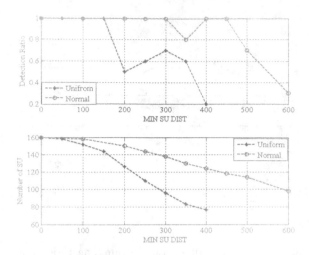

Fig. 3. (a) spread of distributions vs detection ratio (b) spread of distributions vs coherence (c) spread of distributions vs SNR

SUs in a CRN. The two dimensional SU positions are two dimensional random vectors with statistically independent elements. Two cases were considered. In the first case each element is uniformly distributed over $[-X_{max}\sigma_{sd}, X_{max}\sigma_{sd}]$. In the second case, each element is zero mean Gaussian distributed with standard deviation $\{X_{max}\sigma_{sd}\}$.

While keeping the SNR constant and the number of SUs and PUs constant, the σ_{sd} is varied in the range $[1, 6]$. The following simulation is carried on CoSAMP algorithm only. Fig. 4 shows the impact of σ_{sd} on the detection ratio of CoSAMP; coherence of the measurement matrix Φ and average SNR at SUs. When $\sigma_{sd} < 2$ both sets of SUs generate poor detection of PUs. This is because of high coherence of Φ as shown in Fig. 4(b). As $\sigma_{sd} > 2$, coherence of measurement matrix for both plots monotonically decreases which enables CS algorithm to perform efficiently. Fig. 4(a) shows that, each set of SUs from different spatial distribution reaches a maximum detection ratio before dropping to a minimum. This behavior can be explained from the SNR plot in Fig. 4(c). When $\sigma_{sd} > 1$, SUs are being spread out widely across the area causing some SUs to have large distance separation from target PUs. This reduces the RSS and lowers the average SNR at the SUs. The SNR reaches a minimum, where it is difficult for CS algorithm to offer perfect recovery. However the rate at which average SNR decreases is dependent on distribution. Normal distribution having an infinite tail, might push some SUs further away from the area of interest. This influences the performance by having corrupted observations and can reduce the average SNR significantly. From the plots we can observe that, when σ_{sd} is too large, CS fails to perform efficiently in spite of having lower coherence between the columns of Φ.

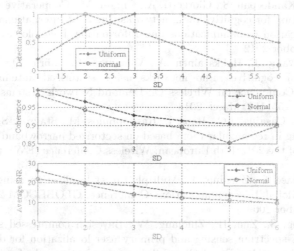

Fig. 4. (a) spread of distributions vs detection ratio (b) spread of distributions vs coherence (c) spread of distributions vs SNR

7 Conclusion

In this paper we formulated a sparse problem to jointly determine the locations and transmission power of target PUs in a CRN using CS algorithm. Useful information about PUs can be extracted with selective positioning of SUs. We proposed a novel approach of pre-selecting a refined set of SUs from a randomly distributed set. A minimum distance separation is used as a constraint to remove closely placed SUs as well as highly correlated observations. This enables CS algorithms to accurately reconstruct a unique sparse vector with location and transmit power level information for transmitting PUs. Simulation results suggest that our approach achieved a reliable determination of PU positions and transmit powers in a practical CRN with a small number of SUs, as sensing stations. Reliable determination was demonstrated when the number of SUs are very close to the theoretical measurement bound of CS. To further test the effectiveness of our method, simulations were run for two spatial probability distributions for SU positions. In both cases our approach achieved the maximum detection ratio with relatively few secondary users performing receive power sensing.

References

1. Wang, S., Yang, Q., Shi, W., Wang, C.: Interference mitigation and resource allocation in cognitive radio-enabled heterogeneous networks. In: 2013 IEEE of Global Communications Conference (GLOBECOM), pp. 4560–4565, December 2013
2. Hu, D., Mao, S.: Co-channel and adjacent channel interference mitigation in cognitive radio networks. In: Military Communications Conference, MILCOM 2011, pp. 13–18, November 2011

3. Mariani, A., Kandeepan, S., Giorgetti, A., Chiani, M.: Cooperative weighted centroid localization for cognitive radio networks. In: 2012 International Symposium on Communication Communications and Information Technologies (ISCIT), pp. 459–464, October 2012
4. Werner, J., Wang, J., Hakkarainen, A., Valkama, M., Cabric, D.: Primary user localization in cognitive radio networks using sectorized antennas. In: 2013 10th Annual Conference on Wireless On-Demand Network Systems and Services (WONS), pp. 155–161, March 2013
5. Arambasic, I., Casajus, J.Q., Raos, I., Raspopoulos, M., Stavrou, S.: Anchor-less self-positioning in rectangular room based on sectorized narrowband antennas. In: Proceedings of the 2013 19th European, Wireless Conference (EW), pp. 1–6, April 2013
6. Chen, F., Valaee, S., Zhenhui, T.: Multiple target localization using compressive sensing. In: Global Telecommunications Conference, GLOBECOM 2009, pp. 1–6. IEEE, November 2009
7. Xue, L., Hong, S., Zhu, H., Zhiqiang, W.: Bayesian compressed sensing based dynamic joint spectrum sensing and primary user localization for dynamic spectrum access. In: 2011 IEEE of Global Telecommunications Conference (GLOBECOM 2011), pp. 1–5, December 2011
8. Jayawickrama, B.A., Dutkiewicz, E., Oppermann, I., Fang, G., Ding, J.: Improved performance of spectrum cartography based on compressive sensing in cognitive radio networks. In: 2013 IEEE International Conference on Communications (ICC), pp. 5657–5661, June 2013
9. Jamali-Rad, H., Ramezani, H., Leus, G.: Sparse multi-target localization using cooperative access points. In: 2012 IEEE 7th Sensor Array and Multichannel Signal Processing Workshop (SAM), pp. 353–356, June 2012
10. Candes, E.J., Wakin, M.B.: An introduction to compressive sampling. IEEE Signal Processing Magazine 25(2), 21–30 (2008)
11. Goldsmith, A.: Wireless Communications. Cambridge University Press, New York (2005)
12. Foucart, S., Rauhut, H.: A mathematical introduction to compressive sensing. Springer (2013)
13. Candes, E.J., Romberg, J.: l1-magic: Recovery of sparse signals via convex programming, 4(14) (2005). www.acm.caltech.edu/l1magic/downloads/l1magic.pdf
14. Tropp, J.A., Gilbert, A.C.: Signal recovery from random measurements via orthogonal matching pursuit. IEEE Transactions on Information Theory 53(12), 4655–4666 (2007)
15. Needell, D., Tropp, J.A.: Cosamp: Iterative signal recovery from incomplete and inaccurate samples. Applied and Computational Harmonic Analysis 26(3), 301–321 (2009)

Mitigation of Primary User Emulation Attacks in Cognitive Radio Networks Using Belief Propagation

Sasa Maric[✉] and Sam Reisenfeld

Department of Engineering, Macquarie University, Sydney, NSW 2109, Australia
sasa.maric@students.mq.edu.au, sam.reisenfeld@mq.edu.au

Abstract. In this paper, we introduce a belief propagation based technique to combat the effects of primary user emulation attacks (PUEA) in Cognitive Radio (CR) Networks. Primary user emulation attacks have been identified as the most serious threat to CR security. In a PUEA, a malicious user emulates the characteristics of a primary user and transmits over idle channels. As a result, secondary users that want to use the channels are tricked into believing that they are occupied and avoid transmitting on those channels. This allows the malicious user to use the channels uncontested. To moderate the effects of PUEA, we propose a defence strategy based on belief propagation. In our solution, each secondary user examines the incoming signal and calculates the probability that it was transmitted from a primary user. These probabilities are known as beliefs. The beliefs at secondary users are reconciled to an agreed decision by comparison to a predefined threshold. The decision is made by a secondary user on whether it is believed that received transmission on a channel originated from a legitimate primary user or from a primary user emulation attacker.

Keywords: Cognitive radio networks · Belief propagation · Primary user emulation attacks · Security

1 Introduction

Traditional spectrum allocation methods allocate spectrum over large geographic regions and time spans to primary users (PUs). Primary users are licensed by a government regulatory office, such as the Federal Communications Commission in the United States. Channels in the licensed spectrum bands are allocated exclusively to primary users and are inaccessible to other users [1]. Users, other than primary users who could potentially use these channels, are called secondary users (SUs). It has been shown that the traditional allocation method of fixed channel allocation to primary users is leading to a very low utilisation across the licensed spectrum [2] [3]. Cognitive Radio, a collection of intelligent methods designed to use the radio spectrum in an efficient and dynamic manner, has been proposed as a solution to the frequency spectrum shortage. Cognitive

© Institute for Computer Sciences, Social Informatics and Telecommunications Engineering 2015
M. Weichold et al. (Eds.): CROWNCOM 2015, LNICST 156, pp. 463–476, 2015.
DOI: 10.1007/978-3-319-24540-9_38

Radio proposes to increase the efficiency of radio spectrum use by allowing secondary users to use channels when they are unoccupied by primary users. In this way, the average percentage of time for which the channels are actively carrying communication signals is increased. As a result, the total data throughput for the same bandwidth allocation is also increased. This must be achieved, while bounding the interference to a level which causes negligible degradation to the quality of primary user communications[1].

Despite its tremendous potential, Cognitive Radio is yet to be accepted as the solution to the radio spectrum shortage problem. One of the reasons for this is cognitive radio networks are susceptible to a number of types of jamming attacks. The most exploited area in cognitive radio is the spectrum sensing phase, where secondary users scan the frequency spectrum looking for available channels which are unoccupied by primary users. During this phase, if an attacker is able to mimic the signal properties of a primary user, he would be able to trick secondary users into believing that available channels are being used by primary users. This would result in secondary users vacating channels and leaving them available for malicious users to utilise uncontested. This form of attack is called a Primary User Emulation Attack (PUEA).

The remainder of this paper is organized as follows. In section 2, we introduce our system model. In section 3, our defense strategy based on belief propagation is presented. In section 4, we present our simulation result and analysis. Lastly, In section 5, we conclude the paper.

1.1 Related Work

A number of mitigation techniques have been proposed to combat primary user emulation attacks. The most promising of these use localisation of the transmitter. A number of methods exist for localisation of transmitters. These localisation methods can be classified into two categories: distributed localisation and centralised localisation. The first approach uses secondary user cooperation. This type of method involves secondary users trying to solve the localisation problem individually using information from cooperating nodes. The second approach is the central approach. In this approach nodes are scattered around the network and collect snapshots of the transmitted signal. These measurements are sent to a central node that processes the information and makes a decision on whether the suspect is a legitimate user or an attacker.

Locdef [4] is a localisation method that uses both localisation of the transmitter and signal characteristics to determine if the transmitter is a malicious user or not. The Locdef scheme uses sensor nodes scattered around the network to take snapshots of the incoming Received Signal Strength (RSS) at different locations in the network. These measurements are sent to a central location for processing. By identifying peaks in the RSS, a central node is able to determine the location of the transmitted signal. Locdef uses a three stage verification scheme to determine the validity of the incoming signal. The first stage uses the RSS of the signal to determine if it is coming from a primary user location or not. In the second stage the receiver looks at the energy of the received signal.

The reason for this is that secondary users are not able to transmit at high power levels, whereas primary users often are. If a suspect passes the first two stages, the scheme moves on to the last stage where it compares the signal characteristics of the incoming signal with the known characteristics of the idle primary user. If the characteristics of the incoming signal do not match the known signal characteristics of the primary user, the transmitter is deemed to be a malicious user.

Papers [5] and [6] present two primary user emulation attack mitigation schemes based on authentication and encryption. In [6] the author outlines a centralised scheme in which each primary user is given a unique ID number and a random variable (HM) by a centralised base station. Every time a suspect becomes active, the base station goes through a two-step authentication process to insure that the suspect is a valid primary user. Before a primary user can access the network, the user must send their ID number to the BS for authentication. The primary user ID is compared to a pool of identification numbers that correspond to all primary users in the area. If the ID number corresponds to one of the ID numbers in the pool, the scheme moves on to step two of the authentication process. If it does not, the user is treated as a malicious user and is ignored. The second step of the process is called the information displacement step. In this step the HM variable is multiplied by an encryption matrix which returns a value M that is compared to a set of expected values. If the value corresponds to the expected values, the transmitter is authenticated as a primary user. If it does not, the transmitter is treated as a malicious user and is ignored.

In [1] the author presents a technique based on belief propagation. This technique uses cooperation between secondary users to localise a transmitter. Comparing this to the known location of a primary user each secondary user is able to determine with a certain probability whether the transmitter is a primary user. The author denotes this probability as a belief. Secondary users in the network calculate their own local belief and exchange them to their neighbours. Then, each secondary user calculates a final belief using its own beliefs and all the beliefs from its neighbours. This paper modifies the algorithm described in [1] and suggests a useful procedure for determining whether the received signal originates from an attacker or not. Our paper presents substantial improvements to the algorithm described in [1].

2   System Model and Assumptions

In this section, we describe the basic system model that is used throughout this paper. To model the relationship between the transmit signal power and the received signal power, the author in [1] considers both path loss and log normal shadowing of the channel. Using these assumptions, we define an equation for the received signal strength from a primary user k as:

$$P_{r(PU_k)} = P_{t(PU_k)} d_{PU_k}^{-\alpha} h, \tag{1}$$

where, $P_{r(PU_k)}$ represents the received signal power from primary user k, $P_{t(PU_k)}$ represents the transmit power of the primary user k, d_{PU_k} represents the distance between a secondary user and a primary user k, h is the shadow fading constant defined as $h = e^{ab}$ where $a = \frac{ln10}{10}$, b is defined as a random Gaussian variable with a mean 0 and variance σ^2, and α is a propagation loss exponent. From Eq. (1) we are able to derive a similar equation to define the received signal power from an attacker as:

$$P_{r(attacker)} = P_{t(attacker)} d_{attacker}^{-\alpha} h_{attacker}, \qquad (2)$$

where, $P_{r(attacker)}$ represents the received signal power from the attacker, $P_{t(attacker)}$ represents the transmit power of the attacker, $d_{attacker}$ represents the distance between the attacker and a secondary node and $h_{attacker}$ is a shadowing constant similar to the one used in Eq. (1).

3   Detecting PUEA Using Belief Propagation

3.1   Original Belief Propagation Method

Belief propagation provides high accuracy detection of primary user emulation attacks. In belief propagation, each secondary user performs local observations and calculates the probability that an incoming signal belongs to a primary user. To accurately detect the presence of a malicious user, neighbouring nodes must communicate with each other and exchange local observations. Local observations are exchanged in the form of messages. Each secondary user computes a belief about whether the suspect is a primary user or an attacker according to its own local observations and the sum of all incoming messages from all its neighbours. A final belief is calculated using the sum of all beliefs of all SUs. This final belief is compared to a predetermined threshold. If the final belief is above the threshold, the suspect is deemed to be a primary user. If it is below, the suspect is considered to be a malicious user. The belief propagation framework is based on pairwise Markov Random Fields (MRF)[7].

Relative power observations of secondary users represent a pattern of receive powers generated by the location of the transmit station. The exchange of information between secondary users enables recognition of patterns for the purposes of determining whether or not the transmission originates at a known primary user location. In MRF we define Y_i as the local power observation at secondary user i, and X_i as the state of the suspect observed at user i. If $X_i=1$ the suspect is a primary user, if $X_i=0$ the suspect is a malicious user. The local function at user i is defined as $\phi_i(X_i, Y_i)$. The local function represents the observations made by a secondary user i about whether the suspect is a primary user or not. The compatibility function $\psi_{ij}(X_i, Y_j)$ is used to model the relationship between secondary users. The higher the compatibility function between two users is the more relevant the local observations of the two users become to each other. For example, if SU_1 is 1m away from SU_2 and SU_1 is 30m away from SU_3, then local observations that come from SU_2 to SU_1 will contribute more to the final belief of SU_1 than local observations that come from SU_3. The joint probability distribution of unknown variable X_i is given by:

$$P(\{X_i\}, \{Y_i\}) = \prod_{i=1}^{I} \phi_i(X_i, Y_i) \prod_{i \neq j} \psi_{ij}(X_i, Y_j), \tag{3}$$

where, I denotes the number of SUs in the network. We aim to compute the marginal probability at secondary user i, which we denote as the belief. The belief at a secondary user i is given in Eq. (4). It is the product of the local function at user i and all messages coming into user i from all the neighbours of i:

$$b_i(X_i) = k\phi_i(X_i, Y_i) \prod_{i \neq j} m_{ij}(X_i), \tag{4}$$

where, m_{ij} is a message from a secondary user i to a secondary user j and, k is a normalisation constant that insures that the beliefs sum to 1. Therefore:

$$k = \frac{1}{\prod_{i \neq j} m_{ij}(1)}. \tag{5}$$

In order to compute the belief at each user, we introduce a message exchange equation that is used to iteratively update the belief at each secondary user. In the l_{th} iteration a secondary user i sends a message $m_{ij}^l(X_i)$ to secondary user j which is updated by:

$$m_{ij}^l(X_i) = C \sum_{X_i} \psi_{ij}(X_i, Y_j)\phi_i(X_i, Y_i) \prod_{k \neq i,j} m_{ki}^{l-1}(X_i), \tag{6}$$

C is another normalisation constant such that $m_{ij}(1) + m_{ij}(0) = 0$, and therefore:

$$C = \frac{1}{\prod_{k \neq ij} m_{ij}^{l-1}(1)(\psi_{ij}(1,0) + \psi_{ij}(1,1))}. \tag{7}$$

Finally, after all secondary users finish computing their beliefs, these beliefs are added up and averaged to derive a final belief. The final belief is then compared to a predefined threshold. If the final belief is higher than the threshold, the suspect is believed to be a primary user. If the final belief is lower than the threshold the suspect is believed to be a malicious user:

$$Honest, \qquad \frac{1}{M} \sum_{i=1}^{M} b_i \geq b_\tau$$

$$Malicious, \qquad \frac{1}{M} \sum_{i=1}^{M} b_i < b_\tau, \tag{8}$$

where, M is the total number of secondary users in the network, $\sum_{i=1}^{M} b_i$ denotes the sum of all the beliefs of all the secondary users on the network and b_τ denotes the pre-set threshold. It is possible that some users would relay false information to other users in the network. However, false information by a small number of nodes would not influence the final belief value significantly.

Local Function. The local function represents the local observations at a single secondary user. Each secondary user calculates its own local function which corresponds to a probability of a suspect being a primary user. To calculate the local function we must compute two probability density functions (PDFs). The first PDF is computed using the RSS measurements that are acquired from the primary user and is denoted by PDF_{pu_k}. The second is a PDF that is computed using RSS measurements acquired from the attacker and is denoted by $PDF_{attacker}$. The local function corresponds to the similarity between the two PDFs. If the PDFs are the same the local function returns a probability equal to 1, which indicates that the suspect is transmitting from a primary user location. The further apart the distributions are the lower the local function and the higher the probability that the suspect is an attacker. The received signal from the primary user can be obtained using the following equation:

$$\frac{P_{r1(PU_k)}}{P_{r2(PU_k)}} = \left(\frac{d_{1(PU_k)}}{d_{2(PU_k)}}\right)^{-\alpha} \left(\frac{h_{1(PU_k)}}{h_{2(PU_k)}}\right), \tag{9}$$

where, $P_{r1(PU_k)}$ and $P_{r2(PU_k)}$ are the RSS values from a primary user(PU_k) to SU_1 and SU_2, $d_{1(PU_k)}$ and $d_{2(PU_k)}$ are the distances between PU_k and SU_1 and SU_2. $h_{1(PU_k)}$ and $h_{2(PU_k)}$ represent the shadow fading between PU_k and secondary users SU_1 and SU_2. It is assumed that the channel response is a circular Gaussian variable $\mathcal{CN}(0,1)$. If we define q as:

$$q = \frac{h_{1(PU_k)}}{h_{2(PU_k)}}, \tag{10}$$

we can then define $B_{i,j}$ as:

$$B_{i,j} = \left(\frac{d_{1(PU_k)}}{d_{2(PU_k)}}\right)^{-\alpha}, \tag{11}$$

therefore, the primary user's PDF of q can be written as follows:

$$PDF_{PU_k}(q) = \frac{1}{\mid B_{i,j}\mid} \frac{2\frac{q}{B_{i,j}}}{((\frac{q}{B_{i,j}})^2 + 1)^2}. \tag{12}$$

The PDF for the attacker is defined in a very similar way to the PDF of a primary user. SUs collect RSS measurements which they then exchanged with their neighbours. We define $P_{r1(attacker)}$ and $P_{r2(attacker)}$ as the received signal strength from the attacker to SU_1 and SU_2 respectively, and the distances between SU_1 an SU_2 and the attacker as $d_{1(attacker)}$ and $d_{2(attacker)}$ respectively. We can then define the value of A as follows:

$$A_{i,j} = \left(\frac{d_{1(attacker)}}{d_{2(attacker)}}\right)^{-\alpha} = \frac{P_{r1(attacker)}/P_{r2(attacker)}}{\pi}, \tag{13}$$

therefore, the attackers PDF can be written as follows:

$$PDF_{attacker}(q) = \frac{1}{\mid A_{i,j} \mid} \frac{2\frac{q}{A_{i,j}}}{((\frac{q}{A_{i,j}})^2 + 1)^2}. \tag{14}$$

To compare the two PDFs we use the Kullback Leibler distance. The Kullback Leibler distance is defined as:

$$KL_{(PDF_{PU_k}, PDF_{attacker})} = \int_0^\infty PDF_{PU_k} log \frac{PDF_{PU_k}}{PDF_{attacker}} dq. \tag{15}$$

The Kullback Leiber (KL) distance calculates the difference between the two PDFs. If the difference between the PDFs is large the KL formula will return a large number and if the distance is small the KL formula will return a small number. To obtain the local function from the KL distance we use the following formula:

$$\phi = \exp(-\min_k KL_{(PDF_{PU_k}, PDF_{attacker})}). \tag{16}$$

The local function returns a probability that a suspect is a primary user. The higher the probability the more likely the suspect is a primary user, the lower the probability the less likely it is that the suspect is a primary user.

Compatibility Function. The compatibility function is essential for cooperation between secondary users. In the belief propagation framework, the compatibility function is a scalar. The higher the compatibility function between two SUs the more relevant the two SUs are to each other. A reasonable compatibility function may be defined by the following expression:

$$\psi_{i,j}(X_i, Y_j) = \exp(-Cd^\beta_{X_i, Y_j}), \tag{17}$$

where, C and β are constants and, d_{X_i, Y_j} represents the distance between secondary users i and j. The compatibility function is heavily dependent on the distance between the two secondary users. If the distance between the secondary users is large then the compatibility function tends to zero. If the distance between secondary users is small the compatibility function tents to 1.

The compatibility function is used to insure that SUs that are far away do not have a large contribution to a particular SUs beliefs. The reason for this is that secondary users at different locations suffer from different shadow fading and the more distant users are the less likely to make a significant contribution to the accuracy of a SUs belief. It also insures that closer cooperating SU beliefs have a greater impact on the belief of a SU.

Complete Algorithm. The belief propagation algorithm used in this paper is summarised in Algorithm 1. Each secondary user performs measurements and calculates their PDF_{PU_k} and their $PDF_{attacker}$ using Eq. (11) and Eq. (13). Using these measurements, each secondary user iteratively computes their local and compatibility functions using Eq. (16) and Eq. (17). Each secondary user then computes and exchanges messages with all its neighbouring nodes. The last step of the algorithm is where each secondary user calculates their belief using their own local observations and the product of all the messages from all their neighbours.

After a number of iterations the mean of all the beliefs is calculated and compared to a predefined threshold. If the final belief is lower than the threshold the suspect is thought of as an attacker, if the final belief is greater than the threshold the suspect is deemed a primary user. The algorithm converges when there is no significant change in the final belief from the previous iteration to the current iteration. Therefore, the algorithm terminates when:

$$\frac{|\, f_b^{l-1} - f_b^l \,|}{f_b^{l-1}} < 0.001, \tag{18}$$

where, $f_b^l = \frac{1}{M} \sum_{i=1}^{M} b_i$, for the lth iteration. This insures that Algorithm 1 converges when there is a change corresponding to less than 0.1% between iterations.

Algorithm 1. Complete defence strategy against the PUEA using belief propagation

---

1: Each secondary user performs measurements using Eq. (11) and Eq. (13)

2: While $\frac{|f_b^{l-1} - f_b^l|}{f_b^{l-1}} < 0.001$

3: **for** Each iteration **do**

4:    Compute the local function using Eq. (16) and the compatibility function using Eq. (17)

5:    Compute messages using Eq. (6)

6:    Exchange messages with neighbours

7:    Compute beliefs using Eq, (4)

8: **end for**

9: Break

10: The PUE attacker is detected according to the mean of all final beliefs based on comparison against threshold.

11: Each SU will be notified about the characteristics of the attacker's signal and ignore them in the future.

3.2   New Belief Propagation Method

This section provides an outline of the changes that were made to the original technique presented in [1]. The two most significant improvements made to the old algorithm are the new simplified local function the new compatibility function.

Local Function. The local function that was used in the original technique suffered from being overly complicated and introducing a high level of complexity into the algorithm making it slow to converge. Our key contribution is the identification of a simpler more efficient local function. The new local function is just as accurate as the previous function. However, instead of doing a large number of numerical evaluations of integrals for each secondary user in the network, the new function calculates a simple arithmetic equation that allows the system to grow linearly instead of exponentially. The new local function that exhibits these desirable characteristics is:

$$\phi_{i,j} = \frac{\mid A_{i,j} - B_{i,j} \mid}{A_{i,j} + B_{i,j}}. \tag{19}$$

The local function is a measure of the similarity between the RSS measurements from a PU and the RSS measurements from a suspect. The closer the correlation between the two RSS values the more likely it is that the suspect is a primary user. The method used to obtain the local function in the old algorithm was computation time intense and had large computational complexity. This was primarily due to the fact that the KL distance was used to calculate the difference between the two probability density functions. The problem with the KL distance is that it uses an integral to determine the dissimilarity between two functions. As the number of secondary users on the network increases, we see a significant difference between the two methods. This is primary due to the fact that the local function has to be evaluated for each pair of secondary users in the network. As the number of SUs in the network increases the number of calculations of the local function increase exponentially. In the sections that follow we present results that prove that our new local function achieves results that are more accurate and efficient than those obtained by the old local function.

Compatibility Function. The compatibility function that was presented in the original paper discouraged cooperation between secondary users in the CR network and as a result decreased the accuracy of the final belief. This was primarily due to the fact that the compatibility function returned values that were very close to zero unless secondary users are located in close proximity. To increase cooperation between SUs we propose the following compatibility function:

$$\psi_{i,j}(X_i, Y_j) = \exp\left(-\frac{d_{X_i,Y_j}}{100}\right). \tag{20}$$

This compatibility function insures that secondary users that are close to each other are able to cooperate and share their result effectively to increase the accuracy of the results. The goal of the modified function is to insure that the messages between secondary users on the network are more relevant. We show in the next section that the new compatibility function is able to improve the performance of the algorithm by allowing a greater degree of cooperation.

4   Simulation Results and Analysis

In this section we present the results of the original BP algorithm against the improved BP algorithm. We chose to use similar simulation parameters as those presented by the authors in [1]. We set the path loss exponent α as 2.5, the transmit power of the secondary user is 0.1W (since the malicious user is also using a cognitive radio this is also the transmit power of the malicious user, we assume this corresponds to a transmission range of about 20 meters). There are 30 secondary users, one primary user and one malicious user deployed in a 100m by 100m grid.

4.1   Original BP Results and Analysis

This section outlines the results that were obtained in [1]. The authors went through a number of scenarios where they moved the locations of the primary and malicious users around the grid. They noted that as the distance between the primary user and malicious user increased, the final belief decreases, meaning that it is easy to distinguish between a primary user and a malicious user if they are far apart. Fig. 1 shows the plot that was obtained using the original BP algorithm.

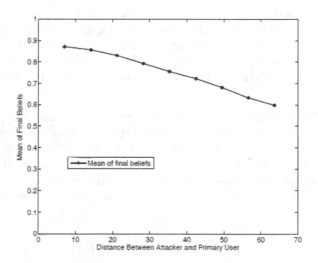

Fig. 1. Final belief Vs Distance (original technique).

We see from the results presented by the authors that the original algorithm is able to distinguish between a legitimate primary user and a malicious user with fairly high accuracy. However, the algorithm that is proposed in the original paper has several deficiencies. The key among these is its high computational complexity. During our simulations we observed an exponential growth in the computational complexity as the number of secondary users in the network is increased. Fig. 2 shows the effects that increasing the number of secondary users has on the computational complexity.

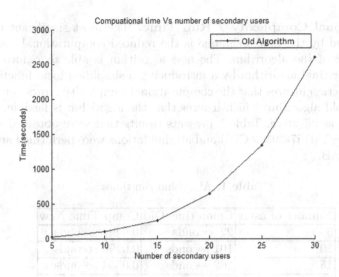

Fig. 2. Computational time of the old technique.

From these results we concluded that although the original algorithm is fairly effective in identifying a malicious user from a primary user, its high computational complexity means that it is not a feasible option for implementation using low power consumption cognitive radio terminals. We identified that the primary reason for the high computational complexity of the original BP algorithm is the computation of the local function. The Kullback Leibler function that is used to evaluate the difference between the primary user probability density function and the attackers probability density function was recognised as the main problem. The reason for this is that the KL function evaluates the difference between two function using an integral expression. If there are n secondary users in the network the KL function has to be evaluated once for each pair of secondary users, which means that it is calculated n^2 times. This is a serious deficiency which makes this algorithm infeasible for practical networks, where the number of users is large.

4.2   New BP Results and Analysis

To combat the deficiencies of the original algorithm, we present a new and improved algorithm that makes two important improvements which increase the accuracy and decease the computational complexity of the original algorithm.

To decrease the computational complexity of the original algorithm we propose a new simplified local function which provides the same level of accuracy with a reduced level of complexity. In addition, we modify the old compatibility function to help increase the level of cooperation between secondary users in the network.

Computational Complexity / Run Time. The most significant improvement obtained by the new technique is the reduced computational complexity and run time of the algorithm. The new algorithm is able to reduce the run time of the original algorithm by a introducing a simplified local function. The new local function insures that the computational complexity grows much slower than in the old algorithm which insures that the algorithm is flexible, scalable and still just as effective. Table 1 presents results that were obtained using an Intel(R) Core(TM) i7-3930k CPU and all simulations were performed and timed using MATLAB.

Table 1. Algorithm run times

Number of users	Comp time Old	Comp Time New
5	22 seconds	0.0491 seconds
10	101 seconds	0.0496 seconds
15	262 seconds	0.0564 seconds
20	648 seconds	0.0682 seconds
25	1337 seconds	0.071 seconds
30	2605 seconds	0.10 seconds

From Table 1 it is clear that the new algorithm is much less computationally complex than the original algorithm. We note that the run times of the new algorithm increase slowly as the number of secondary users in the network is increased. This presents a significant step forward for the algorithm and allows it to be utilised in larger and more complex networks.

Performance and Accuracy. In addition to the reduced computational complexity of the new algorithm, the new algorithm exhibits superior performance to the algorithm presented in [1]. This is primary due to the introduction of a modified compatibility function that allows for a larger degree of cooperation between secondary users. The greater the degree of cooperation between secondary users in the network the lower the chance of false or missed detection of a malicious user. Fig. 3 shows a comparison between the performance of the new algorithm and the performance of the original algorithm.

The perfect BP algorithm would result in a final belief value of 1 when the malicious user and the primary user are at the same location and would result in 0 in all other cases. Through analysis of results we observe that the new algorithm has an average final belief that is smaller than the average of the final belief of the old algorithm. This simple and effective comparison shows that the

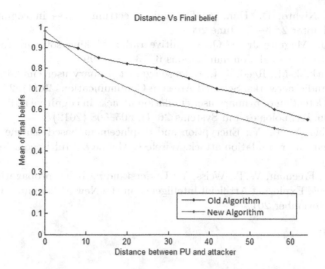

Fig. 3. Comparison of performance between the old and the new techniques.

new algorithm is not just less complicated but also detects PUEA with a higher degree of accuracy.

5   Conclusion

In this paper we present a belief propagation based algorithm to combat the effects of primary user emulation attacks on cognitive radio networks. We introduce key improvements to the algorithm described in [1] in relation to both performance and computational complexity. Through simulation we were able to show that our technique has lower complexity and improved accuracy relative to the technique in [1]. We have shown that the new technique reduces the time of convergence of the BP algorithm from hours to less than a few seconds. Furthermore, despite the simplification of the algorithm we were able to accurately distinguish between primary user and primary user emulation transmissions. These improvements are a direct result of the new local and compatibility functions, which reduce complexity and allow a greater degree of cooperation between secondary users on the CR network. The new algorithm is scalable, efficient, and effective and may be implemented in a low complexity secondary user terminal. The new algorithm provides a significant step forward in the mitigation of primary user emulation attacks in cognitive radio networks using belief propagation.

References

1. Yuan, Z., Niyato, D., Li, H., Song, J.B., Han, Z.: Defeating primary user emulation attacks using belief propagation in cognitive radio networks. Selected Areas in Communications **30**(10), 1850–1860 (2012)

2. Hossain, E., Niyato, D., Han, Z.: Dynamic spectrum access in cognitive radio networks. Chapter 2, 39–72, June 2009
3. Mitola III, J., Maguire Jr., G.Q.: Cognitive radio: Making software radios more personal. IEEE Personal Communications **6**, 13–18 (1999)
4. Chen, R., Park, J.-M., Reed, J.H.: Defense against primary user emulation attacks in cognitive radio networks. Selected Areas in Communications **26**(1), 25–37 (2008)
5. Thanu, M.: Detection of primary user emulation attacks in cognitive radio networks. Collaboration Technologies and Systems **26**(4), 605–608 (2012)
6. Zhou, X., Xiao, Y., Li, Y.: Encryption and displacement based scheme of defense against primary user emulation attack. Wireless, Mobile & Multimedia Networks **4**, 44–49 (2011)
7. Yedidia, J.S., Freeman, W.T., Weiss, Y.: Understanding belief propagation and its generalizations. Exploring Artificial Intelligence in the New Millennium, Chapter 8, 2282–2312, November 2012

Femtocell Collaborative Outage Detection (FCOD) with Built-in Sleeping Mode Recovery (SMR) Technique

Dalia Abouelmaati[1], Arsalan Saeed[1], Oluwakayode Onireti[1], Muhammad Ali Imran[1], and Kamran Arshad[2(✉)]

[1] Institute for Communication Systems (ICS), University of Surrey, Guildford, UK
d.y42011@gmail.com,
{arsalan.saeed,o.s.onireti,m.imran}@surrey.ac.uk
[2] Department of Engineering Science, University of Greenwich, Kent, UK
k.arshad@gre.ac.uk

Abstract. Self-Organizing Networks (SONs) have an important role in the development of the next generation mobile networks by introducing automated schemes. Cell outage detection is one of the main functionalities in self-healing mechanism. Outage detection for small cells has not been discussed in literature with greater emphasis yet. The Femtocell Collaborative Outage Detection (FCOD) algorithm with built-in Sleeping Mode Recovery (SMR) is introduced in this paper. The proposed algorithm is mainly based on the femtocell collaborative detection with incorporated sniffer. It compares the current Femtocell Access Points FAPs' Reference Signal Received Power (RSRP) statistics with a benchmark data. An outage decision is autonomously taken by each FAP depending on a certain threshold value. Moreover, the FCOD algorithm is capable of differentiating between the outage and sleeping cells due to the presence of the built-in SMR technique.

Keywords: Self-organizing networks · Self-healing · Cell outage detection · Heterogeneous cellular network

1 Introduction

Self-Organizing Networks (SONs) have lately been a captivating paradigm for the next-generation cellular networks via standardization bodies [1]-[3]. The aim of SON is to introduce autonomic features such as self-configuration, self-optimization and self-healing functionalities. SON functionalities will therefore enable the automation of certain activities performed by the network operator, thus leading to lower operating expenditure, simplified management and improved efficiency [4]. Self-healing involves automated remote detection of faults and recovery processes to compensate the faults in the network. Cell outage detection is considered as an important stage in the self-healing functionality. The basic function of the detection phase is to automatically detect the cells in outage, i.e. the cells that cannot offer services due to software failures, environmental disasters, technical fault, or component malfunctions [3]. Cell outage causes coverage and capacity gaps, which lead to high user churn rate, as well

© Institute for Computer Sciences, Social Informatics and Telecommunications Engineering 2015
M. Weichold et al. (Eds.): CROWNCOM 2015, LNICST 156, pp. 477–486, 2015.
DOI: 10.1007/978-3-319-24540-9_39

as increased operational costs [5]. In some cases, cell outage can easily be detected by the Operations and Support System (OSS), while some detection might require unplanned site visits, which is a costly task [5], [6].

Cell outage detection algorithms proposed in [7]-[9] are focused on macro-cells. It is expected that future cellular networks will be heterogeneous networks (HetNets), i.e., a mix of macro-cells for ubiquitous user experience and small cells or femto access points (FAPs) for high data rate transmission. Hence, the algorithms proposed in [7]–[9] are not suitable for such networks due to the dense deployment nature of FAPs in the HetNets, as compared to the macro only deployments. Furthermore, there is high possibility of having sparse user statistics in small cells, since they usually support very few users as compared to macro-cells. Recently, [1] proposed a Collaborative Outage Detection (COD) scheme, which is based on the implementation of a distributed outage trigger mechanism and sequential hypothesis testing within a predefined cooperation range. This scheme depends mainly on the Reference Signal Received Power (RSRP) statistics of the users within the cooperative range. Consequently, this approach will fail in detecting cell outage if there are no active users within the cooperation range. Furthermore, the COD and the conventional cell outage detection schemes in literature do not consider sleep mode of FAPs. Therefore, a FAP in idle/sleep mode will be mistakenly taken as in outage, which results in unnecessary compensation procedures and extra costs.

As a solution to the aforementioned challenges, energy efficient Femtocell Collaborative Outage Detection (FCOD) with a built-in Sleeping Mode Recovery (SMR) algorithm is proposed to automatically detect cell outage, by using performance statistics analysis of the collaborative FAPs. The FCOD technique is able to detect cell outage, even in the absence of users and scenarios with low FAPs density within the collaborative range. We consider the energy efficient node controlled mode for the FAPs sleep/wakeup mode. This self-controlled process requires a sniffer and a micro controller to be added to the FAPs to control the sleep and the wakeup cycles [10]. The rest of this paper is organized as follows: In Section 2, we present the network model. In Section 3, we present our proposed FCOD algorithm with SMR. In Section 4, we present extensive simulation based results to substantiate the performance of our proposed algorithm. Finally, we draw the conclusions in Section 5.

2   Network Model

We consider a typical heterogeneous network (HetNet) with *FAPs $\mathcal{F}= \{1,..., F\}$* overlaid on a macrocell. We also consider that one of the femtocells suffers an outage with certain probability in the operational process. The FAP in outage is not able transmit or receive any signal. Furthermore, another femtocell is switched into the sleep mode. The locations of FAPs are assumed to be known to the macrocell base station (MBS). The FAPs transmission powers are assumed to be constant through the outage detection process. In the downlink, FAPs are periodically transmitting reference signals, which assist the channel measurements of the user i.e., RSRP measurement.

These measurements are reported to the FAPs as feedback messages, which help to decide whether there is an outage or not.

We consider that users' positions are unknown to the FAPs and MBS. The users periodically report the neighboring FAPs' RSRP statistics periodically to their serving FAPs, which is used in handover decision and cell reselection process. We assume that the users in a certain area A follow a Poisson point process, $n_A \sim Poi(n;\rho|A|)$, where ρ is the density and n_A is the number of users within a certain area A.

The channel gains of a user u to a FAP f are expressed based on the model described in [11] as:

$$h = \left(\frac{d_o}{d_{u,f}}\right)^a e^{X_{u,f}} e^{Y_{u,f}}, \tag{1}$$

where d_o is the reference distance (1 m), $d_{u,f}$ the distance between FAP f and user u, a is the path loss exponent, while $e^{X_{u,f}}$, and $e^{Y_{u,f}}$ are the shadowing fading factor and multi-path fading factor, respectively. The shadowing fading follows a Gaussian distribution defined by $X_{u,f} \sim N(0,\sigma)$, $\forall u,f$. The multipath fading is Rayleigh fading with zero mean, and therefore $E[e^{Y_{u,f}}] = 0$. We assume that the effects of shadowing-fading are independent over time. According to this hypothesis, the user's RSRP statistics are independent random variables. All the RSRP statistics can be described using (1). Therefore, this distribution can be described according to [12] as follows:

$$r_u \sim \begin{cases} N\left(N_o, \frac{N_o^2}{M}\right) & H_o \\ N\left(P_u + N_o, \frac{(P_u+N_o)^2}{M}\right) & H_1 \end{cases}, \tag{2}$$

where r_u is the RSRP statistics for user u, P_u is the received signal strength for user u, N_o is the noise power, M is the number of samples of the signal (1.4×10^3 /ms for 1.4 MHz bandwidth). H_0 and H_1 denote the outage and normal hypothesis.

3 FCOD-SMR Algorithm

3.1 FCOD with Trigger Stage

Sleep Mode Recovery (SMR) Technique is introduced in this paper to prevent the sleeping FAP from been mistaken as in outage. When a FAP wants to switch to the sleeping mode, it informs the other FAPs within the collaborative range. These FAPs will then replace the current statistics of the sleeping FAP with the benchmark data for this FAP, which represent the normal state, before they start sensing the outage.

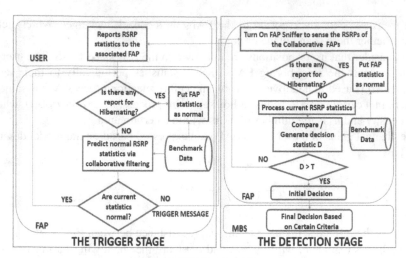

Fig. 1. FCOD Algorithm

When the sleeping FAP becomes active, it informs the FAPs within its collaborative range, in order to be treated as normal. Fig. 1 shows the flowchart for the FCOD algorithm, which includes the SMR approach. The FCOD algorithm involves two stages: the trigger stage and the detection stage.

The energy efficient FCOD technique is based on the node controlled mode, where the FAP detecting outage (i.e. FAP 4 in Fig. 2) uses a sniffer and a micro controller to sense UE activity in order to switch between the sleeping and wakeup mode. When the FAP senses UE activity, it wakes up only if the sensed UE is its subscriber, this avoids the unnecessary activation of the FAP in case of presence of a non-subscriber UE in the vicinity [10]. Once no authorized UE activity is detected the SMR approach will be initiated, FAP 4 will inform the rest of the neighboring FAPs (i.e. FAP 2 and FAP 3) and users within the collaborative range before it switches to the sleep mode as shown in Fig. 2. The collaborative FAPs (i.e. FAP 2 and FAP 3) and users will use the benchmark data (database of normal RSRP statistics) to replace the current RSRP statistics for the sleeping FAP (i.e. FAP 4). Consequently, the sleeping FAPs will not be falsely detected as in outage. After the sleeping FAP becomes active again, it informs the collaborative FAPs in order to be treated normally.

The benchmark data is frequently updated in case new FAPs are introduced into the network or any other changes occur within the collaborative range. The trigger stage, which includes the SMR, is used to check any abnormality (usually an outage) in the FAPs by using the reported user's RSRP statistics to trigger the detection stage. Consequently, the sniffer is not kept on all the time. In the detection stage each FAP within a certain collaborative range (R) uses a sniffer (such as the one used in the node controlled mode to sense the UE activity but with different sensitivity) to sense the neighboring FAPs' current RSRP statistics within a certain collaborative range (the range will be determine according to the sensitivity of the sniffer).

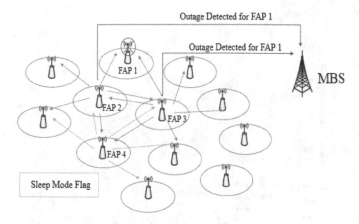

Fig. 2. Conceptual Model for FCOD

The current RSRPs statistics are compared to the benchmark data, which is the previously stored FAPs' RSRP statistics in the normal state. This benchmark data can be stored and exchanged between a group of collaborative FAPs. If the decision statistic (D) for a certain FAP is greater than a certain Threshold value (T), this FAP (i.e. FAP 1) will be initially decided as in outage. The rest of the FAPs within the collaborative range will start sensing using their sniffers. Centralized synchronization is used to manage the initiation of detection for the collaborative FAPs. The initial decision for an outage will be reported to the MBS. The detection stage will always be able to detect the outage regardless of the number of users within the collaborative range. D is determined as follows:

$$D = RSRP - RSRP_0 ,\qquad(3)$$

where $RSRP$ is the normal RSRP statistics from the benchmark data and $RSRP_0$ is the sensed current RSRP statistic for a certain collaborative FAP.

The outage decision is based on the following equation:

$$D > T ,\qquad(4)$$

where T is a heuristically predefined threshold, which is dependent on the false alarm and misdetection rates.

The MBS will check the initial decision reported from the collaborative FAPs (FAP 2 and FAP 3) as shown in Fig. 2. If more than 5% of the FAPs within a certain collaborative range reported an initial decision of an outage for the same FAP, then the MBS will take the final decision that this FAP is in outage. Subsequently, the MBS will start the necessary outage compensation scheme.

The FAPs might need to sense the collaborative users from time to time, in order to avoid delay or undiscovered outage in the trigger stage in case there is no user in the collaborative range. If there are no users sensed, the detection stage will be triggered immediately without waiting for the trigger stage. Another solution for the absence of users is that the detection stage might be triggered randomly from time to time or at a regular time interval.

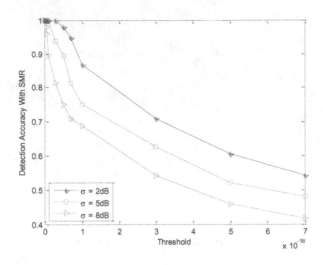

Fig. 3. Detection Accuracy with SMR versus the Threshold for FCOD

3.2   FCOD without Trigger Stage

This scenario is similar to the previous scenario but it only includes the detection stage. The trigger stage is replaced by a timer, to initiate the detection stage in regular time intervals. This eliminates the overhead caused by the trigger stage. However, this scheme might increase the outage detection delay, especially if the outage occurs just after the detection stage, which means the outage won't be detected until the next detection interval.

4   Simulation Results

Simulation Scenario: We consider a two-tier cellular system, which contains multiple femtocells within a macrocell. Femtocells are randomly distributed within the macrocell area (with radius r=1000m). We assume that FAPs transmit with fixed power and the carrier frequency is 2.5GHz with channel bandwidth of 1.4MHz. The users of the femtocell are randomly distributed within the femtocell area (with radius r=50 m). Furthermore, the users are connected with the FAP with the strongest RSRP. The path loss exponent a is set to 4. The number of FAPs and users will vary according to the different scenarios considered. However, the maximum number of FAPs used is 49 and the minimum number of users is 1. The transmission power of the FAP Po = 5 dBm, maximum cooperative range considered R = 600 m, and the shadow fading standard deviation σdB = 8 dB. The FCOD algorithm does not have any restriction on the parametric values (number of users or FAPs).

Fig. 3 illustrates the performance of the FCOD algorithm with SMR. It shows the detection accuracy versus the heuristically set threshold. Furthermore, it illustrates that by using small threshold values, the accuracy is improved significantly. The reason is that if the difference between the normal RSRP statistics from the

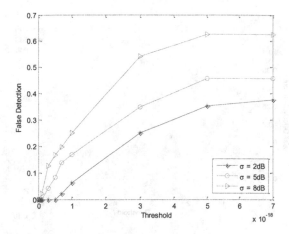

Fig. 4. False Detection versus Threshold for FCOD

benchmark data and the current RSRP is not large, it will be able to detect the outage. Moreover, the figure also shows that when shadowing fading increases (σ=8), the detection stage becomes less accurate than in case of less shadowing (σ=2). This is because of the errors introduced by the shadow fading.

Another algorithm is developed to evaluate the false detection with several threshold values in different channel conditions. Fig. 4 shows the performance of the FCOD algorithm with SMR. It shows the false detection (due to choosing inappropriate threshold value) versus the threshold. Furthermore, it illustrates that by using higher threshold values, the false detection rate increased significantly. The reason is that if the difference between the normal RSRP statistics and the current RSRP is not large, it won't be able to detect the outage. Moreover, the figure also shows that when shadowing fading increases (σ=8), the false detection rate increases during the detection stage compared to the less shadowing (σ=2) case. This is because of the errors introduced by the shadowing fading. Due to the significant importance of differentiating between the outage case and the sleeping mode case, the SMR technique is introduced in this paper to avoid the false detection of the sleeping FAP as an outage. It's also crucial to presents the false detection due to the absence of SMR with several threshold values in different channel conditions.

Fig. 5 demonstrates the performance of the FCOD algorithm without SMR. It shows the false detection versus the threshold. Furthermore, it illustrates that by using small threshold values, the false detection rate increased significantly. As the detection (either false or correct detection) is better with smaller threshold values. The reason is that if the difference between the normal RSRP statics and the current RSRP is not large, it will still be able to detect the outage. However, in this case it's a false detection as it is a sleeping FAP not outage FAP. Moreover, Fig. 5 also shows that when shadowing fading increases (σ=8), the false detection rate decreases compared to the case of less shadowing (σ=2). This is because of the errors introduced by the shadowing fading, which affects the false detection.

Fig. 6 illustrates the performance of the FCOD algorithm with SMR. It shows the detection delay versus FAP transmission power. Furthermore, it shows that the

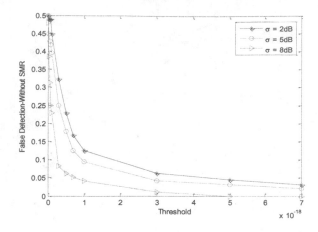

Fig. 5. False Detection-FCOD without SMR versus Threshold

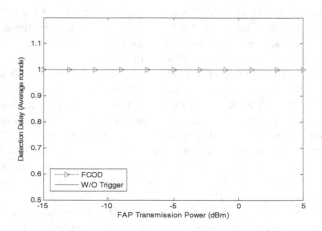

Fig. 6. Detection Delay versus FAP Transmission Power for FCOD

average delay for the FCOD with and without the trigger stage is one round, which means that the trigger stage doesn't affect the delay of the FCOD. The trigger stage function is to optimize the use of the sniffer by not keeping it on all the time. If the trigger stage senses an abnormality of a certain FAP it will initiate the detection stage by turning on the sniffer of the sensing FAP. However, the trigger stage increases the overhead of the FCOD algorithm.

Fig. 7 illustrates the performance of the FCOD algorithm with SMR. It shows the detection accuracy versus FAP transmission power with different threshold values. Furthermore, it demonstrates that by using a lower threshold value it is possible to achieve 100 % accuracy without increasing the FAP transmission power.

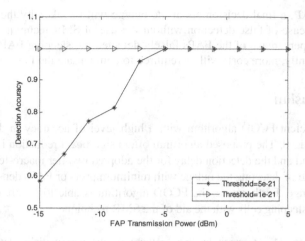

Fig. 7. Detection Accuracy versus FAP Transmission Power

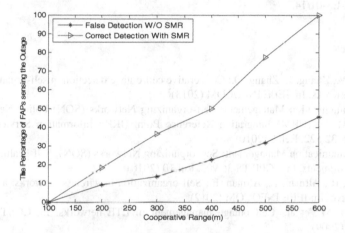

Fig. 8. The Percentage of FAPs Sensing the Outage versus Cooperative Range

However, when a higher threshold value is required due to a limitation of channel condition (power variation will set a limitation for using lower threshold as it might be misleading), a higher accuracy is still be achievable but at a cost of higher transmission power.

Fig. 8 illustrates the performance of the FCOD algorithm with and without SMR. It represents the percentage of FAPs sensing the outage versus the cooperative range, for the case of false and correct detection. Furthermore, it demonstrates that in case of using SMR technique when the collaborative range increases the percentage of FAPs sensing the outage increases, which improve the reliability of the FCOD algorithm. However this collaborative range is limited by the sniffer sensitivity. This figure can be used by the MBS to decide on an outage FAP based on a certain criteria. For example, if the collaborative range R=600m, 100% of the FAPs should detect the outage. The MBS then decide (according to the criteria) that if 5% of these FAPs report an outage for a certain FAP, a final decision about the outage will be taken by the MBS. Fig. 8 also shows the case of false detection without the use of SMR technique. In this case nearly half the percentage of the FAPs falsely detected the sleeping FAP as an outage FAP. Consequently, more costs will be required to compensate the false outage.

5   Conclusion

The energy efficient FCOD algorithm with a high level of accuracy in detecting FAP outage is introduced. The proposed algorithm offers significant reduction in the communication overhead and the detection delay for the adopted two-tier macro-femto scenario. Also it is capable of detecting the outage with minimum users or FAPs density within the collaborative range. Furthermore, the FCOD algorithm is able to differentiate between cell outage and sleeping cells with the aid of the SMR technique.

Acknowledgments. This research leading to these results was partially derived from the University of Greenwich Research & Enterprise Investment Programme grant under agreement number RAE-ES-01/14.

References

1. Wang, W., Zhang, J., Zhang, Q.: Cooperative cell outage detection in self-organizing femtocell networks. In: IEEE INFOCOM (2013)
2. Telecommunication Management: Self-organizing Networks (SON) Policy Network Resource Model (NRM) Integration Reference Point (IRP); Information Service (IS). In: 3GPP TS 32.522, Rel. 9 (2010)
3. Telecommunication Management: Self-organizing Networks (SON); Self-healing concepts and requirements. In: 3GPP TS 36.902, Rel.10 (2010)
4. Combes, R., Altman, Z., Altman, E.: Self-organization in wireless networks: a flow-level perspective. In: IEEE INFOCOM (2012)
5. Amirijoo, M., et al.: Cell outage management in LTE networks. In: COST 2100 TD (09)941 (2009)
6. Self-organizing Networks: NEC's proposals for next-generalization radio network management. In: NEC White Paper (2009)
7. Zoha, A., Saeed, A., Imran, A., Imran, M.A., Abu-Dayya, A.: A SON solution for sleeping cell detection using low-dimensional embedding of MDT measurements. In: IEEE International Symposium on Personal, Indoor and Mobile Radio Communications (2014)
8. Liao, Q., Wiczanowski, M., Stanczak, S.: Toward cell outage detection with composite hypothesis testing. In: IEEE International Conference on Communications (2012)
9. Khanafer, R. Solana, B., Triola, J., Barco, R., Moltsen, L., Altman, Z., Lazaro, P.: Automated Diagnosis for UMTS Networks Using Bayesian Network Approach. IEEE Trans. Veh. Technol. **57**(4) (2008)
10. Navaratnarajah, S., Saeed, A., Dianati, M., Imran, M.: Energy Efficiency in Heterogeneous Wireless Access Networks. IEEE Wireless Communications Magazine **20**(5) (2013)
11. Erceg, V., et al.: An Empirically Based Path Loss Model for Wireless Channels in Suburban Environments. IEEE Journal on Selected Areas in Communications **17**(7) (1999)
12. Shellhammer, S., et al.: Performance of power detector sensors of DTV signals in IEEE 802.22 WRANs. In: Proc. ACM TAPAS (2006)

Resource Allocation for Cognitive Satellite Uplink and Fixed-Service Terrestrial Coexistence in Ka-Band

Eva Lagunas[1]([✉]), Shree Krishna Sharma[1], Sina Maleki[1],
Symeon Chatzinotas[1], Joel Grotz[2], Jens Krause[3], and Björn Ottersten[1]

[1] Interdisciplinary Centre for Security, Reliability and Trust (SnT),
University of Luxembourg, Walferdange, Luxembourg
{eva.lagunas,shree.sharma,sina.maleki,symeon.chatzinotas,
bjorn.ottersten}@uni.lu
[2] Technical Labs, Newtec, Belgium
joel.grotz@newtec.eu
[3] SES, Betzdorf, Luxembourg
jens.krause@ses.com

Abstract. This paper addresses the cognitive Geostationary Orbit (GSO) satellite uplink where satellite terminals reuse frequency bands of Fixed-Service (FS) terrestrial microwave links which are the incumbent users in the Ka 27.5-29.5 GHz band. In the scenario considered herein, the transmitted power of the cognitive satellite user has to ensure that the interference impact on potentially present FS links does not exceed the regulatory interference limitations. In order to satisfy the interference constraint and assuming the existence of a complete and reliable FS database, this paper proposes a Joint Power and Carrier Allocation (JPCA) strategy to enable the cognitive uplink access to GSO Fixed Satellite Service (FSS) terminals. The proposed approach identifies the worst FS link per user in terms of interference and divides the amount of tolerable interference among the maximum number of FSS terminal users that can potentially interfere with it. In so doing, the cognitive system is guaranteed to never exceed the prescribed interference threshold. Subsequently, powers and carriers are jointly allocated so as to maximize the throughput of the FSS system. Supporting results based on numerical simulations are provided. It is shown that the proposed cognitive approach represents a promising solution to significantly boost the performance of conventional satellite systems.

Keywords: Cognitive radio · Satellite communications · Resource allocation

1 Introduction

The Digital Agenda for Europe foresees broadband interactive access as a cornerstone in the recovery and development plan for Europe [1]. The challenging

© Institute for Computer Sciences, Social Informatics and Telecommunications Engineering 2015
M. Weichold et al. (Eds.): CROWNCOM 2015, LNICST 156, pp. 487–498, 2015.
DOI: 10.1007/978-3-319-24540-9_40

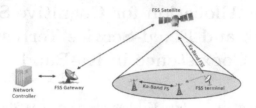

Fig. 1. Spectral coexistence of FSS uplink with the FS terrestrial link in the Ka-band (27.5-29.5 GHz)

objectives set forth by the European Commission are to provide basic broadband access to all Europeans by 2013 (at least 30 Mbps) and to ensure that at least 50% of the households get 100 Mbps by 2020 [2]. In this context, Satellite Communications (SatCom) can play a key role to ensure ubiquitous coverage including remote and rural regions, where terrestrial deployment cannot be guaranteed or its cost is prohibitive. Traditional single-beam satellites operating in C/Ku-band show limitations in supporting the flexible distribution of bandwidth needed for broadband applications [3,4]. Novel multi-beam satellites operating in the Ka-band frequency spectrum have recently gained attention among satellite telecommunication industries because they can significantly enhance the system capacity as a result of higher frequency reuse and multiple narrowly focused beams. Examples of high throughput satellites operating in Ka-band are Eutelsats KaSat [5], VIASAT 1 [6] and SES-12 [7].

The provision of widely available and competitively-priced broadband satellite services critically depends on the availability of radio spectrum resources. Spectrum congestion has been identified as the main limiting factor in providing solutions meeting the user expectations by 2020 [8]. Cognitive Radio (CR) [9] is conceived as an ideal spectrum management tool to solve the spectrum scarcity, as it enables unlicensed systems to opportunistically utilize the underutilized licensed bands. As far as cognitive SatCom is concerned, recently a number of research projects have been undertaken to address the spectrum sharing concept between two satellite systems or between satellite and terrestrial systems [10–13]. Within [10], three scenarios have been identified as potentially positively impacted by the use of CR techniques. In this paper, we consider one of the preselected scenarios in [10]: an uplink Ka-band FSS cognitive system reusing frequency bands of Fixed-Service (FS) terrestrial microwave links (incumbent systems), as depicted in Fig. 1. In this scenario, the FSS terminal users can maximize frequency exploitation by flexible utilization of the FS segment through the adoption of CR techniques in the satellite uplink. Therefore, this scenario falls within the underlay CR paradigm [14], whereby the uplink power density of the FSS system has to ensure that the interference impact on the potentially present FS links does not exceed the regulatory interference limitations. The applicability of CR in the aforementioned scenario was discussed in [15,16], considering realistic channel propagation conditions, concluding that both satellite and terrestrial systems could potentially operate in the same band without degrading each other's performance. Here, we go a step further, and consider designing efficient resource allocation algorithms for this scenario. Our goal is to optimally assign carriers and adjust the transmit power of the cognitive devices so that the individual QoS requirement is satisfied and the interference at each of the FS stations is kept below the given threshold. Assuming a FS database-assisted approach in which the interference caused at the incumbent system can be perfectly determined and assuming free space path loss between satellite terminal and terrestrial stations, the proposed approach identifies the worst FS link per user in terms of received interference and divides the amount of tolerable interference among the maximum number of FSS terminal users that might potentially interfere with it. In so doing, a transmit power limit is obtained per user and carrier level. This simple and conservative strategy guarantees the cognitive system to never exceed the prescribed interference threshold. The severe restriction in transmit powers is compensated by an efficient Joint Power and Carrier Allocation (JPCA) module that allocates the available resources so as to maximize the overall satellite system throughput.

The remainder of this paper is structured as follows. Section 2 reviews the regulatory context as applied to the frequency bands under consideration. Section 3 introduces the signal model. The proposed cognitive exploitation framework is described in detail in Section 4. Section 5 provides supporting results based on numerical data. Finally, conclusions are given in Section 6.

2 Regulatory Context

Satellites featuring Ka-band transponders have existed for more than a decade. At ITU-R level, the frequency allocations to the FSS uplink span across the whole 27.5-30 GHz, with only 500 MHz available exclusively for FSS terminals and the remainder is shared with terrestrial FS links. In Europe, the CEPT framework provides for FSS Earth stations exemption of individual licensing in certain parts of the Ka-band [17]. CEPT sets the band 29.5-30 GHz for exclusive FSS use (same as ITU) and the rest is segmented into specific portions designated for use by uncoordinated FSS Earth stations and for the use by terrestrial FS links. It should be noted, however, that CEPT decisions are not mandatory instruments, and CEPT administrations may choose not to implement them. Examples of european counties not following this decision are UK, Denmark, Austria, Sweden and Lithuania. At the time being, the FSS satellite system is only allowed to operate in the exclusive 29.5-30 GHz band, which is insufficient to meet future demands. Different research projects [10,12] are investigating whether CR techniques could help resolve this specific sharing scenario. In this respect, this paper investigates FSS cognitive satellite terminals operating in the Ka band 27.5-30 GHz, reusing frequency bands of FS links with priority protection.

3   Signal Model

In the Ka band scenario being addressed in this paper, the cognitive FSS terminals might impose interference to the terrestrial FS incumbent system. Let us assume a scenario with L FSS terminal users and N FS microwave links. The aggregated interference power caused by the L cognitive transmitters at the n-th FS microwave station for a particular carrier frequency f_m, $m = 1, \ldots, M$, is given by,

$$I_n(m) = \sum_{l=1}^{L} p_l \cdot g_{l,n}(m), \tag{1}$$

where p_l denotes the transmit power of the l-th FSS terminal and $g_{l,n}(m) = |h_{l,n}(m)|^2$, with $h_{l,n}(m)$ being the average cross-channel coefficient from the l-th FSS terminal to the n-th FS station when the FSS user is transmitting at f_m.

This average cross-channel gains $g_{l,n}(m)$ can be seen as the Cross-Channel State Information (CCSI). Throughout this paper, we assume a FS database-assisted approach in which perfect CCSI is assumed available at the FSS system. Clearly, the accuracy and completeness of the available database determines the quality of the CCSI. In this respect, verification of available database via measurements will be considered in future works. The cross-channel gains $g_{l,n}(m)$ can be expressed as follows,

$$g_{l,n}(m) = G_{\text{Tx}}^{FSS}(\theta_{l,n}) \cdot G_{\text{Rx}}^{\text{FS}}(n, \theta_{n,l}) \cdot L(d_{l,n}, f_m) \tag{2}$$

where,

- $G_{\text{Tx}}^{FSS}(\theta)$: Gain of the FSS transmitting antenna at offset angle θ.
- $\theta_{i,j}$: Offset angle (from the boresight direction) of the i-th station in the direction of the j-th station.
- $G_{\text{Rx}}^{\text{FS}}(n, \theta)$: Gain of the n-th FS station antenna at offset angle θ.
- $L(d, f) = \left(\frac{c}{4\pi df}\right)^2$: Free space path loss with d being the transmitter-receiver distance and f being the carrier frequency.
- $d_{i,j}$: Distance between the i-th transmitter and the j-th receiver.

The radiation patterns $G_{\text{TX}}^{\text{FSS}}(\theta)$ and $G_{\text{Rx}}^{\text{FS}}(n, \theta)$ can be obtained from ITU-R S.465-6 and ITU-R F.1245-2, respectively. Unlike [15, 16], we consider the worst case propagation model that would result from a line-of-sight path through free space, with no obstacles nearby to cause reflection or diffraction. In (2), it is assumed that the interfering signal falls within the victim bandwidth. If the spectra do not overlap completely, then a compensation factor of $B_{\text{overlap}}/B^{\text{FS}}$ is applied, where B_{overlap} stands for the portion of the interfering signal spectral density within the receive filter bandwidth given by B^{FS}.

Satisfying Quality of Service (QoS) requirements of the incumbent FS link requires guaranteeing received interference less than a tolerable level $I_{\text{thr},n}$. This protection condition is expressed mathematically as,

$$I_n(m) \leq I_{\text{thr},n} \tag{3}$$

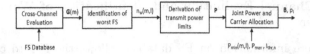

Fig. 2. Block diagram of the cognitive exploitation framework.

Such limitations are defined by the regulatory authorities. Typical reference limitations are given by ITU recommendations such as ITU-R F.758, where the interference level is recommended to be -10 dB below the receiver noise.

Apart from the transmit power limits given by the interference constraints, each FSS terminal has its own power limits. According to CEPT, the maximum EIRP of FSS terminals shall not exceed a value in a range from 55 dBW to 60 dBW. As far as cognitive QoS is concerned, a minimum signal-to-interference and noise ratio (SINR) shall be guaranteed for correct decoding. Therefore, $P_{\min}(m, l) \leq p_l \leq P_{\max}$ where $P_{\max}$ denotes the maximum allowable transmission power due to regulatory issues and $P_{\min}(m, l)$ denotes the minimum required power for l-th FSS user to close the link over the m-th carrier.

4   Cognitive Exploitation Framework

One of the major challenges for cognitive uplink satellite communications is, thus, how to optimally assign carriers and adjust the transmitted power so that the individual QoS requirement is satisfied and the aggregate interference at each of the FS stations is kept below the given limit. The level of received interference at each FS station depends both on the transmitted powers p_l, $l = 1, \ldots, L$, and also on the carrier assignment. Therefore, power and carrier allocation should be considered jointly, which lead to cumbersome optimization problems [18].

In this paper, we follow a simple, but efficient approach in which the transmit power and the assigned carrier of the FSS terminals are determined based on the worst cross-channel condition. The proposed approach first evaluates the cross-channel gains at the carrier level based on the available information at the FS database. With this information, the Network Controller (NC) of the FSS systems is able to identify the worst FS link in terms of received interference per user and per carrier. Next, the amount of tolerable interference of the worst incumbent FS link is divided among the maximum number of FSS terminals that can potentially contribute to its aggregate interference. Subsequently, the maximum transmitted power of each FSS station is derived per each carrier and user. The resulting powers are fed to the JPCA module in order to allocate the resources by maximizing the overall throughput of the FSS system. The schematic diagram of the cognitive exploitation framework is depicted in Fig. 2. Next sections are devoted to describe in detail the proposed cognitive exploitation blocks.

4.1 Cross-Channel Evaluation and Identification of Worst FS Receiver

Having complete information on FS database allows the NC to compute the cross-channel gains $g_{l,n}(m)$, $l = 1, \ldots, L$, $n = 1, \ldots, N$, $m = 1, \ldots, M$, as in (2). Therefore, for each carrier frequency, the cross-channel matrix $\mathbf{G}(m) \in \mathbb{R}^{L \times N}$ can be described as follows,

$$\mathbf{G}(m) = \begin{bmatrix} g_{1,1}(m) & \cdots & g_{1,N}(m) \\ \vdots & \ddots & \vdots \\ g_{L,1}(m) & \cdots & g_{L,N}(m) \end{bmatrix}. \tag{4}$$

For each l-th FSS user, the identification of the worst FS station in terms of interference consists in determining the one with maximum cross-channel gain, $n_{\mathrm{w}}(m, l) = \max_n [\mathbf{G}(m)]_l$, where $[\mathbf{G}(m)]_l$ denotes the l-th row of matrix $\mathbf{G}(m)$ and $n_{\mathrm{w}}(m, l)$ indicates the worst FS station of user l operating in carrier m. These worst FS stations in terms of received interference will determine the maximum allowable transmit power per user.

4.2 Transmit Power Limits

The first step is to obtain the interference power limit for each FSS terminal user at a carrier level. For a particular l FSS user, the interference limit of the worst FS receiver, $I_{\mathrm{thr},n_{\mathrm{w}}(m,l)}$ [W], is broken into different portions according to the maximum number of FSS users that can potentially interfere with it. This is,

$$I_{\mathrm{w}}(m, l) = I_{\mathrm{thr},n_{\mathrm{w}}(m,l)} \left(\frac{B^{\mathrm{FS}}}{B^{\mathrm{FSS}}} \right)^{-1}. \tag{5}$$

where the fraction $\frac{B^{\mathrm{FS}}}{B^{\mathrm{FSS}}}$ gives the maximum number of FSS terminals that fit in the FS frequency band. Therefore, the transmit power limit is established to ensure that the following individual interference constraint is satisfied,

$$I_{\mathrm{w}}(m, l) \leq p_l \cdot G_{\mathrm{Tx}}^{FSS}(\theta_{l,n}) \cdot G_{\mathrm{Rx}}^{\mathrm{FS}}(n, \theta_{n,l}) \cdot L(d_{l,n}, f_m). \tag{6}$$

As a consequence, we can obtain the maximum transmission power that FSS terminals should not exceed to guarantee the incumbent system protection in the following way,

$$p_{\max}(m, l) = \frac{I_{\mathrm{w}}(m, l)}{G_{\mathrm{Tx}}^{FSS}(\theta_{l,n}) \cdot G_{\mathrm{Rx}}^{\mathrm{FS}}(n, \theta_{n,l}) \cdot L(d_{l,n}, f_m)}. \tag{7}$$

Note that there could be some frequencies where no FS is deployed leading to $p_{\max}(m, l) \to \infty$ or very good conditions in which $p_{\max}(m, l) > P_{\max}$. Moreover, we might face the opposite situation in which the interference constraint is too strong and the value of $p_{\max}(m, l)$ is below the minimum required power.

To overcome this infeasibility conditions, the resulting $p_{\max}(m, l)$ are subject to the following adjustments,

$$
p(m, l) = \begin{cases} P_{\max} & \text{if } p_{\max}(m, l) > P_{\max} \\ p_{\max}(m, l) & \text{if } P_{\min}(m, l) \leq p_{\max}(m, l) < P_{\max} \\ 0 & \text{otherwise} \end{cases} \tag{8}
$$

For notational convenience, we define $\mathbf{P} \in \mathbb{R}^{M \times L}$ as the matrix containing the maximum allowable powers due to interference constraint,

$$
\mathbf{P} = \begin{bmatrix} p(1, 1) & \cdots & p(1, L) \\ \vdots & \ddots & \vdots \\ p(M, 1) & \cdots & p(M, L) \end{bmatrix}. \tag{9}
$$

Assuming the worst-case interference in which the complete FS receiver bandwidth is shared with cognitive transmitters is a very conservative assumption which, although protects FS terrestrial system to the maximum extent, it might cause undesirable reductions in the FSS system throughput. This loss in performance, however, can be efficiently compensated with proper carrier allocation by favoring uplink transmissions on carriers with better cross-channel conditions.

The conservative power derivation, thus, ensures that any combination of the powers contained in $\mathbf{P}$ never results in aggregate interference above the acceptable threshold $I_{\mathrm{thr}, n_w(m,l)}$. Next section is devoted to optimally choose a power and carrier combination from $\mathbf{P}$ that maximizes the sum-rate of the FSS system.

4.3   Joint Power and Carrier Allocation (JPCA)

Having tackled the problem of incumbent terrestrial system protection, this section is devoted to optimally allocating the carriers and powers among FSS terminal users by maximizing the total FSS system throughput. We denote $\mathbf{b}_l \in \mathbb{R}^{M \times 1}$ the carrier assignment of l-th FSS user. The elements of $\mathbf{b}_l$ work as an indicator function: "1" if m-th carrier is assigned to the l-th user and "0" otherwise. For notational convenience, we stack all the carrier assignments in the matrix $\mathbf{B} = \begin{bmatrix} \mathbf{b}_1 & \cdots & \mathbf{b}_L \end{bmatrix}$. Here, we assume that the FSS users are allocated within the same beam, and each carrier includes a frame which can accommodate a maximum number of users. However, for simplifying the description of the resource allocation module, we assume herein that each carrier can be assigned to only one user. Extension to the case where a given number of users share the same carrier using time division multiplexing (MF-TDMA or Mx-DMA) is then straightforward. Since we assume that at each time, only one user can use a carrier, thus for each carrier m, we have $\forall m$: $\sum_{l=1}^{L} b(m, l) = 1$. Note that having obtained the carrier allocation matrix $\mathbf{B}$, it is straightforward to compute the corresponding power allocation as $p_l = \mathbf{b}_l^H \mathbf{p}_l$, where $\mathbf{p}_l$ stands for the l-th column of $\mathbf{P}$.

As discussed in the previous section, the information regarding the maximum allowable power for each user over each available carrier is available in the NC.

This information is used to determine the value of $b(m, l)$ to be zero or one by maximizing the overall throughput of the system according to the following optimization problem,

$$\max_{\mathbf{B}} \quad \|\mathrm{vec}(\mathbf{B} \odot \mathbf{R}(\mathbf{SINR}))\|_{l_1} \quad \mathrm{s.t.} \quad \sum_{l=1}^{L} b(m, l) = 1, \tag{10}$$

where $\odot$ denotes the Hadamard product, $\mathrm{vec}(\cdot)$ denotes the vectorization operator, $\|\cdot\|_{l_1}$ denotes the l_1-norm and $\mathbf{R}(\mathbf{SINR})$ denotes the rate matrix with $r_{l,m}$, $l = 1, \ldots, L$, $m = 1, \ldots, M$, elements indicating the DVB-S2X rate [19] associated with the corresponding SINR value. In (10), $\mathbf{SINR} \in \mathbb{R}^{M \times L}$ denotes the SNR values derived from $\mathbf{P}$ as follows,

$$\mathrm{SNR}(m, l) = \frac{p(m, l) \cdot G_{\mathrm{Tx}}^{\mathrm{FSS}}(0) \cdot [G/T]_{\mathrm{Rx}}^{\mathrm{SAT}}(l) \cdot L(D, f_m)}{k B^{\mathrm{FSS}}}, \tag{11}$$

where $[G/T]_{\mathrm{Rx}}^{\mathrm{SAT}}(l)$ is the satellite gain over noise temperature for the l-th FSS terminal user, $G_{\mathrm{Tx}}^{\mathrm{FSS}}(0)$ denotes the FSS terminal antenna gain in the boresight direction $(\theta = 0°)$ and D is the distance between the FSS terminal and the satellite. On the denominator, k is the Boltzmann constant. The SINR values are obtained considering a $[C/I]_{\mathrm{Rx}}^{\mathrm{SAT}}$ term which accounts for the co-channel interference.

We solve the optimization problem in (10) using the Hungarian algorithm [20], which provides an efficient and low complexity method to solve the one-to-one assignment problem in polynomial time.

5   Numerical Evaluation

In this section, we present some numerical results to demonstrate the performance of the proposed JPCA technique to give Ka-band cognitive uplink access to the FSS terminals reusing frequency bands of FS terrestrial microwave links with priority protection.

5.1   Simulation Setup

The FS database is extracted from the ITU-R BR International Frequency Information Circular (BR IFIC) database [21]. In particular, we focus on the database related to Slovenia with more than 3,300 records, which is one of the most complete database available for this scenario in BR IFIC. We consider an FSS multibeam satellite system and focus on a representative beam of 155 km radius with its center located in the capital and largest city of Slovenia, Ljubljana (46.0553°N and 14.5144°E). The beam gain $[G/T]_{\mathrm{Rx}}^{\mathrm{SAT}}(l)$ is computed as [22],

$$[G/T]_{\mathrm{Rx}}^{\mathrm{SAT}}(l) = [G/T]_{\mathrm{Rx,max}}^{\mathrm{SAT}} \cdot \left(\frac{J_1(u(\varphi_l))}{2u(\varphi_l)} + 36 \frac{J_3(u(\varphi_l))}{u(\varphi_l)^3} \right)^2 \tag{12}$$

where $[G/T]_{\mathrm{Rx,max}}^{\mathrm{SAT}}$ is the maximum satellite antenna gain, J_p is the first kind of Bessel's function of order p and $u(\varphi_l) = 2.07123 \cdot \sin(\varphi_l)/\sin(\varphi_{3\mathrm{dB}})$, with φ_l and $\varphi_{3\mathrm{dB}} = 0.2°$ being the satellite nadir angle corresponding to user l and the half-power beamwidth, respectively.

The performance was evaluated by averaging over 50 independent FSS terminal geographical distributions, which were selected uniformly at random for each realization within the considered beam footprint according to the population density database obtained from the NASA Socioeconomic Data and Applications Center (SEDAC) [23]. The number of FSS terminals L is set to be 356, which coincides with the number of carriers to be assigned. A summary of the most relevant parameters and the FSS link budget details are presented in Table 1. An example of FSS terminal users distribution together with the location of the FS stations and the beam pattern of the FSS satellite obtained with (12) is depicted in Fig. 3.

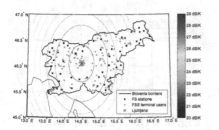

Fig. 3. Beam pattern of the FSS satellite over Slovenia.

Table 1. Simulation Parameters

Parameter	Value
B^{FSS}	7 MHz
Shared band	$27.5 - 29.5$ GHz (285 carriers)
Exclusive band	$29.5 - 30$ GHz (71 carriers)
Parameters for FSS system	
Reuse pattern	4 color (freq./pol.)
Satellite location	13°E
$[G/T]_{\mathrm{Rx,max}}^{\mathrm{SAT}}$	29.3 dB/k
EIRP	50 dBW
$[C/I]_{\mathrm{Rx}}^{\mathrm{SAT}}$	10 dB
$G_{\mathrm{Tx}}^{\mathrm{FSS}}(0)$	42.1 dBi
Antenna pattern	ITU-R S.465
Terminal height	15 m
Altitudes above the sea level	From [24]
D	35,786 km
Parameters for FS system	From database
B^{FS}	7 or 28 MHz
$G_{\mathrm{Rx}}^{\mathrm{FS}}(n,0)$ $\forall n$	34 dBi
Antenna pattern	ITU-R F.1245-2
Antenna height	10 m
$I_{\mathrm{thr,9}}$	-137.55 dBW @ 7 MHz
	-131.53 dBW @ 28 MHz

5.2   Numerical Results

Fig. 4 shows the aggregate interference caused by the cognitive FSS system and received at the FS stations in terms of Cumulative Distribution Function (CDF) with and without the proposed JPCA module. For the latter, a random carrier assignment is considered and $p_l = P_{\max}$, $\forall l$. For the JPCA case, the optimal and one sub-optimal combination of the powers contained in $\mathbf{P}$ are plotted in Fig. 4. The minimum aggregate interference threshold is depicted as well in the figure for comparison purposes. It can be observed that the interference generated by the FSS system at the incumbent system exceeds the acceptable threshold when no optimal resource assignment is employed and it is kept always below the threshold when using the proposed JPCA technique.

The effect of the transmit power limit given by the interference constraint (3) is evident by a glance of Fig. 5, where the SINR of the FSS terminal users is shown in terms of CDF with optimal, suboptimal and without JPCA. An immediate observation is that the SINR degrades when transmit power limitations

Table 2. Throughput per beam

Case	Technique	Value (Mbps)
Exclusive only	w/ JPCA (subopt)	699.5136
	w/ JPCA (opt)	699.5291
Shared+Excl. w/o FS	w/ JPCA (subopt)	3538.0503
	w/ JPCA (opt)	3538.5299
Shared+Excl. w/ FS	w/ JPCA (subopt)	3347.6373
	w/ JPCA (opt)	3538.1431

apply, i.e., when FS terrestrial microwave links are protected. In particular, 35% of FSS users experience SINR values below 9.8 dB in the FSS-FS coexistence case when resources are not allocated optimally, which decreases down to 22.5% if the proposed JPCA is employed, and to 9.3% when the transmit power is not limited (blue line).

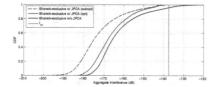

Fig. 4. CDF of aggregate interference

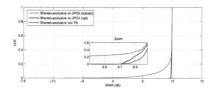

Fig. 5. CDF of SINR distribution

Similarly, Fig. 6 presents the comparison of rate distribution in the considered scenario. From the figure, it can be noted that the beam availability in the presence of the incumbent FS system is less than the beam availability when no power constraints apply (blue line). Further, it is clear that employing the proposed JPCA attains a throughput per user that is very close to that attained in the absence of transmit power limitations.

Fig. 7 illustrates the achieved per beam throughput. To obtain greater insight, Table 2 summarizes the results shown in Fig. 7. It is worth to point out that the additional spectrum together with the optimal JPCA module provides around 405.8% improvement over the conventional exclusive band (29.5-30 GHz) case, which is almost the same that can be achieved in the absence of FS microwave links. This gain reduces to 378.6% when resources are not optimally distributed. The application of optimal JPCA in the non-shared spectrum scenarios does not provide much benefit and this is because, in this particular study, all users and carriers experience similar channel conditions.

6   Conclusion

In this paper, we developed a novel spectrum exploitation framework for cognitive uplink FSS terminals in the band 27.5 − 29.5 GHz, where the incumbent

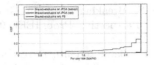

Fig. 6. Per user rate

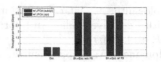

Fig. 7. Throughput per beam

users are Fixed-Service (FS) microwave links. Results based on computer simulations were presented, which showed that the FSS system throughput can be significantly improved by using the proposed cognitive exploitation framework while guaranteeing the sufficient protection of incumbent FS systems.

Acknowledgment. This work was partially supported by the European Commission in the framework of the FP7 Cognitive Radio for SATellite communications - CoRaSat (Grant agreement no. 316779) and by National Research Fund, Luxembourg, under CORE project SpEctrum Management and Interference mitiGation in cognitive raDio satellite networks - SeMIGod, and CORE project SATellite SEnsor NeTworks - SATSENT.

References

1. EU: Digital agenda for europe. http://ec.europa.eu/digital-agenda (accessed December 8, 2014)
2. European Commission: A digital agenda for Europe. In: COM (2010) 245, Brussels, Belgium, May 2010
3. Brandel, D., Watson, W., Weinberg, A.: NASA's Advanced Tracking and Data Relay Satellite System for the Years 2000 and Beyond. Proceedings of the IEEE **78**(7), 1141–1151 (1990)
4. Alegre-Godoy, R., Alagha, N., Vazquez-Castro, M.: Offered capacity optimization mechanisms for multi-beam satellite systems. In: IEEE Int. Conf. on Communications (ICC), Ottawa, Canada, June 2012
5. EUTELSAT. http://www.eutelsat.com (accessed December 9, 2014)
6. VIASAT. https://www.viasat.com (accessed December 9, 2014)
7. SES. https://www.ses.com (accessed January 7, 2015)
8. Aloisio, M., Angeletti, P., Coromina, F., Mignolo, D., Petrolati, D., Re, E.: A system study to investigate the feasibility of terabit/s satellites. In: IEEE Int. Vacuum Electronics Conf. (IVEC), Monterey, CA, USA, April 2012
9. Haykin, S.: Cognitive radio: brain-empowered wireless communications. IEEE Journal on Selected Areas in Communications **23**(2), 201–220 (2005)
10. COgnitive RAdio for SATellite Communications - CoRaSat. European Commission FP7, October 2012
11. Cooperative and Cognitive Architectures for Satellite Networks - CO2Sat. Fonds National de la Recherche Luxembourg (FNR) (2011)
12. Spectrum Management and Interference Mitigation in Cognitive Radio Satellite Networks - SeMiGod. Fonds National de la Recherche Luxembourg (FNR) (2014)
13. Antennas and Signal Processing Techniques for Interference Mitigation in Next Generation Ka Band High Throughput Satellites - ASPIM. European Space Agency, September 2014

14. Goldsmith, A., Jafar, S.A., Maric, I., Srinivasa, S.: Breaking Spectrum Gridlock with Cognitive Radio: An Information Theoretic Perspective. Proc. IEEE **97**(5), 894–914 (2009)
15. Mohamed, A., Lopez-Benitez, M., Evans, B.: Ka band satellite terrestrial co-existence: a statistical modelling approach. Ka and Broadband Communications, Navigation and Earth Observation Conf., Salerno, Italy, October 2014
16. Maleki, S., Chatzinotas, S., Evans, B., Liolis, K., Grotz, J., Vanelli-Coralli, A., Chuberre, N.: Cognitive Spectrum Utilization in Ka Band Multibeam Satellite Communications. IEEE Communication Magazine (to appear)
17. The use of the band 27.5-29.5 GHz by the Fixed Service and uncoordinated Earth stations of the Fixed-Satellite Service (Earth-to-space). ECC Decision ECC/DEC/(05)01, March 2005
18. Hong, M., Luo, Z.: Signal processing and optimal resource allocation for the interference channels. In: Chellappa, R., Theorodiris, S. (eds.) Academic Press Library in Signal Processing: Communications and Radar Signal Processing. Elsevier, Oxford (2014)
19. Digital Video Broadcasting (DVB): DVB-S2X Standard. https://www.dvb.org/standards/dvb-s2x (accessed October 7, 2014)
20. Kuhn, H.: The Hungarian Method for the Assignment Problem. Naval Research Logistics Quarterly **2**, 83–97 (1955)
21. ITU: Br ific for terrestrial services. http://www.itu.int/en/ITU-R/terrestrial/brific (accessed October 7, 2014)
22. Caini, C., Corazza, G., Falciasecca, G., Ruggieri, M., Vatalaro, F.: A Spectrum and Power Efficient EHF Mobile Satellite System to be Integrated with Terrestrial Cellular Systems. IEEE J. Sel. Areas Commun. **10**(8), 1315–1325 (1992)
23. NASA: Socioeconomic Data and Applications Center (SEDAC). http://sedac.ciesin.columbia.edu (accessed February 27, 2015)
24. USA: US National Geospatial-Intelligence Agency (NGA). http://geoengine.nga.mil/muse-cgi-bin/rast_roam.cgi (accessed February 18, 2015)

SHARF: A Single Beacon Hybrid Acoustic and RF Indoor Localization Scheme

Ahmed Zubair[✉], Zaid Bin Tariq, Ijaz Haider Naqvi, and Momin Uppal

School of Science and Engineering,
Lahore University of Management Sciences, Lahore, Pakistan
ahmedzubair2108@gmail.com, {ijaznaqvi,momin.uppal}@lums.edu.pk

Abstract. The inability of GPS (Global Positioning System) to provide accurate position in an indoor environment has resulted in global efforts for a precise indoor position system throughout the last decade. The current state of the art of localization and tracking estimates the position of the mobile node based on attributes like received signal strength (RSS), angle of arrival (AoA) etc. from at least three anchor nodes. This paper presents SHARF; a *single beacon* hybrid acoustic and RF localization scheme in an indoor environment. It combines the RF RSS information for ranging with the angle of azimuth from acoustic localization system based on beacon signals from only *one* target node to *one* anchor node. The experimental results show an improved localization accuracy in comparison to trilateration scheme. All these features, i.e. single beacon, hybrid approach and outlier rejection, posit the superiority of this technique over the existing systems.

Keywords: Indoor localization · Wireless acoustic sensor networks (WASN) · Received signal strength (RSS) · TIme difference of arrival (TDOA) · Angle of arrival (AOA)

1 Introduction

The developments in Global Positioning System (GPS) in the past two decades have revolutionized the way people commute and navigate. GPS, however, does not provide a reasonable position estimate in the indoor or cluttered environments due to high penetration losses, which stymies the reception of signals from at least 4 satellites. The success and utility of GPS has invoked quite an interest in the area of indoor localization among researchers. A right mix of accurate and robust localization would allow the development of several new indoor location based services. Localization of mobile nodes is a key requirement in navigating, tracking and various other applications. Indoor positioning systems (IPS) are the systems where location of a mobile node is estimated based on different attributes inferred from stationary nodes. The applications of an accurate IPS encompass a wide range of applications. For instance, huge shopping malls, airport, supermarkets, museums, storage facilities and parking plazas are few

© Institute for Computer Sciences, Social Informatics and Telecommunications Engineering 2015
M. Weichold et al. (Eds.): CROWNCOM 2015, LNICST 156, pp. 499–510, 2015.
DOI: 10.1007/978-3-319-24540-9_41

examples where IPS can revolutionize the domain of location based services. In addition, robotic mail delivery, context aware robotics, smart rooms and interactive humanoid robots etc. are also amongst possible applications out of the myriads of possibilities, yet to be explored.

Indoor localization and tracking systems may be broadly classified into two types; infrastructure free (IF) localization and infrastructure based (IB) localization [1]. In IF localization, no new infrastructure is required for localization of mobile nodes. Such schemes assume universal WiFi availability (indoors) and thus localization can be implemented on smart phones with no requirement of additional infrastructure. IB localization techniques require additional hardware (other than a smart phone) and careful planning (e.g. deployment of sensors, beacons, routers, antennas, development of maps, etc.) to carry out localization. Although IF localization is gaining genuine interest from research and industrial community, there are areas where IB localization is preferred over IF localization. For instance, for localization of miners in an underground mine, for robot navigation in warehouses and localization of expensive equipment in hospitals or museums, designing a system based on Wi-Fi and smart phone devices may not be a suitable choice.

Localization schemes (be it IF or IB) in general are classified in many ways; range-free or range-based, distributed or centralized, geometric or non geometric, etc. Range-based methods include exploitation of quantities such as TOA (Time of Arrival), TDOA (Time Difference of Arrival), AOA (Angle of Arrival) or RSS (Received Signal Strength) [2] whereas range free localization rely on attributes like number of hops, node density etc [3]. Range-based approaches give better estimates of localization as compared to range-free methods as they are relatively more accurate with providing the information about the direction of the signal source. They exploit the TOA and TDOA information to find out the AOA. On the other hand, the RSS information provides an estimate about the range/distance of the signal source with careful modeling of the environment.

This paper presents SHARF; a hybrid approach for localizing a source based on TDOA and RSS information for acoustic and RF signals and requiring only *one* beacon node and *one* anchor node. Hybrid localization has been previously investigated but most of the previous work is generally concentrated on the hybrid algorithms[4] to find a sweet spot of trade-off between computational complexity and localization accuracy. Some systems exploit the special sensor array structures [5] for better localization accuracy. Unlike these approaches, the proposed system combines the strengths of the two methods of localization to provide a new insight into the localization and accurate position estimation. The TDOA based acoustic localization approach is very accurate if the sensor and beacon nodes satisfy Fraunhofer plane wave condition. However, when it comes to ranging a source in the far field, the acoustic localization system presents problems with a single node since the distance of sensor and beacon node does not remain comparable to the distance between microphone sensors in the array and hence, erroneous estimates appear. To overcome this problem, at least three nodes are required for location estimation through trilateration; which presents its own challenges like clock synchronization and time stamping of events etc. Although acoustic localization has been attempted with 2 sensor nodes with reflective surface [6], the cost of complexity is high. Similarly, the RF localization based on RSS does not determine the direction of arrival with a single node and hence three nodes are required for trilateration in this case too [7]. Moreover, finding the AOA with wireless signals by exploiting TDOA requires at least two antennas and very high speed and high resolution Analog to Digital Converter (ADC) units which are very expensive. The scheme presented in this paper brings down the number of sensor nodes to just one; a single RF-acoustic sensor node for localization and ranging. To the best of authors' knowledge, location estimation with such precision and accuracy has never been achieved with a single node before. Thus SHARF is first of its kind. The scheme presented in [8] is, however, somehow akin to it. Furthermore, in order to cater for the outliers of RF based ranging, we employ a moving average filter and a clustering algorithm which rejects the anomalies in the RSS values. This takes care of rapid fluctuations due to received multi-paths.

The rest of the paper is organized as follows. Section 2 presents the system model and the proposed approach for hybrid localization. The hardware implementation is given in section 3. Section 4 presents the experimental results of the proposed localization algorithm based on implementation on actual hardware platform. Finally, section 5 concludes this paper and highlights some future work.

2   System Model and Proposed Approach

The proposed hybrid wireless-acoustic localization system consists of a single wireless-acoustic anchor node and a wireless-acoustic source beacon as the mobile node. The reduction in the number of anchor nodes from (at least) 3 to 1 shows a three-folds decrease in the cost of the entire setup. In this paper, we assume that we are localizing a single target node. However, extension can be done for multiple targets.

Following are the key features of our measurement setup:

1. The deployed anchor node has complete information about its location and orientation in the environment.
2. The beacon node is present in the far field and behaves as a point source.
3. The nature of acoustic signals from the source are discontinuous and impulsive.
4. Localization is being done in 2D i.e. elevation information is not incorporated in the system. Moreover, the localization is carried out in the coverage area of the anchor node.
5. It is assumed that the link between anchor node and the target node follows the following path loss model

$$P_r = P_r(d_0) + K + 10\gamma log_{10}(d_0/d) + \zeta \qquad (1)$$

where P_r represents the received power, $P_r(d_0)$ is the received signal strength at reference distance d_0 which is usually taken to be equal to $1m$, K (in dB)

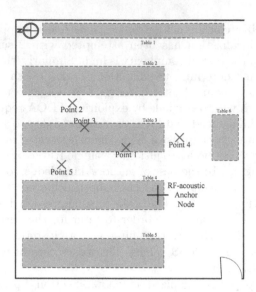

Fig. 1. Experimental setup for wireless-acoustic hybrid localization system. The cross points (X) are the target positions and the origin has been drawn where wireless-acoustic anchor node was placed.

is a unit-less constant depending on channel and antenna characteristics and ζ is a random variable which causes fluctuations in RSS values due to shadowing and changing multipaths. The distribution of ζ is unknown in case of non-stationary indoor environments. $P_r(d_0)$ can be calculated using *Friis* equation or acquired through empirical measurements [9].

6. The RSS values of the beacon signals are averaged over multiple wavelengths through a Moving Average (MA) filter. This averages out the fluctuations due to multi path components and averages out the zero mean noise as well. The MA filter's window size of $M = 15$ is used in our experiments which means that if the target is moving with a speed of 0.3 m/s, with $M = 15$ we roughly average out the RSS values over more than 7 wavelengths of the carrier signal of frequency 433MHz.

Our proposed approach goes through the three stages detailed below before combining the data to reach a final source estimate.

2.1 Path Loss Calculation

Path loss exponent (PLE) is computed through a training phase. Fig. 1 shows the environment of the lab in which the experiments have been conducted. To compute the PLE, RSSI values were recorded at different distances where the anchor node was placed at the coordinates of (0,0). Table 1 shows RSSI values recorded in dBm for multiple distances between the anchor and mobile nodes.

The PLE is computed such that it minimizes the mean squared error between the model and received power during the training phase [10]. Let d_i be the

Table 1. Received Signal Strength Information vs Distance values.

RSSI (dBm)	Distance (m)
-27	1.0
-36	3.17
-43	3.24
-44	3.33
-46	3.9
-49	5.2

distance between the anchor node and the mobile node at the ith iteration, then the received signal power based the model above can be written as:

$$M_{model}(d_i) = K - 10\gamma log_{10}(d_i) \tag{2}$$

where $K \cong 20log_{10}\frac{\lambda}{4\pi d_o}$ is the unit less constant that depends upon the characteristics of the antenna used, d_o is the reference distance in the far field region of the antenna and λ is the wavelength at which the signal is transmitted. In our case the frequency is $433Mhz$ so λ is nearly $0.693m$.

The Mean Square Error (MSE) is than calculated using as under:

$$F(\gamma) = \sum_{i=1}^{n}[M_{measured}(d_i) - M_{model}(d_i)]^2 \tag{3}$$

After taking the first derivative of the above equation and equating that to zero, PLE exponent that minimizes the MSE in the given environment turn out to be 3.26. We make use of this PLE for ranging computations.

2.2 Rejection of Anomalous RSS Values

To reject the anomalous RSS values, we use hyper-ellipsoidal clustering model for outlier detection. Let $R_k = \{r_1, r_2, \cdots r_k\}$ be the first k samples of RSS values collected at the target mobile node in a WSN. Each sample r_i is a $d \times 1$ vector in $\Re^d$, where d is the number of anchor nodes participating in localization. Hyper ellipsoidal outlier detection clusters the normal data points and the points lying outside the clusters are declared as outlier. The boundary of the cluster (an hyper-ellipsoid in this case) is related to a distance metric which typically is a function of mean $m_{R,k}$ and covariance S_k of the incoming RSS data R_k. One example of the distance metric is *Mahalanobis* distance, D_i [11], for which the cluster can be characterized by the following equation

$$e_k(m_R, S_k^{-1}, t) = \{r_i \epsilon \Re^d| \tag{4}$$

$$\underbrace{\sqrt{(r_i - m_{R,k})^T S_k^{-1}(r_i - m_{R,k})}}_{D_i = Mahalanobis\ distance\ of\ x_i} \leq t\}$$

where e_k is the set of normal data points whose Mahalanobis distance, $D_i < t$ and t is the *effective* radius of the hyper-ellipsoid. The choice of t depends on the distribution of the normal data points. If the normal data follows a *chi-squared* distribution, it has been shown that up to 98% of the incoming normal data can be enclosed by the boundary of an hyper-ellipsoid, if the effective radius t is chosen such that $t^2 = (\chi_d^2)_{0.98}^{-1}$ [11].

2.3   Azimuthal Angle Calculation

The angle of azimuth of the acoustic beacon signal from the target node is found using the conventional TDOA technique. For the simplest case in 2D, the DOA can be computed by finding out the time difference of arrival of the sound signal on two spatially distributed microphones where the distance between them is known, as shown in Fig. 2.

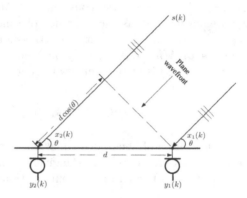

Fig. 2. DOA estimation in 2-D with identical microphones. The source is located in the far field, the incident angle is θ and the spacing between two microphones is d[12].

If microphone $y_1(k)$ is taken as a reference, signal at the second sensor $y_2(k)$ is the delayed version of the same signal at $y_1(k)$, having a delay equal to the time required for the plane wave to travel an extra distance $d \cos \theta$.

Therefore, the TDOA of sound signal between the two sensors is given by:

$$t_d = (d \cos \theta)/c$$

where c is the speed of sound in meters per seconds and t_d is the time delay between two signals .

Also

$$\cos \theta = \frac{c \times t_d}{d} \qquad (5)$$

Here $t_d = n_d \times t_s$ where delay index $n_d = n_p - n_{mean}$ is the difference of peak index from mean index and $t_s = 1/f_s$ is the sampling interval if f_s is the sampling frequency.

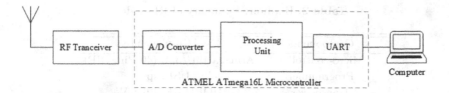

Fig. 3. A block diagram of RF-based localization system.

The value of n_p can be determined by using the cross correlation method [13]. Let the acoustic signal from mics $y_1(k)$ and $y_2(k)$ are digitized and the samples are stored in two arrays $x_1(m)$ and $x_2(m)$ respectively. The correlation between the signals x_1 and x_2 is then given by:

$$R_{x_1 x_2}(n) = \sum_{m=\infty}^{-\infty} x_1(m) x_2(m+n) \qquad (6)$$

The peak index n_p will be then:

$$\arg \max_{n} R_{x_1 x_2}(n) \qquad (7)$$

If the angle ranges between 0 and 180 and t_d is known, then angle θ can be uniquely determined. Therefore, estimating the incident angle θ is essentially identical to calculating the TDOA of the two signals. By extension of the above, to get the range of 0 to 360, three microphones are used to cater for the ambiguity in the direction of source as we have done in our case. In this way, the entire 2-D planes is covered.

3 Hardware Implementation

3.1 Hardware for RF Measurements

RSS-based localization is tested on low cost, DASH7 compatible, prototype wireless sensor nodes which have been designed and manufactured in-house using off-the-shelf components (COTS). We employed Carrier Sense Multiple Access-Collision Avoidance (CSMA-CA) as our channel access protocol. Atmega 16/32L has been used for processing the data from the wireless sensors. The block diagram of wireless localization system is show in Fig. 3. Table 2 summarizes the specification of RF hardware node.

3.2 Hardware for Acoustics Measurements

The acoustic hardware platform is developed using 32-bit ARM-based M4F-Cortex microcontroller that is capable of working upto 168MHz clock frequency, 3 ADC modules with 12-bits resolution and a maximum sampling rate of 2.4

Table 2. Hardware Summary of RF-node

Components	Used/Implemented
Micro-controller	Atmega 16L/32L (40 Pin, DIP)
Programmer	USBasp
Transceiver	HOPE RFM 69CW
Sensors	Illumination, Temperature
Channel Access	CSMA-CA
Networking Protocol	SNAP
Programming Language	C
Debugging Interface	UART

million samples per second. A built-in floating point unit (FPU) offers more flexibility and computing power to the hardware platform. Such a processing speed and fast ADC is a basic requirement to resolve very small differences in the time of arrival of the acoustic signal on spatially displaced microphones array. The sampling rate we used is $100kHz$ but it can be increased further to improve the resolution of the azimuthal angle. However, to meet the Nyquist criterion of sampling, $44.1kHz$ sampling rate standard as used in audio industry will suffice. The accuracy of the estimated angle of arrival would be lesser though. Condenser type, omni-directional analogue microphones with a sensitivity of $-44 \pm 2dB$ and signal to noise ratio of 60dBA are used with a preamplifier. The preamplifier MAX9814 is a low-cost, high-quality microphone amplifier from Maxim Integrated with built-in Automatic Gain Control (AGC). The reason of using the preamplifier with the microphones is that in order to maintain a particular distance between microphones, they need to be spread apart and the signal is carried through wires. This faint signal if not amplified at the microphone will result in accumulation of noise over the wire. This Variable Gain Amplifier (VGA) provides the ability to pick up weak signals too. An acoustic sensor node consists of three microphones 120° apart in 2-D i.e. at the vertices of a right angled triangle. The measurement setup and the hardware specifications have been presented and listed in the Fig. 4 and Table 3 respectively. The distance between each microphone is 30cm. The source localization or direction estimation is implemented in 2-D and works on the principle of TDOA to find the Direction of Arrival (DOA) of sound from the beacon. The spatial positioning of microphones translates into the signal reaching the microphones at slightly different time intervals. These time delays then indicate the DOA to give the angle of azimuth of the source.

4   Experimental Results

The measurement layout for SHARF with position of the anchor node and test points with mobile node has been shown in Fig. 1. The experiments are conducted in AdCom research lab with an area of 80 m^2. It has six wooden tables

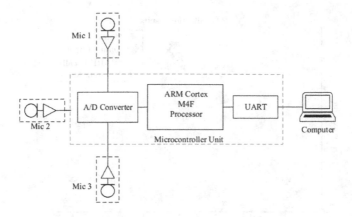

Fig. 4. A block diagram of acoustic localization system.

Table 3. Hardware Summary of acoustic-node

Components	Used/Implemented
Micro-controller	STM32F407VG
Programmer	ST-Link Programmer/Debugger
Microphones sensitivity	$-44 \pm 2dB$
Preamplifier	MAX9814
- Amplifier gain	Variable Gain Adjustment
Programming Language	C
Debugging Interface	ST-Link Debugger

placed in it and two of its walls are of concrete. Additionally there is a plenty of measurement equipment present on the tables to generate multipaths. We assume that the anchor node which had complete information of its position and orientation. It continuously receives the wireless-acoustic beacon signals from the test points marked as X's in the Fig. 1. The RF hardware platform estimates range using RSS values and the acoustic hardware platform provides the azimuthal angle of the target. The results of the localization are then monitored on the PC through a UART interface on the anchor node.

SHARF achieves very good localization accuracy. The mean and standard deviation of location estimation error has been shown in the Fig. 8 over multiple measurements. The results with RF trilateration has also been plotted for alongside to establish the superiority of the proposed scheme.

For the sake of comparison, the error histograms for frequency of error distances for RSS-based trilateration and SHARF at test point 5 for 20 readings have been shown in Fig. 6 and Fig. 7 respectively. The error histograms on this particular point indicate an improvement of around 1.5m while using SHARF localization scheme versus only RF-based trilateration scheme.

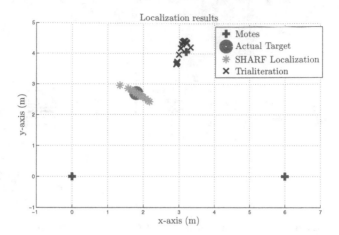

Fig. 5. Ground truth based localization results for SHARF localization and Trilateration systems at point 5.

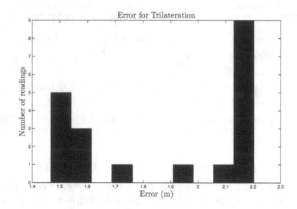

Fig. 6. Distance error histogram for RSS-based Trilateration system at point 5.

Similarly, Fig. 8 shows the anchor node positions, target location and localization results using trilateration and SHARF. We can see for most points, the proposed SHARF scheme outperforms trilateration while using only one beacon node and one anchor node. Please note that for a fair comparison, we use rejection of anomalous RSS values both for trilateration as well as the proposed scheme or otherwise the gain over trilateration can be shown to be much higher. Thus the proposed scheme gives a double advantage; one that it requires only one anchor node and second that it outperforms trilateration-based localization scheme which requires at least three anchor nodes in order to estimate the location of mobile node. Thus SHARF is a localization scheme which not only reduces the size complexity and cost of the system, but also improves the localization accuracy.

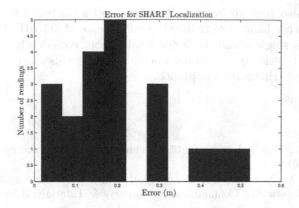

Fig. 7. Distance error histogram for SHARF localization system at point 5.

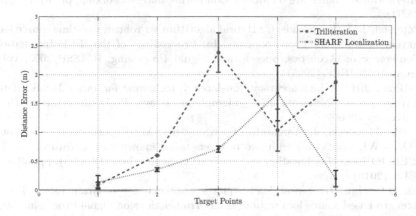

Fig. 8. The mean and standard deviation error bars for SHARF localization and Trilateration systems.

5   Conclusion

This paper proposes and implements SHARF; a single beacon hybrid acoustic-RSS based target localization system. Our solution exploits the strength of TDOA in finding the direction of source from acoustic signals and RSS information in finding the range of the target and combines these information together to mitigate the weakness of acoustic-based localization in ranging and RSS-based localization in direction of the source. RSS-based trilatertion and acoustic-based trilateration systems alone present complexities in synchronization and problem formulation. Moreover, slight error in values of one of the three nodes taking part in trilateration results in a substantial error in the localization error. Cost of the system also goes up by using 3 anchor nodes (at least) for trilateration which, on the other hand, gets reduced to one in case of SHARF. In addition to requiring only a single anchor node for localization, considerable improvement (94%) in mean position error has been achieved using this scheme as compared to the traditional trilateration approach.

References

1. Schneider, D.: New indoor navigation technologies work where gps can't. IEEE Spectrum (2013)
2. Gu, Y., Lo, A., Niemegeers, I.: A survey of indoor positioning systems for wireless personal networks. Communications Surveys & Tutorials, IEEE **11**(1), 13–32 (2009)
3. He, T., Huang, C., Blum, B.M., Stankovic, J.A., Abdelzaher, T.: Range-free localization schemes for large scale sensor networks. In: Proceedings of the 9th Annual International Conference on Mobile Computing and Networking, pp. 81–95. ACM (2003)
4. Peterson, J.M., Kyriakakis, C.: Hybrid algorithm for robust, real-time source localization in reverberant environments. In: Proceedings of the IEEE International Conference on Acoustics, Speech, and Signal Processing, ICASSP 2005, vol. 4, pp. iv-1053. IEEE (2005)
5. DiBiase, J.H.: A high-accuracy, low-latency technique for talker localization in reverberant environments using microphone arrays, Ph.D. dissertation, Brown University (2000)
6. Kietlinski-Zaleski, J., Yamazato, T., Katayama, M.: Experimental validation of TOA UWB positioning with two receivers using known indoor features. In: 2010 IEEE/ION Position Location and Navigation Symposium (PLANS), pp. 505–509. IEEE (2010)
7. So, H.C., Lin, L.: Linear least squares approach for accurate received signal strength based source localization. IEEE Transactions on Signal Processing **59**(8), 4035–4040 (2011)
8. Geng, X., Wang, Y., Feng, H., Chen, Z.: Hybrid radio-map for noise tolerant wireless indoor localization. In: 2014 IEEE 11th International Conference on Networking, Sensing and Control (ICNSC), pp. 233–238. IEEE (2014)
9. Mao, G., Anderson, B., Fidan, B.: Path loss exponent estimation for wireless sensor network localization. Computer Networks **51**(10), 2467–2483 (2007)
10. Goldsmith, A.: Wireless communications. Cambridge University Press (2005)
11. Moshtaghi, M., et. al.: Incremental elliptical boundary estimation for anomaly detection in wireless sensor networks. In: IEEE International Conference on Data Mining (ICDM), pp. 467–476. IEEE (2011)
12. Benesty, J., Chen, J., Huang, Y.: Microphone array signal processing, vol. 1. Springer (2008)
13. Knapp, C., Carter, G.C.: The generalized correlation method for estimation of time delay. IEEE Transactions on Acoustics, Speech and Signal Processing **24**(4), 320–327 (1976)

Massive MIMO and Femto Cells for Energy Efficient Cognitive Radio Networks

S.D. Barnes[1]([✉]), S. Joshi[1], B.T. Maharaj[1], and A.S. Alfa[2]

[1] Department of Electrical, Electronic and Computer Engineering,
University of Pretoria, Pretoria, South Africa
simonbarnes@ieee.org, j.shital84@gmail.com, sunil.maharaj@up.ac.za
[2] Department of Electrical and Computer Engineering,
University of Manitoba, Winnipeg, Canada
alfa@ee.umanitoba.ca

Abstract. In this paper, energy efficiency (EE) was investigated for a cognitive radio network (CRN) where massive multiple-input multiple-output (MIMO) was combined with femto cells. A heterogeneous network was considered to maximise EE, while massive MIMO was implemented to increase spatial reuse and focus energy into smaller spatial regions. Cooperation between femto cell based relay stations (RS) and a micro cell based secondary base station (SBS) allowed for secondary user (SU) quality-of-service (QoS) requirements to be met and also for control over primary user (PU) interference. Two key issues were addressed by the proposed model, namely improved CRN EE and reduced PU interference. The formulated EE optimisation problem was non-convex and thus converted into a semi-definite problem. Simulation results show that combining massive MIMO at the micro cell level with femto cells, lead to a CRN EE improvement. The addition of femto cells also lead to a reduction in PU interference.

Keywords: Cognitive radio network · Energy efficiency · Femto cells · Massive MIMO

1 Introduction

In recent years wireless communication networks have experienced tremendous growth due to an increase in the demand for services such as mobile data communications, high speed internet and live video conferencing. At first glance, it would appear that the spectrum available to operators is limited and would thus not be able to cope with this ever increasing demand for high speed data communication and the associated bandwidth. However, a closer look suggests that spectrum utilisation is more of a problem, rather than the actual availability thereof. Large portions of the usable spectrum are either largely underutilised or partially occupied [1]. Furthermore, the practice of anywhere-anytime access to the network has triggered a dramatic expansion of network infrastructure. It has become essential for mobile network operators to maintain sustainable capacity growth.

© Institute for Computer Sciences, Social Informatics and Telecommunications Engineering 2015
M. Weichold et al. (Eds.): CROWNCOM 2015, LNICST 156, pp. 511–522, 2015.
DOI: 10.1007/978-3-319-24540-9_42

This subsequently leads to a drastic increase in the energy consumption of the network. Given that energy costs (diesel, electricity) are constantly increasing and since this energy expenditure constitutes a major portion (20% - 30%) of the operators's total energy expenditure [2], it is necessary to consider schemes that can increase the energy efficiency (EE) of the network, while maintaining required quality of service (QoS) levels for secondary users (SUs).

Cognitive radio (CR) [11] has been identified as an efficient technology, that allows secondary users (SUs) to make use of underutilised or partially used spectrum that is in fact licensed for use by primary users (PUs). Using licensed channels means that SUs must always consider the interference that they may inflict on PUs. Therefore, it is very important to keep the SU interference level below the level that is acceptable to the PU. Since there is a relationship between the EE of the network and the amount of power required to guarantee the level of QoS experienced by SUs, techniques should be investigated that minimise base station (BS) power consumption while maintaining the level of QoS required by SUs. As suggested in [6], there are many techniques through which the power consumption of the base station can be minimised. Of these techniques, energy aware cooperation between BSs, the integration of femto cells within micro cells and the use of multiple-input multiple-output (MIMO) are considered to be most beneficial. These techniques have already been incorporated into existing wireless communication networks.

"Massive MIMO"[7] is currently seen as a promising cellular network architecture offering several attractive features over regular MIMO. It has been shown that even simple linear pre-coders and detectors are optimal when the ratio of antennas per BS to the number of user terminals per cell is very large [10]. In [9], cooperation between the primary BS (PBS) and the SBS was compared with the case when there is no cooperation between them. It was shown that the signal-to-interference-plus-noise-ratio (SINR) requirements of all SUs can be met when there is cooperation between the secondary base station (SBS) and the PBS. In [12], an EE improvement was demonstrated for a heterogeneous network (HetNet) deployment when compared to a single cell scenario. While HetNets have been considered for cognitive radio networks (CRNs), traditional wireless networks and massive MIMO for traditional wireless networks, the unique contribution in this research work is the combination of both massive MIMO and HetNets in a CRN.

In this paper, femto cells are deployed within the micro cell and a massive MIMO system is used at the secondary BS (SBS). Coordination between an amplify-and-forward relay station (RS), in a femto cell, and the SBS, in the micro cell, is considered. It is assumed that both the RS and the SBS can serve the SUs within their coverage area, depending on the efficiency of the network. These secondary micro cells are assumed to lie within the primary cell. Since the RS has a smaller coverage area, the total transmission power as well as the static power of the RS is less than that of the micro SBS. An optimization problem is formulated where the objective is to maximize the energy efficiency (EE) of the overall CRN while satisfying the QoS requirement at each SU as well as adhering

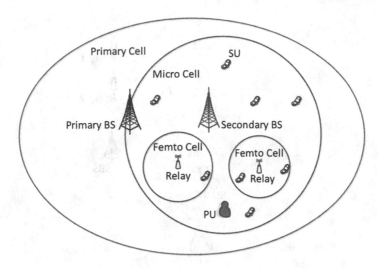

Fig. 1. System model for an underlay CRN with a primary cell covering multiple secondary cells.

to the interference constraints for each PU (with a per antenna power constraint at both the SBS and the RS). Since this optimization problem is non-convex, it is converted into a convex problem. Maximisation of EE for the overall network is then solved. From the results it is shown that the introduction of massive MIMO at the micro cell level together with the addition of femto cells, improves the EE of the network. The addition of femto cells within the micro cells is shown to reduce the interference experienced by the PUs.

Notation: The subscript zero is used to denote micro cell SBSs and PBSs. Any other subscript values denote RSs.

2   System Model

An underlay CRN, as illustrated in Fig. 1, is considered where the SBS is equipped with K_{BS} antennas that are serving N number of SUs ($K_{BS} >> N$). Each of these SUs have a single antenna. Two femto cells are considered within the coverage area of the SBS. Each of these femto cells are equipped with a single RS which has K_R number of antennas, where $K_R \in \{1, 2, 3\}$. The primary cell surrounds the secondary cells. A single PU can be located anywhere within the primary cell. If the PU lies outside the coverage zone of the SBS then the model reduces to the one shown in [5], which is similar to the case of traditional communication network. The PBS is assumed to have a single antenna. A single antenna PU is assumed to be located randomly within the coverage area of the SBS.

It is assumed that the SBS and RSs at each of the femto cells are coordinated in such a way that either a RS or a BS or both of them, can provide service to the SUs. Since the transmission power of the RS is much less than that of the

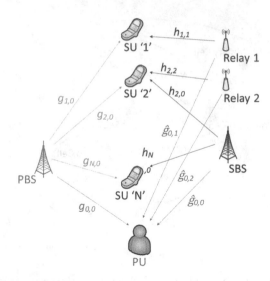

Fig. 2. Conceptual diagram for the proposed model.

BS, it would be preferable to serve the SUs using the RS. However due to the small coverage area of the RS, it may not be able to serve all of the SUs that lie within the secondary micro cell. Hence, if it is energy efficient, those SUs that are not being served by the RS would be served by a combination of the RS and the SBS. Otherwise they would be served by the SBS alone. This decision is taken based on the solution to the optimisation problem for the maximisation of the EE of the overall CRN.

As shown in Fig. 2, $\mathbf{h}_{t,0} \in C^{1 \times K_{BS}}$ and $\mathbf{h}_{t,j} \in C^{1 \times K_R}$ represent the channel gains from the SBS to the tth SU and from the jth RS to the tth SU respectively. The terms $g_{t,0}$ and $g_{0,0}$ represent the channel gains from the PBS to the tth SU and from the PBS to the PU respectively. The terms $\hat{g}_{0,0}$ and $\hat{g}_{0,j}$ represent the channel gains from the SBS to the PU and from the jth RS to the PU respectively. A flat fading sub-carrier is considered, as in [5], and the base band channels are assumed to be perfectly know at both sides of each channel [4]. The received signal power at the tth SU is given by,

$$y_t = \mathbf{h}_{t,0}^H \mathbf{x}_0 + \sum_{j=1}^{R} \mathbf{h}_{t,j}^H \mathbf{x}_j + g_{t,0}^H x_p + n_t, \tag{1}$$

where n_t is the circularly symmetric complex Gaussian noise with zero mean and variance σ_k^2, R is the number of femto cells (i.e. number of RSs), $\mathbf{x}_0 \in C^{K_{BS} \times 1}$, $\mathbf{x}_j \in C^{K_R \times 1}$ and x_p are the transmitted signals from the SBS, the RS and the PBS respectively. The transmitted signal can be written as,

$$\mathbf{x}_j = \sum_{t=1}^{N} \mathbf{w}_{t,j} x_{t,j}, \quad j = 0, 1, ... R \tag{2}$$

$$SINR_t = \frac{|\mathbf{h}_{t,0}^H \mathbf{w}_{t,0}|^2 + \sum_{j=1}^{R} |\mathbf{h}_{t,j}^H \mathbf{w}_{t,j}|^2}{\sum_{i=1, i \neq t}^{N}(|\mathbf{h}_{t,0}^H \mathbf{w}_{i,0}|^2 + \sum_{j=1}^{R} |\mathbf{h}_{t,j}^H \mathbf{w}_{i,j}|^2) + |g_{t,0}^H|^2 + \sigma_t^2} \geq \hat{\gamma}_t \quad (3)$$

where $x_{t,j}$ is the information transmitted from the SBS or the RS to the tth SU, $\mathbf{w}_{t,j}$ is a beam forming vector where j is the transmitter serving user t with $\mathbf{w}_{t,0} \in C^{K_{BS} \times 1}$ and $\mathbf{w}_{t,j} \in C^{K_R \times 1}$

3 Problem Formulation

The objective is to maximize the EE of the overall CRN while satisfying the QoS requirement of the SUs and the interference constraint of the PU. The QoS requirement specifies the information rate, in bit/s/Hz, that each user should achieve. Therefore, it is related to SINR as follows,

$$\log_2(1 + SINR_t) \geq \gamma_t, \quad (4)$$
$$\text{or,} \quad SINR_t \geq 2^{\gamma_t} - 1 = \hat{\gamma}_t,$$

where γ_t is the required QoS of tth SU.

From Eq. (1), the SINR for the tth SU is given by Eq. (3), given at the top of the page. This is the QoS constraint for the SUs. The interference constraint for a PU is the sum of the interference from the SBS and the RSs, and is given by,

$$\zeta_0 = \sum_{i=1}^{N} |\hat{\mathbf{g}}_{0,0}^H \mathbf{w}_{i,0}|^2 + \sum_{i=1}^{N} \sum_{j=1}^{R} |\hat{\mathbf{g}}_{0,j}^H \mathbf{w}_{i,j}|^2. \quad (5)$$

With the SU interference limit denoted by ζ, this can be rewritten as,

$$\zeta_0 = \sum_{i=1}^{N} \sum_{j=0}^{R} |\hat{\mathbf{g}}_{0,j}^H \mathbf{w}_{i,j}|^2 \leq \zeta. \quad (6)$$

The total power consumption per sub-carrier C is the sum of the dynamic power and static power, as shown in Eq. (7),

$$P_{tot} = P_{dynamic} + P_{static}. \quad (7)$$

The dynamic power depends on the transmitted power and the amplifiers's inefficiency ρ. The static power depends on the number of the antennas (K_{BS} or K_R) which are used for serving the SUs and the total power dissipation η in each antenna [5]. The total power is thus given as,

$$P_{tot} = (\sum_{j=0}^{R} \rho_j \sum_{i=1}^{N} ||\mathbf{w}_{i,j}||^2) + (\frac{\eta_0}{C} K_{BS} + \sum_{j=1}^{R} \frac{\eta_j}{C} K_R), \quad (8)$$

where $\| \mathbf{w}_{i,j} \|^2$ is the norm-square of $\mathbf{w}_{i,j}$.

The maximum transmitted power has to be bounded. Since each antenna has its own power amplifier, it is common practice to use the per-antenna constraint on both the SBS and the RS [3]. Each SBS and RS is thus subject to L_j power constraints, such that,

$$\sum_{i=1}^{N} \mathbf{w}_{i,j}^{H} \mathbf{Q}_{j,l} \mathbf{w}_{i,j} \leq q_{j,l}, \quad l = 1, 2, .., L_j \tag{9}$$

where $\mathbf{Q}_{0,l} \in C^{K_{BS} \times K_{BS}}$; $\mathbf{Q}_{j,l} \in C^{K_R \times K_R}$ are the weighting matrices defining the per antenna power constraints for $j = 1, ..., R$ and are positive semi-definite. Each antenna is limited by a limit $q_{j,l} \geq 0$, where $q_{0,l} \gg q_{j,l}$ for $j = 1, ..., R$. The numerical simulation considers $L_0 = K_{BS}$, $L_j = K_R$ and $q_{j,l} = q_j \forall \ell$. $\mathbf{Q}_{j,l} = 1$ at the lth diagonal element and zero elsewhere [5].

The energy efficiency of the network is defined as the QoS requirement of each SU per unit of power consumed by the SBS and the RSs. Thus for user t it is defined as,

$$EE_t = \frac{B \log_2(1 + SINR_t)}{P_{tot}}. \tag{10}$$

The normalised energy efficiency (with respect to bandwidth B) for each user t is given by,

$$EE_t = \frac{\log_2(1 + SINR_t)}{P_{tot}} = \frac{\log_2(1 + \hat{\gamma}_t)}{P_{tot}}. \tag{11}$$

The objective of this paper is to maximise the EE of the overall network, thus the total EE of the network is given by,

$$EE = EE_1 + EE_2 + ... + EE_N, \tag{12}$$

$$EE = \frac{N \cdot \log_2(1 + \hat{\gamma})}{P_{tot}}, \tag{13}$$

where all SUs are assumed to have the same QoS requirement i.e. $\hat{\gamma}_1 = \hat{\gamma}_2 = ... = \hat{\gamma}_N = \hat{\gamma}$.

Thus, since it can be inferred that $\mathbf{w}^* \mathbf{R} \mathbf{w} = Tr[\mathbf{R} \mathbf{w} \mathbf{w}^*] = Tr[\mathbf{R} \mathbf{W}]$ and using Eq. (13) where $\mathbf{w}_{i,j}^H$ is replaced with $\mathbf{W}_{i,j}$ [3], the final optimisation problem is defined as follows,

$$\max_{\forall \mathbf{w}} \quad EE = \frac{N \cdot \log_2(1 + \hat{\gamma})}{\sum_{j=0}^{R} \rho_j \sum_{i=1}^{N} Tr(\mathbf{W}_{i,j}) + (\frac{\eta_0}{C} K_{BS} + \sum_{j=1}^{R} \frac{\eta_j}{C} K_R)}, \tag{14}$$

s.t. $\quad \log_2(1 + SINR_t) \geq \gamma_t, \quad \forall t$

$$\sum_{i=1}^{N} \sum_{j=0}^{R} | \hat{g}_{0,j}^H \mathbf{w}_{i,j} |^2 \leq \zeta,$$

$$\sum_{i=1}^{N} \mathbf{w}_{i,j}^H \mathbf{Q}_{j,l} \mathbf{w}_{i,j} \leq q_{j,l}. \quad \forall j, l$$

The objective function can be expressed as,

$$\max_{\forall \mathbf{w}} \quad EE = \frac{1}{\psi} = \frac{\log_2(1+\hat{\gamma})}{P_{tot}}, \tag{15}$$

where ψ is the new objective function used to maximize the EE.

Eq. (14) above is a non-convex problem and thus needs to be converted into a convex problem in order to solve it. Since the matrix $\mathbf{W}_{i,j}$ is a correlation matrix, it is thus necessary to add a constraint to guarantee that it is Hermitian and positive semi-definite. Thus Eq. (14) reduces to,

$$\min_{\forall \mathbf{w}} \quad \psi, \tag{16}$$

$$\text{s.t.} \quad \psi \cdot \log_2(1+\hat{\gamma}) \geq \sum_{j=0}^{R} \rho_j \sum_{i=1}^{N} Tr(\mathbf{W}_{i,j}) + (\frac{\eta_0}{C}K_{BS} + \sum_{j=1}^{R} \frac{\eta_j}{C}K_R),$$

$$\sum_{j=0}^{R} \mathbf{h}_{t,j}^H ((1+\frac{1}{\hat{\gamma}_t})\mathbf{W}_{t,j} - \sum_{i=1}^{N} \mathbf{W}_{i,j})\mathbf{h}_{t,j} \geq | \, \mathbf{g}_{t,0} \, |^2 + \sigma^2, \quad \forall t$$

$$\sum_{i=1}^{N} \sum_{j=0}^{R} \hat{g}_{0,j}^H \mathbf{W}_{i,j} \hat{g}_{0,j} \leq \zeta,$$

$$\sum_{i=1}^{N} Tr\left[\mathbf{Q}_{j,l}\mathbf{W}_{i,j}\right] \leq q_{j,l},$$

$$\mathbf{W}_{i,j} \succeq 0,$$

$$\mathbf{W}_{i,j} = \mathbf{W}_{i,j}^*.$$

$\mathbf{W}_{i,j}$ is positive semi-definite with rank $[\mathbf{W}_{i,j}] \leq 1$. With an additional constraint of rank $[\mathbf{W}_{i,j}] = 1$ [8], Eq. (16) is an exact equivalent of Eq. (14). Since that constraint is not included in Eq. (16), it is a relaxed solution for Eq. (14).

4   Numerical Results

Numerical results have been obtained for the scenario shown in Fig. 1. SUs were randomly distributed in the given area with at least one SU present in each femto cell. For the sake of simplicity, only one antenna was considered at the PBS (the results can easily be extended to the multi-antenna case at the PBS).

The solution to the optimisation problem was solved so as to determine the EE of the system (while adhering to the constraints shown in Eq. (15)). The channel model from [7] was adopted for $\mathbf{h}_{t,j}$. The parameter values used when performing the simulations are listed in Table 1 [5].

Simulation results for the case where the number of SUs in the network is small, i.e. 5, are shown in Fig. 3. The EE when femto cells are employed is compared to the EE when only a BS is employed (no RS). The EEs of two

Table 1. Channel Parameters for the numerical results [5].

Parameters	Values
Micro cell radius	500 m
Femto cell radius	50 m
Minimum distance between SUs and BS/relays	35m/3.5m
Number of SUs	5 and 12
Number of PU	1
Number of relays	2
Number of antennas at SBS	20 - 100
Number of antennas at relays	0, 1, 2, 3
QoS constraint of SU	2 - 3 bit/s/Hz
Interference Temperature (IT)	0.1 mW
Carrier frequency	2 GHz
Number of sub-carriers	600
Total bandwidth	10 MHz
Sub-carrier bandwidth	15 kHz
Standard deviation of log-normal shadowing	7 dB
Path and penetration loss, distance d (km), (Micro Cell)	$148.1 + 37.6\log_{10}(d)$ dB
Path and penetration loss, distance d (km), (Femto Cell)	$127 + 30\log_{10}(d)$ dB
Noise variance	-127 dBm
Noise figure	5 dB
Power amplifier efficiency	
- Micro BS	0.388
- Relay	0.052
Circuit power per antenna	
- Micro BS	66 mW
- Relay	0.08 mW
Per antenna constraints	
- micro BS	189 mW
- Relay	25.6 mW

different levels of QoS are also compared. When the number of SUs is small and their QoS requirements are low it is not very beneficial to use more antennas for the femto cells. When the QoS requirement of the SUs is 2 bit/s/Hz, the improvement in EE for 1 antenna per RS, 2 antennas per RS and 3 antennas per RS is almost identical. The case when the QoS requirement is 3 bit/s/Hz, is similar.

From the plot it is evident that when the number of antennas at the SBS is small ($K_{BS} < 30$) the EE increases with the number of SBS antennas and is also higher when femto cells are employed. The extent to which EE is improved by the use of femto cells is dependant on the QoS requirement of the SUs. For the scenario where $K_{BS} = 30$ antennas, a femto cell with three antennas is employed and the QoS requirement of the SUs is 2 bit/s/Hz (shown by the dashed lines), there is a 0.149 b/s/Hz/W improvement in EE. However, there is a 0.425 b/s/Hz/W improvement when the QoS requirement is 3 bit/s/Hz

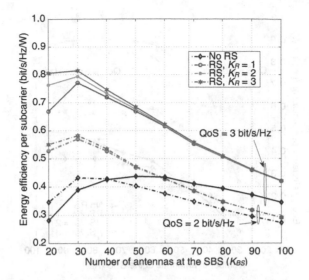

Fig. 3. Total energy efficiency of each BS when 5 SUs were considered (solid line: QoS of 3 bit/s/Hz; dashed line: QoS of 2 bit/s/Hz).

(shown by the solid lines). This improvement in EE is due to improved energy focusing, as a consequence of the large number of antennas used at the SBS, and also because the power gain due to a decrease in dynamic power is larger than the gain due to an increase in static power (caused by the addition of antennas).

When the number of antennas at the SBS becomes large ($K_{BS} > 30$) the improvement in EE begins to decrease and there is no further improvement to be gained by the addition of more antennas at the SBS, since the EE improvement steadily decreases due to an increase in the static power. Furthermore, the gain obtained from the deployment of femto cells, when compared to the case where no femto cells were deployed, becomes more distinguished as the QoS requirement of the SUs increases.

When no femto cells are employed, there is also an initial improvement in EE with the addition of antennas at the SBS. However, the improvement begins to decrease when $K_{BS} > 30$ for a QoS of 2 bit/s/Hz and when $K_{BS} > 50$ for a QoS of 3 bit/s/Hz.

Fig. 4 shows the energy efficiency of the system when the number of SUs is increased to 12. When no femto cells are deployed and when the SU QoS requirement is 2 bit/s/Hz (shown by the dashed line), then there is an improvement in EE when $K_{BS} < 80$ antennas, after which the gain in EE remains almost constant. When the QoS requirement increases to 3 bit/s/Hz, there is a continuous improvement in EE with an increment in the number of antennas at the SBS. When femto cells are deployed for the same scenario, the gain in EE increases with an increase in the number of femto cell antennas. This is due to a decrease in the propagation loss as a result of the use of femto cells, as well as an improvement in energy focusing due to the larger number of antennas.

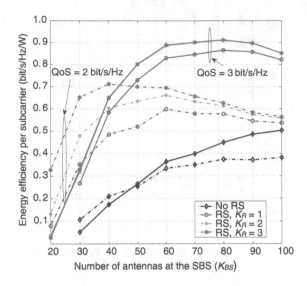

Fig. 4. Total energy efficiency of each station when 12 SUs are considered (solid line: Qos of 3 bit/s/Hz; dashed line: Qos of 2 bit/s/Hz).

The rate at which EE decreases, after the addition of a certain number of SBS antennas, varies according the SU QoS requirement. Similar to the 5 SUs case, the improvement in EE is larger for the higher QoS requirement.

The corresponding interferences per sub-carrier are listed in Table 2. It can be seen that the interference experienced by the PU, due to the SUs, decreases remarkably (by 22.2 %, from -10 dBm to -12.22 dBm) with the addition of femto cells. The result shows constant interference irrespective of the number of antennas used at either micro BS or femto cell. This is due to the fact that in the case of a micro BS, the problem is formulated according to the interference temperature (IT) limit. Thus, for a given IT (in this case 0.1 mW) the micro BS minimises its transmit power. For the case where femto cells are added to the network there is neither an increase nor a decrease in the interference experienced by the PU, since the PUs are located outside of the coverage of area of the RS. However, since the addition of femto cells divides the SUs into a micro BS region and a RS region, the number of users to be served by the BS is decreased; thereby decreasing the total interference to the PU. This result suggests that there would be a drastic decrease in the interference experienced by the PU if more users are served by the RS.

Table 2. Total interference to PU from each station (12 SUs with IT=0.1 mW).

Scenario	No RS	$K_R = 1$	$K_R = 2$	$K_R = 3$
ζ_0 (dBm)	-10.00	-12.22	-12.22	-12.22

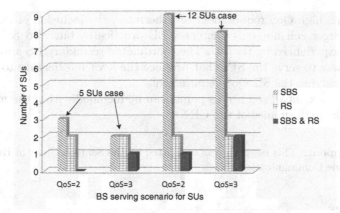

Fig. 5. Number of SUs served by different base stations and QoS requirements.

Fig 5 shows how the SUs are served by the different base stations. It explains the nature of the curves in Fig. 3 and Fig. 4. For the case of five SUs, when the QoS requirement is 2 bit/s/Hz, three SUs are served by the micro SBS and two SUs are served by the RS. When the QoS requirement is 3 bit/s/Hz, two SUs are served by the micro SBS, two SUs by the femto RS and the rest of the SUs by a combination of both. This suggests that when the SU QoS requirement increases it would be beneficial to serve the SUs using a combination of the SBS and the RS. Since the SUs are served in an efficient way, the rate at which EE decreases, after 40 antennas are added to the SBS, is similar for both QoS requirement scenarios. For the 12 SU case, when the QoS requirement is 2 bit/s/Hz, nine SUs are served by the micro BS, two SUs are served by the RS and one SU is served by both. When the QoS requirement of the SUs is 3 bit/s/Hz, eight SUs are served by the micro BS, two SUs are served by RS and two SUs are served by both.

5   Conclusion

From the results obtained it can be concluded that the gain in EE, obtained from the deployment of femto cells, decreases as the number of antennas at the SBS increases (when the number of SUs in the network is small). Thus, when the SU QoS requirement is small, the gain in EE is almost negligible when a large number of antennas is used at the SBS. The benefit of deploying femto cells becomes more pronounced as the number of SUs in the network increases, as well as when the SU QoS requirements are increased. Similarly, when the number of SUs and/or the network SU QoS requirements increase, it becomes more beneficial to increase the number of antennas at the femto cell RS. Irrespective of the number of SUs, the EE increases when the QoS requirement of the SUs increases. Massive MIMO is thus an effective approach for increasing the EE at the BS, for the case when the CRN needs to support a large number of users with a high QoS requirement. Furthermore, the inclusion of femto cells within the larger cell not only increases EE drastically, but also reduces the interference experienced by the PU. The coordinated secondary BS and RSs act in a cordial way to serve the SUs that produce the least interference to the PU, while still satisfying the SU QoS requirements.

In future work, small cell access points can be included within the micro cell to improve the performance of the CRN.

Acknowledgment. This research was supported by the Sentech Chair in Broadband Wireless Mobile Communication.

References

1. Report of the spectrum efficiency working group. Spectrum Policy Task Force (2002)
2. Bayhan, S., Alagoz, F.: Scheduling in centralized cognitive radio networks for energy efficiency. IEEE Trans. Veh. Technol. **62**(2), 582–595 (2013)
3. Bengtsson, M., Ottersten, B.: Optimal and suboptimal transmit beamforming. In: Handbook of Antennas in Wireless Communications (2001)
4. Björnson, E., Jorswieck, E.: Optimal resource allocation in coordinated multi-cell systems. Foundations Trends Commun. Inform. Theory **9**(2–3), 113–381 (2012)
5. Björnson, E., Kountouris, M., Debbah, M.: Massive MIMO and small cells: improving energy efficiency by optimal soft-cell coordination. In: 20th Int. Conf. Telecommun., Casablanca, Morocco, pp. 1–5 (2013)
6. Hasan, Z., Boostanimehr, H., Bhargava, V.K.: Green cellular networks: A survey, some research issues and challenges. IEEE Commun. Surveys Tuts. **13**(4), 524–540 (2011)
7. Hoydis, J., ten Brink, S., Debbah, M.: Massive MIMO in the UL/DL of cellular networks: How many antennas do we need? IEEE J. Sel. Areas Commun. **31**(2), 160–171 (2013)
8. Huang, Y., Palomar, D.P.: Rank-constrained separable semidefinite programming with applications to optimal beamforming. IEEE Trans. Signal Process. **58**(2), 664–678 (2010)
9. Islam, H., Liang, Y.C., Hoang, A.T.: Joint beamforming and power control in the downlink of cognitive radio networks. In: Proc. IEEE Wireless Commun. Netw. Conf., Kowloon, Hong Kong, pp. 21–26 (2007)
10. Marzetta, T.L.: Noncooperative cellular wireless with unlimited numbers of base station antennas. IEEE Trans. Wireless Commun. **9**(11), 3590–3600 (2010)
11. Mitola III, J., Maguire Jr, G.Q.: Cognitive radio: Making software radios more personal. IEEE Pers. Commun. **6**(4), 13–18 (1999)
12. Saker, L., Elayoubi, S.E., Chahed, T., Gati, A.: Energy efficiency and capacity of heterogeneous network deployment in LTE-advanced. In: Proc. 18th European Wireless Conf., Poznan, Poland, pp. 1–7 (2012)

Hybrid Cognitive Satellite Terrestrial Coverage

A Case Study for 5G Deployment Strategies

Theodoros Spathopoulos[1], Oluwakayode Onireti[1], Ammar H. Khan[2], Muhammad A. Imran[1], and Kamran Arshad[3(✉)]

[1] Institute for Communication Systems (ICS), University of Surrey, Guildford, UK
th.spathopoulos@gmail.com, {o.s.onireti,m.imran}@surrey.ac.uk
[2] Inmarsat Global Limited, London, UK
ammar.khan@inmarsat.com
[3] Department of Engineering Science, University of Greenwich, Kent, UK
k.arshad@gre.ac.uk

Abstract. The explosion of mobile applications, wireless data traffic and their increasing integration in many aspects of everyday life has raised the need of deploying mobile networks that can support exponentially increasing wireless data traffic. In this paper, we present a Hybrid Satellite Terrestrial network, which achieves higher data rate and lower power consumption in comparison with the current LTE and LTE-Advanced cellular architectures. Furthermore, we present a feasibility study of the proposed architecture, in terms of its compliance with the technical specifications in the current standards.

Keywords: Hybrid satellite terrestrial · 5G · Control and user plane separated

1 Introduction

The increasing demand for data in mobile communication networks has resulted in the need for developing sufficient and advanced network infrastructures to support higher capacity and data rate. The forecasts in [1] shows that by 2018 the mobile data traffic will be 6.3 times higher than it was in 2013. In addition to this, the global CO_2 emissions of the mobile communications sector are expected to rise to 178 Megatons in 2020 [2]. Consequently, alternative approaches in the design and operation of future mobile networks are being investigated. The concept under investigation in this paper is the separation of the control (C)-plane and the user (U)-plane in the Radio Access Network (RAN). The C-plane provides ubiquitous coverage via the macro cells at low frequency band. On the other hand, the U-plane functionality is provided by the small/data cells at a higher frequency band, such as 3.5, 5, 10 GHz, where new licensed spectrum is expected to be available for future use. The use of such bands for small cells can lead to a significant increase in capacity, since they can offer bandwidth up to 100 MHz [3]. Likewise, cross-tier interference is avoided by operating the macro and small cells on separate frequency bands, thus leading to improvement in spectral efficiency. The C-plane and U-plane are not necessarily handled by the same

© Institute for Computer Sciences, Social Informatics and Telecommunications Engineering 2015
M. Weichold et al. (Eds.): CROWNCOM 2015, LNICST 156, pp. 523–533, 2015.
DOI: 10.1007/978-3-319-24540-9_43

node and are separated. Consequently, this gives the network operators more flexibility, since the C-plane (control/macro cells) manages UEs connectivity and mobility [4]. The separated plane architecture also enables reduction in energy consumption as it leads to longer data cell sleep periods due to their on demand activation [5], [6]. Furthermore, base station (BS) cooperation in the U-plane can be done more effectively since control signalling can be performed through a separate wireless path.

In this paper, a hybrid satellite terrestrial network architecture is presented, where a satellite is deployed to provide C-plane functionality, while femtocells are deployed to provide U-plane functionality. The operating frequency band for the satellite is considered to be L-band (1-2 GHz), as proposed in Inmarsat's BGAN system [7]. Satellites have cognitive capability, i.e. real-time intelligence which can be used to maximise the utilisation of available radio resources and to improve link performance. Such intelligence includes knowledge of the location of UEs and femtocells within its coverage, which enables associating UEs to the most suitable femtocells. In general, satellites offer much wider spatial coverage compared to macro BSs. A typical satellite can offer control signalling to a whole country, thus leading to significant reduction in physical infrastructure and maintenance cost, when compared with using the latter for control signalling. The feasibility of the proposed network architecture is based on the "dual connectivity" feature, which enables the simultaneous transmission of the U-plane and the C-plane by different nodes.

The purpose of this paper is to present the hybrid architecture, and compare it with existing cellular technologies, for a variety of scenarios, as well as to examine the compliance of its simulation results with the state of the art cellular standards. The rest of the paper is organised as follows. In section 2, the definition of the hybrid satellite terrestrial network architecture is presented, by defining the functions of the network elements. Section 3 describes the techniques applied in the hybrid network to achieve an effective resource utilisation at the terrestrial and satellite parts. In section 4, the details and the assumptions of the network simulations are presented. In section 5, a performance comparison between LTE, LTE-Advanced and the Hybrid architecture is made for different scenarios. In section 6, the compliance of the performance results with the current 4G standards is investigated and the suggestions to be taken into consideration in the promising 5G cellular standards are also presented. Finally section 7 concludes this paper.

2   Network Architecture

The International Telecommunication Union (ITU) defines a "hybrid satellite terrestrial system" as the one that employs satellite and terrestrial components that are interconnected, but operate independently of each other [8]. In such systems, the satellite and terrestrial components use separate network management systems and can operate in different frequency bands. An illustration of the proposed Hybrid network is shown in Figure 1, where the UE is operating in dual mode, communicating simultaneously both with the satellite and the eNBs.

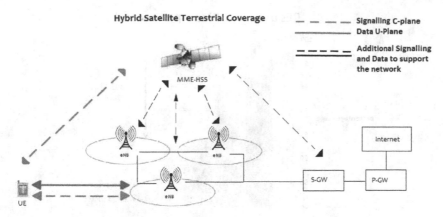

Fig. 1. Hybrid Satellite Terrestrial Architecture

From a higher level architectural point of view, as the satellite can provide coverage to the whole terrestrial network, it is used as a Home Subscriber Server (HSS) entity, carrying detailed information about the subscribers. In addition, since the satellite can communicate with the backbone network, such as the Serving Gateway (S-GW), for both data and signalling purposes, it can also serve as a Mobility Management Entity (MME), which is responsible for the mobility management of the users.

The terrestrial part of the network consists of femtocells/eNBs that are interconnected with fibre optics network. In addition, fibre optics is also used for the connections between the eNBs and the backbone network. This assumption enables reliable and fast data transfer among the terrestrial network elements, which minimises transmission errors and latency.

The reason for having two paths for the C-plane communication is that for some User Equipment (UE) activities, signalling from both the U-plane (eNB) and the C-plane (satellite) are required for successful operation. For example, power coordination and handover procedures require accurate measurement, which cannot be provided through the satellite channel due to high latency. Hence, cooperation of both data and signalling planes is essential for the smooth UE operation.

3   Resource Utilisation

The main advantage of separating the C-plane from the U-plane in cellular networks is the ability to replace part of the resources reserved for the signalling of the U-plane, with actual data. In general, the complete separation of the two planes is not possible, due to the fact that some of the C-plane functionalities have to be in the U-plane to support the reliability of the actual data transmission. In that sense, part of the Downlink Control Information (DCI) needs to occupy some of the available physical resources reserved for data transmission.

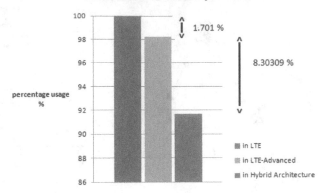

Fig. 2. Percentage usage of REs in the U-plane for the 3 architectures

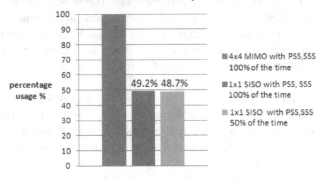

Fig. 3. Percentage usage of REs in the C-plane

The information that each physical channel needs to carry, is related to the occupied physical resources in the Orthogonal Frequency Division Multiplexing (OFDM) resource grid. These physical resources are called Resource Elements (REs).

In general, from the total available resources in an OFDM resource grid, 25% is occupied by the C-plane and 75% by the U-plane [9]. Regarding the U-plane (terrestrial) communication of the hybrid network, one of the C-plane signalling channels that must be used to support data transfer is the Physical Hybrid ARQ Indicator Channel (PHICH), which is responsible for providing ARQ acknowledgements [10]. Since the reliability of the useful data transfer is also based on a variety of upper layer protocols, it is possible to loosen the acknowledgment restrictions and reduce the resources reserved for the PHICH by a factor of 1/6 (from what was suggested in Release 8 LTE Resource Grid). By doing so, it is possible to substitute the rest 5/6 of the REs used for PHICH with actual data. In Figure 2, a comparison between the number of REs used for the U-plane in LTE, LTE-Advanced and the hybrid architecture is presented. The figure shows that about 1.7% reduction in the U-plane control signalling is achieved by separating the C-plane from the U-plane in LTE-Advanced as compared in LTE. Furthermore, the hybrid architecture can offer about 8.3% reduction in the U-plane control signalling, as compared to LTE, due to the reduction in the number of REs used for PHICH.

The following assumptions are made regarding the resources reserved for the control signalling of the C-plane in the hybrid architecture: a) Since the Reference Signals (RSs) are closely related with the number of antennas used in the system, and since the C-plane is responsible for low data rate communication, by deploying a single beam (single antenna) satellite it is possible to reduce the number of REs reserved for the RSs. b) In addition to that, since the serving satellite is used as an HSS/MME, it contains information about all the UEs. Furthermore, since the UEs communicate with the same satellite, part of the transmitted control information remains the same. Consequently, it is possible to reduce the transmission of the Primary and Secondary Synchronisation Channels (PSS and SSS) by 50% of the time. By doing so, as it can be seen in Figure 3, the resources reserved for the control information of the C-plane are further reduced thus, occupying less bandwidth on the satellite.

4   Network Simulations

In order to provide the performance results of the proposed network, a case study of providing high speed data coverage to the whole UK area was simulated in Matlab. For the calculation of the satellite's power consumption, the formula of the Friis equation was used,

$$P_t = \frac{P_r}{G_t G_r \left(\dfrac{\lambda}{4\pi d}\right)^2} \qquad (1)$$

considering $G_t = 10dB$ and $G_r = 1.5dB$ as the typical antenna gains of the transmitter and receiver respectively, $P_r = -80\ dBm$ for the minimum receive power, λ the wavelength, and d the distance of the satellite orbit from the earth user (36,000 km for GEO and 800 km for LEO). In addition, the same equation was used to calculate the power consumption of the terrestrial part of the network, assuming as total eNBs' power needs, the sum of power required for pure wireless transmission needs between each active eNB and its edge serving user. The assumption made for the terrestrial part was that each femtocell/eNB could serve an area with radius $R_{femto} = 10m$ and each Macro BS (used for signaling in LTE-Advanced), could serve an area with radius $R_{macro} = 5km$. Moreover, for the calculation of the capacity provided per km^2, the Shannon's capacity law was used

$$C_{per\_user} = B \log_2 (1 + \frac{P_r}{N_0 B + I_0}),$$

(2)

where $B = 20\ MHz$ the available bandwidth per cell, $No = -174\ dBm/Hz$ the noise spectral density and I_o the interference produced by the neighboring active eNBs assuming that the requirement of having and existing line-of-sight (LOS) path between the UE and the satellite is satisfied.

5   Network Performance and Comparison Between Different Scenarios

In this section, a comparison of the proposed architecture with LTE and LTE-Advanced has been done for different scenarios. Initially, since the satellite is exclusively responsible for providing the C-plane functions, the overall performance of the hybrid network is mainly based on the selection of satellite orbit. The performance specifications of a Geostationary Earth Orbit (GEO) satellite and a Low Earth Orbit (LEO) satellite are presented in Table 1. As it can be seen, a LEO satellite constellation presents better performance in terms of power consumption and latency of signal transmission, compared with a GEO satellite. However, the latter offers wider coverage and less capital and operating expenditures (CAPEX and OPEX), which is a topic beyond the scope of this paper, since a single GEO satellite can provide coverage even to a whole continent. Furthermore, the inter-satellite handovers that occur in LEO constellations increase the C-plane complexity and may also introduce further delay in the control functionality, which are however factors that are not taken into consideration in this paper.

For purely power reduction objectives and in order to substantiate the superiority of the hybrid network, a LEO satellite constellation is assumed to be deployed for the C-plane functions. The different scenarios simulated, represent three different case studies. In i) all UEs are active and all eNBs are switched on, in ii) all UEs are active and 2 out of 25 eNBs/km^2 are cooperating to enhance local performance, and in iii) when 13 out of 25 eNBs/km^2 are considered to serve idle users and are switched off.

In the simulations, 5 UEs per eNB was considered on average and the available bandwidth was 20MHz per eNB. In addition, the coverage radius of each eNBs was considered to be 10m. The performance results regarding the U-plane capacity achieved per architecture are shown in Table 2 and are illustrated in Figure 4. As it can be seen, the hybrid architecture achieves the highest capacity in all scenarios. This is a result of the reduction in the resources reserved for the PHICH, as discussed in section 3. At this point it is worthy to mention that it is impossible to switch off any of the unused or underused eNBs in LTE, because the desired "always connected" behaviour of the UEs will be interrupted.

Regarding the power consumption of the C-plane, only the performance results of LTE-A and the Hybrid architecture are presented. Power consumption of LTE is omitted due to the fact that it is a non-separated architecture, and the corresponding

C and U-planes are transmitted simultaneously, by the same eNB. As a result, the power consumption of the C-plane and U-plane are the same. In Table 3, the C-plane power consumption for the separation architectures is presented. As it can be seen, for all the scenarios, the hybrid architecture consumes the least power for wireless signal transmission in terms of mW/km². In Table 4, the power consumption for wireless transmission purposes of the network as a whole is illustrated (C-plane and U-plane) and a comparison of the different scenarios is presented in Figure 5. As it was expected, the power consumption of a separated architecture is higher than the power consumption of a non-separated architecture. This is due to the fact that in a separated architecture, umbrella coverage network elements are set on top of the already existing network infrastructure for providing the C-plane functions and thus, their power consumption has to be added to the network's total power consumption. However, the results in scenario iii, which represents the non- peak traffic hours of the network, show that the power consumption of the hybrid network can be less, compared with both LTE-Advanced and LTE. This shows that the proposed architecture represents a strong candidate for future mobile energy efficient technologies.

Table 1. Specifications of Different Satellite Deployment Scenarios.

Satellite Orbit	Power Consumption [mW/km²]	Earth to satellite transmission dalay [ms]	RRC_IDLE to RRC_CONNECTE D delay [ms]
GEO at 36,000 km	119.71	120	800
LEO at 800 km	0.059116	2.6	280.8

Table 2. U-Plane Capacity per Architecture.

Technology	U-plane capacity [Gbps/km²]		
	Scenario i	Scenario ii	Scenario iii
LTE	357.95	359.06	N/A
LTE-A	358.81	359.92	357.94
Hybrid	363.3	364.42	362.42

Table 3. Power Consumption of the C-Plane

C-plane	C-plane power consumption [mW/km²]		
	Scenario i	Scenario ii	Scenario iii
LTE-A deploying 5km macro BSs	0.061459	0.061534	0.062457
LEO satellite (800km)	0.059116	0.059116	0.059116
GEO satellite (36,000km)	119.71	119.71	119.71

Table 4. Power Consumption of the Whole Network

Technology	Total network's power consumption [mW/km^2]		
	Scenario i	Scenario ii	Scenario iii
LTE	12.112	12.163	N/A
LTE-A	12.173	12.224	11.988
Hybrid with LEO satellite	12.171	12.222	11.986
Hybrid with GEO satellite	131.822	131.873	131.673

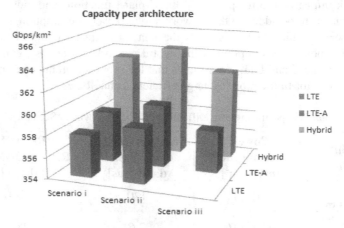

Fig. 4. Capacity per architecture

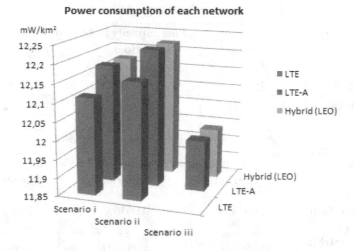

Fig. 5. Power consumption of each network

6   Compliance of the Proposed Hybrid Architecture with 4G Cellular Standards

The performance requirements for a mobile technology to be considered as 4G or beyond 4G, must comply with the requirements of the International Mobile Tele-communication (IMT) Advanced standard. These requirements suggest that the average spectral efficiency must be greater than 2.2 bits/s/Hz, and also the C-plane latency for the transmission from RRC_IDLE to RRC_CONNECTED state, must be less than 100msec [11].

Through simulating the network with the specifications described in section 4, the Hybrid network architecture achieves 2.85 bits/s/Hz, considering femtocells with 10m radius that serve on average 5 UEs per cell. Regarding the C-plane latency, as it was also discussed in Section 5, the deployment of a GEO satellite results in C-plane latency of 800 ms and the deployment of LEO satellite in 280.8 ms. Of course both results are not compliant with the IMT-Advanced requirements however, they can be considered as a suggestion in the development of future cellular technologies, such as the promising 5G. In that sense, in order to allow the deployment of such technologies in future mobile standards, it is suggested to loosen the above C-plane restriction to 800 ms for GEO or 300ms for LEO satellites. It is worthy to mention, that such a delay in the states' transition, occurs due to the fact that the state of the art satellite network architectures may not be capable of processing large amounts of data (as the ones discussed in this paper), retransmitting them to the backhaul network for further process. However, extensive research is being made on advanced satellite network architectures that will be capable of high speed data processing without retransmission, fact that will enable future network architectures, as the one proposed in this paper, to be implemented offering gigabit end user services. Integrating such advanced satellite payloads in the proposed architecture, it will definitely meet the IMT-Advanced latency specifications.

In addition to the above mentioned requirements, it is also useful, to present a comparison of the hybrid network's performance regarding the LTE-Timers. The most important of them are: T300; T301 and T310, which indicate the maximum delay for a connection establishment and re-establishment request, as well as for physical layer problems. The possible values according to LTE-Advanced are within the ranges [400-8000] ms for T300 and T301, and [50-2000] ms for T310. Hence, according to the limitations in the wireless signal processing, the single return through the satellite signal transmission has an average delay of 500ms for GEO and 25 ms for LEO satellites, which fit within the LTE-Timer range. Furthermore, the values also imply that even if a transmission fails, it is possible to retransmit the desired signal before the expiration of the timer.

7   Conclusion

The proposed Hybrid Satellite Terrestrial architecture gives encouraging results towards its consideration for possible deployment in future mobile networks.

The hybrid architecture, compared with state of the art technologies, gives the highest capacity per square meter and the lowest power consumption per square meter for wireless transmission purposes. Moreover, the technical specifications of the proposed architecture complies with the 4G standards. The spectral efficiency and the transmission delay meet the requirements of IMT-Advanced and the LTE-Advanced timers, respectively. The delay that occurs in the state transmission between the RRC_IDLE and the RRC_CONNECTED state, does not meet the C-plane delay requirements suggested from IMT-Advanced. However, this issue provides a design drive for satellites to minimize the latency beyond the theoretical bound as much as possible and enable such hybrid architectures to be deployed in future mobile standards.

Regarding the technical specifications of the satellite part to meet the bandwidth and data rate specifications for the transition from the current existing technologies to the suggested network architecture, there are already deployed mobile satellite systems can provide enhanced broadband capabilities and services. One of such is Inmarsat's Global Xpress system, which offers seamless worldwide coverage with advanced data rates up to 50Mbps [12]. In that sense, the UE convenience will be easier to be achieved.

As a final comment, the feasibility of the proposed architecture was based on the technical specification derived from the simulations made. Of course the issues of CAPEX and OPEX definitely play an important role for the realistic implementation of the Hybrid network, as well as for its comparison with the existing technologies. Assuming that for the U-plane, the same optical fiber network is going to be used for the interconnection among the femtocells/eNBs for each separation architecture, the investigation of the network's cost mainly focuses on the C-plane implementation. However, in case of such a study, the results have to be derived considering the whole lifecycle of the network, since by deploying a satellite, the maintenance cost of the C-plane is nowhere near the maintenance cost of the macro BSs network used in LTE-Advance.

Acknowledgments. This research leading to these results was partially derived from the University of Greenwich Research & Enterprise Investment Programme grant under agreement number RAE-ES-01/14.

References

1. Rupert, W.: Mobile data will grow 6.3 times between 2013 and 2018 and growth will be strongest outside Europe and North America, October 21, 2013. http://www.analysysmason.com/AboutUs/News/Insight/Mobile-data-Oct2013/ (accessed August 18, 2014)
2. Gruber, M., Blume, O., Ferling, D., Zeller, D., Imran, M.A., Strinati, E.C.: EARTH—energy aware radio and network technologies. In: IEEE International Symposium on Personal, Indoor and Mobile Radio Communications, pp. 1–5, September 2009
3. Ishii, H., Kishiyama, Y., Takahashi, H.: A novel architecture for LTE-B: C-plane/U-plane split and phantom cell concept. In: IEEE Globecom Workshops (GC Wkshps), pp. 624–630, December 2012

4. Zakrzewska, A., Lopez-Perez, D., Kucera, S., Claussen, H.: Dual connectivity in LTE HetNets with split control- and user-plane. In: IEEE Globecom Workshops, pp. 391–396, December 2013
5. Capone, A., Filippini, I., Gloss, B., Barth, U.: Rethinking cellular system architecture for breaking current energy efficiency limits. In: Sustainable Internet and ICT for Sustainability (SustainIT), pp. 1–5, October 2012
6. Mohamed, A., Onireti, O., Qi, Y., Imran, A., Imran, M., Tafazolli, R.: Physical layer frame in signalling-data separation architecture: overhead and performance evaluation. In: European Wireless, May 2014
7. Inmarsat Global Limited: BGAN. http://www.inmarsat.com/service/bgan-link
8. Sooyoung, K., Heewook, K., Do Seob, A.: A Cooperative Transmit Diversity Scheme for Mobile Satellite Broadcasting Systems. Advanced Satellite Mobile Systems, ASMS, 72–75, August 2008
9. Cui, D.: LTE peak rates analysis. In: IEEE Wireless and Optical Communications Conference (WOCC), pp. 1-3, May 2009
10. GPP: Ts 36.300: Evolved Universal Terrestrial Radio Access (E-UTRA) and Evolved Universal Terrestrial Radio Access Network (E-UTRAN); Overall description; Stage 2, ETSI (2010)
11. ITU Radiocommunication Sector (ITU-R): ITU global standard for international mobile telecommunications "IMT-Advanced". http://www.itu.int/ITU-R/index.asp?category= information&rlink=imt-advanced&lang=en
12. Inmarsat Global Limited: Global Xpress. http://www.inmarsat.com/service-group/global-xpress

Energy-Efficient Resource Allocation Based on Interference Alignment in MIMO-OFDM Cognitive Radio Networks

Mohammed El-Absi$^{(\boxtimes)}$, Ali Ali, Mohamed El-Hadidy, and Thomas Kaiser

Institute of Digital Signal Processing, Duisburg-Essen University, Essen, Germany
mohammed.el-absi@uni-due.de

Abstract. In this paper, we propose an energy-efficient interference alignment (IA) based resource management algorithm for multi-input multi-output (MIMO) orthogonal frequency division multiplexing (OFDM) cognitive radio (CR) systems. The proposed algorithm provides the secondary users (SUs) with the opportunity for underlay sharing of the primary system spectrum. The proposed algorithm ensures the quality-of-service (QoS) of the primary system by guaranteeing the minimum transmission rate. The problem is formulated as a mixed-integer non-convex optimization problem, in which the objective is to maximize the energy efficiency, and the constraints are the per-user power budget and QoS demand of the primary system. To tackle mixed-integer and non-convexity nature of the problem, we propose a sub-optimal energy-efficient algorithm through two successive steps. The first step schedules the subcarriers among the SUs based on IA while the second step iteratively allocates the power based on Dinkelbach's scheme. Simulations reveal that the proposed algorithm achieves significant improvement in the energy efficiency compared to the traditional spectrum-efficient algorithm.

Keywords: Cognitive radio · Interference alignment · Resource allocation · Energy efficiency · MIMO · OFDM

1 Introduction

Cognitive radio (CR) is considered as a promising technology that overcomes the scarcity and the underutilization of the spectrum [1]. In this context, the secondary users (SUs) are allowed to share the same spectrum bands with the primary users (PUs) provided that the quality of service (QoS) of the PUs is guaranteed. Recently, interference alinement (IA) is merged with CR as an efficient interference management technique in order to achieve optimal utilization of the system resources. IA is a cooperative transmission approach that achieves an optimal sum-rate for K-user interference channels at high signal-to-noise-ratio (SNR). IA is performed by aligning the interference signals from the undesired transmitters in certain subspaces, termed as interference subspaces, and the desired signal in the other subspaces, termed as interference-free subspaces [2],[3].

© Institute for Computer Sciences, Social Informatics and Telecommunications Engineering 2015
M. Weichold et al. (Eds.): CROWNCOM 2015, LNICST 156, pp. 534–546, 2015.
DOI: 10.1007/978-3-319-24540-9_44

Energy efficiency gains much of interest nowadays due to the rapidly increasing cost of energy [4,5]. The ever-increasing data-rate demands require energy-efficient transmissions in order to prolong the battery lifetime of wireless devices. However, energy-efficient based IA resource allocation in multi-input multi-output (MIMO) CR systems is rarely addressed in the literature, where many research works considered the problem of IA based resource allocation in MIMO CR systems aiming at maximizing the spectral efficiency of CR systems [6–9]. In this regards, the work in [10] proposed energy-efficient resource allocation based on IA in CR systems. Nevertheless, this work is focused only on narrow-band CR systems in addition to that it is restricted to a limited number of SUs.

In this work, we investigate energy-efficient resource allocation algorithm based on IA technique in dense MIMO orthogonal frequency division multiplexing (OFDM) CR systems. Resource management problem is formulated on the base of IA in order to enable the SUs to underlay the primary system spectrum. The proposed algorithm satisfies the QoS of the PU by guaranteeing the minimum transmission rate of the PU. In problem formulation, each subcarrier is assigned to a feasible number of SUs in order to meet IA feasibility conditions. The resource allocation problem is formulated as a mixed-integer non-convex optimization problem. To tackle the mixed-integer and non-convexity nature of the problem, a sub-optimal energy-efficient algorithm is proposed through two steps. First step assigns each subcarrier to a feasible number of SUs while the second step allocates the power among all subcarriers and all users.

2 System Model

In this model, K transmitter-receiver pairs are assumed, where a cognitive radio system with $K-1$ pairs of SUs is coexisted with a single-user broadband primary system. All the K nodes are equipped with M_T transmit antennas and M_R receive antennas. User 1 refers to the PU that occupies a bandwidth of B Hertz divided into N subcarriers. Each subcarrier has a bandwidth $W = B/N$ Hertz. Underlay spectrum sharing is assumed through this work, where the SUs guarantee the QoS of the PU. In order to accomplish co-channel interference free transmission, IA is applied to give the opportunity for the different SUs to share the CR spectrum with optimal interference management. Due to the frequency orthogonality of OFDM systems, MIMO IA can be applied independently on each subcarrier as a combination of linear precoder at the transmitter and interference suppression decoder at the receiver [3]. Therefore, we model the system focusing on a specific subcarrier n. For the nth subcarrier, the D symbol data streams $\mathbf{x}_k^n$ are precoded at the kth transmitter using a unitary matrix $\mathbf{V}_k^n \in \mathbb{C}^{M_T \times D}$. This precoder aligns the desired data at its own receiver in the interference-free subspace while the interference signals from other SU transmitters are aligned at the interference subspace [2,11]. With perfect channel knowledge, the received signal at the kth receiver on the nth subcarrier is written as

$$\mathbf{y}_k^n = \mathbf{U}_k^{nH}\mathbf{H}_{kk}^n\mathbf{V}_k^n\mathbf{x}_k^n + \sum_{j=1,j\neq k}^{K} \mathbf{U}_k^{nH}\mathbf{H}_{kj}^n\mathbf{V}_j^n\mathbf{x}_j^n + \mathbf{U}_k^{nH}\mathbf{z}_k^n, \qquad (1)$$

where $\mathbf{U}_k^n \in \mathbb{C}^{M_R \times D}$ is a unitary linear interference suppression matrix applied at the kth receiver, and $\mathbf{H}_{kj}^n \in \mathbb{C}^{M_R \times M_T}$ denotes the channel frequency response between j^{th} transmitter and k^{th} receiver. $\mathbf{z}_k^n \in \mathbb{C}^{M_R \times 1}$ is the zero mean unit variance circularly symmetric additive white Gaussian noise (AWGN) vector with variance σ^2 at the kth receiver.

In IA, the interference can be totally nullified when the condition $M_T + M_R - (K+1)D \geq 0$ is achieved [12]. The precoder and decoder matrices can be designed to achieve IA using closed-form solution or other algorithmic methods as presented in the literature for many cases (e.g. [2,13,14] and references therein). In feasible IA systems, the interference is concentrated in the interference subspace, and hence the leakage interference in the desired subspace is trivial [15]. Accordingly, the received signal in (1) becomes

$$\mathbf{y}_k^n = \mathbf{U}_k^{nH}\mathbf{H}_{kk}^n\mathbf{V}_k^n\mathbf{x}_k^n + \mathbf{U}_k^{nH}\mathbf{z}_k^n. \qquad (2)$$

The total sum-rate of the CR system in addition to the PU is expressed as [13]

$$R = \sum_{n=1}^{N}\sum_{k=1}^{K} \log_2 \left| \mathbf{I}_D + \frac{1}{\sigma^2}\mathbf{U}_k^{nH}\mathbf{H}_{kk}^n\mathbf{V}_k^n\mathbf{S}_k^n\mathbf{U}_k^n\mathbf{H}_{kk}^{n\,H}\mathbf{V}_k^{nH} \right|, \qquad (3)$$

where $\mathbf{S}_k^n \in \mathbb{R}^{D \times D}$ is the input covariance matrix of the kth user on the nth subcarrier, and hence the transmitted power by the kth user over the nth subcarrier is $P_k^n = \mathrm{Tr}\,(\mathbf{S}_k^n)$. Since $\mathbf{U}_k^{nH}\mathbf{H}_{kk}^n\mathbf{V}_k^n$ is considered as the effective channel and has a rank of D, the sum-rate in (3) can be formulated using spectral decomposition into

$$R = \sum_{n=1}^{N}\sum_{k=1}^{K}\sum_{d=1}^{D} \log_2 \left(1 + \frac{P_{k,d}^n \lambda_d \left(\mathbf{U}_k^{nH}\mathbf{H}_{kk}^n\mathbf{V}_k^n\right)}{\sigma^2} \right), \qquad (4)$$

where $P_{k,d}^n$ is the allocated power to the dth data stream at the kth user on the nth subcarrier and $\lambda_d \left(\mathbf{U}_k^{nH}\mathbf{H}_{kk}^n\mathbf{V}_k^n\right)$ is the dth eigenvalue of $\mathbf{U}_k^{nH}\mathbf{H}_{kk}^n\mathbf{V}_k^n$. Further, we denote $\lambda_d \left(\mathbf{U}_k^{nH}\mathbf{H}_{kk}^n\mathbf{V}_k^n\right)$ as $\lambda_{k,d}^n$.

3 Problem Formulation

The energy efficiency is defined as the amount of information being transmitted in one Hertz per Joule energy consumption (bits/Hz/Joule). Our objective is to maximize the energy efficiency of the system while the QoS of the PU is guaranteed. The QoS of the PU is guaranteed as the minimum transmission rate, which is described as

$$\sum_{n=1}^{N}\sum_{d=1}^{D} \log_2 \left(1 + \frac{P_{1,d}^n \lambda_{1,d}^n}{\sigma^2} \right) \geq R_Q, \qquad (5)$$

where R_Q is the minimum transmission rate that should be guaranteed to achieve the required QoS.

In this work, the overall power consumption is expressed as

$$\mathcal{E} = \sum_{k=1}^{K}\sum_{n=1}^{N}\sum_{d=1}^{D} P_{k,d}^{n} + \sum_{k=1}^{K}\left(P_{ct}^{k} + P_{cr}^{k}\right), \tag{6}$$

where P_{ct}^{k} and P_{cr}^{k} are the transmitter-circuit and the receiver-circuit power consumption for the kth user, respectively [16].

IA allows the SUs to share the spectrum resources simultaneously with the PU, which increases the degrees-of-freedom of the CR system. Nevertheless, according to IA feasibility conditions, the number of SUs that is allowed to share the PU on a given subcarrier is restricted up to a certain number of SUs written as

$$K_f = \frac{M_T + M_R}{D} - 2. \tag{7}$$

Therefore, the formulation of IA based resource management problem should consider this limitation by scheduling only K_f SUs on a given subcarrier. The problem can be formulated as

$$P1: \underset{\mathcal{P},\mathcal{W}}{\arg\max}\, \frac{R(\mathcal{P},\mathcal{W})}{\mathcal{E}(\mathcal{P},\mathcal{W})} = \frac{\sum\limits_{n=1}^{N}\sum\limits_{k=1}^{K}\sum\limits_{d=1}^{D} w_k^n \log_2\left(1 + \frac{P_{k,d}^n \lambda_{k,d}^n}{\sigma^2}\right)}{\sum\limits_{k=1}^{K}\sum\limits_{n=1}^{N}\sum\limits_{d=1}^{D}\left(w_k^n P_{k,d}^n\right) + \sum\limits_{k=1}^{K}\left(P_{ct}^k + P_{cr}^k\right)} \tag{8a}$$

$$\text{s.t.}: \sum_{n=1}^{N} w_1^n = N \tag{8b}$$

$$\sum_{n=1}^{N}\sum_{d=1}^{D} w_k^n P_{k,d}^n \leq P_k^{\max} \quad \forall k \tag{8c}$$

$$P_{k,d}^n \succeq 0, \qquad \forall n, k, d \tag{8d}$$

$$\sum_{n=1}^{N}\sum_{d=1}^{D} \log_2\left(1 + \frac{P_{1,d}^n \lambda_{1,d}^n}{\sigma^2}\right) \geq R_Q \tag{8e}$$

$$w_k^n \in \{0,1\}\ \forall k,n \tag{8f}$$

$$\sum_{k=2}^{K} w_k^n = K_f\ \forall n, \tag{8g}$$

where $\mathcal{P} = \{P_{k,d}^n, \forall k, n, d\}$ and $\mathcal{W} = \{w_k^n = \{0,1\}, \forall k, n\}$ are the power allocation and user selection indicators, respectively. w_k^n is a binary variable that indicates whether the kth SU is allowed to access the nth subcarrier, where $w_k^n = 1$ if and only if the nth subcarrier is allocated to the kth SU and 0 implies otherwise. w_1^n is always 1 since the PU is guaranteed to access all the spectrum which is satisfied by the constraint (8b). The constraint (8c) represents the kth user total power constraint $P_k^{\max}$, while a positive transmission power at each antenna is guaranteed by (8d). The constraint (8e) ensures that the QoS of the

PU as stated in (5). The equality condition $\sum_{k=2}^{K} w_k^n = K_f$ ensures that any given subcarrier can be shared by K_f SUs in addition to the PU, where IA feasibility is accomplished.

4 Sub-optimal Energy-Efficient Algorithm

The optimization problem of $P1$ is a non-convex and mixed-integer optimization problem, which is mostly prohibitive to solve. The non-convexity nature is a result of the objective function which is the ratio of two functions, and the mixed-integer nature comes from the integer constraint that is used for SUs scheduling. Therefore, we propose a sub-optimal scheme in order to solve Problem $P1$ efficiently with low computational complexity. Firstly, we avoid the mixed-integer nature of Problem $P1$ by finding the indicators $\mathcal{W}$ using frequency scheduling. After that, the objective function is simplified using techniques from nonlinear fractional programming in order to allocate the power among users and subcarriers aiming at maximizing the energy efficiency of the system. The detailed description of the sub-optimal algorithm is provided in the next section.

4.1 Frequency Scheduling

The integer constraint, that is used for user scheduling in (8f), is an obstacle in tackling the optimization problem. Therefore, frequency scheduling needs to be performed in case of having a dense CR system, where the number of SUs is greater than K_f, in order to find $\mathcal{W}$. In this step, we schedule K_f SUs to share the PU a given subcarrier. This step can overcome the IA feasibility constraint and guarantees feasible and perfect IA on each subcarrier [17]. The scheduling operation chooses the SUs with strong direct effective channel since this provides more power gain to save extra energy.

The description of the scheduling step can be commenced by defining $\mathcal{N}$ and $\mathcal{B} = \{2, .., K\}$ to be the sets contain all the non-assigned subcarriers and all the SUs, respectively. Furthermore, define $\mathcal{C} = \{c(1), .., c(N_\mathcal{C})\}$ to be the sets of all possible combinations of K_f SUs, where $N_\mathcal{C}$ denotes the number of combinations while $c(i) \in \mathcal{C}$ refers to the group of users inside the ith combination. The first element in each group is the PU in addition to the K_f SUs. Each combination must satisfy that $c(i) \subseteq \{1, \mathcal{B}\}$ and $c(i) \neq c(j); \forall (i \neq j)$. For the nth subcarrier, the combination selection can be formulated mathematically as

$$c_n^* = \arg\max_{c(i)} \sum_{k \in c(i)} \left\| \mathbf{U}_k^{n\mathrm{H}} \mathbf{H}_{kk}^n \mathbf{V}_k^n \right\|_F, \tag{9}$$

where the SUs inside this cluster are the only allowed to transmit over that subcarrier in addition to the PU.

At the beginning of the scheduling algorithm, all the possible combinations are generated to form $\mathcal{C}$ using the SUs in the set $\mathcal{B}$. Afterwards, the subcarriers are assigned sequentially to groups, where a given subcarrier, e.g the nth subcarrier, is allocated to the combination c_n^* that achieves the scheduling criterion in (9). After finding c_n^*, the indicator w_k^n is set to be 1 for all the SUs in c_n^* and 0 otherwise. The scheme is repeated until allocating all subcarriers among the clusters. The scheduling procedures are included in Algorithm 1.

4.2   Power Allocation

By means of frequency scheduling, the subcarrier indicators $\mathcal{W}$ are already determined. Therefore, the power allocation problem can be formulated as follows

$$P2 : \arg\max_{\mathcal{P}} \frac{R(\mathcal{P})}{\mathcal{E}(\mathcal{P})} = \frac{\sum_{n=1}^{N} \sum_{k \in c_n^*} \sum_{d=1}^{D} \log_2\left(1 + \frac{P_{k,d}^n \lambda_{k,d}^n}{\sigma^2}\right)}{\sum_{n=1}^{N} \sum_{k \in c_n^*} \sum_{d=1}^{D} \left(P_{k,d}^n\right) + \sum_{k=1}^{K} \left(P_{ct}^k + P_{cr}^k\right)} \tag{10a}$$

$$\text{s.t.} : \sum_{n=1}^{N} \sum_{d=1}^{D} P_{k,d}^n \leq P_k^{\max} \quad \forall k \tag{10b}$$

$$P_{k,d}^n \succeq 0, \qquad \forall n, k, d \tag{10c}$$

$$\sum_{n=1}^{N} \sum_{d=1}^{D} \log_2\left(1 + \frac{P_{1,d}^n \lambda_{1,d}^n}{\sigma^2}\right) \geq R_Q. \tag{10d}$$

Hence, the optimization problem $P2$ is now non-convex quasiconcave fractional program, where the numerator is concave in $P_{k,d}^n$ and the denominator is affine [18]. Since quasiconcave fractional programs share some important properties with concave programs [19], it is possible to solve concave-convex fractional programs with many of the standard methods for concave programs.

In this work, the iterative Dinkelbach's method [20] is deployed to solve the quasiconcave problem of $P2$ in a parameterized concave form. Let χ is a compact set of feasible solutions of the optimization problem, where $\mathcal{P} \in \chi$. The following objective function

$$\arg\max_{\mathcal{P} \in \chi} \frac{R(\mathcal{P})}{\mathcal{E}(\mathcal{P})}$$

can be associated using Dinkelbach's method [20] with the following parametric concave program

$$\mathcal{F}(\lambda) = \arg\max_{\mathcal{P} \in \chi} R(\mathcal{P}) - \lambda\mathcal{E}(\mathcal{P}), \tag{11}$$

where $\lambda \in \mathbb{R}$ is treated as a parameter. It can be shown that $\mathcal{F}(\lambda)$ is convex, continuous and strictly decreasing in λ [20]. We define λ^* as the maximum energy efficiency of the considered system which is given by

$$\lambda^* = \frac{R(\mathcal{P}^*)}{\mathcal{E}(\mathcal{P}^*)} = \arg\max_{\mathcal{P} \in \chi} \frac{R(\mathcal{P})}{\mathcal{E}(\mathcal{P})}. \tag{12}$$

According to Dinkelbach's method [20], we can achieve the maximum energy efficiency λ^* when

$$\underset{\mathcal{P} \in \chi}{\arg\max} \; R(\mathcal{P}) - \lambda^* \mathcal{E}(\mathcal{P}) = R(\mathcal{P}^*) - \lambda^* \mathcal{E}(\mathcal{P}^*) = 0 \qquad (13)$$

for $R(\mathcal{P}) \geq 0$ and $\mathcal{E}(\mathcal{P}) > 0$ [20,21].

In summary, Dinkelbach proposes an iterative method to find increasing values of feasible λ by solving the parameterized problem

$$\mathcal{F}(\lambda_l) = \underset{\mathcal{P} \in \chi}{\arg\max} \; R(\mathcal{P}) - \lambda_l \mathcal{E}(\mathcal{P}), \qquad (14)$$

where λ_l denotes the lth iteration. The iterative process continues until the absolute difference value $|\mathcal{F}(\lambda_l)|$ becomes as small as a pre-specified ϵ.

Accordingly, Problem $P2$ is turned into solving a group of convex problems, which is definitely more manageable. Therefore, for a given λ, Problem $P2$ becomes

$$P3 : \underset{\mathcal{P}}{\arg\max} \left(\sum_{n=1}^{N} \sum_{k \in c_n^*} \sum_{d=1}^{D} \log_2 \left(1 + \frac{P_{k,d}^n \lambda_{k,d}^n}{\sigma^2} \right) \right) - \qquad (15a)$$

$$\lambda \times \left(\sum_{n=1}^{N} \sum_{k \in c_n^*} \sum_{d=1}^{D} P_{k,d}^n + \sum_{k=1}^{K} \left(P_{ct}^k + P_{cr}^k \right) \right)$$

$$\text{s.t.} : \sum_{n=1}^{N} \sum_{d=1}^{D} P_{k,d}^n \leq P_k^{\max} \quad \forall k \qquad (15b)$$

$$P_{k,d}^n \succeq 0, \qquad \forall n, k, d \qquad (15c)$$

$$\sum_{n=1}^{N} \sum_{d=1}^{D} \log_2 \left(1 + \frac{P_{1,d}^n \lambda_{1,d}^n}{\sigma^2} \right) \geq R_Q. \qquad (15d)$$

Problem $P3$ is convex, where the Lagrangian can be written as

$$\mathcal{L} = \sum_{n=1}^{N} \sum_{k \in c_n^*} \sum_{d=1}^{D} \log_2 \left(1 + \frac{P_{k,d}^n \lambda_{k,d}^n}{\sigma^2} \right) - \lambda \times \left(\sum_{n=1}^{N} \sum_{k \in c_n^*} \sum_{d=1}^{D} P_{k,d}^n + \sum_{k=1}^{K} \left(P_{ct}^k + P_{cr}^k \right) \right)$$

$$(16)$$

$$+ \alpha \left(\sum_{n=1}^{N} \sum_{d=1}^{D} \log_2 \left(1 + \frac{P_{1,d}^n \lambda_{1,d}^n}{\sigma^2} \right) - R_Q \right)$$

$$+ \sum_{k=1}^{K} \sum_{n=1}^{N} \sum_{d=1}^{D} P_{k,d}^n \mu_{k,d}^n - \sum_{k=1}^{K} \beta_k \left(\sum_{n=1}^{N} \sum_{d=1}^{D} P_{k,d}^n - P_k^{\max} \right),$$

where α is the non-negative Lagrange multiplier corresponding to the minimum PU QoS rate in (15d). The Lagrange multiplier vector $\boldsymbol{\mu}$, which has non-negative elements $\mu_{k,d}^n \forall n, k, d$, considers the positive power transmission in (15c). $\boldsymbol{\beta}$ is the Lagrange multiplier vector corresponding to the maximum power budget for each user in the system as in (15b), which has non-negative elements $\beta_k, \forall k$. After rearranging the Karush-Kuhn-Tucker (KKT) conditions, we get

$$P_{1,d}^n = \left[\frac{1+\alpha}{\lambda + \sum\limits_{k=1}^{K} \beta_k} - \frac{\sigma^2}{\lambda_{k,d}^n} \right]^+ \tag{17}$$

$$P_{k,d}^n = \left[\frac{1}{\lambda + \sum\limits_{k=1}^{K} \beta_k} - \frac{\sigma^2}{\lambda_{k,d}^n} \right]^+, \tag{18}$$

where $[y]^+ = \max(0, y)$. These Lagrange multipliers can be solved numerically using ellipsoid or interior point method.

Remark: At low SNR, Problem $P3$ may have no solution since the constraint in (15d) is not feasible to be achieved. To avoid this case, we firstly check if the constraint in (15d) is feasible or not [10]. This can be satisfied by switching the SUs into sleep mode and performing power allocation aiming at maximizing the throughput of the PU as follow

$$P4 : \max_{P_{1,d}^n} \sum_{n=1}^{N} \sum_{d=1}^{D} \log_2 \left(1 + \frac{P_{1,d}^n \lambda_{1,d}^n}{\sigma^2} \right) \tag{19a}$$

$$\text{s.t.} : \sum_{n=1}^{N} \sum_{d=1}^{D} P_{1,d}^n \leq P_1 \tag{19b}$$

$$P_{1,d}^n \geq 0. \tag{19c}$$

This problem can be efficiently solved using a successive application of the conventional waterfilling concept as follows [22]

$$\hat{P}_{1,d}^n = \left[\nu - \frac{\sigma^2}{\lambda_{1,d}^n} \right]^+, \tag{20}$$

where ν is the waterfilling level. The constraint in (15d) is feasible if and only if $\sum\limits_{n=1}^{N} \sum\limits_{d=1}^{D} \log_2 \left(1 + \frac{\hat{P}_{1,d}^n \lambda_{1,d}^n}{\sigma^2} \right) \geq R_Q$. Otherwise, the transmission mode is changed from IA into single user PU MIMO system as in [10] in order to provide the PU with the full resources to achieve the maximum throughout.

4.3 The Proposed Algorithm

The proposed energy-efficient IA algorithm for MIMO-OFDM CR systems is summarized in Algorithm 1. As discussed before, frequency scheduling is performed in order to obtain $\mathcal{W}^*$ as in the steps 1-14. After that, we check whether

---

Algorithm 1. Sub-Optimal Energy-Efficient Algorithm

1: Initialize $\mathcal{N} = \{1, 2, .., N\}$, $\mathcal{B} = \{2, .., K\}$, the maximum number of iterations L
   and the maximum tolerance ϵ
2: Set $\lambda = 0$ and iteration index $l = 0$
3: Find $\mathcal{C}$
4: $n = \mathcal{N}(1)$; (the first element in $\mathcal{A}$)
5: **while** $\mathcal{N}$ is not empty **do**
6:     **for all** $c(i) \in \mathcal{C}$ **do**
7:         Find $\mathbf{V}_k^n$ and $\mathbf{U}_k^n$; $\forall k \in c(i)$
8:         Evaluate $\psi^n = \sum_{k \in c(i)} \left\| \mathbf{U}_k^{n\,\mathrm{H}} \mathbf{H}_{kk}^n \mathbf{V}_k^n \right\|_F$
9:     **end for**
10:     Choose the set c_n^* such that ψ^n is maximized
11:     Set $w_k^n = 1$ $\forall k \in c_n^*$ and 0 otherwise
12:     Remove n from $\mathcal{N}$ and Set $n = n + 1$
13: **end while**
14: **Output** $\mathcal{W}^*$
15: Switch SUs into sleep mode and solve Problem $P4$ using (20)
16: **if** $\sum_{n=1}^{N} \sum_{d=1}^{D} \log_2 \left(1 + \frac{\hat{P}_{1,d}^n \lambda_{1,d}^n}{\sigma^2} \right) \geq R_Q$ **then**
17:     Switch on SUs
18:     **while Convergence** = False and $l < L$ **do**
19:         Solve Problem $P3$ as in (17) and (18) and obtain $\acute{\mathcal{P}}$
20:         **if** $R(\acute{\mathcal{P}}) - \lambda_l \mathcal{E}(\acute{\mathcal{P}}) < \epsilon$ **then**
21:             **Convergence** = True
22:             **Return** $\mathcal{P}^* = \acute{\mathcal{P}}$ and $\lambda^* = \frac{R(\mathcal{P}^*)}{\mathcal{E}(\mathcal{P}^*)}$
23:         **else if ; then**
24:             Set $l = l + 1$ and $\lambda_l = \frac{R(\acute{\mathcal{P}})}{\mathcal{E}(\acute{\mathcal{P}})}$
25:             **Convergence** = False
26:         **end if**
27:     **end while**
28: **else if ; then**
29:     Change transmission mode of the PU into single user MIMO and Switch SUs
       into sleep mode.
30: **end if**

---

the available resources are sufficient to guarantee the minimum QoS rate. When QoS is guaranteed, the power is allocated by solving a group of convex problems aiming at finding $\mathcal{P}^*$ as in the steps 17-26. Otherwise, the PU utilizes the full resources in order to maximize the throughput of the primary system by changing the transmission mode into single user MIMO system.

5   Simulation Results

In this section, we evaluate the performance of the proposed energy-efficient resource allocation algorithm using numerical simulations. A PU that occupies 5 MHz bandwidth is assumed, where the number of subcarriers is $N = 64$. Each

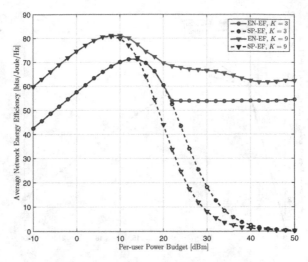

Fig. 1. Network energy efficiency versus maximum per-user power budget $P_k^{\max}$ for different numbers of users.

subcarrier has a bandwidth of 78.128 kHz, and the noise variance is $\sigma^2 = -60$ dBm. A CR system is assumed to share the PU spectrum based on IA technique. For all nodes in this scenario, the PU and SUs, are equipped with 2 antennas $M_T = M_R = 2$, and each node sends one data stream. Channel realizations have been drawn from independent and identically distributed Gaussian distribution with zero mean and unit variance. The circuit power consumption of the transmit circuit and receive circuit is assumed to be $P_{ct}^k = P_{cr}^k = 32$ dBm for all users. The minimum data-rate requirement for the PU is $R_Q = 25$ Mbits/s. For the purpose of performance comparison, the following algorithms are considered in the simulation:

1. **EN-EF**: Resource management is performed according to the proposed energy-efficient method as described in Algorithm 1.
2. **SP-EF**: The resources are allocated aiming at maximizing the spectral effeceincy as described in [9].

Fig. 1 depicts the average system energy efficiency versus the maximum per-user transmit power budget $P_k^{\max}$. At low SNR regime, it can be observed that the energy efficiency of *EN-EF* algorithm increases as the maximum per-user transmit power budget increases until reaching the maximum energy efficiency. After that, this scheme slightly decreases and converges to a specific energy efficiency value, where any additional increase in the transmitted power is not beneficial from energy efficiency point of view. It is noted for *EN-EF* algorithm that as the number of users increases the energy efficiency performance gets more benefit from the multiuser diversity, which is translated to commence an additional power gain to the system and save energy. On the other side, the energy efficiency of *SP-EF* algorithm behaves identical to *EN-EF* at low SNR

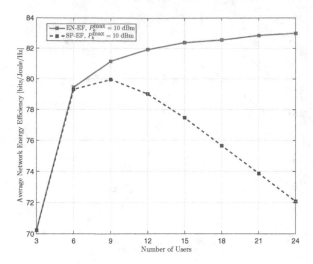

Fig. 2. Network energy efficiency versus the number of users when per-user power budget $P_k^{\mathrm{max}} = 10$ dBm.

regime while its energy efficiency performance dramatically decreases with the increase of the maximum per-user transmit power budget since each user uses the maximum power budget to maximize the sum-rate of the system. It is noted that the energy efficiency of *SP-EF* scheme at middle and high SNR regimes decreases as the number of SUs increases since each user uses its power budget and, hence, as the number of SUs increases the used power increases. This result is more clarified in Fig. 2 where this figure presents the energy efficiency of both schemes with the number of users when per-user power budget $P_k^{\mathrm{max}} = 10$ dBm. It is noted that the energy efficiency of *EN-EF* scheme increases with the number of users while *SP-EF* scheme decreases.

6   Conclusion

In this paper, we propose an energy-efficient resource allocation algorithm for MIMO-OFDM CR systems that underly a PU. The optimization problem is formulated as a non-convex mixed-integer problem, in which the per-user power budget and the QoS of the PU are considered. The problem is handled through two steps. In the first step, frequency scheduling is performed to allocate the sub-carriers among the SUs. In the second step, the power allocation is considered by exploiting Dinkelbach's method, where an iterative power allocation algorithm is proposed for maximizing the system energy efficiency. Simulations show that the proposed scheme provides considerable gains on energy efficiency with ensuring the QoS of the PU.

References

1. Haykin, S.: Cognitive radio: brain-empowered wireless communications. IEEE Journal on Selected Areas in Communications 201–220 (February 2005)
2. Cadambe, V.R., Jafar, S.: Interference alignment and degrees of freedom of the K-user interference channel. IEEE Transactions on Information Theory 3425–3441 (2008)
3. Cadambe, V.R., Jafar, S.: Reflections on interference alignment and the degrees of freedom of the K-user MIMO interference channel. IEEE Information Theory Society Newsletter 5–8 (2009)
4. Feng, D., Jiang, C., Lim, G., Cimini, L.J., Feng, G., Li, G.Y.: A survey of energy-efficient wireless communications. IEEE Communications Surveys Tutorials 15(1), 167–178 (2013)
5. Han, C., Harrold, T., Armour, S., Krikidis, I., Videv, S., Grant, P., Haas, H., Thompson, J.S., Ku, I., Wang, C., Le, T.A., Nakhai, M.R., Zhang, J., Hanzo, L.: Green radio: radio techniques to enable energy-efficient wireless networks. IEEE Communications Magazine 49(6), 46–54 (2011)
6. Perlaza, S.M., Fawaz, N., Lasaulce, S., Debbah, M.: From spectrum pooling to space pooling: Opportunistic interference alignment in MIMO cognitive networks. IEEE Transactions on Signal Processing 58(7), 3728–3741 (2010)
7. Sboui, L., Ghazzai, H., Rezki, Z., Alouini, M.-S.: Achievable rate of cognitive radio spectrum sharing MIMO channel with space alignment and interference temperature precoding. In: IEEE International Conference on Communications (ICC), pp. 2656–2660 (2013)
8. El-Absi, M., Shaat, M., Bader, F., Kaiser, T.: Interference alignment based resource management in MIMO cognitive radio systems. In: Proceedings of 20th European Wireless Conference, pp. 1–6, May 2014
9. El-Absi, M., Shaat, M., Bader, F., Kaiser, T.: Interference alignment with frequency-clustering for efficient resource allocation in cognitive radio networks. In: IEEE Global Communications Conf. (Globecom), December 8–12, 2014
10. Zhao, N., Yu, F.R., Sun, H.: Power allocation for interference alignment based cognitive radio networks. In: 2014 IEEE Conference on Computer Communications Workshops (INFOCOM WKSHPS), pp. 742–746, April 2014
11. Jafar, S., Fakhereddin, M.J.: Degrees of freedom for the MIMO interference channel. IEEE Transactions on Information Theory 53(7), 2637–2642 (2007)
12. Yetis, C., Gou, T., Jafar, S., Kayran, A.: On feasibility of interference alignment in MIMO interference networks. IEEE Transactions on Signal Processing 58, 4771–4782 (2010)
13. Gomadam, K., Cadambe, V.R., Jafar, S.: A distributed numerical approach to interference alignment and applications to wireless interference networks. IEEE Transactions on Information Theory 57(6), 3309–3322 (2011)
14. El-Absi, M., El-Hadidy, M., Kaiser, T.: A distributed interference alignment algorithm using min-maxing strategy. Transactions on Emerging Telecommunications Technologies (2014)
15. Zhao, N., Yu, F.R., Sun, H.: Adaptive energy-efficient power allocation in green interference alignment wireless networks. IEEE Transactions on Vehicular Technology PP(99), 1 (2014)
16. Cui, S., Goldsmith, A., Bahai, A.: Energy-constrained modulation optimization. IEEE Transactions on Wireless Communications 4(5), 2349–2360 (2005)

17. Zhao, N., Qu, T., Sun, H., Nallanathan, A., Yin, H.: Frequency scheduling based interference alignment for cognitive radio networks. In: 2013 IEEE Global Communications Conference (GLOBECOM), pp. 3447–3451, December 2013
18. Boyd, S., Vandenberghe, L.: Convex Optimization. Cambridge University Press, New York (2004)
19. Schaible, S.: Fractional programming. In: Handbook of global optimization, vol. 2 of Nonconvex Optim. Appl. Kluwer Acad. Publ., Dordrecht, pp. 495–608 (1995)
20. Schaible, S.: Fractional programming. ii, on Dinkelbach's algorithm. Management Science **22**(8), 868–873 (1976)
21. Isheden, C., Chong, Z., Jorswieck, E., Fettweis, G.: Framework for link-level energy efficiency optimization with informed transmitter. IEEE Transactions on Wireless Communications **11**(8), 2946–2957 (2012)
22. Tse, D., Viswanath, P.: Fundamentals of Wireless Communication. Cambridge University Press, Wiley series in telecommunications (2005)

Standards and Business Models

Receiving More than Data - A Signal Model and Theory of a Cognitive IEEE 802.15.4 Receiver

Tim Esemann$^{(\boxtimes)}$ and Horst Hellbrück

Department of Electrical Engineering and Computer Science,
CoSA Center of Excellence, Lübeck University of Applied Sciences, Lubeck, Germany
{tim.esemann,horst.hellbrueck}@fh-luebeck.de

Abstract. In standard medium access, transmitters perform spectrum sensing. Information about concurrent interferers is gained mainly during this sensing period. Especially during transmission respectively reception there is a blind gap where transmitter and receiver have limited capabilities to detect interferer. Standard radio receiver devices for IEEE 802.15.4 provide solely data output and no cognitive capabilities. Particularly mobile interferer create problems when moving gradually into reception range. First, they create small interference before actually causing collision later, when approaching. However, small interference is not yet detectable by todays transceivers. As a solution, we provide a signal model and an architecture for an extended cognitive IEEE 802.15.4 receiver as a basis for advanced signal processing for interference detection. The results of our theoretical analysis verify that the received signal contains signal marks of the interferer and therefore holds more information than transmitted data. Our theory is evaluated by simulations and experiments with a pair of IEEE 802.15.4 transmitter and an extended cognitive receiver.

Keywords: Spectrum sensing · Interference · Signal model · IEEE 802.15.4

1 Introduction

The number of devices with wireless interfaces increases continuously with numerous devices operating in the scarce spectrum available for unlicensed ISM bands. Spectrum utilization in ISM bands is very heterogeneous with many standards competing within the same frequency range. For 2.4 GHz we have IEEE 802.11 (WLAN), IEEE 802.15.1 (Bluetooth), and IEEE 802.15.4 (in some literature named Zigbee) suitable for low power and also for mobile devices. In addition, several proprietary wireless transmissions operate in that frequency range. Therefore, concurrent transmission with interference occurs regularly. Concurrent transmission takes place when at least two wireless transmitters utilize the same or parts of frequency spectrum and a receiver is in reception range. During concurrent transmissions signals interfere with each other on the receiver side.

© Institute for Computer Sciences, Social Informatics and Telecommunications Engineering 2015
M. Weichold et al. (Eds.): CROWNCOM 2015, LNICST 156, pp. 549–561, 2015.
DOI: 10.1007/978-3-319-24540-9_45

Strong interference degrades the performance of a wireless system as transmission errors occur. With competing devices using the same standard the term collision is used preferably. Wireless standards like IEEE 802.15.4 apply schemes like carrier sensing (CS) or collision avoidance (CA) to avoid collisions and interference [1]. CS is performed prior start of transmission but the transceiver is "blind" during the transmission itself.

Although this is a general problem for wireless transmissions, we will focus on a solution for IEEE 802.15.4 within this paper. A pair of standard IEEE 802.15.4 transceivers is not able to detect and identify reliably interference during transmission and reception. To the best of our knowledge we are the first to propose a cognitive receiver for IEEE 802.15.4 to enable spectrum awareness during transmission. The contributions are as follows: We provide a theoretical analysis of quadrature demodulated signals and interferences. We introduce a new model for an extension of the physical layer (PHY) of an IEEE 802.15.4 transceiver towards a cognitive receiver. It serves as a basis for future signal processing capabilities e.g. to enable comprehensive interference detection. We provide both, simulation and experimental results of our implementation with GNU Radio. The results of the evaluation show that we have reached a further step towards a cognitive receiver.

The rest of this article is organized as follows: Section 2 will introduce related work and demonstrate the need for new approaches. We will analyze the problem of concurrent interference and its impact as signal marks in the received signal in Section 3. Section 4 evaluates the theory by simulations. The paper concludes with a short summary and presents future work in Section 5.

2   Related Work

The goal of our approach is to increase the spectrum awareness during transmission. We will introduce recently published approaches for advanced spectrum sensing prior to and during an ongoing transmission and discuss it in relation to our work.

Akyildiz et al. describe in [2] that spectrum sensing is an important requirement to exploit unused frequencies. The authors distinguish between in-band and out-of-band sensing. In contrast to our work in-band sensing in [2] is only considered prior to transmission. Therefore, a trade-off between sensing and transmission has to be found in order to gain reasonable interference avoidance and transmission period. On the other hand out-of-band sensing is able to sense other frequency bands during an ongoing transmission, but not in the band that is currently utilized for transmission. A comprehensive summary of spectrum sensing schemes is given by Yücek and Arslan in [3] and Ariananda et al. in [4]. The sensing schemes under investigation achieve a variety in performance and accuracy. Schemes providing more detailed spectrum awareness are usually more complex and time-consuming. The three most prevalent schemes are energy detection, cyclostationary feature detection and matched filters. All schemes perform sensing prior to a transmission and not during the transmission. To ensure spectrum awareness during transmission in all these solutions a third radio for sensing is required whereas in our approach one pair of transceivers is sufficient.

Another solution to perform spectrum sensing during transmission is cooperative spectrums sensing. A survey on cooperative spectrum sensing in order to increase spectrum awareness is given by Akyildiz et al. in [5]. With cooperation of multiple and spatially separated sensing devices the spectrum awareness can be significantly improved. On the other hand this yields in more operational effort due to multiple sensing devices and additional overhead caused by exchange of sensing information. Cooperative sensing cannot be implemented with a single pair of transmitter and receiver.

In the past new approaches for spectrum sensing even during transmission were introduced. In [6] the authors propose to divide the transmission band into subbands, whereas a redundant subband is continuously used for spectrum sensing. This reduces bandwidth efficiency as a redundant frequency range with no data transmission being required. Another approach to achieve spectrum awareness during ongoing transmission is to utilize multicarrier waveforms and to analyze subcarriers at the receiver. In [7] Farhang-Boroujeny suggests to measure and to compare the energy of each received subcarrier in order to detect concurrent transmissions. It allows in-band concurrent transmitter detection even during ongoing transmissions, but requires wideband multicarrier transmission which is not available for IEEE 802.15.4 devices. As energy detection is proposed, again it is not possible to identify any specific signal marks from other interferer. With recent advances in full-duplex wireless communication [8] schemes like simultaneous transmit-and-sense seem to be achievable in the future. However, to the best of our knowledge current results have not yet exceeded the status of preliminary experiments [9] and analytical examination of the advantages [10]. Furthermore, our approach does not require any additional and complex antenna configuration within transmit and receive path.

In conclusion, several techniques and schemes to provide spectrum awareness have been introduced in the past. *Spectrum sensing schemes prior transmission* provide information about signal marks from interferer only during execution, but not during subsequent transmission. *Spectrum sensing schemes during transmission* either require redundant subbands for sensing, multicarrier waveforms or complex antenna circuitry and configuration. In the following sections, we will describe how small signal interference changes the received signal and how to build a radio receiver for IEEE 802.15.4 that receives more than data.

3 Problem and Analysis

As introduced in Section 1 the increasing utilization of wireless systems will result in a heterogeneous and dynamic radio environment. In such a radio environment concurrent wireless transmissions using the same frequency range will interfere with each other. Many of these radio transceivers are mobile today. A mobile and transmitting transceiver appearing in the scene interferes with small power first and with closer distance it finally disrupts the transmission of other systems and causes collisions. Hence, it is important to detect such interferer reliably in advance before collisions occur.

Today's wireless systems like IEEE 802.15.4 [1] transceiver use Carrier Sense Multiple Access Collision Avoidance (CSMA/CA) among each other in heterogeneous radio environment. Transceivers perform carrier sensing (a simple energy detection) immediately prior to transmission. If no other transmission is detected during spectrum sensing the transmitter starts its own transmission. After the receiver has decoded the data frame, it is checked for transmission errors by calculating the cyclic redundancy check (CRC). Occurring bit errors are detected reliably with CRC but the reason cannot be identified. In conclusion, spectrum awareness can only be provided during the spectrum sensing period (SS) as illustrated in Figure 1 by the white background. During the transmission and reception there is a "blind gap" illustrated as grey background that we will quantify in the following. In the IEEE 802.15.4 standard the measurement duration for carrier sensing is specified to be $128us$ (measurement duration of 8 symbols [1] p.54). With maximum transmission duration of 4.2 ms this yields in a spectrum awareness of only 3% of the total time interval. With minimum transmission duration (by sending acknowledgement frames) spectrum awareness is increased to not more than 25% of the total time interval.

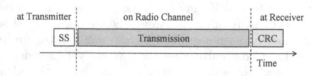

Fig. 1. Limited Spectrum Awareness during Transmission

The aim of our work is to show that it is possible to receive more than just data in order to improve spectrum awareness during transmission. We propose to analyze the received and demodulated signal of single received frames or even parts of these frames, for marks of another concurrent interfering signal. Our theoretical analysis shows that the received signal includes information about interference occurred during an ongoing transmission. Therefore, we propose to add cognitive capabilities to the receiver. This is the basis for future work on signal processing of the received signal. First preliminary results from experiments analyzing the received signal at the receiver and a conceptual hardware setup are published in [11,12]. Our previous work shows an implementation with a basic modulation scheme (MSK) and a preliminary study to integrate it into a standard receiver. The work did not provide a signal model and no theoretical analysis of the received and demodulated signal.

In our approach we assume that the interfering signal is still not large enough to cause a collision and transmission errors. This assumption is reasonable as especially in a heterogeneous environment with mobile devices a radio transceiver needs to be very sensitive to concurrent radios to avoid interfering with their transmissions. It is important to detect the signal of a concurrent radio on an overlapping frequency band as soon as possible to adapt transmission parameters accordingly in advance of a collision. After this initial explanation we will describe the digital demodulation process and provide mathematical expressions for an Offset Quadrature Phase Shift Keying (OQPSK) modulated signal. The theoretical analysis and model were validated by simulations and experiments with a real IEEE 802.15.4 radio link. Furthermore, our concept was adapted to IEEE 802.15.4 transmission without affecting standard compliant data transfer. For simplicity, the presented mathematical analysis does not consider noise in the environment. However, the experimental results in 4 show that the analytic results hold also for noisy signals.

3.1   Quadrature Demodulation of an OQPSK Modulated Signal with Interference

The Offset Quadrature Phase Shift Keying (OQPSK) signal can be written as [14]:

$$s_{OQPSK}(t) = a_c[m_I(t)\cos(\omega_c t) + m_Q(t)\sin(\omega_c t)] \tag{1}$$

OQPSK utilizes half-sine pulse shaping, $m_I(t)$ and $m_Q(t)$. Where in-phase (I) and quadrature component (Q) are misaligned by half a symbol duration. The demodulation of such a OQPSK signal with a quadrature demodulator follows several stages as depicted in Figure 2. Equation (2) to (4) show the result of each stage in detail. First, the received OQPSK modulated and real signal $s_{recO}(t)$ is converted with Hilbert transform into a complex signal $S_{recO}(t)$.

$$S_{recO}(t) = a_c[m_I(t)\cos(\omega_o t) + j m_Q(t)\sin(\omega_c t)] \tag{2}$$

Second, the complex signal is quadrature demodulated, resulting in $S_O(t)$.

$$
\begin{aligned}
S_O(t) &= S_{rec}(t) \times e^{-j\omega_c t} \\
&= a_c[m_I(t)\cos(\omega_c t) + j m_Q(t)\sin(\omega_c t)] \\
&\quad \times [\cos(\omega_c t) - j\sin(\omega_c t)] \\
&= a_c[\underbrace{m_Q(t) + (m_I(t) - m_Q(t))\cos^2(\omega_c t)}_{I_{so}(t)} \\
&\quad + j\underbrace{(m_Q(t) - m_I(t))\cos(\omega_c t)\sin(\omega_c t)]}_{Q_{so}(t)}
\end{aligned}
\tag{3}
$$

Third, the phase angle of the demodulated signal $\varphi(t)$ is determined with *arc tangent* function. Finally, bit decision is made based on the determined phase angle $\varphi(t)$.

$$\varphi(t) = \arctan\left(\frac{Q_s(t)}{I_s(t)}\right) \tag{4}$$

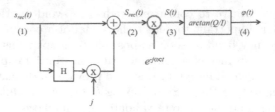

Fig. 2. Limited Spectrum Awareness during Transmission

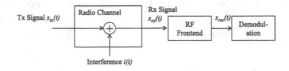

Fig. 3. Reception of a transmitted signal with superimposed interference from a concurrent transmitter

If another concurrent radio signal $i(t)$ interferes with the transmitted OQPSK modulated signal $s_{OQPSK}(t)$, it is superimposed as shown in (5) and Figure 3.

$$s_{recOI}(t) = s_{OQPSK}(t) + i(t) \tag{5}$$

With concurrent transmission, (2) and (3) are extended by additional components ($Interference$) as shown in (6) and (7). $\hat{S}$ and $\hat{I}$ are the Hilbert transformed signal components of the OQPSK and the interfering signal.

$$S_{recOI}(t) = [S_{OQPSK}(t) + I(t)] + j[\hat{S}_{OQPSK}(t) + \hat{I}(t)] \tag{6}$$

$$
\begin{aligned}
S_{OI}(t) &= S_{rec}(t) \times e^{-j\omega_c t} \\
&= S_{OQPSK}(t) \times e^{-j\omega_c t} + I(t) \times e^{-j\omega_c t} \\
&= \underbrace{\left[\underbrace{I_{Soqpsk}(t)}_{OQPSK\ only} + \underbrace{I(t)\cos(\omega_c t) + \hat{I}(t)\sin(\omega_c t)}_{Interference} \right]}_{I_{SOI}(t)} \\
&\quad + j\underbrace{\left[Q_{Soqpsk}(t) - I(t)\sin(\omega_c t) + \hat{I}(t)\cos(\omega_c t) \right]}_{Q_{SOI}(t)}
\end{aligned}
\tag{7}
$$

Finally, inserting the corresponding $I_s(t)$ and $Q_s(t)$ component into (4) results in the phase angle of the demodulated signal that additionally contains signal marks of the interfering signal. In order to extract the influence of interference, we introduce an extension of a traditional receiver which is presented in the next section.

3.2   Interference Extraction out of Received OQPSK Modulated Signal

To extract the influence of the interference signal we apply a method which is known from interference cancellation technique [14]. But, here we apply it the other way around. We extract the interference signal components from the demodulated signal as shown in Figure 4 and Equation (8).

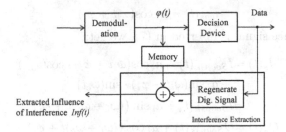

Fig. 4. Interference Extraction

$$\varphi_{int}(t) = \underbrace{\varphi_s(t)}_{received} - \underbrace{\varphi_{OQPSK}(t)}_{regenerated} = Inf(t) \tag{8}$$

Remember that we consider cases where the interference is still not large enough to cause transmission errors. Demodulated and decoded data is used to regenerate the demodulated signal $\varphi_{OQPSK}(t)$ as it supposed to be like without impact of interference. This regenerated signal $\varphi_{OQPSK}(t)$ is subtracted from the actual received and demodulated signal $\varphi_s(t)$ including the interference. Inserting the in-phase $I(t)$ and quadrature $Q(t)$ components from (7) (*received*) and (3) (*regenerated*) into *arc tangent* of (4) and successively into (8) results in a rather more complex expression. Corresponding signal marks from the interfering signal are hardly observable within this complex term. Therefore, we further simplify this expression by applying an approximation. Considering an interfering signal with signal strength that is much smaller than our actually transmitted and received signal, we use the approximation that:

$$\lim_{x \to 0} \tan x \approx x \tag{9}$$

Instead of (8) the approximated influence of interference $\widetilde{Inf}(t)$ is expressed as:

$$\varphi_{int}(t) \approx \tan\left(\varphi_{int}(t)\right) = \widetilde{Inf}(t) \tag{10}$$

This approximation and (8) results in the following term:

$$\tan\left(\varphi_{int}(t)\right) = \tan\left(\varphi_s(t) - \varphi_{OQPSK}(t)\right)$$
$$= \frac{\tan\left(\varphi_s(t)\right) \cdot \tan\left(\varphi_{OQPSK}(t)\right)}{1 + \tan\left(\varphi_s(t)\right) \cdot \tan\left(\varphi_{OQPSK}(t)\right)} \tag{11a}$$

At first glance this does not seem to be a true simplification, but the *tangent* suspends the *arc tangent* from (4). As shown in the following section these assumptions will simplify the expression of $Inf(t)$.

3.3 Influence of Interference

We consider a sinusoidal signal to show the influence of the superimposed interfering signal.

$$i_{cos}(t) = a_i \cos(\omega_i t + \varphi_i) \tag{12}$$

If this interference signal is inserted in (7), we get:

$$
\begin{aligned}
I_s(t) =& I_{Soqpsk}(t) + a_i \cos(\omega_i t + \varphi_i) \cdot \cos(\omega_c t) \\
& + a_i \sin(\omega_i t + \varphi_i) \cdot \sin(\omega_c t) \\
=& I_{Soqpsk}(t) + a_i \sin\left((\omega_i - \omega_c)t + \phi_i\right)
\end{aligned} \tag{13a}
$$

$$Q_s(t) = Q_{Soqpsk}(t) + a_i \cos\left((\omega_i - \omega_c)t + \phi_i\right) \tag{13b}$$

With (11) and further trigonometric identities and successive simplifications this yields in (14).

$$
\begin{aligned}
\widetilde{Inf}(t) =& \frac{c_4 c_3 \cos\left(\beta\right) + c_4(c_2 - c_3) \cos\left(\alpha\right) \cos\left(\alpha - \beta\right)}{c_4 c_3 \sin\left(\beta\right) + c_4(c_2 - c_3) \cos\left(\alpha\right) \sin\left(\alpha - \beta\right) + c_1 \left(c_2 \cos(\alpha)\right)^2 + c_1 \left(c_3 \sin(\alpha)\right)^2} \\
c_1 =& a_c \ , \ c_4 = a_i \ , \ c_2 = m_I(t) \ , \ c_3 = m_Q(t) \\
\alpha =& \omega_c t \ , \ \beta = \left((\omega_i - \omega_c)t + \phi_i\right)
\end{aligned}
$$
$$\tag{14}$$

The resulting term of $\widetilde{Inf}(t)$ includes signal components and therefore marks of the superimposed sinusoidal interference. It is influenced by its amplitude a_i, frequency ω_i and phase ϕ_i. Corresponding examples of such signals are depicted in Figure 5. The upper signal presents the demodulated signal with interference. The second signal presents the demodulated signal without interference, respectively the regenerated demodulated signal. The example of an extracted influence of interference in the third graph is a result of a sinusoidal interference with a SIR of 14 dB and a frequency of 50kHz. The extracted influence of interference shows significant signal marks caused by the interfering signal. The width of the sinusoidal cycles is dependent on the frequencies of the transmitted and interfering signal, ω_c and ω_i. Amplitude of the interfering signal determines the amplitude values of the extracted influence of interference. This is because the amplitude of the interference a_i respectively c_4 is not part of the last two sinusoidal terms of denominator of Equation (14) and these components stay constant if c_4 varies. Whereas the transmitted symbols corresponding to $m_I(t)$ and $m_Q(t)$ and the phase of interfering ϕ_i determines the phase shifts.

The results in this section show that the extracted influence of an interfering signal after demodulation contains signal marks corresponding to the interfering signal. The presented signal model is validated in the next Section 4 with baseband simulation and experiments with an OQPSK modulated signal and a superimposed OQPSK modulated interfering signal.

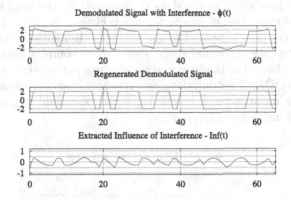

Fig. 5. Interference Extraction with sinusoidal interference for 64 demodulated bits

4 Evaluation

We have implemented an extended IEEE 802.15.4 receiver with software defined radios (SDR) composed of an USRP2 [15] and signal processing with GNU Radio [16]. USRP2 is a hardware frontend for GNU Radio applications responsible for up- and down-conversion of RF signals and furthermore for digital-to-analog and analog-to-digital conversion. Our extended IEEE 802.15.4 receiver is completely implemented in GNU Radio. Signal processing relevant to IEEE 802.15.4, i.e. demodulation and decision device, is based on the work of Schmid, presented in [17]. The interference extraction is implemented according to Figure 4. The received IEEE 802.15.4 signal is A/D-converted with a sampling frequency of 4 MS/s. After demodulation including clock recovery the sample rate of the digital signal is 2 MS/s corresponding to the chip rate 2 MChips/s of a standard IEEE 802.15.4 transmission. An IEEE 802.15.4 transmitter is set up accordingly.

4.1 Baseband Simulation with OQPSK

First, GNU Radio simulation was employed to show that even an OQPSK modulated interfering signal generates significant signal marks in the extracted influence of interference $Inf(t)$. Therefore, a second interfering OQPSK modulated signal was generated and superimposed in baseband on the original signal. Considering (5) this yields in:

$$s_{rec}(t) = s_{OQPSK-Tx}(t) + s_{OQPSK-Interferer}(t) \tag{15}$$

A carrier frequency offset of 50kHz compared to the original Tx-signal was chosen to simulate another concurrent OQPSK transmitting radio device. The resulting signals are depicted in Figure 6, again with an SIR of 14dB. The occurring

signal marks caused by the interfering OQPSK signal are dependent on the
transmitted data of the original transmitter and the interferer. For this simulation the interfering transmitter signal is modulated with random data. If the
data is incidentally similar to the data transmitted by the original transmitter
the amplitude of the influence of interference is close to zero, see the start of the
depicted signal $Inf(t)$. Compared to the extracted influence of interference of a
sinusoidal interference the signal shape shows more complex variations. This is
due to the dependency the in-phase and quadrature part of the interfering signal
are varied by its OQPSK modulation. Nevertheless, baseband simulation showed
that even with a more complex interfering signal observable signal marks occur
in the extracted influence of interference.

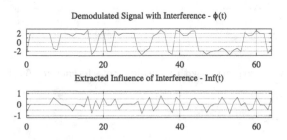

Fig. 6. Interference Extraction with OQPSK interference for 64 demodulated bits

4.2   Measurement with an Extended IEEE 802.15.4 Receiver

Finally, the extended IEEE 802.15.4 receiver was evaluated in a real and therefore
noisy radio environment with mobile IEEE 802.15.4 interferer as depicted in
Figure 7. Distance between IEEE 802.15.4 transmitter and extended receiver
was fixed to 3 m. Distance between the concurrent and interfering IEEE 802.15.4
transmitter and the extended receiver was varied from 5 to 1.5 m. At a distance
of 1.5 m between receiver and interferer single chip errors start to occur and
therefore risk of an upcoming collision arises.

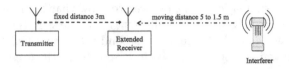

Fig. 7. Measurement setup for moving interferer

Short frames comparable to an acknowledgement frame were transmitted
within the experiment. A section of the extracted influence of interference for

256 chips is depicted in Figure 8. An initial measurement without interfering signal was conducted first. No significant signal marks are present except noisy variations of the amplitude. Subsequently the distance between interferer and extended receiver was shortened from 5 m to 1.5 m. At the beginning of each section depicted in Figure 8 no interference is present. Next to the 50th chip in both plots the interferer starts its transmission and therefore superimposes its signal. At this point in time the amplitude of the extracted influence of interference increases by approximately 10dB (5m) and 20dB (1.5m) respectively. Even if the interferer is 5 m away from the extended receiver the occurring signal marks are observable in $Inf(t)$.

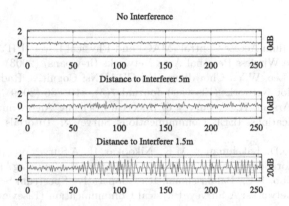

Fig. 8. Extracted Influence of interference $Inf(t)$ for 256 demodulated chips for different distances between cognitive receiver and interferer

The conducted simulations and experiments show that concurrent and interfering transmission generate signal marks within the received and demodulated signal. With an extended receiver we will be able to receive more than data. Note again, that no additional third radio is required in our approach.

5   Conclusion

In this paper we have motivated the need of spectrum awareness during transmission. We have shown with theoretical analysis, simulation and experiments that signal marks from concurrent interfering signals are observable in the received and demodulated signal. Once the receiver observes these signal marks and is able to assign them to a corresponding interfering source, the receiver will be able to inform the transmitter that concurrent transmission occurred. This information might be transmitted within a corresponding acknowledgement frame. Implementing such protocol will be part of our future work. Our proposed signal model and theory of an extended cognitive receiver is the basis for advanced signal processing to detect and identify interfering transmitter. Next, we will implement such signal processing and investigate its performance to detect and identify different kind of sources of interference. We plan to analyze the extracted influence of interference with signal processing like performing an FFT, analyzing the distribution of the amplitude values (i.e. determing m-order moment) or others. Additionally, we will evaluate our approach in terms of the occurrence of single bit errors and in case of multiple interferer. Finally, we will implement the interference extraction into a mobile IEEE 802.15.4 transceiver with a small-scale SDR extension. An additional RF-frontend performs the down-conversion into baseband and ADC, whereas the signal processing in Figure 4 is implemented into a small FPGA. The conceptual setup is described in [12].

References

1. Wireless Medium Access Control (MAC) and Physical Layer (PHY) Specifications for Low-Rate Wireless Personal Area Networks (lr-wpans) (2003)
2. Akyildiz, I., Lee, W.Y., Chowdhury, K.: CRAHNs: Cognitive Radio Ad Hoc Networks. Ad Hoc Networks (Elsevier) Journal **7**(5), 810–836 (2009)
3. Yücek, T., Arslan, H.: A Survey of Spectrum Sensing Algorithms for Cognitive Radio Applications. IEEE Communications Survey & Tutorials **11**(1), 116–130 (2009)
4. Ariananda, D.D., Lakshmanan, M.K., Nikookar, H.: A Survey on Spectrum Sensing Techniques for Cognitive Radio. In: Proceedings of the 2nd CogART (2009)
5. Akyildiz, I., Lo, B.F., Balakrishnan, R.: Cooperative Spectrum Sensing in Cognitive Radio Networks: A Survey. Physical Communication (Elsevier) Journal **4**(1), 40–62 (2011)
6. Yin, W., Ren, P., Du, Q., Wang, Y.: Delay and Throughput Oriented Continuous Spectrum Sensing Schemes in Cognitive Radio Networks. IEEE Transactions on Wireless Communications **11**(6), 2148–2159 (2012)
7. Farhang-Boroujeny, B.: Filter Bank Spectrum Sensing for Cognitive Radios. IEEE Transactions on Signal Processing **56**(5), 1801–1811 (2008)
8. Choiand, J.I., Jain, M., Srinivasan, K., Levis, P., Katti, S.: Achieving Single Channel, Full Duplex Wireless Communication. In: Proceedings of the Sixteenth Annual International Conference on Mobile Computing and Networking. MobiCom 2010, pp. 1–12. ACM, New York (2010)
9. Ahmed, E., Eltawil, A. Sabharwal, A.: Simultaneous Transmit and Sense for Cognitive Radios using Full-Duplex: A first Study. In: IEEE Antennas and Propagation Society International Symposium (APSURSI), pp. 1–12 (2012)
10. Afifi, W., Krunz, M.: Exploiting Self-Interference Suppression for Improved Spectrum Awareness/Efficiency in Cognitive Radio Systems. In: Proceedings of the IEEE INFOCOM 2013 Main-Conference (2013)
11. Esemann, T., Hellbrück, H.: CSOR - Carrier Sensing On Reception. In: Proceedings of the 4th CogART, Barcelona, Spain (2011)
12. Esemann, T., Hellbrück, H.: Integrated Low-Power SDR enabling Cognitive IEEE 802.15.4 Sensor Nodes. In: Proc. of the 8th Karlsruhe Workshop on Software Radios, Karlsruhe, Germany (2014)
13. Proakis, J.G., Salehi, M.: Digital Communications, 5th edn. McGraw Hill (2008)

14. Patel, P., Holtzman, K.: Analysis of a Simple Successive Interference Cancellation Scheme in a DS/CDMA System. IEEE Journal on Selected Areas in Communications **12**(5), 796–807 (1994)
15. Ettus Research LLC: USRP2 Universal Software Radio Peripheral (2009)
16. GNU Radio. http://gnuradio.org
17. Schmid, T.: GNU Radio 802.15. 4 En-and Decoding. Networked & Embedded Systems Laboratory, UCLA, Technical Report TR-UCLANESL-200609-06 (2006)

Prototype of Smart Phone Supporting TV White-Spaces LTE System

Takeshi Matsumura[1](✉), Kazuo Ibuka[1], Kentaro Ishizu[1], Homare Murakami[1,2],
Fumihide Kojima[1,2], Hiroyuki Yano[1], and Hiroshi Harada[2,3]

[1] NICT, Wireless Network Research Institute, Smart Wireless Laboratory,
3-4, Hikarino-Oka, Yokosuka, Kanagawa 239-0847, Japan
matsumura@nict.go.jp
[2] NICT, Social ICT Research Center, 4-2-1, Nukui-Kitamachi,
Koganei, Tokyo 184-8795, Japan
[3] Department of Communications and Computer Engineering,
Graduate School of Infomatics, Kyoto University,
Yoshida-Konoe-cho, Sakyo-ku, Kyoto 606-8501, Japan

Abstract. Recently, secondary use of white-spaces has been expected as one of the technologies to mitigate spectrum resource shortage problem. In particular, secondary utilization of white-spaces in the TV band (TVWS) has attracted attention and some standardizing activities such as IEEE802.11af and IEEE802.22 have been promoted. For the mobile communication systems, it is possible to secure more channels for traffic balancing by utilizing the TVWS, while there are still some difficulties in miniaturization and low power consumption for TVWS utilization. In this study, the authors prototyped a smart phone type TV band device, which supports a TVWS LTE system, by applying frequency conversion technology. This prototyped smart phone is fully operated by an internal battery and can connect to the TVWS eNB previously developed by the authors. Commercial LTE Band 1 is also supported and the phone can select and smoothly switch between two bands. This is the world's first prototyped smart phone supporting the TVWS that demonstrates not only the feasibility of miniaturization and low power consumption, but also the possibility of spectrum expansion in mobile communication system towards next generation.

Keywords: LTE · Spectrum sharing · TV white-spaces · Frequency conversion · Smart phone

1 Introduction

Recently, a variety of attractive applications for mobile terminals such as smart phones and tablet terminals have been widely used and demand of communication traffic has been burgeoning rapidly. This trend will continue over the next decade and International Telecommunications Union (ITU) estimates that a required frequency band for mobile communication will become 1,280–1,720 MHz by 2020 [1]. On the other hand, since spectrum resource is limited, it is urgently required to achieve the

© Institute for Computer Sciences, Social Informatics and Telecommunications Engineering 2015
M. Weichold et al. (Eds.): CROWNCOM 2015, LNICST 156, pp. 562–572, 2015.
DOI: 10.1007/978-3-319-24540-9_46

practical use of spectrum sharing technology for improving efficiency of spectrum usage. In such a situation, communication system utilizing white-spaces (WS), which is allocated to existing communication systems but not used "temporally" or "specially," has been expected as one of spectrum sharing technologies. In particular, exploitation of WS in the TV band (TVWS, TV White-spaces) has been studied worldwide due to its attractive propagation property and penetrability, and some communication systems such as IEEE802.11af for WiFi system [2] and IEEE802.22 for WRAN (Wireless Regional Area Network) system [3] have been standardized as TVWS communication systems.

For the mobile communication systems, it is possible to secure more channels for traffic balancing by utilizing the TVWS. However, the realization of miniaturization and low power consumption is a major challenge for TVWS utilization by mobile communication systems. Since the allocated frequency to TV broadcasting is lower and considerably wider than that of existing mobile communication systems and further downsizing of conventional RF (Radio Frequency) circuit and component are difficult, especially in development of small-size and wide-band antenna and filter. Furthermore, adjacent channel leakage power of secondary systems in the TV bands is severely limited in comparison with conventional standards, to protect primary users, i.e. TV broadcasters [4],[5]. On the other hand, demand of portable-size TVWS devices has been increased for further investigation of interference with/from primary users under mobile environment and also the vertical-handover based on spectrum sharing technology towards next generation of 5G.

To develop communication devices supporting heterogeneous network access including the TVWS, the frequency conversion system is one of the feasible technologies due to its versatility that enables the deployment of any communication systems to desired frequency bands. In the mobile communication system, some prototypes in the TVWS by utilizing the frequency conversion technology have been reported [6]-[8]. However, further miniaturization is still necessary for development of actual mobile communication environment in the TVWS. In this study, we prototyped a smart phone type TV band device which supports a TVWS LTE (Long Term Evolution) system, by applying the frequency conversion technology. This prototyped smart phone is fully operated by an internal battery and can connect to the TVWS eNB (enhanced Node B) developed by the authors in the previous work [8]. In addition, commercial LTE Band 1 is also supported and the phone can select and smoothly switch between bands according to an available channel list in the TVWS provided by a TV white-space database (WSDB) which is also developed by NICT [9],[10]. This is the world's first prototyped smart phone supporting the TVWS LTE system that demonstrates feasibility of miniaturization and low power consumption.

The rest of this paper is organized as follows. In Section 2, physical features and hardware design of the prototyped smart phone are described. Measured RF performance and spurious response are described in Section 3. Finally, we conclude this paper with Section 4.

2   Smart Phone Prototype

2.1   Physical Features

Fig. 1 shows a prototyped smart phone supporting the TVWS LTE system. This prototype is based on the off-the-shelf smart phone and its physical features are substantially the same as commercial ones, as summarized in Table 1. Frequency conversion circuits are additionally incorporated to support the TVWS LTE system instead of the existing TDD-LTE system which the original smart phone accommodates. Two microSIM card slots are incorporated in the phone; one is for the developed TVWS LTE system and the other one is for commercial LTE networks. The LTE system for both bands is based on the 3GPP standard release 8. Using original application, the prototyped smart phone selects and smoothly switches communication bands between the TVWS and commercial LTE Band 1.

Fig. 2 shows channel assignment of the TVWS LTE system with a signal bandwidth (BW) of 5 MHz, 10 MHz and 20 MHz to TV channels with a channel BW of 6 MHz. A center frequency of the TVWS LTE with a BW of 5 MHz is assigned to a center frequency of a single TV channel with a BW of 6 MHz. Correspondingly, center frequencies of the TVWS LTE with a BW of 10 MHz and 20 MHz are assigned to center frequencies of two and four consecutive TV channels, respectively.

Size and weight are also virtually the same as original ones, since only the RF circuits are replaced from existing TDD-LTE system to TVWS LTE system. This prototype is fully operated by an internal battery with a current capacity of 2,600 mAh. Since additional circuits for frequency conversion consume extra current, the standby time for the TVWS LTE system is shortened to 290 hours in comparison with 440 hours for commercial LTE network in Band 1.

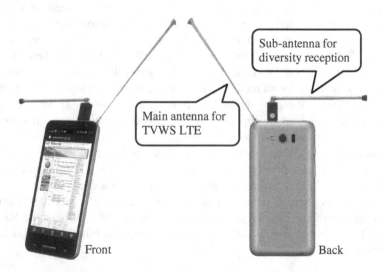

Sub-antenna for diversity reception

Main antenna for TVWS LTE

Front

Back

Fig. 1. Prototyped smart phone supporting TVWS LTE system.

Table 1. Physical Features.

Item	Description
LTE/3G	3GPP release 8
Wi-Fi	IEEE802.11a/b/g/n (2.4GHz/5GHz)
Bluetooth	Version 4.0
GPS	Bulit-in
CPU	APQ8064T 1.7GHz (Quad core)
Platform	Android 4.2 (JellyBean)
Internal Memory	RAM : 2GB, ROM : 32GB
SIM card slot	microSIM × 2
External Memory	microSD/microSDHC/microSDXC
Display	4.7 inch TFT panel, 1,920 x 1,080 full-HD Touch panel : electrostatic
Camera	1,340 MegaPixel
Battery	2,600 mAh
Battery Life (Standby time)	Up to 490 hours for 3G Up to 440 hours for LTE Band 1 Up to 290 hours for TVWS LTE
Battery Life (Talk time)	Up to 10 hours for 3G
Battery charge time	190 min with rapid charge adapter 270 min with wireless charger
External Antenna Port	SMA-P (for sub-antenna in UHF band)
Size	132 mm x 65 mm x 10.9 mm
Weight	146 g

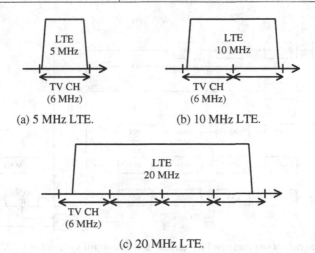

(a) 5 MHz LTE.   (b) 10 MHz LTE.

(c) 20 MHz LTE.

Fig. 2. LTE channel allocation to TV channels (based on 6 MHz BW).

2.2   Access to the White-Space Database

In the rule established by the FCC, a TV band device without any sensing function for TV broadcasting must access to the WSDB and acquire an available channel list according to its geo-location, before starting wireless communication in the TV band [4]. Therefore, this prototyped smart phone accesses to the WSDB and receives an available channel list via commercial network or internet access with a built-in WiFi system or the existing LTE system first, by sending its geo-location acquired by the built-in GPS (Global Positioning System) module. Here, by enhancing capabilities of the WSDB in accordance with the standard of the LTE communication system, the available channel list returned from the WSDB takes into account the operational status of neighbor TVWS eNBs [10]. Users can confirm the list of available channels on the display and switch the communication band arbitrarily by using original application.

2.3   Hardware Design

Fig. 3 shows a brief block diagram of implemented RF circuit and its control system allowing the TVWS communication in the prototyped smart phone. A customized RF IC, which supports LTE Band 1 and Band 38, is mounted on the RF board and connected to the modem IC. Here, Band 1 is a commercial band in Japan and its frequency ranges are 1,920–1,980 MHz. and 2,110–2,170 MHz for an uplink and a downlink, respectively. Band 38 with a frequency range of 2,570–2,620 MHz is not used in Japan so far. Duplex modes of Band 1 and Band 38 are FDD (Frequency Division Duplex) and TDD (Time Division Duplex), respectively. For communication in the TV band, communication frequency of Band 38 is converted to the UHF (Ultra High Frequency) band by using the frequency conversion technology. In this prototype, frequency conversion circuits are implemented instead of the RF circuits for Band 38, and thus, FDD-LTE system in the commercial band and TDD-LTE system in the TVWS are supported. Table 2 summarizes supported LTE communication systems in the phone.

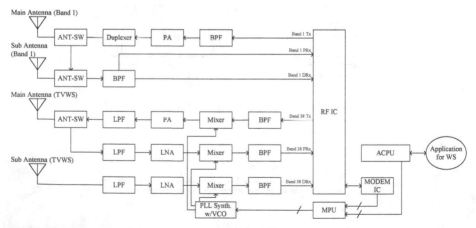

Fig. 3. Block diagram of implemented RF circuit and its control system for TVWS communication on prototyped smart phone.

Table 2. Supported LTE Communication Systems.

Item	Description
Supporting Frequency	UHF : 470-710 MHz (TVWS) Band 1 : 1,920-1,980 MHz (uplink) / 2,110-2,170 MHz (downlink)
Bandwidth	5, 10, 20 MHz
Transmission Power	Up to 20 dBm
Modulation	BPSK / QPSK / 16QAM / 64QAM BPSK : Control channel 64QAM : Downlink data channel
Duplex system	UHF : TDD Band 1 : FDD
Multiple Access	OFDMA (downlink) SC-FDMA (uplink)

For the frequency conversion, an RF signal of 2.6 GHz is converted to the UHF band with an upper local signal. Supporting frequency range in the UHF band is 470–710 MHz, which is the TV band in Japan. Since the same architecture is applied in our previously prototyped TVWS eNB [8], this prototyped smart phone enables connection to this eNB. To alleviate deterioration of frequency accuracy caused by the frequency conversion system, a high accurate TCXO (Temperature Compensated Crystal Oscillator) is used as a reference clock for the PLL (Phase Locked Loop) synthesizer with a VCO (Voltage Controlled Oscillator), which generates the upper local signal.

Additional MPU (Micro-Processing Unit) is implemented to control the communication frequency in the TVWS. The MPU is informed use channel information from the ACPU (CPU for Application) and generates control signals for the PLL synthesizer IC.

2.4   Antenna for TVWS

The prototyped smart phone supports diversity reception, as shown in Fig. 3. A built-in rod antenna originally for one-segment broadcasting reception is diverted to a main antenna for the TVWS LTE system and internally connected to an antenna port on the RF board. This antenna is retractable and used for both transmission and reception. A radiation efficiency of more than –2.7 dB is obtained in free space by measurement. For diversity reception, an additional antenna can be connected via an SMA connector, which is optionally attached to the phone. For the existing LTE system in Band 1, main and sub antennas are implemented internally.

3   RF Performance

3.1   Measured RF Characteristics

RF performance of the TVWS LTE system is measured based on the standard of the LTE communication system by using a radio communication tester (MT8820C,

Anritsu). Fig. 4 shows a measurement system for TVWS LTE. Since the upper local is used to convert a communication frequency of 2.6 GHz in the prototyped smart phone, IQ (Inphase and Quadrature) polarity in the TV band is reversed. MT8820C does not support reversed IQ polarity, and thus, a mixer is inserted between MT8820C and the prototyped smart phone to reverse IQ polarity in the TV band. After the frequency of the RF signal is converted back to 2.6 GHz by an upper local with a frequency range of 3.0–3.3 GHz, IQ polarity is reversed once again to original polarity. To suppress unwanted signals generated by the mixer, LPF (Low Pass Filter) with a cut-off frequency of 780 MHz and BPF (Band Pass Filter) with a pass-band of 2,570–2,620 MHz are inserted in UHF and 2.6 GHz parts, respectively. In addition, attenuator is inserted between the LPF and the phone to avoid saturation of the mixer during transmission measurement. Note that all losses generated from the additional circuits in Fig. 4 are taken into account to measurement results described below.

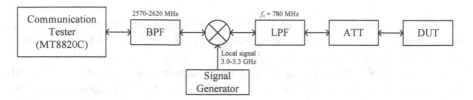

Fig. 4. Measurement system for RF performance of TVWS LTE.

Table 3. Measured RF Characteristics for TVWS LTE.

ITEM	Unit	RF Characteristics		
		Lch 473 MHz	Mch 593 MHz	Hch 707 MHz
Tx power	dBm	20.27	19.36	19.34
Frequency error	ppm	−0.041	-0062	−0.021
EVM (Data)	% rms	7.35	5.95	8.15
EVM (RS)	% rms	6.59	6.00	7.19
OBW	MHz	4.4775	4.455	4.4775
ACLR (Lower)	dBc	−28.27	−39.16	−27.07
ACLR (Upper)	dBc	−27.53	−39.70	−27.45
Rx Sensitivity	dBm	−92.6	−90.2	−87.0

Table 3 summarizes measured RF characteristics. Transmission power and signal BW are set to +20 dBm and 5 MHz, respectively, for measurement of all transmission characteristics. All the measurement was performed at a low channel (Lch) of 473 MHz, a middle channel (Mch) of 593 MHz and a high channel (Hch) of 707 MHz. Although the output power at Hch is slightly lower than that at Lch, frequency charac-teristics of the output power is sufficiently flat within 1 dB. Frequency error with a high accuracy of less than ±0.1 ppm is achieved in all channels and without deterioration due to the frequency conversion. An occupied BW (OBW) of about 4.5 MHz and an error vector magnitude (EVM) of less than 12.5 % in all channels are adequate in comparison with the requirement in the 3GPP standard.

A sufficient result of an adjacent channel leakage ratio (ACLR) at the Mch is obtained, while ACLR at both Lch and Hch is about −27 dBc and does not satisfy with the 3GPP standard. Fig. 5 shows typical frequency characteristics of ACLR and transmission power of the phone. Both ACLR in upper and lower sides indicates almost same frequency characteristics. ACLR is sufficiently low in a frequency range of 520–690 MHz but remarkably deteriorated in the edge of the TV band, i.e. 470 and 710 MHz. On the other hand, the transmission power is almost flat in the entire TV band. In addition, ACLR of a power amplifier (PA) device is not remarkably deteriorated even in the edge of the TV band in comparison with that in Fig. 5. Therefore, deterioration of ACLR is mainly caused by impedance mismatch of the PA with the mixer and LPF, due to wide frequency range of the TVWS. This is one of challenges in component development and circuit design for further prototype of portable size TV band devices.

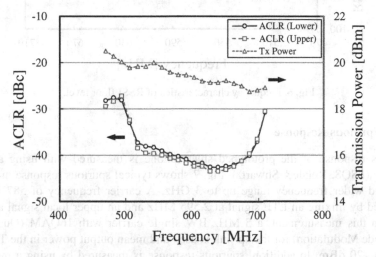

Fig. 5. Frequency characteristics of ACLR and transmission power.

Receiver sensitivity is defined as minimum received power with a throughput of more than 95 % in QPSK modulation, as following the 3GPP standard. As indicated in Table 3, a receiver sensitivity of −92.6 dBm, −90.2 dBm, −87.0 dBm is obtained at the Lch, the Mch and the Hch, respectively. This result is inadequate in comparison with a requirement in the 3GPP standard but sufficient in comparison with other TV white-space communication standards such as IEEE802.11af [2]. Fig. 6 shows frequency characteristics of RSSI (Received Signal Strength Indicator) floor level that indicates received power at which RSSI does not change due to lower received signal level than receiver noise level. This RSSI floor level almost agrees with receiver sensitivity. From this result, higher RSSI floor level is observed in higher frequency range, resulting in deterioration of the receiver sensitivity. This deterioration might be caused by insufficient frequency characteristics of gain and NF (Noise Figure) of the mixer, and also noise or spurious in a local signal. For further improvement of RF performance, it is important to flatten frequency characteristics of components constituting the RF circuit and to implement design techniques for noise reduction in wide frequency range of the TV band.

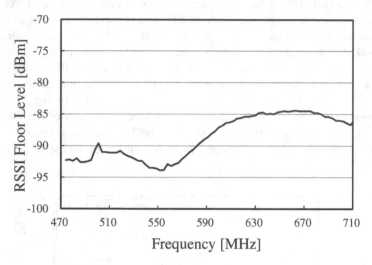

Fig. 6. Frequency characteristics of RSSI floor level.

3.2   Spurious Response

Spurious response of the prototyped smart phone is measured with using a signal analyzer (FSQ8, Rohde& Shwartz). Fig. 7 shows typical spurious response in the TV band and wider frequency range up to 3 GHz. A carrier frequency of 587 MHz is generated by mixing an LTE signal at 2,595 MHz and an upper local signal at 3,182 MHz. In this measurement, a 5 MHz BW single carrier with 16QAM (Quadrature Amplitude Modulation) for the up-link is used and mean output power in the TV band is set to +20 dBm. In addition, spurious response is measured by using a max hold function with a resolution BW (RBW) of 100 kHz.

No remarkable spurious is observed and floor noise level is sufficiently low about less than –60 dBm over the entire TV band, as shown in Fig. 7(a). This unwanted signal level outside using channel complies with criteria in rules regulated by the FCC and the Ofcom, except for slopes of spectrum mask. On the other hand, 2nd harmonics with a signal level of –53.8 dBm is observed at 1,174 MHz, as shown in Fig. 7(b). This spurious level is sufficiently low and no other remarkable spurious is observed in the wide frequency range.

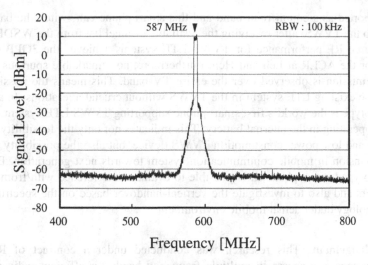

(a) In wide frequency range.

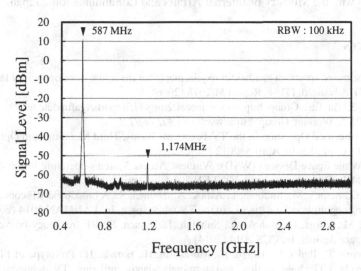

(b) In wide frequency range.

Fig. 7. Prototyped smart phone supporting TVWS LTE system.

4   Conclusion

In this study, the prototyped smart phone enabling TVWS communication with the LTE system is described and its physical features are summarized. In this prototyped smart phone, communication frequency of the existing LTE Band 38 is converted to the UHF band by utilizing frequency conversion technology. Existing LTE Band 1 is also supported as a commercial band and this smart phone can switch the band from Band 1 to the TVWS, after receiving the available channel list from the WSDB.

Measured RF performance for TVWS LTE systems achieves the 3GPP standard except for the ACLR at Lch and Hch. Furthermore, no remarkable spurious and unwanted emission is observed over the entire TV band. This means the possibility to deploy the existing LTE system in the TVWS without crucial alterations. In addition, this prototype is the world's first smart phone supporting TVWS LTE system, capable of fully operating in the internal battery. This indicates not only the feasibility of portable size and low power consumption TVWS device, but also the possibility of spectrum expansion in mobile communication system towards next generation. By using this prototyped smart phone, it is possible to evaluate interference with/from the primary users and also to investigate the vertical-handover based on the spectrum sharing technology under actual mobile environment.

Acknowledgement. This research was conducted under a contract of R&D on "dynamic spectrum access in multiple frequency bands for efficient radio resource utilization" with the Ministry of Internal Affairs and Communications, Japan.

References

1. ITU: Estimated spectrum bandwidth requirements for the future development of IMT-2000 and IMT-Advanced. ITU-R Report M.2078 (2006)
2. IEEE802.11af Task Group. http://www.ieee802.org/11/Reports/tgafupdate.htm
3. IEEE802.22 Working Group. http://www.ieee802.org/22/
4. FCC: Unlicensed Operation in the TV Broadcast Bands, Third Memorandum Opinion and Order. FCC, pp. 12–36, April 5, 2012
5. ETSI: White Space Devices (WSD); Wireless Access Systems operating in the 470 MHz to 790 MHz TV broadcast band. ETSI EN 301 598, April 2014
6. Varga, G., Schrey, M., Subbiah, I., Ashok, A., Heinen, S.: A broadband RF converter empowering cognitive radio networks in the TV white space. In: LANMAN 2014 (2014)
7. Schrey, M., Varga, G., Ashok, A., Subbiah, I., Heinen, S.: RF frequency converters for white space devices. In: CCS 2014 (2014)
8. Matsumura, T., Ibuka, K., Ishizu, K., Murakami, H., Harada, H.: Prototype of FDD/TDD dual mode LTE base station and terminal adaptor utilizing TV white-spaces. In: CROWNCOM 2014, pp. 317–322 (2014)
9. Ishizu, K., Murakami, H., Harada, H.: TV white space database for coexistence of primary-secondary and secondary-secondary systems in mesh networking. In: WPMC 2012, pp. 118–122 (2012)
10. Ibuka, K., et al.: Development and field experiment of white-spaces LTE communication system in UK digital terrestrial TV band. In: Proc. IEEE 81st VTC (2015)

Strategic Choices for Mobile Network Operators in Future Flexible UHF Spectrum Concepts?

Seppo Yrjölä[1(✉)], Petri Ahokangas[2], Jarkko Paavola[3], and Pekka Talmola[4]

[1] Nokia Networks, Oulu, Finland
seppo.yrjola@nokia.com
[2] Oulu Business School, Oulu, Finland
petri.ahokangas@oulu.fi
[3] Turku University of Applied Sciences, Turku, Finland
jarkko.paavola@turkuamk.fi
[4] Nokia Technologies, Salo, Finland
pekka.hk.talmola@nokia.com

Abstract. This paper seeks to identify and discuss for mobile network operators business opportunities and strategic choices in the new flexible hybrid use concept of the Ultra High Frequency spectrum (470–790 MHz) by Digital Terrestrial TV and Mobile Broadband. More flexible use of the band aims to increase the efficiency of spectrum use in delivering fast growing and converging MBB, media and TV content to meet changing consumer needs. The framed opportunities and created simple rules indicate that the MNOs could benefit significantly from the new UHF bands enabling to cope with increasing asymmetric media data traffic and to offer differentiation through personalized broadcasting and new media services. As a collaborative benefit concept opens up new business opportunities in delivering TV and media content using MBB network. Furthermore, it had potential to transform the business ecosystem around both the broadcasting and the MBB by introducing new convergence opportunities.

Keywords: Strategy · Simple rules · Business opportunity · Mobile network operator · Broadcasting · Mobile broadband · UHF · Spectrum sharing · 5G

1   Introduction

The mobile broadband industry is starting to suffer from the scarcity of radio spectrum with the increasing data traffic and rapidly changing user habits [1]. As the downstream media content, video in particular, is the biggest and fastest growing part of the traffic [2], asymmetry in MBB networks is increasing with average downlink to uplink ratio in the new 4th generation LTE networks being approximately 10:1 and growing. Latest changes in consumption characteristics with ubiquitous high data speed demand, has put mobile network operators against a disruptive change.

Thus, in the broadcasting industry the importance of DTT platform providing audiovisual media and traditional free-to-air services have been challenged by competing delivery platforms, Over the Top (OTT) media delivery over the Internet, bypassing operators and higher general regulatory UHF spectrum fees. While the customers' TV

© Institute for Computer Sciences, Social Informatics and Telecommunications Engineering 2015
M. Weichold et al. (Eds.): CROWNCOM 2015, LNICST 156, pp. 573–584, 2015.
DOI: 10.1007/978-3-319-24540-9_47

type of media content consumption has been on steady growth, the media content delivery and consumption mechanism will and have already started to change. Consumers are changing their consumption habits from linear real time to non-linear usage with the growing demand for more personalized longtail content ranging from commercial on-demand services to user generated content and channels with interactivity [3]. Reception of the TV content is happening more via cable, satellite, fixed broadband and, particularly, via MBB. Especially the broadband delivery is meeting the requirements of personalized content better than traditional broadcast methods.

The exclusive spectrum availability through auctions has been limited and even the largest MNOs face the risk of running out of spectrum in the future provided that the predicted data rate growth continues as estimated. Making new exclusive spectrum available for MBB networks is difficult due to the lack of unallocated spectrum and the costly and lengthy traditional 'command & control' spectrum auctioning & refarming process. This is becoming increasingly complex in the future due to difficulties in finding unused exclusive spectrum and high costs and time needed for the reallocation process.

As it is well-known that many spectrum bands are currently only lightly occupied in time and space, more flexible ways of allocating spectrum, e.g., spectrum sharing, has lately received growing interest among regulators considering new ways of fulfilling the different spectrum demands to meet the mobile traffic growth while maintaining the rights of the original incumbent systems operating in the bands. Currently regulation can be regarded as the key driver for speeding up spectrum sharing, see e.g. [4] and [5].

With these views to the future, spectrum regulators are on one hand considering responding to the changing environment by gradually compressing and withdrawing some DTT licenses of lower demand and repurposing these for MBB. On the other hand, in order to continue fulfilling the national public social service obligations the most used and in particular national broadcasters' DTT licenses will continue to seize part of the UHF spectrum for the foreseeable future on a non market determined basis. The DTT technology evolution will improve the efficiency of the spectrum utilization through evolution from DTT to Terrestrial Digital Video Broadcasting DVB-T2 technology which is more spectrum efficient than DVB-T and also better supports wide area single frequency networks (SFN).

All discussed trends and drivers are transforming broadcasting business environment, in particular for the UHF broadcasting spectrum holders, and opens up new business opportunities as well as risks due to increasing pressure for innovative flexibility and sharing in spectrum usage. For the present, it may be observed that DTT operators have not been offered incentives for changing their spectrum usage. Instead, they have seen unilateral acts from regulators and MNOs towards further compressing their DTT spectrum to give room for additional MBB capacity.

CEPT recently set up Task Group 6 (TG6) "Long term vision for the UHF broadcasting band" [4], to identify and analyze possible scenarios for the development of the band taking account technology and service development. Accordingly, in the European Commission's Radio Spectrum Policy Groups' (RSPG) published report [6] on a long-term strategy on the future use of the UHF band (470-790 MHz) in the EU states as follows [6]: *"The RSPG recommends that member states should have the flexibility to use the 470-694 MHz band for WBB downlink, provided that such use is compatible with the broadcasting needs in the relevant Member State and does not create a constraint on the operations of DTT in this band, including for neighboring countries."*

LTE for broadcasting as a mobile centric broadcast solution, it is not entirely "new." Precursors such as Digital Video Broadcasting–Handheld (DVB-H) or Qualcomm developed Media Forward Link Only (MediaFLO) were less than successful suffering costly investments in dedicated infrastructure and devices and the lack of scale and harmonization in spectrum and devices. Favorable national regulation and standardization supported more successful deployments in China with China Multimedia Mobile Broadcasting (CMMB) system and South Korea with Digital Mobile Broadcasting (DMB) both currently considering ways to scale up with global 3GPP based ecosystem. Although the underlying technical concepts, in particular LTE-Advanced Carrier Aggregation (CA) and evolved Multimedia Broadcast Multicast Service Broadcast (eMBMS) are known and have been standardized in 3GPP [7] and [8], whereas technical flexible use concept with DTT TV broadcasting has not been validated. In addition, there is no work on the business impacts of the concept related to flexible UHF use.

Previous works on business analysis for DTT MBB – UHF spectrum hybrid use or sharing was limited as focus has been on TVWS concept [9]. The general business drivers, enablers and potential impacts of the spectrum sharing on the MBB market were described in [10] and incentives and strategic dynamic capabilities for the key stakeholders in the flexible use of the UHF were discussed in [11]. In this paper we focus on analyzing the flexible use of the UHF spectrum by DTT and MBB. In the development of new flexible spectrum usage or sharing models, it is important to consider the underlying business opportunities and strategic choices to create business models that are sound for all the key stakeholders. This paper investigates:

1) What are the business opportunities and how are they framed for MNOs exploring the flexible UHF concept?
2) What kind of strategic choices do MNOs have to make regarding flexible use?

The anticipatory action learning in a future-oriented mode research methodology [12] was applied in this paper utilizing the capacity and expertise of the policy, business and technology research communities. Simple Rules strategic framework [13] was used in analyzing MNOs strategic choices. The rest of this paper is organized as follows. First, the flexible DTT MBB usage concept is presented in Section 2. Theoretical background for strategic framework is introduced in Section 3. The research methodology applied and the business opportunities and Simple Rules strategy for MNOs in using flexible UHF concept are derived in Section 4. Finally, conclusions are drawn in Section 5.

2   Overview of the Flexible Use of UHF Spectrum

The DTT broadcast has traditionally operated on spectrum bands from 470 to 862 MHz. The 800 MHz band (790-862 MHz) is under deployment for MBB use throughout Europe and World Radiocommunication Conference (WRC) 2012 made a decision on 700 MHz band re-allocation after the WRC-15. In the coming WRC-15 the new IMT spectrum identification agenda point will address co-primary allocation with mobile of the lower UHF band (470–694 MHz) that currently has a primary allocation to broadcasting. Further in the 2012 FCC in USA published the notice of proposed rulemaking on 600 MHZ BC television spectrum incentive auction [14].

TG6 in their long term vision for the UHF broadcasting Band developed the following scenarios how to accommodate both delivery of TV content as well as additional capacity for MBB [4]:

1) Class A: Primary usage of the band by existing and future DVB terrestrial networks.
2) Class B: Hybrid usage of the band by DVB and/or downlink LTE terrestrial networks.
3) Class C: Hybrid usage of the band by DVB and/or LTE (including uplink) terrestrial networks.
4) Class D: Usage of the band by future communication technologies.

In the following analysis we will focus on the flexible hybrid scenario Class B in which the LTE Supplemental downlink (SDL) CA technology introduces a flexible way of how to take freed TV channels to mobile use while maintaining capability to deliver TV content both in conventional living room large screen use cases as well as in new mobile use cases on smart phones and tablets. Supplemental downlink combines traditional paired FDD spectrum pair with additional downlink channel [7].

Although the consumers' interest in the traditional TV programs remains or even increases, the ways how TV content is delivered and consumed is radically changing. Users are more and more receiving the TV media content via cable, satellite, fixed broadband and, especially, via MBB. Furthermore real time linear one way usage is gradually changing into non-linear usage, location independent consumption with increasing demand for interactivity. This reduced demand of DTT as the main delivery mechanism impacted negatively on the value of the service spectrum use [3]. With these insights and foresights it could be further hypothesized that some 'underutilized' and lower valued TV frequencies could and will be reassigned and/or shared with mobile use.

As the availability of the freed TV channels can vary largely between different geographical areas and countries, the scenario b) above proposes to assign them first for the MBB downlink use only. Compared with widely deployed Frequency Division Duplex (FDD) or Time Division Duplex (TDD) access methods, the SDL technique offers better interference free compatibility with the remaining DTT use in the country and across the national borders: FDD operation requires more harmonized conditions with a wider spectrum and also TDD utilizes uplink which is less compatible with the DTT use. Additional flexibility, if needed, to hybrid use of the UHF band allowing different deployment schedules in different regions and countries could be gained through utilizing functionalities that are already developed for shared spectrum access like e.g. recently widely discussed Licensed Shared Access (LSA) concept [15]. SDL allows both the unicast and multicast uses in a flexible way based on demand with eMBMS [8] technique providing tools for cell capacity optimization to cope with present large traffic asymmetry as well as future converged broadcasting services. Additionally, as the SDL base station radios start utilizing freed DTT frequencies one by one locally, there will be no impact on interleaved spectrum used by Program Making and Special Events (PMSE) services.

In the flexible hybrid use concept the evolution of the UHF spectrum can follow the market demand within regulatory frame. Potential evolution of flexible use of the UHF band for Europe is illustrated in the Fig. 1. Already in the first phase hybrid SDL CA concept [7] could be utilized in the deployment of the 700MHz band after WRC-15 through better co-existence characteristics with potential across the border TV transmitters. In the Flexible DTT-MBB scenario the amount of SDL MBB and DTT in the lower UHF band are determined by the market needs and in the long term future integrated UHF Multimedia network vision DTT technologies could even be completely replaced by converged LTE based delivery platform using either SDL and/or eMBMS to deliver TV media content [11].

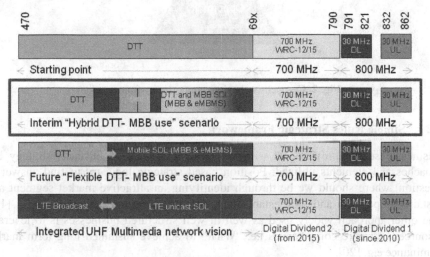

Fig. 1. Evolution of the flexible use of the UHF band with MBB [11].

3   Business Opportunity Based Simple Rules Strategic Framework

3.1   Co-opetitive Business Opportunity Framework

An opportunity has been generally defined in the business literature as the possibility to serve customers better and differently [16] framed by enablers, limiting factors as well as challenges caused by the business context. In the flexible UHF context business opportunities are made to create and deliver value for the stakeholders, value that is co-created among various actors from converging MBB and broadcast (BC) ecosystems as a joint effort. In addition to value co-creation, an equally important aspect of value is the ability to capture value by the stakeholders, i.e., obtain profits [17] which in the context of this research can be called value co-capture. The term co-opetition, defined as the coexistence of competition and cooperation within the value creating business context, illustrates the increased complexity of the UHF co-primary business environment, where companies simultaneously compete and cooperate with each other not only over spectrum but also over customers. Value co-creation could be seen as a cooperative and the parallel value co-capture as a competitive process [18]. Fig. 2 below illustrates the analysis frame used in this paper to develop and frame the business opportunities for MNOs.

Fig. 2. Co-opetitive business opportunity framework

3.2   Simple Rules Strategic Framework

Business research provides us with numerous examples of business strategy approaches and elements utilized. Position based strategic logic try to find answer to question: where should we be through identifying an attractive market segment and sustainable position and then establishing, strengthening and defending it e.g. [19]. The other approach widely used as well in well structured businesses is to leverage resources and core competences i.e. "What" to achieve sustained long term market dominance e.g. [20].

Traditional approaches, however, include several limitations in rapidly changing complex markets: they do not build around the business opportunity, have only weak linkages to the key business processes, depict resources rather than activities, and lack needed flexibility to seize fast changing opportunities. In this paper we adopt the business strategy approach presented in [13] that partly helps to answer to the concerns discussed above.

In emerging, dynamic and systemic environments this novel "Simple Rules" approach sees business strategies as built around the business opportunity and the key processes needed to seize them flexibly and timely. A simple rule provides guidelines within which opportunities could be pursued with selected key processes. The proposed framework consists of five categories:

1) How-to rules for conducting business in an unique way
2) Boundary rules for defining the boundaries of the business opportunities of the stakeholders,
3) Priority rules that help to identify and rank the criteria for opportunity decision making,
4) Timing rules that help in synchronizing, coordinating and pacing emerging opportunities, and
5) Exit rules that help in identifying basis for exit or selecting initiatives to be stopped.

4   Analysis of the Simples Rules for MNO's Flexible UHF Business

The research methodology applied, business opportunities and strategic choices as Simple Rules created and their analysis are summarized in this section.

4.1   Methodology

Business opportunities and strategic choices as simple rules were created utilizing the Anticipatory Action Learning (AAL) methodology, in a future-oriented mode [12]. In developing foresight the methodology represents a unique, reflexive, and iterative process of questioning and creating the future from transformational point of view. In this interactive and collaborative approach conversation and dialog among cross-disciplinary participants, from multiple domains concerned with the research project is essential.

The elements of business opportunity and strategy analysis discussed in this paper were created in a series of future oriented project planning workshops in April-October 2014 organized by the Finnish FUHF research consortium consisting of end to end Finnish UHF ecosystem with expertise in the areas of policy, business and technology.

4.2   Business Opportunities

In the analysis for the business opportunity elements of flexible UHF use, five key ecosystem roles are identified: the National Regulator (NRA), MNOs, BC Network Operators (BNO), TV media content providers and device and infrastructure vendors. As far as flexible UHF concept is concerned, the roles of the regulator and both the broadcasting and mobile broadband operators are vital in adopting of novel UHF concept and spectrum sharing technologies in general. In addition when developing and analyzing the opportunity frame authors argue that three domains; policy, business, and technology, affecting flexible spectrum usage concepts should proceed in tandem. Enabling, limiting and challenging elements framing the business opportunities for the MNO are listed in Table 1.

Business and technology elements can be identified as enablers for value co-creation. Fast growing demand and lack of exclusive spectrum combined with the radical changes in the TV media consumption habits will urge the adoption of novel more flexible and efficient spectrum management concepts. Furthermore different spectrum sharing schemes are high in regulators agenda. Utilization of the LTE ecosystem scale and harmonization will reduce risk related technology maturity. High adjacent collaborative new business potential with media content players for MNOs could emerge with broadcasting content delivery to variety of smart devices. However, at the same time with lowered entry barrier to UHF spectrum BNOs and new types of operators could consider entering the MBB business.

Regarding limiting factors, sound, sustainable and harmonized regulatory environment can be the limiter that needs to be addressed before MNO can co-create and co-capture value from it with broadcasting & media partners. The limited spectrum availability with potential national restriction and obligations may negatively influence the MNOs outlook on flexible use and the spectrum valuation. A specific technology item to be considered is the need to relocate PMSE services essential for the media program making. In addition to MNO opportunities it is essential to consider reciprocal incentives for the current BC spectrum holders to further transition to flexible use.

Regulatory risk and uncertainty are the main elements of the co-opetitive challenges in the competitive domain. First, the complexity of the flexible spectrum framework and the license and transaction cost might impact the value of the spectrum and the required time of recovering the network investments. Secondly, in their regulatory strategy MNOs have to balance between exclusive spectrum and flexible/shared spectrum options and their interdependencies. On the technology domain MNOs need to pay attention to dynamic capabilities needed to deploy, manage and optimize multilayered unicast-multicast network under flexible sharing conditions.

In summary, in order to realize the business potential and opportunities of flexible UHF spectrum use, MNO have occasion to simultaneously co-create and co-capture value with broadcasting media players in a co-opetitive business environment where co-operation (spectrum) and competition (customers & services) exist parallel to each other.

MNOs are in unique position to leverage additional downlink capacity flexible UHF concept offers. Faster access to QoS licensed UHF spectrum without mandatory coverage obligations will help them to timely cope with booming asymmetric data needs. Additional capacity combined with scalable and flexible unicast-multicast solution will enable MNOs to better retain and grow existing customer base with changing demand and consumer habits. Furthermore, personalized converged mobile broadband and media broadcasting services offer opportunity for differentiation.

Table 1. Elements framing business opportunities

	Business opportunity framing elements
Enablers	Lack of exclusive spectrum triggers new spectrum access approaches
	Consumers TV and media consumption habits are changing towards unlinear, mobile and multi-device usage
	Commercial TV service providers and national BS are already offering streaming services
	Additional potentially lower cost capacity to cope with asymmetric traffic
	UHF spectrum offers superior coverage and in building data penetration
	New DTT technologies improves UHF spectrum utilization efficiency, DVB-T2 transition on the other hand might delay opening
	Co-primary allocation improves overall spectrum use efficiency
	Lower entry barrier to broadcasting/video-on-demand business
	Potential to extend MNO's business to broadcasting content delivery
	Spectrum sharing in general on regulatory agenda
	Harmonized LTE technology base offering scale
Limiters	Need for global and national spectrum regulation may slow down entry - Harmonization is a pre-condition to enable potential benefit fully.
	Lack of BCs willing to discuss flexible co-primary use. Seen as a threat.
	Limited spectrum availability limits MNO business opportunities
	Regulatory framework restrictions may reduce the economic value
	Other UHF incumbent like PMSE or TV White Space might delay and constrain introduction
	National broadcasting policy and regulatory requirement e.g. coverage, reliability, free-to-air and must carry rules, consumer data
Challenges	Regulatory risk and uncertainty related to timing, term, licenses and flexibility
	Impact on the further availability of traditional exclusive spectrum
	Spectrum license cost with potentially higher transaction costs associated with shared use.
	May change the competitive environment with BC interest in deploying their own LTE networks.
	Increased technical and operational complexity with related capital and operational costs
	New capabilities needed for network management and optimization
	Timely availability of terminals and potential impact on cost and complexity

4.3   Simple Rules

Using the above summarized future-oriented action research method; we created a strategy as Simple Rules for mobile network operators deploying the flexible UHF concept applying the Simple Rules strategy approach from [13]. The developed MNOs' strategic rules are summarized in Table 2.

How to reinforce customer retention and acquisition while further strengthen dominant market position are key strategic elements of MNOs. Fundamental means to achieve these is to obtain all available spectrum, prioritizing exclusive, and to manage and optimize it across all the spectrum resources. In addition to network parameter based load balancing, novel traffic steering concepts considering as well QoE view enables MNOs to best match the personalized user demand with the network capacity supply. Collaborating with the TV and media domain could enhance the utilization of the dominant market position in MBB as well as to explore growth pockets in broadcasting.

Regarding opportunity boundaries, MNOs should exploit their existing infrastructure assets and 3GPP ecosystem with available LTE technologies to ensure early use and economies of scale. Active participation to policy and regulation processes is needed to educate the regulator about converging technology and business opportunities in UHF and the long term investment nature of MBB business.

MNOs could prioritize emerging opportunities through retaining control over spectrum and the network enabling to enhance QoS and QoE for the current mobile services e.g. video streaming that offers new revenue opportunities. As an option at early phase of flexible UHF spectrum businesses, MNOs could value average revenue per user (ARPU) over operational efficiency to utilize their customer base. In the future as potential MBB broadcasting convergence proceeds MNOs could consider acquire BC network assets to gain spectrum and infra.

Timing rules are essential in synchronizing opportunities across the company. High efficiency scalable data offload could be implemented first in order to optimize the use of the spectrum assets. Next improved capacity and QoS enables to personalize mobile broadband data to different customer segments. Broadcasting business opportunities exploration in confined areas e.g. live events could follow after the internal asset leverage. In collaborative set up with media content players complementary TV and broadcast content delivery could be next with evolution to potential future wide area TV distribution replacement by LTE broadcast technologies.

Regarding mandatory go / no-go opportunity exit rules MNOs should defend their "bloodline" scarce and finite exclusive spectrum. Another source of differentiation in entering more personalized "unicast" services is the detailed network data. The subscriber data management and customer billing relationship will be a unique asset in the design of new services and the service level differentiations.

Table 2. Summary of Developed Simple Rules

Opportunities	How to	Boundary	Priority	Timing	Exit
Utilize and grow existing customer base with changing demand Gain faster access to QoS licensed UHF spectrum without mandatory coverage obligations Offer personalized mobile broadband data and "ubicast" media delivery services for differentiation	Advance customer retention and acquisition Invest in scale and dominant market position Always prioritize exclusive spectrum Optimize usage of all spectrum assets Partner with the TV and media industry in the future	Exploit existing infrastructure assets Leverage available LTE technologies to ensure early use and economies of scale Actively participate in the policy and regulation process Turn media ICT convergence a source of competitive advantage	Retain control over spectrum and the network Enhance QoS and QoE for the current mobile services e.g. video streaming first New revenue opportunities, higher ARPU Acquire BC network assets to gain spectrum and infra	High efficiency scalable data offload first Personalized mobile broadband data next Explore BC business opportunities in confined areas e.g. live events then Complement TV and broadcast content delivery Future wide area TV distribution replacement	If defending exclusive spectrum becomes impossible If withholding customer data inside company becomes impossible If one cannot keep detailed network information to oneself

5   Conclusions

This paper discuss the transformative role of flexible co-primary UHF area spectrum concept in the future mobile broadband and broadcasting networks as an endeavor to meet the growing traffic demand and changing consumption characteristics of the customers.

We utilized co-opetitive business opportunity framework for understanding mobile network operator's opportunities and how they are framed from policy, technology, and business perspectives in future flexible co-primary UHF spectrum networks. Opportunity analysis was used in creating and discussing strategic choices as simple rules. In developing foresight the Anticipatory Action Learning and in particular action research in a future-oriented mode was used.

We argue that policy and regulation will be on the one hand the key enabler in the path toward flexible use of UHF spectrum and on the other hand play key role in removing limiting and challenging elements critical in the first steps of that path. Ongoing transformative change in media and broadcasting business lower the barrier for change supported by mobile technology development in particular related to 3GPP LTE evolution.

The proposed opportunities and related simple rules could help operators to retain existing customers, strengthen market position and win over new customers by offering personalized mobile broadband data and "ubicast" media delivery services. With MBB broadcast concept on flexible UHF spectrum, linear, traditional TV broadcast can be extended to smart devices providing the scalability and flexibility to combine linear and non-linear TV, on-demand and interactive TV. This can significantly reshape the business ecosystem around the mobile broadband and media and open up new converging and co-operative business opportunities with media and TV industry. MNOs are optimally positioned to explore new business opportunities in parallel with traditional business model.

The strategic choices as simple rules provide a dynamic framework for MNOs for exploring and exploiting emerging opportunities, developing dynamic capabilities to respond transforming environment and building business models to leverage new flexible UHF spectrum access approaches.

In the future, flexible UHF usage concept business studies will need to be expanded to cover also other key stakeholders. In particular, co-operative business model with broadcast domain will be an important aspect to study.

Acknowledgments. This work has been performed as a part of Future of UHF project preparation phase. The authors would like to acknowledge the project consortium: Digita Networks, Elisa, Finnish Communications Regulatory Authority, Nokia, RFtuote, Schneider Finland, Telia Sonera, Turku University of Applied Sciences, University of Turku, VTT Technical Research Centre of Finland, YLE, Åbo Akademi University and Tekes Finnish Innovation Fund.

References

1. Report ITU-R M.2243: Assessment of the global mobile broadband deployments and forecasts for International Mobile Telecommunications (2011)
2. Cisco white paper: Cisco Visual Networking Index: Global Mobile Data Traffic Forecast Update, 2013–2018 (2014). http://www.cisco.com/c/en/us/solutions/collateral/serviceprovider/visual-networking-index-vni/white_paper_c11-520862.pdf
3. Lewin, D., Marks, P., Nicoletti, S.: Valuing the use of spectrum in the EU an independent assessment for GSMA (2013). http://plumconsulting.co.uk/pdfs/Plum_June2013_Economic_Value_of_spectrum_use_in_Europe.pdf
4. ECC Report 224: Long Term Vision for the UHF broadcasting band (2014)
5. The White House, President's Council of Advisors on Science and Technology (PCAST) Report: Realizing the Full Potential of Government-Held Spectrum to Spur Economic Growth (2012)
6. Draft RSPG report 14-585(rev1): RSPG Opinion on a long-term strategy on the future use of the UHF band (470-790 MHz) in the European Union (2014)
7. 3GPP technical report TR 36.808. Evolved Universal Terrestrial Radio Access (E-UTRA); Carrier Aggregation; Base Station (BS) radio transmission and reception (2012)
8. 3GPP technical specification TS 25.346: Introduction of the Multimedia Broadcast/Multicast Service (MBMS) in the Radio Access Network (RAN); Stage 2 (2011)
9. Mwangoka, P., Marques, P., Rodriguez, J.: Exploiting TV White Spaces in Europe: The COGEU Approach (2011). http://www.ictcogeu.eu/pdf/publications/Y2/IEEE%20DySPAN2011_COGEU_paper.pdf
10. Chapin, J., Lehr, W.: Cognitive radios for dynamic spectrum access – The path to market success for dynamic spectrum access technology. IEEE Commun. Mag. **45**(5), 96–103 (2007)
11. Yrjölä, S., Ahokangas, P., Matinmikko, M., Talmola, P.; Incentives for the key stakeholders in the hybrid use of the UHF broadcasting spectrum utilizing supplemental downlink: a dynamic capabilities view. In: International Conference on 5G for Ubiquitous Connectivity (2014)
12. Inayatullah, S.: Anticipatory action learning: Theory and practice. Futures **38**, 656–666 (2006)
13. Eisenhardt, K.M., Sull, D.M.: Strategy as simple rules. Harvard Business Review **79**(1), 107–116 (2001)
14. FCC Report 12-118: Broadcast Television Spectrum Incentive Auction NPRM (2012). http://www.fcc.gov/document/broadcast-televisionspectrum-incentive-auction-nprm
15. Draft ECC Report 205: Licensed Shared Access (2013)
16. Hansen, D., Shrader, R., Monllor, J.: Defragmenting Definitions of Entrepreneurial Opportunity. Journal of Small Business Mgmnt. **49**(2), 283–304 (2011)
17. West, J.: Value capture and value networks in open source vendor strategies. In: Proceedings of the 40th Annual Hawaii International Conference on System Sciences (2007)
18. Brandenburger, A., Nalebuff, B.: Co-opetition. Doubleday, New York (1998)
19. Porter, M.: The Five Competitive Forces That Shape Strategy. Harvard business Review (2008)
20. Prahalad, C.K., Hamel, G.: The core competence of the corporation. Harvard Business Review **68**(3), 79–91 (1990)

Spatial Spectrum Holes in TV Band: A Measurement in Beijing

Sai Huang$^{(\boxtimes)}$, Yajian Huang, Hao zhou, Zhiyong Feng, Yifan Zhang, and Ping Zhang

Key Laboratory of Universal Wireless Communications Ministry of Education, Wireless Technology Innovation Institute (WTI), Beijing University of Posts and Telecommunications, Beijing, China
huangsai@bupt.edu.cn

Abstract. Spatial spectrum holes are areas where TV signal strength falls below a certain threshold and TV frequency can be utilized without license. In our measurement, we prove the existence of spatial spectrum holes considering shadowing and building penetration. To evaluate the influence of shadowing, two dimensional radio environment mapping (REM) is constructed for a $500\,m \times 530\,m$ area in the downtown. According to the REM, a maximum attenuation of $30\,dB$ can be caused by building blockage and shadowing. To measure the loss of wall penetration, a three dimensional measurement is conducted in the outer and inner area of a 12-floor building. It is found that the wall attenuation approximately follows a normal distribution with a mean of $24.31\,dB$. The distribution of spatial spectrum holes is then plotted indicating spatial spectrum holes are abundant especially in the outskirts of Beijing.

Keywords: TV white space · Spectrum measurement · Radio environment mapping · Spatial spectrum access opportunities

1 Introduction

TV band is considered most suitable for spectrum relocation and measurements were conducted all over the world to check the feasibility of CR (cognitive radio) in TVWS (TV white space). Islam *et al.* in [1] reported the utilization of TV band is only 52.35% in Singapore while Bao *et al.* in [2] found the utilization of TV band is 54.78% in Vietnam. In these measurements, spectrum holes are considered as the time and frequency on which TV tower is not transmitting. However apart from time and frequency domain spectrum holes, there are also spectrum holes in the spatial domain. According to FCC regulation, devices can utilize the TV band in areas where TV signal strength falls below -114 dBm even if TV tower is transmitting [3]. Since the aforementioned measurements are

This work was supported by the National Natural Science Foundation of China (61227801, 61421061), the National Key Technology R&D Program of China (2014ZX03001027-003).

© Institute for Computer Sciences, Social Informatics and Telecommunications Engineering 2015
M. Weichold et al. (Eds.): CROWNCOM 2015, LNICST 156, pp. 585–592, 2015.
DOI: 10.1007/978-3-319-24540-9_48

all conducted at fixed locations with good TV signal coverage, they are unable to reveal the geographical aspects of spectrum holes.

Chen et al. in [4] conducted a mobile measurement in Beijing and predicted the signal strength with large-scale propagation model. They found the TV signal strength is above -85 dBm all over Beijing. Although there are no large spatial spectrum holes according to Chen, there may be small outdoor spatial spectrum holes because of shadowing and indoor spatial spectrum holes caused by wall attenuation.

This paper studies the existence of spatial spectrum holes in Beijing by measuring the typical shadowing and wall attenuation condition in the city. To evaluate the influence of shadowing, radio environment mapping (REM) is constructed for a $500m \times 530m$ area in the downtown. To measure the loss of wall penetration, three dimensional measurement is conducted in the outer and inner area of a 12-floor building. The wall attenuation is then calculated and approximated as a normal distribution. Utilizing the calculated value of shadowing and wall attenuation, the geographical distribution of spatial spectrum holes is plotted on the basis of the large-scale fading result of Chen.

The rest of the paper is organized as follows. Section 2 introduces the measurement of shadowing. The measurement of wall attenuation is presented in Section 3. Section 4 demonstrates the geographical distribution of spectrum holes and Section 5 concludes the paper.

2   Measurement of Shadowing

Signal strength may attenuate a great deal due to shadowing caused by building blockage. To investigate the effect of shadowing in downtown Beijing, REM is constructed for a typical area of the city.

2.1   Measurement Setting

The measurement region is a university located on the 3rd ring of Beijing and the coordinate is 116.348345E and 39.967108N. The shape of the area is almost rectangular with a length of 530 meters and width of 500 meters as illustrated in Fig. 1. The buildings in the area are mostly between 6-floor and 15-floor in height, which is typical of downtown Beijing. The university has two main landscapes, i.e., dense residential area in the north and open playground in the south (see Fig. 1).

In China, the $470 - 806$ MHz spectrum band is allocated for terrestrial TV broadcasting and the bandwidth of each TV channel is 8 MHz. As measurement needs to be taken at a large number of sampling points in REM, only channel 22 ($478 - 486$MHz) was measured. Channel 22 is utilized by digital TV broadcasting and is transmitting in the entire measurement period.

Measurement equipment includes Anritsu MS2720T handheld spectrum analyzer and omnidirectional broadband antenna BOGER DA753G. The antenna is connected to the spectrum analyzer by a low-loss cable and kept three meters above the ground in the measurement. The resolution bandwidth (RBW) of the spectrum analyzer is set as 200 kHz. Signal strength is measured at each position 200 times and averaged to eliminate the fluctuations caused over time. Moreover, the Global Positioning System (GPS) module inside MS2720T is utilized to ensure that measurement is taken at the pre-selected positions.

2.2 Simulated Annealing Assisted Electron Repulsion

Generally, REM is conducted in two steps, i.e., sampling and interpolation. Signal strength is first measured at a number of sampling points and then interpolation is utilized to estimate signal strength of the entire area. The most commonly utilized sampling algorithm is symmetric sampling which places sampling points on a uniform grid. In our scenario, symmetric sampling usually put sample points in inaccessible areas, for example inside buildings, as illustrated by the yellow dots in Fig. 1. Therefore a novel sampling algorithm named simulated annealing assisted electron repulsion (SAER) is proposed.

In SAER, the sampling points and boundaries of inaccessible areas are treated as electrons such that the repulsive force between them can ensure the uniform distribution of sampling points and keep them out of inaccessible areas simultaneously. Initially, M fixed electrons are put uniformly on the boundaries of inaccessible areas and N free electrons are deployed in the sampling area at random positions. There are $M + N - 1$ Coulombic forces on each electron. The resultant of Coulombic forces $\mathbf{F}_i$ on the ith electron e_i (at position $\mathbf{r_i}$) is

$$\mathbf{F}_i\left(\mathbf{r}\right) = \xi \cdot Q_i \cdot \sum_{j=1, j \neq i}^{N+M} Q_j \frac{(\mathbf{r_i} - \mathbf{r_j})}{\|\mathbf{r_i} - \mathbf{r_j}\|^3}, \tag{1}$$

where ξ is the Coulombic constant, Q_i and Q_j are the quantities of e_i and e_j respectively. Assuming the Coulombic forces on each electron is constant in a short interval Δt, the displacement of the electron is

$$\mathbf{L_i} = \mathbf{v}_i \cdot \Delta t + \frac{1}{2} \cdot \mathbf{a}_i \cdot \Delta t^2, \tag{2}$$

where $\mathbf{a}_i$ is the acceleration caused by the Coulombic forces. Utilizing (2), the position of the ith electron after Δt can be calculated. The displacement process can be carried out iteratively until the resultant forces approach zero, which suggests the sampling positions reach a relatively stable distribution.

In addition, simulated annealing is adopted to prevent the sampling position from converging to local optimal. The objective function is chosen as

$$\Phi(S) = \Delta F_i^{k+1} = \left\|\mathbf{F}_i^{k+1}\right\| - \left\|\mathbf{F}_i^k\right\|, \tag{3}$$

where $\mathbf{F}_i^k$ and $\mathbf{F}_i^{k+1}$ are the resultant of Coulombic forces on i^{th} electron in the kth and $(k+1)^{th}$ iteration. The probability of accepting a new state is given by

$$P\left(S_i \rightarrow S_{i+1}\right) = \begin{cases} 1 & \Phi(S_{i+1}) \leq \Phi(S_i) \\ e^{\frac{\Phi(S_i) - \Phi(S_{i+1})}{T}} & \Phi(S_{i+1}) > \Phi(S_i), \end{cases} \tag{4}$$

where S_i and S_{i+1} denote the former and current positions of the electron respectively. T is the system temperature which goes to zero. In an iteration, if the new state is accepted, the electron is moved to position S_{i+1}. Otherwise, the position of the electron stays unchanged.

In Fig. 1, the sampling points generated using SAER is marked by red dots. It can be clearly recognized that all the positions are placed outside the inaccessible area and the distribution is approximately uniform.

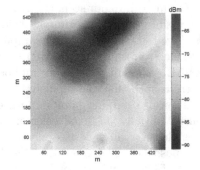

Fig. 1. Sampling Points Distribution in REM   **Fig. 2.** REM of the Selected Area

2.3 Results Analysis

After measuring signal power at 64 sampling points, Kriging interpolation is used to estimate the signal strength of the entire region. To validate the accuracy of our REM, signal strength of 80 randomly chosen positions (verification points) are measured. Mean relatively error (MRE) is adopted to assess the accuracy and can be defined as follows.

$$MRE = \frac{1}{N} \sum_{n=1}^{N} \frac{|P_n^* - P_n|}{P_n}, \tag{5}$$

where P_n^* is estimated signal strength at the the nth verification point and P_n is the measured value. The MREs of SAER and symmetric sampling are 4.72% and 6.58% respectively. Since some sampling points generated by symmetric sampling are inaccessible, we replaced them with their nearest accessible points.

The resultant REM is plotted in Fig. 2. The maximum signal strength comes from the southeast corner of the university (−60 dBm), where a playground locates. The northern part has many tall buildings and experiences the minimum signal power (approximately −95 dBm). It is shown that the signal strength fluctuates between −95 dBm and −60 dBm. If there is no shadowing, signal strength of the entire area would be approximately the maximum value (−60 dBm). Because of shadowing of different degrees, signal strength is distributed in a wide range. With strong shadowing, the variation of signal strength can be more than 30 dB even in a small area.

3   Measurement of Building Penetration Loss

Although signal strength estimated by large-scale measurement is much higher than the threshold defined by FCC, indoor spatial spectrum holes may appear owing to building penetration loss. To figure out the influence of building penetration loss, a three dimensional measurement is conducted for a typical building in Beijing.

3.1   Measurement Settings

The measurement is conducted in a 12-floor masonry building located in the university mentioned in the previous measurement. The building mostly consists of concrete, brick and coated windows, which is typical for northern China. The shape of the building is irregular as illustrated in Fig. 3.

Since a number of positions need to be measured, only channel 22 is measured. Measurement equipment and their connection are also similar to the previous section. Since GPS fails to work in the inner area of the building, the measurement positions are acquired with the help of a tachometer. The measurement equipment is kept in a chart to move around.

To limit the complexity of the measurement, we choose the 4th, 7th, 9th and 12th floor and the surface of building to conduct our measurement. The surface of building has the strongest signal strength since it is free from wall attenuation. The 4th, 7th, 9th and 12th floors are the inner area of the building.

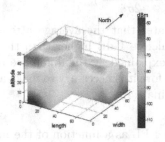   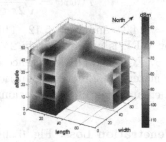

Fig. 3. Outdoor Signal Strength   **Fig. 4.** Indoor Signal Strength

3.2   Results Analysis

With the recorded signal strength at the sampling points, Kriging interpolation is utilized to estimate signal strength in entire area. Another 180 randomly chosen locations is also measured to verify the accuracy of our estimation.

The estimated signal strength for the outer and inner area of the building are plotted in Fig. 3 and Fig. 4. The MREs for both the inner and outer area are all 5.9%, indicating our estimation is reliable.

Fig. 3 shows that the signal strength of the southwest part is higher than the northeast. This is because the southwest part is facing radiation and it further validates that shadowing aggravates the attenuation of signal strength. The signal strength on the roof (−60 dBm) is 45 dB higher than the bottom.

From the vertical view, signal strength is strongly related with the height or floor of the measurement positions. Generally speaking, signal strength increases with height. From the horizontal view, signal strength of innermost spots (−120 dBm) is 50 dBm lower than the outermost ones.

3.3   Analysis of the Penetration Loss

The existence of spatial spectrum holes is strongly related to the penetration loss. Therefore We analyze two kinds of penetration loss, i.e., wall penetration loss (WPL) and floor penetration loss (FPL) and denote them as l_{wpl} and l_{fpl}.

Wall Penetration Loss. Fig. 5 shows the statistical histogram of the WPL, which resembles a normal distribution. To assess the normality of the WPL, Kolmogorov-Smirnov(K-S) test is conducted at significance level ($\alpha = 0.02$), wherein the empirical PDF is compared to a normal PDF with the mean and standard deviation estimated from the measured data. As the fitting curve indicates, the WPL can be well approximated by a Gaussian distribution expressed by (6).

$$p(l_{wpl}) = \frac{1}{\sqrt{2\pi\sigma^2}} \exp(-\frac{(l_{wpl} - u)^2}{2\sigma^2}), \tag{6}$$

where $u = 24.31$ and $\sigma = 5.12$. This indicates that the average attenuation caused by WPL is approximately 24 dB, which is 1 dB higher than the value reported by FCC [3]. This difference can be explained by the difference in construction material in different countries.Therefore it is necessary to make adjustments when applying FCC regulations in China.

Floor Penetration Loss. Fig. 6 shows the FPL as a function of the number of ceilings between the measurement site and the roof. It can be seen that the FPL increases almost linearly with the number of floors. Hence a linear model for FPL is assumed as follows.

$$l_{fpl} = l_o + \kappa \cdot n_{FL} + \xi, \tag{7}$$

where l_o is the initial value and κ is the increase in loss per floor, n_{FL} is the number of ceilings signal passes through and ξ is the statistical variation.

Linear least-square regression is utilized to calculate l_o and κ. The resultant l_o and κ are 5.31 dB and 15.49 dB respectively. The goodness of fit measured by Root Mean Square Error (RMSE) is 0.4413. From our analysis, a 20.8 dB loss exists when signal penetrates the ceiling between two floors. Our result is 2.5 dB higher compared to COST231 [5], which can be explained by the fact that the interlayer may be thicker in Beijing.

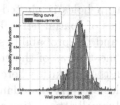

Fig. 5. Cumulative Distribution Function of WPL

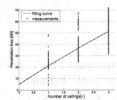

Signal strength (dBm)	Probability distribution of spatial holes
(-∞,-90)	100%
(-90,-85)	100%
(-85,-80)	100%
(-80,-75)	97%-100%
(-75,-70)	80%-97%
(-70,-65)	46%-80%
(-65,+∞)	0%-46%

Fig. 6. Model of FPL for Different Number of Floors   **Fig. 7.** The probability distribution of spatial holes in Beijing

4   Distribution of Spatial Spectrum Holes

Chen *et al.* conducted a mobile measurement to check the geographical distribution of TV signal strength. However, their results consider only large scale fading. As shadowing and wall attenuation also contribute to loss in signal strength, spatial spectrum holes may appear. Since shadowing and wall attenuation loss are all random variables, the probability that spatial holes exist can be calculated as follows.

$$P_{holes} = \int_{p-\varepsilon-l_s}^{+\infty} \int_0^{+\infty} f(l_{wpl}, l_s)dl_{wpl}dl_s, \tag{8}$$

where $f(l_{wpl}, l_s)$ is the joint probability density function of the WPL and shadowing, while p is the signal strength predicted by large scale fading and ε is the threshold for spatial spectrum hole. According to FCC regulation, ε is -114

dBm. However, the minimum working power level for digital TV is −90 dBm according to Chinese standard [6]. Analog TV makes even looser requirement about working power level. The −114 dBm threshold required by FCC should be relaxed in China.

Since there is no official regulation about the threshold in China, a −95 dBm threshold is assumed. As shadowing is tightly coupled with the blockage condition and its distribution is relatively complex, a fixed shadowing loss of 5dB is adopted. In this case, the probability of spatial spectrum holes is determined by the distribution of wall attenuation.

The probability distribution of spatial spectrum holes is plotted in Fig. 7, with different colors denoting different probabilities of spatial holes. In the central area of the city, TV signal strength is strong and the probabilities of spatial spectrum hole is low (the blue areas). However, most areas of the city have at least 80% probability of spatial spectrum holes. In the outskirt, where the signal strength is below −75 dBm, spatial spectrum holes are very abundant.

The spatial spectrum holes are mostly caused by wall attenuation and appear in indoor areas. Therefore we call them indoor spatial spectrum holes, which can be used by the short-range communication such as Wi-Fi or dense small cell deployment in 5G.

5   Conclusion

In this paper, we conducted a comprehensive measurement in TV band in Beijing to validate the existence of spectrum holes in the spatial domain. With radio environment mapping, we show that a variation of 30 dB in signal strength can be caused by building blockage and shadowing. By measuring the inner and outer area of a typical building, we find that the wall penetration loss obeys normal distribution with a mean of 24.31 dB. As a result, the geographical distribution of spatial spectrum holes is plotted indicating spatial spectrum holes are abundant in the outskirts of city. Spatial spectrum holes is very important since it can be utilized by short-range communication such as Wi-Fi or dense small cell deployment in 5G.

References

1. Islam, M.H., et al.: Spectrum survey in Singapore: occupancy measurements and analyses. In: Proc. 3rd Int. Conf. Crowncom, pp. 1–7 (2008)
2. Bao, V.N.Q., et al.: Vietnam spectrum occupancy measurements and analysis for cognitive radio applications. In: ATC, pp. 135–143 (2011)
3. FCC: Second report and order and memorandum opinion and order, pp. 8–260 (2008)
4. Chen, K., et al.: Spectrum Survey for TV Band in Beijing. In: 21st ICT, pp. 267–271 (2014)
5. COST 231 Final Report: Digital Mobile Radio Towards Future Generation Systems. COST Telecom Secertariat, Brussels (1999)
6. GB/T 26686–2011: General specification for digital terrestrial television receiver. Chinese Standards Press (2011)

TV White Space Availability in Libya

Anas Abognah$^{(\boxtimes)}$ and Otman Basir

University of Waterloo, Waterloo, ON N2L3G1, Canada
{aabognah,obasir}@uwaterloo.ca

Abstract. TV White Space (TVWS) has gained much attention recently as a potential new frontier for spectrum-hungry wireless applications. The impact of TVWS technology depends on the availability and status of TVWS spectrum. Multiple studies have been conducted to assess the availability of TVWS spectrum in many countries and regions. In this paper, we present the results obtained on assessing the availability of TVWS spectrum in Libya. Radio propagation analysis was conducted to predicted coverage of TV stations and a threshold-based method based on the minimum field strength was adopted to determine TVWS availability. A minimum of 58% of channels in the UHF TV spectrum were found to be available on average in the most populated region of the country. These results promise great potentials for TVWS technology in Libya.

Keywords: TV White Space · Spectrum management · Spectrum sharing

1 Introduction

The communication spectrum is a scarce national resource of which effective management and regulation is crucial for the development of any nation's economy. Sound long-term spectrum planning and management policies are key to maximize the benefits to the national economy and ensures healthy growth of the ICT sector and infrastructure. In this context, TVWS represents a new challenge to spectrum management while at the same time promising to enable an array of new technologies and applications to support better utilization of the communication spectrum and achieve positive economic return on the national scale.

1.1 What is TVWS?

TV White Space refers to the underutilized frequency bands in the VHF and UHF frequency rage that are traditionally allocated to terrestrial TV broadcast services. The size and the state of these bands vary widely from country to country and from region to region. In most regions of the world much of the white space is in the range of 47–790MHz. The term white space originally referred to the guard bands deliberately left blank between analogue TV channels to avoid interference. Applications utilizing these bands (called secondary users) must not interferer with the existing TV broadcasts and other applications licensed in the area (primary users).

1.2 Why TVWS?

Harvesting unused TVWS is of great importance because of the superior properties of the VHF and UHF bands. Devices transmitting in the VHF and UHF range require lower power and the signal can travel longer ranges compared to the higher frequencies such as the 2.5 GHz used for WiMax for instance. Current standards implementing TVWS and Cognitive Radio such as the IEEE802.22 WRAN promise to provide coverage from a single base station for up to 30-100km [11]. In addition, VHF and UHF frequencies have superior propagation properties both indoor and outdoor where no line-of-sight is required for good reception. These characteristics make TVWS technologies a good candidate to provide wireless information services to rural areas with minimum infrastructure investments.

1.3 TVWS in Libya

Libyan regulators are yet to adopt regulations governing the use and deployment of TVWS technology and devices in the country. Radio regulation in Libya is closely aligned with the ITU global and regional policies and regulations. Libya is in ITU-R Region-1 encompassing Europe, Africa, and much of Asia. Therefore, any future TVWS regulations and policies will be closely similar to that of ITU-R Region-1 countries. In 2013, the Libyan Ministry of Communications and Informatics (MCI) released the Libyan National Frequency Plan (LNFP) for public consultations and comments on future directions for national spectrum planning including the topics of spectrum sharing, cognitive radio, and TVWS [4]. According to the LNFP, the spectrum allocations in the VHF range from 47MHz to 230MHZ and the UHF range from 470MHz to 790MHz assigned for broadcasting services are shown in figure 1.

It can be seen from this figure that TV broadcast bands are more abundant in the UHF range than VHF range. Moreover, the bands in the VHF are more fragmented which adds more complexity to spectrum management and decreases the potential benefits of these bands. This analysis is consistent with ITU-R region 1 countries including the EU countries where more focus is given to the UHF bands as the primary bands for TVWS. As such, any future implementation of TVWS systems in Libya will be similar to that in Europe and will most likely focus on exploiting white space in the UHF bands. However, the large geographical area of Libya and the low population density may allow for VHF bands utilization in TVWS thus increasing the potentials of TVWS technology in providing wireless access to rural areas throughout the country and help in closing the digital divide. Extensive studies are needed to assess the actual availability of TVWS bands and the potentials of TVWS technologies. This work is a contribution in this direction.

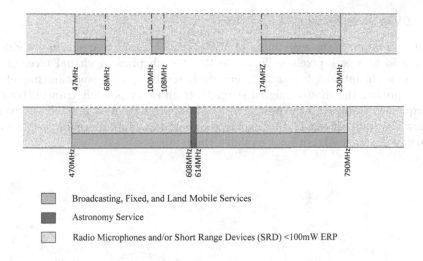

Fig. 1. LNFP in the 47-230MHz VHD and the 470-790MHz UHF ranges

2   Related Work

TVWS estimation studies have been conducted in various parts of the world. In [3] the authors study the availability of TVWS in 11 European countries using the ITU-R P.1456 and the Longley-Rice (ITM) models. They concluded that almost 56% of the 470MHz-790MHz UHF bands are available with only co-channel protection. This number drops to 25% if adjacent channel restrictions are applied. The study only considered transmissions from TV stations. In [2] the authors present their results for TVWS availability in Italy and conclude that TVWS in Italy is available mostly in areas with light population density. A threshold-based approach is used in this study. Further studies are conducted for TVWS availability in India [9], Australia [5], and Japan [10]. All of these studies confirm the potential of TVWS technology in providing additional spectrum for existing and new applications but with varying degrees in each country.

To the author's best of knowledge, there has been no previous attempt to estimate the availability of TVWS in Libya. This work aims to close this gap by estimating TVWS availability in various test regions in Libya using a threshold-based approach.

3   TVWS Estimation

Estimating the availability of TVWS depends heavily in regulatory-set parameters and rules that define the protection regions for TV transmitters and received power threshold for acceptable TV reception for example. Different methods are adopted by different regulators. For this study, we adopt a threshold-based model and design an algorithm to estimate the availability of TVWS.

3.1 TVWS Estimation Algorithm

In order to compute the availability of TVWS the designated area is first divided into a grid of k x k pixels as in figure 2. At each pixel, a virtual receiver is assumed with antenna height h_{rx}. For each receiver, a propagation model is used to predict the received signal strength from every possible transmitter on all frequency channels. If the predicted Field Strength from all possible receivers is below a specific threshold, the pixel is designated as white space for the current frequency channel. Algorithm 1 explains these steps.

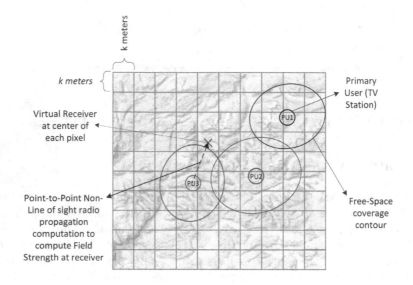

Fig. 2. TVWS Estimation

3.2 Point-to-Point Radio Propagation Computation

The Irregular Terrain Model (also called the Longley Rice model) [8] and The ITU-P.1546 [6] propagation models where both used to predict the received Field Strength (dBuV/m) at each receiver in the test area from each possible transmitter. The ITM model is a radio propagation module for frequencies from 20MHz-20GHz developed by A.G. Longley and P. L. Rice in the 1960s. The module incorporates statistical signal analysis with terrain information to predict median field strength (dBuV/m). An-open source tool called SPALT [7] was used to perform the path loss computation based on the Irregular Terrain Model (ITM).

The second model used is the ITU-P.1546 Propagation Model. This model is published by the International Telecommunications Union (ITU) and is recommended for use for terrestrial TV broadcast coverage prediction for frequencies

---

Algorithm 1. TVWS Estimation Algorithm

Data: Receivers, Transmitters, Frequencies, DTM maps
Result: TVWS Availability
initialization;
for *All Frequencies* do
   for *all Receivers* do
      for *all Transmitters* do
         Calculate received power at current receiver location;
         if *Field Strength $>=$ Threshold* then
            Not white space;
            Go to next receiver;
         end
         Designate pixel as white space;
      end
   end
end

---

in the range of 30MHz-3GHz. The model is based on a set of field-strength curves for different frequencies, antenna height, and distances with equations for interpolation and extrapolation for a range of these values. A MATLAB program was used for the ITU model calculation.

3.3 Field Strength Threshold

The Field Strength at the receiver is compared to a thresholds based on the minimum acceptable location probability. Location probability in TV broadcast planning represents the probability of successful signal reception in a small area (usually 100m by 100m). This probability is critical in determining the coverage area of each transmitter and thus defines the area of service. Table 1 shows different location probabilities for different service types with corresponding minimum field strength values of the received signal based on 2006 ITU Geneva agreement (GE06)[1].

Table 1. Location probabilities and field strengths for different service types

Location Probability	Minimum Field Strength	Service Type
99%	60 dBuV/m	Mobile TV
95%	56 dBuV/m	Fixed Digital TV
50%	48 dBuV/m	Fixed Analogue TV

4 Results

TVWS availability in Libya was estimated using TV transmitters' data obtained from the ITU databases and the regulator in Libya. The data included 343 TV

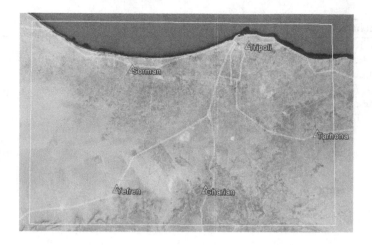

Fig. 3. Test Area 1 (5 TV stations)

stations in the UHF range from 470MHz to 790MHz providing TV coverage for the towns and cities of Libya. The assessment was conducted on two test areas:

1. Test Area 1

    A smaller area encompassing the capital city of Tripoli and the western mountains of "Jabal Nafosa". This area is of the most populated in the country and features interesting mix of terrain from coastal urban environment to mountainous rural terrain. The area contains five sites for primary analogue TV stations each consisting of multiple transmitters in the UHF range (470MHz-79MHz). Figure 3 shows a map of this area.

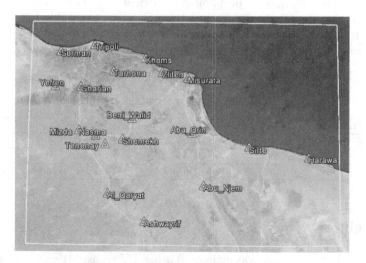

Fig. 4. Test Area 2 (19 TV stations)

2. Test Area 2

A larger area covering the entire western coastal area of Libya. This area covers multiple large cities (Tripoli, Misurata, Azzawia, ...) were existing TV coverage may limit the availability of TVWS but also includes rural areas and smaller towns were TVWS may be abundant. Figure 4 shows a map of this area.

The test areas were divided into a grid of pixels of 1000m by 1000m each. A receiver is assumed at each pixel and the received signal strength in term of Field Strength (dBuV/m) was measured from each transmitter in the area. If the maximum signal strength is below a certain threshold, the pixel is deemed as white space. The process is repeated for each frequency. The Irregular Terrain Model (ITM) and the ITU-P.1546 propagation models were used to predict the received field strength at each location in the test area. Below are the results obtained using each of the two models.

4.1 Irregular Terrain Model Results

Table 2 shows the percentage of total available channels in the UHF range of 470MHz to 790MHz in test area 1 at different threshold level. Each threshold level corresponds to the minimum field strength for different primary service below which reception of the service becomes unacceptable. For instance, field

Table 2. TVWS availability results with Longley-Rice (ITM) propagation model

Primary service type	Threshold (dBuV/m)	TVWS availability (%)	Average spectrum available (MHz)
Fixed Analog TV	48	58.37	163.27
Fixed Digital TV	56	62.89	176.28
Mobile TV	60	65.05	183.17

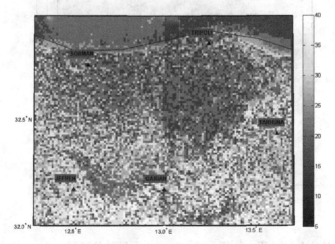

Fig. 5. TVWS heatmap of Tripoli area using Longley-Rice (ITM) model

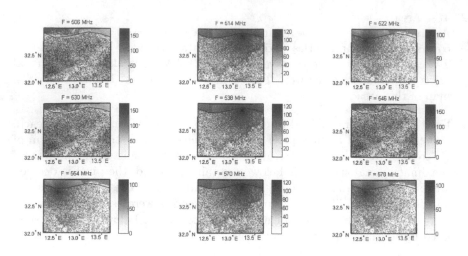

Fig. 6. Heatmaps of field strength values for different channels (ITM model)

Table 3. TVWS availability results with ITU-P.1546 propagation model

Primary service type	Threshold (dBuV/m)	TVWS availability (%)	Average spectrum available (MHz)
Fixed Analog TV	48	60.63	194.02
Fixed Digital TV	56	69.49	222.35
Mobile TV	60	74.22	237.50

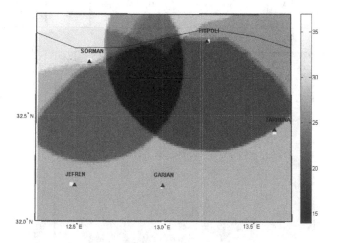

Fig. 7. TVWS Availability in Tripoli area using ITU-P.1546 propagation model

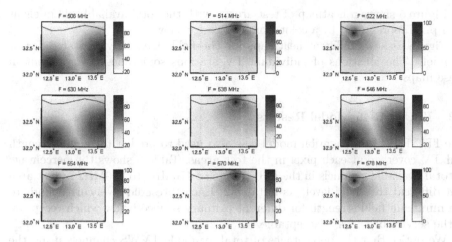

Fig. 8. Heatmaps of field strength values for different channels (ITU-P.1546 model)

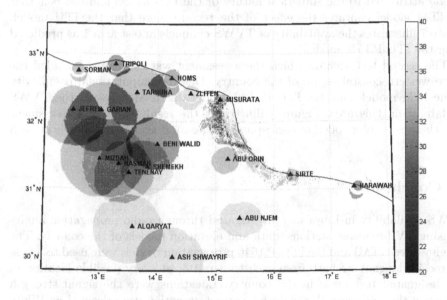

Fig. 9. TVWS availability in the western coastal area of Libya (ITU-P.1546 model used)

strength at a receiver for analog TV below 48 dBuV/m renders the service unacceptable. As the threshold level increases (when going from fixed to mobile service for instance), the percentage of available TV channels for secondary use will also increase. This is a direct result of the fact that increasing the threshold level corresponds to decreasing the protection margin of the primary service, thus resulting in more channels being available for secondary use.

Figure 5 shows a heatmap of test area 1 with the total available channels at each pixel represented by a color from the range shown.

Figure 6 shows different heatmaps for the field strength values at different channels. The locations of individual TV stations sometimes are apparent in these maps.

4.2 ITU-P.1546 Model Results

The ITU-P.1546 propagation model was also used to predict the signal trength and TV coverage at each pixel in the test areas. Table 3 shows the percentage of total available channels in the UHF range of 470MHz to 790MHz in test area 1 at different threshold level. As in table 2, Each threshold level corresponds to the minimum field strength for different primary service below which reception of the service becomes unacceptable.

We notice that the percentages of total available TVWS channels using the ITU-P.1546 model are higher than those obtained with the ITM model. This can be attributed to the statistical nature of the ITU model and the fact that the ITM model captures the effect of the terrain more that the ITU model. Figure 7 illustrates the availability of TVWS channels in test area 1 as predicted using the ITU-P.1546 model.

The second test area on which the assessment was conducted included the entire western coastal region of the country. Due to computational complexity of the ITM model, only the ITU-P.1546 model was used to assess the TVWS availability in this area. Figure 9 illustrates the results from the assessment with the pixels color coded to correspond to the total available channels at each location.

5 Conclusion

TVWS availability in Libya was investigated through radio propagation analysis using TV broadcast stations' data and elevation models of the country. The Longley-Rice (ITM) and the ITU-P.1546 propagation models were used to assess the received signal strength from each transmitter at every possible location in the designated test areas in the country. Locations were the signal strength fell below the minimum threshold for correct reception were deemed as White Space and were designated for use by secondary applications. For all the tests performed in the most populated areas of the country no less than an average 58.37% of total channels were available over all locations with minimum average available spectrum of 163.27MHz. These results are highly encouraging for future implementations of TVWS technology in Libya. While these results do not consider neighboring channels protection and protection of other primary services in the TV spectrum such as wireless microphones, it is worth mentioning that the TV transmitters data used are highly conservative and include all stations registered by the Libyan regulator with the ITU notification database according to the GE06 agreement. This data does not include operational status information for these stations, and since Libya is going through the transmission from analog to digital TV, most of these analog stations are off-line. in addition, satellite TV remains the most popular TV service in the country and a small portion of the population (if any) rely on terrestrial TV services. All of these factors combined with the obtained results in this study assert the potential of future TVWS applications in Libya.

References

1. ITU, GE06 Agreement. http://www.itu.int/ITU-R/terrestrial/broadcast/plans/ge06/
2. Barbiroli, M., Carciofi, C., Guiducci, D., Petrini, V.: White spaces potentially available in Italian scenarios based on the geo-location database approach. In: 2012 IEEE International Symposium on Dynamic Spectrum Access Networks (DYSPAN), pp. 416–421, October 2012
3. Van de Beek, J., Riihijarvi, J., Achtzehn, A., Mahonen, P.: Tv White Space in Europe. IEEE Transactions on Mobile Computing 11(2), 178–188 (2012)
4. The Libyan Ministry of Communications and Informatics: Libyan National Frequency Plan (LNFP). Tech. rep., Tripoli, Libya (2013). http://cim.gov.ly/uploads/LNFP_PC_english.pdf
5. Freyens, B., Loney, M.: Opportunities for white space usage in Australia. In: 2011 2nd International Conference on Wireless Communication, Vehicular Technology, Information Theory and Aerospace Electronic Systems Technology (Wireless VITAE), pp. 1–5, February 2011
6. ITU: Method for point-to-area predictions for terrestrial services in the frequency range 30 MHz to 3000 MHz. Tech. rep. (2013). http://www.itu.int/rec/R-REC-P.1546/en
7. Magliacane, J.A.: K: SPLAT. http://www.qsl.net/kd2bd/splat.html
8. Longley, A.G., Rice, P.L.: Prediction of tropospheric radio transmission loss over irregular terrain. A computer method. Tech. rep., Institute for Telecommunication Sciences (1968)
9. Shendkar, T., Simunic, D., Prasad, R.: TVWS opportunities and regulatory aspects in India. In: 2011 14th International Symposium on Wireless Personal Multimedia Communications (WPMC), pp. 1–5, October 2011
10. Shimomura, T., Oyama, T., Seki, H.: Analysis of TV white space availability in Japan. In: 2012 IEEE Vehicular Technology Conference (VTC Fall), pp. 1–5, September 2012
11. Stevenson, C., Chouinard, G., Lei, Z., Hu, W., Shellhammer, S., Caldwell, W.: IEEE 802.22: The first cognitive radio wireless regional area network standard. IEEE Communications Magazine 47(1), 130–138 (2009)

Emerging Applications for Cognitive Networks

Cognitive Aware Interference Mitigation Scheme for LTE Femtocells

Ismail AlQerm$^{(\boxtimes)}$ and Basem Shihada

King Abdullah University of Science and Technology, Thuwal, Saudi Arabia
{ismail.qerm,basem.shihada}@kaust.edu.sa

Abstract. Femto-cells deployment in today's cellular networks came into practice to fulfill the increasing demand for data services. However, interference to other femto and macro-cells users remains an unresolved challenge. In this paper, we propose an interference mitigation scheme to control the cross-tier interference caused by femto-cells to the macro users and the co-tier interference among femtocells. Cognitive radio spectrum sensing capability is utilized to determine the non-occupied channels or the ones that cause minimal interference to the macro users. An awareness based channel allocation scheme is developed with the assistance of the graph-coloring algorithm to assign channels to the femto-cells base stations with power optimization, minimal interference, maximum throughput, and maximum spectrum efficiency. In addition, the scheme exploits negotiation capability to match traffic load and QoS with the channel capacity, and to maintain efficient utilization of the available channels.

Keywords: Cognitive radio · Femtocells · Macro users · Radio channels · Cross-tier interference · Co-tier interference

1 Introduction

The lack of resources of the macro-cells networks makes them unable to fulfill the data services demand in the indoor areas. An efficient solution is to deploy femto-cells base stations (FBSs) which are capable of communicating users over a broadband wire-line connection [1]. FBSs are short-range, low-cost/low power and can be easily installed by the users in addition to the fact that they reduce the load of the Macro Base Stations (MBSs). However, interference is considered as a technical challenge that affects the femto-cells deployment [1]. There are two types of interference: cross-tier and co-tier. Cross tier is the interference to the macro users caused by the FBSs installed within the same sub-band (SB). Co-tier interference is the one among the deployed femto-cells contending for the same channel. These types of interference lead to service disruption, throughput degradation and connection droppings.

There are several proposed schemes for resource allocation in femto-cells deployments with interference consideration. For example, the schemes proposed

© Institute for Computer Sciences, Social Informatics and Telecommunications Engineering 2015
M. Weichold et al. (Eds.): CROWNCOM 2015, LNICST 156, pp. 607–619, 2015.
DOI: 10.1007/978-3-319-24540-9_50

in [2] and [3] aim to handle both types of interference using uncoordinated and coordinated resource assignment algorithms as in [2], and Q-learning based interference coordination as in [3]. However, the coordination between MBSs and FBSs is difficult due to the requirements of scalability, security, and the availability of backhaul bandwidth in addition to the fact that the number of the deployed FBSs is not fixed. The authors in [4] and [5] propose a scheme that assigns dedicated channels for the communication of FBSs over the up-link (UL) and the down-link (DL). This goes against the idea of improving spectrum utilization by accessing the macro-cells spectrum opportunistically. Femto-cells resource allocation mechanisms are investigated in [6] and [7] to mitigate interference. These mechanisms use cognitive radio and game theory to support their resource allocation methodologies. However, both schemes are limited to channel allocation without consideration of QoS requirements and the scheme in [7] considers cross-tier interference only. The work in [8] aims to maximize the weighted sum rate of the femto-macro network in a delay tolerant scenario. However, this requires high information overhead among MBSs and FBSs. Fractional Frequency Reuse (FFR) technique proposed in [9] shows capability to mitigate interference in multiple cells deployments. However, the cell edge users suffer from lower data rates because of the increase in path-loss and interference [10]. The FFR strategy has been widely used in multi-macrocell environments for suppressing interference as in [11]. The use of cognitive radio [12] in femto-macro deployment effectively contributes in solving the cross-tier interference problem by exploiting its spectrum sensing capability to allocate under-utilized channels. In addition, it is considered for spectrum assignment in order to increase the flexibility and the autonomy of the network in addition to interference mitigation [13].

In this paper, we propose an interference mitigation scheme that aims to mitigate cross-tier interference caused by FBSs to the macro users and the co-tier interference that affects the FBSs that contend to access the free sub-channels in the DL. The scheme enhances spectrum sensing and improves detection capability to find free sub-channels for FBSs to access in which the cross-tier interference is minimal. An adaptive power graph coloring spectrum assignment algorithm is used in conjunction with environment awareness to allocate sub-channels for FBSs that mitigate the co-tier interference. In addition to interference control, the scheme ensures that the selected sub-channel satisfies QoS requirements and matches with the traffic load. Other advantages of the proposed scheme include considering traffic priorities, and maintaining efficient utilization of the available sub-channels which enhances the spectrum efficiency.

The paper is organized as follows, section 2 presents the system model and the interference sources. The interference mitigation scheme and the femto-cells sub-channel allocation mechanism are described in Section 3, while section 4 presents the performance evaluation of our scheme compared to others. Finally, the paper is concluded in section 5.

2   System Model

We consider the interference problem of the DL in a network that consists of macro-cells and femto-cells where the priority of sub-channel access is for the macro-cell users. FBSs can access the sub-channel but with minimal interference to the macro-cell users. Spectrum sensing is employed to detect the available sub-channels that FBSs can access. Spectrum occupancy information is used to allocate the free sub-channels according to the awareness based algorithm to be discussed in the next section. The considered deployment employs Orthogonal Frequency Division Multiple Access (OFDMA) as a channel access technique for the DL with M hexagonal grid macro-cells and F femto-cells in range of each macro-cell. The bandwidth allocated for each MBS is divided into 6 SBs using FFR. Each SB is composed of N_c sub-channels. The MBS can access any of these sub-channels at any time instant. However, the sub-channels are not utilized most of the time.

The femto-cells considered deployment is depicted in Fig. 1 where they are distributed randomly and uniformly in each SB. Each femto-cell is assumed to have variable number of users active at any time instant. The sub-channels are assumed to be almost static with minor variations and follow Rayleigh multi-path fading distribution. Femto-cells deal with two types of connections which are the link between femto-cells and macro-cells and the link between FBSs and their associated users. There are three types of gains considered in the signal propagation model and they contribute to the total channel gain calculated in (1). These gains include antenna's gain (A), shadowing gain (S) and path loss gain (G).

$$H = A + S + G \qquad (1)$$

The received signal to interference and noise ratio (SINR) of a macro-cell user k over sub-channel n is calculate as,

$$SINR_{k,n} = \frac{P_{k,n}H_{k,n}}{I_1 + I_2 + N_{n,k}} \qquad (2)$$

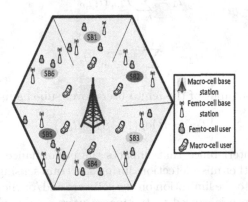

Fig. 1. Femto-cells network

where $P_{k,n}$ is the received power of the macro-user k over sub-channel n, I_1 and I_2 are the two interference imposed by the other MBSs and FBSs respectively and $N_{n,k}$ is the additive white Gaussian noise (AWGN) power. The two types of experienced interference by the macro user I_1 and I_2 are calculated according to (3) and (4) respectively.

$$I_1 = \sum_{l=1}^{M} P_{l,n} H_{k,l,n} \tag{3}$$

$$I_2 = \sum_{j=1}^{F} z^* P_{j,n} H_{k,j,n} \tag{4}$$

where $P_{l,n}$ and $P_{j,n}$ are the transmission powers of the other MBS and FBS over the nth sub-channel respectively and z^* is the factor that indicates if the sub-channel is assigned to a certain femto-cell. It takes a value of 1 if the sub-channel is assigned and 0 otherwise. l and j are the indexes of the MBSs and FBSs respectively. The achievable throughput by the macro user k over sub-channel n is given by,

$$T_{k,n} = B \log(1 + SINR_{k,n}) \tag{5}$$

where B is the sub-channel bandwidth.

Following a similar process, the SINR for a femto user i served over sub-channel n is calculated as,

$$SINR_{i,n} = \frac{P_{i,n} H_{i,n}}{I_1 + I_2 + N_{n,i}} \tag{6}$$

The interference imposed by MBS I_1 and the other FBSs interference I_2 are calculated according to (7) and (8) respectively.

$$I_1 = \sum_{l=1}^{M} P_{l,n} H_{i,l,n} \tag{7}$$

$$I_2 = \sum_{j=1, j \neq i}^{F} z^* P_{j,n} H_{i,j,n} \tag{8}$$

The achievable throughput of the femto user i over sub-channel n is given by,

$$T_{i,n} = B \log(1 + SINR_{i,n}) \tag{9}$$

The cross-tier interference that impacts the performance of the macro-cell users is caused by either miss detection during spectrum sensing or hidden macro users problem. Due to the limitation on the software and/or the hardware sensing capability, interference is caused to the macro users as a result of an incorrect detection. Note that the probability of miss detection depends on the sensing methods, (e.g., the energy detector, the cyclostationarity-feature sensing, and the matched-filtering sensing). Matched-filtering is known as the best approach for spectrum sensing as it maximizes the received SINR [14]. However, it is difficult since it requires dedicated receiver for each signal. The performance of energy detector is limited by the energy threshold and the types of signals. Besides, it fails when the noise becomes non-stationary because of the presence of the cross-tier interference. However, energy detector is the easiest to implement in actual systems. The hidden macro-cell users problem is similar to the hidden node problem in carrier sense multiple access (CSMA). It is caused by many factors including severe multi-path fading, shadowing, and high penetration loss in the areas sensed by femto-cells.

Contending between FBSs for channel access especially in the dense femto-cells deployment is the main reason to encounter co-tier interference that affects the performance of the femto users. Other problems like hidden terminal and exposed terminal problems also contribute to this interference as the femto-cells network is similar to other wireless network once free channels are detected. Adjacent channel interference is another type of interference that affects the macro users on the edge of cell. It is caused if different but adjacent channels are occupied by the macro users and FBSs respectively. However, this interference can be leased by reasonable layout of base stations deployment.

3   The Interference Mitigation Scheme

In this section, we describe the interference mitigation scheme to control both cross-tier and co-tier interference that impact femto-cells operate under the coverage of macro-cells. The scheme also aims to maximize throughput, allocate sub-channels that satisfy QoS requirements by assigning priorities for different types of traffic, and ensure efficient spectrum utilization. Spectrum sensing is exploited to support this scheme in order to mitigate cross-tier interference. Both interference mitigation mechanisms are detailed in the following sections.

3.1   Cognitive Based Cross-tier Interference Mitigation

Cognitive spectrum sensing is employed by the FBSs to determine whether certain SB include free sub-channels. This forces the FBSs to cease their channel access if the sub-channel is busy with macro user transmission. If all SBs are busy, the FBS tries to access the SB with the minimal interference to macro users. The presence of MBS transmissions is detected in the DL signal.

Our scheme implements an enhanced energy detection based spectrum sensing that effectively explores the interference range and maximizes the detection sensitivity. According to the energy detection approach, the signal observed by the FBS is expressed as,

$$y(x) = h(x)s(x) + w(x) \tag{10}$$

where $s(x)$ is the signal transmitted by the MBS, $h(x)$ is the channels gain from the MBS to FBS, $w(x)$ is the AWGN sample, and x is the sample index. The average received energy is given by,

$$Y(X) = \frac{1}{X} \sum_{x=0}^{X-1} |y(x)|^2 \tag{11}$$

where X is the total number of samples. Spectrum sensing aims to distinguish between the following two hypotheses,

$$H_0 : y(x) = w(x) \tag{12}$$

$$H_1 : y(x) = h(x)s(x) + w(x) \tag{13}$$

The hypothesis H_0 is for miss-detection and $H1$ is for correct detection. Energy detection is defined by two probabilities, the probability of detection P_D and the probability of false alarm P_F. The occupancy of sub-channels by macro-cell users can be determined by comparing the metric Y against a threshold λ. Therefore, the P_D is calculated as follows,

$$P_D = Pr(Y > \lambda|H_1) = Q_m(\sqrt{2 * SINR}, \sqrt{\lambda}) \tag{14}$$

where m is the product of time and bandwidth and $Q_m(.,.)$ is the generalized Marcum Q-function [15]. The P_F is calculated as follows,

$$P_F = Pr(Y > \lambda|H_0) = \frac{\Gamma(m, \lambda/2)}{\Gamma(m)} \tag{15}$$

where $\Gamma(.)$ and $\Gamma(.,.)$ are the complete and incomplete gamma functions, respectively [16]. Both probabilities are calculated for each SB as the product function of all sub-channels contained in each SB.

We enhance the normal energy detection procedure to improve the detection capability. The decision regrading channel access by FBS is determined by quantifying how harmful is the interference caused to the macro receivers if the FBS uses the sub-channel. The interference to the macro users is deemed to be harmful if it causes the signal-to-interference ratio (SIR) at the macro receiver to fall below a threshold SIR^*. This threshold depends on the macro receiver robustness toward interference and varies from one service to another. In addition, it may depend on the characteristics of the interfering signal (e.g., signal waveform, continuous versus intermittent interference, etc.) [17]. From the above definitions, we define the interference range of the FBS as the maximum distance from a macro receiver at which the incurred interference is still considered harmful. Consequently, the interference range depends on the macro user interference tolerance not just the FBS transmission power. Let P_m and P_f denote the transmission power of the MBS and the FBS respectively. The distance between the macro-cell transmitter and receiver is denoted as R. The interference range of the FBS (D) is determined according to,

$$\frac{P_m R^{-\alpha}}{P_f D^{-\alpha}} = SIR^* \tag{16}$$

where α is the path loss factor. We deduce from (16) that the macro receiver can tolerate the interference caused by FBS as long as the distance between them is greater than D. As a result, we can define the detection sensitivity (DS) as the minimum SNR of the MBS at which an FBS should be still capable of detecting the macro signal. The FBS should be able to detect active macro transmission within a radius of $R+D$ according to (16). Therefore, the sensitivity is calculated as follows,

$$DS = \frac{P_m(D+R)^{-\alpha}}{N_0} \qquad (17)$$

The spectrum sensing is conducted periodically to stay aware of any MBS starts to transmit. During the sensing period, the QoS degradation incurred by the macro users in accessing the band is determined. The choice of the sensing period depends on the type of the service running on the macro user terminal and has to be set for each SB. For example, the sensing period is less for services that vary over a much larger time scale.

3.2 Awareness Based Co-tier Interference Mitigation

In this section, we develop an awareness based algorithm to mitigate co-tier interference between femto-cells that are using the same SB. The algorithm aims to maximize system throughput, improve spectrum efficiency, and adapt transmission power according to FBS SINR requirements. It is assumed that each group of FBSs are assigned to certain SB and able to access its sub-channels. Each FBS is aware of its interference profile which is characterized by certain interference weight W_i. This weight is exploited to label the interference over the link between any two FBSs and is calculated as,

$$W_i = \sum_{e_i} 10^{I_i/10} \qquad (18)$$

where I_i is I_2 that is calculated using (4). Due to the relatively low distance between the FBS and its femto-cell associated user, the throughput maximization problem for a femtocell can be written as follows,

$$max \sum_{i \in V} \sum_{j \in V} \sum_{n \in N_c} z_i^* z_j^* B log(1 + \frac{P_i}{\sum_{j \neq i} P_j H_{ji} + N}) \qquad (19)$$

where V is the group of FBSs that share the same SB and $z*$ is the assignment indicator of the sub-channel, P_i is the transmission power of the FBS, H_{ji} is the total gain between FBS j and FBS i, and N is the corresponding noise power. Note that i and j are the indexes of the interfering FBSs.

The FBSs channel allocation problem can be modeled as a graph coloring problem with support of an awareness based mechanism between the FBSs sharing the same SB. The awareness mechanism aims to share the interference weight, data rate requirement, traffic type and traffic load information for each FBS among other FBSs sharing the same SB. Consequently, each FBS is aware of its network environment. Interference weight is the basic metric for the graph coloring channel assignment while data rate requirement, traffic type and traffic load are exploited to ensure QoS, assign traffic priority, and improve spectrum utilization. The sub-channel assignment process as in Algorithm 1 starts by mapping the network into a bidirectional graph $GR = (V, E, W)$ with a group of vertices $V = \{v1, v2...\}$ where each vertex v represents an FBS and a group of edges $E = \{e1, e2...\}$ where an edge e is the link connecting two vertices with interference weight w. A larger weight implies having a larger sum of the path loss and shadowing values. The problem is equivalent to coloring each vertex with one color from $C = \{c1, c2...\}$ and assign proper power level to the respective vertex in order to maximize the throughput and mitigate interference. The color of the vertices represents the available sub-channel, and a color pool of the interference graph relies on which particular SBs can be used by that graph. Once the graph is established, we label each FBS with the total interference

---

Algorithm 1.. Co-tier Channel Assignment

Require: $GR = (V, E, W)$, $C = \{c1, c2...\}$, FBS data rate, traffic load and traffic type
Ensure: Sub-channel assignment with maximum throughput T , maximum spectral
efficiency and minimum interference I
BEGIN
Define V_0 as the number of vertices in GR
Define i as the index for the FBS
for $(a = 1$ to $a = V_0)$ **do**
    Calculate W_i, Tp_i, $\forall V \in GR$
    if $(W_i = W_{max})$ **then**
        Select FBSs with maximum W
    end if
    if FBS with W_{max} is not unique **then**
        find FBS with $(Tp_i = Tp_{max})$
    end if
    Color the vertex v with color c (Assign the sub-channel to the FBS)
    Adapt transmission power as in (21)
end for
while (1) **do**
    Check U for each FBS
    if $(U < Th_{min})$ **then**
        Switch the users associated the FBS to other FBSs with condition that $(U \leq Th_{max})$ for the target FBSs
    end if
end while
END

---

weight calculated in (18). At this moment, the vertex with the highest interference weight is colored with an appropriate color with condition that the selected sub-channel satisfies application QoS. If there are more than one FBS with the same interference weight, we check the traffic priority. For example, if one FBS has real-time traffic or it is loaded more than others, it will have the priority to access the sub-channel. Traffic load and application data rate are used to quantify the traffic priority. An indicator Tp is assigned to the FBS to refer to the priority of its traffic. The indicator is ordered in ascending order as the traffic priority increases. The coloring process to maximize throughput can be characterized as follows,

$$argmax \sum_{i \in V} \sum_{n \in N_c} z^* B log(1 + SINR_{i,n}) \qquad (20)$$

$$s.t. \sum_{n \in N_c} z^* \leq 1, \forall i \in V$$

where N_c denotes the specific set of sub-channels, which are used by the vertices involved, $SINR_{i,n}$ is the SINR of femtocell i over the sub-channel n and V is a set of vertices in the graph. Then, the power is adapted for the FBS according to $SINR_{target}$ and the current transmission power as follows,

$$P_i^* = P_i \, SINR_{target} \frac{I_2 + N(i)}{H_{i,n}} \qquad (21)$$

where P_i is the current transmission power, $H_{i,n}$ is the channel gain experienced by FBS i while accessing sub-channel n and $N(i)$ is the corresponding noise power. Finally, the FBS is removed from the graph GR and the process repeated again until the set V is empty. In addition, the awareness mechanism manages to improve spectrum utilization by sharing the number of users associated with each FBS (U). If this number falls below certain threshold Th_{min}, all the users associated with this FBS are switched into another FBSs in the same domain with a condition that U for these FBSs is not exceeding certain threshold Th_{max} and QoS of the switched users is guaranteed. Consequently, the sub-channel is released for other FBSs and this improves spectrum utilization and reduces contention of other FBSs.

4   Performance Evaluation

In this section, the performance of our proposed cognitive aware interference mitigation scheme is evaluated through simulation. The simulation environment parameters are presented in Table 1. Note that d is the distance between the user and the base station.

We compare our scheme with the two schemes proposed in [18] and [19] for interference mitigation in femto-cell macro-cell deployment. In addition, we compare it with the standard scheme that does not implement any interference mitigation mechanism. The scheme proposed in [18] is cognitive based (CR-based) and aims to allocate resources in femto-cell networks to maximize the throughput while minimizing interference to macro-cell users nearby only. However, the scheme (femto-macro) proposed in [19] considers mitigating the

Table 1. Simulation environment parameters

Parameter	Value
Carrier frequency	2 GHz
Cellular layout	Hexagonal grid
Macro-cell radius	500 m
Femto-cell radius	10 m
Path loss MBS user	$L = 15.3 + 37.6 log(d)$
Path loss FBS user	$L = 38.46 + 20 log(d) + 0.7(d)$
Lognormal shadowing	0 mean, 8 dB standard deviation
MBS transmission power	45 dBm
FBS transmission power	20 dBm
White noise power density	- 174 (dBm.Hz^{-1})
Number of SBs	6
Number of sub-channels per SB	10
Macro-user per macro-cell	30
The penetration loss of walls Lw	10 dB

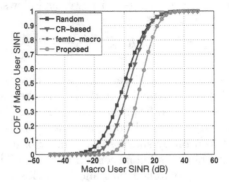

Fig. 2. CDF of macro-cell user's SINR

interference among femto-cells in addition to macro users interference control. The Cumulative Distribution Function (CDF) of the macro user's SINR is presented in Fig. 2. The proposed scheme achieves the highest macro-user's SINR in contrast to the standard random resource allocation, the CR-based and the femto-macro schemes. The reason is that both the CR scheme and the femto-macro scheme employ simple energy detectors to detect the vacant sub-channel. However, our scheme considers accurate detection by better evaluating the interference range and improving the detection sensitivity.

The CDF of the femto-user's SINR is shown in Fig. 3. The proposed scheme achieves significantly better performance than all the other schemes owing to the interference coordination between FBSs with the awareness based spectrum allocation mechanism. The CR-based scheme has no capability to mitigate interference between FBSs while the femto-macro scheme is based on clustering which limits each cluster of femto-cells to access only one sub-channel. This increases the probability of collisions between the contending FBSs. On the other hand, the proposed scheme is not limited to one sub-channel and it comprises awareness and information exchange between the FBSs for channels allocation which does not only mitigate interference but also improves spectrum utilization and maximizes throughput.

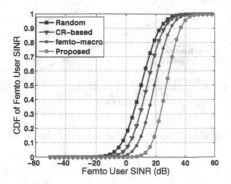

Fig. 3. CDF of femto-cell user's SINR

Fig. 4 presents the spectrum efficiency achieved by all the schemes. It can be noticed that the proposed scheme recorded the highest spectrum efficiency as it considers efficient spectrum utilization by switching users from under-utilized FBSs and free more sub-channels. Fig. 5 presents the throughput achieved by

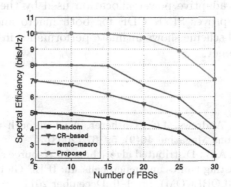

Fig. 4. Spectral efficiency comparison for all schemes

the femto-cell as a function of various number of FBSs. The transmitted traffic considered here is real-time traffic. Our proposed scheme achieved the highest throughput compared to other schemes as it employs adaptive power alloca-tion, considers users QoS requirements and traffic priority which enhances the throughput. Moreover, we notice that the throughput decreases as the number of FBSs increases. It is mainly due to the increase in the probability of collision as the number of FBSs grows.

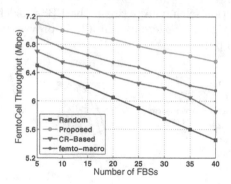

Fig. 5. Femto-cell throughput as a function of the number of FBSs

5   Conclusion

In this paper, we proposed a novel interference mitigation scheme for both cross-tier and co-tier interference in LTE femto-cell and macro-cell deployment. An improved version of cognitive radio spectrum sensing with better detection capability was exploited to mitigate the cross-tier interference. Moreover, an awareness based co-tier interference mitigation mechanism with the aid of graph coloring algorithm was proposed. The mechanism also ensures QoS requirements, supports traffic priority, and improves spectral efficiency by using smart user-FBS association. The adaptive power allocation used by the interference mitigation mechanism improves SINR CDF for both macro and femto users. In addition, the proposed scheme shows ultimate performance in terms of throughput and spectral efficiency.

References

1. Chandrasekhar, V., Andrews, J.G., Gatherer, A.: Femtocell networks: a survey. IEEE Communications Magazine **46**(9), 59–67 (2008)
2. Shi, Y., MacKenzie, A.B.: Distributed algorithms for resource allocation in cellular networks with coexisting femto- and macrocells. In: IEEE Global Telecommunications Conference (GLOBECOM), pp. 1–6, December 2011
3. Bennis, M., Niyato, D.: A q-learning based approach to interference avoidance in self-organized femtocell networks. In: IEEE GLOBECOM Workshops (GC Wkshps)E, pp. 706–710, December 2010
4. Shi, Y., MacKenzie, A.B., DaSilva, L.A., Ghaboosi, K., Latva-aho, M.: On resource reuse for cellular networks with femto- and macrocell coexistence. In: IEEE Global Telecommunications Conference (GLOBECOM), pp. 1–6, December 2010
5. Sun, Y., Jover, R.P., Wang, X.: Uplink interference mitigation for ofdma femtocell networks. IEEE Transactions on Wireless Communications **11**(2), 614–625 (2012)
6. Kaimaletu, S., Krishnan, R., Kalyani, S., Akhtar, N., Ramamurthi, B.: Cognitive interference management in heterogeneous femto-macro cell networks. In: IEEE International Conference on Communications (ICC), pp. 1–6, June 2011

7. Bennis, M., Perlaza, S.M.: Decentralized cross-tier interference mitigation in cognitive femtocell networks. In: IEEE International Conference on Communications (ICC), pp. 1–5, June 2011
8. Agustin, A., Vidal, J., Munoz-Medina, O., Fonollosa, J.R.: Decentralized weighted sum rate maximization in mimo-ofdma femtocell networks. In: IEEE GLOBECOM Workshops (GC Wkshps), pp. 270–274, December 2011
9. Boudreau, G., Panicker, J., Guo, N., Chang, R., Wang, N., Vrzic, S.: Interference coordination and cancellation for 4g networks. IEEE Communications Magazine 47(4), 74–81 (2009)
10. Ali, S.H., Leung, V.C.M.: Dynamic frequency allocation in fractional frequency reused ofdma networks. IEEE Transactions on Wireless Communications 8(8), 4286–4295 (2009)
11. Talwar, S., Lee, S.-C., Kim, H.: Wimax femtocells: a perspective on network architecture, capacity, and coverage. IEEE Communications Magazine 46(10), 58–65 (2008)
12. AlQerm, I., Shihada, B., Shin, K.G.: Cogwnet: a resource management architecture for cognitive wireless networks. In: 22nd International Conference Computer Communications and Networks (ICCCN), pp. 1–7, July 2013
13. AlQerm, I., Shihada, B., Shin, K.G.: Enhanced cognitive radio resource management for lte systems. In: IEEE 9th International Conference on Wireless and Mobile Computing, Networking and Communications (WiMob), pp. 565–570, October 2013
14. Cabric, D., Mishra, S.M., Brodersen, R.W.: Implementation issues in spectrum sensing for cognitive radios. In: Conference Record of the Thirty-Eighth Asilomar Conference on Signals, Systems and Computers, 2004, vol. 1, pp. 772–776, November 2004
15. András, S., Baricz, Á., Sun, Y.: The generalized Marcum Q-function: an orthogonal polynomial approach, ArXiv e-prints, October 2010
16. Arfken, G.: The incomplete gamma function and related functions. Mathematical Methods for Physicists, 3rd edn, pp. 565–572. Academic Press, Orlando (1985)
17. National Telecommunications and Information Administration (NTIA). Interference protection criteria, phase 1: Compilation from existing sources (2005)
18. Oh, D.-C., Lee, H.-C., Lee, Y.-H.: Cognitive radio based femtocell resource allocation. In: International Conference on Information and Communication Technology Convergence (ICTC), pp. 274–279, November 2010
19. Ning, G., Yang, Q., Kwak, K.S., Hanzo, L.: Macro- and femtocell interference mitigation in ofdma wireless systems. In: IEEE Global Communications Conference (GLOBECOM), pp. 5068–5073, December 2012

Packet Loss Rate Analysis of Wireless Sensor Transmission with RF Energy Harvesting

Tian-Qing Wu and Hong-Chuan Yang$^{(\boxtimes)}$

Department of Electrical and Computer Engineering,
University of Victoria, Victoria, Canada
{twu,hy}@uvic.ca

Abstract. RF energy harvesting is a promising potential solution for providing convenient and perpetual energy supply to low-power wireless sensor networks. In this paper, we investigate the performance of overlaid wireless sensor transmission powered by RF energy harvesting from existing wireless system for delay sensitive traffic. We derive the exact closed-form expression for the distribution function of harvested energy over a certain number of coherence time over Rayleigh fading channels with the consideration of hardware limitation, such as energy harvesting sensitivity and efficiency. We further analyze the packet loss probability of sensor transmission subject to interference from existing system.

Keywords: RF energy harvesting · Sensitivity · Wireless sensor transmission

1 Introduction

Wireless sensor networks (WSNs) are used in a wide range of applications, such as environment monitoring, surveillance, health care, intelligent buildings and battle field control [1]. The sensor nodes of WSN are usually powered by batteries with finite life time, which manifests as an important limiting factor to the functionality of WSN. Replacing or charging the batteries may either incur high costs for human labor or be impractical for certain application scenarios (e.g. applications that require sensors to be embedded into structures). Powering sensor nodes through ambient energy harvesting has therefore received a lot of attentions in both academia and industrial communities [2,3]. Various techniques have been developed to harvest energy from conventional ambient energy sources, such as solar power, wind power, thermoelectricity, and vibrational excitations [4–7].

RF energy is another promising candidate ambient energy source for powering sensor nodes. Recently, there has been a growing interest in RF energy harvesting due to the intensive deployment of cellular/WiFi wireless systems in addition to traditional radio/TV broadcasting systems [8]. It has been experimentally proved that RF energy harvesting is feasible from the hardware implementation viewpoint. In [9], the authors developed prototypes for devices that communicate with each other using ambient RF signals from TV/cellular systems as the

© Institute for Computer Sciences, Social Informatics and Telecommunications Engineering 2015
M. Weichold et al. (Eds.): CROWNCOM 2015, LNICST 156, pp. 620–630, 2015.
DOI: 10.1007/978-3-319-24540-9_51

only power source. In [10], the authors present the experimental performance (e.g., charging time of the sensor and received signal power at the sink) of RF energy harvesting using PowerCast energy harvesters [11]. Although these previous works have proved a visible future for the wireless application based on RF energy harvesting, most performance results are obtained through laboratory experiments. There is still a lack of effective theoretical models that can analytically predict the performance of WSNs powered by RF energy harvesting.

Previous literature on RF energy harvesting can be summarized as following. The fundamental performance limits of simultaneous wireless information and energy transfer systems over point-to-point link were studied in [12,13]. In [14], the authors consider a three-node multiple-input multiple-output (MIMO) wireless system, where one receiver harvests energy and another receiver decodes information from the signal transmitted by a common transmitter. A cognitive network that can harvest RF energy from the primary system is considered in [15]. The authors propose an optimal mode selection policy for sensor nodes to decide whether to transmit information or to harvest RF energy based on Markov modelling. In [16], the authors investigate mode switching between information decoding and energy harvesting, based on the instantaneous signal channel and interference condition over a point-to-point link. In most of these works, it is generally assumed that the channel gain remains constant during the whole energy harvesting circle, including obtaining channel state information, making decision accordingly, and then harvesting energy or decoding information. It worths to point out that wireless fading channels are in general time varying with channel coherence time in the order of milliseconds. The harvested energy over one channel coherence time may not be sufficient for channel estimation alone, not to mention information transmission/decoding.

With these observations in mind, we consider an overlaid sensor transmission scenario where a sensor-to-sink communication link operates in the coverage of an existing wireless system over the same frequency. We assume that the sink has a constant power source and that the sensor needs to harvest RF energy from the transmission of existing wireless system. Specifically, the sensor node can only harvest RF energy when its received signal power is larger than a certain sensitivity level [14]. As such, the existing system, being either cellular, WiFi or TV broadcasting systems, serves as the ambient source for sensor energy harvesting and as interference source during sensor transmission. Such an overlaid implementation strategy of RF-energy powered WSN has the potential to offer attractive and green solutions to a wide range of sensing applications, particularly in view of the increasingly severe spectrum scarcity. We consider delay sensitive traffic scenario, where the sensor needs to periodically transmit a new packet to the sink. We investigate the packet transmission performance of the sensor-to-sink link over Rayleigh fading wireless channels over multiple channel coherence time. The statistical distribution of the amount of energy that can be harvested over a fixed number of channel coherence time is derived with the consideration of harvesting sensitivity and efficiency. We study the packet loss probability of delay sensitive traffic, which is dependent on the amount of harvested energy as well as interference amount experienced during packet transmission. We also examine the effect of traffic intensity and the energy storage capacity at the sensor on the packet loss probability based on the exact analytical results. These analytical results will help determine what type of sensing applications that the proposed overlaid implementation strategy can effectively support.

The remainder of the paper is organized as follows. In Section 2, we introduce the system and channel model under consideration. The performance of the proposed sensing implementation for delay sensitive traffic is evaluated in Section 3. Concluding remarks are given in Section 4.

2   System and Channel Model

2.1   System Model

We consider the point-to-point packet transmission from a single-antenna wireless sensor to its sink over a flat Rayleigh fading channel. The sink and the sensor are deployed in the coverage area of an existing wireless system, which could be cellular, WiFi or TV broadcasting systems. We assume that the sensor can harvest RF energy from the transmitted signal of the existing system, and use it as its sole energy source for transmission, as illustrated in Fig. 1.

In the energy harvesting stage, the sensor harvests RF energy from the radio transmission of existing wireless systems over multiple channel coherence time. Typically, the sensor can harvest RF energy only when the received signal power is larger than a power threshold, denoted by P_{th} [14]. In general, P_{th} should be greater than the receiver sensitivity for information reception.

During the packet transmission stage, the sensor will transmit its collected information to the sink using harvested energy. We assume that the energy consumed for information collection is negligible compared with the energy used for transmission [17]. Then the energy that can be used for transmission is approximately equal to the harvested energy. Also note that the sensor transmission will suffer interference from the existing system in this stage, the effect of which will be further discussed in the following sections. Due to the low transmission power and short transmission duration, we ignore the interference that the sensor transmission may generate to the existing system.

2.2   Channel Model

We adopt a log-distance path loss plus Rayleigh block fading channel models for the operating environment [19] while ignoring the shadowing effect for the sake of presentation clarity. In particular, the channel gain between the BS and the sensor remains constant over one channel coherence time, denoted by T_c, and changes to an independent value afterwards. Let h_n denote the fading channel gain over the nth coherence time, where $h_n \in \mathcal{CN}(0, 1)$. For notational conciseness, we use α_n

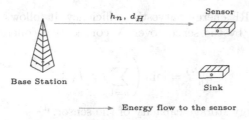

(a) Energy Harvesting Stage

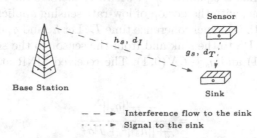

(b) Packet Transmission Stage

Fig. 1. System model for two-stage sensor transmission with RF energy harvesting.

to denote its amplitude square, i.e. $\alpha_n = ||h_n||^2$, whose PDF for Rayleigh fading channel under consideration is given by

$$f_{\alpha_n}(x) = e^{-x}. \tag{1}$$

Then the instantaneous received signal power at the sensor over the nth coherence time is given by $P_n = \overline{P}\alpha_n$, where $\overline{P}$ is the average received power at the sensor due to path loss, given by

$$\overline{P} = \frac{P_T}{\Gamma d_H^\lambda}, \tag{2}$$

where P_T is the constant transmission power of BS, d_H is the distance from BS to the sensor, λ is the path loss exponent of the environment, ranging from 2 to 5, and Γ is a constant parameter of the log-distance path loss model. Specifically, $\Gamma = \frac{PL(d_0)}{d_0^\lambda}$, where d_0 is a reference distance of the antenna far field, and $PL(d_0)$ is linear path loss at distance d_0, depending on the propagation environment.

We assume, as is the case in real world systems [10][11], the sensor can only harvest energy when the instantaneous received signal power P_n is greater than the sensitivity level P_{th} and the harvested energy is proportional to $P_n - P_{th}$. Consequently, the amount of energy that the sensor can harvest during the nth coherence time can be represented as [14]

$$E_n = \begin{cases} \eta T_c(P_n - P_{th}), & P_n \geq P_{th}; \\ 0, & P_n < P_{th}, \end{cases} \tag{3}$$

where $0 \leq \eta \leq 1$ is RF energy harvesting efficiency. It follows that the amount of energy harvested by the sensor over N consecutive coherence time can be given by

$$E_h^{(N)} = \min \left(\sum_{n=1}^{N} E_n, E_c \right), \tag{4}$$

where E_c is the energy storage capacity of the sensor. [1]

The transmission power of the sensor when it uses the harvested energy over N coherence time is equal to $\frac{E_h^{(N)}}{T_s}$, where T_s denotes the transmission time duration. We assume, with the notion of low-rate sensing applications, that T_s is much smaller than the channel coherence time T_c. Let h_s and g_s denote the fading channel gains from BS to the sink and from the sensor to the sink, respectively, where $h_s \in \mathcal{CN}(0,1)$ and $g_s \in \mathcal{CN}(0,1)$. The received SINR at the sink can be calculated as

$$\gamma_s = \frac{\frac{E_h^{(N)}}{T_s d_T^\lambda} ||g_s||^2}{\frac{P_T}{d_I^\lambda} ||h_s||^2 + \Gamma \sigma^2}, \tag{5}$$

where d_T is the distance from the sensor to the sink, d_I is the distance from BS to the sink, and σ^2 is the variance of the additive noise at the sink. In general, the sensor and the sink are very close to each other, i.e. $d_T \ll d_H \approx d_I$. In the following, we study the performance of such overlaid sensor transmission when it is used to support low-rate data traffics.

3   Performance Analysis for Delay Sensitive Traffic

For certain sensing applications, such as smart metering and environment monitoring, the sensor node needs to periodically send their collected information (e.g. energy usage, temperature, humid information) to the sink. Any delay in the delivery of these information may render them useless. Therefore, the goal is to successfully transmit these information packet within a fixed time duration. As such, an important performance metric for such application is the packet loss probability, i.e. the percent of packets that could not be delivered to the sink in time. We now analyze the packet loss probability of the proposed overlaid sensing implementation with RF energy harvesting. An accurate quantification of this metric will help determine the sensing applications that could be supported with the proposed implementation.

3.1   Distribution of Harvested Energy over N Coherence Time

We are interested in the distribution fuction of the harvested energy of the sensor over N channel coherence time, which will be used for packet loss probability analysis.

---

[1] E_c can also be viewed as the energy threshold, above which the sensor can carry out packet transmission.

We first consider the one coherence time case, i.e. $N = 1$. The CDF of the harvested energy can be simply represented as

$$F_{E_h^{(1)}}(x) = \Pr[E_h^{(1)} < x] = \Pr[E_1 < x], \; x \le E_c. \tag{6}$$

After substituing (3) into (6) and some manipulation, we have

$$F_{E_h^{(1)}}(x) = 1 - e^{-\frac{x}{\eta T_c \overline{P}} - \frac{P_{th}}{\overline{P}}}, \; x \le E_c. \tag{7}$$

For the multiple channel coherence time case, i.e. $N > 1$, we denote the number of channel coherence time, in which the sensor can harvest energy, by N_a. According to the total probability theorem, the CDF of the harvested energy is shown as

$$F_{E_h^{(N)}}(x) = \Pr[E_h^{(N)} < x] = \sum_{i=0}^{N} \Pr[\sum_{n=1}^{N} E_n < x, N_a = i]. \tag{8}$$

When the ith largest received power is larger than P_{th} and the $(i+1)$th largest one is lower than P_{th}, the number of coherence time that the sensor can harvest energy is $N_a = i$. We denote the ordered version of N i.i.d. random variables α_n as $\alpha_{1:N} \ge \alpha_{2:N} \ge \cdots \ge \alpha_{N:N}$, and the sum of the $i-1$ largest variables as $\beta_i = \sum_{j=1}^{i-1} \alpha_{j:N}$. We can show that $N_a = i$ if and only if $\alpha_{i:N} \ge \frac{\Gamma d_H^\lambda P_{th}}{P_T}$ and $\alpha_{i+1:N} < \frac{\Gamma d_H^\lambda P_{th}}{P_T}$. Therefore, $F_{E_h}(x)$ can be calculated as,

$$F_{E_h^{(N)}}(x) = \sum_{i=2}^{N-1} \Pr[\beta_i + \alpha_{i:N} < \frac{x}{\eta T_c \overline{P}} + \frac{i P_{th}}{\overline{P}}, \alpha_{i:N} \ge \frac{P_{th}}{\overline{P}}, \alpha_{i+1:N} < \frac{P_{th}}{\overline{P}}] \tag{9}$$

$$+ \Pr[\alpha_{1:N} < \frac{P_{th}}{\overline{P}}] + \Pr[\frac{P_{th}}{\overline{P}} \le \alpha_{1:N} < \frac{x}{\eta T_c \overline{P}} + \frac{P_{th}}{\overline{P}}, \alpha_{2:N} < \frac{P_{th}}{\overline{P}}]$$

$$+ \Pr[\beta_N + \alpha_{N:N} < \frac{x}{\eta T_c \overline{P}} + \frac{N P_{th}}{\overline{P}}, \alpha_{N:N} \ge \frac{P_{th}}{\overline{P}}]$$

$$= \sum_{i=2}^{N-1} \int_{\frac{P_{th}}{\overline{P}}}^{\frac{x}{i\eta T_c \overline{P}} + \frac{P_{th}}{\overline{P}}} \int_{(i-1)y}^{\frac{x}{\eta T_c \overline{P}} + \frac{i P_{th}}{\overline{P}} - y} \int_0^{\frac{P_{th}}{\overline{P}}} f_{\beta_i,\alpha_{i:N},\alpha_{i+1:N}}(t,y,z) dt dy dz$$

$$+ \int_0^{\frac{P_{th}}{\overline{P}}} f_{\alpha_{1:N}}(t) dt + \int_0^{\frac{P_{th}}{\overline{P}}} \int_{\frac{P_{th}}{\overline{P}}}^{\frac{x}{\eta T_c \overline{P}} + \frac{P_{th}}{\overline{P}}} f_{\alpha_{1:N},\alpha_{2:N}}(t,y) dt dy$$

$$+ \int_{\frac{P_{th}}{\overline{P}}}^{\frac{x}{N\eta T_c \overline{P}} + \frac{P_{th}}{\overline{P}}} \int_{(N-1)y}^{\frac{x}{\eta T_c \overline{P}} + \frac{N P_{th}}{\overline{P}} - y} f_{\beta_N,\alpha_{N:N}}(t,y) dt dy,$$

where $f_{\alpha_{1:N}}(x,y)$, $f_{\alpha_{1:N},\alpha_{2:N}}(x,y)$, $f_{\beta_N,\alpha_{N:N}}(x,y)$, and $f_{\beta_i,\alpha_{i:N},\alpha_{i+1:N}}(x,y,z)$ are the marginal and joint PDFs of $\alpha_{i:N}$ and β_i, whose closed-form expression can be obtained in [18]. By properly substituting the closed-form expression of $f_{\alpha_{1:N}}(x,y)$, $f_{\alpha_{1:N},\alpha_{2:N}}(x,y)$, $f_{\beta_N,\alpha_{N:N}}(x,y)$, and $f_{\beta_i,\alpha_{i:N},\alpha_{i+1:N}}(x,y,z)$ into (9) and carrying out integration, the close form expression of the CDF of harvested energy is obtained as

$$
F_{E_h^{(N)}}(x) = \begin{cases}
\sum_{i=2}^{N} \frac{N!(1-e^{-\frac{P_{th}}{P}})^{N-i}e^{-\frac{iP_{th}}{P}}}{(i-1)!(i-2)!(N-i)!} \sum_{m=0}^{i-2}(1-i)^{i-2-m}\binom{i-2}{m}\sum_{j=0}^{m}\frac{m!}{(m-j)!} \\
\left\{(i-1)^{m-j}\sum_{k=0}^{i-2-j}\frac{(i-2-j)!}{(i-2-j-k)!i^{k+1}}\left[\left(\frac{P_{th}}{P}\right)^{i-2-j-k}\right.\right. \\
\left.-e^{-\frac{x}{\eta T_c P}}\left(\frac{x}{i\eta T_c P}+\frac{P_{th}}{P}\right)^{i-2-j-k}\right] - e^{-\frac{x}{\eta T_c P}}\sum_{s=0}^{m-j}(-1)^{m-j-s}\binom{m-j}{s} \\
\left.\frac{\left(\frac{x}{\eta T_c P}+\frac{iP_{th}}{P}\right)^s\left(\frac{x}{i\eta T_c P}+\frac{P_{th}}{P}\right)^{i-1-j-s}-\left(\frac{P_{th}}{P}\right)^{i-1-j-s}}{i-1-j-s}\right\} + (1-e^{-\frac{P_{th}}{P}})^N \\
+N(1-e^{-\frac{P_{th}}{P}})^{N-1}(e^{-\frac{P_{th}}{P}}-e^{-\frac{x}{\eta T_c P}-\frac{P_{th}}{P}}), & x \le E_c; \\
1, & x > E_c.
\end{cases}
\tag{10}
$$

After taking derivative with respect to x, the PDF of $F_{E_h}(x)$ is derived and given as

$$
f_{E_h^{(N)}}(x) = \begin{cases}
\sum_{i=2}^{N}\frac{N!(1-e^{-\frac{P_{th}}{P}})^{N-i}e^{-\frac{x}{\eta T_c P}-\frac{iP_{th}}{P}}}{(i-1)!(i-2)!(N-i)!}\sum_{m=0}^{i-2}(1-i)^{i-2-m}\binom{i-2}{m} \\
\sum_{j=0}^{m}\frac{m!}{(m-j)!}\left\{(i-1)^{m-j}\sum_{k=0}^{i-2-j}\frac{(i-2-j)!}{(i-2-j-k)!i^{k+1}}\left(\frac{x}{i\eta T_c P}+\frac{P_{th}}{P}\right)^{i-2-j-k-1}\right. \\
\left(\frac{x}{i\eta^2 T_c^2 P^2}+\frac{P_{th}}{\eta T_c P^2}-\frac{i-2-j-k}{i\eta T_c P}\right) + \sum_{s=0}^{m-j}(-1)^{m-j-s}\binom{m-j}{s}\frac{i^s}{i-1-j-s} \\
\{(\frac{x}{i\eta T_c P}+\frac{P_{th}}{P})^{i-2-j}(\frac{x}{i\eta^2 T_c^2 P^2}+\frac{P_{th}}{\eta T_c P^2}-\frac{i-1-j}{i\eta T_c P}) \\
\left.\left.+(\frac{P_{th}}{P})^{i-1-j-s}(\frac{x}{i\eta T_c P}+\frac{P_{th}}{P})^{s-1}(-\frac{x}{i\eta^2 T_c^2 P^2}-\frac{P_{th}}{\eta T_c P^2}+\frac{s}{i\eta T_c P})\}\right\} \\
+\frac{N}{\eta T_c P}(1-e^{-\frac{P_{th}}{P}})^{N-1}e^{-\frac{x}{\eta T_c P}-\frac{P_{th}}{P}} + (1-e^{-\frac{P_{th}}{P}})^N\delta(x) \\
+[1-F_{E_h}(E_c)]\delta(x-E_c), & N>1; \\
\frac{1}{P\eta T_c}e^{-\frac{x}{P\eta T_c}-\frac{P_{th}}{P}}+(1-e^{-\frac{P_{th}}{P}})^N\delta(x)+[1-F_{E_h}(E_c)]\delta(x-E_c), & N=1,
\end{cases}
\tag{11}
$$

where $\delta(\cdot)$ denotes the impulse function. Note that the PDF involves two impulse function at 0 and E_c due to the capacity constraints.

3.2   Packet Loss Probability Analysis

We assume that the sensor must collect and transmit one packet to the sink over a fixed time duration T_F. The number of coherence time in T_F, denoted by N, is approximately equal to $\lfloor\frac{T_F}{T_c}\rfloor$. The sensor will first harvest RF energy for N channel coherence time and then transmit the packet to the sink using the harvested energy. We focus on low rate sensing application and ignore the potential packet collision with other sensors. We also assume that, with adoption of certain error correction coding scheme, the packet can be successfully received by the sink if the received SINR at the sink during packet transmission is above γ_T. As such, packet loss will occurs if and only if the received SINR at the sink during packet transmission is below the threshold γ_T. This may be due to insufficient harvested energy, poor sensor to sink channel quality, as well as strong interference from BS. Mathematically, the packet loss probability of the sensor transmission is given by

$$
P_{PL} = \Pr[\gamma_s < \gamma_T] = \Pr\left[\frac{\frac{E_h^{(N)}}{T_s d_T^\lambda}\|g_s\|^2}{\frac{P_T}{d_I^\lambda}\|h_s\|^2 + \Gamma\sigma^2} < \gamma_T\right].
\tag{12}
$$

Conditioning on $E_h^{(N)}$, the packet loss probability can rewritten in terms of the PDFs of $E_h^{(N)}$, $||g_s||^2$, and $||h_s||^2$, denoted by $f_{E_h^{(N)}}(\cdot)$, $f_{||g_s||^2}(\cdot)$, and $f_{||h_s||^2}(\cdot)$, respectively, as

$$P_{PL} = \int_0^{E_c} \int_0^{\infty} F_{||g_s||^2} \left(\frac{T_s \gamma_T d_T^\lambda (\frac{P_T y}{d_I^\lambda} + \Gamma \sigma^2)}{z} \right) f_{||h_s||^2}(y) f_{E_h^{(N)}}(z) dy dz. \tag{13}$$

The PDF of $||h_s||^2$ and $||g_s||^2$ for the Rayleigh fading channel model under consideration are commonly given by

$$f_{||h_s||^2}(x) = f_{||g_s||^2}(x) = e^{-x}. \tag{14}$$

After proper substitution and some manipulations, we can rewrite P_{PL} as

$$P_{PL} = \int_0^{E_c} \left(1 - \frac{ze^{-\frac{T_s \gamma_T \Gamma d_T^\lambda \sigma^2}{z}}}{z + \frac{P_T}{d_I^\lambda} T_s \gamma_T d_T^\lambda} \right) f_{E_h^{(N)}}(z) dz. \tag{15}$$

Finally, the packet loss probability for delay sensitive traffic can be calculated by substituting (11) into (15) and carrying out numerical integration. Note that only finite integration of some basic functions are involved in the calculation.

3.3 Numerical Results

We assume the same parameters for RF energy harvesting system as in [9]. In particular, the transmission power of BS is $P_T = 10kW$. The distance from BS to the sensor, BS to the sink and the sensor to the sink are set as $d_H = 100$ meters, $d_I = 100$ meters and $d_T = 1$ meter, respectively. The pass loss exponent λ is assumed to be 3, the channel coherence time T_c be $100ms$, and the transmission time of the sensor T_s be $1ms$. The sensitivity of the sensor is assumed to be $P_{th} = -10dBm = 0.1mW$ [10]. For simplicity, we assume harvesting efficiency $\eta = 1$ and packet loss constant $\Gamma = 1$.

In Fig. 2, we plot the PDF of harvested energy over $N = 1, 2, 3$ coherence time. We can see that $E_h^{(N)}$ follows a mixed distribution with impulse at $x = 0$ and $x = E_c$, which represents the probability that the sensor can not harvest any energy over N coherence time and the probability that the sensor will be fully charged after N coherence time. With the increase of the number of coherence time N, the continuous portion of probability mass moves towards right, with the distribution of harvested energy spreads more widely along the energy axis. This is because when N increases, the sensor has larger probability to harvest more energy.

In Fig. 3, we plot the packet loss probability at the sink as a function of the SINR threshold for different energy capacity E_c with $N = 3$. We can see when γ_T is small, the packet loss probability shows approximately linear degradation. We also observe that larger energy capacity E_c leads to smaller packet loss probability. However, the benefit of lowing packet loss probability shrinks with

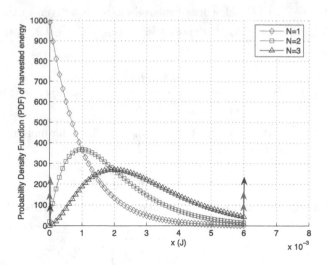

Fig. 2. Distribution of harvested energy over N channel coherence time ($E_c = 0.006J$).

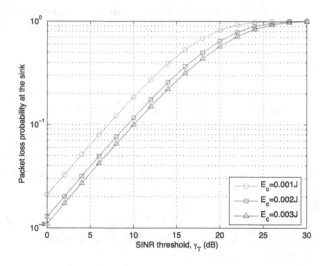

Fig. 3. Packet loss probability at the sink for different energy storage capacity.

the increase of the energy capacity E_c. This is because when E_c gets larger, the sensor has smaller probability to be fully charged, such that the effect of the energy capacity on the packet loss probability gradually reduces.

In Fig. 4, we plot the packet loss probability at the sink as a function of the number of the channel coherence time before each packet transmission. We can see the packet loss probability at the sink gradually reduces as N increases, and converge to a constant value when N is very large. This is due to the existence

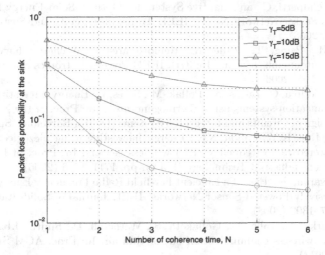

Fig. 4. Packet loss probability at the sink over N coherence time.

of energy storage capacity E_c, which limits the total harvested energy and in turn the transmission power. Moreover, we notice that higher SINR threshold leads to higher packet loss probability, as expected by intuition.

4   Conclusion

In this paper, we investigated the packet transmission performance of wireless sensor nodes powered through harvesting RF energy from existing wireless systems. We derive the exact closed-form expression for the distribution function of harvested energy over a certain number of coherence time over Rayleigh fading channels, based on which we further analyze the packet loss probability of sensor transmission with the consideration of hardware limitation, such as harvesting sensitivity and energy storage capacity, and interference from existing system. The analytical results will greatly facilitate the design and optimization of such sensor system powered by RF energy harvesting for the appropriate target sensing applications.

References

1. Akyildiz, I., Su, W., Sankarasubramaniam, Y., Cayirci, E.: A Survey on Sensor Networks. IEEE Communications Magazine **40**(8), 102–114 (2002)
2. Seah, W., Eu, Z.A., Tan, H.P.: Wireless sensor networks powered by ambient energy harvesting (wsn-heap) - survey and challenges. In: 1st International Conference on Wireless Communication, Vehicular Technology, Information Theory and Aerospace Electronic Systems Technology, Wireless VITAE 2009, pp. 1–5 (2009)
3. Sudevalayam, S., Kulkarni, P.: Energy Harvesting Sensor Nodes: Survey and Implications. IEEE Communications Surveys Tutorials **13**(3), 443–461 (2011)

4. Alippi, C., Galperti, C.: an Adaptive System for Optimal Solar Energy Harvesting in Wireless Sensor Network Nodes. IEEE Trans. on Circuits and Systems **55**(6), 1742–1750 (2008)
5. Weimer, M., Paing, T., Zane, R.: Remote area wind energy harvesting for low-power autonomous sensors. In: 37th IEEE Power Electronics Specialists Conference, pp. 1–5 (2006)
6. Mateu, L., Codrea, C., Lucas, N., Pollak, M., Spies, P.: Energy harvesting for wireless communication systems using thermogenerators. In: Proc. of the XXI Conference on Design of Circuits and Integrated Systems (DCIS), Barcelona, Spain (2006)
7. Tan, Y.K., Hoe, K.Y., Panda, S.K.: Energy harvesting using piezoelectric igniter for self-powered radio frequency (RF) wireless sensors. In: Proc. of IEEE Intel. Conference on Industrial Technology (ICIT), pp. 1711–1716 (2006)
8. Le, T., Mayaram, K., Fiez, T.: Efficient Far-field Radio Frequency Energy Harvesting for Passively Powered Sensor Networks. IEEE Journal of Solid-State Circuits **43**(5), 1287–1302 (2008)
9. Liu, V., Parks, A., Talla, V., Gollakota, S., Wetherall, D., Smith, J.R.: Ambient backscatter: wireless communication out of thin air. In: Proc. ACM SIGCOMM, pp. 1–13 (2013)
10. Baroudi, U., Qureshi, A., Talla, V., Gollakota, S., Mekid, S., Bouhraoua, A.: Radio frequency energy harvesting characterization: an experimental study. In Proc. IEEE TSPCC, pp. 1976–1981 (2012)
11. Powercast Corporation. http://www.powercastco.com
12. Varshney, L.R.: Transporting information and energy simultaneously. In: Proc. IEEE Int. Symp. Inf. Theory (ISIT), pp. 1612–1616 (2008)
13. Grover, P., Sahai, A.: Shannon meets tesla: wireless information and power transfer. In: Proc. IEEE Int. Symp. Inf. Theory (ISIT), pp. 2363–2367 (2010)
14. Zhang, R., Ho, C.K.: MIMO Broadcasting for Simultaneous Wireless Information and Power Transfer. IEEE Trans. Wireless Commun. **12**(5), 1989–2001 (2013)
15. Park, S., Heo, J., Kim, B., Chung, W., Wang, H., Hong, D.: Optimal mode selection for cognitive radio sensor networks with RF energy harvesting. In: Proc. IEEE PIMRC, pp. 2155–2159 (2012)
16. Liu, L., Zhang, R., Chua, K.: Wireless Information Transfer with Opportunistic Energy Harvesting. IEEE Trans. Wireless Commun. **12**(1), 288–300 (2013)
17. Raghunathan, V., Ganeriwal, S., Srivastava, M.: Emerging Techniques for Long Lived Wireless Sensor Networks. IEEE Commun. Mag. **44**(4), 108–114 (2006)
18. Yang, H.C., Alouini, M.S.: Order Statistics in Wireless Communications. Cambridge University Press (2011)
19. Goldsmith, A.: Wireless Communications. Cambridge University Press (2005)

Distributed Fair Spectrum Assignment for Large-Scale Wireless DSA Networks

Bassem Khalfi[1], Mahdi Ben Ghorbel[2]([✉]), Bechir Hamdaoui[1],
and Mohsen Guizani[2]

[1] Oregon State University, Corvallis, OR, USA
{khalfib,hamdaoui}@eecs.orst.edu
[2] Qatar University, Doha, Qatar
mahdi.benghorbel@qu.edu.qa, mguizani@ieee.org

Abstract. This paper proposes a distributed and fair resource allocation scheme for large-scale wireless dynamic spectrum access networks
based on particle filtering theory. We introduce a proportionally fair
global objective function to maximize the total network throughput while
ensuring fairness among users. We rely on particle filtering theory to
enable distributed access and allocation of spectrum resources without
compromising the overall achievable throughput. Through intensive simulation, we show that our proposed approach performs well by achieving
high overall throughput while also improving fairness between users.

Keywords: Particle filtering · Dynamic spectrum access · Distributed
spectrum assignment · User fairness

1 Introduction

The increasingly growing number of wireless devices, along with the continually
rising demand for wireless bandwidth, has created a serious shortage problem in
the wireless spectrum supply. This foreseen spectrum shortage is shown to be due
to the lack of efficient spectrum allocation and regulation methods rather than
due to the scarcity of spectrum resources [1]. As a result, Dynamic Spectrum
Access (DSA) has been promoted as a potential candidate for addressing this
shortage problem, which essentially allows spectrum users to locate spectrum
opportunities and use them efficiently without harming legacy users [2]. Many
research attempts have been conducted to enable effective DSA. While many of
them have focused on spectrum sensing related challenges, others have focused on
developing resource allocation techniques that help access and utilize spectrum
resources efficiently [3].

Enabling DSA while maximizing the total throughput has been one of the
key challenges for resource allocation in DSA systems [3]. Many researchers have
proposed centralized approaches aiming to maximize the total throughput [4].
Although these methods achieve optimal or near-optimal performances, they
have limitations when it comes to scalability and computational complexity,

© Institute for Computer Sciences, Social Informatics and Telecommunications Engineering 2015
M. Weichold et al. (Eds.): CROWNCOM 2015, LNICST 156, pp. 631–642, 2015.
DOI: 10.1007/978-3-319-24540-9_52

especially when applied to large-scale systems. Therefore, distributed approaches are more attractive, and can be more effective when applied to DSA. This is because the decision will be taken locally by each user instead of being taken centrally, making each user send its information to the central agent so as to allow it to make such a decision.

There have been many resource allocation approaches proposed in the literature for enabling distributed DSA [3,5,6]. For example, the authors in [6] proposed Q-learning for distributed multiband spectrum access and power allocation. The authors in [7] proposed objective functions that rely on Q-learning to allocate spectrum resources in a distributed manner, where the focus was on the throughput maximization. In [8], particle filtering theory was used also for promoting distributed resource allocation in DSA systems, where it was shown that it can achieve higher throughput than what the technique proposed in [7] achieves when using the same objective functions.

One common concern with most of these distributed approaches is that they aim to increase throughput but without taking into account any fairness consideration. Even though throughput maximization-based approaches maximize the overall network throughput, they may lead to starvation of some Secondary Users (SU)s, resulting thus, in not treating all users equally fairly. This means that some users may get very limited amounts of throughput when compared to others. It is therefore important that fairness should be taken into account when designing these distributed allocation techniques. In the literature, fairness has been proposed with centralized approaches [8–10]. The authors in [8] considered the maximization of the minimum objective function to address user fairness. Although, the proposed objective function achieves better fairness, proportionally fair methods [11] are anticipated to achieve higher fairness. They target to balance between two conflicting behaviors: the cooperative behavior using the sum maximization and the minimum maximization which penalizes the users with high throughput.

With all of this in mind, this paper proposes a distributed allocation technique that jointly combines proportional fairness with particle filtering to assign spectrum in large-scale DSA networks. Since the global spectrum assignment optimization problem suffers from a high computational complexity and does not scale well, our technique aims to achieve suboptimal allocation while ensuring fairness among the different users. Using simulation, we compare the throughput performance of the proposed technique with that of the minimum fairness technique, proposed in [8].

The remainder of this article is organized as follows. In Section 2, we describe our system and channel model. In Section 3, we formulate the resource allocation problem for large-scale DSA systems, and discuss the issues related to the derivation of the optimal solution with respect to the used objective function. We apply particle filtering for distributed spectrum allocation in Section 4. Evaluations are provided in Section 5. Finally, we conclude the paper in Section 6.

2   Large-Scale DSA System Model

We consider a DSA system composed of n DSA agents competing to communicate over m non-overlapping bands, where a DSA agent represents a transmitter-receiver pair of SUs. The n agents are uniformly distributed within a cell where a primary system is communicating as illustrated in Fig. 1. We assume that the m bands have been perfectly sensed and declared as available using spectrum sensing technique (we will not discuss this technique as it is beyond the scope of this paper). As we are considering a large-scale system, the number of users is assumed to be very high compared to the number of the available bands ($n >> m$).

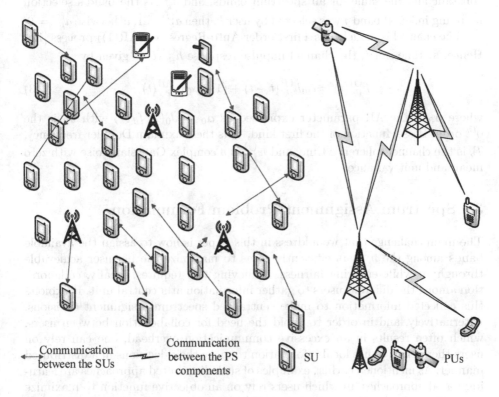

<table>
<tr><td>◄——►</td><td>Communication
between the SUs</td><td>⚡</td><td>Communication
between the PS
components</td><td>SU</td><td>PUs</td></tr>
</table>

Fig. 1. Large scale DSA system.

At each time slot t, user (agent) i tries to communicate with its correspondent receiver by selecting one band from the pool of the available bands. Each user aims to achieve the maximum possible throughput R_i with respect to its allowed power budget P_i. If we assume that user i selects band j, then the achieved throughput can be expressed as

$$R_i(t) = B^{(j)} \log_2(1 + \gamma_i^{(j)}(t)), \tag{1}$$

where $B^{(j)}$ is the jth channel bandwidth and $\gamma_i^{(j)}(t)$ is the received Signal to Interference plus Noise Ratio (SINR), expressed as

$$\gamma_i^{(j)}(t) = \frac{P_i^{(j)}|h_{ii}^{(j)}(t)|^2}{\sum_{\substack{k=1 \\ k \neq i}}^{n} a_k^{(j)} P_k^{(j)}|h_{ik}^{(j)}(t)|^2 + N_0 B^{(j)}}. \tag{2}$$

Here $h_{ik}^{(j)}(t)$ is the jth channel impulse response from the kth transmitter to the ith receiver, N_0 is the power spectral density of the noise which is assumed to be constant and the same for all spectrum bands, and $a_k^{(j)}$ is the band's selection mapping index. If band j was selected by user k, then $a_k^{(j)} = 1$, otherwise $a_k^{(j)} = 0$.

The channel is modeled as a first order Auto-Regressive (AR(1)) process [12]. Hence, at time slot t, the channel impulse response $h_{ik}^{(j)}(t)$ is given by

$$h_{ik}^{(j)}(t) = \alpha_0 h_{ik}^{(j)}(t-l) + (1-\alpha_0) w_i^{(j)}(t), \tag{3}$$

where α_0 is the AR parameter expressed as $\alpha_0 = J_0(2\pi f_d T_b)$ with J_0 is the 0th order Bessel function of the first kind, f_d is the maximum Doppler frequency, T_b is the channel coherence time, and $w_i^{(j)}$ is a complex Gaussian noise with zero mean and unit variance.

3   Spectrum Assignment Problem Formulation

The main challenge that we address in this paper is how to assign the available bands among the n users efficiently so as to maximize the per-user achievable throughput while ensuring fairness. Achieving this requires, ideally, collaboration among the different users to gather information at a central unit. It exploits this collected information to make centralized spectrum assignment decisions. Alternatively, and in order to avoid the need for collaboration between users, which often results in an excessive communication overhead, one can rely on users themselves to use local information to make their decisions in a distributed manner. As mentioned earlier, examples of such distributed approaches are learning based approaches, in which users rely on an objective function to maximize their achieved throughput. Authors in [7] showed that while the use of intrinsic objective functions results in fluctuating behaviors, the use of global objective functions, which take into account other users' decisions, though improve the overall system performance, are slow in doing so. The sum objective function is then defined as

$$\mathcal{O}_i^{sum}(t) = \sum_{k=1}^{n} R_k(t). \tag{4}$$

A common problem with the above functions is that they do not ensure fairness among users. In an attempt to address fairness, using a common global objective function known as bottleneck optimality, *max-min* approach has been proposed in [13] and is expressed as

$$\mathcal{O}_i^{\min}(t) = \min_{1 \leq k \leq n} R_k(t). \tag{5}$$

This objective function is more suitable for users having the same requirements, which is generally not the case in wireless communications. Although max-min solves the problem of starvation, users with high requirements will be penalized while users with low requirements will get more service than what they need. For a more efficient fair allocation, proportional fair [11], is shown to strike a good balance between two conflicting objectives: the maximization of the total throughput and the max-min fairness which may penalize users with high requirements. It is defined as

$$\mathcal{O}_i^{\mathrm{PF}}(t) = \sum_{k=1}^{n} \log_2(R_k(t)). \tag{6}$$

Using the proportional fair global objective function, we formulate our optimization for each user i as follows

$$\max \quad \mathcal{O}_i^{\mathrm{PF}}(t) \qquad \forall\, t \tag{7a}$$

$$\text{s.t.} \quad \sum_{j=1}^{m} a_i^{(j)}(t) = 1 \quad \forall\, t. \tag{7b}$$

This is a non-linear integer programming problem of the allocation index $\mathbf{a}_i^{(j)}(t) = [a_i^{(1)}(t), a_i^{(1)}(t), ..., a_i^{(m)}(t)]$. The constraint (7b) is used to control the number of the bands that each user could select at each time. This is behind the combinatorial nature of the problem where each user is allowed to select one single band.

Optimally allocating the m bands among the n users requires relaying all the network information such as the channels' fading and the users' power budgets to a central processing unit. By doing so, not only the system suffers from a huge network overhead, but it also incurs high computational processing time. The computational complexity is m^n and thus, it increases exponentially with the network scale. Hence, the allocation problem is NP-hard. Therefore, applying a distributed approach is more appealing to reduce the exchange overhead. In this case, each user has to take its own decision, $\mathbf{a}_i(t)$, and exchange its measured throughput and allocated channel to other users such that the global system evolves towards an optimum spectrum allocation $\mathbf{a}(t)$.

One of our main contribution in this paper is to consider fair distributed resource allocation. To the best of our knowledge, fairness has been addressed with centralized spectrum allocation so far and without any focus on the system scalability [3].

4   Fair Distributed Spectrum Assignment for Large-Scale DSA

One key merit that distributed resource allocation schemes possess is low signaling overhead. Local decisions are made following the exchange of some information (e.g. the achieved throughput and the selected band) among users and tracking the system evolution over time. In this context, particle filter based approaches are known to have strong tracking capabilities and can be adapted to non-linear and non-Gaussian estimation problems [14]. However, since the problem of spectrum assignment comes down to an estimation problem, we propose distributed particle filtering to estimate at each time slot the best spectrum allocation that achieves the fairness goal.

The concept of distributed particle filtering is derived from the sequential estimation and importance sampling techniques. Each user needs to interact with some or all other users in order to get the best estimation of the unknown. We model the evolution of the estimation of the best spectrum allocation as a discrete-time state-space model given by

$$\mathbf{a}(t) = \mathcal{X}(\mathbf{a}(t-1)) + \mathbf{u}(t), \tag{8a}$$

$$R_i(t) = \Psi_i(\mathbf{a}(t)) + v_i(t), \tag{8b}$$

where $\mathcal{X}(.)$ is a known function that describes the state's change. $\Psi_i(.)$ is the function that links the global state $\mathbf{a}(t)$ to the local observation $R_i(t)$. It is a non-linear function of the state $\mathbf{a}(t)$. $\mathbf{u}$ and $v_i(t)$ are two stochastic noises of the state and the observation models, respectively. The noises are assumed to be white and independent of the past and the present states. Equation (8a) describes the relation between the state at instants t and $t-1$. Note that $R_i(t)$ is seen as the measurement to be observed locally by user i.

The two equations (8a) and (8b) provide a probabilistic model of our problem formulation. The goal of distributed particle filtering is to get the channel assignment matrix $\mathbf{a}(t)$ sequentially using all the local measurements $R_i(t)$ of all the users i up until the current time t.

Since the channels' fading changes over time for the whole system, this affects the spectrum selection for each user at each time slot. Fortunately, with the presence of an inherent correlation between the channel realizations, the channel state at time t could be estimated from the previous spectrum assignment; i.e., at time $t-1$. We assume that each user relays its band selection, denoted as $\mathbf{a}_i(t) = (a_i^1, ..., a_i^m)$, along with its measured observation, $R_i(t)$, to the other users. This information allows the other users to estimate their best selections during the next time slot. Denoting the other users' band selections by $\mathbf{a}_{-i}(t-1)$, the global function that governs the state change and executed by each user could be expressed as

$$\mathcal{X}(t) = \arg\max_{\mathbf{a}_i(t)} \mathcal{O}_i^{\mathrm{PF}}(t)|\{\mathbf{a}_{-i}(t) = \mathbf{a}_{-i}(t-1), \tilde{h}(t)\}, \tag{9}$$

where $\tilde{h}$ is the estimate of the channel according to (3).

With conventional Bayesian approaches, to estimate $\mathbf{a}(t)$, we should compute the posterior $f(\mathbf{a}(t)|R_{1:n}(0:t))$, where f denotes a probability density function and $R_{1:n}(0:t)$ is the vector that contains the observed throughput from $t' = 0$ until $t' = t$. The state can be sequentially estimated in two steps: a prediction phase given by Equation (10a) and an update phase using Equation (10b) [14].

$$f(\mathbf{a}(t)|R_{1:n}(0:t-1)) = \int f(\mathbf{a}(t)|\mathbf{a}(t-1))f(\mathbf{a}(t-1)|R_{1:n}(0:t-1)), \quad (10a)$$

$$f(\mathbf{a}(t)|R_{1:n}(0:t)) = \frac{f(R_{1:n}(t)|\mathbf{a}(t))f(\mathbf{a}(t)|R_{1:n}(0:t-1))}{f(R_{1:n}(t))|R_{1:n}(0:t-1))}. \quad (10b)$$

Although the recursion can simplify the derivation of $f(\mathbf{a}(t)|R_{1:n}(0:t))$, it could not be straightforwardly computed due to the non-linearity and the involvement of an integral quantity.

Particle filtering theory provides an interesting tool to overcome this issue. Instead of computing the posterior $f(\mathbf{a}(t)|R_{1:n}(0:t))$, it is sufficient to consider a large number of samples from this distribution. These samples should be carefully drawn to reflect the original probability density function. Hence, it could be approximated by

$$f(\mathbf{a}(t)|R_{1:n}(0:t)) = \sum_{k=1}^{N_s} w^k(t)\delta(\mathbf{a}(t) - \mathbf{a}^k(t)), \quad (11)$$

where N_s is the number of samples, $\mathbf{a}^k(t)$ is the kth sample and $w^k(t)$ is the correspondent weight. But, since we will apply the particle filter distributively, instead of estimating $\mathbf{a}(t)$, user i estimates only its channel selection $\mathbf{a}_i(t)$ by considering a local density function known as importance density $f(\mathbf{a}_i(t)|R_i(0:t), \mathbf{a}_i(t-1), \mathbf{a}_{-i}(t))$. In this case, the particles, $\mathbf{a}_i^k(t)$, are composed by the other users' selections, $\mathbf{a}_{-i}(t)$, and a possible selection of user i. User i forwards its optimal selection $\mathbf{a}_i(t)$ to the other users to be considered in their particles. Although this importance density is optimal [15], its implementation is challenging, and hence, we instead consider the following

$$\pi(\mathbf{a}_i(t)|\mathbf{a}(t-1)) = f(\mathbf{a}_i(t)|\mathbf{a}^k(t-1), \mathbf{a}_{-i}(t)). \quad (12)$$

The weight at each sample is deduced from the previous weight and by taking into consideration the new observation. From the importance function in (12), it follows that

$$w_i^k(t) = w_i^k(t-1)f(R_i(t)|\mathbf{a}^k). \quad (13)$$

These weights are then normalized.

Over time, the weights of the different particles at each user become negligible, i.e., $w_i^k(t) \approx 0 \ \forall \ k$ except for a few particles whose weights become very large. This problem is often known as the *samples degeneracy*. This implies that huge computations will be dedicated to update particles with very minor contributions. The idea of re-sampling is to make the particles with large weights more dominant while rejecting the particles with small weights [16]. This results in Algorithm 1.

Algorithm 1. Distributed particle filtering for fair spectrum assignment in large-scale DSA systems.

INPUT: The available bands: m.

OUTPUT: The assigned spectrum for each user: $\{\mathbf{a}_i\}_{1 \leq i \leq n}$

Initialization At the first time slot $t = 0$

for all DSA user i **do**

- Generate random samples of the possible channel assignment $\{\mathbf{a}_i^k(0)\}_{k=1}^{N_s}$;
- Set the weights to be equal $\{w_i^k\}_{k=1}^{N_s} = \frac{1}{N_s}$;
- Select the best band;
- Exchange the received throughput and the selected band with other users;

end for

for all time slot t **do**

    for all DSA user i **do**

    1.   **Prediction:** Compute possible particles using (12);

    2.   **Decision:** Select the band of the particle giving the highest reward;

    3.   Start the transmission on the selected bands;

    4.   Update the channels estimation;

    5.   **Weighting:** Compute possible particles using (13);

    6.   **Normalizing the weight:**

    7.   **Re-sampling:** Apply re-sampling to avoid degeneracy;

    8.   Exchange the received throughput and the selected band;

    end for

end for

---

5   Simulation Results

We consider a DSA system with $n = 100$ agents communicating over $m = 10$ bands. We assume that at the beginning of each time episode, the sensing process is performed and the available bands are determined. The channels between the transmitter and its correspondent receiver as well as the other receivers are assumed to be Rayleigh fading channels with an average channel gain $\left[\frac{d}{d_{ki}}\right]^{\eta}$ where $d = 1Km$ is a reference distance, d_{ki} is the distance between the ith transmitter and the kth receiver and η is the path-loss exponent that is set to 3. We set the average gain of the direct channel link to be 3 dB stronger than the average gains of the interference channels. The number of particles at each user is set to $N_s = 20$ particles. We assume that each user uses an elastic traffic model [6]. In this model, each user i has its own throughput requirement threshold, $R_i^{\text{th}}(t)$, which is uniformly distributed in the interval $[0, 10kbit/s]$. The power budget for all the users is set to $4dBm$. We rely on this model to allow each user to specify its own QoS requirements, which can be different across different users. Hence, instead of considering $R_i(t)$ in the observation, we consider the reward $r_i(R_i(t))$ that follows the elastic model.

To study the performance of our scheme, we investigate the per-agent achieved throughput at each time slot. Fig. 2 shows that the distributed particle filtering approach achieves better per-agent throughput, on the average, when compared to the minimum throughput approach. This could be explained by the fact that the minimum scheme tends to penalize the users with good channels at the expense of favoring those users with poor channels to achieve the same level of throughput. On the other hand, the sum objective approach achieves, as expected, the highest throughput among the other approaches.

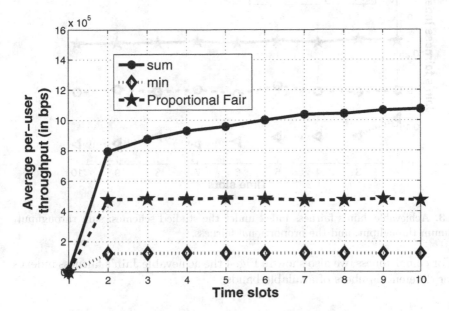

Fig. 2. Achievable throughput under the studied schemes: sum throughput, minimum throughput, and the proportional fairness.

Now, in order to evaluate the fairness of the proposed distributed approach, we use and measure the Jain's fairness index [17], defined in Eq. (14), under each of the studied approaches, and compare that achieved under our scheme with those achieved under the other ones.

$$J(t) = \frac{\left(\sum_{i=1}^{n} R_i(t) \right)^2}{n \sum_{i=1}^{n} R_i^2(t)}. \tag{14}$$

In Fig. 3, we plot the Jain's index $J(t)$. We first notice that our proposed fair distributed approach achieves better fairness than the two other approaches. The total sum throughput has the lowest fairness index since the objective is to select the best channels that allow to reach the highest total throughput rather than accounting for every user's satisfaction. Recalling Fig. 2, we conclude that ensuring fairness comes at the expense of lowering the total throughput that the network as a whole can achieve. The figure also illustrates that our approach outperforms the minimum fairness when different users have different QoS requirements. Although the latter achieves better fairness performance when users have the same requirements, for non-homogeneous environment, our approach is more suitable.

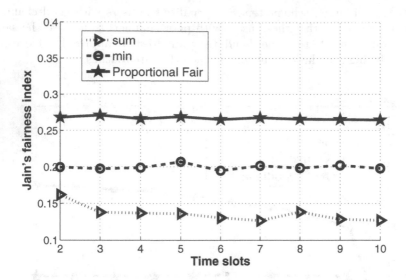

Fig. 3. Achievable Jain's fairness index under the studied schemes: sum throughput, minimum throughput, and the proportional fairness.

For completeness, we also show in Fig. 4 the achievable Jain's fairness indexes under different numbers of available bands.

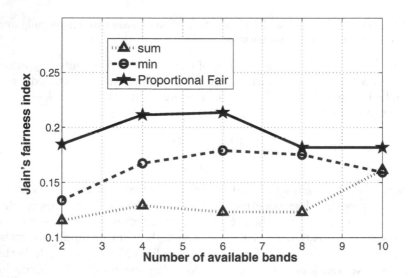

Fig. 4. Achievable Jain's fairness index when varying the number of bands.

6   Conclusions

This paper proposes a particle filtering-based technique for fair and distributed spectrum allocation in large-scale DSA systems. When compared with other approaches, the technique is shown to ensure the best fairness among users while still achieving a reasonably high network throughput.

Acknowledgment. This work was made possible by NPRP grant # NPRP 5- 319-2-121 from the Qatar National Research Fund (a member of Qatar Foundation). The statements made herein are solely the responsibility of the authors.

References

1. Notice of proposed rule making and order, Federal Communications Commission (FCC), Tech. Rep. Et docket no. 03–322, December 2003
2. Zhao, Q., Sadler, B.: A survey of dynamic spectrum access. IEEE Signal Processing Magazine **24**(3), 79–89 (2007)
3. Tragos, E., Zeadally, S., Fragkiadakis, A., Siris, V.: Spectrum assignment in cognitive radio networks: A comprehensive survey. IEEE Comm. Surveys Tutorials **15**(3), 1108–1135 (2013)
4. Wang, H., Ren, J., Li, T.: Resource allocation with load balancing for cognitive radio networks. In: Proceedings of IEEE Global Telecommunications Conference (GLOBECOM 2010), pp. 1–5, December 2010
5. Liu, K., Zhao, Q.: Distributed learning in cognitive radio networks: multi-armed bandit with distributed multiple players. In: Proceedings of IEEE International Conference on Acoustics Speech and Signal Processing (ICASSP), pp. 3010–3013, March 2010
6. Mehdi, B.G., Bechir, H., Rami, H., Mohsen, G., MohammadJavad, N.: Distributed dynamic spectrum access with adaptive power allocation: energy efficiency and cross-layer awareness. In: Proceedings of IEEE Conference on Computer Communications Workshops (INFOCOM WKSHPS), pp. 694–699, April 2014
7. MohammadJavad, N., Bechir, H., Kagan, T.: Efficient objective functions for coordinated learning in large-scale distributed OSA systems. IEEE Tran. on Mobile Computing **12**(5), May 2013
8. Ben Ghorbel, M., Khalfi, B., Hamdaoui, B., Hamdi, R., Guizani, M.: Resource allocation for large-scale dynamic spectrum access system using particle filtering. In: IEEE Global Telecommunications Conference (GLOBECOM 2014), December 2014
9. Byun, S.-S., Balasingham, I., Liang, X.: Dynamic spectrum allocation in wireless cognitive sensor networks: Improving fairness and energy efficiency. In: IEEE Veh. Technology Conf. (VTC Fall), September 2008
10. Peng, C., Zheng, H., Zhao, B.Y.: Utilization and fairness in spectrum assignment for opportunistic spectrum access. Mobile Networks and Applications **11**(4), 555–576 (2006)
11. Bertsimas, D., Farias, V.F., Trichakis, N.: The price of fairness. Operations Research **59**(1), 17–31 (2011)
12. Zhang, D., Tian, Z.: Adaptive games for agile spectrum access based on extended Kalman filtering. IEEE Journal of Selected Topics in Signal Processing **1**(1), 79–90 (2007)

13. Shi, H., Prasad, R., Onur, E., Niemegeers, I.: Fairness in wireless networks: Issues, measures and challenges. IEEE Communications Surveys Tutorials **16**(1), 5–24 (2014)
14. Hlinka, O., Hlawatsch, F., Djuric, P.: Distributed particle filtering in agent networks: A survey, classification, and comparison. IEEE Signal Processing Magazine **30**(1), 61–81 (2013)
15. Doucet, A., Godsill, S., Andrieu, C.: On sequential monte carlo sampling methods for Bayesian filtering. Statistics and Computing **10**(3), 197–208 (2000)
16. Liu, J.S., Chen, R.: Blind deconvolution via sequential imputations. J. of the American Statistical Asso. **90**, 567–576 (1995)
17. Jain, R., Chiu, D.-M., Hawe, W.R.: A quantitative measure of fairness and discrimination for resource allocation in shared computer system. Eastern Research Laboratory, Digital Equipment Corporation (1984)

Multiple Description Video Coding for Underlay Cognitive Radio Network

Hezerul Abdul Karim[1(✉)], Hafizal Mohamad[2], Nordin Ramli[2], and Aduwati Sali[3]

[1] Faculty of Engineering, Multimedia University, 63100 Cyberjaya, Selangor, Malaysia
hezerul@mmu.edu.my
[2] Wireless Communication Cluster, MIMOS Berhad,
Technology Park Malaysia, 57000 Kuala Lumpur, Malaysia
{hafizal.mohamad,nordin.ramli}@mimos.my
[3] Department of Computer & Communication, Faculty Engineering,
Universiti Putra Malaysia, 43400 Serdang, Selangor, Malaysia
aduwati@eng.upm.edu.my

Abstract. Cognitive radio (CR) with spectrum sharing allows new or secondary devices to co-exist with primary (licensed) users (PU) in accessing the spectrum. This is known as underlay CR. It allows the secondary users (SU) to transmit multimedia data services (video transmission) at low power and low data rate when the PU is using the spectrum. Hence SU can still enjoy uninterrupted video services with minimum tolerable quality. However, problem arises when SUs are subjected to interferences mainly from PU and other SUs. The objective of this paper is to provide error-free video transmission to SU in the underlay CR transmission by using an error resilience method, namely Multiple Description Coding (MDC). Since the underlay mode CR is characterized by low power, low data rate and possibly high packet loss rate, base layer video streaming of a Scalable Video Coding (SVC) with MDC is a feasible solution. The base layer video is coded using MDC with even and odd frames generating two descriptions. Simulation results show that transmitting video in the underlay CR using MDC perform better objectively and subjectively than using a single description coding (SDC).

Keywords: Cognitive radio · Underlay · Multiple description video coding · Scalable video coding · Error resilience

1 Introduction

Cognitive Radio (CR) is a technology that can address the problem of inefficient usage of radio spectrum through enhanced spectrum management capabilities [1]. The spectrum management includes opportunistic access to the spectrum and spectrum sharing. In opportunistic access, CR system allows new or secondary devices to opportunistically access a portion of spectrum, which belongs to PU as in the overlay CR [2]. In spectrum sharing, the low power SU share the spectrum with the PU as in the case of underlay CR. In underlay CR technique, the SU spreads it bandwidth large enough to ensure tolerable amount of interference to the PU.

© Institute for Computer Sciences, Social Informatics and Telecommunications Engineering 2015
M. Weichold et al. (Eds.): CROWNCOM 2015, LNICST 156, pp. 643–652, 2015.
DOI: 10.1007/978-3-319-24540-9_53

Video transmission over CR networks has received many attentions from the research community to fully utilize the available bandwidth resulting from the usage of CR, for example in [3]-[8]. Most of the papers use cross-layer technique to optimize video transmission over CR. Recently, video transmission over underlay CR has been investigated in [7] and [8]. It is a great challenge to obtain high quality video transmission in underlay CR [7]. The challenges are due to the need for the SU to use low transmission power and low transmission rate so that they do not generate interference to the PU. Moreover, SU will suffer interferences from both the PUs and other SUs. In [8], uninterrupted video services are provided to the SU by allowing the SU to receive video in both overlay and underlay mode of CR. The base layer of the scalable video coding (SVC) is transmitted with I-frame error resilience technique during the underlay mode for acceptable video quality.

The underlay mode can be characterized as having low data throughput due to the low transmission power and high packet loss due to interferences from both the PUs and other SUs [7],[9]-[11]. Hence, an error resilience video transmission in the underlay CR network is needed. One of the ways to provide the error resilience is by using multiple description coding (MDC) technique [12].

MDC has been used in video coding community to provide error resilience 2D and 3D video transmission ([12] and [13]). MDC allows a source to be encoded into two or more correlated redundant descriptions. If one of the descriptions is received, acceptable video quality can be achieved. More descriptions will provide higher video quality.

MDC for CR has been studied in [14], [15] and [16]. In [14], MDC is used to transmit images in single and two cognitive radio channels with optimized coding rates within certain amount of time. MDC is employed to cope with the packet losses caused by PU traffic interruptions. However, application of MDC for video transmission in the underlay CR is not considered in this paper. In [15], MDC framework based on Priority Encoding Transmission packetization technique is employed to cope with both primary interruptions and secondary collisions. Different amount of Forward Error Correction (FEC) is assigned to the different descriptions. Again, only image data is used as the source in this paper. Furthermore, underlay transmission is not considered in [15]. MDC multicast (MDCM) in orthogonal frequency division multiplexing (OFDM-based) cognitive radio network (CRN) is studied in [16]. The paper shows that the system throughput using the MDCM scheme with optimized resource allocation is much higher than using the conventional multicast (CM) scheme. However, the proposed system in [16] does not consider video transmission using underlay CR.

The problem of video transmission using MDC in the underlay CR is investigated in this paper. The underlay CR can be subjected to video packet loss as described in [7] and [8]. Specifically, in this paper, we propose technique of using MDC for the transmission of the SVC's base layer to enhance its error resilience property. The remainders of this paper will be organized as follows. Section 2 of the paper presents an overview of MDC followed by the proposed system model in Section 3. Section 4 presents the simulation results and discussion. The paper is concluded in Section 5.

2   Overview of MDC

MDC is one of the error resilience coding techniques that generate multiple/separate descriptions which are correlated and equally important. The descriptions can be in the form of multimedia data such as audio, video and text. In this paper, the concentration is on video data. Any descriptions provide low but at acceptable video quality. Additional descriptions provide incremental improvements to the video quality. The multiple descriptions of the video data may be sent through either the same or separate physical channels. Acceptable video quality can be maintained as long as the two video descriptions are not simultaneously (in terms of the spatial location and time) affected by packet losses [12].

Commonly, two descriptions are generated by MDC encoder. In general more than two descriptions are possible. MDC is popular because it can provide adequate video quality without retransmission and it is suitable for low delay application, hence suited for real time application such as video communication. It also simplifies network design where feedback is not needed and all MDC packets can be equally treated. When MDC is combined with   Multi Path Transport (MPT), traffic dispersion and load balancing in network can be achieved [12]. However, due to the redundancies generated, MDC has less coding efficiency than a single description coder (SDC).

Fig. 1 shows the general block diagram of MDC encoder and decoder for two descriptions. In general, it can be extended to more than two descriptions. In Fig. 1, the two descriptions created by the encoder are sent separately across two channels. The total bit rate is $R = R_1 + R_2$, where R_1 and R_2 are the bit rates used to send each description. In the case of two channels, if $E1$ and $E2$ are received, the decoder invokes Decoder 0 to decode $E1$ and $E2$, and produces a high-quality reconstruction with central distortion D_0. If only $E1$ is received, Decoder 1 is invoked to decode signal from Channel 1 producing a lower, but still acceptable quality reconstructions with side distortions D_{11}. Decoder 2 is invoked when only $E2$ is received, producing acceptable quality reconstructions with side distortions D_{12}. A balanced design is achieved when $R_1=R_2$ and $D_1=D_{11}=D_{12}$.

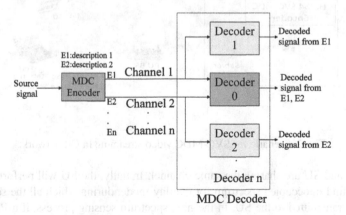

Fig. 1. General block diagram of MDC [12].

A single description coder (SDC) minimises D_0 for a fixed total rate R, and rate-distortion function $R(D_0)$ is used to measure the performance. MDC has contradictory requirements to simultaneously minimise both D_0 and D_1. At one extreme, minimising D_0 by simply alternating the R bits of an SDC bit stream into each description will have unacceptably high D_1. At the other extreme, minimising D_1 by simply duplicating the SDC bit stream with rate R into each description will have a large D_0 because it uses $2R$ bits to achieve D_0. The redundancy r is defined as the additional bit rate required by the MDC coder, $r = R-R^*$, where R is the total bit rate of the two streams in MDC, and R^* is the reference bit rate from (SDC).

In Fig. 1, the mismatch occurs when only $E1$ or $E2$ is received and only Decoder 0 is available. Decoder 0 is designed to decode both $E1$ and $E2$ together. It fails to decode when only $E1$ or $E2$ is received. The mismatch can be eliminated by having Decoder 1 and 2 specially designed to decode $E1$ and $E2$ respectively [12].

3   System Models

The scenario for SVC-MDC video streaming over the CR wireless network is illustrated in Fig. 2. The H.264/SVC-MDC encoder ([17] and [18]) is used to generate the multiple descriptions. The H.264/SVC-MDC encoder is based on even and odd frames MDC [19]. The video sequence is decomposed into even and odd frames. H264/SVC-MDC will encode the video once and many layers of video that can support spatial, temporal or quality scalability will be generated at the server. It also includes the two MDC layers, $E1$ and $E2$.

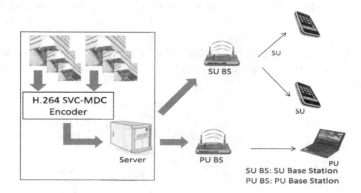

Fig. 2. Scenario for SVC-MDC video streaming in CR network.

The PU and SU are sharing the same channel. Initially, the SU will perform spectrum sensing to find unoccupied spectrum for overlay mode, during which all the scalable layers can be transmitted to the SU. In the next spectrum sensing process, if a PU is found, the SU Base Station (BS) system will switch to underlay mode and transmit the base layer only with MDC capabilities to SU. Fig. 3 shows how the MDC layers, *E1* and *E2*, are generated at the server in Fig. 2. The *E1* and *E2* MDC layers can be selected by the SU BS to be delivered to SU under the error prone underlay CR mode.

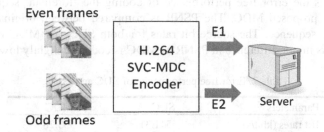

Fig. 3. E1 and E2 MDC layer generation at the server.

The even and odd frames based MDC [19] are used because of simpler implementation and theoretically does not involve modification to the existing encoder, hence less extra processing. It also can avoid mismatch coding due to each frame in *E1* and *E2* is predicted from the previous even frame and odd frame respectively. Due to this configuration, no side information is needed to prevent the mismatch coding. Furthermore, the even and odd frames based MDC can be implemented on various video encoder such as MPEG2, MPEG4 and H.264. The method can also be easily extended to more than two descriptions.

The underlay model for CR used in this paper is as described in [8]. The model produces high packet error rate of about 20%. This high packet error rate is due to the interference from PU and other SU.

4 Simulation Results

For video simulation, H.264/SVC-MDC codec [18] is used to code and decode the video sequence. The main simulation parameters for the codec are shown in Table1. The original video sequence is decomposed into even and odd frames video sequence. H.264/SVC-MDC is used to encode the decomposed video sequence to generate two descriptions, namely *E1* and *E2* as shown in Fig. 3.

Table 1. SVC Parameters.

Parameters	Value
Input video	foreman_cif.yuv
Video width	352
Video height	288
Quantisation parameters	27.5 for SDC
	30 for MDC
Frame format	IPPP
No of I-frame	One I-frame for every 60 P-frames (SDC)
	One I-frame for every 30 P-frames (MDC)

4.1   Error Free Environment

The coding performance of the proposed H.264/SVC-MDC codec in underlay condition is first evaluated under error free environment where there are no packet losses. Table 2 shows the error free performance of coding the 'foreman' sequence using SDC and the proposed MDC. The PSNR is compared between original and compressed video sequence. The source bit rates for both SDC and MDC is about 540 kbit/s. There is minor difference in PSNR as MDC is coded at slightly lower bit rates.

Table 2. Error free performance of SDC and MDC.

Parameters	SDC	MDC
Bit rates (kbit/s)	541.31	533.28
Average PSNR (dB)	37.69	36.35
Quantisation parameter	27.5	30

4.2   Underlay - Ideal MDC Channel

Ideal MDC channel assumes only one of the MDC streams are received by the decoder and the other stream is completely lost [19]. In this experiment, the two descriptions will be sent by the SU base station to the SU in the underlay mode. It was found in [8] that the video Packet Error Rate (PER) is about 20%. The video frames within the two MDC descriptions ($E1$ and $E2$) are dropped according to the PER. For fair comparison, the video frames from the SDC are also dropped using the same PER.

In the simulation, when $E1$ is subjected to 20% PER, $E1$ description is considered lost completely due to the very high PER. The decoded video sequence is reconstructed from just $E2$ utilizing frame interpolation to reconstruct the full sequence. Similarly, when $E2$ is subjected to 20% PER, $E2$ description is considered lost completely and video sequence is reconstructed from just $E1$.

Table 3 shows the error prone performance of coding the same sequence using SDC and MDC in the underlay mode that is subjected to 20% PER. It is assumed that when the $E1$ or $E2$ MDC description is subjected to 20% PER, it is considered lost completely. The results show that even though E2 MDC stream is completely lost by the 20% PER, a good video quality of PSNR 32.53dB can be achieved by just using $E1$ MDC stream. Similarly, when $E1$ MDC stream is completely lost, an acceptable video quality of PSNR 32.55dB can be achieved by just using E2 MDC stream.

Table 3. Performance of SDC and MDC in ideal MDC channel for underlay CR.

Average PSNR (dB)	SDC	MDC
Error free	37.69	36.35
At 20% packet loss	26.86	32.53 (using $E1$)
(underlay mode)		32.55 (using $E2$)

Fig. 4 shows the subjective quality comparison of the video sequence with SDC and MDC during transmission in the underlay mode. It can be seen that MDC perform better than SDC subjectively. However, due to the average frame interpolation MDC results in ghost or double image artifact as shown in Fig. 4(c) and Fig. 4(d) that can hardly be noticed when the sequence is played. Improved results can be obtained by using frame interpolation that considers motion information between frames.

(a) Original   (b) SDC, 20% PER

(c) MDC, 20% PER on E2   (d) MDC, 20% PER on E1

Fig. 4. Subjective quality of video transmission in ideal MDC channel for underlay CR.

4.3   Underlay – MC-CDMA

In this experiment, the SU is modeled as Multi Carrier CDMA (MC-CDMA) with BPSK modulation ([8] and [9]). The simulation results are based upon Additive White Gaussian Noise (AWGN) channel and the video packet size $L = 1000$ bits. Fig. 5 shows the theoretical AWGN and simulation performance for underlay CR mode for two SUs. As an example, the video packet for both SDC and the proposed MDC is dropped according to the BER at 4dB and 5dB E_b/N_o. There is more video packet dropped at 4dB E_b/N_o compared to 5dB E_b/N_o. In this experiment, any of the two of the MDC streams can be subjected to the video packet loss. The PSNR of the proposed MDC and the SDC is shown in Table 4.

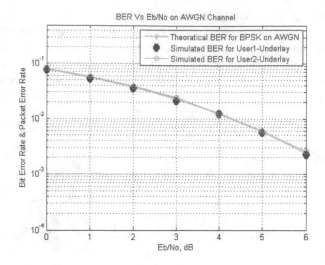

Fig. 5. BER performance for SU in underlay CR.

Table 4. Average PSNR Performance of SDC and MDC in underlay CR.

Channel Condition	Average PSNR (dB)	
	SDC	MDC
Error-free	37.69	36.35
5 dB E_b/N_o	35.36	35.29
4 dB E_b/N_o	32.96	35.15

It can be seen from Table 4 that the proposed MDC outperforms SDC at 4dB E_b/N_o with average PSNR of 35.15dB compared to 32.96dB. At 5dB E_b/N_o, where there is less packet loss, average PSNR of the proposed MDC is slightly lower than SDC due to the redundancies generated by the MDC. However, the redundancies from the proposed MDC can be used to mitigate the error (high packet loss) in the underlay CR channel. Fig. 6 shows the subjective quality improvement obtain with the proposed MDC during transmission in the underlay mode. Improved results can be obtained with proper error concealment at the MDC decoder and is subjected to future studies.

(a) SDC at 4dB E_b/N_o (b) MDC at 4dB E_b/N_o

Fig. 6. Subjective quality of video transmission in simulated underlay CR.

5   Conclusion

In this paper, the problem of video streaming using SVC in underlay mode of wireless CR has been investigated to allow uninterrupted multimedia services to the SU. In the underlay mode, interferences from PU and other SU results in video packet loss to the existing SU. Hence, some form of error resilience need to be embedded during the scalable video encoding process. In this paper, even and odd frame MDC is used as the error resilience technique to mitigate the packet losses in the underlay CR. Simulation results shows PSNR improvement of 7 dB for 'foreman' sequence is achieved by the proposed MDC at 20% PER in ideal MDC channel for underlay CR. The results also shows PSNR improvement of about 2 dB by the proposed MDC for the same sequence in simulated underlay CR channel at 4dB E_b/N_o. Future works include using channel coding to improve performance at the physical layer and using MDC with proper error concealment at the application/video.

Acknowledgments. The work presented was supported by the FRGS scheme at Multimedia University under the project account MMUE/140083.

References

1. Ramli, N., Hashim, W., Mohamad, H., Abbas, M.: Cognitive Radio Technology: A Survey. My Convergence **5**, 36–42 (2011)
2. Mohamedou, A., Sali, A., Ali, B.M., Othman, M.: Genetic fuzzy scheduler for spectrum sensing in cognitive radio networks. In: 12th International Symposium on Communication and Information Technologies, pp. 770–775. IEEE Press, Gold Coast Australia (2012)
3. Luo, H., Ci, S., Wu, D.: A Cross-layer Design for the Performance Improvement of Real-time Video Transmission of Secondary Users over Cognitive Radio Networks. IEEE Transactions on Circuits and Systems for Video Technology **21**, 1040–1048 (2011)
4. Hu, D., Mao, S., Hou, Y.T., Reed, J.H.: Scalable Video Multicast in Cognitive Radio Networks. IEEE Jounal on Selected Areas in Communications **28**, 434–444 (2010)
5. Hu, D., Mao, S.: Streaming Scalable Videos over Multi-Hop Cognitive Radio Networks. IEEE Transactions on Wireless Communications **9**, 3501–3511 (2010)
6. Bocus, M.Z., Coon, J.P., Canagarajah, C.N., McGeehan, J.P., Armour, S.M.D., Doufexi, A.: Resource Allocation for OFDMA-Based Cognitive Radio Networks with Application to H.264 Scalable Video Transmission. EURASIP Journal on Wireless Communications and Networking **245673**, 10 (2011)
7. Guan, B., He, Y.: Optimal resource allocation for video streamingover cognitive radio networks. In: IEEE International Workshop on Multimedia Signal Processing, pp. 1–6. IEEE Press, Hangzhou (2011)
8. Abdul Karim, H., Mohamad, H., Ramli, N., Sali, A.: Scalable video over overlay/underlay cognitive radio network. In: 12th International Symposium on Communication and Information Technologies, pp. 668–672. IEEE Press Gold Coast, Australia (2012)
9. Chakravarthy, V., Li, X., Wu, Z., Temple, M.A., Garber, F., Kannan, R., Vasilakos, A.: Novel Overlay/Underlay Cognitive Radio Waveforms using SD-SMSE Framework to Enhance Spectrum Efficiency—Part I: Theoretical Framework and Analysis in AWGN Channel. IEEE Trans. On Communication **57**, 3794–3804 (2009)

10. Oh, J., Choi, W.: A hybrid cognitive radio system: a combination of underlay and overlay approaches. In: IEEE Vehicular Technology Conference Fall, pp. 1–5. IEEE Press, Ottawa (2010)

11. Srinivasa, S., Jaafar, S.A.: The throughput potential of cognitive radio: a theoretical perspective. In: Asilomar Conference on Signals, Systems and Computers, pp. 221–225. Asilomar, CA (2006)

12. Wang, Y., Reibman, A.R., Lin, S.: Multiple Description Coding for Video Communications. IEEE Proceedings 93, 57–70 (2005)

13. Abdul Karim, H.: Multiple Description Coding for 3D Video. PhD Thesis. University of Surrey, UK (2008)

14. Li, H.: Multiple description source coding for cognitive radio systems. In: Fifth International Conference on Cognitive Radio Oriented Wireless Networks and Communications, pp. 1–5, ICST, Cannes (2010)

15. Chaoub, A., Ibn-Elhaj, E.: Multiple description coding for cognitive radio networks under secondary collision errors. In: 16th IEEE Mediterranean Electrotechnical Conference (MELECON), pp. 27–30. IEEE Press, Yasmine Hammamet (2012)

16. Sheng-yu, L., Wen-jun, X., Yan, G., Kai, N., Jia-ru, L.: Resource Allocation for Multiple Description Coding Multicast in OFDM-Based Cognitive Radio Networks. The Journal of China Universities of Posts and Telecommunication 19, 51–57 (2012)

17. Schwarz, H., Marpe, D., Wiegand, T.: Overview of the Scalable Video Coding Extension of The H.264/AVC Standard. IEEE Transactions on Circuits and Systems for Video Technology 17, 1103–1120 (2007)

18. Karim, H.A., Hewage, C.T.E.R., Worrall, S., Kondoz, A.M.: Scalable Multiple Description Video Coding for Stereoscopic 3D. IEEE Transaction on Consumer Electronics 54,745–752 (2008)

19. Apostolopoulos, J.G.: Error-resilient video compression via multiple state streams. In: International Workshop on Very Low Bitrate Video Coding, Kyoto, Japan, pp. 168–171 (1999)

Device-Relaying in Cellular D2D Networks: A Fairness Perspective

Anas Chaaban[1] (✉) and Aydin Sezgin[2]

[1] Electrical Engineering Program, KAUST, Thuwal, KSA
anas.chaaban@kaust.edu.sa
[2] Institute of Digital Communication Systems, RUB, Bochum, Germany

Abstract. Device-to-Device (D2D) communication is envisioned to play a key role in 5G networks as a technique for meeting the demand for high data rates. In a cellular network, D2D allows not only direct communication between users, but also device relaying. In this paper, a simple instance of device-relaying is investigated, and its impact on fairness among users is studied. Namely, a cellular network consisting of two D2D-enabled users and a base-station (BS) is considered. Thus, the users who want to establish communication with the BS can act as relays for each other's signals. While this problem is traditionally considered in the literature as a multiple-access channel with cooperation in the uplink, and a broadcast channel with cooperation in the downlink, we propose a different treatment of the problem as a multi-way channel. A simple communication scheme is proposed, and is shown to achieve significant gain in terms of fairness (measured by the symmetric rate supported) in comparison to the aforementioned traditional treatment.

Keywords: Device-relaying · D2D · Multi-way · Symmetric rate

1 Introduction

A cellular network consisting of multiple users who want to communicate with a base-station (BS) is typically studied in the literature as a combination of a multiple-access channel (MAC) [1] in the uplink (UL), and a broadcast channel (BC) [2] in the downlink (DL). The demand for higher rates of communication has motivated researchers to seek methods to improve the performance of transmission schemes for the MAC and the BC. This research has taken many different directions in the past (code design, MIMO, etc.). Although these directions will remain of great importance, a new promising direction that will be of high importance in the future is user cooperation.

User cooperation is possible in a cellular network by enabling Device-to-Device communication (D2D). D2D is a technique that is envisioned as a potential solution for meeting the high data-rate demand in future cellular networks [3]. The main idea of D2D is that devices within close proximity communicate with each other directly, without involving a BS in their communication [4],

© Institute for Computer Sciences, Social Informatics and Telecommunications Engineering 2015
M. Weichold et al. (Eds.): CROWNCOM 2015, LNICST 156, pp. 653–664, 2015.
DOI: 10.1007/978-3-319-24540-9_54

thus offloading some traffic from the BS. D2D is not only useful for this scenario where the devices act as either a source or a destination, but also for scenarios where devices can act as relays that support other devices' communication with the BS [5]. The latter variant, which we term device-relaying cellular network (DRCN), enables user cooperation. It also meets the goal of D2D, which is mainly offloading traffic from the BS. For example, a user with bad channel quality who requires T resource units (time/frequency) to send data directly to the BS, requires $T' < T$ resource units to send the same data through a user with a better channel quality (acting as a relay), which in turn frees $T - T'$ resource units at the BS. Therefore, this enables the BS to support more users; the ultimate goal of designing a network.

User cooperation has been studied earlier in the context of the MAC and the BC [6,7,8,9]. In this paper, we would like to shed light on the following aspect: Interpreting the DRCN as a cooperative MAC or BC is restrictive, in the sense that it restricts communication over a given channel resource (time/frequency) to either UL or DL but not both. Enabling both UL and DL simultaneously over a given channel resource can lead to better performance in terms of achievable rates since the coding scheme can be designed jointly for the UL and DL. Therefore, we propose approaching this problem from another perspective. In particular, we approach the problem from a multi-way channel perspective, where the multi-way channel is an extension of the two-way channel [10] to multiple users, and has been studied in [11,12,13].

New opportunities arise when we treat the DRCN as a multi-way channel, namely, multi-way communication and relaying. Multi-way communication refers to simultaneous direct communication between multiple nodes over the same channel resource. In the simplest case with two users, multi-way communication can double the channel capacity as shown in [14]. Multi-way relaying on the other hand basically refers to the idea of having a relay node compute a function of the received codewords [15] which is simultaneously useful at multiple destinations, and is based on physical-layer network coding [16]. In its simplest form, multi-way relaying in a network with two users can double the capacity of the network in comparison to one-way relaying [17,18]. Using these opportunities can boost the performance of a DRCN in comparison to the traditional cooperative MAC/BC treatment of the problem.

To show the aforementioned gains, we consider a simple DRCN consisting of two D2D-enabled users and a BS. We propose a simple transmission scheme for the network based on a time sharing combination between three components: two-way communication between user 1 and the BS, two-way communication between user 2 and the BS, and two-way relaying through user 1 (we assume that user 1 is the stronger user). While the sum-capacity of this channel has been characterized in [11] within a constant gap, the scheme proposed in [11] is not fair, since the weaker user is switched off. In practice, it is interesting to maximize the throughput of the network, under a fairness constraint among the users. This corresponds to maximizing the minimum achievable rate, known as max-min fairness. Thus, we focus here on the fair DRCN where rates are allocated to the users equally (symmetric rate). We write the achievable symmetric rate of the proposed scheme which is based on simultaneous UL/DL, and an alternative scheme based on separate UL/DL. Then we compare the achievable rates numerically as a function of the strength of the D2D channel and of the transmit power available. The comparison shows that as the quality of the D2D channel becomes better, the gain achieved by device-relaying increases. Furthermore, the simultaneous UL/DL scheme is simpler than separate UL/DL (which involves block-Markov encoding and backward decoding) and better performing in terms of symmetric rate. This comparison gives guidelines for future cellular network design, regarding the switching on/off of device-relaying capabilities. Namely, if the gain per unit cost (power, complexity, etc.) obtained by device-relaying is higher that a given target value, then the device-relaying functionality is switched on. Otherwise, the DRCN is operated as a combination of a MAC/BC.

The paper is organized as follows. We introduce the system model in Section 2. Then, we propose a simultaneous UL/DL scheme based on two-way communication/relaying in Section 3. Next, a separation based scheme which combines cooperative MAC and BC schemes is introduced in Section 4. Finally, the paper is concluded with a numerical evaluation and discussion in Section 5. Throughout the paper, we use bold-face letters to denote vectors and normal-face letters to denote scalars. The function $C(x)$ is used to denote $\frac{1}{2}\log(1 + x)$ for $x \geq 0$, and $C^+(x) = \max\{0, C(x)\}$. For $x \in [0, 1]$, $\bar{x}$ denotes $1 - x$.

2   System Model

Consider a cellular network consisting of a base station (BS) and two users as shown in Figure 1. Users 1 and 2 want to communicate with the BS (node 3) in both directions, UL and DL. In the UL, user $i \in \{1, 2\}$ wants to deliver x_{3i} to the BS which decodes $\hat{x}_{3i}$, while in the DL, the BS wants to deliver x_{i3} to user i, which decodes $\hat{x}_{i3}$. We assume that all nodes have full-duplex capability[1], and that the users can establish D2D communication through the channel indicated by h_3 in Fig. 1. Note that since the users do not have signals intended to each other, the D2D channel serves relaying purposes only, leading to a DRCN.

Node $i \in \{1, 2, 3\}$ sends the signal x_i of length n, with an average power constraint $\|x_i\|^2 \leq nP_i$. The received signals at the three nodes can be written as

$$y_i = h_j x_k + h_k x_j + z_i, \tag{1}$$

for distinct $i, j, k \in \{1, 2, 3\}$. Here x_i is the transmit signal of node i, which can in general contain a combination of fresh information and relayed information. That is, x_i depends on both the information originating at node i ($x_{ji}, j \neq i$), and the past received signals at this node. The variable $h_i \in \mathbb{R}$ denotes the

---

[1] Although the main idea can be extended to the half-duplex case, discussing the full-duplex case is more suitable here since the main concern is future networks which are expected to have sufficiently good full-duplex capabilities.

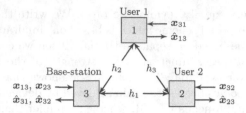

Fig. 1. In a device-relaying cellular network, users 1 and 2 want to communicate with the base station in the uplink, which in turn wants to communicate with the users in the downlink. The users can relay information to each other via the D2D channel h_3.

coefficient of the channel between nodes j and k. The channel coefficients are assumed to be static, and globally known at all nodes. The noise signal $z_i \in \mathbb{R}$ is assumed i.i.d. Gaussian with zero mean and unit variance. Note that since all nodes operate in a full-duplex mode (same time and frequency), the channels are reciprocal. We assume without loss of generality that $h_2^2 \geq h_1^2$, while the D2D channel h_3 is arbitrary.

The goal of this paper is to study the impact of the D2D channel on the achievable communication rate, subject to a fairness constraint so that the rates of all signals appearing in Fig. 1 are equal. We consider two variants of communication. In the first variant, the UL and DL phases take place simultaneously, while in the second variant, they are separated. The first variant is interesting by itself due to the potential gain arising from the use of multi-way communication and relaying. The second variant is interesting since it reflects the gain achieved by employing device-relaying in a traditional cellular system with a separated UL/DL. Moreover, the second variant serves as a benchmark for comparing the performance of modern techniques in a DRCN. We start by presenting the transmission scheme for the simultaneous UL/DL variant.

3   Simultaneous Uplink/Downlink

In a traditional cellular network without device-relaying capabilities, the users would communicate with the BS in a uni-directional manner. That is, users 1 and 2 would send the signals $x_1 = x_{31}$ and $x_2 = x_{32}$ e.g. to the BS, respectively, forming a MAC. Similarly, the BS would send the signals $x_3 = x_{13} + x_{23}$ to users 1 and 2, respectively, forming a BC. The availability of the D2D link h_3 between users 1 and 2 enables enhancing this scheme by allowing cooperation between the users [8,9]. So far, user cooperation was also applied in a uni-directional manner. That is, a user relays the other user's signal to the BS, or relays the BS signal to the other user. But cooperation can be established in a better way, by making a user relay a signal which is simultaneously useful at the BS and the other user. This can be achieved by physical-layer network coding. Furthermore, if the users and the BS can send information at the same time, then, a channel between a user and the BS can be operated as a two-way channel. With these

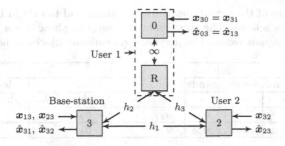

Fig. 2. The DRCN transformed into a multi-way relay channel with a direct link between two of the users (here user 2 and the BS).

ideas in mind, several schemes can be designed for the DRCN as described in the following paragraph.

A transmission scheme for the DRCN can be obtained from the multi-way relay channel with 3 users [19,20], where user i wants to send an independent signal x_{ji} to user j via a relay node by using the following idea. First, the stronger node, say user 1, is split into two nodes, a relay (R) and a virtual user (user 0). Since these two nodes are co-located, we can model them as two separated nodes connected by a channel with infinite capacity, leading to the representation in Fig. 2. If $h_1 = 0$, the problem reduces to a multi-way relay channel, where users 2 and 0 do not send signals to each other, i.e., $x_{20} = x_{02} = 0$. Thus, a transmission scheme for this case is readily obtained from the multi-way relay channel. If $h_1 \neq 0$, then from a multi-way relay channel point-of-view, we have interference between user 2 and 3 (BS) in Fig. 2. Even in this case, interference can be dealt with by using backward decoding and interference neutralization as shown in [13]. Another possibility for communicating over this channel is by switching user 2 off, and operating the DRCN as a two-way channel between user 1 and the BS. Although this scheme achieves the sum-capacity of the channel within a constant gap [11], it is not a fair scheme. In this paper, we propose a fair and simple scheme, which combines schemes for the two-way channel and the two-way relay channel in a TDMA fashion. We present the achievable rate, and then describe its achievability in detail.

Proposition 1. *The rates $R_{ij} > 0$ satisfying $R_{13} \leq \bar{R}_{13}$, $R_{23} \leq \bar{R}_b + \bar{R}_{u3}$, $R_{31} \leq \bar{R}_{31}$, and $R_{32} \leq \bar{R}_b + \bar{R}_{u2}$, are achievable in a DRCN, where*

$$\bar{R}_{13} \triangleq \alpha C(h_2^2 P_3), \qquad \bar{R}_{31} \triangleq \alpha C\left(\frac{h_2^2 P_1}{\bar{\beta}}\right), \tag{2}$$

$$\bar{R}_{u2} \triangleq \beta C\left(\frac{h_1^2 P_2}{\bar{\alpha}}\right), \qquad \bar{R}_{u3} \triangleq \beta C\left(h_1^2 P_3\right), \tag{3}$$

$$\bar{R}_b \triangleq \gamma \min\left\{C^+\left(\frac{h_3^2 P_2}{\bar{\alpha}} - \frac{1}{2}\right), C^+\left(h_3^2 P_3 - \frac{1}{2}\right), C\left(\frac{h_2^2 P_1}{\bar{\beta}}\right), C\left(\frac{h_3^2 P_1}{\bar{\beta}}\right)\right\}, \tag{4}$$

and $\alpha + \beta + \gamma = 1$.

Table 1. Summary of the operations at users 1 and 2 and the BS in the three phases of the transmission scheme. The signal sent by user 1 in phase 3 is a function $f(\cdot)$ of two-way relaying signals received in previous transmission blocks, denoted b_2' and b_3'.

Phase	1		2		3	
	sends	decodes	sends	decode	sends	decodes
User 1	u_{31}	u_{13}	-	-	$b_1 = f(h_2 b_3' + h_3 b_2')$	$h_2 b_3 + h_3 b_2$
User 2	-	-	u_{32}	u_{23}	b_2	b_1
BS	u_{13}	u_{31}	u_{23}	u_{32}	b_3	b_1

Proposition 2. *Simultaneous UL/DL in the DRCN achieves a symmetric rate* $R_{31} = R_{32} = R_{13} = R_{23} = R_{sim}$, *where*

$$R_{sim} = \max_{\alpha,\beta,\gamma} \min\{\bar{R}_{13}, \bar{R}_{31}, \bar{R}_b + \bar{R}_{u3}, \bar{R}_b + \bar{R}_{u2}\}.$$

To achieve R_{sim}, communication over the DRCN is established using the following component schemes. The first component is two-way communication [14], which is used for sending signals directly from user 1 to the BS and vice versa, and from user 2 to the BS and vice versa. The second component is two-way relaying through user 1, which is used for sending signals from user 2 to the BS and vice versa, via compute-forward relaying [17] at user 1. These components are combined by using TDMA. Thus, we divide the communication session into three phases:

1. Phase 1 spans a fraction α of the total transmission duration, and is reserved for two-way communication between user 1 and the BS,
2. phase 2 spans a fraction β of the total transmission duration, and is reserved for two-way communication between user 2 and the BS, and
3. phase 3 spans a fraction γ of the total transmission duration, and is reserved for two-way relaying between user 2 and the BS via user 1.

Since user 1 is active in phases 1 and 3, then this user can transmit at a power of $P_1/\bar{\beta}$ without violating the power constraint. Similarly, user 2 can transmit in phases 2 and 3 at a power of $P_2/\bar{\alpha}$. The BS always transmits at a power P_3. Next, we describe the transmission procedure in each of the three phases (a summary is given in Table 1).

Phases 1 and 2: In phase 1, user 1 and the BS communicate as in a two-way channel [14], while user 2 remains silent. That is, user 1 sends u_{31} and the BS sends u_{13} with powers $P_1/\bar{\beta}$ and P_3, respectively. At the end of the transmission, user 1 decodes u_{13} and the BS decodes u_{31}. The achievable rate of this phase is as given in the constraints in (2) [14], where the factor α accounts for the duration of phase 1.

Similar transmission is used between user 2 and the BS in phase 2, where user 2 sends u_{32} and the BS sends u_{32} with powers $P_2/\bar{\alpha}$ and P_3, respectively. This achieves the rates given in (3), where the factor β accounts for the duration of phase 2.

Phase 3: In phase 3, all three nodes are active. Namely, user 2 and the BS communicate through user 1 which acts as a bi-directional relay. Thus, user 2 sends a signal b_2 with power $P_2/\bar{\alpha}$, and the base station sends a signal b_3 with power P_3. The two signals have rate R_b, and are constructed using a nested-lattice code [15]. User 1 thus is able to decode $h_2 b_3 + h_3 b_2$ [17,21], which is possible if the rate R_b satisfies

$$R_b \leq \min \left\{ C^+ \left(h_2^2 P_3 - \frac{1}{2} \right), C^+ \left(\frac{h_3^2 P_2}{\bar{\alpha}} - \frac{1}{2} \right) \right\}. \tag{5}$$

Then, user 1 maps the decoded sum to a signal b_1 with power $P_1/\bar{\beta}$ and rate R_b, and sends this signal to both user 2 and the BS in the next transmission block. User 2 and the BS can decode this signal if

$$R_b \leq \min \left\{ C \left(\frac{h_2^2 P_1}{\bar{\beta}} \right), C \left(\frac{h_3^2 P_1}{\bar{\beta}} \right) \right\}. \tag{6}$$

After decoding b_1, user 2 and the BS obtain $h_2 b_3 + h_3 b_2$. User 2 then extracts b_3 by subtracting b_2, and the BS extracts b_2 by subtracting b_3. This leads to the rate constraint (4), where the factor γ accounts for the duration of phase 3.

The combination of these three phases achieves the rates given in Proposition 1. Since we are seeking a symmetric rate (fair scheme), we calculate the minimum between the achievable rates for a given α, β, γ, and optimize the outcome over all time sharing parameters satisfying $\alpha + \beta + \gamma = 1$. This leads to the rate R_{sim} given in Proposition 2. In order to evaluate the performance of the given scheme, we compare it with a scheme based on uplink/downlink separation. This scheme is described briefly next.

4 Uplink/Downlink Separation

The DRCN can be interpreted as a MAC with cooperation (MAC-C) between the transmitters in the UL, and as a BC with cooperation (BC-C) between the receives in the DL. As such, in addition to the simultaneous UL/DL operation mode explained above, the network can be operated as a combination of a MAC-C and a BC-C by separating the UL and DL in time. That is, communication is divided into two phases, an UL MAC-C phase and a DL BC-C phase. These two models have been studied in literature. The following paragraphs review results on the achievable rate regions in the MAC-C and the BC-C.

4.1 MAC-C

The capacity of the UL phase has been studied in [7,8], where a transmission scheme based on Willems' results on the MAC with generalized feedback [6] was proposed. Shortly, in transmission block t, user i sends $x_i(t) = x_{ji}(t) + x_{3i}(t) + \alpha_i x_c(t)$, where x_{ji} is the cooperation signal form user i to user $j \neq i$ with power p_{ji}, x_{3i} is the signal intended to the BS with power p_{3i}, and x_c is a common signal sent by both users to the BS, where the power of $\alpha_i \boldsymbol{x}_c$ is p_{ci}. The signal $\boldsymbol{x}_c$ is available at both users as a result of cooperation using $\boldsymbol{x}_{ji}$. In other words, $\boldsymbol{x}_c(t)$ is a function of the signals $\boldsymbol{x}_{ji}(t-1)$ and $\boldsymbol{x}_{ij}(t-1)$ where the latter has been decoded by user i in transmission block $t-1$. The power constraint of user i is satisfied if $p_{ji} + p_{3i} + p_{ci} \leq P_i$. The decoding of cooperation signals is done by decoding $\boldsymbol{x}_{ji}$ while treating $\boldsymbol{x}_{3i}$ as noise at user i, while using the common known signal $\boldsymbol{x}_c$ as side-information. The BS decodes both the intended signals $\boldsymbol{x}_{31}$ and $\boldsymbol{x}_{32}$, and the cooperation signals $\boldsymbol{x}_{21}$, $\boldsymbol{x}_{12}$, and $\boldsymbol{x}_c$. The achievable symmetric rate of this scheme is given as follows [8].

Theorem 1. *The symmetric uplink rate $R_{31} = R_{32} = R^{\mathrm{u}}(P_1, P_2)$ is achievable in the MAC-C, where $R^{\mathrm{u}}(P_1, P_2) = \max \min \left\{ \bar{R}_1^{\mathrm{u}}, \bar{R}_2^{\mathrm{u}}, \frac{1}{2}\bar{R}_{\Sigma,1}^{\mathrm{u}}, \frac{1}{2}\bar{R}_{\Sigma,2}^{\mathrm{u}} \right\}$, the maximization is over all power allocations satisfying $p_{ji} + p_{3i} + p_{ci} \leq P_i$, and where*

$$\bar{R}_i^{\mathrm{u}} \triangleq C \left(\frac{h_3^2 p_{ji}}{1 + h_3^2 p_{3i}} \right) + C \left(h_j^2 p_{3i} \right), \quad i, j \in \{1, 2\}, \ i \neq j \tag{7}$$

$$\bar{R}_{\Sigma,1}^{\mathrm{u}} \triangleq C \left(h_2^2 P_1 + h_1^2 P_2 + 2\sqrt{h_1^2 h_2^2 p_{c1} p_{c2}} \right) \tag{8}$$

$$\bar{R}_{\Sigma,2}^{\mathrm{u}} \triangleq C \left(h_2^2 p_{31} + h_1^2 p_{32} \right) + C \left(\frac{h_3^2 p_{21}}{1 + h_3^2 p_{31}} \right) + C \left(\frac{h_3^2 p_{12}}{1 + h_3^2 p_{32}} \right). \tag{9}$$

4.2   BC-C

The BC-C has been studied in [9], where the capacity of the channel was studied. A transmission scheme for the BC-C was proposed based on a combination of superposition block-Markov encoding, decode-forward at user 1 (the stronger user), and compress-forward at user 2 (the weaker user). Namely, the BS sends $\boldsymbol{x}_3 = \boldsymbol{x}_{c3} + \boldsymbol{x}_{23} + \boldsymbol{x}_{13}$, where $\boldsymbol{x}_{c3}$ has power p_{c3}, $\boldsymbol{x}_{23}$ has power p_{23}, and $\boldsymbol{x}_{13}$ has power p_{13}. The signal $\boldsymbol{x}_{c3}$ is a cooperation signal, desired at user 2, but also decoded by user 1 for relaying in subsequent transmissions. That is, user 1 decodes $\boldsymbol{x}_{c3}$ and uses it to generate $\boldsymbol{x}_1$ with power P_1. The signals $\boldsymbol{x}_{23}$ and $\boldsymbol{x}_1$ are scaled versions of each other, i.e., $\boldsymbol{x}_1 = \alpha \boldsymbol{x}_{23}$, which allows increased rates of decoding $\boldsymbol{x}_{23}$ at user 2. Finally, the signal $\boldsymbol{x}_{13}$ is dedicated to user 1. User 2 helps user 1 in decoding this latter signal by compressing its received signal $\boldsymbol{y}_2$, and sending the compressed signal as $\boldsymbol{x}_2$ with power p_2. The power constraints are satisfied if $p_{c3} + p_{23} + p_{13} \leq P_3$, and $p_2 \leq P_2$. User 1 decodes $\boldsymbol{x}_{c3}$ first, then it decodes $\boldsymbol{x}_2$ and decompresses $\boldsymbol{y}_2$, and next combines this decompressed $\boldsymbol{y}_2$ with $\boldsymbol{y}_1$ for decoding $\boldsymbol{x}_{13}$. User 2 decodes $\boldsymbol{x}_{c3}$ and $\boldsymbol{x}_{23}$. This results in the following achievable symmetric rate [9].

Theorem 2. *The symmetric downlink rate $R_{13} = R_{23} = R^{\mathrm{d}}(P_1, P_2, P_3)$ is achievable in a BC-C, where $R^{\mathrm{d}}(P_1, P_2, P_3) = \max \min\{ \bar{R}_1^{\mathrm{d}}, \bar{R}_{2,1}^{\mathrm{d}}, \bar{R}_{2,2}^{\mathrm{d}} \}$, the maximization is over all power allocations satisfying $p_2 \leq P_2$ and $p_{c3} + p_{23} + p_{13} \leq P_3$, and where*

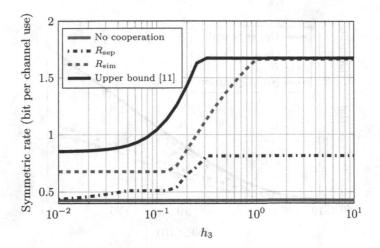

Fig. 3. Symmetric rates R_{sim} and R_{sep} as a function of the D2D channel h_3 for a DRCN with $P_1 = P_2 = P_3 = 100$, $h_2 = 1$, and $h_1 = 0.15$.

$$\bar{R}_1^{\mathsf{d}} \triangleq C \left(h_2^2 p_{13} + \frac{h_1^2 h_3^3 p_{13} p_2}{1 + h_3^2 p_2 + (h_1^2 + h_2^2) p_{13}} \right) \tag{10}$$

$$\bar{R}_{2,1}^{\mathsf{d}} \triangleq C \left(\frac{h_1^2 (p_{c3} + p_{23}) + h_3^2 P_1 + 2\sqrt{h_1^2 h_3^2 p_{23} P_1}}{1 + h_1^2 p_{13}} \right) \tag{11}$$

$$\bar{R}_{2,2}^{\mathsf{d}} \triangleq C \left(\frac{h_2^2 p_{c3}}{1 + h_2^2 p_{13}} \right). \tag{12}$$

4.3   MAC-BC-C Scheme

From Theorems 1 and 2, we can design a transmission strategy for the DRCN. Namely, we can state the following achievable symmetric rate.

Proposition 3. *Separate UL/DL in the DRCN achieves a symmetric rate* $R_{31} = R_{32} = R_{13} = R_{23} = R_{sep}$, *where* $R_{sep} = \max \min \left\{ \tau R^{\mathsf{u}}(p_1^{\mathsf{u}}, p_2^{\mathsf{u}}), \bar{\tau} R^{\mathsf{d}} \left(p_1^{\mathsf{d}}, p_2^{\mathsf{d}}, \frac{P_3}{\bar{\tau}} \right) \right\}$, *the maximization is over* $\tau \in [0,1]$, $p_i^{\mathsf{u}} \leq P_i$, *and* $p_i^{\mathsf{d}} = (P_i - \tau p_i^{\mathsf{u}})/\bar{\tau}$, $i = 1, 2$.

The transmission scheme achieving the symmetric rate R_{sep} operates as follows. First, the transmission is divided into two phases, an UL phase whose duration is a fraction $\tau \in [0,1]$ of the overall transmission time, and a DL phase during the remaining transmission time. During the UL phase, users 1 and 2 communicate with the BS using the MAC-C scheme of [8], with powers p_1^{u} and p_2^{u}. Thus, the uplink symmetric rate given by $R^{\mathsf{u}}(p_1^{\mathsf{u}}, p_2^{\mathsf{u}})$ is achievable. During the DL phase, the BS communicates with users 1 and 2 with power $P_3/\bar{\tau}$, which in turn cooperate with powers p_1^{d}, and p_2^{d}, respectively. Thus, the downlink symmetric rate given by

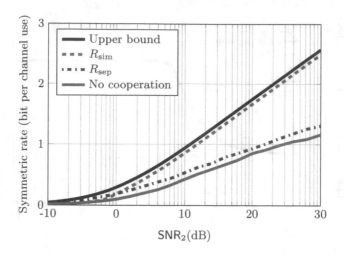

Fig. 4. Symmetric rates R_{sim} and R_{sep} as a function of SNR_2 for a DRCN with $P_1 = P_2 = P_3 = P$ and $h_1 = 0.5$, $h_2 = 1$, and $h_3 = 2$.

$R^{\text{d}}\left(p_1^{\text{d}}, p_2^{\text{d}}, \frac{P_3}{\bar{\tau}}\right)$ is achievable. The overall symmetric rate is the minimum between $\tau R^{\text{u}}(p_1^{\text{u}}, p_2^{\text{u}})$ and $\bar{\tau} R^{\text{d}}\left(p_1^{\text{d}}, p_2^{\text{d}}, \frac{P_3}{\bar{\tau}}\right)$, maximized over the parameters τ, p_1^{u}, and p_2^{u}. It remains to guarantee that the power constraints are satisfied. Since the BS is active for a fraction $\bar{\tau}$ of the time, the power constraint at the BS is satisfied. The power constraints at the users are satisfied by the choice of p_i^{d} given $p_i^{\text{u}} \leq P_i$.

5   Comparison and Discussion

In order to compare the performance of the simultaneous UL/DL and the separate UL/DL schemes, we choose a setup where $h_2 = 1$, $h_1 = 0.15$, and $P_1 = P_2 = P_3 = P = 100$, i.e., the signal-to-noise ratio of h_2 is $\text{SNR}_2 = h_2^2 P = 20\text{dB}$ and that of h_1 is $\text{SNR}_1 = h_1^2 P = 3.5\text{dB}$. We calculate the achievable symmetric rates R_{sim} and R_{sep} numerically, for $h_3 \in [0.01, 10]$. As we can see in Fig. 3, both schemes achieve cooperation gain. This gain increases when h_3 becomes larger than the weaker channel h_1. Furthermore, the simultaneous UL/DL scheme outperforms the separate UL/DL scheme.

The behaviour of R_{sim} can be interpreted as follows. If h_3 is small, then relaying through user 1 does not help since the resulting rate (4) will be close to zero. In this case, γ is set to zero, and α and β are set so that $\alpha C(h_2^2 P) = \beta C(h_1^2 P)$ and $\alpha + \beta = 1$, leading to a rate given by half the harmonic mean of $C(h_2^2 P)$ and $C(h_1^2 P)$. As h_3 increases beyond h_1 but is still smaller than h_2, the rate $\bar{R}_b$ becomes larger than $C(h_1^2 P)$ and it becomes better for user 2 to communicate with the BS through user 1. In this case, β is set to zero, and we need to choose α and γ such that $\min\{\bar{R}_b, \alpha C(h_2^2 P)\}$ is maximized. Since $\bar{R}_b$ increases with h_3, so does R_{sim}. This increase continues until h_3 becomes larger than h_2, at which $\bar{R}_b$ stops to increase since h_2 becomes the bottleneck.

At this point, setting $\alpha = \gamma = \frac{1}{2}$ achieves R_{sim} close to optimal. On the other hand, if h_3 is small, the MAC-BC-C performs similar to a MAC/BC without cooperation, and its performance improves slowly as h_3 increases. Then, the rate R_{sep} saturates due to (12). The gain achieved by simultaneous UL/DL is significant especially at large h_3 (nearly two-fold), and can be interpreted as a multi-way communication/relaying gain.

Simultaneous UL/DL achieves lower gain for lower values of h_3 in comparison to MAC/BC without cooperation. This observation is interesting for future cellular network design, since it indicates when the relaying functionality of a device should be switched on. Namely, if the rate-gain is beyond a certain value that is acceptable for a given price (power, complexity, etc.), then the functionality is switched on, otherwise, the network is operated in a MAC/BC mode.

As a conclusion, in a cellular network with D2D communication, where users are allowed to relay information to/from other users, simultaneous UL/DL can achieve higher symmetric rates than separate UL/DL. The achievable symmetric rates are plotted in Fig. 4 for a setup with $h_1 = 0.5$, $h_2 = 1$, $h_3 = 2$, and $P = P_1 = P_2 = P_3$, as a function of $\text{SNR}_2 = h_2^2 P$. A significant gain is achieved by simultaneous UL/DL in the moderate/high SNR regime. This multi-way communication/relaying gain can be a potential solution for high data-rate demand in future communication systems, especially that D2D communication is envisioned to be part of those systems.

Acknowledgment. This work is supported by the German Research Foundation, Deutsche Forschungsgemeinschaft (DFG), Germany, under grant SE 1697/5.

References

1. Ahlswede, R.: Multi-way communication channels. In: Proc. of 2nd International Symposium on Info. Theory, Tsahkadsor, Armenian S.S.R., pp. 23–52, September 1971
2. Cover, T.M.: Broadcast channels. IEEE Trans. on Info. Theory **IT–18**(1), 2–14 (1972)
3. Asadi, A., Wang, Q., Mancuso, V.: A survey on device-to-device communication in cellular networks. IEEE Communications Surveys and Tutorials (99), April 2014
4. Laya, A., Wang, K., Widaa, A.A., Alonso-Zarate, J., Markendahl, J., Alonso, L.: Device-to-device communications and small cells: enabling spectrum reuse for dense networks. IEEE Wireless Communications **21**(4), 98–105 (2014)
5. Tehrani, M.N., Uysal, M., Yanikomeroglu, H.: Device-to-device communications in 5G cellular netowrks: Challenges, solutions, and future directions. IEEE Communications Magazine, 86–92, May 2014
6. Willems, F.M.J.: Informationtheoretical results for the discrete memoryless multiple access channel, Ph.D. dissertation, Katholieke Univ. Leuven, Leuven, Belgium, October 1982
7. Sendonaris, A., Erkip, E., Aazhang, B.: User cooperation diversity part I: System description. IEEE Trans. on Communications **51**(11), 1927–1938 (2003)
8. Kaya, O., Ulukus, S.: Power control for fading cooperative multiple access channels. IEEE Trans. on Wireless Comm. **6**(8), 2915–2923 (2007)

9. Liang, Y., Veeravalli, V.V.: Cooperative Relay Broadcast Channels. IEEE Trans. on Info. Theory **53**(3), 900–928 (2007)
10. Shannon, C.E.: Two-way communication channels. In: Proc. of Fourth Berkeley Symposium on Mathematics, Statistics, and Probability, vol. 1, pp. 611–644 (1961)
11. Chaaban, A., Maier, H., Sezgin, A.: The degrees-of-freedom of multi-way device-to-device communications is limited by 2. In: Proc. of IEEE International Symposium on Info. Theory (ISIT), Honolulu, HI, pp. 361–365, June 2014
12. Ong, L.: Capacity results for two classes of three-way channels. In: Proc. of the: International Symposium on Communications and Information Technologies (ISCIT), Gold Coast, QLD, October 2012
13. Maier, H., Chaaban, A., Mathar, R., Sezgin, A.: Capacity region of the reciprocal deterministic 3-way channel via delta-y transformation. In: Proc. of the 52nd Annual Allerton Conference on Communication, Control, and Computing, Monticello, Illinois, October 2014
14. Han, T.S.: A general coding scheme for the two-way channel. IEEE Trans. Info. Theory **30**(1), 35–44 (1984)
15. Nazer, B., Gastpar, M.: Compute-and-forward: Harnessing interference through structured codes. IEEE Trans. on Info. Theory **57**(10), 6463–6486 (2011)
16. Wilson, M.P., Narayanan, K., Pfister, H.D., Sprintson, A.: Joint physical layer coding and network coding for bidirectional relaying. IEEE Trans. on Info. Theory **56**(11), 5641–5654 (2010)
17. Nam, W., Chung, S.-Y., Lee, Y.H.: Capacity of the Gaussian two-way relay channel to within 1/2 bit. IEEE Trans. on Info. Theory **56**(11), 5488–5494 (2010)
18. Avestimehr, A.S., Sezgin, A., Tse, D.: Capacity of the two-way relay channel within a constant gap. European Trans. in Telecommunications **21**(4), 363–374 (2010)
19. Chaaban, A., Sezgin, A.: Signal space alignment for the Gaussian Y-channel. In: Proc. of IEEE International Symposium on Info. Theory (ISIT), Cambridge, MA, pp. 2087–2091, July 2012
20. Gao, F., Cui, T., Jiang, B., Gao, X.: On communication protocol and beamforming design for amplify-and-forward N-Way relay networks. In: Proc. of IEEE International Workshop on Computational Advances in Multi-Sensor Adaptive Processing (CAMSAP), Aruba, December 2009
21. Nazer, B.: Successive compute-and-forward. In: Proc. of the 22nd International Zurich Seminar on Communication (IZS 2012), Zurich, Switzerland, March 2012

Interference Mitigation and Coexistence Strategies in IEEE 802.15.6 Based Wearable Body-to-Body Networks

Muhammad Mahtab Alam$^{(\boxtimes)}$ and Elyes Ben Hamida

Qatar Mobility Innovations Center (QMIC), Qatar Science and Technology Park
(QSTP), PO. Box. 210531, Doha, Qatar
{mahtaba,elyesb}@qmic.com

Abstract. This paper is focused on understanding the impact of interference in wearable wireless body-to-body networks (BBN). We have presented and compared two *non-collaborative* schemes (i.e., Time-shared
and channel hopping) and one *collaborative* technique (i.e., CSMA/CA).
For the performance evaluation, different metrics such as packet error
rate (PER), packet reception ratio (PRR), energy consumption and
latency are considered. In order to have accurate evaluation, a comprehensive and realistic simulation framework and cross-layered based
system models are developed in a network simulator. Finally, the results
show that, for *non-collaborative* channel hopping approach outperforms
the time shared scheme in all the metrics especially even at lowest transmission power. Whereas, CSMA/CA approach performs much better in
terms of delay as well as PRR, however, it is costly in terms of energy
consumption.

Keywords: Wearable body-to-body networks · Interference mitigation ·
Coexistence · IEEE 802.15.6 · Performance evaluation

1 Introduction

A Wearable Wireless Sensor Networks is a self-organized network at the human
body scale. It consists of heterogeneous smart devices which are low-power,
miniaturized, hardware-constrained (with limited processing and storage capabilities), and attached to (or implanted inside) a human body. These devices
can be sensors (to sense, transmit and receive data), actuators (to react according to the perceived data) or coordinators (to act as a gateway for the external
network). Typically sensors are connected to monitor physiological signs (e.g.
heartbeat, temperature, etc.), movement and activity (e.g. acceleration, orientation etc.) and surrounding environments (e.g. temperature, toxic gases, etc.).
Wearable wireless sensor networks have gained significant attention in daily life
applications. In health-care sector, remote and mobile monitoring of patients
from physician or hospitals is a reality, self monitoring and early diagnosis is
also possible. Athletes and players uses various wearable devices to maintain

© Institute for Computer Sciences, Social Informatics and Telecommunications Engineering 2015
M. Weichold et al. (Eds.): CROWNCOM 2015, LNICST 156, pp. 665–677, 2015.
DOI: 10.1007/978-3-319-24540-9_55

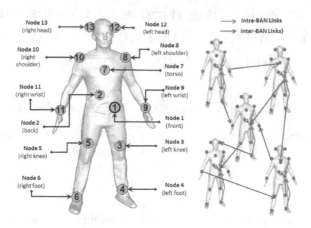

Fig. 1. Intra-BAN and Inter-BAN Networks and Interference Scenarios.

their fitness. Further the concept of augmented reality is getting mature due to convergence of technologies, data and computing.

In this work, we will emphasize on our on-going research with an application scenario of wearable wireless sensor networks for rescue and critical operations for emergency and disaster management [1]-[2]. In the given application context, most often, the existing infrastructure is either damaged or over-saturated, therefore, body area networks will create a new network for wireless communication. Further, multiple human bodies will enable coordination and communication through wireless Body-to-Body Networks (BBN). Figure 1, shows an example in which number of nodes are placed on a body for the intra-body communication and multiple bodies are closely located to effectively coordinate and communicate with each other in rescue and disaster operation. The coordinating nodes are responsible and controlling the communication on the body as well as between the bodies. However, one of the fundamental problem while being close to each other is that the sensors connected on one body interfere with the sensors connected on the other bodies and therefore, can interrupt and interfere the intra and inter body communications. In this regard, most of the literature focus on adjacent channel interference (i.e., interference from other standards), whereas interference mitigation and coexistence schemes for BBN is very limited. The focus in this paper is to ensure effective communication within and between multiple bodies by applying suitable coexistence strategies. In this context, recently released IEEE 802.15.6 standard proposed three methods including *beacon shifting*, *channel hopping* and *active superframe interleaving*. The performance of these strategies are yet to be evaluated especially in the context of BBN. Concerning that, we have considered a simulation-based approach mainly because of lack of commercially available IEEE 802.15.6 compliant radio transceiver for prototyping and experimentation. Further, BBN is very complex networks to analyze analytically because there are many parameters with huge set of possible combinations.

The contributions of this paper are as follows. First, we proposed modified, simpler and more efficient versions of coexistence schemes. For non-collaborative approach, *time shared* mechanism is implemented which is a simplified version of *beacon shifting* since it does not has to maintain big table with beacon shifting indexes. For the case of *channel hopping* technique a random channel selection method is adapted. Further, a modified and simpler IEEE 802.15.6 compliant *carrier sense multiple access/collision avoidance CSMA/CA* based *collaborative* coexistence strategy is implemented and evaluated. Second, as the reliability and quality-of-service are key performance constraints for the given applications, therefore, the performance of the coexistence strategies are evaluated in terms of packet error rate, packet delivery ratio, packet latency and energy consumption. The evaluation is achieved under realistic environment including accurate intra and inter BAN mobility and radio link modeling, realistic pathloss and channel models, IEEE 802.15.6 proposed MAC models (i.e., CSMA/CA and scheduled access), which are developed for BBN systems.

2 Related Works

Generally the interference mitigation is classified into *collaborative* and *non − collaborative* coexistence techniques. In collaborative methods multiple nodes interact with each other to manage coexistence, whereas, in non-collaborative multiple nodes manage coexistence without any interaction. The initial research studies on WBAN interference mainly concentrate on the impact from other technologies (aka., adjacent channel interference) such as IEEE 802.11, IEEE 802.15.1, *etc*. It is clear from the previous research works such as [3–6], that there is a dominant interference from other networks in WBAN. These approaches of interference analysis are only enough for intra-BAN communication, where each node is synchronized with its coordinator and are configured at the same transmit power. However, with an advent of body-to-body communications, inter-BAN interference and its mitigation is a new problem. Merely a few studies targeted this issue, for example, [7] focused the study on the measurement of the coupling between 10 bodies in a room at (2400-to-2500) MHz. An average pathloss of -67.9 dB and standard deviation of 5 dB was observed. These bodies were separated by 1-to-5 meters in hospital environment. Though the measurements conducted are interesting, however it is only limited to static case without any mobility considerations. Further, it does not show how much application performance loss is expected with evaluated interference. Finally, the coexistence strategy results are limited to only packet delivery. Three co-located WBANs were configured to operate at different transmission power levels, with chirp and duty cycling sampling receivers in [8]. The results showed that under high traffic density the chirp receiver is more immune to interference than the sampling receiver where the PLR was around 1% for up to 10 co-located users. However, the transceivers operate at *ultra wide band* (UWB), whereas the impact of interference in *narrow band* is much more stronger and evident.

More recently, the authors addressed in [9] the issue of co-channel interference between co-located multiple BANs. Two uncoordinated approaches are presented, first, a semi-random strategy is used to re-allocate the slots in TDMA mode. A coordinating node checks if the total interference experienced by all the receiver nodes (based on the random slot assignment) is less then the current slot, the slot assignment for the next frame changes otherwise, it remains unchanged. In the second approach, a minimum interference slot assignment algorithm is chosen instead of assigning random slot. These proposed approaches are limited due to number of un-realistic assumptions. First in random slot assignment, the performance of actual throughput and delay suffers especially due to lack of realistic mobility, low-to-high traffic and nodes density. Further, the actual interference is not calculated or estimated instead it is based on assumptions. Finally no coexistence method of IEEE 802.15.6 standard is evaluated. With reference to inter-BAN interference mitigation, the recently released IEEE 802.15.6 standard (targeted for WBAN), has proposed several methods for coexistence. These include, *beacon shifting*, *channel hopping* and *active superframe interleaving* [10]. To best of our knowledge, the performance of these methods in particular with inter-BANs context is yet to be evaluated. Further, all these methods are non-collaborative and are based on pre-defined strategies. In addition to that, in this paper we will also analyze the impact of collaborative technique using IEEE 802.15.6 compliant CSMA/CA method. This can be considered as implicit collaborative technique in which nodes do not share any specific information to each other but the interference can be minimized through only proper channel sensing.

3   System Models

Wireless Body-to-Body Networks (BBN) is relatively a new dimension of WBAN in which multiple bodies interact and share certain information. Fig. 1, shows an overview of on-body links and body-to-body links. In this section, we will explain various cross-layer components of the BBN system.

3.1   IEEE 802.15.6 MAC Models

The IEEE 802.15.6 standard provides a great flexibility to the researchers and developers to adapt the medium access as their requirements. In classical healthcare WBAN systems, *time division multiple access* (TDMA) based medium access control is most often considered. Every sensor node has a dedicated slot to transfer its data to the other sensors or coordinator. Moreover, works such as [11]-[12] can further help to optimize the slot scheduling based on the traffic load. Historically, limited attention has been given to CSMA/CA, however, very-low duty cycle CSMA/CA based protocols such as [13] seems very attractive. IEEE 802.15.6 MAC can be implemented through CSMA/CA, TDMA, slotted aloha, scheduled access as well as polling and posting mechanisms. The MAC layer can operate in three different modes. In beacon mode with superframe boundary,

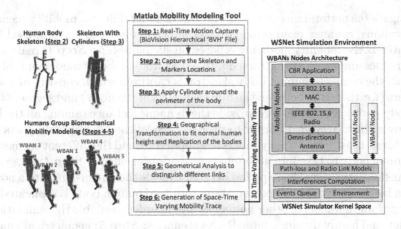

Fig. 2. Joint Biomechanical, Group Mobility and Radio Link Modeling for BANs and BBNs.

the higher priority and emergency data transfer can execute in exclusive access phase (EAP) including both EAP1 and EAP2. For regular non-emergency traffic two random access phase (i.e., RAP1 and RAP2) can be considered. Both EAP and RAP can use only CSMA/CA or slotted aloha channel access schemes. Further, managed access phase (MAP) can be scheduled both in beacon enabled and non-beacon modes. Application-specific optimal MAC configurations are presented for intra-BAN in one of our earlier work [14]. Those configurations are evaluated through physical and MAC parameters for the scheduled access. In this paper, we have extended the work for body-to-body communications through extensive simulations. IEEE 802.15.6 proposed CSMA/CA MAC and scheduled access MAC are implemented in a packet-oriented network simulator as explained later in section 4.1 for an inter-BAN analysis.

3.2 Realistic Mobility and Physical Layers Modeling for BBNs

The accurate mobility, path-loss and radio link modeling is a key requirement in order to get more insight into the performance of wireless communication stacks under real deployment and operating assumptions [15–17]. This is especially true in the context of BANs and BBNs, whose radio channels might undergo harsh multi-path fast fading and time-varying slow fading due to human body shadowing effects [18]. To that end, we consider in this work the Intra-BAN biomechanical mobility and radio link models which we recently introduced in [15], and we extend these to handle the inter-BANs case.

Intra/Inter-BANs Biomechanical Mobility and PathLoss Modeling. Modeling the mobility and posture behaviors of real human bodies is a complex task. One solution consists in exploiting real-time motion capture data and to couple them with geometrical transformation and analysis techniques to properly investigate the performance of BANs and BBNs under different mobility scenarios (*e.g.* walking, running, exercising, *etc*). As shown in Figure 2, our proposed Intra and Inter-BANs mobility modeling is based on six main steps: *Step 1*: real motion capture measurements are extracted into our Matlab mobility modeling tool [15]; *Step 2*: the complete human body skeleton is captured which consists in a set of markers (*i.e.* the joints between the different parts of the body) and segments (*i.e.* the body parts). These markers provide the dynamic distances among all the locations over time; *Step 3*: In order to properly model the human body parts (*e.g.* arms, torso, head, legs, *etc*.), cylinders are applied around the different segments of the human body to take into account body shadowing effects; *Step 4*: geographical transformations are then applied in order to scale the dimensions into a normal human height and width. Moreover, the determined human body is replicated into a configurable numbers of other human bodies in order to enable the simulation of complex and highly dynamic inter-BANs scenarios; *Step 5*: geometrical analysis is thus applied in order to determine the types of all the available links (*e.g.* LOS or NLOS, Intra or Inter BANs) and during the whole trace duration. Exact link types during mobility are evaluated by checking the intersection of the cylinders between all the links. If a link intersects with a cylinder, then the link is declared as NLOS, otherwise it is in LOS state; *Step 6*: finally, space-time varying links and mobility traces are generated and stored in an external file, which ultimately can be fed into the **WSNet** packet-oriented simulation environment [16] to enable the realistic performance evaluation of high level communication protocols. More details about the six steps can be found in [14]. Once the space-time varying links and mobility traces are properly generated for a given mobility scenario, channel models can be applied in order to assess the performance of radio-links. The IEEE 802.15.6 standard has already proposed various channel models, including the *CM3* (*body surface to body surface*) and *CM4* (*body surface to external*) models. However, it was shown that these models provide only basic distance-based path-loss without any time varying effects and correlations features [15]. Due to these limitations, the enhanced IEEE 802.15.6 path-loss models are used as presented in [15] and [19].

Interference Modeling. In order to correctly model the interference which might disturbs the reception of packets at the physical layer, one common way consists in replacing the SNR (*signal-to-noise-ratio*) [15] by a SINR (*signal-to-interference-plus-noise-ratio*). Sources of interference include Intra-BAN and/or Inter-BAN nodes operating in the same frequency band, *i.e.*, *co-channel interference*, or in different frequencies bands, *i.e.*, *adjacent channel interference*. The proper calculation of the SINR value for a given radio link, between the two nodes i (transmitter) and j (receiver), requires the knowledge of all the signals which are currently and concurrently being received at the receiver j. At any time instant t, the current SINR value can be computed as follows:

$$SINR_{ij}^t[mW] = \frac{P_i^{TX} \cdot PL(d_{ij})}{N_j + \sum_{k \neq i,j} \alpha_{ik} \cdot P_k^{TX} \cdot PL(d_{kj})}, \tag{1}$$

where P_i^{TX} stands for the transmission power of the transmitter node i; N_j is the power of the thermal background noise at the receiver node j; α_{ik} the rejection factor between the channels associated with the nodes i and k ($\alpha_{ik} = 1$ in this work); P_k^{TX} is the transmission power of the interfering node k. We consider a full interference model where any node k can potentially generate interference at a given receiver j.

Radio Link Modeling. Finally, in order to determine if a given transmission was successful (despite of interference), it is important to evaluate the corresponding *packet-error-rate* (PER), as: $PER_{ij} = 1 - (1 - BER_{ij}^t)^n$; where n is the packet length in bits, and BER_{ij}^t is the corresponding *bit-error-rate* which is computed based on the current SINR level at time t (*i.e.* $SINR_{ij}^t$), and the considered physical layer characteristics (*e.g.* data rates and modulation schema), as follows:

$$BER_{ij}^t = \begin{cases} 0.5 \times e^{-Eb/No} & \text{DBPSK} \\ Q(\sqrt{4 \times Eb/No} \times sin(\pi/4 \times \sqrt{(2)})) & \text{DQPSK} \end{cases} \tag{2}$$

Where, Eb/No is the energy per bit to noise power spectral density ratio in dB which is computed based on the current SINR level, as: $Eb/No[dB] = SINR_{ij}^t[dB] + 10 \times log_{10}(BW/R)$; where BW is the bandwidth in Hz, and R is the data rate in bps.

IEEE 802.15.6 Compliant Interference Mitigation and Coexistence Strategies. The IEEE 802.15.6 standard proposed three techniques for coexistence as briefly mentioned earlier in sec. 2. With reference to beacon shifting technique and in general a beacon packet (transmitted by a coordinator) contains number of important information. It includes timings of the superframe including beacon period, nodes slot duration, number of the slots assignments, sleep duration, coexistence methods, etc. The beacon shifting is important and required to avoid the collisions of the beacons between multiple BANs. This is achieved by having a different pseudo random sequence at each BAN coordinator which helps randomize the start of the superframe. However, this method alone does not guarantee the interference avoidance between multiple BANs. To have more reliable coexistence mechanism, in this paper, we adapted beacon shifting technique as time-shared approach. In this approach, during the active duration of one BAN, all the other BANs will be in sleep mode and the body-to-body interference can be avoided. This technique does not require to manage any random sequence and is more simple to implement especially under static network where each superframe period is selected according to number of BANs in the surroundings. Channel hopping is another coexistence approach proposed in IEEE 802.15.6 standard which can be applied in scheduled MAC.

In this method the coordinator, generate a channel hopping sequence based on 16-bits Galois linear feedback shift register (LFSR) with a generator polynomial function: $g(x) = X^{16} + X^{14} + X^{13} + X^{11} + 1$. More details on the channel separation and exact calculation of channel hop can be found in [10]. In channel hoping technique, we used a random channel mechanism with every channel has equal probability to be selected. Each BAN operate in one fixed channel for its intra-BAN communication. In narrow band spectrum, there are 79 channels which can be used within the frequency range of $[2400 - 2483.5]$MHz, having center frequency as $fc = 2402.00 + 1.00 \times nc(MHz)$, where $nc = 0, 1, ...77, 78$. Finally we have implemented CSMA/CA medium access method, which can be considered as a implicit collaborative technique for coexistence.

4   Performance Evaluation

In order to understand the impact of body-to-body interference, first, a reference scenario is considered in which multiple bodies are located in close vicinity to communicate without any coexistence strategy. Second, three coexistence strategies (as explained in previous section) are implemented and their results are compared and presented in the following sections.

4.1   Simulation Setup

A packet-oriented network simulator called **WSNet** [16], is used as shown in Fig. 2. It contains various models for wireless sensor networks, wireless local area network and adhoc networks. However, previously it does not contain WBAN specific modules. Therefore, we have developed WBAN specific modules which are explained in section 3 with focus on IEEE 802.15.6 standard compliance. An overview of the developed frame work is shown in Fig. 2. Following are the brief details of the development. The simulation setup is based on version 3.0, which is an up-to-date version of **WSNet**. We consider 5 human bodies, each of them having one coordinating node and 11 sensor nodes as shown in Fig. 1. Five co-located BANs are moving altogether within a distance of 3 meters apart (please note that, this is in compliance with the IEEE 802.15.6 standard in which upto 10 BANs can co-locate in volume of $(6 * 6 * 6)m^3$. At the application layer, consistent packets of 50 bytes of payload, are generated using CBR (constant bit rate) model. The packets are generated at a rate of 100 ms (which satisfy most of the medical signals requirements (i.e., upto 4 Kb/s as effective throughput) [14]). From the application layer, every packet is parsed into the MAC layer. CSMA/CA and scheduled access MAC protocols are developed based on the IEEE 802.15.6 standard. At the PHY layer, *differential quadrature phase shift keying* (DQPSK) modulation model is developed for the *narrowband* (2450 MHz), using the formulas of EbNo, BER and PER as shown in Sec. 3.2. Enhanced IEEE 802.15.6 pathloss models (cf. Sec. 3.2) are implemented. Finally, the real-time motion captured-based inter-BAN mobility traces are imported in **WSNet** which provides accurate space and time variations. By using all the above explained models, the WSNet's XML configuration files (i.e., **xml**) are generated as follows: the number of BAN varies from

1 to 5, transmit power varies between 0 dBm, -10 dBm, -20 dBm and -25 dBm. The coexistence schemes varies from the reference scheme (i.e., without any coexistence) to time-shared, channel hopping and CSMA/CA schemes for 50 iterations and with 95% confidence intervals. The simulations are executed for walking, sitting/standing and running mobility patterns for a duration of 63 sec.

4.2 Results

After having accurate simulating environment, in this section, four performance metrics i.e., PER, PRR, Energy Consumption and Packet Latency, are considered for the evaluation of both *collaborative* and *non − collaborative* schemes under the given application context. At first, average PER distribution is computed as shown in Fig. 3, using accurate radio link model (i.e, explained in Sec. 3.2). It can be noticed that, in a reference scenario (i.e., Fig. 3-a), as the number of BANs increases from 2 to 3, the PER starts increasing sharply and reaches to 1. In comparison, all the co-existence schemes perform much better, only CSMA/CA based approach suffers marginally when number of BAN reaches beyond 3. For the case of PRR, the worse case under lowest transmission power is presented in Fig 4-a. PRR for reference scenarios for 1 BAN is 94.24%, however, as the BAN increases from 2 to 3 the PRR reduces to 0%. It can be seen that, both channel hopping and time-shared perform much better with PRR is above 95% even under -25 dBm. For the case of CSMA/CA, it performs within 95% requirement as long as the transmission power is -20 dBm, for the case of -25 dBm, its performance degrades significantly as can be seen in Fig 4-a. Further, more detailed results of PRR are presented in Tab. 1. Concerning the average packet delay of a single transmission at the lowest transmit power, for the reference scenario as the number of BAN increases from 2, all the packets starts colliding and the coordinator does not receive any packet. For the coexistence schemes, channel hopping and CSMA/CA has a consistent delay, whereas time shared has gradual increase with the increase of BANs as shown in Fig 4-b. More details can be seen in the Tab. 1. Finally for the energy consumption, different current consumption values are considered from TI's cc2420 radio transceiver. For example, for transmission, [17.4 11 9.2 8] mA is used against the power levels (i.e., [0 -10 -20 -25] dBm). For the reception and idle modes, 19.7 mA is used, whereas for the sleep 0.9 mA is used. The energy consumption is estimated by considering a battery of 3 volts. The results are shown in Fig 4-c and Tab. 1. It can be observed that, for the reference scenario, the energy consumption increases nearly 10 times as the number of BANs increases upto 2 and then it matches with CSMA/CA which consumes maximum energy as the nodes are always in active state. Channel hoping and time-shared schemes perform more energy efficient even under higher number of BANs. To conclude, there is a trade-off between *collaborative* and *non − collaborative* coexistence techniques. CSMA/CA performs well for PER/PRR until -20 dBm, however, the performance degrades significantly at -25 dBm, whereas, both time-shared and channel hopping schemes performs much better. The main advantage of collaborative approach is that it has minimum delay which could be important

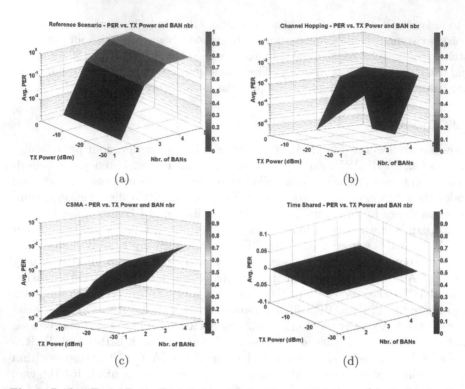

Fig. 3. Packet Error Ratio Distribution of Coexistence Schemes, (a): Reference Scenario, (b): Channel Hopping, (c): CSMA/CA and (d): Time shared

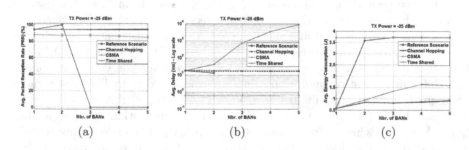

Fig. 4. (a): Average packet reception ratio for multiple BANs in various coexistence schemes. (b): Average packet delay for multiple BANs under coexistence schemes. (c): Average energy consumption for multiple BANs in different coexistence schemes.

for time critical applications. however, it has much higher energy consumption as the nodes are active all the time which could be optimized in the future by applying low power listening protocols. Finally, channel hopping is appeared as the best scheme for *non − collaborative* approach, in which all the performance metrics are optimized under lower transmission power.

Table 1. PRR, energy consumption and latency under varying TX power and BANs for coexistence schemes.

Performance Metrics	TX Power (dBm)	BAN (nbr)	Reference Scenario	Channel Hopping	CSMA/CA	Time Shared
PRR (%)	-20	1	99.78	99.76	97.20	99.77
		3	0	99.77	94.55	99.76
		5	0	99.77	93.84	99.76
	0	1	100	100	99.76	100
		3	0	100	99.01	99.99
		5	0	100	98.65	100
Latency (ms.)	-20	1	12.7	12.7	0.52	12.5
		3	Inf.	12.6	0.54	39.2
		5	Inf.	12.6	0.54	6460
	0	1	12.5	12.7	0.51	12.5
		3	Inf.	12.4	0.52	38.5
		5	Inf.	12.5	0.52	6349
Energy(J)	-20	1	0.55	0.54	3.71	0.54
		3	3.72	0.84	3.71	1.31
		5	3.72	0.92	3.71	1.63
	0	1	0.55	0.55	3.72	0.55
		3	3.72	0.84	3.72	1.31
		5	3.72	0.82	3.72	1.63

5 Conclusion

In this paper we have analyzed the impact of interference in wearable wireless body-to-body networks. First of all, rescue and critical application based scenario is considered and corresponding system models are developed. The models are developed around IEEE 802.15.6 standard in a network simulator called WSNet. The standard's proposed channel models for narrow band are enhanced to have space and time variations as well as dynamic distances among all the nodes on the body. Accurate radio-link and mobility models are developed for both, on-body (i.e., by using bio mechanical approach), and body-to-body networks (i.e., by using group mobility model). Further, IEEE 802.15.6 compliant scheduled access and CSMA/CA MAC protocols are implemented. Two *non – collaborative* coexistence techniques (i.e., Time-shared and channel hopping) are evaluated and one *collaborative* (i.e., CSMA/CA) approach is explored. The performance is evaluated against several metrics such as PER, PRR, latency and energy consumption. It is found that for *non – collaborative* case, channel hopping scheme performs much better under lower transmission power and should be selected for inter-body interference mitigation. Whereas, for the *collaborative* case, CSMA/CA performs very well for both delay and PRR, however, it consume energy, which could be optimized in the future by applying low power medium access approaches.

Acknowledgment. This publication was made possible by NPRP grant #[6 − 1508 − 2 − 616] from the Qatar National Research Fund (a member of Qatar Foundation). The statements made herein are solely the responsibility of the authors.

References

1. Hamida, E.B., Alam, M.M., Maman, M., Denis, B., D'Errico, R.: Wearable body-to-body networks for critical and rescue operations the crow^2 Project. In: IEEE PIMRC 2014 - Workshop on The Convergence of Wireless Technologies for Personalized Healthcare, pp. 2145–2149, September 2014
2. Alam, M.M., Hamida, E.B.: Surveying wearable human assistive technology for life and safety critical applications: Standards, challenges and opportunities. Sensors **14**(5), 9153–9209 (2014)
3. Martelli, F., Verdone, R.: Coexistence issues for wireless body area networks at 2.45 ghz. In: 18th EW Conference, pp. 1–6, April 2012
4. Davenport, D.M., Ross, F., Deb, B.: Coexistence of wban and wlan in medical environments. In: 70th VTC Conference, pp. 1–5, September 2009
5. Hayajneh, T., Almashaqbeh, G., Ullah, S., Vasilakos, A.: A survey of wireless technologies coexistence in wban: analysis and open research issues. Wireless Networks **20**(8), 2165–2199 (2014)
6. Jie, D., Smith, D.: Coexistence and interference mitigation for wireless body area networks: Improvements using on-body opportunistic relaying, CoRR, vol. abs/1305.6992 (2013)
7. Davenport, D.M., Ross, F., Deb, B.: Wireless propagation and coexistence of medical body sensor networks for ambulatory patient monitoring. In: Proceedings of IEEE BSN, pp. 41–45, April 2009
8. Dotlic, I.: Interference performance of ieee 802.15.6 impulse-radio ultra-wideband physical layer. In: 22nd International PIMRC Conference, pp. 2148–2152, September 2011
9. Alasti, M., Barbi, M., Syrafian, K.: Uncoordinated strategies for inter-ban interference mitigation. In: 25th International PIMRC Conference, pp. 1–5, September 2014
10. Ieee standard for local and metropolitan area networks - part 15.6: Wireless body area networks, pp. 1–271 (2012)
11. Marinkovic, S., Popovici, E., Spagnol, C., Faul, S., Marnane, W.: Energy-efficient low duty cycle mac protocol for wban. IEEE Trans. on Info. Tech. in Biomed. **13**(6), 915–925 (2009)
12. Omeni, A.J.B.O., Wong, A.C.W., Toumazou, C.: Energy efficient medium access protocol for wireless medica basn. IEEE Trans. on Biomed. Circuits and Syst. **2**(4), 251–259 (2008)
13. Alam, M.M., Berder, O., Menard, D., Sentieysr, O.: Tad-mac: Traffic-aware dynamic mac protocol for wbasn. IEEE JETCAS Journal **43**(1), 109–119 (2012)
14. Alam, M.M., Hamida, E.B.: Performance evaluation of ieee 802.15.6 mac for wbsn using a space-time dependent radio link model. In: 11th AICCSA Conference, pp. 1–8, November 2014

15. Alam, M.M., Ben Hamida, E.: Towards accurate mobility and radio link modeling for ieee 802.15.6 wearable body sensor networks. In: 10th WiMob Conference, pp. 298–305, October 2014

16. Ben Hamida, E., Chelius, G., Gorce, J.M.: Impact of the physical layer modeling on the accuracy and scalability of wireless network simulation. Simulation **85**(9), 574–588 (2009)

17. Ben Hamida, E., Chelius, G.: Investigating the impact of human activity on the performance of wireless networks: An experimental approach. In: WoWMoM 2010 Conference, pp. 1–8, June 2010

18. Hamida, E., D'Errico, R., Denis, B.: Topology dynamics and network architecture performance in wbsn. In: 4th NTMS Conference, pp. 1–6, February 2011

19. Wiserban - smart miniature low-power wireless microsystem for ban, Tech. Rep. WP3 - D3.1v5, August 2011. http://cordis.europa.eu/docs/projects/cnect/4/257454/080/deliverables/001-WiserBANWP3D31Final20111007.pdf (Online)

Workshop Cognitive Radio for 5G Networks

Distributed Power Control for Carrier Aggregation in Cognitive Heterogeneous 5G Cellular Networks

Fotis Foukalas$^{(\boxtimes)}$ and Tamer Khattab

Electrical Engineering, College of Engineering, Qatar University, Doha, Qatar
{foukalas,tkhattab}@qu.edu.qa

Abstract. In this paper, we study the distributed optimal power allocation for the carrier aggregation in next generation (5G) cognitive radio networks. The presented study relies on the power control and carrier aggregation principles of wireless communication systems. Our approach differs from the conventional well-known water filling (WF) algorithm in the sense that we provide decentralized solution, wherein all of the Lagrange multipliers are not handled equally over the heterogeneous fading channels. This is accomplished in order to provide distributed power control over the heterogeneous fading channels that are considered non-identically distributed and non-identical Nakagami-m channels. To this end, we first formulate the optimization problem and in the sequel, we solve it using the alternating direction method of multipliers (ADMM), which provides to our solution the required decomposition for each channel and the robustness through the augmented Lagrangian. For benchmarking, we provide comparison to other prominent decomposition methods like dual decomposition method (DDM). Simulation results highlight the performance gain of ADMM in terms of number of iterations. The achievable sum rates are also depicted for different network setups. Comparison to the WF is also provided that reveals the gain of the applied decomposition methods (i.e. ADMM and DDM) to the cognitive heterogeneous 5G cellular networks.

Keywords: Carrier aggregation · Optimal power allocation · Heterogeneous fading channels · Alternating direction method of multipliers · Decomposition methods

1 Introduction

5G technologies include the cognitive cellular networks concept relying on the heterogeneous networks deployment of macro, pico and other different size of cells. The difference among those cells is their carrier frequency that can be used either to provide higher capacity or coverage [1]. An additional cognitive radio aspect in 5G wireless communication systems (beyond LTE-Advanced) is the carrier aggregation (CA). Although CA first introduced in Rel.8 of the LTE system for a static

© Institute for Computer Sciences, Social Informatics and Telecommunications Engineering 2015
M. Weichold et al. (Eds.): CROWNCOM 2015, LNICST 156, pp. 681–695, 2015.
DOI: 10.1007/978-3-319-24540-9_56

implementation within particular bands, today, the application of CA in Het-Nets is already implemented (Rel.12) and it is being further extended to Rel.13 within a multi-band context [3]. In this way, the aggregation of heterogeneous dispersed bands is carried out giving the freedom to operators for better spectrum exploitation including the bandwidth expansion [2]. Such an heterogeneous wireless medium should be taken into account in the design of known techniques of wireless communications such as power control [4].

Power control relies on the channel state information (CSI) that is sent from the user equipment (UE) back to the base station (BS) through feedback channel resulted in the well-known water filling (WF) algorithm [4]. Power control with CA in cognitive cellular networks with HetNets deployment should be revisited due to the heterogeneity of the multiple channels that can be aggregated. In such a dynamic wireless medium, wherein the channel gains are not considered identically distributed and non identical random variables, the conventional WF algorithm is not practical anymore. Thereby, we need to devise new solutions to manipulate more than one Lagrange multipliers and on the other hand, to not provide a global solution (i.e. one Lagrange multiplier) due to the heterogeneity. Towards this end, several works have been dealt with the multiple Lagrange multipliers issue as discussed below.

In [5], authors studied the problem of power allocation for interference channels. For the solution of a 2-channel system model, they use the augmented Lagrangian method in conjunction with the steepest descent method to optimize the augmented Lagrange function to each iteration. Augmented Lagrangian method is a modified dual-ascent method with an additional penalty condition bringing thereby robustness and yielding convergence without assumptions like strict convexity of the objective function [6]. In [7], the authors studied power control in a cognitive radio network application, wherein the interference originated from the secondary network to the primary is considered. Looking into their solution for the problem of power allocation, they proposed a Gauss-Seidel sequential iterative method. Gauss-Seidel is used for the calculation of the power allocation vector including the transmit power of other secondary base stations contributing to the interference at the primary receivers. In [8], the author proposed a dynamic power control algorithm that allows each femtocell user to adapt its outage probability specification to minimize the total energy consumption in the system and guarantees a minmax fairness in terms of worst outage probability to all the femtocell users. In [9], authors studied a cognitive radio model for the power control with constraint on the transmit power and the interference power resulting in a Lagrange dual function with two Lagrange multipliers. They proposed a dual decomposition method (DDM) for solving this problem dealing with the two multipliers. Such solution adopts the decomposability for a given network resource allocation problem providing architectural alternatives for a more modularized network design [10].

Obviously, decomposition methods is one of the powerful tool that naturally looks for parallel optimization algorithms [6]. For example, in [11], the authors provide an optimal design of multiuser DSL spectrum using DDM to manage an exponential complexity that increases with the number of DSL channels. In particular, the power constraints are imposed through the use of Lagrangian multipliers, which can be chosen correctly in order to achieve the optimisation objective across different tones. The DDM has been also applied to the power control for spectrum sharing cognitive radio networks for decoupling the problems of the transmit power and interference power calculating thereby the corresponding Lagrange multipliers [12].

In this paper, we study the power control problem with CA in heterogeneous networks, wherein the optimal power allocation is accomplished over heterogeneous channel conditions. In order to model such heterogeneous system, we assume a channel model with independent but not identically distributed channels [13]. Additionally, we assume non-identical Nakagami-m fading channels that gives the system model more heterogeneous characteristics [14]. The problem is formulated with separate power constraints for each channel assuming optimal power allocation policy through Lagrange multiplier for each one. Since we don't look for a global solution (i.e. one unique Lagrange multiplier for all channels), we devise an algorithm that can provide on one hand decomposition and on the other hand local information exchange at each iteration leading to a smaller number of iterations as compared to the other state of the art decomposition method as the DDM. The proposed algorithm follows the principles of the alternating direction method of multiplier (ADMM) that represents an advanced DDM that combines the idea of DDM and the augmented Lagrangian method [6]. ADMM has been recently used to solve several problems in wireless communications; we mention here for example the need for a distributed multi-cell coordinated beamforming solution, wherein multiple base stations (BSs) collaborate with each other in the beamforming design to mitigate the intercell interference as presented in [15]. Finally, in order to establish a benchmark of the proposed ADMM algorithm, we solve our problem using the DDM providing a practical algorithm. The obtained simulation results indicate the advantage of using ADMM compared with DDM in terms of convergence and number of iterations.

The rest of this paper is organised as follows. Section 2 give details about the system model and the channel model. Section 3 provides the theory for the CA in HetNets assuming optimal power allocation and heterogeneous fading channels. Section 4 provides the details on the ADMM based solution and the Section 5 the details on the DDM approach. Simulation results and useful insights are provided in Section 6 and the paper summary is provides in Section 7.

2   System and Channel Models

2.1   System Model

The proposed system model is considered for an heterogeneous network (HetNet), wherein the large and small cells are separated through the use of different frequencies. The considered HetNet consists of cells of different sizes that are called macro-, micro-, pico- and femto-cells.

Fig. 1 depicts the system model of our macro/micro/pico and femto cells HetNet deployment, in which all cells use different frequency channels and fading impairments as explained below in the channel model description. Although, the system model shows three cells, it will be expended to more generic case using the derived analysis below. Under this premise, there is no interference problem and the throughput gain for this option will be the highest one. We also assume that the HetNet is able to provide carrier aggregation (CA) among the heterogeneous bands. Each band within each cell can provide one or multiple component carriers, i.e. channels, for aggregation offering the highest rate to the end-user, whereby the CA in heterogeneous cognitive cellular networks can be realized. In the next section, we give the details about the channel model of the considered system model.

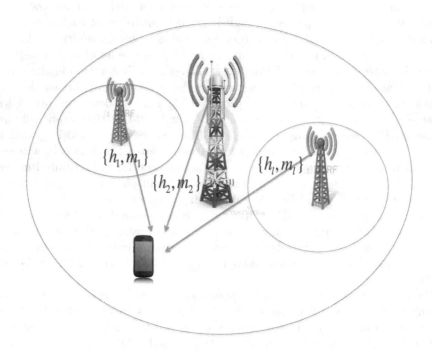

Fig. 1. Cognitive Heterogeneous Cellular Networks with Carrier Aggregation.

2.2 Channel Model

We assume that the CA over the considered HetNet system model can be modeled as L parallel channels with heterogeneous fading channel characteristics. In particular, each component carrier (i.e. channel) with $l \in 1, .., L$ can be aggregated by the transmitter (Tx) using the channel state information (CSI) received by the receiver (Rx) for each channel. Considering L channels in our channel model, it is identical to assume L parallel channels [16]. We assume that the feedback for each channel is provided by the Rx to Tx in an efficient way either per channel or over the whole bandwidth [17]. The input–output (X, Y) relationship for each channel of the L parallel channels with CA is described as follows:

$$Y_l = h_l X_l + n_l, \ \forall l \in L \tag{1}$$

where h_l is the channel gain of the $l - th$ channel and n_l the noise that is a zero-mean unit-variance complex Gaussian random variable independent of the noise on the other channels.

Based on these assumptions, the average power of the $l - th$ channel is given by:

$$g_l = E[|\ h_l\ |^2], \ \forall l \in L \tag{2}$$

under the following constraint:

$$\Sigma_{l=0}^{L-1} h_l = 1. \tag{3}$$

The signal-to-noise-ratio (SNR) for the $l - th$ channel is equal thereby to:

$$\gamma_l = \frac{h_l p_l}{\sigma_l{}^2 B_l}, \ \forall l \in L \tag{4}$$

where p_l is the transmit power of the $l - th$ channel, the σ_l the variance of the noise and the B_l is the bandwidth of each channel.

Moreover, the following assumptions apply about our channel model:

- The bandwidth of each channel B_l is equal and fixed.
- The number of channels should provide the following rule $L = B_l T_d$ in respect to the delay spread T_d having assumed a multi-carrier system.
- Each channel has a channel gain denoted as $\{H_l\}_{l=0}^{L-1}$.
- Each channel is considered invariant within a coherence period T_c and thereby the number of symbols per channel is equal to $K_l = [B_l T_c]$.

Having defined the system and channel models, the aim of this work is to provide the most efficient power control scheme for CA over HetNets by maximizing the sum achievable rate. To this end, we first model the carrier aggregation over heterogeneous fading channels defining the required performance analysis, and next, we explain the problem under consideration.

3   Carrier Aggregation over Heterogeneous Fading Channels

CA in HetNets can be assumed as the CA over heterogeneous fading channels, where the latter can be analysed as the sum achievable rate over independent and non-identically distributed (i.n.d.) channels in terms of power and non-identical

Nakagami-m channels in terms of fading impairments. In this way, the different channels to be aggregated expose heterogeneous conditions. Under this premise, the sum achievable rate in a CA system is defined as follows.

First, we assume that for each channel the power control is employed for the adaptation over the fading channel conditions through the channel feedback. Thereby, an optimal power allocation is carried out. For bench-marking purpose, we assume that each channel performs the well known water-filling (WF) algorithm and thus, the optimal power allocation for each l-th channel is given as follows [4]:

$$P_l(h_l) = \left(\lambda_l - \frac{\sigma_l{}^2}{h_l} \right). \tag{5}$$

The corresponding achievable average rate over the fading channel is obtained as follows:

$$C_l(h_l) = \int \log_2 \left(\frac{h_l p_l}{\lambda_l \sigma^2 B_l} \right) f(g_l) dg_l. \tag{6}$$

The performance of the CA system over heterogeneous fading channels is considered as the sum rate as follows:

$$C_{tot} = \sum_{l=1}^{L} C_l = \sum_{l=1}^{L} C(h_l). \tag{7}$$

In order to model the HetNets environment, we assume that the channel gains are heterogeneous, i.e. independent and non-identically distributed (i.n.d) in terms of power and non-identical Nakagami-m in terms of fading impairments. In this case, the instantaneous SNR γ_l of each channel is considered a gamma distributed random variable with probability density function (PDF) given by [14]:

$$f_{\gamma_l}(\gamma) = \frac{m_l{}^{m_l} \gamma^{m_l-1} \exp^{-m_l \gamma / \bar{\gamma}_l}}{\bar{\gamma}_l{}^{m_l} \Gamma(m_l)} \tag{8}$$

where the fading parameter is considered different for each l-th channel denoted as m_l as well as the average SNR $\bar{\gamma}_l$. Thus, heterogeneous fading channels can be assumed as also pointed out in [13], wherein the fading impairments are modeled with different PDFs. In our case, we assume the generalized case of Nakagami-m for changing the factor m accordingly, since our main focus is on the power control scheme for CA in HetNets. More specifically, we look for the power control scheme that does not give a global solution for the i.n.d and non-identical fading channel gains. The contribution of this paper is to find the overall optimal power allocation $P(h_1, .., h_l)$ of the proposed CA over heterogeneous fading channels. The considered optimal power allocation is accomplished using the alternating direction method of multipliers (ADMM) providing thereby a more efficient and robust decomposition and learning among the fading channels with heterogeneous characteristics. Our future work on this topic, we will incorporate more sophisticated fading channel formula including scheduling among the channels with different bandwidth options for each channel [13].

4 Power Control for Carrier Aggregation in Heterogeneous Fading Channels

We formulate the problem of maximizing the sum rate over the transmit power of each $l - th$ component channel. The problem is formulated for $l \in L$ channels as follows:

$$\max_{p_1,\ldots,p_L} C_{tot} = \sum_{l=1}^{L} C_l \tag{9}$$

$$s.t. \quad \sum_{l=1}^{L} P_l(h_l) \leq \bar{P}_l \tag{10}$$

where the two constraints guarantee that each channel $l - th$ follows each one optimal power allocation policy.

It is evident from the problem defined in (9) that a solution using decomposition principles could provide the mathematical framework to build an analytic foundation for the design of requested distributed power control. For example, assuming the WF algorithm, each subproblem can be solved isolated resulting in the individual sum achievable rate that can not give an efficient distributed and coordinated solution. We look for a solution that can be achieved by exchanging information about the channels conditions in order to provide solutions on their separate problems at local level leading to the efficient overall solution at distributed level. One known solution of such a problem is the dual decomposition that can solve the problems separately and update the optimal values using the subgradient method. Nevertheless, there is a more powerful method that relies on the decomposition principles providing more robustness in such distributed problems. This method is known as the Alternating Direction Method of Multipliers (ADMM).

In particular, the ADMM combines the principles of the dual decomposition using also the augmented Lagrangian tool for gradually learning. In particular, the ADMM method consists of the following steps in order to solve our problem:

- To formulate the augmented Lagrangian function:

$$L(p_1,..,p_L,\lambda_1,..,\lambda_L) = \sum_{l=1}^{L} C_l$$
$$+\lambda_1 \left(P_1(h_1) - \bar{P}_1\right) + ... + \lambda_L \left(P_L(h_L) - \bar{P}_L\right)$$
$$+\frac{\rho}{2} \left(\left(P_1(h_1) - \bar{P}_1\right)^2 + ... + \left(P_L(h_L) - \bar{P}_L\right)^2\right)$$

$$\tag{11}$$

688   F. Foukalas and T. Khattab

- To designate the dual decomposition as follows:

$$\min_{\lambda_1,..,\lambda_L} \{ \max_{p_1,..,p_L} C_1 + ... + C_L$$
$$+\lambda_1 \left(P_1(h_1) - \bar{P}_1\right) + ... + \lambda_L \left(P_L(h_L) - \bar{P}_L\right)$$
$$+\frac{\rho}{2} \left(\left(P_1(h_1) - \bar{P}_1\right)^2 + ... + \left(P_L(h_L) - \bar{P}_L\right)^2 \right) \}$$

(12)

where the $\lambda_1,..,\lambda_L$ are the dual variables for each of L channels.
- To solve the inner subproblems through optimization decomposition solution using Gauss-Seidel or block-coordinate descent method:

$$P_1(h_1)^{k+1} = \arg\min_{p_1} L(p_1,..,p_L^k, \lambda_1^k,..,\lambda_L^k)$$

(13)

$$= C(1) + \lambda_1 \left(P_1^k(h_1) - \bar{P}_1\right)$$
$$+\rho \left(\left(P_1^k(h_1) - \bar{P}_1\right)^2 + ... + \left(P_L^k(h_L) - \bar{P}_L\right)^2 \right)$$

(14)

$$\vdots$$

$$P_2(h_L)^{k+1} = \arg\min_{p_L} L(p_1^{k+1},..,p_L, \lambda_1^k,..,\lambda_L^k)$$

(15)

$$= C(L) + \lambda_L \left(P_L^k(h_L) - \bar{P}_L\right)$$
$$+\rho \left(\left(P_1^{k+1}(h_1) - \bar{P}_1\right)^2 + ... + \left(P_L^k(h_L) - \bar{P}_L\right)^2 \right)$$

(16)

- To solve the outer problem using the subgradient updates:

$$\lambda_1^{k+1} = \lambda_1^k + \rho(P_1(h_1)^{k+1} - \bar{P}_1)$$

(17)

$$\vdots$$

$$\lambda_L^{k+1} = \lambda_L^k + \rho(P_L(h_2)^{k+1} - \bar{P}_L)$$

(18)

Instead of the dual decomposition method (DDM), ADMM, as its name suggests, alternatively performs one iteration of the Gauss-Seidel step $(13-16)$ and one step of the outer subgradient update for speeding up its convergence. Notably, the augmented Lagrangian is minimized jointly with respect to the

L primal variables. The optimal power allocation $P_l(h_l)$ with L variables are updated in an alternating or sequential fashion. Separating the maximization for the optimal power allocation of two channels into two steps is precisely what allows for decomposition [6]. In order to benchmark the ADMM performance, we also devise the DDM for our problem and we describe the algorithm in details below.

5   Benchmarking with the Dual Decomposition Method

DDM is a powerful tool that can be used for decomposing a problem to sub-problems applying the separation principles in networking systems [18]. We will provide the solution of the proposed optimization problem using DDM n order to establish a benchmark to the proposed ADMM-based algorithm for comparison purposes. In this way, we can also highlight the architectural differences among the two decomposition approaches.

We first formulate the Lagrangian function of the optimization problem as follows:

$$L(p_1, .., p_L, \lambda_1, .., \lambda_L) =$$
$$\sum_{l=1}^{L} C_l + \lambda_1(P_1(\gamma_1) - \bar{P}_1) + ... + \lambda_L(P_L(\gamma_L) - \bar{P}_L)$$

(19)

and the Lagrangian dual function is given as follows:

$$g(\lambda_1, .., \lambda_L) =$$
$$\max_{p_1, .., p_L} \sum_{l=1}^{L} C_l + \lambda_1(P_1(\gamma_1) - \bar{P}_1) + ... + \lambda_L(P_L(\gamma_L) - \bar{P}_L)$$

(20)

where the $\lambda_1, .., \lambda_L$ Lagrange multipliers are considered the link price [10].

The dual function can be minimized to obtain an upper bound on the optimal value of the original optimization problem:

$$\min_{\lambda_1, .., \lambda_L} \quad g(\lambda_1, .., \lambda_L)$$

(21)

$$s.t. \ \lambda_1 > 0$$
$$\vdots$$
$$\lambda_L > 0$$

where the optimal dual objective g^* forms the duality gap $C_{tot}^* - g^*$, which is indeed zero since the Karush-Kuhn-Tucker (KKT) conditions are satisfied.

The DDM algorithm used for the problem solution is described in the algorithm below. The DDM is simulated in parallel with the ADMM and useful insights are discussed in the section below.

Algorithm 1. Dual Decomposition algorithm with L component channels.

- Parameters: constant step size α and constant convergence value ϵ.
- Initialize: variables $\lambda_1^k = 1, .., \lambda_L^k = 1$ for all L channels.

1. The Lagrangian dual problem is solved locally by the BS, which aggregates the L channels and then send the feedback the solutions to the corresponding channels.
2. The BS updates its prices for each component channel $l \in L$ using the subgradient as follows:

$$\lambda_1^{k+1} = \lambda_1^k - \alpha(\sum_1 \bar{P}_1 - P_1(\lambda_1)) \tag{22}$$

$$\vdots$$

$$\lambda_L^{k+1} = \lambda_L^k - \alpha(\sum_L \bar{P}_L - P_L(\lambda_L)) \tag{23}$$

and broadcasts the new prices $\lambda_1^{k+1}, .., \lambda_L^{k+1}$.
3. Set $k \longrightarrow k + 1$ and go to step 1) until satisfying termination criterion.

- Stop once $| \lambda_l^{k+1} - \lambda_l^k | \le \varepsilon, \forall l \in L$ simultaneously, where ε is the convergence rule.

---

6   Simulation Results and Useful Insights

In this section, simulation results are presented and useful insights are discussed. We opt to provide the outage probability for each channel that reveals better the impact of different Lagrange multiplier values resulted by the different applied methods. The outage probability formula can be found to several references, e.g. [4].

Fig. 2 depicts the outage probability using two component carriers (CCs) assuming the following heterogeneous channel conditions: $CC - 1$: $5dB, m = 1, CC - 2 : 15dB, m = 2$ where the first term denotes the average SNR of the specific CC and the second term the fading m parameter. The simulation has been carried out using ADMM, DDM and WF algorithms. Focusing on the first carrier, i.e. CC1, it is inferred that the ADMM outperforms the DDM in terms of required number of iterations. In particular, the ADMM requires 18 iterations in order to converge to the capacity solution and the DDM requires 23

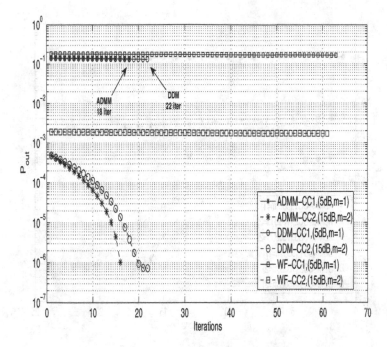

Fig. 2. Outage probability of 2 CCs of $5dB$ and $m = 1$, $15dB$ and $m = 2$ using ADMM and DDM. Benchmark with the WF algorithm is also provided.

iterations. This is provided in the ADMM by the parallel solutions of the inner subproblems that is not taken into account in the DDM. For benchmarking purpose, the results using WF algorithm are depicted that requires 63 iterations. In this way, the advantage of decomposition methods for handing the heterogeneity of the channels is manifested. The same outcome can be observed for the second carrier, i.e. CC2, with an additional interesting performance gain that is the outage probability, which shows lower values than those required for the first channel. This could be explained as the results of having better channel conditions for the CC2 compared to the CC1's ones.

Fig. 3 below depicts the performance of the ADMM in comparison to DDM using three CCs, i.e. CC1, CC2 and CC3. The results also corroborate the advantage of using ADMM instead of DDM having a better number of iterations as long as the channel conditions are better. The outage probability for better channel conditions is improved as well. It is also observed that the higher average SNR for a particular fading channel condition, e.g. $m = 2$ does not have an impact on the number of iterations for the two prominent decomposition methods that applied in this paper and corroborate the benefit of seperation principles in wireless communications through the dual decomposition [18]. Finally, it should be noted that for better channel conditions, e.g. CC2, the outage proba-

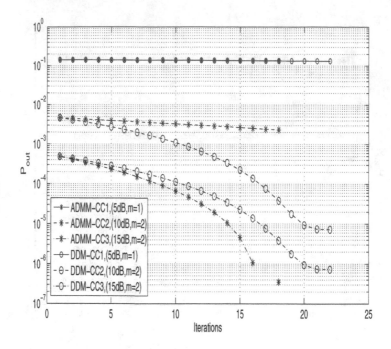

Fig. 3. Outage probability of 3 CCs of $5dB$ and $m = 1$, $10dB$ and $m = 2$, $15dB$ and $m = 2$ using ADMM and DDM.

bility is low since the Lagrange multiplier decreases significantly almost to zero. This behavior is observed for both ADMM and DDM verifying the fact that the decomposition provides achievable rates without power control. However, the WF requires power control for achieving the capacity with higher impact on the performance in terms of outage probability.

Fig. 4 depicts the achievable sum rate in bits per sec over average SNR in dB at the CC-1 assuming fading channel with $m = 1$, average SNR equal to $10dB$ and fading channel $m = 2$ at the CC-2 and finally average SNR equal to $15dB$ and fading channel $m = 2$ at the CC-3. The results are depicted using ADMM, DDM and WF algorithms respectively. It is evident from the results that both decomposition methods, i.e. ADMM and DDM outperforms the WF and there is a gain of using the DDM at the low power regime although in terms of iterations ADMM performs better as have been discussed above. Most importantly, for more than 2 CCs, the performance gain of using decomposition methods is significant due to the provided coordinated solution for each particular link comparing to the isolated WF solution, which does not deal with the distributed nature of the problem at hand.

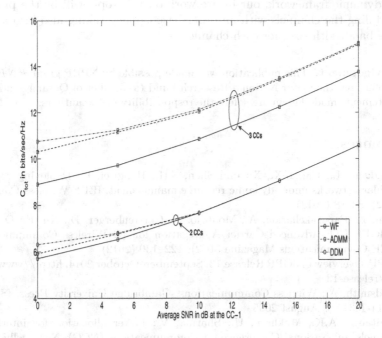

Fig. 4. Sum rate (bits/sec) over the average SNR in dB at the CC-1 with $m = 1$, assuming $10dB$ average SNR and $m = 2$ at the CC-2, and $15dB$ average SNR and $m = 2$ at the CC-3, using ADMM, DDM and WF algorithms.

7   Conclusion and Future Work

In this paper, we study the power control problem when carrier aggregation in heterogeneous networks is deployed in future cognitive 5G cellular networks. Our study assumes the optimal power allocation over fading channels with heterogeneous characteristics in terms of power and fading impairments. To this end, we model the channel gains with heterogeneous characteristics that is carried out assuming independent and non-identically distributed (i.n.d.) in terms of power and non-identical Nakagami-m in terms of fading impairments. Under this premise, we formulate the problem of maximizing the achievable rate of the CA over the transmit power constraints of the channels. The problem solution is carried out in a distributed and coordinated fashion employing the alternating direction method of multipliers (ADMM) as a powerful tool for providing decomposition. The particular method is devised, applied and presented in this paper. In order to benchmark the proposed method, we provide a problem solution using the dual decomposition method (DDM) too. Simulation results are obtained and illustrated, which corroborate the fact that ADMM converges faster than the DDM. In terms of sum rate, the decomposition methods for such a distributed problem provides better result than the classical WF. Having defined such a dynamic framework, our future work on this topic will be the provision of scheduling the channels with an order-based policy taking also into account variable bandwidth sizes for each channel.

Acknowledgment. This publication was made possible by NPRP grant $\#NPRP$ 6-1326-2-532 from the Qatar National Research Fund (a member of Qatar Foundation). The statements made herein are solely the responsibility of the authors.

References

1. Yongkang, L., Cai, L.X., Xuemin Shen, S.H., Hongwei, L.: Deploying cognitive cellular networks under dynamic resource management. IEEE Wireless Communi. **20**(2), 82–88 (2013)
2. Shen, Z., Papasakellariou, A., Montojo, J., Gerstenberger, D., Xu, F.: Overview of 3GPP LTE-Advanced Carrier Aggregation for 4G Wireless Communications. IEEE Communications Magazine **50**(2), 122–130 (2012)
3. 3GPP, Overview of 3GPP Release 13, September–October 2014. http://www.3gpp.org/release-13
4. Goldsmith, A.: Wireless Communications. Gambridge University Press. ISBN-IO 0–511-13315-4, August 2005
5. Chaitanya, A.K., Mukherji, U., Sharma, V.: Power allocation for interference channels. In: National Conference on Communications (NCC), New Delhi, India, pp. 1–5, February 2013
6. Boyd, S., Parikh, N., Chu, E., Peleato, B., Eckstein, J.: Distributed Optimization and Statistical Learning via the Alternating Direction Method of Multipliers. Foundations and Trends in Machine Learning, Now Publ **3**(1), 1–122. doi:10.1561/2200000016
7. Gong, S., Wang, P., Niyato, D.: Optimal power control in interference-limited cognitive radio networks. In: IEEE Int. Conf. Communication Systems (ICCS), pp. 82–86, Non. 2010
8. Ten, C.W.: Optimal power control in Rayleigh-fading heterogeneous networks. In: IEEE Proc. INFOCOM 2011, pp. 2552–2560, April 2011
9. Zhang, E., Yin, S., Li, S., Yin, L.: Optimal power allocation for OFDM-based cognitive radios with imperfect channel sensing. In: 8th Intern. Conf. WiCOM 2012, pp. 1–5, September 2012
10. Palomar, D.P., Chiang, M.: A Tutorial on Decomposition Methods for Network Utility Maximization. IEEE Journal on Sel. areas in Comm. **24**(8), August 2006
11. Cendrillon, R., Yu, W., Moonen, M., Verlinden, J., Bostoen, T.: Optimal Multiuser Spectrum Balancing for Digital Subscriber Lines. IEEE Trans. Communi. **54**(5), 922–933 (2006)
12. Zhang, L., Xin, Y., Liang, Y.-C., Poor, H.V.: Cognitive multiple access channels: optimal power allocation for weighted sum rate maximization. IEEE Trans. Communi. **57**(9), 2754–2762 (2009)
13. Morsi, R., Michalopoulos, D., Schober, R.: Multi-user Scheduling Schemes for Simultaneous Wireless Information and Power Transfer Over Heterogeneous Fading Channels, January 2014. arXiv:1401.1943

14. Suraweera, H., Karagiannidis, G.K.: Closed-Form Error Analysis of the Non-Identical Nakagami-m Relay Fading Channel. IEEE Comm. Letters **12**(4), April 2008
15. Shen, C., Chang, T.-H., Wang, K.-Y., Qiu, Z., Chi, C.-Y.: Distributed Robust Multicell Coordinated Beamforming With Imperfect CSI: An ADMM Approach. IEEE Trans. Sign. Proc. **60**(6), 2988–3003 (2012)
16. Lozano, A., Tulino, A.M., Verdu, S.: Optimum power allocation for parallel Gaussian channels with arbitrary input distributions. IEEE Trans. Inf. Theory **52**(7), 3033–3051 (2006)
17. Berardinelli, G., Sorensen, T.B., Mogensen, P., Pajukoski, K.: Transmission over multiple component carriers in LTE-A uplink. IEEE Wirel. Commun. **18**(4), 63–67 (2011)
18. Ribeiro, A., Giannakis, G.B.: Separation Principles in Wireless Networking. IEEE Wirel. Commun. **56**(9), 4488–4505 (2010)

Design of Probabilistic Random Access
in Cognitive Radio Networks

Rana Abbas[✉], Mahyar Shirvanimoghaddam, Yonghui Li, and Branka Vucetic

The University of Sydney, Sydney, NSW, Australia
{rana.abbas,mahyar.shirvanimoghaddam,yonghui.li,
branka.vucetic}@sydney.edu.au

Abstract. In this paper, we focus on the design of probabilistic random access (PRA) for a cognitive radio network (CRN). The cognitive base station (CBS) allows the secondary users (SUs) to reuse the sub-channels of the primary users (PUs) provided that the interference of the SUs to the PUs is below a predetermined threshold. PUs transmit over a fixed set of channels with fixed transmission powers that are scheduled by the CBS. With this prior information, CBS optimizes the probabilistic random transmissions of the SUs. In each time slot, SUs transmit over a random number of channels d, chosen uniformly at random, according to a certain degree distribution function, optimized by the CBS. Once the signals of the SUs and PUs are received, CBS then implements successive interference cancellation (SIC) to recover both the SUs' and PUs' signals. In the signal recovery, we assume that the PUs' signals can be recovered if the interference power (IP) of the SUs to the PUs is below a predetermined threshold. On the other hand, we assume the SUs' signals can be recovered if its received SINR is above a predetermined threshold. We formulate a new optimization problem to find the optimal degree distribution function that maximizes the probability of successfully recovering the signals of an SU in the SIC process under the SINR constraints of the SUs while satisfying the IP constraints of the PUs. Simulation results show that our proposed design can achieve higher success probabilities and a lower number of transmissions in comparison with conventional schemes, thus, significantly improving signal recovery performance and reducing energy consumption.

Keywords: Probabilistic random access · Cognitive radio · SIC · IP · SINR · Degree distribution

1 Introduction

Cognitive radio (CR) has been known to be a promising technology to achieve the efficient utilization of the radio spectrum. In CR networks (CRNs), unlicensed secondary users (SUs), are allowed access to the radio spectrum owned by the licensed primary users (PUs), provided that the PUs are guaranteed a certain level of protection. Optimal resource allocation algorithms i.e. channel and power

© Institute for Computer Sciences, Social Informatics and Telecommunications Engineering 2015
M. Weichold et al. (Eds.): CROWNCOM 2015, LNICST 156, pp. 696–707, 2015.
DOI: 10.1007/978-3-319-24540-9_57

allocation, among the SUs that maximize their data rates or minimize their transmit power requirements have been well-investigated for multiple scenarios and are known to be NP-hard. Accordingly, numerous sub-optimal algorithms for resource allocation have been proposed for both downlink and uplink CR transmissions [1,2]. However, these approaches do not scale well as the number of users in the network increases and their activity becomes more dynamic.

To overcome these problems, random access protocols provide a simple solution that significantly reduces processing and signalling overhead. A commonly used approach is to employ random access over the control channels. That is, users perform contentions for channel access request. Once access request is granted, data will be transmitted over the allocated channels. Commonly used contention based schemes in CRNs include ALOHA, slotted-ALOHA and carrier sense multiple access (CSMA) [3,4,5]. These models assume that the SUs contend to access the channels only when the PUs are inactive.

In [6], authors proposed another random access approach where the CBS predetermines a certain transmission probability and makes it known to all the SUs. The PUs' transmissions are fixed whereas the SUs transmissions are randomized according to the assigned transmission probability. It is shown that such a simple random transmission can offer significant improvements in performance, in certain cases, for both the PUs and SUs, compared to fixed transmissions. It is argued that from a design point of view, controlling the probabilities is easier than controlling the power. However, the paper only considered a very single case of single channel and no analysis was done to derive the design criteria for choosing the optimal transmission probability.

In [7,8,9], some probabilistic random access (PRA) schemes were proposed where each user transmits over a subset of sub-channels, which are selected uniformly at random, according to a degree distribution, predetermined by the base station. The PRA can then be represented by a bipartite graph, and a message passing algorithm can be implemented at the base station to recover the users' signals. Optimization is then carried out by using the conventional analytical tools of codes-on-graph for binary erasure channels to maximize the probability of having an interference-free clean packet in each iteration. That is mainly because it is assumed that successful signal recovery is only possible when a 'clean packet' is available.

In this paper, we extend the work in [6,7,8,9] to an uplink CRN. We assume the PUs' channel and power allocations are scheduled, and thus, known priori at the CBS. The CBS performs maximal-ratio combining (MRC) to combine the multiple copies of each user's signals over its respective sub-channels and implements successive interference cancellation (SIC) to recover the SUs' and PUs' signals. Under the conventional interference power (IP) constraint, a PU's signal can be successfully recovered if the IP caused by the SUs to that PU is below a predetermined threshold. Moreover, under the conventional SINR constraint, an SU's signal can be successfully recovered if its SINR is above a predetermined threshold. Due to the IP and SINR constraints, the 'clean packet' model becomes sub-optimal. Accordingly, we formulate a new optimization problem to find the optimal degree distribution that maximizes the success probability of the SIC process of the SUs while satisfying the IP constraints of the PUs. This is equivalent to maximizing the probability of having a received SINR of an SU greater than or equal to a predetermined threshold in each iteration of the SIC. Simulation results show that our proposed design can achieve significantly lower error probabilities and requires a lower number of transmissions, in comparison with the conventional approach.

The rest of this paper is organized as follows. Section 2 presents the system model. In section 3, we describe the probabilistic random transmission scheme and the SIC process. In section 4, we analyze the system performance in an asymptotic setting and formulate our degree distribution optimization problem. Numerical results are provided in section 5. Finally, section 6 concludes the paper.

2 System Model

We consider an uplink CRN, including a CBS, a set of K_p active PUs, denoted by $\mathcal{K}_p$, and a set of K_s active SUs denoted by $\mathcal{K}_s$. There are in total N orthogonal sub-channels of equal bandwidth in the network. Channels are assumed to be reciprocal and block fading; that is, we assume the channel coefficients remain constant for the whole transmission block but vary independently from one block to the other. Let y_n denote the received signal vector at the CBS over the nth sub-channel, where $1 \leq n \leq N$. Then, it can be expressed as follows:

$$y_n = \sum_{k \in \mathcal{K}_p} g_{k,n} x_{k,n} + \sum_{i \in \mathcal{K}_s} h_{i,n} u_{i,n} + e_n, \tag{1}$$

where $g_{k,n}$ is the channel gain between PU_k and the CBS over the nth sub-channel, and $h_{k,n}$ is the channel gain between SU_k and the CBS over the nth sub-channel. $x_{k,n}$ and $u_{i,n}$ are the transmitted signals of each of the PUs and SUs to the CBS, over the nth sub-channel. e_n is the additive white Gaussian noise (AWGN) random variable with zero mean and variance σ_e^2.

Each PU is allocated one distinct set of sub-channels. We denote by $\mathcal{N}_p^{(k)}$ the set of $N_p^{(k)}$ sub-channels allocated to PU_k. We denote by $\mathcal{N}_s^{(k)}$ the set of $N_s^{(k)}$ sub-channels chosen by SU_k. Then, $x_{k,n} = 0$ for $n \notin \mathcal{N}_p^{(k)}$, and $u_{k,n} = 0$ for $n \notin \mathcal{N}_s^{(k)}$. Moreover, we denote by $Q_{k,n}$ and $P_{k,n}$ the power of $x_{k,n}$ and $u_{k,n}$, respectively.

3 Random Transmission Scheme

In this section, we describe the random transmission scheme for the previously described system.

3.1   Channel Access

For a given transmission block, each SU chooses a random degree d obtained from a predefined degree distribution $\Omega(x) = \sum_i \Omega_i x^i$, where Ω_i is the probability that $d = i$. Then, the SU chooses d sub-channels uniformly at random to transmit over. We define a $K_s \times N$ random channel access matrix $\mathbf{A}$ with integer elements $a_{k,n} \in \{0,1\}$, where $a_{k,n} = 1$ means SU_k is transmitting in sub-channel n, and $a_{k,n} = 0$ means SU_k is not transmitting in sub-channel n. Thus, it is easy to show that the elements of $\mathbf{A}$ are independent identically distributed (i.i.d.) Bernoulli random variables with a success probability of $\frac{1}{N}\bar{\Omega}$, where $\bar{\Omega}$ is the average degree and is given by $\sum_i i\Omega_i$. Then, we can represent the probabilistic random

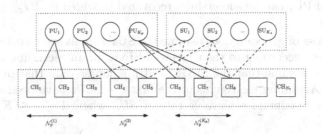

Fig. 1. Bipartite Graph Illustration of the Random Transmission Scheme

transmission scheme by a bipartite graph as shown in Fig.1. The PUs and SUs are shown by circles and referred to as variable nodes while the sub-channels $[\mathrm{CH}_i]_1 \leq i \leq N$ are shown by squares and referred to as check nodes. The number of edges connected to each variable node corresponds to the number of sub-channels it is allocated, and it is called the degree of the respective variable node. The solid edges represent the transmissions of the PUs whose number is assumed to be fixed e.g. PU_1 is of degree 2 in Fig. 1. On the other hand, the dashed edges represent the transmissions of the SUs whose number is a random variable with a distribution pre-determined by the CBS.

3.2   Successive Interference Cancellation

The CBS employs SIC to recover each user's signals. Each user is assumed to transmit the same signals over its respective sub-channels. The CBS can, then, combine the received transmissions of each user over all respective sub-channels using MRC, and the overall received SINR at the CBS can be represented as the sum of all individual SINRs. Note that the CBS is assumed to have the perfect knowledge of the PUs' channel state, transmit power and allocated sub-channels. We also assume that the CBS first attempts to recover the signals of the PUs. The maximum achievable rate of PU_i, where $1 \leq i \leq K_p$, is shown below:

$$R_p^{(i)} = \frac{N_p^{(i)}}{N} \log \left(1 + \sum_{n \in \mathcal{N}_p^{(i)}} \gamma_{p,n}^{(i)} \right), \tag{2}$$

where

$$\gamma_{p,n}^{(i)} = \frac{|g_{i,n}|^2 Q_{i,n}}{\sum_{k=1}^{K_s} a_{k,n} |h_{k,n}|^2 P_{k,n} + \sigma_e^2}. \tag{3}$$

We denote by $I_{p,n}^{(i)} = \sum_{k \in \mathcal{K}_s} a_{k,n} |h_{k,n}|^2 P_{k,n}$ the interference power caused to PU_i's transmission over the nth sub-channel, where $n \in \mathcal{N}_p^{(i)}$. Thus, the total interference caused by the SUs to PU_i can be expressed as $I_p^{(i)} = \sum_{n \in \mathcal{N}_p^{(i)}} I_{p,n}^{(i)}$. The signals of PU_i can be successfully recovered provided that $I_p^{(i)}$ is below the threshold $I_{th}^{(i)}$.

Without loss of generality, we assume the SUs' signals are recovered through the SIC process according to their received SINR, in an ascending order. More specifically, we assume that the SINR of SU_k is larger than that of SU_{k-1}, for $1 \leq k \leq K_s$. Assuming the signals of the first $i-1$ SUs have been successfully recovered, the maximum achievable rate of SU_i, where $1 \leq i \leq K_s$, is shown below:

$$R_s^{(i)} = \frac{N_s^{(i)}}{N} \log \left(1 + \sum_{n \in \mathcal{N}_s^{(i)}} \gamma_{s,n}^{(i)} \right), \tag{4}$$

where

$$\gamma_{s,n}^{(i)} = \frac{a_{i,n} |h_{i,n}|^2 P_{i,n}}{\sum_{k=i+1}^{K_s} a_{k,n} |h_{k,n}|^2 P_{k,n} + \sigma_e^2}. \tag{5}$$

Thus, we can express the total SINR of SU_i as $\gamma_s^{(i)} = \sum_{n \in \mathcal{N}_s^{(i)}} \gamma_{s,n}^{(i)}$. The signals of SU_i can be successfully recovered provided that their received SINR $\gamma_s^{(i)}$ at the ith iteration of SIC is above the threshold $\gamma_{th}^{(i)}$.

It will be shown later that the design problem is dependent on the SUs' received power rather than transmit power. Assuming that the SUs are able to estimate their channel gains from the downlink given the reciprocity of the channel, the CBS needs to broadcast the received power constraints only, imposed on each sub-channel on a per user basis. The SUs can, then, adaptively tune their power as necessary. Accordingly, we define a power vector $\mathbf{p} = [P_n]_{1 \leq n \leq N}$, where P_n is the received power constraint imposed on the nth sub-channel on a per user basis.

4   Asymptotic Performance Analysis of PRA in CRNs

In this section, we analyze the relationship between the system constraints (I_{th} and γ_{th}) and the different system metrics (N, K_p and K_s). We formulate an optimization problem to find the degree distribution that can maximize this probability of successfully recovering the signals of the SUs for a given setup.

4.1 Probability Density Function of the IP

Let us first calculate the power of interference introduced to the PUs.

Lemma 1. *In an asymptotically large network ($N \to \infty$, $K_s \to \infty$), the probability density function of the total interference power induced over the subchannels of PU_k, $\forall k \in \mathcal{K}_p$, follows the Poisson distribution below:*

$$Pr(I_p^{(k)} = iP_o^{(k)}) = e^{-\alpha N_p^{(k)}} \frac{\left(\alpha N_p^{(k)}\right)^i}{i!}, \tag{6}$$

where $\alpha = \frac{K_s}{N}\bar{\Omega}$, and $P_n = P_o^{(k)}$ $\forall n \in \mathcal{N}_p^{(k)}$. Its average and standard deviation are given below:

$$\mathbb{E}[I_p^{(k)}] = \alpha N_p^{(k)} P_o^{(k)}, \quad \sigma_{I_p^{(k)}} = \alpha N_p^{(k)} P_o^{(k)}. \tag{7}$$

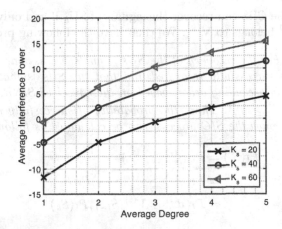

Fig. 2. The average interference power for a total of $N = 128$ sub-channels assigned equally to $K_p = 60$ PUs and shared by K_s SUs.

The proof of this lemma is provided in Appendix 7. In Fig. 2, the average IP is shown as a function of the average degree $\bar{\Omega}$ and the number of users K_s. $P_o^{(k)}$ is set to 0 dB $\forall k \in \mathcal{K}_p$. The average IP per PU is shown to increase with the number of K_s, as expected from Lemma 1. It is worthy of noting that N is fixed for all three simulations and that the increase in the average IP in fact corresponds to the increase in the ratio $\frac{K_s}{N}$ rather than K_s itself.

4.2 Probability of Success of the SUs

As in Section III-B, we assume the SUs' signals are recovered in an ascending order, based on their received SINR, with $\gamma_s^{(i)} \leq \gamma_s^{(i-1)}$ for $1 \leq i \leq K_s$. Given that the signals of the first $i - 1$ SUs have been successfully recovered, we can rewrite (5) and express the total SINR of SU$_i$ as follows:

$$\gamma_s^{(i)} = \sum_{n \in \mathcal{N}_s^{(i)}} \frac{P_n}{d_n^{(i)} P_n + \sigma_e^2}, \tag{8}$$

where $d_n^{(i)}$ is a random variable that represents the number of users, other than SU$_i$, transmitting in the nth sub-channel and whose signals have not been recovered yet. We define $\mathbf{d}^{(i)} = [d_n^{(i)}]_{1 \le n \le N_s^{(i)}}$ and refer to it as the observation vector. The CBS can then recover the signals of SU$_i$, if and only if, $\gamma_s^{(i)} \ge \gamma_{th}^{(i)}$, which will happen for certain values of $\mathbf{d}^{(i)}$. Let $\mathbf{V}^{(k)}$ denote the set of all vectors $\mathbf{v}$ that can satisfy the SINR constraint for SU$_k$. It can then be found that:

$$\mathbf{V}^{(k)} = \{(v_1, v_2, ..., v_{N_s^{(k)}}) | \sum_{n \in \mathcal{N}_s^{(k)}} \frac{P_n}{v_n P_n + \sigma_e^2} \ge \gamma_{th}^{(k)}\} \tag{9}$$

In other words, the CBS can recover the signals of SU$_k$ if and only if the observation vector $\mathbf{d}^{(k)}$ belongs to $\mathbf{V}^{(k)}$. We then have the following proposition:

Proposition 1. *For the recovery of the SUs' signals, we assume that the PUs' signals have been successfully recovered and that the SUs' signals are ordered and recovered in an ascending order, based on their received SINR. Let S_i be the event of having $\gamma_s^{(i)} \ge \gamma_{th}^{(i)}$. Then, the probability of successfully recovering the signals of SU$_i$, through the SIC process, can be calculated as follows:*

$$\begin{aligned} Pr(S_i) &= Pr(\gamma_s^{(i)} \ge \gamma_{th}^{(i)}) \\ &= Pr(\gamma_s^{(i)} \ge \gamma_{th}^{(i)} | S_{i-1}) Pr(S_k) \\ &= Pr(\mathbf{d}^{(i)} \in \mathbf{V}^{(i)} | S_{i-1}) Pr(S_k), \end{aligned}$$

for $1 \le i \le K_s$.

4.3   Clean Packet Model

As mentioned before, authors in [7,8,9] have implemented the iterative recovery process of codes-on-graph for the binary erasure channel (BEC) in PRA schemes. As in Fig. 1, the system is mapped onto a bipartite graph and the signal recovery is visualized as a message passing algorithm [10]. However, at the receiver side, successful signal recovery can only take place if an interference-free clean packet has been received at the destination.

From Section III-B, the observation vector of the 'clean packet' model must have the following form for successful signal recovery:

$$\{\mathbf{d}^{(i)} | \exists j, d_j^{(i)} = 0, 1 \le j \le i\}.$$

Let us consider the case where the received power of an SU's signal is less than or equal to its SINR threshold. Then, if the received signal is interference-free, it can be successfully recovered in our design. For such a case, the observation vectors of both designs are the same for $d_m = 1$.

On the other hand, from (9), we can see that the set of observation vectors that ensure successful recovery will generally be larger for our design; thus, it is expected to provide a higher probability of success. The 'clean packet' model can be seen as a special case of our design. Interestingly, when the SINR threshold is higher than that of the received power per signal for an SU, the 'clean packet' model fails to service any SUs at all. However, for sufficiently high degrees, our approach can still service a significant fraction of the SUs.

4.4   Optimization of the Degree Distribution

Given a CRN system of K_p PUs and K_s SUs transmitting over a set of N sub-channels, we formulate an optimization problem to find the degree distribution that maximizes the probability of successfully recovering the SUs' signals through the SIC process, while satisfying the IP constraints of the PUs. The CBS has the perfect knowledge of the PUs channel allocation, power, and respective IP constraints. It also has knowledge of the number of active SUs and their respective SINR constraints. Accordingly, the optimization problem can be formulated as follows:

$$\max_{\mathbf{p}, \Omega(x)} \sum_{k=1}^{K_s} \Pr(S_k)$$

$$\text{s.t.} \qquad \text{(i)} \sum_{i=1}^{d_m} \Omega_i = 1, \ \Omega_i \geq 0, \qquad \forall 1 \leq i \leq d_m$$

$$\text{(ii)} \mathbb{E}\left[\sum_n I_{p,n}^{(k)}\right] \leq I_{th}^{(k)}, \qquad \forall k \in \mathcal{K}_p.$$

Condition (i) ensures the sum of all probabilities is equal to 1. Condition (ii) ensures that the PUs are protected by the IP constraint on a per user basis. With reference to Lemma 1, it can easily be seen that this condition determines the value of $\bar{\Omega}$ and $\mathbf{p}$. Optimization is carried out using the covariance matrix adaptation evolution strategy (CMA-ES)[11] and can be easily modified for different IP and SINR thresholds.

5   Numerical Results

In this section, we investigate the system performance for different setups. Results are averaged over 10000 samples. The received power constraint per sub-channel P_o is taken to be 0 dB, the number of sub-channels N is set to 128 [12].

For ease of analysis, we now assume that $P_n = P_o$, where $1 \leq n \leq N$. This condition dictates that all sub-channels have the same received power constraint.

For a practical system, this also reduces the signalling overhead. This can be easily justified for the case where $N_p^{(i)} = N_p$ and $I_{th}^{(i)} = I_{th}$, for $1 \leq i \leq K_p$. We adopt this assumption in our simulations. We also assume that $\gamma_{th}^{(i)} = \gamma_{th}$, for $1 \leq i \leq K_s$.

Table 1. Results of CMA-ES optimization for $\frac{K_s}{N} = \frac{60}{128}$

$\log_{10} \gamma_{th}/P_o$	0 dB		1 dB	
d_m	4	8	4	8
Ω_1	0.0002	0.0003	0.0001	0.0002
Ω_2	0.5072	0.0831	0.1108	0.1295
Ω_3	0.0041	0.1619	0.1727	0.1529
Ω_4	0.4885	0.1744	0.7163	0.2112
Ω_5		0.2255		0.0674
Ω_6		0.0382		0.2406
Ω_7		0.1758		0.1188
Ω_8		0.1408		0.0794
ϵ	3.00e-03	3.75e-04	2.67e-04	4.05e-04

In Table 1, we show the results of CMA-ES for $\frac{K_s}{N} = \frac{60}{128}$ and a maximum degree of 4 and 8. Using the results of [7], we proceed to compare the achievable error probabilities of both designs; the error probability is denoted by ϵ and defined as $1 - \frac{1}{K_s} \sum_{k-1}^{K_s} \mathbb{P}_{s,k}$. Results are shown in Fig. 3. As predicted, the proposed design outperforms the 'clean packet' model even for $\frac{\gamma_{th}}{P_o} = 1$. As our proposed design relies on the overall received SINR, the sum of all individual SINRs, it makes use of all transmissions over the different sub-channels rather than interference-free transmissions only, thus, achieving better performance for the same power requirements. Finally, in Fig. 4, we consider the probability of having the IP caused by the SUs to the PUs below a given threshold. We use the results from Table 1, for $d_m = 8$ and $\log_{10} \frac{\gamma_{th}}{P_o} = 0$dB. We find $\bar{\Omega}$ to be around 5.23. Interestingly, for $I_{th} \leq$ -5dB, the probability of successfully recovering a PU's signals becomes independent of the threshold and solely dependent on the number of SUs supported in the network. Even more so, for thresholds as high as 10 dB, the probability of successfully recovering a PU's signals is almost one for any number of SUs. It is worthy of noting that Condition (ii) in Section IV-D can be easily modified to limit this probability by restricting $\Pr \left(\sum_n I_{p,n}^{(k)} \leq I_{th}^{(k)} \right) \leq \delta$, where δ is a predefined threshold.

6   Practical Considerations

In our system, the CBS is assumed to have the perfect knowledge of the PUs' activity and channel conditions, their respective IP constraints, the number of

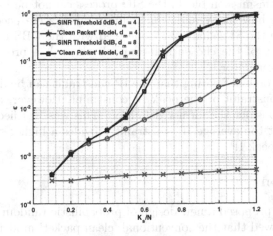

Fig. 3. Error probability of proposed scheme in comparison to the 'Clean Packet' model for different ratios of $\frac{K_s}{N}$

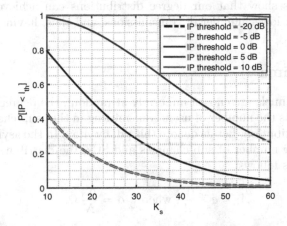

Fig. 4. Probability of IP being below the threshold for different values of K_s

active SUs and their respective SINR threshold. We assume this is made known to the CBS over the control channel, where transmissions are deterministic in duration and nature. Accordingly, the CBS can find the received power constraints necessary and the optimal degree distribution function to meet the system constraints. Then, the control channel can also be used to make the degree distribution known to the SUs. As the SUs are assumed to be able to estimate their channel gains from the downlink given the reciprocity of the channel, the signalling overhead is significantly reduced in comparison to fixed resource allocation.

For a given transmission block, the SIC process cannot be initiated without the knowledge of how many and which sub-channels were accessed by which users. We assume the SUs share the same seed with the CBS to determine the number and index of the chosen sub-channels through a pre-defined pseudo-random number generator [7].

Finally, it is worthy of noting that the IP constraint can be defined as either the average IP constraint or the peak IP constraint. However, throughout this paper, we only consider the former definition. This was justified in [13], where it was shown that the average IP constraint provides a higher system capacity than that of the peak IP.

7   Conclusion

In this paper, we proposed a new design of probabilistic random access schemes in CRNs. We showed that the conventional 'clean packet' model is sub-optimal under the IP and SINR constraints. We formulated a new optimization problem, based on CMA-ES, to maximize the probability of successful recovering the SUs' signals in the SIC process while satisfying the IP constraints of the PUs. Numerical results show that our degree distributions can achieve lower error probabilities with lower number of transmissions, and thus, having lower power requirements.

Proof of Lemma 1

Since the sub-channels are chosen uniformly at random, the degree of each sub-channel, defined as the number of users transmitting in that sub-channel, follows the binomial distribution. Let us denote this pdf by $\Lambda(x)$. In the asymptotic case, that is for a large number of sub-channels and SUs, the distribution converges to Poisson [14], as follows:

$$\Lambda_i = e^{-\alpha}\frac{\alpha^i}{i!}, \text{ where } \alpha = \frac{K_s}{N}\bar{\Omega}.$$

From (2), the IP constraint for PU_k was defined as: $I_p^{(i)} = \sum_{n=1}^{N_p^{(k)}} \sum_{k\in\mathcal{K}_s} a_{k,n}P_n = \sum_{n=1}^{N_p^{(k)}} u_n P_n$, where u_n is a random variable representing the number of SUs transmitting in the nth sub-channel. Intuitively, the probability of having u_n SUs transmitting in a sub-channel n is the same for all $1 \le n \le N$, and $\Pr(I_{p,n}^{(k)} = u_n P_n)$ is simply Λ_{u_n}. Assuming equal power allocation, that is $P_n = P_o^{(k)} \forall n \in \mathcal{N}_p^{(k)}$, we can express the pdf of $I_p^{(k)}$ as follows:

$$\Pr(I_p^{(k)} = i) = \Pr(u_1 P_o^{(k)} + u_2 P_o^{(k)} + ... + u_{N_p^{(k)}} P_o^{(k)} = i) = \bigotimes_{n=1}^{N_p^{(k)}} \Lambda_z\big|_{z=\frac{i}{P_o^{(k)}}}, \forall k \in \mathcal{K}_p,$$

where $\bigotimes$ denotes the convolution operation. As $\Lambda(x)$ was shown to be poisson distributed, and as the sum of poisson distributed random variables is also a poisson random variable, we arrive at (6).

References

1. Liang, Y.C., Chen, K.C., Li, Y., Mahonen, P.: Cognitive Radio Networking and Communications: An Overview. IEEE Transactions on Vehicular Technology **60**(7), 3386–3407 (2011)
2. Domenico, A.D., Strinati, E., Benedetto, M.D.: A Survey on MAC Strategies for Cognitive Radio Networks. IEEE Commununications Surveys & Tutorials **14**(1), 21–44 (2012)
3. Yang, L., Kim, H., Zhang, J., Chiang, M., Tan, C.W.: Pricing-based spectrum access control in cognitive radio networks with random access. In: Proceedings of IEEE INFOCOM, pp. 2228–2236 (2011)
4. Wang, S., Zhang, J., Tong, L.: Delay analysis for cognitive radio networks with random access: a fluid queue view. In: Proceedings of IEEE INFOCOM, pp. 1–9 (2010)
5. Chen, T., Zhang, H., Maggio, G.M., Chlamtac, I.: Cogmesh: a cluster-based cognitive radio network. In: Proceedings of IEEE International Symposium on New Frontiers in Dynamic Spectrum Access Networks, pp. 168–178 (April 2007)
6. Barman, S.R., Merchant, S.N., Madhukumar, A.: Random transmission in cognitive uplink network. In: International Conference on Mobile Services, Resources, and Users (MOBILITY), pp. 94–98 (2013)
7. Liva, G.: Graph-based Analysis and Optimization of Contention Resolution Diversity Slotted ALOHA. IEEE Transactions on Communications **59**(2), 477–487 (2011)
8. Paolini, E., Liva, G., Chiani, M.: High throughput random access via codes on graphs: coded slotted ALOHA. In: Proceedings of IEEE International Conference on Communications (ICC), pp. 1–6 (April 2011)
9. Liva, G., Paolini, E., Lentmaier, M., Chiani, M.: Spatially-coupled random access on graphs. In: IEEE International Symposium on Information Theory (ISIT), pp. 478–482 (July 2012)
10. Luby, M., Mitzenmacher, M., Shokrollahi, M.A.: Analysis of Random Processes via AND-OR Tree Evaluation. SODA **98**, 364–373 (1998)
11. Hansen, N., Ostermeier, A.: Adapting arbitrary normal mutation distributions in evolution strategies: the covariance matrix adaptation. In: Proceedings of IEEE International Conference on Evolutionary Computation, pp. 312–317 (1996)
12. Xu, H., Li, B.: Efficient resource allocation with flexible channel cooperation in OFDMA cognitive radio networks. In: Proceedings of IEEE INFOCOM, pp. 1–9 (2010)
13. Zhang, R.: On Peak versus Average Interference Power Constraints for Protecting Primary Users in Cognitive Radio Networks. IEEE Transactions on Wireless Communications **8**(4), 2112–2120 (2009)
14. Shokrollahi, M.A.: Raptor codes. IEEE Transactions on Information Theory **52**(6), 2551–2567 (2006)

On the Way to Massive Access in 5G: Challenges and Solutions for Massive Machine Communications

(Invited Paper)

Konstantinos Chatzikokolakis[1], Alexandros Kaloxylos[2], Panagiotis Spapis[2], Nancy Alonistioti[1], Chan Zhou[2], Josef Eichinger[2], and Ömer Bulakci[2(✉)]

[1] Department of Informatics and Telecommunications, National and Kapodistrian University of Athens, Athens, Greece
{kchatzi,nancy}@di.uoa.gr
[2] Huawei European Research Center, Future Wireless Technologies, Munich, Germany
{alexandros.kaloxylos,panagiotis.spapis,chan.zhou,
joseph.eichinger,oemer.bulakci}@huawei.com

Abstract. Machine Type Communication (MTC) is expected to play a significant role in fifth generation (5G) wireless and mobile communication systems. The requirements of such type of communication mainly focus on scalability (i.e., number of supported end-devices) and timing issues. Since existing cellular systems were not designed to support such vast number of devices, it is expected that they will throttle the limited network resources. In this paper, we introduce an effective solution for handling the signalling bottlenecks caused by massive machine communications in future 5G systems. The proposed approach is based on a device classification scheme using the devices' requirements and position for forming groups of devices with the same or similar device characteristics. Our scheme is analysed, and the evaluation results indicate that the proposed solution yields significant reduction in collisions compared to the standard when MTC devices attempt to access the Random Access CHannel (RACH).

Keywords: 5G · Group-based communications · Machine type communication · Massive connectivity · Random access channel

1   Introduction

According to the available predictions, the number of simple end-devices (e.g., sensors) that transmit short messages in periodic or asynchronous mode will grow considerably between 2013 and 2018, reaching 2,013 million in number, compared to 341 million reported in 2013, making thus, the need to be supported by 5G systems imperative [1]. The communication of such devices with the core network is commonly referred as Machine-to-Machine (M2M) Communication or Machine Type Communication (MTC) [2]. To efficiently support M2M communication, it is required to design new schemes that will lead to the reduction of signalling messages both in downlink and in uplink communication and avoid potential communication bottle-

© Institute for Computer Sciences, Social Informatics and Telecommunications Engineering 2015
M. Weichold et al. (Eds.): CROWNCOM 2015, LNICST 156, pp. 708–717, 2015.
DOI: 10.1007/978-3-319-24540-9_58

necks for a 5G operator in channels such as the random access and the paging. In this paper, we focus primarily on the uplink communication and, more specifically, on how a vast number of devices can be supported through a 5G Random Access CHannel (RACH).

Massive MTC scenarios require the development of innovative solutions so as to avoid signalling congestion in future mobile communication networks. The current design of mobile networks is tailored for human-to-human (H2H) communications (i.e., Telephony, SMS, Streaming services, etc.) [3]. In such deployments, for accessing the network, the user equipments (UEs) follow the contention-based random access procedure, which occurs in LTE networks in every Random Access Opportunity (RAO). However, such network designs are unlikely to be able to handle the MTC applications, where a large number of machines will attempt to transmit simultaneously small amounts of data.

For massive MTC, the collision rate is large for the RACH access procedure, as the number of devices is very high for a single cell; whereas, the number of the RACH preambles is very limited. For example, assuming a cell with 1000 users, 64 RACH preambles and 30ms packet arrival interval, the collision probability is almost certain (99.97%) [4]. Furthermore, the MTC network characteristics impose additional requirements for mobile networks. Specifically, deployment of large number of devices will lead to collisions that will increase the battery consumption for the MTC devices. Periodic network access may lead to increased latency and QoS degradation, and poor spectrum efficiency may occur due to allocation of resources for small data transmissions.

This paper is organized as follows. In the next section, we provide a thorough analysis of the state of the art approaches for improving the access to RACH, and we assess their suitability for 5G communication networks. In Section 3, we present a novel solution for reducing the collisions of MTC devices during the RACH access and, in Section 4, we provide evaluation results that show resiliency and robustness of the proposed solution against the demanding conditions of a 5G network. Finally, Section 5 concludes the paper.

2   State of the Art Analysis

Numerous solutions for handling the RACH procedure in wireless networks with large number of devices have been proposed in the literature. Those solutions can be classified into *pull-based* and *push-based schemes* depending on the signalling process.

The *pull-based schemes* may be further classified into several categories. In Access Class Barring (ACB) schemes [5]-[8], the UEs are categorized to access classes, and based on their access class, the UEs determine whether their RACH procedure should be delayed. Physical RACH (PRACH) resource separation between H2H and M2M schemes [9] suggest that the separation of resources can be achieved by separating either the preambles or the time-frequency RACH resources (i.e., Resource Blocks) into two groups so as to increase the PRACH slots reserved for MTC and reduce collisions. Solutions on Dynamic allocation of RACH Resources [10]-[12] suggest that base station (BS), referred to as evolved Node Bs (eNBs) in long-term evolution (LTE) networks, dynamically allocate PRACH resources based on PRACH overload and, thus, coordinating the overall procedure and reducing the failed attempts. In back-off schemes [6][8][13], the random access attempts for MTC devices and typical user terminals are treated separately using different delay schemes. Additionally, for reducing the collision rate, MTC devices could be classified further based on their delay requirements, and different back-off procedures could be applied for the various classes [6]. In addition, these solutions could be combined with other approaches (e.g., group-based random access procedure in case of collisions [13]). Other approaches focus on slotted access [4] where M2M terminals are allowed to transmit preambles in specific random access slots. In such schemes, each eNB broadcasts the random access cycle, and MTC terminals calculate random access slots based on their identity and the received random access cycle. However, preamble collision is unavoidable if several M2M devices share the same random access slot. Finally, in group (aka, cluster) based solutions MTC devices can be grouped according to QoS requirements [14] or geographical location [13][15]. In these approaches, a group head is selected to communicate with the eNB on behalf of the group. The group head receives requests from group members and relays them to the eNB.

The *push-based schemes* are paging-based approaches in which the RACH procedure is triggered by the eNB rather than the UE. All MTC devices in idle mode listen to the paging message, and the devices initialize random access procedure when their IDs are included in the paging message [16].

However, the above-mentioned solutions are not targeting 5G networks and are not sufficient to fulfil the stringent requirements for future cellular communication networks. As described afore, random access resources are extremely limited for the considered number of devices, and the collisions during the random access procedure lead to unacceptable latency levels. Although some of the solutions are applied in MTC scenarios, they are only evaluated for a small number of devices ranging from tens to few hundreds, making their scalability questionable.

A solution that follows a form of slotted access scheme seems appropriate so as to avoid extra collisions. Furthermore, as the network environments will be much denser in the future, it is reasonable to assume that the devices can compose groups with direct communication so as to alleviate the load of the BS. Thus, the combination of clustering mechanisms, slotted access scheme, and possibly device-to-device (D2D) communication between the Cluster Members (CMs) seems as a promising solution.

3   Classification and Cluster-Based RACH Access of MTC Devices

In the previous sections, we have provided a brief analysis of the available solutions in the literature related to MTC in cellular networks, and we have argued that, in 5G communication scenarios, the existence of huge number of MTC devices will lead to signalling congestion when accessing the RACH. In this section, we provide a description of our solution. The proposed scheme aims at optimizing the random access process using:

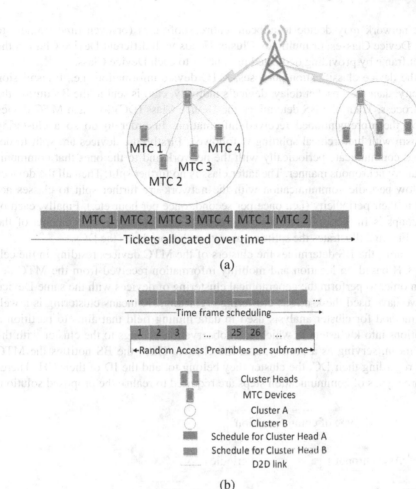

Fig. 1. MTC RACH access scheme.

a) device classification and clustering mechanisms,
b) D2D communication inside each cluster, and,
c) an appropriate slotted access scheme for the communication of the Cluster Heads (CHs) with the network.

Figure 1 captures a conceptual view of the considered environment with huge number of MTC devices operating and attempting to access the RACH. The devices should be classified based on their communication needs and should form clusters according to their location and mobility patterns. As illustrated in Figure 1 (a) the Cluster Members will send their data to their Cluster Head with D2D links following a time-slotted scheme for intra-cluster communication. The Cluster Heads are responsible to send the aggregated data to the network. In order to do so, Cluster Heads follow the scheduling information received from the network. In Figure 1 (b) is shown that the network may decide to allocate entire subframes (or even time frames) to specific Device Classes, or multiplex Cluster Heads with different Device Class in the same subframe by providing dedicated preambles to each Device Class.

For the device classification purposes, MTC device information (i.e., transmission periodicity, data size, packet delay, device's mobility, etc.) is sent to the BS during the attach process. Then, the BS determines the Device Class (DC) for each MTC device based on the aforementioned received information. In order to do so a clustering mechanism with hierarchical splitting is followed. Firstly, the devices are split to devices that communicate periodically with the network and to the ones that communicate in an asynchronous manner. The latter class is no further split. Then all the devices that follow periodic communication with the network are further split to classes according to their periodicity (i.e., once per second, once per hour etc.). Finally, each of these groups is further split based on the communication delay requirements of the devices. Figure 2 illustrates the splitting operation that produces the DCs.

Afterwards, the BS determines the clusters of the MTC devices residing in the cell and the CH based on location and mobility information received from the MTC devices. In order to perform the geographical clustering of devices with the same Device Class, we have used the k-means clustering algorithm. K-means clustering is a well known method for cluster analysis used in data mining field that aims to partition n observations into k clusters in which each observation belongs to the cluster with the nearest mean, serving as a prototype of the cluster. Then, the BS notifies the MTC devices regarding their DC, the cluster they belong to, and the ID of their CH. Thereafter, three types of communication steps are required to realise the proposed solution.

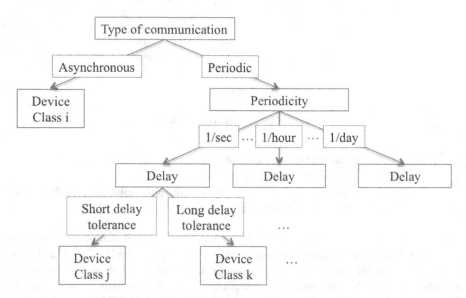

Fig. 2. Hierarchical splitting operation for Device Classes

The first type comprises a set of messages exchanged between the BS and the CHs. Firstly, the BS broadcasts RACH access scheduling information to CHs. The CHs are scheduled to be able to transmit periodically every N time slots with the BS. N may vary for different CHs depending on their DC, which determines the communication needs of the DC. Then, CHs attempt to access RACH during the specified RAOs, as scheduled. On successful communication, the CH transmits the aggregated monitoring data of its cluster to the BS. Although the CHs that transmit periodically are multiplexed and, thus, scheduled on specific time frames (specified by a periodicity factor), collision may still occur either due to having too many CHs scheduled in a frame or, due to having RACH access attempts triggered from devices with asynchronous communication. In case of collisions, the BS reschedules the CHs and broadcasts the new schedule to the CHs.

Table 1. Classification and Cluster-based RACH access algorithm

Step	Description
0	MTC devices are attached to the network and they send their device information
1	eNB determines the Device Class of the MTC devices through a clustering mechanism with hierarchical splitting
1.1	$\forall$ MTC device perform classification based on Type of Communication and produce two classes of MTC devices; the ones with periodic transmissions and the ones with the asynchronous transmissions. Devices with asynchronous transmissions are no further classified.
1.2	$\forall$ MTC device with periodic transmissions perform classification based on the devices periodicity and give label s to the device, $s \in$ ClassLabels $= \{class^1_1, class^1_2, ..., class^1_i, ..., class^1_N)$
1.3	$\forall$ i $\in$ ClassLabels, $\forall$ MTC device perform classification based on communication delay and add label t to the device, $t \in$ ClassLabels$^2 = \{class^2_1, class^2_2, ... class^2_i, ... , class^2_N)$. Each device has an s, t label now
2	Based on the Device Class of each device perform geographical clustering using k-means. The value of k determines the number of clusters and the maximum geographical distance between the CH and a CM.
3	eNB notifies the MTC devices regarding their DC, the cluster they belong to, and the ID of their CH. CHs also receive scheduling information for accessing the RACH
4	CH performs time-division scheduling, and broadcasts reservation token information to the CMs. D2D communication sessions with the CMs according to the produced schedule are established
5	CH performs RACH access on the RAOs defined by the network
6	Steps 2-5 are repeated so as to perform cluster merging particularly for non-stationary devices

The second type of communication comprises a set of signalling steps for intra-cluster communication between the CH and each CM. Initially, the CH produces a time-division schedule, distributes reservation token information to the CMs, and establishes D2D communication sessions with the CMs according to the produced schedule. Then, the CMs communicate with the CH according to the schedule received. In case CMs leave a cluster, they notify first the CH over specific tokens reserved by the CH for control information, and they wait for a reply from the CH. Once the device has left the cluster, the CH calculates, if necessary, a new schedule and disseminates it to the CMs.

Finally, the third type of communication is related to the operation of merging clusters. In this case, the CH needs to communicate with MTC devices that are not members of the cluster. Such operation is executed by the BS, if needed. More specifically, BS may decide merging of clusters based on their position and their mobility patterns. In that case, the BS will send a notification to the CHs to merge and indicate which device will be the new CH of the merged cluster. Table 1 summarizes the proposed mechanism steps.

4   Evaluation Results

For the assessment of the proposed scheme, we have utilized a simulation scenario comprising several machine types with various service requirements and mobility characteristics. The purpose is to assess the proposed scheme's capability to reduce the overhead of RACH due to massive deployment of MTC devices in a small geographical area. In our simulation scenario, we have assumed the deployment of 1 BS in the area, having 1 RAO per time frame (i.e., maximum 64 devices per time frame may successfully access the RACH due to 64 random access preambles available). Table 2 summarizes the simulation setup in terms of numbers of devices, traffic characteristics, and devices' mobility for a massive deployment of MTC devices in a 5G scenario, see Test Case 11 *Massive deployment of sensors and actuators* in [17]. The devices have been uniformly distributed to those characteristics.

Table 2. Requirements for MTC scenario in 5G networks.

Type	Value
Number of devices	[30.000, 300.000]
Transmission Periodicity	Periodic (1/minute, 1/hour, 1/day) or asynchronous
Data Transmission size	20, 75 or 125 bytes
Packet Delay	Small delay [5,10] msec or Larger delay [1, 5] sec
Device's Mobility	Stationary or Low mobility (i.e., <3km/h)

In the evaluated scenario, we have dropped uniformly the considered number of devices in the 387 m x 552 m grid area covered by the BS [18] and have measured the number of collisions and the collision rate of the devices over a time window of 1 hour when accessing the RACH. Figure 3 highlights the merits of our work compared to the current standard and shows significant gains regarding the collision rate. More specifically, applying our solution to the network reduces 2 times the collision rate for 30.000 devices and as the number of the devices increases we end up having up to a 2.8-time collision rate reduction for 300.000 devices.

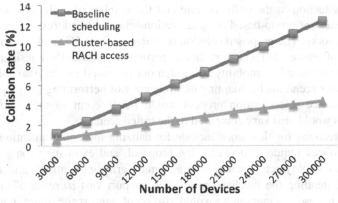

Fig. 3. Collision rate

Figure 4 highlights the benefits of our solution regarding the total number of collisions. The collisions are 4 times reduced for 30.000 devices, while this reduction reached up to 7.3 times for 300.000 devices. Overall, we observe that the developed scheme enables the deployment of a large number of devices in a small deployment area, since we take advantage of the similar traffic characteristics of co-located devices.

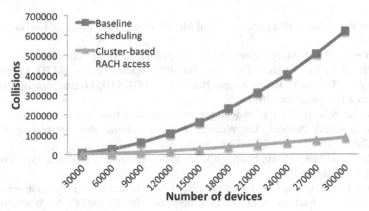

Fig. 4. Total number of collisions

5 Conclusion

The vast increase of the MTC devices that is expected for 5G networks will lead to exhaustive usage of limited network resources, such as the RACH. Additionally, such devices may have strict latency requirements and limited battery lifetime, making, thus, the collision rate of their requests a critical point that should be carefully studied. In this paper, we have assessed how currently available solutions may fail address these problems adequately and, we have presented a cluster-based solution, which exploits the traffic and mobility characteristics of each MTC device. The results show significant reduction in the collision rate and the number of collisions, and indicate the effectiveness of group-based communication schemes to address the requirements of future networks. The proposed solution is evaluated for a 5G scenario with massive deployment of sensors and actuators. In this particular scenario the considered devices are stationary or have low mobility. Although our proposed mechanism could also be applied in other scenarios by adapting the frequency of performing geographical clustering, the device classification process would be different as the communication requirements would also vary compared to the tested scenario.

Future directions for this work include formalizing the communication messages and the interfaces required to realize the proposal, and evaluating the performance under various deployment scenarios (e.g., smart grid applications and emergency scenarios). Extending the mechanism so as to support vast increase of unscheduled events (e.g., in case of emergency) would also be of interest for future work. Finally, the effect in the power/battery consumption could be considered as well, since the machines grouping and coordination seems to increase the battery durability of the sensors, due to the lower transmission power required for communicating via the cluster head.

References

1. Cisco Visual Networking Index: Global Mobile Data Traffic Forecast Update, 2013–2018 (2014)
2. Emmerson, B.: M2M: the Internet of 50 billion devices. M2M Magazine (2010)
3. 3GPP TS 36.300, Evolved Universal Terrestrial Radio Access (E-UTRA) and Evolved Universal Terrestrial Radio Access Network (E-UTRAN); Overall description; Stage 2 (Release 12), December 2014
4. Zhou, K., Nikaein, N.: Packet aggregation for machine type communications in LTE with random access channel. In: Wireless Communications and Networking Conference (WCNC), 2013, pp. 262–267. IEEE, April 2013
5. 3GPP TR 37.868 v11.0.0, Study on RAN Improvements for Machine Type Communications, September 2011
6. Cheng, J.-P., Lee, C., Lin, T.-M.: Prioritized random access with dynamic access barring for RAN overload in 3Gpp. LTE-A networks. In: GLOBECOM Workshops, 2011, pp. 368–372. IEEE, December 2011
7. Zheng, K., Hu, F., Wang, W., Xiang, W., Dohler, M.: Cooperative Access Class Barring for Machine-to-Machine Communications. IEEE Transactions on Wireless Communications 11(1), 27–32 (2012)
8. Cheng, M.-Y., Lin, G.-Y., Wei, H.-Y., Hsu, C.-C.: Performance evaluation of radio access network overloading from machine type communications in LTE-A networks. In: WCNC, 2012, pp. 248–252. IEEE, April 1–1, 2012
9. Lee, K.-D., Kim, S., Yi, B.: Throughput comparison of random access methods for M2M service over LTE networks. In: GLOBECOM Workshops, 2011, pp. 373–377. IEEE, December 2011
10. Choi, S., et al.: Automatic configuration of random access channel parameters in LTE systems. In: Proc. 2011 IFIP Wireless Days, pp. 1–6, October 2011

11. Lo, A., Law, Y.W., Jacobsson, M.: Enhanced LTE advanced random-access mechanism for massive machine-to-machine (M2M) communications. In: Proc. 27th Meeting of Wireless World Research Forum, October 2011

12. Zhou, K., Nikaein, N., Knopp, R.: Dynamic resource allocation for machine-type communications in LTE/LTE-A with contention-based access. In: Wireless Communications and Networking Conference (WCNC), 2013, pp. 256–261. IEEE, April 7–10, 2013

13. Kookjin, L., JaeSheung, S., Yongwoo, C., Ko, K.S., Sung, D.K., Shin, H.: A group-based communication scheme based on the location information of MTC devices in cellular networks. In: 2012 IEEE International Conference on Communications (ICC), pp. 4899–4903, June 10–15, 2012

14. Lien, S.-Y., Chen, K.-C.: Massive Access Management for QoS Guarantees in 3GPP Machine-to-Machine Communications. IEEE Communications Letters 15(3), 311–313 (2011)

15. Wang, S.-H., Su, H.-J., Hsieh, H.-Y., Yeh, S.-P., Ho, M.: Random access design for clustered wireless machine to machine networks. In: 2013 First International Black Sea Conference on Communications and Networking (BlackSeaCom), pp. 107–111, July 3–5, 2013

16. Wei, C.-H., Cheng, R.-G., Tsao, S.-L.: Performance Analysis of Group Paging for Machine-Type Communications in LTE Networks. IEEE Transactions on Vehicular Technology 62(7), 3371–3382 (2013)

17. Fallgren, M., Timus, B. (eds.): Future radio access scenarios, requirements and KPIs, METIS deliverable D1.1, March 2013. https://www.metis2020.com/documents/deliverables/

18. Monserrat, J.F., Fallgren, M. (eds.): Simulation guidelines, METIS deliverable D6.1, October 2013. https://www.metis2020.com/documents/deliverables/

An Evolutionary Approach to Resource Allocation in Wireless Small Cell Networks

Shahriar Etemadi Tajbakhsh[✉], Tapabrata Ray, and Mark C. Reed

University of New South Wales, Canberra, Australia
{s.etemaditajbakhsh,t.ray,mark.reed}@unsw.edu.au

Abstract. In this paper we consider the problem of joint resource allocation, user association and power control for optimizing the network utility which is a function of users' data rates in a wireless heterogeneous network. This problem is shown to be NP-hard and non-convex. We propose an evolutionary algorithm to solve the problem. We show the gain of joint optimization over the scenarios with fixed power is considerable. Also in terms of computation time, our algorithm is substantially improved over the previously proposed algorithm by the authors.

Keywords: Small cell networks · Resource allocation · Optimization · Evolutionary algorithms

1 Introduction

As a promising technology to cope with ever growing demand for bandwidth intensive applications by the wireless users, small cells are supposed to bring the base stations to the close proximity of the users maybe in their homes or offices. Thousands or even more cells might be exploited within an area which was previously covered by a small number of cells. As a consequence, highly scalable algorithms are required in such densified networks for load balancing, power control and interference management, and channel (or time slot) allocation. For various reasons, the algorithms currently in use for conventional cellular networks should be revisited. For instance, unlike current cellular networks where statistically speaking the load variation of the cells is relatively minor (due to the law of large numbers), in small cells the load associated to a cell (particularly if conventional algorithms for user association are applied) may dramatically change over time. Also because of shorter distances between the transmitters interference mitigation and power management is more sensitive.

In [1,2], the optimization problem of joint resource allocation and user association for maximizing the network utility (which is a function of user data rates typically with proportional fairness consideration) is solved using some relaxations. Recently, the authors in [3] have studied a more general scenario where the power allocated for each resource at each base station is also an optimization variable and can be tuned. This problem is shown to be NP-hard and non-convex. In [3] a greedy algorithm is proposed to explore for the optimal solution. At each

© Institute for Computer Sciences, Social Informatics and Telecommunications Engineering 2015
M. Weichold et al. (Eds.): CROWNCOM 2015, LNICST 156, pp. 718–724, 2015.
DOI: 10.1007/978-3-319-24540-9_59

round of the algorithm a perturbation operator manipulates the solution found in the last iteration and if the new solution was better it updates the solution. Despite the considerable gain over the fixed power scenario in [1,2], the algorithm is very slow and hardly practical for a real time optimization of network parameters.

In this paper, an evolutionary algorithm to the optimization problem of joint resource allocation, user association and power control is introduced. This algorithm is substantially faster than the previously proposed algorithm in [3]. Not only the proposed algorithm converges faster but also it is possible to be parallelized over several processors as it is a population based algorithm where each solution within the population can be evaluated simultaneously on a distinct processor.

It should be noted that the problems of load balancing (see [4] for a comprehensive survey) and power control (see for instance [5,6,7] for a given set of transmitters and receivers) have been separately studied. However, unifying these problems within an identical framework is not straightforward at all. In the following the optimization problem and the proposed solution are described and evaluated via numerical experiments.

2   System Model and Problem Formulation

We study a wireless cellular network consisting of N users, M base stations and a set of K channels[1] available at each of the base stations. Cells are categorized to L tiers according to the range of their transmission power. We consider the downlink between the base stations and the users. The set of all users, all base stations, and all the channels which are identically available for association at each base station for association) are denoted by $\mathcal{U}$, $\mathcal{B}$ and $\mathcal{F}$, respectively. The subset of base stations belonging to tier l is denoted by $\mathcal{B}^l$.

In this paper we assume the power allocated to each frequency channel j from each base station k is an optimization variable and is denoted by P_k^j. The achievable rate for user i associated to base station k on channel j, denoted by c_{ik}^j, is typically a logarithmic function of signal to noise and interference ratio (SINR). In this paper we assume:

$$c_{ik}^j = \log(1 + \frac{P_k^j g_{ik}}{N_0 + \sum_{\ell:\ell\in\mathcal{B},\ell\neq k} P_\ell^j g_{i\ell}}) \qquad (1)$$

where g_{ik} is the channel gain between base station k and user i, which includes path loss, shadowing and antenna gains and N_0 is the thermal noise power. We consider a snapshot of the network i.e. g_{ik} is assumed to be fixed which is similar to the assumption of [1,2].

We denote association of user i to base station k on channel j by x_{ik}^j where $x_{ik}^j = 1$ indicates the user is associated and j by $x_{ik}^j = 0$ otherwise. We use

---

[1] In this paper we consider an OFDM based system, however the algorithm proposed here can be extended to a time division multiple access system as well.

proportional fairness criterion in defining our objective function to maintain a balance between providing high data rates to the users and fairness in allocating network resources. It is shown in [8] that by maximizing the sum of logarithms of data rates the proportional fairness is achieved. Therefore, the optimization problem of joint resource allocation, user association and power control is formulated as follows.

$$\underset{P,X}{\text{maximize}} \quad \sum_{i=1}^{N} \log c_i$$

s.t.

$$c_i = \sum_{k=1}^{M} \sum_{j=1}^{K} x_{ik}^j c_{ik}^j, \ \forall i \in \mathcal{U}$$

$$c_{ik}^j = \log \left(1 + \frac{P_k^j g_{ik}}{N_0 + \sum_{\ell:\ell \in \mathcal{B}, \ell \neq k} P_\ell^j g_{i\ell}} \right), \forall i \in \mathcal{U}, k \in \mathcal{B}, j \in \mathcal{F} \quad (2)$$

$$x_{ik}^j \in \{0,1\}, \ \ \forall i \in \mathcal{U}, \ k \in \mathcal{B}, \ j \in \mathcal{F}$$

$$\sum_{i=1}^{N} x_{ik}^j \leq 1, \ \ \forall k \in \mathcal{B}, \ j \in \mathcal{F}$$

$$\sum_{k=1}^{M} x_{ik}^j \leq 1, \ \ \forall i \in \mathcal{U}, \ j \in \mathcal{F}$$

$$P_\ell^j \leq p_0^l, \ \ \forall \ell \in \mathcal{B}^l, \ j \in \mathcal{F}$$

where P and X are the set of all P_k^j's and x_{ik}^j's, respectively. The first and the second constraint identify the data received by each user i. The fourth constraint indicates that each channel at each base station can be allocated to one user at most. The fifth constraint ensures that a user is not associated to the same channel in more than one base stations. Finally, the last constraint applies an upper bound p_0^l to the transmission power in each tier l (which is an important constraint in real systems). As mentioned before this problem is non-convex, mixed integer and NP-hard as it is reduced to the problem of [1,2] if the power is fixed. In the next section we describe the proposed evolutionary algorithm to solve the formulated problem.

3   Evolutionary Algorithm

The algorithm evolves over a ceratin number of iterations and for a population $\mathcal{S}$ of size S. At each iteration for each member $s \in \mathcal{S}$ of the population one of the three perturbation operators $\mathcal{O} = \{O_1, O_2, O_3\}$ is chosen to manipulate s and generate $s' = O_i(s)$. Each member of the population results in a value for objective function $f(s) = \sum_s \log c_i$. If $f(s') \geq f(s)$, s is replaced with s', otherwise s remains in the population. It should be noted that any $s \in S$ is a solution satisfying the constraints in optimization problem 2. In the second phase of each iteration any member of the updated population is combined with a partner from the same population using a reproduction operator to generate a population of children who are the population for starting next iteration. In the following, the three operators, the reproduction operator, switching mechanism between operators for each member and the initialization of the algorithm are described in detail.

- **Operator 1 ($\mathcal{O}_1$):** This operator randomly chooses a channel from a random base station, say P_k^j and adds a random value from a Gaussian distribution subject to a ceiling for the transmission (if the new value exceeds the ceiling, the transmission power is updated to the ceiling value (*Power Control*).

- **Operator 2 ($\mathcal{O}_2$) :** Two frequency channels, say j_1 and j_2, from the same base station are chosen randomly and the users associated to these two channels (say i_1 and i_2, respectively) are swapped. In other words, j_1 is allocated to i_2 and i_1 is associated to j_2. It should be noted that when the power is fixed the schedule of resource allocation (the order which the frequency channels are allocated to users associated to the same base station) does not matter. However, in our work this schedule changes the SINR observed by the users and hence the network utility (*Resource Scheduling*).

- **Operator 3 ($\mathcal{O}_3$):** The third operator randomly chooses a user i_1, a channel j_1 at base station ℓ_1 where currently is allocated to a user i_2. If channel j_1 has not been allocated to user i_1 from any other base station, i_2 is replaced with i_1. Otherwise, if channel j_1 has been allocated to i_1 at some other base station say base station ℓ_2, then channel j_1 is allocated to user i_2 at base station ℓ_2 and channel j_1 is allocated to ℓ_1 (*User Association*).

- **Operator Switching:** For any member of the population, we start from $\mathcal{O}_1$. If an operator $\mathcal{O}_i$ cannot improve the solution for τ successive iterations operator $\mathcal{O}_j$ is replaced where $j = (i+1) \mod 3$.

- **Reproduction Operator:** A fraction $\frac{1}{r}$ of the best solutions $\mathcal{S}' \subseteq \mathcal{S}$ (with highest utility function value) in the current iteration are selected and r replicas of each $s \in S'$ are placed in a new $\mathcal{S}$ (i.e. all the old members of S are replaced with these replicas). Any member $s \in \mathcal{S}$ is matched with a randomly chosen member $s' \in \mathcal{S}$. To combine s and s' to generate a child $s"$ we use a crossover operator where each pair $< i : x_{ik}^j = 1, P_k^j >$ is chosen randomly from either s or s'. Therefore the child s'' is a random mix of the resource allocations in s and s'. If $f(s'') > f(s), f(s')$, s'' is replace with s.

- **Initialization:** We have chosen the solution found in [1,2] as the start point. This is reasonable as this solution is optimal for fixed power on all the channels within the same base station. All the members of the first iteration are set to the same solution.

As it can be inferred from the description of algorithm, it is different from standard evolutionary algorithms as it incorporates a greedy sub-algorithm within the main algorithm where at each round the solution selected for each member of the solution cannot be worse than the solution in the last generation. In other words, this algorithm can be named as a greedy-evolutionary algorithm.

In the next section we evaluate the performance of the algorithm via numerical experiments.

4   Numerical Experiments

In our simulations we evaluate the performance of the proposed algorithm and also we compare it to the algorithm proposed in [3]. Particularly we show that the proposed algorithm in the current paper is fundamentally faster both due to its faster convergence and also the possibility of running the algorithm over several processors. A set of $N = 100$ users and $M = 22$ base stations including $|\mathcal{B}^1| = 2$ macro base stations (first tier) and $|\mathcal{B}^2| = 20$ femtocell base stations (second tier) and $K = 20$ frequency channels available at each base station are considered. The users are spread over an $1000m \times 1000m$.

We assume the transmission power by the macro base station and femtocell base stations are upper limited by $P_k^j \leq 52dBm, \forall k \in \mathcal{B}^1, j \in \mathcal{F}$ and $P_k^j \leq 34dBm, \forall k \in \mathcal{B}^2, j \in \mathcal{F}$, respectively. To initialize the algorithm we assume all the transmission power for all the channels at each macro base stations and all the channels at each femtocell are set to be $P_k^j = 46dBm, \forall k \in \mathcal{B}^1, j \in \mathcal{F}$ and $P_k^j = 34dBm, \forall j, k \in \mathcal{B}^2$, respectively. We model the path loss as $L(d) = 30 + 37\log(d)$ and $40+34\log(d)$ for the macro and femtocell base stations, respectively where d is the distance between the transmitter and receiver. The thermal noise power is $\sigma_{noise}^2 = -104dB$ and the shadowing is assumed to be log normal random variable S with standard deviation $\sigma_{shadow} = 8dB$. Antenna gains for the first and second tiers are $g_A^1 = -15dBi$ and $g_A^2 = -5dBi$, respectively. The channel gain g_{ik} is assumed to be $g_A^l - L(d) + S, l = 1, 2$.

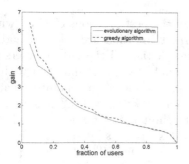

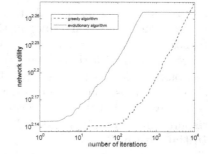

Fig. 1. Distribution of gain for the variants of the algorithm

Fig. 2. Evolution of the network utility over iterations

In Fig. 1 we represent the distribution of the gain of the greedy algorithm in [3] and the proposed evolutionary algorithm in this paper over the optimization solution with fixed power assumption in [1,2]. As it can be observed the performance of the two algorithms are completely close. The average gain for

Table 1. Transmission power characteristics in the best solution found by the algorithm

	max power	min power	mean power
Macro cells	48.36	29.31	37.97
Femto cells	34.00	3.04	29.21

the evolutionary algorithm is 2.08. However, it should be noted that the current algorithm can be processed on 50 processors while the greedy algorithm should be only processed on a single machine as it is a sequential algorithm. However, not only because of the possibility of parallel computing but also because of the diverse pool of solutions generated at each round and reproduction the algorithm also converges in a dramatically less number of iterations. Fig. 2 shows the evolution of network utility in the evolutionary algorithm over 800 iterations and greedy algorithm over 10000 for iterations of the algorithm (we stopped the evolutionary algorithm after 800 iterations and fixed the solution for comparison). The evolutionary algorithm converges after nearly 400 iterations while the greedy algorithm finds a slightly better solution in about 10000 iterations. Therefore the difference in computation time is enormous and makes the evolutionary algorithm highly convenient in responding to the dynamics of the network.

Finally the maximum, minimum and average power in the solution found by the evolutionary algorithm are represented in table 1. The maximum power for the femtocells is matching with the upper limit on the power. Therefore, one can conclude that if the upper limit is removed it would be possible to obtain better solutions. However it should be noted for various reasons and according to radio communication regulations the transmission power cannot exceed a certain limit.

5   Conclusion

In this paper we proposed an evolutionary algorithm to the optimization problem of joint resource allocation, user association and power management in wireless small cell networks. The problem is NP-hard, non-convex and mixed integer. The proposed algorithm is substantially more efficient in terms of its computational time over the previously proposed algorithm by the authors. As a population based algorithm, the evolutionary algorithm in this paper can be run on several processors at the same time. Moreover, the algorithm itself converges dramatically faster because of it larger pool of solutions and reproduction.

References

1. Fooladivanda, D., Rosenberg, C.: Joint resource allocation and user association for heterogeneous wireless cellular networks. IEEE Transactions on Wireless Communications **12**(1), 248–257 (2013)

2. Ye, Q., Rong, B., Chen, Y., Al-Shalash, M., Caramanis, C., Andrews, J.G.: User association for load balancing in heterogeneous cellular networks. IEEE Transactions on Wireless Communications **12**(6), 2706–2716 (2013)
3. Tajbakhsh, S.E., Ray, T., Reed, M.C.: Joint power control and resource schedulingin wireless heterogeneous networks. In: International Conference on Telecommunications, ICT, Sydney, Australia (2015)
4. Andrews, J.G., Singh, S., Ye, Q., Lin, X., Dhillon, H.S.: An overview of load balancing in HetNets: Old myths and open problems. IEEE Wireless Communications **21**(2), 18–25 (2014)
5. Qian, L.P., Zhang, Y.J.A., Huang, J.: Mapel: Achieving global optimality for a non-convex wireless power control problem. IEEE Transactions on Wireless Communications **8**(3), 1553–1563 (2009)
6. Sung, C.W.: Log-convexity property of the feasible SIR region in power-controlled cellular systems. IEEE Communications Letters **6**(6), 248–249 (2002)
7. Chiang, M., Tan, C.W., Palomar, D.P., O'Neill, D., Julian, D.: Power control by geometric programming. IEEE Transactions on Wireless Communications **6**(7), 2640–2651 (2007)
8. Kelly, F.: Charging and rate control for elastic traffic. European transactions on Telecommunications **8**(1), 33–37 (1997)

Coexistence of LTE and WLAN in Unlicensed Bands: Full-Duplex Spectrum Sensing

Ville Syrjälä[✉] and Mikko Valkama

Department of Electronics and Communications Engineering,
Tampere University of Technology, Tampere, Finland
`ville.syrjala@tut.fi`

Abstract. Problem of opportunistic use of the unlicensed 5-GHz band for LTE carrier aggregation (LTE unlicensed) is studied from the point-of-view of cognitive full-duplex transceivers. In this paper, an initial study of the impact of self-interference on the performance of cyclostationary spectrum sensing algorithm is given, in case where a full-duplex transceiver tries to opportunistically use parts of the band used by a WLAN signal. Effective sensing while transmitting is natively possible, because WLAN and LTE signals have different cyclic properties. The evaluation of the impact focuses on extensive system simulations and simulation analysis. It is concluded that the self-interference can indeed interfere with the cyclostationary spectrum sensing. However, the effect can be lowered by lowering the bandwidth of the aggregated signal and instead using higher spectral density for the lower bandwidth signal.

Keywords: LTE unlicensed · Full-duplex radio · Spectrum sensing · Cyclostationary spectrum sensing

1 Introduction

Ever-increasing growth in the use of mobile data requires that the available spectrum is used as efficiently as possible. Another solution to answer for the growth is to allocate more spectra for mobile data. One proposed solution to combine the both of these solutions is so called LTE-Unlicensed (LTE-U), in which Long Term Evolution (LTE) downlink is aggregated to the unlicensed 5-GHz band. Wireless LAN (WLAN) signals are already allocated to the unlicensed 5-GHz band, but aggregating LTE downlink is much simpler for LTE transceivers, than utilizing the WLAN simultaneously with licensed LTE for additional throughput. Furthermore, LTE provides higher spectral efficiency compared to WLAN, as well as longer range. However, the use of unlicensed spectrum for carrier aggregation is more complicated than the use of licensed bands, because other systems already exist in the band. Therefore the coexistence methods need to be considered to enable LTE-U [1], [2].

One solution for the coexistence of aggregated LTE downlink and WLAN in the unlicensed bands is the cognitive radio technology based on opportunistic spectrum access [3], [4]. Unfortunately, even though opportunistic spectrum access has received

© Institute for Computer Sciences, Social Informatics and Telecommunications Engineering 2015
M. Weichold et al. (Eds.): CROWNCOM 2015, LNICST 156, pp. 725–734, 2015.
DOI: 10.1007/978-3-319-24540-9_60

huge amounts of research input in recent years, there have not been many practical implementations. This is natural, because the licensed users do not wish any additional interference to the bands allocated for them. In LTE-U, however, the idea is that open spectrum is used for carrier aggregation. This might still be fairly problematic, because the co-existence methods should be designed so that the existing systems are minimally interfered. To help this, this paper proposes use of wireless full-duplex technology to be used for opportunistic spectrum access in LTE-U. This basically means that the opportunistic user (LTE user) can sense for the primary signal (WLAN), while it transmits the signals itself. This makes the whole opportunistic spectrum access more attractive, because the secondary user can react to the primary signal faster, since the sensing can be carried out all the time.

In the literature, it has already been shown that utilizing full-duplex technology for cognitive radio application offers many benefits over the traditional receiver architectures [5], [6], [7]. This paper carries out an initial study on what is the effect of the self-interference on the performance of cyclostationary spectrum sensing algorithm in LTE-U, and proposes the use of full-duplex radio technology for LTE-U. This enables listen-whilst-talking, instead of the de facto listen-before-talk. This is promising, because WLAN and LTE signals have significantly different cyclic properties. This has not been considered in the existing literature. Special emphasis is given to study the effect of the bandwidth of the aggregated LTE signal on the performance of cyclostationary spectrum sensing. More specific coexistence strategies are not yet discussed in this paper.

The outline of this paper after this section is as follows. The second section shortly presents the general level ideas of full-duplex radio architecture and cyclostationary spectrum sensing, and shortly discusses the use of cognitive full-duplex radio technology in LTE-U. The third section then describes the simulator and simulation parameters. The simulation results and the corresponding analysis are given in the fourth section. Finally, the fifth section concludes the work.

2   Cognitive Full-Duplex Transceiver Utilizing Cyclostationary Spectrum Sensing

This section shortly describes the cognitive full-duplex radio transceiver and the idea of using full-duplex cognitive radio transceivers combined with cyclostationary spectrum sensing in LTE-U. Finally, cyclostationary spectrum sensing is presented in detail.

2.1   Cognitive Full-Duplex Radio Transceiver

Principal illustration of a cognitive full-duplex transceiver is given in Fig. 1. The basic principle is like in any modern direct conversion transceiver, but since the transmitter transmits at the same center frequency as the receiver receives, the own signal (called self-interference from now on) needs to be cancelled at the receiver. In the full-duplex transceiver structure of Fig. 1, this is done in two stages. First, an analog filter is tuned to match the self-interference channel as closely as possible in analog domain. Then, the signal to be transmitted is fed through the filter and subtracted from the received signal. The second stage of the cancellation is done in digital domain, where a digital filter mimics the remaining self-interference channel, and the cancellation is carried out using the transmitted samples. The digital cancellation is very important part even though the analog cancellation is done already in the analog domain, because of the limitations in analog-domain filtering. [8]

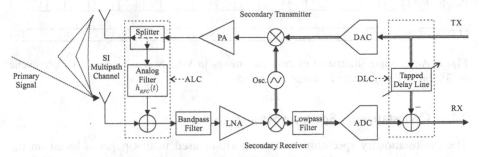

Fig. 1. Principal illustration of a cognitive full-duplex transceiver with analog linear cancellation (ALC) and digital linear cancellation (DLC) of the self interference.

The cyclostationary spectrum sensing is done in the digital domain after both of the self-interference cancellation stages have been carried out.

2.2   Use of Full-Duplex Transceivers to Enable LTE-U

Usual solutions for coexistence in LTE-U are based on so-called listen-before-talk principle [1], [2]. The base station (secondary user) basically listens for the primary user and transmits if it does not exist, and then again stops transmission to listen. This is very inefficient, because of the pauses, but it also potentially causes interference to the existing systems, if the primary system begins transmission right or shortly after the secondary user begins transmission.

Using full-duplex radio technology in LTE-U offers two key benefits over the de facto listen-before-talk principle. With full-duplex radio technology so-called listen-whilst-talk principle can be utilized. First key benefit is that the LTE base station does not need to stop its transmission while sensing, because the sensing and transmission can be carried out simultaneously. The second key benefit is that when the sensing is done simultaneously with the transmission, the secondary user can react to the primary user transmission instantaneously and discontinue or change the transmission so that it does not interfere with the primary system.

The different cyclic properties of the LTE and WLAN signals enable efficient use of the listen-whilst-talk principle. In LTE-U, cyclic frequencies of the primary and secondary signals are very different. The primary signal is IEEE 802.11 family signal with 64 subcarriers and 8 or 16 sample cyclic prefix. The length of the OFDM symbol without cyclic prefix (cyclic delay) is 3.2 μs and with cyclic prefix 3.6 μs or 4 μs. The secondary signal on the other hand has totally different properties, e.g., the 10 MHz mode has 1024 subcarriers with 72 or 256 sample cyclic prefix. The length of the

OFDM symbol without cyclic prefix (cyclic delay) is 66.67 µs and with cyclic prefix 71.35 µs or 83.33 µs. The cyclic frequencies, and more importantly, the cyclic delays are totally different. The timings of these signals and their relationship to the sensing period are roughly illustrated in Fig. 2.

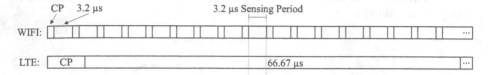

Fig. 2. An example illustration of different timings in WLAN signal with 16 sample cyclic prefix and LTE signal with 72-length cyclic prefix.

2.3   Cyclostationary Spectrum Sensing

The cyclostationary spectrum sensing algorithm used in this paper is based on the statistical tests proposed in [9]. For an OFDM signal $x(n)$, its conjugate cyclic auto-correlation function

$$R_x^{f_k} = \hat{R}_x^{f_k} - \varepsilon(f_k) \tag{1}$$

is non-zero for the cyclic frequencies of the OFDM signal $f_k \in A$ (A is a set of cyclic frequencies of the OFDM signal $x(n)$), and zero when $f_k \notin A$. In (1), $\varepsilon(f_k)$ is the estimation error in the sample estimate of the conjugate cyclic autocorrelation function

$$\hat{R}_x^{f_k} = \frac{1}{N} \sum_{n=0}^{N-1} x(n)x^*(n-\tau)e^{\frac{-j2\pi f_k n}{N}} . \tag{2}$$

Here, τ is the autocorrelation delay parameter. The statistical test is based on the assumption that the estimation error is asymptotically Gaussian distributed zero-mean complex random variable. The test can therefore be formulated for known cyclic frequencies $f_k \in A$ as a hypothesis test

$$
\begin{aligned}
H_0 &: \forall f_k \in A : \hat{R}_x^{f_k} = \varepsilon(f_k) \\
H_1 &: \exists f_k \in A : \hat{R}_x^{f_k} = R_x^{f_k} + \varepsilon(f_k),
\end{aligned}
\tag{3}
$$

where H_0 is a null hypothesis (the cyclostationary signal is not present) and H_1 is one-hypothesis (the cyclostationary signal exists). This is a very simple binary classi-fication task, so the test can be formulated as a simple threshold test [9]. Notice that because of the motivation in the previous subsection, the residual self-interference is considered to be noise in this test, and therefore included in the estimation error.

Following the test proposed in [10], the threshold test for the presence of cyclosta-tionarity, when the noise is assumed to be zero-mean Gaussian distributed, can be formulated into a simple form

$$F_{\chi^2}(T) > 1 - p , \tag{4}$$

where p is the desired false-alarm rate, F_{χ^2} denotes the cumulative distribution function of the well-known χ^2 distribution, and the χ^2 distributed test statistic can be computed as

$$T = \hat{R}_{x,v}^{f_k} \hat{\Sigma}_{2c}^{-1} (\hat{R}_{x,v}^{f_k})^T , \tag{5}$$

where vector

$$\hat{R}_{x,v}^{f_k} = \left[\mathrm{Re}\left\{ \hat{R}_x^{f_k} \right\} \quad \mathrm{Im}\left[\hat{R}_x^{f_k} \right] \right] . \tag{6}$$

In the computation of the test statistic T, the estimate of the covariance matrix of $\hat{R}_{x,v}^{f_k}$ can be written as

$$\hat{\Sigma}_{2c} = \begin{bmatrix} E\left[\mathrm{Re}\left\{ \hat{R}_x^{f_k} \right\}^2 \right] & E\left[\mathrm{Re}\left\{ \hat{R}_x^{f_k} \right\} \mathrm{Im}\left\{ \hat{R}_x^{f_k} \right\} \right] \\ E\left[\mathrm{Re}\left\{ \hat{R}_x^{f_k} \right\} \mathrm{Im}\left\{ \hat{R}_x^{f_k} \right\} \right] & E\left[\mathrm{Im}\left\{ \hat{R}_x^{f_k} \right\}^2 \right] \end{bmatrix} . \tag{7}$$

For computational simplicity this test assumes that the estimation error is zero-mean Gaussian distributed. In practice, this is not strictly the truth, because the residual self-interference contributes to the total noise of the system, and it only resembles a Gaussian distributed signal without strictly being one.

3   Simulator

3.1   Simulation Routine

First, OFDM modulated signal waveforms are generated for the own transmitter (self-interference) and primary-user transmitter. The signals are not in any way synchronized to each other. Both signals are then put through independent multipath channels. Additive white Gaussian noise is generated and summed to the sum of the two signals. The total self-interference cancellation is modelled in two stages. First, the analog-domain self-interference cancellation is modelled, so that only the first multipath component of the self-interference signal is suppressed to the desired level. In this process, white Gaussian noise is used as an error to model the estimate of the first multipath component in the cancellation. Then, digital domain self-interference cancellation is modelled, which aims to cancel the other multipath components of the self-interference signal as well as to improve the cancellation of the first multipath component. Once again, white Gaussian noise is used as an estimation error in the digital self-interference cancellation algorithm. Therefore, the self-interference cancellation is not only modelled as a simple attenuation, but as a more realistic process.

The signal with self-interference partially cancelled is then fed to the cyclostationary spectrum sensing algorithm set to detect the primary user signal.

3.2   Simulation Parameters

The simulator is using 20 MHz sampling rate. The primary-user signal is an OFDM signal with 312.5 kHz subcarrier spacing and with 64 subcarriers of which 52 around the center subcarrier are active and the remaining subcarriers are nulled. 16QAM subcarrier modulation is used, and cyclic prefix length is set to 8 samples per OFDM symbol. The self-interference signal (own signal) is an OFDM signal with 15 kHz subcarrier spacing and with 1024 subcarriers of which varying amount of subcarriers are active. The active subcarriers are always evenly around the center subcarrier and all the other subcarriers are nulled. 16QAM subcarrier modulation is also used for the self-interference signal. The cyclic prefix is 72 samples long. The signal is modelled by first generating the native LTE-signal samples with 15.36 MHz sampling rate. The signal is then 4-times oversampled, and then linear interpolation and filtering are then used to the signal samples, to get the samples from the correct positions to model the signal more accurately at the 20 MHz sampling rate. Linear interpolation gives rather accurate model since, the signal is relatively narrow-bandwidth compared to the sampling rate after the oversampling.

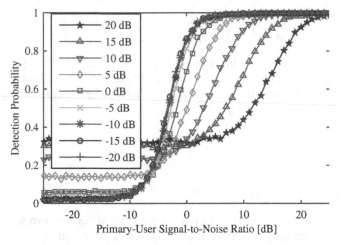

Fig. 3. Detection probability as a function of the primary-user signal-to-noise-ration when the total power of the self-interference after the self-interference cancellation stages is set the amount (in dB) denoted in the legend above the total additive white Gaussian noise power. The amount of active subcarriers (bandwidth) of the self-interference signal is 600 (full).

The cyclostationary spectrum sensing algorithm is set so that it gives 2 % false-alarm rate in the case that the only interferer is the white-Gaussian noise. The non-Gaussian statistical properties of the self-interference are the only reason for false-alarm rates that are not 2 % in the results, and without the self-interference the false-alarm rate is always on average 2 %.

4   Simulation Results and Analysis

In Fig. 3, the detection probability is given as a function of the primary-user signal-to-noise ratio. Different curves denote the different total power differences (in dB) between the self- interference signal and the additive white Gaussian noise. For example the legend entry 20 dB means that the total power of the self-interference signal is 20 dB above the total power of the additive white Gaussian noise. In practice, lower legend entry means better self-interference cancellation. In these results, the self-interference signal has 600 active subcarriers. In the curves, we can see that when the self-interference is suppressed to around a level of the white Gaussian noise, we get quite near to the performance level of when there is no self-interference at all. However, it seems that the self-interference still has a small effect on the detection. Even with good self-interference cancellation levels, it seems that self-interference has clear effect on the detection results. This is natural, because first of all, self-interference might have similar cyclic frequencies, but also, statistical properties of the self-interference are clearly different from those of the white Gaussian noise. The cyclostationary spectrum sensing algorithm after all is derived for the case, where the noise is pure white Gaussian noise.

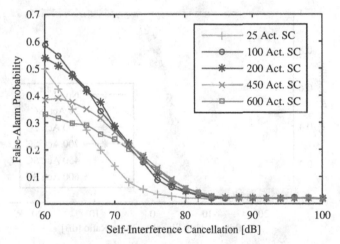

Fig. 4. False-alarm probability as a function of self-interference cancellation level. The additive white Gaussian noise power equals the self-interference power when self-interference cancellation is set to 80 dB (in the x-axis of the figure). The amount of active subcarriers (bandwidth) of the self-interference signal is varied.

In Fig. 4, Fig. 5 and Fig. 6, the results are given for different amount of active subcarriers (denoted in the legend), which practically means different bandwidths.

In Fig. 4, the false-alarm probability is given as a function of self-interference cancellation when level of the additive white Gaussian noise is set to 80 dB below the self-interference before the self-interference cancellation. This basically means that the total self-interference power after the cancellation is at the same power level with the additive white Gaussian noise, when self-interference cancellation is 80 dB. From the figure we can see that when the power of the self-interference is very high, lower-bandwidth signal increases the false-alarm probability the most. This is because its statistical properties are less and less similar to those of the white Gaussian noise when its bandwidth is made narrower. Very high-power self-interference is however not very interesting, because in that case the self-interference has already very huge effect on the detection performance no matter what bandwidth is used. More interesting levels of self-interference cancellation are around 80 dB or less, because there the effect of the self-interference is relatively small, and it would be attractive if the self-interference cancellation would not need to suppress the self-interference below the noise floor, and we are still able to get good detection results. We can see that by lowering the used amount of subcarriers, the false-alarm probability gets nearer and nearer to the case without the self-interference. The detection algorithm does not suffer much from the narrowband interferer. This is interesting result, because the amount of active subcarriers can be varied based on need during the primary signal detection. It should also be kept on mind, that even whilst the amount of active subcarriers is lowered, the power allocated per subcarrier increases in relation, so it is possible to get more throughput per subcarrier.

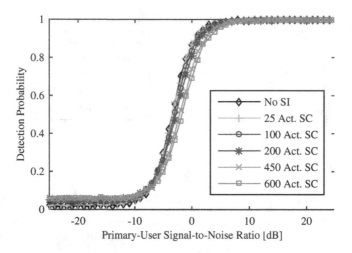

Fig. 5. Detection probability as a function of the primary-user signal-to-noise-ration when total additive white Gaussian noise power is set to the same level as the total power of the self-interference after the self-interference cancellation stages. The amount of active subcarriers (bandwidth) of the self-interference signal is varied.

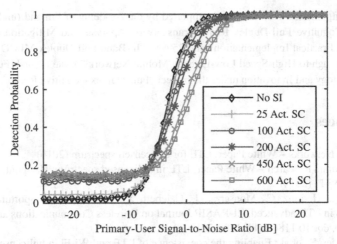

Fig. 6. Detection probability as a function of the primary-user signal-to-noise-ration when total additive white Gaussian noise power is set to 5 dB lower level than the total power of the self-interference after the self-interference cancellation stages. The amount of active subcarriers (bandwidth) of the self-interference signal is varied.

In Fig. 5 and Fig. 6, the detection probability is studied for the cases where the power of the self-interference is at the same level as and 5 dB above of, respectively, the total power of the additive white Gaussian noise. We can see that if the self-interference can be suppressed to the same level as the additive white Gaussian noise, the performance impact on the detection is relatively small, even with higher band-width signal. However, when the self-interference can only be suppressed 5 dB above the additive white Gaussian noise level, the self-interference has still quite clear impact on the detection result especially with higher bandwidth signals. However, the impact can be lowered by lowering the bandwidth of the secondary signal.

5   Conclusions

In this paper, the use of full-duplex radio technology was proposed for LTE unlicensed application to allow simultaneous transmission and sensing in opportunistic spectrum access. More specifically this problem was studied from the point of view of a cognitive full-duplex radio transceiver utilizing cyclostationary spectrum sensing. It was shown that the residual self-interference has some impact on the spectrum sensing algorithm performance if it cannot be suppressed to around the level of the noise floor of the receiver. However, if this is not possible, the impact is still relatively small even when the secondary signal is 5 dB above the noise floor. Also, the impact can be lowered by lowering the bandwidth of the secondary signal while keeping the same total transmission power.

734   V. Syrjälä and M. Valkama

Acknowledgements. This work was supported by the Academy of Finland (under the projects 276378 "Cognitive Full-Duplex Radio Transceivers: Analysis and Mitigation of RF Impairments, and Practical Implementation" and 259915 "In-Band Full-Duplex MIMO Transmission: A Breakthrough to High-Speed Low-Latency Mobile Networks"), the Finnish Funding Agency for Technology and Innovation under the project "Full-Duplex Cognitive Radio".

References

1. Nokia Networks – White Paper, LTE for unlicensed spectrum (2014)
2. Qualcomm Research – White Paper, LTE in unlicensed spectrum: Harmonious coexistence with Wi-Fi (2014)
3. Osa, V., Herranz, C., Monserrat, J., Gelebert, X.: Implementing opportunistic spectrum access in LTE-advanced. EURASIP Journal on Wireless Communications and Networking (2012). doi:10.1186/1687-1499-2012-99
4. Abinader, F., et al.: Enabling the coexistence of LTE and Wi-Fi in unlicensed bands. IEEE Communications Magazine **52**(11), November 2014
5. Tsakalaki, E., Alrabadi, O., Tatomirescu, A., de Carvalho, E., Pedersen, G.: Concurrent communication and sensing in cognitive radio devices: Challenges and enabling solution. IEEE Trans. on Antennas and Propagation **62**(3), March 2014
6. Cheng, W., Zhang, X., Zhang, H.: Full duplex spectrum sensing in non-time-slotted cognitive radio networks. In: Proc. Military Communications Conference 2011, Baltimore, MD, November 2011
7. Riihonen, T., Wichman, R.: Energy detection in full-duplex cognitive radios under residual self-interference. In: Proc. CROWNCOM 2014, Oulu, Finland, June 2014
8. Bharadia, D., McMilin, E., Katti, S.: Full duplex radios. In: Proc. SIGCOMM 2013, Hong Kong, August 2013
9. Dandawate, A., Giannakis, G.: Statistical tests for presence of cyclostationarity. IEEE Trans. Signal Processing **42**(9), 2355–2369 (1994)
10. Turunen, V., et al.: Implementation of cylostationary feature detector for cognitive radios. In: Proc. International Conference on Cognitive Radio Oriented Wireless Networks, Hannover, Germany, June 2009

Research Trends in Multi-standard Device-to-Device Communication in Wearable Wireless Networks

Muhammad Mahtab Alam[1]([✉]), Dhafer Ben Arbia[1,2], and Elyes Ben Hamida[1]

[1] Qatar Mobility Innovations Center (QMIC), Qatar Science and Technology Park
(QSTP), PO. Box. 210531, Doha, Qatar
{mahtaba,dhafera,elyesb}@qmic.com

[2] Polytecnic School of Tunisia, SERCOM Laboratory, University of Carthage,
B.P. 743, 2078 La Marsa Tunisie, La Marsa, Tunisia

Abstract. Wearable Wireless Networks (WWN) aim to provide attractive alternate for conventional medical care system. It is an effective way of monitoring patients within clinics, hospitals and remotely from home, offices etc. In this paper we extend the classical envisioned applications from medical health-care to rescue and critical applications for disaster and emergency management using WWN. There are number of challenges to effectively realize this application and several of those are presented in this paper along with various opportunities. We review multi-standard and multiple technologies based wearable wireless cognitive system for Device-to-Device (D2D) communication. Coexistence and inter-operability is one of the important challenges which are discussed along with utilization of possible technologies for on-body, body-to-body and off-body communications.

Keywords: Wearable wireless networks · Heterogeneous networks · Coexistence · Interoperability · Device to device communication · Cognitive radio

1 Introduction

With the revolution and emergence of tremendous amount of growth in various technologies, it is predicted that in the next five years there will be about fifty billion devices world-wide, which means on-an-average every person on the planet will be equipped with about six to seven devices. This massive influx of devices will deluge with huge amount of data which has to deal with disruptive technologies and powerful but smart computing. Further these devices are 'heterogeneous' which are based on multiple technologies and standards to achieve specific applications data rates, reliability and quality-of-service and so on.

In cellular networks Device-to-Device (D2D) communication exploit direct communication between nearby mobile devices to improve the spectrum utilization, overall throughput, and energy consumption, while enabling new peer-to-peer and location-based applications and services. D2D-enabled LTE devices can

© Institute for Computer Sciences, Social Informatics and Telecommunications Engineering 2015
M. Weichold et al. (Eds.): CROWNCOM 2015, LNICST 156, pp. 735–746, 2015.
DOI: 10.1007/978-3-319-24540-9_61

also become competitive for fallback public safety networks, which must function when cellular networks are not available or fail [1]. In additions to that, the efficient usage of frequency spectrum is very vital and therefore the role of software defined cognitive radios in the future technologies is very important [2].

In this paper we focus on one of the emerging technology called wearable to provide an additional ad-hoc network to support public safety networks for emergency management. Typically wearable body sensor networks (WBSN) consists of tiny, smart, low-power, and self-organized sensors to observe physiological signals of a human body. However in our research we extend the communication from on-body to body-to-body and off-body networks to effectively realize the applications such as rescue and critical for disaster and emergency management [3].

We address several challenges as well as opportunities for such heterogeneous wearable wireless networks (WWN). This includes standardization and compliance, effective coexistence and interoperability among multiple technologies, and how to ensure end-to-end network routing and connectivity especially in heterogeneous networks. First several WBSN applications and their requirements are discussed followed by the suitable architecture to realize wide range of applications. Second, with regards to number of different standards which are currently used for WWN, a comparative analysis and utilization of these standards are discussed. Third, several coexistence schemes are explored to ensure effective coexist among multiple technologies and the issues related to interoperability are discussed. Further, the impact of multiple technologies and standards-based applications on network, medium access control (MAC) and physical (PHY) layers are presented. Followed by the software-defined cognitive radio to coordinate and control multiple standards. Finally we present a conclusion and future research directions.

2 Applications and Architecture

Wearable technology is one of the most upcoming and emerging technology which can be seen in many dimension of daily-life as explain below. The wearable computer is defined as a mean of personal empowerment through human-computer interaction of smart devices [4]. The key characteristics of these devices are that they are always on and always available or ready for the users interaction. Unlike portable or other smart devices, wearable devices do not need to turn on and we can augment the reality of the physical world instantly and more powerfully from our surroundings as a result the intelligence can be significantly enhanced.

Wearable wireless network is an enabling ubiquitous monitoring and communication system. Typical envisioned applications range from the medical field (e.g., vital sign monitoring, automated drug delivery, etc.), to sports and fitness, entertainment and augmented reality, gaming and ambient intelligence and so on. However, with regards to applications such as disaster, rescue and critical missions, workers safety in harsh environments (e.g., oil and gas fields, refineries, petro chemical and mining industries) as well as roadside and building workers, wearable WBSN technology can also play a vital role to not only save human

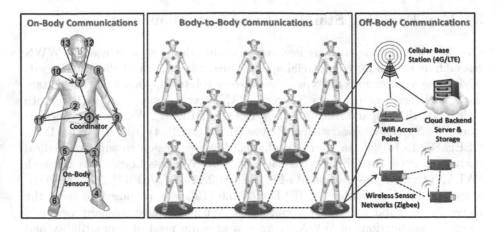

Fig. 1. A Generic Architecture for Wearable Wireless Networks.

lives but also to protect critical and valuable assets [5]. In this paper, we will emphasize on these applications as most of the other applications and their corresponding requirement can be inclusive as well.

The specific application characteristics can help to specify and confined many requirements, however, in general, there are number of parameters which can impact. For example, the devices can be only coordinators, or it can be based on the combinations of both sensors and coordinator, further, actuators can also be included. For our applications context we require all these devices, further we require various types of these devices such as source, sink, gateways and multistandard nodes. Traffic patterns can be periodic, event driven as well as burst, which includes audio, video and data. The network can be centralized for the on-body communication and distributed for the body-to-body networks.

The generic architecture of wearable wireless networks (WWN) consists of inbody, On-Body, Body-to-Body, and off-Body communication networks as shown in Fig. 1. In-body networks is mainly based on implant devices inside the human body such as heart, kidney, ear, birth control and back pain, etc. On body often called as Wireless Body Sensor Networks (WBSN) typically contains multiple sensors (to sense the physiological signals), actuators (to react according to the perceived signal) and a coordinator which control and coordinate the other sensors (or nodes) within WBSN. Often a coordinator is much more powerful in terms of out-reach, resources and control, which can interconnect the WBSN to remote/external network infrastructures using beyond-WBSN communications (e.g., 4G/LTE/5G, WiFi etc.). Therefore, we require multi-standard based D2D communication which is necessary for emergency management especially in the context where either existing infrastructure is either completely damaged or over saturated to improve end-to-end network connectivity and latency. In order to achieve this vision there are many challenges that are discussed in the following section, including coexistence and interoperability, multi-standards, cross-layer etc.

3   Key Enabling Standards and Compliance

Over the last decade, various low power standards have been used in WWN research as well as for commercial applications, where most of them partly satisfying the requirements for typical health-care related applications. These standards includes Personal Area Network (PAN) technologies, such as Bluetooth (IEEE 802.15.1) [6] and Bluetooth Low Energy (BLE) [7], Wireless Sensors Network (WSN) technologies, such as Zigbee (IEEE 802.15.4) [8], Ultra Wide Band (IEEE 802.15.4a) [9], an alternate physical layer extension to support medical body area networks (IEEE 802.15.4j) [10], and Wireless Local Area Network (WLAN) technologies, such as Wi-Fi (IEEE 802.11a/b/g/n) [11]. More recently, a specific BAN standard, i.e. IEEE 802.15.6 [12], was proposed to meet the increasing demand for WWN applications. With reference to many new and emerging applications of WWN, there is a growing need of compatibility and compliance among multiple standards.

Tab. 1 presents number of different standards and their compliance against various parameters and constraints. In particular with reference to rescue and critical applications for emergency management multiple-standards are required. For example, for on-body communication, low-power WPAN standards such as IEEE 802.15.6 is more suitable, however, it is not designed for body-to-body communication, for that matter, using IEEE 802.15.4, IEEE 802.11, 4G/LTE D2D are required which can extend the networks connectivity in an effective manner. Whereas, for off-body, one of the end-device of BBN should be able to communicate through cellular networks or infrastructure-based networks such as 4G/LTE. To conclude, existing devices, such as smartphones already supports many standards, but existing protocol stack are not smart enough to provide connectivity or routing between different network technologies and this is one of the important challenge for the future wireless networks.

4   Coexistence and Interoperability

Most of the WWN related standards and technologies operate on the same frequency ISM bands which results in significant interruption to each other. In this regard, the initial research studies on WWN interference mainly concentrate on the impact from other technologies (aka., adjacent channel interference) such as IEEE 802.11, IEEE 802.15.1, etc. It is clear from the previous research works that there is a dominant interference from other networks in WBSN [13–16]. Therefore, to coexist in harmony, certain information needs to be shared, hence interoperability is very important. In this context, coexistence strategies can be used which are often categorized as collaborative and non-collaborative. Several non-collaborative schemes are proposed in IEEE 802.15.6 standard such as beacon shifting, channel hopping and active superframe interleaving. However, the performance of all these schemes is yet to be evaluated especially in the context of heterogeneous networks. Moreover, these approaches of interference analysis are only enough for intra-BSN communication; where each node is synchronized

Table 1. Comparison of the Key enabling Standards for Wearable Wireless Sensor Networks.

Parameters	IEEE 802.11 a/b/g/n (Wi-Fi)	IEEE 802.15.1 (Bluetooth)	IEEE 802.15.1 (BLE)	IEEE 802.15.4 (Zigbee)	3GPP LTE/4G	IEEE 802.15.4j (MBAN)	IEEE 802.15.6 (WBAN)
Modes of Operation	Adhoc, Infrastucture	Adhoc	Adhoc	Adhoc	Infrastucture	Adhoc	Adhoc
Physical (PHY) Layers *	NB	NB	NB	NB	NB	NB	NB, UWB, HBC
Radio Frequencies (MHz)	2400, 5000	2400	2400	868/915/2400	700, 750, 800, 850, 900,1900, 1700/2100	2360-2390/2390-2400	402-405, 420-450, 863-870, 902-928, 950-956, 2360-2400, 2400-2483.5
Power Consumption	High (∼ 800mW)	Medium(∼ 100mW)	Low (∼ 10mW)	Low(∼ 60mW)	NA	Low(∼ 50mW)	Ultralow Power (∼ 1mW at 1m distance)
Maximal Signal Rate	Up to 150 Mb/s	Up to 3 Mb/s	Up to 1 Mb/s	Up to 250 Kb/s	Up to 300 Mb/s	Up to 250 Kb/s	10 Kb/s to 10 Mb/s
Communication Range	Up to 250 m (802.11n)	100 m (class 1 device)	> 100 m	Up to 75 m	Up to 100 Km	Up to 75 m	Up to 10 m
Networking Topology	Infrastructure-based	Adhoc very small networks	Adhoc very small networks	Infrastructure-based	Adhoc, Peer-to-Peer, Star, Mesh	Adhoc, Peer-to-Peer, Star	Intra-WBAN: 1 or 2-hop star. Inter-WBAN: non-standardized
Topology Size	2007 devices for structured Wi-Fi BSS	Up to 8 devices per Piconet	Up to 8 devices per Piconet	Up to 65536 devices per network	NA	Up to 65536 devices per network	Up to 256 devices per body, and up to 10 WBANs in 6m^3
Target Applications	Data Networks	Voice Links	Healthcare, Fitness, beacon, security, etc.	Sensor Networks, home automation, etc.	Data Networks and Voice Links	Short range Medical Body Area Networks	Body Centric applications
Target BAN Architectures	Off-body	On-body	On-body	Body-to-Body, Off-Body	Body-to-Body, Off-Body	On-Body	On-Body

* NB: Narrowband, UWB: ultra Wide Band, HBC: Human Body Communication

with its coordinator and are configured at the same transmit power. However, with an advent of body-to-body communications, inter-BSN interference and its mitigation is a new problem. On the other hand, collaborative strategies are necessary for viable inter-operability (i.e., to share key information between other standards), furthermore, they can also help for inter-BSN interference. A little effort has been done so far and there are many opportunities for research in this area.

In one of our recent work [17], several coexistence (both including collaborative and non-collaborative) strategies were evaluated for body-to-body communication. IEEE 802.15.6 standard has proposed beacon shifting, channel hopping and superframe interleaving as coexistence schemes which are all considered as non-collaborative. We have considered scheduled access MAC protocol of IEEE 802.15.6 standard as a reference case (i.e., without any coexistence), then a time shared approach and channel hopping are used for comparisons. In addition to that, we also considered CSMA/CA (carrier sense multiple access/collision avoidance), which can be considered as implicitly collaborative scheme in which multiple nodes sense the channel to avoid collisions and interference. In this

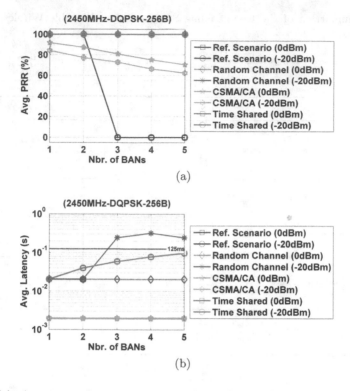

(a)

(b)

Fig. 2. (a): Average packet reception ratio for multiple BANs in various coexistence schemes. (b): Average packet delay for multiple BANs under coexistence schemes.

section due to shortage of space we only presents part of our findings, though some more details can be found in [17]. As an example, we analyze the performance of average packet reception ratio (PRR) and average delay as performance metrics under varying transmission power for three coexistence schemes and one reference scenario and only one of the configurations (i.e., 2450 MHz with highest data rates or in other words differential quadrature phase shift keying modulation (DQPSK) and maximum payload size of 256 bytes) is shown in Fig. 2. These results are conducted in a packet oriented network simulator and all the detail regarding setup and configurations can be found in [17].

From the presented results, it is found that the PRR for reference scenario for 1 BSN is 94.24%, however, as the BSN increases from 2 to 3 the PRR reduces sharply to 0% and since the packets are not received at all therefore we don't have any delay values in Fig. 2-b. Whereas, it can be noticed that, both channel hopping and time-shared perform much better with PRR being above 95% even under -20 dBm, though their delay performance is not as good as PRR and especially as the number of BANs reaches 3 or more channel hopping with -20dBm does not satisfy the upper bound of latency requirements of IEEE 802.15.6 standard. The time-shared scheme is almost independent of the transmit power as far as the PRR and delay is concerned it is only dependent on the number of BANs in which latency increases linearly with increase in number of BANs. For the case of CSMA/CA, the PRR reaches to almost 80% just with 2 BANs and it continuously degrade with increase in number of BANs. With reference to its delay requirements, it fully satisfy all the constraints of IEEE 802.15.6 standard. To sum up, all coexistence schemes have pros and cons depending upon the specific configurations which can satisfy the applications requirements such as latency, PRR, energy efficiency etc., and the best scheme can be selected based on the specific constraints.

5   Cross Layer Networking Protocols

The interoperability of communication technologies in critical and public safety context is important to analyze and the impact on the upper layers in particular the routing layer needs to be investigated. A mechanism between the MAC/PHY and the routing should be in charge of reacting, switching and coexisting with multiple and simultaneous technologies. Below we will present an overview on existing investigations of heterogeneous networks with specific interest on the MAC/PHY and routing layers.

Issued in 2012, the IEEE 802.15.6 norm document details requirements for WBSN [12]. This norm covered many points in particular the communication range of the WBSN nodes, the ability to reconnect dynamically the disconnected nodes. The WWNs are supposed to gather body physical measurements and forward it to a distant monitoring system. Referring to the possible network tactical architecture [18], a BSN may interface with various network technologies mentioned in Section 4. In order to have effective BSN deployment there are several issues and challenges ranging from the hardware to the application layer. Below we will highlight some of the most important aspects for cross layers.

Physical layer must deal with unpredictable topology and network changes. Body sensors must be able to operate with wireless networks and low power [18] in such conditions. The media access control (MAC) layer needs to minimize the packets collisions and allow fair channel access. MAC protocols are also mandatory to increase network capacity, energy efficiency and guarantee a better quality-of-service (QoS). The hidden nodes phenomenon [19] for example is highly considerable, due to the NLOS (Non-Line-Of-Sight) between some nodes which can be caused by mobility and unpredictable topology changes. NLOS depends on the selection of the appropriate MAC techniques to adopt: Carrier Sense Multiplexing Access (CSMA) or Time Division Multiplexing Access (TDMA).

In higher layer, routing in WWN has to handle frequent nodes disconnections and reconnections, which will influence on the capabilities required for the routing protocols to adopt and implement. This should be ensured without causing excessive traffic overhead or computational burden on the power constrained devices [20]. To meet requirements detailed in [12], and referring to the effective networking model presented in [5], a variety of Ad hoc networks are compared in [21]. Mobile Adhoc Networks (MANETs) are the classical approach regarding the implementation of public safety networks, broadcasting communications with multi-hop communications based on reactive and proactive routing protocols. A study on the evaluation of MANETs in emergency and rescue scenario is investigated in [22]. The assessment of the MANETs routing protocols referring to the classes proactive, reactive, hybrid and hierarchical routing protocols. We will discuss first whether this class of the routing protocol is appropriate to the application of WWN or not.

Proactive routing protocols, such as Optimized Link State Routing protocol (OLSR) and Destination Sequenced Distance Vector (DSDV), exchange continuously information to keep up-to-date routes to all network nodes, and is important in case of victim's evacuation or rescue missions. However, this may affect negatively the bandwidth utilization. On the other hand, reactive routing protocols such as Ad hoc On-demand Distance Vector (AODV) and Dynamic Source Routing (DSR) are characterized by their two mechanism components: route discovery and route maintenance. The latency to initiate the communication and the delay to detect network changes exclude reactive routing protocols from tactical and rescue context. Besides these flat routing protocols are not appropriate to critical and rescue context. The hybrid and hierarchical protocols combine two or more proactive and reactive routing protocols and divide the network into zones or clusters. This leads to manage environment sectors through a cluster head which reduce the power consumption and increase network performance, e.g. Cluster Based Routing Protocol (CBRP).

For public safety networks, an ad-hoc congnitive radio (CR) based spectrum-aware routing protocol is proposed in [23]. This routing protocol comes up with the use of the white spaces spectrum resources in TVWS (Television White Space). It presents a specific adoption of ADOV routing protocol due to the unpredictable availability of the TVWS which requires a hop-by-hop routing. The contribution lies in the fact that during the route request (RREQ), the proposed routing protocol includes the TVWS availability of nodes. Each RREQ will inform about the source nodes and TVWS availability. It evaluates the performance in a specific scenario, and validate the adoption of AODV under the controlled simulation conditions [23].

Delay Tolerant Network (DTN) is a network based on nodes encountering. Where a node waits until it encounters another node to deliver the packets. Characterized by its latency, DTN is suitable for low density networks [21], but not in critical and rescue missions. All forwarding mechanisms of DTN are based on opportunistic communications where routes are built dynamically through any encountered nodes (e.g. Epidemic forwarding, PRoPHET forwarding, MaxProp and TTR) [24].

The future routing protocols, are required to consider latency, reliability, mobility, thermal-effects and energy consumption [25]. These challenges leads to the need of a cross-layer networking solution which will be in charge of selecting and calling the communication technology needed in each connection.

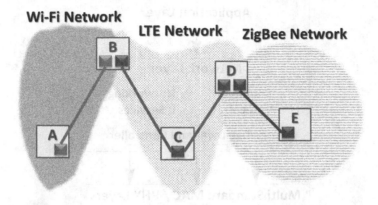

Fig. 3. Multi-Standards Compliant Heterogeneous Wearable Wireless Networks.

6   Software Defined-Based Multi-standard Cognitive Radio

The specific selection among the above presented networking protocols depends upon the specific application. An appropriate routing protocol for the WWN under the given application case, will consequently affect the selection of suitable MAC/PHY layers. Therefore, the heterogeneous technologies adoption in the communication layer is inevitable. Cognitive Radio thus is a promising technology that may fulfill the requirements, imposed by the environment of the WWN (healthcare, emergency and rescue, military, etc.) applications and then by the routing layer. Looking into the heterogeneity aspect in the communication layer, regarding the existing multiple standards (Wi-Fi 802.11, LTE/5G 802.16, ZigBee 802.15.4, Bluetooth 802.15.1, WSN 802.15.4, UWB 802.15.3a), there are many issues to investigate. For example, power consumption, which differ from one technology to another and influence the communication interface. Storage capabilities regarding the packet size and the buffering capacity that has an important role to reduce the processing overload in multiple layers. The various technologies radio range and the bandwidth support are also critical parameters that should be considered for a suitable selection of technology.

A multi-standard node uses more than one technology in communication layer and has the ability to operate with different routing protocols. This node could switch from a communication technology to another (i.e., to keep using different MAC/PHY interfaces), in some prospective conditions. For example the parameters such as, the remaining battery power, the signal strength, the unreachable destinations, the bandwidth specified by the standard may induce the node to be complemented by another available technology interface to get over these issues. In addition, several nodes may also have the ability to communicate with more than one node using different radio technologies as shown

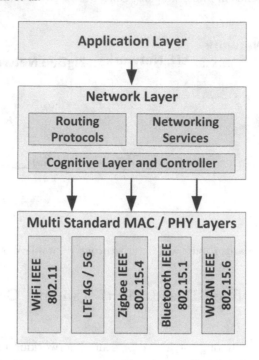

Fig. 4. Cross layer-based Cognitive Protocol Stack for WWN.

in Fig. 3 (e.g. connected to a distant node over LTE/5G interface, and with a close node over Wi-Fi interface).

The basic idea shown in Fig. 3, is that a node-A would like to communicate with a node-E, and based on the best available networks, the transmitted messages could be forwarded through the nodes-B, C and D, using different network technologies, until reaching node-E. In this particular example, node-A and node-B are using Wi-Fi connection, and since Node-E is out of reach of the Wi-Fi range of Node-B, so they are connected through LTE network (i.e., node-C and node-D). Consequently, the routing protocol in the up-layer will be adapted with the requirements imposed by the communication layer.

Fig. 4 shows the cross layer based cognitive radio controlled protocol stack for future emerging WWNs. It can be noticed that a cognitive layer is added to select the suitable technologies (i.e., multi-standards) based on the applications requirements, data rates, channel conditions, radio link quality and many other parameters. The appropriate routing and navigation is also selected based on the specific MAC/PHY technologies through cross-layer interaction.

7   Conclusion

To conclude, a wearable wireless networks is an upcoming and emerging technology. It can be applied to number of applications other than classical health-care.

In paper we emphasize in our on-going research work for rescue and critical applications. We have presented several important issues pertaining to effective realization of this specific application. In particular we have highlighted existing standards and the need to have multi-standards compliant devices. Further, coexistence and interoperability challenges and the possible solutions are explored. In this aspect, one of the key limitations (for testing various coexistence strategies), is the lack of IEEE 802.15.6 standard compliant devices. Further, a cross-layer existing state of the art is presented and the importance of cognitive radio is highlighted and how it can influence the future WWN. Finally software-defined based multi-standard cognitive radio is presented which can control, configure, select and switch between multiple technologies based on the specific requirements.

Acknowledgment. This publication was made possible by NPRP grant #[6 − 1508 − 2 − 616] from the Qatar National Research Fund (a member of Qatar Foundation). The statements made herein are solely the responsibility of the authors.

References

1. Lin, X., Andrews, J.G., Ghosh, A., Ratasuk, R.: An overview on 3g pp. device-to-device proximity services. CoRR, vol. abs/1310.0116 (2013). http://arxiv.org/abs/1310.0116
2. Rusek, F., Persson, D., Lau, B.K., Larsson, E., Marzetta, T., Edfors, O., Tufvesson, F.: Scaling up mimo: Opportunities and challenges with very large arrays. IEEE Signal Processing Magazine **30**(1), 40–60 (2013)
3. Hamida, E.B., Alam, M.M., Maman, M., Denis, B., D'Errico, R.: Wearable body-to-body networks for critical and rescue operations the crow^2: Project. In: IEEE PIMRC 2014 - Workshop on The Convergence of Wireless Technologies for Personalized Healthcare, pp. 2145–2149, September 2014
4. U. of Toronto. Definition of wearable computer (1998). http://wearcomp.org/wearcompdef.html
5. Alam, M.M., Hamida, E.B.: Surveying wearable human assistive technology for life and safety critical applications: Standards, challenges and opportunities. Sensors **14**(5), 9153–9209 (2014)
6. IEEE standard for local and metropolitan area networks - part 15.1:-part 15.1: Wireless medium access control (MAC) and physical layer (PHY)specifications for wireless personal area networks (WPANS) (2005)
7. Bluetooth.org. Sig, bluetooth (2015). http://www.bluetooth.com/Pages/Bluetooth-Smart.aspx
8. IEEE standard for local and metropolitan area networks - part 15.4: Lowrate wireless personal area networks (LR-WPANS), pp. 1–314 (2012)
9. Amendment to 802.15.4-2006: Wireless medium access control(MAC) and physical layer (phy) specifications for low-rate wireless personal area networks (LR-WPANS) (2007)
10. Bluetooth.org. part 15.4: Low-rate wireless personal area networks (LR-WPAN) amendment 4: Alternative physical layer extension to support medical body area network (mban) services operating in the 2360 mhz 2400 mhzband (2013). http://standards.ieee.org/finndstds/standard/802.15.4j-013.html802.15.4j-2013.html

11. Wireless LAN medium access control (mac) and physical layer (phy) specifications (2012)
12. IEEE standard for local and metropolitan area networks - part 15.6: Wireless body area networks, pp. 1–271 (2012)
13. Martelli, F., Verdone, R.: Coexistence issues for wireless body area networks at 2.45 ghz. In: 18th EW Conference, pp. 1–6, April 2012
14. Davenport, D.M., Ross, F., Deb, B.: Coexistence of wban and wlan in medical environments. In: 70th VTC Conference, pp. 1–5, September 2009
15. Hayajneh, T., Almashaqbeh, G., Ullah, S., Vasilakos, A.: A survey of wireless technologies coexistence in wban: analysis and open research issues. Wireless Networks **20**(8), 2165–2199 (2014)
16. Jie, D., Smith, D.: Coexistence and interference mitigation for wireless body area networks: Improvements using on-body opportunistic relaying. CoRR, abs/1305.6992 (2013)
17. Alam, M.M., Hamida, E.B.: Interference mitigation and coexistence strategies in ieee 802.15.6 based wearable body-to-body networks. CROWNCOM 2015, LNICST **156**, pp. 1–13, 2015
18. Chen, M., Gonzalez, S., Vasilakos, A., Cao, H., Leung, V.: Body area networks: A survey. Mobile Networks and Applications **16**(2), 171–193 (2011). doi:10.1007/s11036-010-0260-8
19. Cheung, L.Y., Chia, W.Y.: Designing tactical networks perspectives from a practitioner, DSTA horizons, Tech. Rep., June 2013
20. Hoebeke, J., Moerman, I., Dhoedt, B., Demeester, P.: An overview of mobile ad hoc networks: applications and challenges. Journal of the Communications Networks **3**(3), 60–66 (2004)
21. Reina, D.G., Askalani, M., Len-Coca, J.M., Toral, S.L., Barrero, F.: A survey on ad hoc networks for disaster scenarios. In: 6th INCoS Conference, September 2014
22. Quispe, L.E., Galan, L.M.: Review: Behavior of ad hoc routing protocols, analyzed for emergency and rescue scenarios, on a real urban area. Expert Systems Applications **41**(5), 2565–2573 (2014). doi:10.1016/j.eswa.2013.10.004
23. Bourdena, A., Mastorakis, G., Pallis, E., Arvanitis, A., Kormentzas, G.: A spectrum aware routing protocol for public safety applications over cognitive radio networks. In: 2012 International Conference on Telecommunications and Multimedia (TEMU), pp. 7–12, July 2012
24. Martn-Campillo, A., Crowcroft, J., Yoneki, E., Mart, R.: Evaluating opportunistic networks in disaster scenarios. Journal of Network and Computer Applications 36(2), 870–880 (2013). http://www.sciencedirect.com/science/article/pii/S1084804512002275
25. Bangash, J.I., Abdullah, A.H., Anisi, M.H., Khan, A.W.: A survey of routing protocols in wireless body sensor networks. Sensors 14(1), 1322–1357 (2014). http://www.mdpi.com/1424-8220/14/1/1322

Implementation Aspects of a DSP-Based LTE Cognitive Radio Testbed

Ammar Kabbani[✉], Ali Ramadan Ali, Hanwen Cao, Asim Burak Güven,
Yuan Gao, Sundar Peethala, and Thomas Kaiser

Institute of Digital Signal Processing, Duisburg-Essen University, Essen, Germany
{ammar.kabbani,ali.ali,hanwen.cao,asim.gueven,yuan.gao,
sundar.peethala,thomas.kaiser}@uni-due.de

Abstract. One of the key issues of designing radio communication systems is to enhance the efficiency and flexibility of the available radio spectrum. In this context, reconfigurable implementation of the different system layers such as Media Access Control (MAC), Physical (PHY), and RF layers with optimal cross-layer design is driving nowadays research on designing reliable and robust mobile communication systems. This paper presents the software and hardware implementation aspects of a DSP-Based cognitive LTE testbed and addresses the design, the implementation and the realistic verification of the key components of the system.

Keywords: Cognitive Radio · LTE · Testbed · DSP

1 Introduction

In order to meet the increasing requirements of data throughput and the number of interconnected devices, the cellular communication are developing mainly in two directions. First, the spectral efficiency has been improved from 2G era's about 1 bps/Hz to today's 4G LTE Advanced (LTE-A)'s over 20 bps/Hz. The large spectral efficiency requires sophisticated and power-consuming signal processing which seems to be approaching a practical limit when considering the feasibility in practical implementations, such as the number of active antennas in mobile terminal, the interference mitigation techniques within both single device and from network perspective. The limitation in achievable spectral efficiency has pushed the research/development (R&D) and regulation into the other direction, thus enabling the utilization of more radio spectrum with dynamic sharing with the cognitive radio (CR) [1] technology. During the past 15 years, several CR standards have been defined, such as the IEEE 802.22, IEEE 802.11af and ECMA-392. However, they are alternatives to the mainstream cellular technologies and lack of the endorsement by both large network operators and vendors. As a result, despite of the promising concept of CR, it is still far from commercial success. In order to bridge the gap between CR and the cellular technologies, in

© Institute for Computer Sciences, Social Informatics and Telecommunications Engineering 2015
M. Weichold et al. (Eds.): CROWNCOM 2015, LNICST 156, pp. 747–758, 2015.
DOI: 10.1007/978-3-319-24540-9_62

the project kogLTE [2], we are enhancing the LTE system with the CR functionalities to enable the flexible dynamic spectrum access, especially for accessing the
significant underutilized TV band. In this paper, the flexible architecture of the
cognitive LTE system is presented in Fig. 1. The main component in this architecture is the dual TI C6670 DSP platform which is used for LTE physical layer
(PHY) base band processing, spectrum sensing (SSM) and to run the cognitive
engine (CE). The ARM processor runs a full LTE protocol stack (PS) implementation and communicates with LTE PHY through an ethernet interface. The
baseband platform communicates with the RF front-end (RF-FE) through the
common public radio interface (CPRI). The key implementation aspects of spectrum sensing, cognitive engine, RF control and the integration of hardware and
software components are introduced in this paper. The architecture and implementation of the testbed makes it particularly suitable for the fast deployable
eNodeB system based on tethered aerial platform which is proposed in the FP7
ABSOLUTE project [3]. The organization of this paper is as follows: The Spectrum Sensing Module (SSM) is described in Section 2. The Cognitive Engine
Module is illustrated in Section 3. Section 4 gives a brief description of the RF
Control Module, and Section 5 describes the integration of the complete system,
while Section 6 concludes the paper.

2   Spectrum Sensing

The *spectrum sensing module (SSM)* is an essential component to enable the cognitive functionality of the LTE cognitive testbed. It performs periodic and event
based spectrum sensing and provides the cognitive engine (CE) with the knowledge about the spectrum environment. Based on the cognitive engine requirements, SSM can perform either blind detection of the signal (e.g. energy detection) or it can also classify the signal (e.g. cyclostationarity-based sensing) [4].
Signal classification is important to distinguish primary and secondary users
which helps the cognitive engine to allocate available spectrum resources properly. SSM is implemented on one core of the Texas Instruments (TI) Digital
Signal Processor (DSP) (TMS320C6670) [5] and communicates with the cognitive engine which runs on the same core through shared memory. There are three
procedures defined for the spectrum sensing module (SSM). The initial procedure
is the initial sensing where a simple energy detector based on a TI DSP library
function is implemented. This function will return the calculated power for each
of the channels to be sensed (the channels to be sensed are provided by the cognitive engine) as shown in Fig. 2 below. The power measurement can be repeated
several times to get the mean power of the sensed channel. After the sensing is
done a task status flag is set to inform the cognitive engine that the sensing task
is finished. The other two procedures are the in-band sensing and out-of-band
sensing. The in-band sensing procedure is triggered periodically (each 1-160 ms
[6]) to ensure that the operating channel is still free. Energy detection is used to
detect any interference in the operating channel. If an interference is detected,
the interference flag is set to alarm the cognitive engine that shall select a new

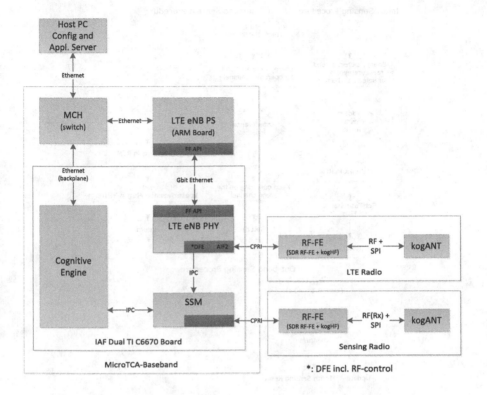

Fig. 1. Block diagram of the cognitive LTE Small Cell Base Station

operating channel based on the information in its data base or *radio environment map (REM)* which contains sensing results of other possible operating channels. The out-of-band sensing procedure is also carried out periodically (each 0.1 - 60 sec [6]) to sense one or several channels around the operating channel. If the classification flag is set, the SSM performs an advanced sensing algorithm to classify the type of the signal (DVB-T, LTE etc.), otherwise a simple sensing algorithm like energy detection is sufficient. If an increased interference in one of the sensed channels is detected the interference flag is set to alarm the cognitive engine.

3 Cognitive Engine

In this section we describe the SW architecture of the cognitive engine which is depicted in Fig. 3. From the figure it can be seen that the cognitive engine and the spectrum sensing module constitute the cognitive extension. The overall goal in designing the architecture in the logical sense was to keep the cognitive extension as modular as possible to attach it to a commercially available 3GPP LTE Protocol Stack. In the following the cognitive engine modules are highlighted.

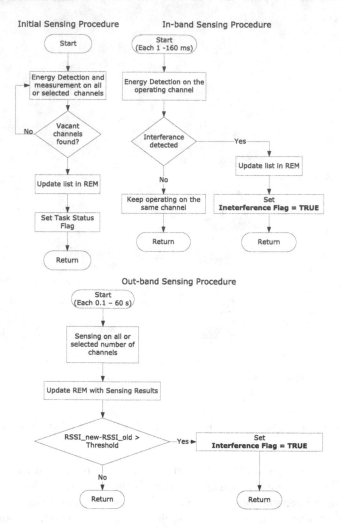

Fig. 2. Sensing Procedures

3.1 Cognitive Manager

The *Cognitive Manager - CM* is the central part of the architecture and is responsible for the information exchange between the internal and external entities. In particular, it acts as a master to manage the procedures carried out by the slave modules such as the Optimization & Decision Unit or Cognition Unit which perform specific functions. Further, inspired by the architecture in [6] we have defined for the cognitive manager the state machine shown in Fig. 4. When the cognitive BS is turned on, it goes first into the initialization state to find an operating channel. During this procedure, the cognitive devices (BS and UEs) perform spectrum sensing to gather information about the radio environment. The process of gathering information about the radio environment can be further assisted by a database access or in other words with a *REM - Radio Environment Map*. The retrieved information is then stored in the *Cognitive Database*. Next, this information is then used by the *Optimization & Decision Unit - ODU* to execute the core channel selection algorithms. After a suitable channel is found, the CM configures the LTE protocol stack, to be more specific the carrier frequency and the maximum transmit power for the DL and UL are passed to the *Radio Resource Control - RRC* layer. Then, the CM goes into the operating state and is ready for data transmission. During the operating state, the CM carries out two main periodic tasks. One of them is the regular updating of the local REM residing in the cognitive database controlled by a timer $T_{refresh\_REM}$. The other one characterized as background procedures in Fig. 4 is the periodic execution of spectrum sensing tasks. Both of these procedures serve the purpose of keeping the radio environmental information up-to-date. Moreover, in case the interference in the used channel increases above a predefined threshold the *CM_channel_move* flag is set to TRUE and the channel evacuation procedure is executed. This procedure involves notifying the cognitive UEs to stop any transmission and wait for a new channel assignment.

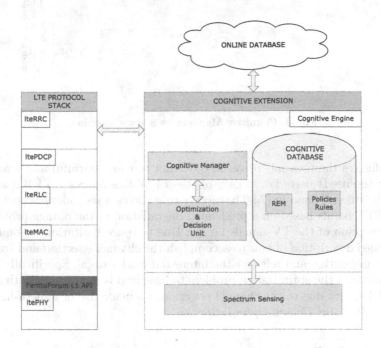

Fig. 3. SW Architecture of the Cognitive Extension

3.2 Optimization and Decision Unit - ODU

The ODU is as already mentioned responsible for dynamically allocating resources for the cognitive devices. In particular it selects an unused channel

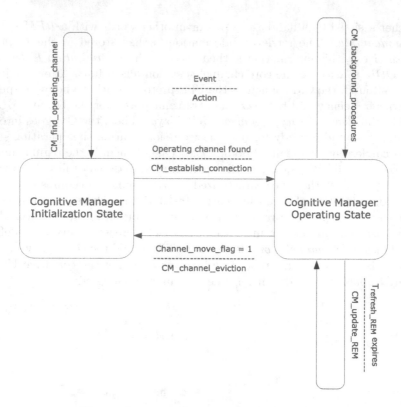

Fig. 4. Cognitive Manager as a state machine

and configures the transmit power in order not to cause harmful interference to primary (licensed) users i.e. in case of the *TV White Spaces - TVWS* according to the rules defined in [7]. This access rule allows a secondary access to the licensed TV bands based on a predefined degradation of the outage probability in the reception of the TV signals i.e 1%. Due to space limitation we omit here the detailed description of how to accomplish the channel selection and transmit power configuration and refer for the interested reader to [8]. Specifically, based on the fact that the optimization problem to be solved is non-convex with a combinatorial aspect due to the channel selection, we made use of a meta-heuristic method called *Ant Colony Optimization*.

3.3 Cognitive Database

The cognitive database has the task of storing the spectrum sensing results and a location specific copy of the REM retrieved through an online database. Within the external database access procedure, the CM sends to the online database its geolocation coordinates and the external database calculates the received primary signal levels at the position of the cognitive device with the usage of path loss models and information about the TV transmitters which can be gathered from the national institute for broadcasting services in Germany [9]. We used the *Okumura-Hata* path loss model for urban areas and implemented the database on a PC running the C code. The target platform from TI (TMS320C6670) communicates with the database running on the PC through an ethernet interface.

4   RF Control

The RF control module (part of the digital front-end DFE in Fig. 1 above) is implemented to control the RF front-end (RF-FE), including RF parameter configurations (e.g. carrier frequency, bandwidth, automatic gain control (AGC), etc.). To enable the reliable transmission of RF control commands between the baseband DSP (the master node) and the RF front-end (the slave node), a two-layer protocol is defined, namely data link layer (DLL) and physical layer (PHY). In the DLL layer, the basic RF control commands and procedures are defined. In order to adapt to the flexible RF control commands in cognitive radio, a frame structure as shown in Fig. 5 with variable length is applied, the end of which is marked by the special symbol "END" for the purpose of frame synchronization. Additionally, a finite state machine with three states, i.e. "RFctrl Initial", "RFctrl Config" and "RFctrl Failed", is defined as shown in Fig. 6. The PHY layer of the RF control can apply the existing RS232 or the common public radio interface (CPRI) in which the fast control and management (C&M) channel is used to convey the RF control commands. The parity check in the PHY layer is used to detect errors in the received command frame based on which the slave node can configure the RF units successfully and reply the acknowledgement (ACK) to the master node or just reply the negative acknowledgement (NACK).

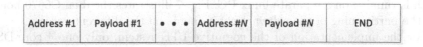

Fig. 5. Command frame structure

5   Integration Aspects of the Cognitive LTE Testbed

In this section, the hardware and software implementation aspects of the cognitive small cell base station testbed will be described. The software partitioning is planned such that the PHY, CE, and Spectrum sensing are running on a dual

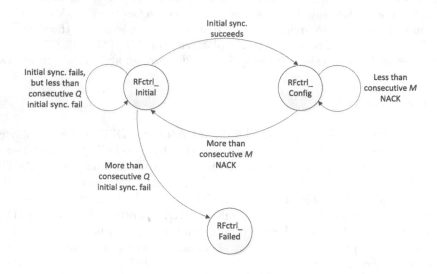

Fig. 6. Finite state machine in the master node

C6670 board which has two TI (TMS320C6670) DSPs with 4 cores each [10]. The board has optical connectors (SFPs) towards the RF boards and a Gigabit Ethernet interface (GbE) to the ARM board. The physical layer implementation is following LTE 3GPP specification and is adapted for small cell operation. The protocol stack (PS) including Media Access Control (MAC) is running on a Qseven (Q7) compatible i.MX6 CPU-Module ARM board [11]. The ARM board has two Gigabit Ethernet interfaces for high bandwidth connection to the dual C6670 baseband processing platform and to the Evolved Packet Core (EPC) which is running on a separate linux PC. Fig. 7 illustrates the dual C6670 board and the partitioning of the software components on the different cores.

For the implementation of the cognitive LTE system, only one 4 core-DSP of the dual TI C6670 board is used (The other DSP might be used if more complex algorithms are needed in future). Core0 is used for initialization of the different software modules and also the resource management of the system. Cognitive Engine and Spectrum sensing modules are running on Core1. Core2 is used for the complete LTE PHY baseband processing, while Core3 is used as an interface to the PS running on the ARM board. This interface is used to allow the communication between PHY and PS based on a Gigabit Ethernet (GbE) communication. It follows the Small Cell Forum specification [12], and provides an API towards both Layer 1 and Layer 2 that allows both to communicate with each other through a platform specific transport layer. In this case, the transport layer is the TI Ethernet driver and network coprocessor (NETCP) [13]. The RF boards used include two components: the converter module (kogHF) to allow the

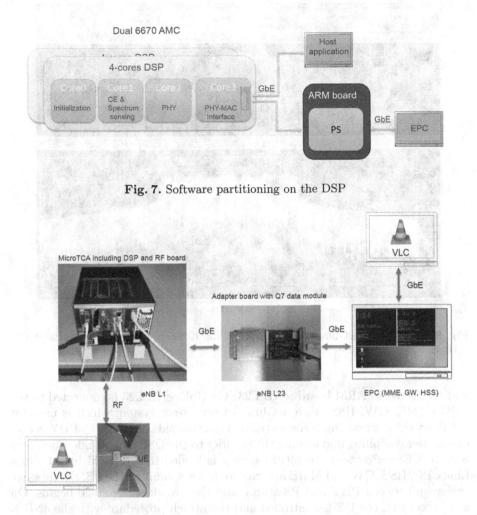

Fig. 7. Software partitioning on the DSP

Fig. 8. The different hardware components of the cognitive LTE testbed

system to work in TV white space. It is used as an up/down converter between TV whitespace frequency range (470 - 790 MHz) and LTE radio-frequency (e.g. 2.6 GHz). The second component is the software defined radio RF front-end (SDR RF-FE) [14], which is used as a reconfigurable LTE RF front-end. It has a common public radio interface (CPRI) [15] towards the baseband DSP board with a data rate of 2457.6 Mbit/s, Xilinx Spartan-6 FPGA and 2-antenna duplex operation with variable RF signal bandwidth. Fig. 8 shows a picture of the testbed and its different components.

A commercial UE dongle with a test SIM was used for the test. It communicates with the eNB via antennas. At the eNB side, the DSP (PHY) is connected to the SDR RF board via CPRI interface to exchange the baseband IQ signal,

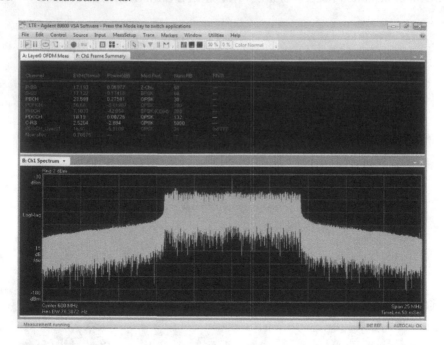

Fig. 9. The LTE spectrum and the EVM of the different downlink channels at 600 MHz carrier frequency

and connected to ARM board (eNB L23) via GbE. eNB L23 is connected to the EPC (MME, GW, HSS) with a GbE. An access-net system which is used for data and video streaming is connected on the other side of the EPC. PHY is executed after compiling and loading the binaries to the DSP using Code Composer Studio (CCS). Protocol Stack (PS) binary is loaded to the ARM board. On a Linux PC, HSS, GW and MME are executed. After running the PS, the message exchange between PHY and PS starts, and the broadcasting stage begins. On a separate PC, the UE is controlled and the attach procedure with the eNB is initiated. After RACH, connection and authentication stages, UE is attached to the eNB. Further more, the responsibility of the cognitive extension which is running on the DSP is to sense the environment and choose the right channel available for transmission without affecting the primary users by controlling the RF unit and the PS configuration.

Measurement results of the downlink decoded information at 600 MHz carrier frequency are shown in Fig. 9 and Fig. 10. Agilent MXA spectrum and signal analyzer was used for capturing and decoding the downlink signal.

For video streaming test, an access net machine has been attached to the GW on which a VLC software runs as a media server to stream video to the UE. A DL video streaming with about 8 Mbps throughput has been verified at the UE side with 10 MHz bandwidth. Fig. 11 shows a screen shot of the video received at the UE side.

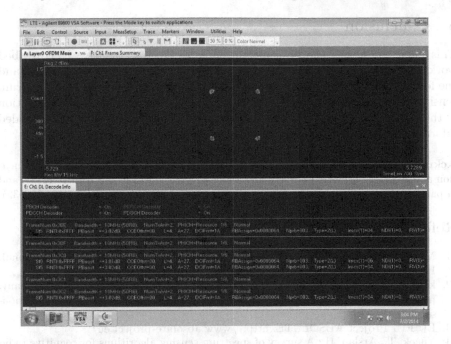

Fig. 10. Constellation and downlink decoded information

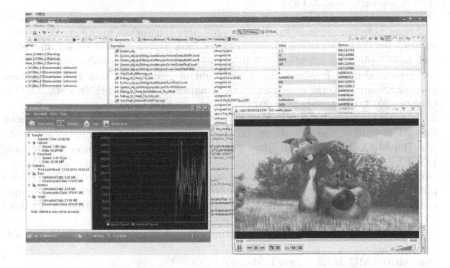

Fig. 11. Screen shot of video streaming

6 Conclusion

This paper presented the implementation aspects of a cognitive LTE testbed working in TV white space spectrum. It focuses on the implementation and the features of the main components of the system such as LTE PHY, spectrum sensing, cognitive engine, and RF control. The software and hardware integration of the testbed was discussed, and measurement results of the downlink decoded information in TV white space spectrum were presented.

Acknowledgments. This work was supported in part by Federal Ministry of Education and Research (BMBF) of Germany under Project kogLTE (FKZ 16BU1213), and in part by the European Commission under Project ABSOLUTE (FP7-ICT- 318632).

References

1. Mitola, J., Maguire, G.: Cognitive Radio: Making Software Radios More Personal. IEEE Pers. Commun. **6**(4), 13–18 (1999)
2. kogLTE project of federal ministry of education and research (BMBF) of germany. http://www.vdivde-it.de/KIS/vernetzt-leben/kognitive-drahtlose-kommunikation-ssysteme/koglte
3. EU FP7 Project ABSOLUTE. http://www.absolute-project.eu/
4. Yucek, T., Arslan, H.: A survey of spectrum sensing algorithms for cognitive radio applications. In: IEEE Communications Surveys and Tutorials, pp. 116–130 (2009)
5. TMS320C6670 Multicore Fixed and Floating-Point System-on-Chip. http://www.ti.com/product/tms320c6670
6. IEEE standard for information technologytelecommunication and information exchange between systems wireless regional area networks (wran)specific requirements - part 22: Cognitive wireless ran medium access control (mac) and physical layer (phy) specifications: Policies and procedures foroperation in the tv bands amendment 1: Management and control plane interfaces and procedures and enhancement to the management information base (mib). IEEE P802.22a/D2, October 2013, pp. 1–551, May 2014
7. ECC/CEPT: Technical and operational requirements for the possible operation of cognitive radio systems in the white spaces of the frequency band 470–790 MHz, January 2011. http://www.erodocdb.dk/Docs/doc98/official/pdf/ECCREP159.PDF
8. Selen, Y., Kronander, J.: Optimizing power limits for white space devices under a probability constraint on aggregated interference. In: IEEE International. Symposium on Dynamic Spectrum Access Networks (DYSPAN), pp. 201–211, October 2012
9. IRT: DVB-T Transmitters in North-Rine Westphalia, November 2013. http://www.irt.de/
10. Dual 6670 AMC. http://www.iaf-bs.de/
11. eDM-QMX6. http://www.data-modul.com/de/home.html
12. Small Cell Forum: LTE eNB L1 API definition v1.1, October 2010. http://www.smallcellforum.org
13. Texas Instruments: Keystone architecture, network coprocessor(NETCP)user guide, November 2010
14. SDR Radio Front-End. http://www.hhi.fraunhofer.de/
15. Common Public Radio Interface (CPRI). http://www.cpri.info/

Construction of a Robust Clustering Algorithm for Cognitive Radio Ad-Hoc Network

Nafees Mansoor, A.K.M. Muzahidul Islam$^{(\boxtimes)}$, Mahdi Zareei,
Sabariah Baharun, and Shozo Komaki

Malaysia-Japan International Institute of Technology (MJIIT),
Universiti Teknologi Malaysia (UTM), Johor Bahru, Malaysia
nafees@nafees.info,{muzahidul.kl,sabariahb,shozo.kl}@utm.my,
m.zareei@ieee.org

Abstract. With the swift expansion of wireless technologies, the demand for radio spectrum is ever growing. Besides the spectrum scarcity issue, spectrums are also underutilized. Cognitive radio customs an open spectrum allocation technique, which ensures efficient handling of the frequency bands. However, suitable network architecture is must for the implementation of cognitive radio networks. This paper presents a robust cluster-based architecture for cognitive radio ad-hoc network. Considering the spatial variance of the spectrum, the proposed architecture splits the network into groups of cluster. Set of free common channels resides every cluster that enables smooth shifting among control channels. The paper also introduces a parameter called Cluster Head Determining Factor (CHDF) to select cluster-heads. Each cluster comprises of a secondary cluster-head to combat the re-clustering issue for mobile nodes. Conclusively, to evaluate the performance of the proposed architecture, simulation is conducted and comparative studies are performed. From the simulation result, it is found that the proposed cluster-based architecture outperforms other recently developed clustering approaches by upholding a reduced number of clusters in the network.

Keywords: Cognitive radio networks · Ad-hoc networks · Cluster-based network · Network architecture · Re-clustering

1 Introduction

There is a rapid growth in wireless applications and technologies, which carries an ever-increasing demand for radio spectrums. However, radio spectrum is a limited natural resource and is almost fully distributed. This leads to a spectrum scarcity problem for the forthcoming wireless technologies. On the other hand, due to the current command-and-control based spectrum allocation method, radio spectrum is underutilized with variance of frequency, time and space [1].

J. Mitola III initiates the idea of Cognitive Radio Network (CRN), where utilization of the unused spectrum in an opportunistic manner is the main objective [1, 2]. Cognitive radio network, an intelligent wireless communication system, has the

© Institute for Computer Sciences, Social Informatics and Telecommunications Engineering 2015
M. Weichold et al. (Eds.): CROWNCOM 2015, LNICST 156, pp. 759–766, 2015.
DOI: 10.1007/978-3-319-24540-9_63

ability to adjust itself on the situation and to make relevant changes in operating parameters such as carrier frequency, transmit-power, modulation strategy, etc in runtime. In CRN, licensed users are considered as Primary Users (PUs) and Secondary Users (SUs) are the unlicensed users who use the free spectrum opportunistically.

A decentralized and self-configured wireless network is considered as wireless ad-hoc network, where the network does not depend on preexisting infrastructures. The decentralized feature of wireless ad-hoc networks allows the network to be more scalable than of wireless managed network. Mobile Ad-hoc Networks (MANETs) and Wireless Sensor Networks (WSNs) are the two popular types of wireless ad-hoc networks [3]. Due to the flexible and dynamic spectrum usage behavior over other ad-hoc technologies, CRN has received a profound interest to the network researchers for the last few years. To scale down ad-hoc networks, clustering is a widely practiced scheme, where nodes are logically grouped based on certain criteria.

In this paper, a robust cluster-based spectrum aware network architecture for cognitive radio ad-hoc network is presented. The proposed architecture splits the network into clusters where the spatial variations of spectrum availability are considered for clustering. A parameter named Cluster Head Determining Factor (CHDF) is introduced to select cluster-heads where clusters' operations are coordinated by the cluster-heads [4, 5]. Each cluster comprises of a secondary cluster-head to combat the re-clustering issue for mobile nodes. The cluster components of the proposed architecture are Cluster-Heads (CHs), Secondary Cluster-Heads (SCHs), Cluster Members (CM), and Forwarding Nodes (FNs). Simulation results show that the proposed cluster-based architecture outperforms other recently developed clustering approaches by upholding a reduced number of clusters in the network.

The paper is organized as follows. In section 2, a brief analysis on different architectures for CRN is presented. The network model of the proposed architecture is discussed in section 3. The proposed cluster-based architecture for CRN is described in section 4. In section 5, simulation results of the proposed architecture are presented and compared. Conclusion and future works have been discussed in section 6.

2   Related Works

Cognitive radio network has received an intense interest to network researchers for the last few years. This section discusses various lately proposed network architectures for CRN.

Ad-hoc CRN architectures can be divided into two groups, one is non cluster-based architectures and another one is cluster-based architectures [6]. Both groups of architectures are vital for the concrete deployment of CRN.

Most of the non cluster-based network architectures for CRN suffer from enlarged communication overheads and inefficient to multi-hop scenario [7-9].

To solve the issues associated with non cluster-based architectures, cluster-based architectures are introduced. Spectrum awareness and local control channel assignment are the main two concerns of reviewed architectures. Spectrum aware cluster-based architecture presented in [10] suffers from frequent re-clustering problem as clusters are formed with lesser number of common channel (often equal to 1). Re-grouping is dominant in [11] with the presence of the PU, as the architecture considers global control channel. Latency in intra-cluster communication and re-clustering for mobile nodes are the main limitations of [12], where clusters are formed based on affinity propagation message-passing technique.

Architecture presented in [13] turns out to be unrealistic with fading control channel and suffers from the re-clustering problem for mobile nodes. Degree based clustering method presented in [14] requires extra processing time and re-clustering issue is acute for mobile nodes. Proposed cluster-based architecture in [15] shows up stable performance for varying spectrum availability. However, the architecture adds extra delay in intra-cluster communication because of the larger cluster size and re-clustering is essential for mobility of nodes. One of the widely conversed cluster-based CRN architecture CogMesh [16] constructs clusters around a specific local channel called master channel. CogMesh practices licensed spectrum for control messaging, which may interfere PU transmission. The architecture in [16] has some provisions for nodes' mobility. However, not all the clusters have the mechanism to deal with the re-clustering issue for mobile nodes. Re-clustering issue is dominant in the dynamic clustering scheme presented in [17] both for varying spectrum availability and mobile nodes. Though nodes' mobility is considered in the architecture presented in [18], however, the architecture suffers from frequent re-clustering problem with varying spectrum availability.

Thus, the recent proposed architectures attain several critical topics for concrete development of CRN. However, a stable architecture in terms of varying spectrum and nodes' mobility is still due for CRN [6] .

3   Network Model

Our assumed ad-hoc network comprised of self-organized Cognitive Radios (CRs)/ Secondary Users (SUs), where SUs have the capability to sense and utilize available free spectrums autonomously. Co-existing with PUs, SUs are location aware and have the processing ability to calculate own CHDF value. CRs are also aware of the CHDF values of the neighboring CRs. The spectrum band is distributed over non-overlapping orthogonal channels with distinctive channel ID for each channel. Licensed spectrum of the PU is only available for SUs if the PU's transmission is absent. Subject to the geographical location, channel availability differs for SUs. We consider that a SU detects available spectrum by sensing free frequency bands using methods such as energy detectors, cyclostationary feature extraction, or Eigenvalue-based feature extraction [19].

The proposed clustering mechanism is independent from any specific PU activity model. We consider Semi-Markov ON– OFF model to evaluate the performance of the proposed architecture. Semi-Markov ON-OFF process is modeled on any channel for the PU traffic. Busy (ON) or idle [20] are the two states that we have considered for any channel. The activation period of any channel is assumed to be an independent random variable. This assumption is realistic when spectrum bands are licensed to independently operating PUs (e.g., channels operated by different TV stations).

We consider IEEE 802.22 standard for the operating frequencies of the system, where SU uses a free channel opportunistically and vacates the channels whenever PUs presence is sensed. To avoid interference with PUs, we assume the presence of a simple interference avoidance model in the system.

We also consider there are two transceivers in each CR, where one is used for control and the other one is used for data transmission. With the ability for least switching delay, each transceiver is spectrum aware. Equal transmission range is considered for all the cognitive radios. A link exists between two radios *iff* they are in each other's communication range and share at least one common channel. We also assume that there is a global common control channel exists in the network. Each cluster declares its own control channel once the cluster formulation is completed. Intra-cluster and inter-cluster communications are coordinated by the Cluster Heads (CHs). For inter-cluster communication, Forwarding Nodes (FNs) are used, where FNs are those nodes that are positioned at the edge of two neighboring clusters and can hear beacons from both clusters.

4   Cluster Formation

CRs need to alter different channels to discover the neighbors. Upon the completion of neighbor discovery, nodes share Accessible Channel Lists (ACLs), C_i and neighbors lists N_i among 1-hop neighbors (where $i = 1,2,3,\ldots, n$). Then the cluster formation stage starts, where the proposed clustering scheme is defined as a maximum edge biclique problem [5, 21]. Based on neighbor list N_i and accessible channels list C_i, each CR_i constructs an undirected bipartite graph $G_i(A_i, B_i, E_i)$. Graph G (V, E) is called bipartite if vertices set V can be split into two disjoint sets A and B where $A \cup B = V$, such that all edges in E connect vertices from A to B. Here, $A_i = CR_i \cup N_i$, and $B_i = C_i$. An edge (x, y) exists between vertices $x \in A_i$ and $y \in B_i$ if $y \in C_i$, i.e., channel y is in the channel list of CR_i. Fig. 1 (a) presents the connectivity graph of a CRN with the accessible channel set in the brackets. From the bipartite graph, each node in the network constructs its own maximum edge biclique graph.

From the maximum edge biclique graph, node determines new C_i and N_i values. These two values are vital for the proposed architecture as CHDF of a node is calculated only using these two values. Our clustering scheme aims to allocate maximum number of free common channels per cluster with suitable amount of member nodes.

We introduce a parameter called Cluster Head Determining Factor (CHDF) to select cluster heads. Every CR calculates CHDF based on equation (1) (Fig. 1 (b)).

$$CHDF_i = \sqrt[c_i]{C_i^{N_i}} ; \quad i = 1,2,3 \ldots n \tag{1}$$

where, Ci is the number of free common channels and Ni is the number of neighboring nodes of CRi. A node announces itself as cluster head if its own CHDF value is higher than all its neighbors. Once the CHDF value of a node CRi is lesser than any of its neighbor, CRi joins the neighboring CH and becomes a cluster member (CM).

After the cluster formation, CH selects SCH from the CMs based on the CHDF value. The SCH takes charge of the cluster if current CH moves out, which deceases reclustering possibility. The proposed cluster-based network is presented in Fig. 1 (c), where the solid line denotes the logical link and dotted line denotes physical links. CH defines and upholds operating channels for the cluster. To find the existence of any other clusters in the neighborhood, CMs check their neighbor list for other cluster heads. CM becomes the FN and connects two clusters once it finds other CH in the neighbor list. In our proposed cluster-based architecture, cluster consists of one CH, one SCH and CMs. All cluster members are 1-hop apart from the CH. FN connects two neighboring clusters, where there can be maximum two intermediate FNs between two neighboring CHs. In the proposed architecture, intra-cluster communications are performed using the local common channels.

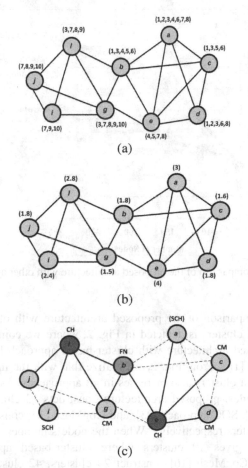

Fig. 1. (a) Connectivity graph of a Cognitive Radio Network with the accessible channels sets in the brackets.(b) CHDF value for each node, (c) Proposed cluster-based network.

5   Simulation Results

We use MATLAB as a simulation tool to evaluate the performance of our proposed cluster-based architecture. Moreover, to perform the comparative study of our proposed architecture, we compare the simulation results of our proposed architecture with three other recently developed approaches, namely cluster-based approach [22], spectrum opportunity-based control channel (SOC) approach [23], and CogMesh [16]. The simulation area is considered as 10000 m^2 (square meters), where the cognitive nodes are positioned randomly in the simulation environment. We consider the communication range for each node to 500 meters and also consider 10 channels in the simulation environment.

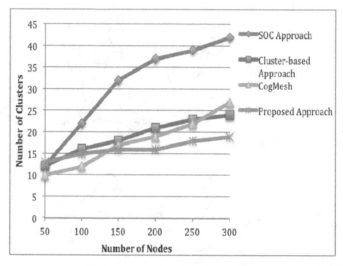

Fig. 2. Performance comparison of the proposed architecture with other approaches (in terms of number of clusters).

Performance comparison of the proposed architecture with other architectures in terms of number of clusters is depicted in Fig. 2, where we compare the simulation result of the proposed architecture with cluster-based approach [22], SOC approach [23], and CogMesh [16]. From Fig. 2, we realize that with the increasing number of nodes, the number of clusters also increases in all approaches. As shown in Fig. 2, in a network of 50 nodes, proposed architecture constructs 13 clusters, where cluster-based approach [22], SOC approach [23], and CogMesh [16] construct 12 clusters, 12 clusters and 10 clusters respectively. When the node's number upturns to 300, proposed architecture gives 19 clusters where cluster-based approach [22], SOC approach [23] and CogMesh [16] construct 24 clusters, 42 clusters and 27 clusters respectively. With the increasing number of nodes, the SOC approach generates higher number of clusters comparing with other three approaches. Meanwhile, comparing with all the approaches, our proposed architecture constructs lesser number of clusters with increasing number of nodes.

6   Conclusion and Future Works

In this paper, we proposed a robust spectrum aware cluster-based architecture that breaks the CRN ad-hoc network into clusters. In the proposed architecture, set of free common channels resides every cluster that enables smooth shifting among control channels. Each cluster comprises with a secondary cluster-head to combat the re-clustering issue for mobile nodes. Conclusively, to evaluate the performance of our method, simulation is conducted and comparative studies are performed. From the simulation, it has been found that our proposed architecture performs better compare with other recently developed architectures. Our next research step is to develop routing and broadcasting protocols for the proposed architecture.

Acknowledgment. This research is partially supported by MJIIT Research Grant with Vote No. 4J044 and GUP TIER 1 the research grant of Universiti Teknologi Malaysia (UTM) with Vote No. 05H61 for the year 2014 to 2015.

References

1. Akyildiz, I.F., Lee, W.Y., Vuran, M.C., Mohanty, S.: NeXt generation/dynamic spectrum access/cognitive radio wireless networks: A survey. Computer Networks **50**, 2127–2159 (2006)
2. Mitola, J., Maguire, G.Q.: Cognitive radio: Making software radios more personal. IEEE Personal Communications **6**, 13–18 (1999)
3. Muzahidul Islam, A.K.M., Wada, K., Chen, W.: Dynamic Cluster-based Architecture and Data Congregation Protocols for Wireless Sensor Network. International Journal of Innovative Computing, Information and Control (IJICIC) **9** (2013)
4. Mansoor, N., Muzahidul Islam, A., Zareei, M., Baharun, S., Komaki, S.: A stable cluster-based architecture for cognitive radio Ad-Hoc networks. In: 2014 Annual International Conference on IEEE Region 10 (TENCON), pp. 1–6, (2014)
5. Mansoor, N., Muzahidul Islam, A., Zareei, M., Baharun, S., Komaki, S.: Cluster modelling for cognitive radio Ad-hoc networks using graph theory. In: 2014 International Conference on Applied Mathematics, Modelling and Simulation (ICAMMS), pp. 1–8 (2014)
6. Mansoor, N., Muzahidul Islam, A.K.M., Zareei, M., Baharun, S., Wakabayashi, T., Komaki, S.: Cognitive Radio Ad-Hoc Network Architectures. A Survey. Wireless Personal Communications **81**, 1117–1142 (2015). 2015/04/01
7. Xin, C.S., Cao, X.J.: A cognitive radio network architecture without control channel. In: 2009 IEEE Global Telecommunications Conference, Globecom 2009, vol. 1–8, pp. 796–801 (2009)
8. Chen, K.-C., Peng, Y.-J., Prasad, N., Liang, Y.-C., Sun, S.: Cognitive radio network architecture: part I – general structure. In: Presented at the Proceedings of the 2nd International Conference on Ubiquitous Information Management and Communication, Suwon, Korea (2008)
9. Hu, Z., Sun, L., Tian, H.: A framework of access network architecture for 4G systems based on cognitive radio. In: 2009 5th International Conference on Wireless Communications, Networking and Mobile Computing, vol. 1–8, pp. 1722–1725 (2009)

10. Zhao, J., Zheng, H., Yang, G.H.: Spectrum sharing through distributed coordination in dynamic spectrum access networks. Wireless Communications and Mobile Computing **7**, 1061–1075 (2007)
11. Zhao, Q., Qin, S., Wu, Z.: Self-organize network architecture for multi-agent cognitive radio systems. In: 2011 International Conference on Cyber-Enabled Distributed Computing and Knowledge Discovery (CyberC), pp. 515–518 (2011)
12. Baddour, K.E., Ureten, O., Willink, T.J.: A Distributed Message-passing Approach for Clustering Cognitive Radio Networks. Wireless Personal Communications **57**, 119–133 (2011)
13. Guo, C., Peng, T., Xu, S.Y., Wang, H.M., Wang, W.B.: Cooperative spectrum sensing with cluster-based architecture in cognitive radio networks. In: 2009 IEEE Vehicular Technology Conference, vol. 1–5, pp. 445–449 (2009)
14. Zhang, J.-Z., Wang, F., Yao, F.-Q., Zhao, H.-S., Li, Y.-S.: Cluster-based distributed topology management in Cognitive Radio Ad Hoc networks. In: 2010 International Conference on Computer Application and System Modeling (ICCASM), pp. V10-544–V10-548 (2010)
15. Asterjadhi, A., Baldo, N., Zorzi, M.: A cluster formation protocol for cognitive radio ad hoc networks. In: 2010 European Wireless Conference (EW), pp. 955–961 (2010)
16. Chen, T., Zhang, H.G., Maggio, G.M., Chlamtac, I.: CogMesh: a cluster-based cognitive radio network. In: 2007 2nd IEEE International Symposium on New Frontiers in Dynamic Spectrum Access Networks, vol. 1 and 2, pp. 168–178, (2007)
17. Mansoor, N., Islam, A.K.M.M., Baharun, S., Komaki, S., Wada, K.: CoAd: a cluster based adhoc cognitive radio networks architecture with broadcasting protocol. In: 2013 International Conference on Informatics, Electronics and Vision (ICIEV), pp. 1–6 (2013)
18. Talay, A.C., Altilar, D.T.: United nodes: cluster-based routing protocol for mobile cognitive radio networks. Iet Communications **5**, 2097–2105 (2011)
19. Yucek, T., Arslan, H.: A survey of spectrum sensing algorithms for cognitive radio applications. IEEE Communications Surveys and Tutorials **11**, 116–130 (2009)
20. Lu, L., Zhou, X., Onunkwo, U., Li, G.Y.: Ten years of research in spectrum sensing and sharing in cognitive radio. EURASIP Journal on Wireless Communications and Networking **2012**, 1–16 (2012)
21. Mansoor, N., Muzahidul Islam, A., Zareei, M., Baharun, S., Komaki, S.: Spectrum aware cluster-based architecture for cognitive radio ad-hoc networks. In: 2013 International Conference on Advances in Electrical Engineering (ICAEE), pp. 181–185 (2013)
22. Li, X., Hu, F., Zhang, H., Zhang, X.: A Cluster-Based MAC Protocol for Cognitive Radio Ad Hoc Networks. Wireless Personal Communications **69**, 937–955 (2013). 2013/03/01
23. Lazos, L., Liu, S., Krunz, M.: Spectrum opportunity-based control channel assignment in cognitive radio networks. In: Presented at the Proceedings of the 6th Annual IEEE Communications Society Conference on Sensor, Mesh and Ad Hoc Communications and Networks, Rome, Italy (2009)

On the Effective Capacity of Delay Constrained Cognitive Radio Networks with Relaying Capability

Ahmed H. Anwar[1]([✉]), Karim G. Seddik[2], Tamer ElBatt[3,4], and Ahmed H. Zahran[4]

[1] Electrical Engineering and Computer Science Department,
University of Central Florida, Orlando 32816, USA
a.h.anwar@knights.ucf.edu
[2] Electronics Engineering Department, American University in Cairo,
AUC Avenue, New Cairo 11835, Egypt
kseddik@aucegypt.edu
[3] Wireless Intelligent Networks Center (WINC),
Nile University, Smart Village, Giza, Egypt
[4] Department of EECE, Faculty of Engineering,
Cairo University, Giza, Egypt
{ahzahran,telbatt}@ieee.org

Abstract. In this paper we analyze the performance of a secondary link in a cognitive radio relaying system operating under a statistical quality of service (QoS) delay constraint. In particular, we quantify analytically the Effective Capacity improvement for the secondary user when it offers a packet relaying service to the primary user packets that are lost under the SINR interference model. Towards this objective, we utilize the concept of Effective Capacity introduced earlier in the literature as a metric to quantify the wireless link throughput under statistical QoS delay constraints, in an attempt to support real-time applications using cognitive radios. We study a two-link network, a single secondary link and a primary network abstracted to a single primary link, with and without relaying capability. We analytically prove that exploiting the packet relaying capability at the secondary transmitter improves the Effective Capacity of the secondary user. Finally, we present numerical results that support our theoretical findings.

1 Introduction

Over the past decade, there has been surge in demand for the wireless spectrum due to the bandwidth-hungry applications, e.g., multimedia communications. Moreover, there has been ample evidence that the wireless spectrum has been significantly underutilized. In [1], the cognitive radio (CR) concept has been first

A.H. Anwar—This work was made possible by grants number NPRP 4-1034-2-385 and NPRP 5-782-2-322 from the Qatar National Research Fund (a member of Qatar Foundation). The statements made herein are solely the responsibility of the authors.

© Institute for Computer Sciences, Social Informatics and Telecommunications Engineering 2015
M. Weichold et al. (Eds.): CROWNCOM 2015, LNICST 156, pp. 767–779, 2015.
DOI: 10.1007/978-3-319-24540-9_64

introduced as a promising technology due to its opportunistic, agile and efficient spectrum utilization merits. Cognitive radios enable secondary users (SUs) to co-exist with the primary (licensed) users (PUs) in the same frequency band without causing harmful interference. Three major cognitive radio paradigms have been introduced in the literature: underlay, overlay, and interweave [2].

Providing quality of service (QoS) guarantees has been a daunting challenge for wireless networks, in general, and for cognitive radio networks, in particular. The Effective Capacity (EC) concept originally proposed in [3] is a throughput performance metric for a wireless link under statistical QoS (delay) constraints. It is considered the wireless dual concept to the notion of "Effective Bandwidth" which was originally coined for wired networks in [4].

Introducing the relay nodes in cognitive networks has been studied in [5], the authors used cooperative relay node to assist the transmission of CRNs. In [6], proposed an adaptive cooperation diversity scheme including best-relay selection while ensuring the QoS of the primary user. In [7], the authors proposed a feedback-based random access channel scheme for cognitive relaying networks. However, delay constraints for opportunistic users with real-time communication requirements were not considered.

In [8], we quantified the EC gains and transmission power reduction attributed to exploiting the primary user feedback at the secondary transmitter. However, the SUs in [8] do not provide a relaying service to the unsuccessful primary packets. Previous work did not studied the effective capacity of the cognitive radios. However,the closest to our work is [9], where the EC for interference and delay constrained cognitive radio relaying channels is characterized. The system model in [9] hinges on the underlay cognitive radio paradigm, whereas our system exhibits the characteristics of interweave cognitive radios which mandates spectrum sensing and allows for SU-PU co-existence as long as the SINR is above an acceptable threshold. In addition, we add a relaying service at the SU rather than using dedicated relaying nodes as in [9].

Our main contribution in this paper is to show that a higher EC, and hence, a higher data rate can be sustained if the secondary user offers a packet relaying service to the primary user. We develop a queuing theoretic analysis to capture the gains of adding relaying capability to the cognitive radio network. We show analytically that adding a relaying capability to cognitive radio networks not only increases its EC but also helps the PU to evacuate its queue faster and, hence, giving more opportunity to the SU to transmit over the shared channel.

The rest of the paper is organized as follows. A background on the EC concept is given in Section 2. The system model and underlying assumptions are presented in Section 3. In Section 4, the EC problem for cognitive relaying networks is formulated and analyzed. Afterwards, the numerical results and discussion are presented in Section 5. Finally, we conclude the paper and point out potential directions for future research in Section 6.

2   Background: Effective Capacity

In [3], Wu and Negi introduced the notion of *effective capacity* (EC) of a wireless
link as the maximum constant arrival rate that can be supported by a given channel service process while satisfying a statistical QoS requirement specified by the
QoS exponent, denoted θ. The EC concept is a link layer modeling abstraction
to incorporate QoS requirements, such as delay, into system performance analysis studies of wireless systems. Using EC as a performance metric enables us to
evaluate the cognitive radio network throughput under statistical QoS constraint
without performing queuing analysis.

If Q is defined as the stationary queue length, then θ is the decay rate of the
tail distribution of the queue length Q, that is

$$\lim_{q \to \infty} \frac{\log \Pr(Q \geq q)}{q} = -\theta. \tag{1}$$

From (1), it is clear that the EC captures a probabilistic QoS constraint.
Practically, θ, which depends on the statistical characterization of the arrival
and service processes, establishes bounds on the delay (or buffer length). It has
been established in [3] that the EC for a given QoS exponent θ is given by

$$-\lim_{t \to \infty} \frac{1}{\theta t} \log_e \mathbb{E}\left\{e^{-\theta S(t)}\right\} = -\frac{\Lambda(-\theta)}{\theta}, \tag{2}$$

where $\Lambda(\theta) = \lim_{t \to \infty} \frac{1}{t} \log_e \mathbb{E}\left\{e^{-\theta S(t))}\right\}$ is a function of the logarithm of the
moment generating function of $S(t)$, $S(t) = \sum_{k=1}^{t} r(k)$ represents the time accu-
mulated service process and $\{r(k), k = 1, 2, \cdots\}$ is the discrete, stationary and
ergodic stochastic service process.

3   System Model

We consider a time slotted system as shown in Fig. 1. Where data is transmitted
in frames of duration T seconds, that fits exactly in one time slot. The primary
network traffic is abstracted to a single primary link. Hence, our analysis is valid
for any number of primary users. Assuming one frequency channel, the primary
transmitter will access the channel whenever it has a packet to send. On the
other hand, the single SU attempts to access the medium with a certain policy,
described later, based on the spectrum sensing outcome. The SU is assumed
to have a packet to send at the beginning of each time slot (i.e. the SU queue
is saturated). We assume that the SU uses the first N seconds out of the slot
duration T for spectrum sensing.

In the rest of this paper, we refer to the system where the SU offers a relay
service to the "undelivered" primary packets, besides sending its own packets,
as the "Cognitive Relay system". On the other hand, the baseline system with
no relaying capability is referred to as "No-relay system". The EC of both systems is analyzed under the SINR interference model. According to the cognitive
radio system adopted in this paper (which is a hybrid between underlay and

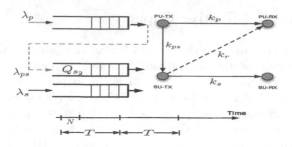

Fig. 1. Cognitive Relay System Model.

interweave), the SU transmits its packets with a lower power level P_1 when the channel is sensed busy. However, if the medium is sensed idle, the SU transmits with a higher power level P_2. These power levels correspond to the SU transmission rates of r_1 and r_2 for busy and idle mediums, respectively. We assume non-perfect spectrum sensing. Hence, a miss-detection event occurs if the PU is active and the medium is sensed idle by the SU. On the other hand, a false alarm occurs when the medium is sensed busy while the primary user is not sending. Simple energy detection [10] is adopted as the spectrum sensing mechanism.

The discrete time secondary link input-output relations for idle and busy channels in the ith symbol duration are given, respectively, by

$$y(i) = h(i)x(i) + n(i) \qquad i = 1, 2, \cdots \tag{3}$$

$$y(i) = h(i)x(i) + s_p(i) + n(i) \qquad i = 1, 2, \cdots, \tag{4}$$

where $x(i)$ and $y(i)$ represent the complex-valued channel input and output, respectively. $h(i)$ denotes the fading coefficient between the cognitive transmitter and receiver, $s_p(i)$ is the interference signal from the primary network on the SU and $n(i)$ is the additive thermal noise at the secondary receiver modeled as a zero-mean, circularly-symmetric complex Gaussian random variable with variance $\mathbb{E}\{|n(i)|^2\} = \sigma_n^2$. The channel bandwidth is denoted by B. The channel input is subject to the following average energy constraints: $\mathbb{E}\{|x(i)|^2\} \leq P_1/B$ or $\mathbb{E}\{|x(i)|^2\} \leq P_2/B$ for all i's, when the channel is sensed to be busy or idle, respectively. The fading coefficients are assumed to have arbitrary marginal distributions with finite variances, that is, $\mathbb{E}\{|h(i)|^2\} = \mathbb{E}\{z(i)\} = \sigma^2 < \infty$, where $|h(i)|^2 = z(i)$. Finally, we consider a block-fading channel model and assume that the fading coefficients stay constant for a block of duration T seconds (i.e., one frame duration) and change independently from one block to another.

In the proposed model, we leverage a perfect error-free primary feedback channel. The primary receiver sends a feedback at the end of each time slot to acknowledge the reception of packets. Typically, the PU receiver sends an ACK if a packet is correctly received, however, a NACK is sent if a packet is lost. Failure of reception is attributed to primary channel outage. In case of an idle slot, no feedback is sent. The SU is assumed to overhear and decode this primary feedback perfectly and to act as follows: if an ACK/no feedback is overheard, the SU behaves normally and starts sensing the channel in the next time slot. On the other hand, if a NACK is overheard by the SU, yet, it can successfully decode the PU's data packet, then the SU stores it in the relay queue and sends an ACK to the PU as explained in the next section.

We consider a cognitive relaying system where the SU plays a role in relaying the "undelivered" primary packets. We recall that the SU has two separate queues; the first queue stores packets to be relayed for the PU (Relay queue). The second queue stores the SU own packets (Secondary Queue). The SU senses the medium and accesses it with either P_1 or P_2 according to the sensing outcome, while giving the advantage to evacuate the relay queue first.

We assume that all four links in the studied system are subject to outage, that is, the outage probability in the primary link is denoted k_p, in the PU-TX and SU-TX link is denoted k_{ps}, in the SU-TX and PU-RX (relaying channel) is denoted k_r and in the secondary link is denoted k_s.

In our model, we assume that the PU occupies the wireless channel with a fixed prior probability ρ_p [9]. The channel sensing can be formulated as a hypothesis testing problem between the additive white Gaussian noise $n(i)$ and the primary signal $s_p(i)$ in noise. Noting that there are NB complex symbols in a duration of N seconds, this can be expressed mathematically as follows:

$$H_0 : y(i) = n(i), \qquad i = 1, \cdots, NB; \tag{5}$$

$$H_1 : y(i) = s_p(i) + n(i), \qquad i = 1, \cdots, NB. \tag{6}$$

Hence, it is straightforward to write down the probabilities of false alarm P_f and detection P_d as follows:

$$P_f = Pr(Y > \omega | H_0) = 1 - P\left(\frac{NB\omega}{\sigma_n^2}, NB\right); \tag{7}$$

$$P_d = Pr(Y > \omega | H_1) = 1 - P\left(\frac{NB\omega}{\sigma_{sp}^2 + \sigma_n^2}, NB\right), \tag{8}$$

where ω is the energy detector threshold, $Y = \frac{1}{NB}\sum_{i=1}^{NB} |y(i)|^2$ and $P(x,a)$ denotes the regularized lower gamma function defined as $P(x,a) = \frac{\gamma(x,a)}{\Gamma(a)}$ where $\gamma(x,a)$ is the lower incomplete gamma function. Note that the test statistic Y is chi-square distributed with $2NB$ degrees of freedom.

4   The Effective Capacity of the Relaying Secondary User under the SINR Model

In order to perform EC analysis for the cognitive radio relaying system, the primary activity has to be analytically quantified. Therefore, a queuing analysis for both, primary queue and relay queue, is conducted in this section. Afterwards, we will develop the system Markov chain that characterizes the cognitive user EC.

4.1 Primary User Queue Analysis

We assume that the primary packets arrive according to a Bernoulli arrival process. At each time slot, a new packet arrives with probability $0 \leq \lambda_p \leq 1$. The PU is assumed to send a packet in a time slot as long as its queue is non-empty. Hence, the PU access probability is expressed as $\Pr\{\text{PU accesses}\} = \rho_p$, where ρ_p denotes the probability of a non-empty primary queue. When the PU sends a packet one of three scenarios may arise:

- The packet is successfully received by the PU-RX. In this case, the packet is dropped from the secondary relay queue, if it was successfully received by the SU-TX.
- The packet is successfully received by the SU-TX but not received by the PU-RX. In this case, the packet is stored in the relay queue and dropped from the primary queue.
- The packet is neither received by the PU-RX nor the SU-TX. Hence, the PU will re-transmit it in the next time slot with probability one.

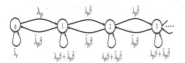

Fig. 2. Primary Queue Markov Chain.   **Fig. 3.** Relay Queue Markov Chain.

The primary queue is modeled as a discrete-time Markov chain, where λ_p is the packets arrival rate at the PU. β is the service rate of this birth-death primary queue. The Markov chain state χ_n represents that there are n packets in the primary queue in this time slot. The events governing the transitions between states can be summarized as follows:

- $\Pr(\chi_{n+1}|\chi_n)$ means that a new packet is added to the queue due to either a new packet arrival while no packet is served within the same time slot.
- $\Pr(\chi_n|\chi_n)$ means that no new arrivals and no packet is serviced or new packet arrived while another one is successfully served by either the primary channel or the relay channel.
- $\Pr(\chi_n|\chi_{n+1})$ means that no new arrivals while a packet is successfully served by either the primary channel or the relay channel.

Applying the global balance equations at the states of the Markov chain we can characterize ρ_p as a function of β and λ_p as shown in the appendix.

4.2 Relay Queue Analysis

In order to characterize the Effective Capacity of the cognitive relaying user and complete the analysis, we need to characterize the non-empty probability of the relay queue. The arrival rate is λ_{ps} and the service rate is β_r. The relay queue can be also modeled by a birth-death queue, hence, we use similar steps to characterize ρ_r (details are given in the Appendix).

4.3 Modeling the Cognitive Radio Channel

Along the lines of [11], we develop a Markov chain capturing the dynamics of the cognitive radio channel where the state represents the sensing outcome (B-B, MD, FA, I-I) and the channel reliability (ON, OFF), as illustrated next.

Not knowing the channel conditions, the secondary transmitter sends at fixed rates. More specifically, the transmission rate is fixed at r_1 bits/s in the presence of active primary users while the transmission rate is r_2 bits/s when the PU is idle. We initially construct a state-transition model for cognitive transmissions by considering the cases in which the fixed transmission rates are smaller or greater than the instantaneous channel capacity values, and also incorporating the sensing decision and its correctness. In particular, if the fixed rate is smaller than the instantaneous channel capacity, we assume that reliable communication is achieved and the channel is in the ON state. Otherwise, we declare that outage has occurred and the channel is in the OFF state. Note that information has to be retransmitted in such a case.

4.4 State Transition Dynamics

The state transition dynamics of the SU are captured in the Markov chain depicted in Fig. 4. It is an eight states' Markov chain. Each state represents the sensing process outcome and the SU link ON or OFF as discussed next. Regarding the decision of channel sensing and its correctness, we have the following four possible cases: the channel is busy and detected busy (B-B), the channel is busy and detected idle (MD), idle and detected busy (FA), and, finally, the channel is idle and detected idle (I-I). In each case, we have two link outage possibilities, namely ON and OFF, depending on whether the transmission rate exceeds the instantaneous channel capacity or not. In order to identify these states, we have to first determine the instantaneous channel capacity in each time slot. Note that if the channel is detected busy, the secondary transmitter sends packets with power P_1. Otherwise, it transmits with a higher power, P_2. Considering the interference σ_{sp} caused by the primary users as additional Gaussian noise, we can express the instantaneous channel capacities in the above four cases as follows:

$$C_l = B \log_2(1 + SNR_l z(i)), \tag{9}$$

where SNR_l denotes the average signal-to-noise ratio (SNR) for each possible scenario l, where $l = 1, 2, 3, 4$. It is straightforward to write the SNRs in these four cases, that is $SNR_1 = \frac{P_1}{B(\sigma_n^2 + \sigma_{sp}^2)}$, $SNR_2 = \frac{P_2}{B(\sigma_n^2 + \sigma_{sp}^2)}$, $SNR_3 = \frac{P_1}{B\sigma_n^2}$ and $SNR_4 = \frac{P_2}{B\sigma_n^2}$. Note that in scenarios 1 and 3, the channel is detected busy and, hence, the transmission rate is r_1 while it is r_2 in scenarios 2 and 4.

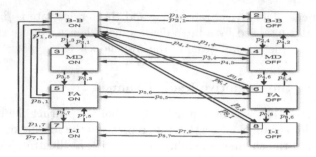

Fig. 4. The Markov chain model for the cognitive radio channel.

If these fixed rates are below the instantaneous capacity values, i.e., if $r_1 < C_1, C_3$ or $r_2 < C_2, C_4$, then the cognitive transmission is considered to be in the ON state where reliable communication is achieved. On the other hand, if $r_1 \geq C_1, C_3$ or $r_2 \geq C_2, C_4$, outage occurs and the secondary user transmission is in the OFF state. In those cases, reliable communication is not attained, and hence, the information has to be resent. It is assumed that a simple automatic repeat request (ARQ) mechanism is incorporated in the communication protocol to acknowledge the reception of data and to ensure that erroneous data is retransmitted. This state-transition model with eight states is depicted in Fig. 4. In states 1, 3, 5, and 7, the cognitive radio channel is in the ON state, and $r_1(T - N)$ bits in states 1 and 5, and $r_2(T - N)$ bits in states 3 and 7 are transmitted and successfully received. On the other hand, the transmission rate is zero in the OFF states.

The above Markov chain is fully characterized by its transition probability matrix $\mathbf{R}_{M \times M}$ defined as:

$$\mathbf{R}_{M \times M} = [p_{i,j}], 1 \leq i, j \leq M. \tag{10}$$

Given the EC expression in (2) and the state transition model in Fig. 4, the EC can be expressed as follows:[1]

$$EC(\theta) = \frac{\Lambda(-\theta)}{-\theta} = \max_{r1, r2} \frac{1}{-\theta} \log_e sp(\mathbf{\Phi}(-\theta)\mathbf{R}), \tag{11}$$

where the matrix $\mathbf{R}$ is the state transition matrix as defined above, and $sp(\mathbf{\Phi}(-\theta)\mathbf{R})$ is the spectral radius of the matrix $\mathbf{\Phi}(-\theta)\mathbf{R}$, that is, the maximum of the absolute of all eigenvalues of the matrix. Therefore, to reach a closed form expression for the EC, we need to get the eigenvalues of the matrix $\mathbf{\Phi}(-\theta)\mathbf{R}$. $\mathbf{\Phi}(-\theta)$ is a diagonal matrix defined as $\mathbf{\Phi}(-\theta) = diag(\phi_1(-\theta), \phi_2(-\theta), \cdots, \phi_M(-\theta))$ whose diagonal elements are the moment generating functions of the Markov process in each of the M states.

---

[1] The proof can be found in [12, Ch.7].

In order to fully characterize the EC, we first characterize the transition probability matrix $\mathbf{R}$ as follows.

$$\begin{aligned} p_{1,1} &= \rho_p P_d \Pr(r_1 < C_1(i + TB)|r_1 < C_1(i)) \\ &= \rho_p P_d \Pr(z(i + TB) > \alpha_1|z(i) > \alpha_1), \end{aligned} \tag{12}$$

where $\alpha_1 = \frac{2^{\frac{r_1}{B}}}{SNR_1}$, the term $\Pr(r_1 < C_1(i + TB)|r_1 < C_1(i))$ represents the probability that the channel is ON (SU not in outage), ρ_p is the prior probability of the primary channel being busy, P_d is the probability of detection as in (8).

Note that $p_{1,1}$ depends, in general, on the joint distribution of $(z(i + TB), z(i))$. However, since fading changes independently from one block to another in the block-fading model, we can further simplify $p_{1,1}$ to

$$p_{1,1} = \rho_p P_d \Pr(z(i + TB) > \alpha_1) = \rho P_d \Pr(z(i) > \alpha_1)$$

Thus, we can immediately see that the transition probability $p_{1,1}$ does not depend on the original state. Hence, due to the block fading assumption, we can express

$$p_{i,1} = \rho_p P_d \Pr(z(i) \geq \alpha_1) \text{ for } i = 1, 2, \cdots, 8.$$

Similarly, $p_{i,2} = p_2 = \rho_p P_d \Pr(z < \alpha_1)$, $p_{i,3} = p_3 = \rho_p(1 - P_d) \Pr(z \geq \alpha_2)$, $p_{i,4} = p_4 = \rho_p(1 - P_d) \Pr(z < \alpha_2)$, $p_{i,5} = p_5 = (1 - \rho_p)P_f \Pr(z \geq \alpha_3)$, $p_{i,6} = p_6 = (1 - \rho_p)P_f \Pr(z < \alpha_3)$, $p_{i,7} = p_7 = (1 - \rho_p)(1 - P_f) \Pr(z \geq \alpha_4)$ and $p_{i,8} = p_8 = (1 - \rho_p)(1 - P_f) \Pr(z < \alpha_4)$, where $\alpha_2 = \frac{2^{\frac{r_1}{B}}}{SNR_2}$, $\alpha_3 = \frac{2^{\frac{r_1}{B}}}{SNR_3}$ and $\alpha_4 = \frac{2^{\frac{r_1}{B}}}{SNR_4}$. Hence, the transition probability matrix is constructed as a unit rank matrix:

$$\mathbf{R} = \begin{bmatrix} p_{1,1} & p_{1,2} & \cdots & p_{1,8} \\ \cdot & \cdot & \cdots & \cdot \\ \cdot & \cdot & \cdots & \cdot \\ p_{8,1} & p_{8,2} & \cdots & p_{8,8} \end{bmatrix} = \begin{bmatrix} p_1 & p_2 & \cdots & p_8 \\ \cdot & \cdot & \cdots & \cdot \\ \cdot & \cdot & \cdots & \cdot \\ p_1 & p_2 & \cdots & p_8 \end{bmatrix}. \tag{13}$$

4.5   Characterizing the Effective Capacity

If we define Q as the stationary queue length, then θ is defined as the decay rate of the tail distribution of the queue length Q.

Hence, we have the following approximation for the buffer violation probability for large queue lengths, denoted by q_{max}

$$P(Q \geq q_{max}) \approx \exp^{-\theta q_{max}}. \tag{14}$$

Therefore, larger θ corresponds to more strict QoS constraints whereas smaller θ implies looser constraints. In certain settings, constraints on the queue length can be mapped to delay-QoS constraints.

In practical applications, the value of θ depends on the statistical characterization of the arrival and service processes, bounds on the delay or buffer lengths, and the target values of the delay or buffer length violation probabilities.

The effective capacity for a given QoS exponent θ is given by equation (1) where $S(t) = \sum_{k=1}^{t} r(k)$ represents the time accumulated service process and $\{r(k),\ k=1,2,\cdots\}$ is the discrete, stationary and ergodic stochastic service process.

Note that the service rate is $r = r_1(T - N)$ if the cognitive system is in state 1 or 5. Similarly, the service rate is $r = r_2(T - N)$ for states 3 and 7.

In OFF states, transmission rates exceed the instantaneous channel capacities and reliable communication is not possible. Hence, their service rates are effectively zero.

The state transition model for both systems under investigation, "Cognitive Relay system" and "No-relay system", is essentially the same. This is attributed to the fact that the cognitive channel (secondary link) has the same dynamics in both systems. However, the EC will be different due to the presence of the secondary relay queue in the cognitive relaying model. Next, we characterize the Effective Capacity of the cognitive relaying system using the state transition model described in the previous subsection.

For the cognitive radio channel with the state transition model described earlier, the spectral radius of $sp(\mathbf{\Phi}(-\theta)\mathbf{R})$ is the rank of this matrix. Hence, the normalized effective capacity in bits/s/Hz is given by

$$
\begin{aligned}
EC_{relay}(SINR, \theta) = \frac{-(1 - \rho_r)}{\theta TB} \log_e((p_1 + p_5) \exp^{-(T-N)\theta r_1} \\
+ (p_3 + p_7) \exp^{-(T-N)\theta r_2} + p_2 + p_4 + p_6 + p_8).
\end{aligned}
\tag{15}
$$

On the other hand, the EC for the baseline "No-Relay system" is derived in [11] and is given by

$$
\begin{aligned}
EC_{no-relay}(SINR, \theta) = \frac{-1}{\theta TB} \log_e((p_1 + p_5) \exp^{-(T-N)\theta r_1} \\
+ (p_3 + p_7) \exp^{-(T-N)\theta r_2} + p_2 + p_4 + p_6 + p_8).
\end{aligned}
\tag{16}
$$

From both equations, (15) and (16), the EC in case of relay system is degraded by the probability of having an empty relay queue. It is obvious since the SU starts to send its own packets only if the relay queue is empty.

5   Numerical Results

In this section, we present numerical results that provide further insights about the effect of relaying on EC of CRNs. We show results for the relaying system and compare it with the baseline system where the SU has no relaying capability [11]. The numerical values used for the system parameters are as follows: $SNR_1 = 6.9$ db, $SNR_2 = 10$ db, $SNR_3 = 30.7$ db, $SNR_4 = 40$ db, $k_p = 0.6$, $k_{ps} = 0.2$, $k_r = 0.4$ if the PU is active, $k_r = 0.2$ if the PU is idle, $r_1 = 1000$ bps, $r_2 = 6000$ bps and $lambda_p = 0.38$. We also set $T = 0.1$ sec, $N = 0.026$ sec, $\lambda = 1.7$ and $B = 1000$ Hz. Note that the optimal values for r_1 and r_2 are obtained by simple numerical search such that, EC is maximized.

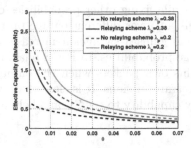

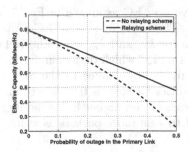

Fig. 5. SU EC of the Relay system and the no-relay system.

Fig. 6. EC of the Relay system and the no-relay system versus primary link outage probability.

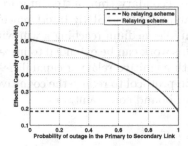

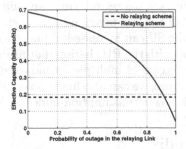

Fig. 7. EC of the Relay system and the no-relay system versus primary-secondary link outage probability

Fig. 8. EC of the Relay system and the no-relay system versus relaying link outage probability

In Fig. 5, we plot the SU EC for the cognitive relaying and the no-relay (baseline) system versus the, statistical QoS constraint, delay exponent θ for a sensing duration of $N = 0.026$ and for two values of the PU packet arrival rate. Clearly, as the delay exponent θ increases (stricter delay requirements), the effective capacity (the maximum rate that the channel can sustain in bit/sec/hertz) decreases. The same result can be easily distilled from the EC definition in (2). Moreover, it is shown that when the SU helps relaying the "unsuccessful" PU packets, for the set of outage probabilities given before, the secondary user attains higher EC. As θ increases, the performance gain decreases since stricter QoS constraints limits the secondary user throughput. Finally, it is intuitive to notice that the SU EC decreases as the primary user become more active in accessing the medium due to higher packet arrival rate, λ_p.

Next, we investigate the system behavior versus different link outage probabilities. It is worth noting that we only plotted the EC versus the outage probability values that preserve the system queues stability as explained in the appendix. In Fig. 6, the EC is plotted versus k_p while fixing other outage probabilities like $k_{ps} = 0.2$ and $k_r = 0.4$. Similarly, we investigated the effect of k_{ps} and k_r on the cognitive user EC as shown in Fig. 7 and 8, while fixing other probabilities of outage. The No relay system EC remains constant over different outage probabilities for k_{ps} and k_r, while the Relay system gains a significant increase in terms of its EC which decays under high outage probabilities. In Fig. 8 relaying is not giving any chance to the SU to send his own packets when the outage probabilities k_r exceeds 0.92 due to multiple retransmissions. It is clear that we can always obtain higher EC by adding a relay capability to the SU and helping the PU to send more packets as well.

6   Conclusion and Future Work

In this paper we study a two-link network, a primary network abstracted to a single primary link, a single secondary link with relaying capability. We show that exploiting the packet relaying capability at the secondary transmitter improves the EC of the secondary user. It is shown that under the "SINR Interference" model the SU can increase its chance to find an idle medium reaching a win-win situation with the PU sharing that medium. This work can be extended to investigate the case of multiple secondary users with cognitive relaying capabilities. In the future, we can also find a power allocation protocol to reduce the cognitive network power consumption under an EC lower bound constraint.

Appendix

Given the Markov chain in Fig. 2, to characterize the non-empty queue probability, we apply the global balance equation (GBE) on each state. Let β is the service rate of Q_p (let χ_i denote $\Pr(\chi_i)$). Applying the GBE at state 0:

$$\chi_0 \lambda_p = \chi_1 \bar{\lambda}_p \beta \ \rightarrow \ \chi_1 = \frac{\lambda_p}{\bar{\lambda}_p \beta} \chi_0. \tag{17}$$

Applying the balance equation at state 1,

$$\chi_1 (\lambda_p \bar{\beta} + \bar{\lambda}_p) = \chi_0 \lambda_p + \chi_2 \bar{\lambda}_p \beta \rightarrow \chi_2 = \frac{1}{\beta} \left(\frac{\lambda_p \bar{\beta}}{\bar{\lambda}_p \beta} \right)^2 \chi_0. \tag{18}$$

Recursively, $\chi_i = \frac{1}{\beta} \left(\frac{\lambda_p \bar{\beta}}{\bar{\lambda}_p \beta} \right)^i \chi_0 \ \ \forall i \geq 1$. Since $\sum_{i=1}^{\infty} \chi_i = 1$, we can calculate χ_0 with some manipulations. To ensure queue stability, we must have $\lambda_P \bar{\beta} \leq \bar{\lambda}_p \beta$, hence,

$$\chi_0 = \left[1 + \frac{\lambda_p}{(\bar{\lambda}_p \beta - \lambda_p \bar{\beta})} \right]^{-1}. \tag{19}$$

After some mathematical manipulations we can express χ_0 as:

$$\chi_0 = 1 - \frac{\lambda_p}{\beta}. \tag{20}$$

Finally we have $\rho_p = 1-\chi_0$. Where the service rate $\beta = 1-k_p \times k_{ps}$. Similarly, one can characterize, ρ_r. Then $\beta_r = (1-k_r|PU active) \times \rho_p + (1-k_r|PU idle) \times (1-\rho_p)$. Hence we can write $\rho_r = \frac{\lambda_{ps}}{\beta_r}$.

References

1. Mitola III, J., Maguire Jr., G.Q.: Cognitive radio: making software radios more personal. IEEE Personal Communications **6**(4), 13–18 (1999)
2. Goldsmith, A., Jafar, S., Maric, I., Srinivasa, S.: Breaking spectrum gridlock with cognitive radios: An information theoretic perspective. Proceedings of the IEEE **97**(5), 894–914 (2009)
3. Wu, D., Negi, R.: Effective capacity: a wireless link model for support of quality of service. IEEE Transactions on Wireless Communications **2**(4), 630–643 (2003)
4. Chang, C.S., Thomas, J.A.: Effective bandwidth in high-speed digital networks. IEEE Journal on Selected Areas in Communications **13**(6), 1091–1100 (1995)
5. Jia, J., Zhang, J., Zhang, Q.: Cooperative relay for cognitive radio networks. In: INFOCOM 2009, pp. 2304–2312. IEEE (2009)
6. Zou, Y., Zhu, J., Zheng, B., Yao, Y.: An adaptive cooperation diversity scheme with best-relay selection in cognitive radio networks. IEEE Transactions on Signal Processing **58**(10), 5438–5445 (2010)
7. Helal, N.M., Seddik, K.G., El-Keyi, A., El Batt, T.: A feedback-based access scheme for cognitive-relaying networks. In: Wireless Communications and Networking Conference (WCNC), pp. 1287–1292. IEEE (2012)
8. Anwar, A.H., Seddik, K.G., ElBatt, T., Zahran, A.H.: Effective capacity of delay constrained cognitive radio links exploiting primary feedback. In: 11th International Symposium on Modeling Optimization in Mobile, Ad Hoc Wireless Networks (WiOpt), pp. 412–419. IEEE, Tsukuba (2013)
9. Musavian, L., Aissa, S., Lambotharan, S.: Effective capacity for interference and delay constrained cognitive radio relay channels. IEEE Transactions on Wireless Communications **9**(5), 1698–1707 (2010)
10. Akyildiz, I.F., Lee, W.Y., Vuran, M.C., Mohanty, S.: NeXt generation/dynamic spectrum access/cognitive radio wireless networks: a survey. Computer Networks **50**(13), 2127–2159 (2006)
11. Akin, S., Gursoy, M.C.: Effective capacity analysis of cognitive radio channels for quality of service provisioning. IEEE Transactions on Wireless Communications **9**(11), 3354–3364 (2010)
12. Chang, C.S.: Performance guarantees in communication networks. Springer Science & Business Media (2000)

Cooperative Spectrum Sharing Using Transmit Antenna Selection for Cognitive Radio Systems

Neha Jain[✉], Shubha Sharma, Ankush Vashistha, Vivek Ashok Bohara, and Naveen Gupta

WiroComm Research Lab, Indraprastha Institute of Information Technology
(IIIT-Delhi), New Delhi, India
{neha13158,shubha12104,ankushv,vivek.b,naveeng}@iiitd.ac.in

Abstract. In this paper, a spectrum sharing scheme that utilizes the two-phase cooperative decode-and-forward relaying protocol is proposed. The cooperation between primary (i.e. licensed) and secondary (i.e. unlicensed) system helps in achieving the desired target rate for the primary system and spectrum access for cognitive (i.e. secondary) system. In the proposed scheme, secondary transmitter which is equipped with multiple antennas uses transmit antenna selection to improve the primary's performance by reducing the interference level of secondary signal at primary receiver, while keeping the performance of secondary system unaffected. Closed form expressions for outage probability have been derived for both systems by varying transmit power level at secondary transmitter. The theoretical results have been compared with simulation results to validate the analysis done in this paper.

Keywords: Spectrum sharing · Transmit antenna selection

1 Introduction

Radio frequency (RF) spectrum, considered as the most limited resource for wireless communications has been congested due to its diversified use. However large portion of the spectrum remains unutilized because of the variation in spectrum utilization with respect to time and location. This unutilized spectrum is termed as "spectrum holes" or white spaces. Cognitive radio has emerged as a solution to address this spectrum scarcity problem [1]-[2]. Moreover, the generation of mobile system is continuously upgrading in every 10 years because of the growing demand of people in communicating as well as in accessing the information. Fifth generation wireless systems, commonly abbreviated as "5G", is the next step in this continuous innovation and evolution of wireless technology. 5G has been envisioned to support 1,000-fold gains in capacity. Cognitive radio can be seen as a promising step towards 5G technology. It can sense and identify the unused frequency bands and use them for its own (unlicensed) transmission. Beside the interweave technique [3]-[4], in which the underutilized spectrum is accessed opportunistically by the cognitive user, cooperative spectrum sharing

M. Weichold et al. (Eds.): CROWNCOM 2015, LNICST 156, pp. 780–789, 2015.
DOI: 10.1007/978-3-319-24540-9_65

[5] has been recently proposed as an alternative framework to realize a cognitive radio network. In cooperative spectrum sharing (CSS), secondary transmitter relays the data of primary system in order to get spectrum access over licensed band of primary user. In this architecture, primary and secondary system consists of transmitter receiver pair known as primary transmitter (PT) - primary receiver (PR) and secondary transmitter (ST) - secondary receiver (SR) respectively, are allowed to coexist in the same frequency band with the assurance that secondary system will improve the performance of primary system.

Substantial amount of literature has demonstrated the performance of conventional CSS protocol under decode and forward relaying [6]-[7]. In these schemes, whenever the instantaneous transmission rate of primary system drops below the target rate, it seeks cooperation from the neighbouring terminals which may help it in achieving the target rate. Secondary transmitter (ST) "disguises" itself as a relay and collaborates with primary system by forwarding its data to the destination. Primary system returns the favour by helping the secondary system with spectrum access. However, the performance of CSS protocols is limited by the interference tolerable at PR from ST. Moreover, most of these schemes have been confined to single antenna system. Recently, some work has also been proposed where multiple antenna CR system have been used to enhance the performance of both systems [8]-[9]. The authors in [8] proposed a scheme with multiple antennas at ST node which utilizes zero-forcing precoding technique in order to cancel the interference at PR caused due to presence of cognitive system. But the application of this precoding technique requires perfect transmit channel state information (CSI) at ST. Assuming that perfect transmit CSI is available at ST may not be practically feasible in the case of fading environment. Moreover, in [8], as ST is working as an amplify and forward relay, therefore while forwarding the data from PT to PR, it will amplify both the required signal as well as noise received from PT. In [9], authors have proposed a CSS scheme in which ST is equipped with two antennas. Both the antennas receive primary's data which is decoded at ST and then forward this data by selecting one of the two antennas randomly. This will improve the performance of primary system when compared to conventional CSS scheme because of increase in probability of successful decoding of primary's data. However it still suffers from the drawback on the amount of interference at PR due to presence of secondary system which is same as conventional CSS system.

In this paper, we have proposed a transmit antenna selection [10] based scheme with multiple antennas at ST node which can alleviate the drawbacks of [6]-[7], [9]. Moreover, unlike [8], proposed scheme doesn't require perfect CSI, it just requires partial CSI feedback to select the best among the set of antennas at ST (that maximizes the post processing SNR at PR). This reduces the transmitter complexity and lowers the feedback bandwidth while preserving the gains from diversity [11]-[12]. In the proposed scheme, once primary and secondary system enter into CSS, PT broadcasts its data in half of the overall time slot (represented as phase 1) which is received by all the present nodes i.e. PR, ST and SR. After receiving primary's data ST will try to decode it. In the remaining half of the time slot (phase 2), ST chooses the antenna having larger instantaneous gain between ST and PR for primary's data transmission and secondary's data is transmitted via other antenna which has comparatively lower gain as shown in Fig. 1. Finally, the data received in both the phases, is decoded using maximum rate combining (MRC) at PR. However, if ST fails to decode primary's data, it will remain silent in phase 2. This technique is advantageous in two ways; first, we can improve the performance of primary system by reducing the interference caused due to secondary's data at PR. Second, the performance of secondary system is unaffected because of interference cancellation at SR. Moreover, when ST works as a pure relay and transfer only primary's data, in such a scenario, ST can also be seen as a selection combiner [13]. Consequently, PR will receive its signal from a selection combiner and a direct link (PT-PR). The performance of primary as well as secondary system has been analysed by deriving the closed form expressions for outage probability. The results demonstrate the considerable improvement in the performance of primary system along with spectrum access for secondary system.

Throughout this paper, a complex Gaussian random variable (RV) Z with mean μ and variance σ^2 is denoted as $Z \sim \mathcal{CN}(\mu, \sigma^2)$. An exponentially distributed RV X with mean $\frac{1}{\lambda}$ is denoted as $X \sim \varepsilon(\lambda)$. $\sim$ is used to indicate "has the distribution of" and i.i.d is used to represent independent and identically distributed. The transpose of a matrix A is denoted by A^T. $f_X(x)$ symbolizes the probability density function (PDF) of RV X and $f_{X,Y}(x,y)$ symbolizes the joint PDF of RVs X and Y. Moreover, $F_X(x)$ symbolizes the cumulative distribution function (CDF) of RV X and $F_{X,Y}(x,y)$ symbolizes the joint CDF of RVs X and Y. The rest of the paper is organized as follows. Section 2 describes the proposed system model and obtains the analytical results for outage probability of primary and secondary systems. Section 3 discusses the simulation results and finally section 4 concludes the paper.

2   Model Description with Performance Analysis

2.1   System Model

The primary and secondary system consists of transmitter receiver pair known as PT-PR and ST-SR respectively. We have considered multiple antennas at ST, named as ST1 and ST2.[1] Channels between the links are modeled as Rayleigh flat fading channels and the channel coefficients between PT-PR, PT-SR, PT-ST(1), PT-ST(2), ST(1)-PR, ST(2)-PR, ST(1)-SR, ST(2)-SR is $h_1, h_2, h_3, h_4, h_5, h_6, h_7, h_8$ respectively. Here, $h_i \sim \mathcal{CN}(0, d_i^{-v})$ where, v is the path loss component and d_i is the normalized distance between the corresponding link. The normalization is done with respect to the distance between PT-PR link therefore, $d_1 = 1$. The instantaneous gain of each channel is given as $\gamma_i = |h_i|^2$ where, $\gamma_i \sim \varepsilon(d_i^v)$.

---

[1] For ease of analysis, we have assumed that ST is equipped with two antennas, however the results obtained can be easily extrapolated to scenarios where ST is equipped with multiple (>2) antennas.

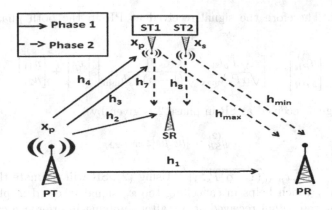

Fig. 1. Transmission Phases

2.2   System Equations

In transmission phase 1, PT broadcasts primary signal i.e. x_p which is received by all the nodes. Therefore, signal received at PR is given as

$$y_{PR}^{(1)} = \sqrt{P_p}x_p h_1 + n_{11} \tag{1}$$

where, P_p is the power assigned to PT and $n_{ij} \sim \mathcal{CN}(0,\sigma^2)$ is the AWGN in i^{th} phase of transmission at j^{th} receiver and j=1,2,3 corresponds to PR, SR, ST respectively. The signal received at SR in phase 1 is given by

$$y_{SR}^{(1)} = \sqrt{P_p}x_p h_2 + n_{12}. \tag{2}$$

Since ST is equipped with two antennas, hence the signal received at ST can be given as

$$\begin{bmatrix} y_{ST}^{(1)} \\ y_{ST}^{(2)} \end{bmatrix} = \sqrt{P_p} \begin{bmatrix} h_3 \\ h_4 \end{bmatrix} x_p + n_{13}. \tag{3}$$

In transmission phase 2, ST decodes the primary signal (i.e. x_p) and transmits it along with its own signal (i.e. x_s). As ST has two antennas, in order to reduce interference at PR, it will transmit x_p and x_s from the antenna which provides maximum and minimum instantaneous gain between ST-PR respectively. Therefore, signal received at PR in phase 2 is given by

$$y_{PR}^{(2)} = \begin{bmatrix} h_{\max} & h_{\min} \end{bmatrix} z + n_{21} \tag{4}$$

where, $h_{\max} = \begin{cases} h_5 \text{ if } \gamma_5 > \gamma_6 \\ h_6 \text{ if } \gamma_5 \leq \gamma_6 \end{cases}$, $h_{\min} = \begin{cases} h_6 \text{ if } \gamma_5 > \gamma_6 \\ h_5 \text{ if } \gamma_5 \leq \gamma_6 \end{cases}$,

$z = \begin{bmatrix} \sqrt{\alpha P_s}x_p & \sqrt{(1-\alpha)P_s}x_s \end{bmatrix}^T$, α and $(1-\alpha)$ is the fraction of power provided by the secondary transmitter to transmit primary signal and secondary signal respectively. Therefore the signal received at PR in the both phases can be written as

$$\begin{bmatrix} y_{PR}^{(1)} \\ y_{PR}^{(2)} \end{bmatrix} = \begin{bmatrix} \sqrt{P_p}h_1 & 0 \\ \sqrt{\alpha P_s}h_{\max} & \sqrt{(1-\alpha)P_s}h_{\min} \end{bmatrix} \begin{bmatrix} x_p \\ x_s \end{bmatrix} + \begin{bmatrix} n_{11} \\ n_{21} \end{bmatrix}. \tag{5}$$

Now, the signal received at SR in phase 2 is given by

$$y_{SR}^{(2)} = [h_7 \ h_8]z + n_{22} \tag{6}$$

where, $z = \left[\sqrt{\alpha P_s}x_p \ \sqrt{(1-\alpha)P_s}x_s\right]^T$. Using (2), SR will estimate the primary signal (i.e $\hat{x}_p$) which helps in cancelling the x_p signal received in phase 2 and hence the overall signal received at SR after applying interference cancellation is given as

$$y_{SR} = \sqrt{(1-\alpha)P_s}h_8 x_s + n_{22}. \tag{7}$$

2.3   Outage Probability of Primary System

Outage at primary system occurs when system fails to achieve the target transmission rate (R_{pt}). There are two such cases: In first case, outage occurs if ST is unable to decode the primary signal in phase 1 and along with this, the link between PT-PR also fails to achieve R_{pt}, or in second case, outage occur if ST successfully decodes x_p but still overall rate achieved at PR is less than R_{pt}. Therefore, the expression for outage probability at primary system is given as

$$P_{out}^{PR} = P[R_{11} < R_{pt}]P[R_{13} < R_{pt}] + P[R_{13} > R_{pt}]P[R_{MRC} < R_{pt}] \tag{8}$$

where, R_{11} is the transmission rate achieved in phase 1 between PT-PR link, R_{13} is the transmission rate achieved between PT-ST in phase 1 and R_{MRC} is the rate achieved at PR after applying MRC of both transmission phases. Solving for (8),

$$R_{11} = \frac{1}{2}\log_2\left(1 + \frac{P_p \gamma_1}{\sigma^2}\right). \tag{9}$$

The factor $\frac{1}{2}$ is due to the fact that the whole transmission is divided into two phases.

$$P[R_{11} < R_{pt}] = P\left[\gamma_1 < \frac{\sigma^2 \rho}{P_p}\right] = 1 - e^{-\frac{\sigma^2 \rho}{P_p}}. \tag{10}$$

as, $\rho = 2^{2R_{pt}} - 1, \gamma_1 \sim \varepsilon(1)$.

$$R_{13} = \frac{1}{2}\log_2\left(1 + \frac{P_p \gamma_3}{\sigma^2} + \frac{P_p \gamma_4}{\sigma^2}\right) \tag{11}$$

and

$$P[R_{13} < R_{pt}] = P\left[\gamma_3 + \gamma_4 < \frac{\sigma^2 \rho}{P_p}\right]. \tag{12}$$

We assume that the distances between the antennas at ST is negligible as compare to distance between the nodes, hence $d_3 = d_4, d_5 = d_6, d_7 = d_8$. Therefore, γ_3 and γ_4 are i.i.d and hence $f_{\gamma_3,\gamma_4}(\gamma_3, \gamma_4) = f_{\gamma_3}(\gamma_3)f_{\gamma_4}(\gamma_4)$ where,

$$f_{\gamma_3} = \begin{cases} d_3^v e^{-d_3^v \gamma_3} & \gamma_3 > 0 \\ 0 & \text{otherwise.} \end{cases}$$

Therefore,

$$P[R_{13} < R_{pt}] = \int_0^{\frac{\sigma^2 \rho}{P_p}} \int_0^{\frac{\sigma^2 \rho}{P_p} - \gamma_4} f_{\gamma_3,\gamma_4}(\gamma_3, \gamma_4) d\gamma_3 d\gamma_4$$

$$= 1 - \left[\left(1 + \frac{\sigma^2 \rho}{P_p} d_3^v \right) e^{-\frac{\sigma^2 \rho}{P_p} d_3^v} \right]. \tag{13}$$

Moreover,

$$P[R_{13} > R_{pt}] = \left[\left(1 + \frac{\sigma^2 \rho}{P_p} d_3^v \right) e^{-\frac{\sigma^2 \rho}{P_p} d_3^v} \right]. \tag{14}$$

The rate at PR after MRC is obtained as

$$R_{\text{MRC}} = \frac{1}{2} \log_2(1 + \text{SNR}_{\text{MRC}}) \tag{15}$$

where, $\text{SNR}_{\text{MRC}} = \frac{P_p \gamma_1}{\sigma^2} + \frac{\alpha P_s \gamma_{\max}}{(1-\alpha)P_s \gamma_{\min} + \sigma^2}$, $\gamma_{\max} = \max(\gamma_5, \gamma_6), \gamma_{\min} = \min(\gamma_5, \gamma_6)$. Therefore,

$$P[R_{MRC} < R_{pt}] = P \left[\frac{P_p \gamma_1}{\sigma^2} + \frac{\alpha P_s \gamma_{\max}}{(1-\alpha)P_s \gamma_{\min} + \sigma^2} < \rho \right]. \tag{16}$$

After solving, we get

$$P[R_{MRC} < R_{pt}] = 1 - e^{-\frac{\sigma^2}{P_p}(\rho - \frac{\alpha}{1-\alpha})} + \frac{2}{n} e^{-\frac{d_5^v \sigma^2 \rho}{\alpha P_s} + \frac{mp}{n}}$$

$$\left(Ei \left[\frac{\rho \sigma^2}{P_p} \left(\rho - \frac{\alpha}{1-\alpha} \right) - \frac{mp}{n} \right] - Ei \left[-\frac{mp}{n} \right] \right) \tag{17}$$

where, $\alpha \leq \frac{\rho}{\rho+1}$, $m = \left(\frac{1-\alpha}{\alpha} \right) \rho + 1$, $n = \left(\frac{1-\alpha}{\alpha} \right) \frac{P_p}{\sigma^2}$, $p = \frac{d_5^v P_p}{\alpha P_s} - 1$ and Ei represents the exponential integral defined as $Ei(x) = -\int_{-x}^{\infty} \frac{e^{-t}}{t} dt$. For detailed derivation of (17), please refer to Appendix A. After substituting (10), (13), (14) and (17) in (8), we get

$$P_{out}^{PR} = (1 - e^{-\frac{\sigma^2 \rho}{P_p}}) \left(1 - \left[\left(1 + \frac{\sigma^2 \rho}{P_p} d_3^v \right) e^{-\frac{\sigma^2 \rho}{P_p} d_3^v} \right] \right) + \left(\left[\left(1 + \frac{\sigma^2 \rho}{P_p} d_3^v \right) e^{-\frac{\sigma^2 \rho}{P_p} d_3^v} \right] \right)$$

$$\left(1 - e^{-\frac{\sigma^2}{P_p}(\rho - \frac{\alpha}{1-\alpha})} + \frac{2}{n} e^{-\frac{d_5^v \sigma^2 \rho}{\alpha P_s} + \frac{mp}{n}} \left(Ei \left[\frac{\rho \sigma^2}{P_p} \left(\rho - \frac{\alpha}{1-\alpha} \right) - \frac{mp}{n} \right] - Ei \left[-\frac{mp}{n} \right] \right) \right) \tag{18}$$

Special case when $\alpha=1$ ST acts as a selection combiner. In such senerio, $\text{SNR}_{\text{MRC}} = \frac{P_p\gamma_1}{\sigma^2} + \frac{P_s\gamma_{\max}}{\sigma^2}$. Therefore (16) reduces to,

$$P[R_{MRC} < R_{pt}] = P\left[\frac{P_p\gamma_1}{\sigma^2} + \frac{P_s\gamma_{\max}}{\sigma^2} < \rho\right] \tag{19}$$

After solving,

$$P[R_{MRC} < R_{pt}] = e^{-g}\left(\frac{e^{((2\mu g)-2\psi)}}{2\mu - 1} - \frac{2e^{((\mu g)-\psi)}}{\mu - 1} - 1\right)$$
$$- \left(\frac{e^{2\psi}}{2\mu - 1} - \frac{2e^{-2\psi}}{\mu - 1} - 1\right) \tag{20}$$

where, $g = \frac{\rho\sigma^2}{P_p}$, $\psi = \frac{d_5^v \rho\sigma^2}{P_s}$ and $\mu = \frac{d_5^v P_p}{P_s}$. For detailed derivation of (20), please refer Appendix B.

2.4 Outage Probability of Secondary System

Outage probability of a secondary system is the probability by which secondary receiver fails to decode secondary signal with the target rate i.e. R_{st}. If in phase 1, links between PT-ST and PT-SR fails in decoding x_p, interference cancellation at SR in phase 2 is not possible and hence outage will be declared for secondary system. The outage probability for secondary system can be given as [6]

$$P_{out}^{SR} = 1 - P[R12 > Rpt]P[R13 > Rpt]P[RSR2 > Rst] \tag{21}$$

where, R_{12} is the transmission rate achieved between PT-SR link in phase 1, R_{13} is the transmission rate achieved at ST in phase 1 (given in (14)) and R_2^{SR} is the rate achieved at SR in phase 2. Solving for (21),

$$R_{12} = \frac{1}{2}\log_2\left(1 + \frac{P_p\gamma_2}{\sigma^2}\right). \tag{22}$$

Therefore,

$$P[R_{12} > R_{pt}] = P\left[\gamma_2 > \frac{\rho\sigma^2}{P_p}\right] = e^{-\frac{d_4^v \rho\sigma^2}{P_p}}. \tag{23}$$

Moreover,

$$R_2^{SR} = \frac{1}{2}\log_2\left(1 + \frac{P_s(1-\alpha)\gamma_7}{\sigma^2}\right). \tag{24}$$

Therefore,

$$P[R_2^{SR} > R_{st}] = P\left[\gamma_7 > \frac{\rho_s\sigma^2}{P_s(1-\alpha)}\right] = e^{-\frac{d_7^v \rho_s\sigma^2}{P_s(1-\alpha)}} \tag{25}$$

where, $\rho_s = 2^{2R_{st}} - 1$.

After substituting (23), (14) and (25) in (21), we get

$$P_{out}^{SR} = 1 - \left[\left(\left(1 + \frac{\sigma^2\rho}{P_p}d_3^v\right)e^{-\frac{\sigma^2\rho}{P_p}d_3^v}\right)e^{-\frac{d_2^v \rho\sigma^2}{P_p}}e^{-\frac{d_7^v \rho_s\sigma^2}{P_s(1-\alpha)}}\right] \tag{26}$$

3   Simulation Results and Discussion

In this section, we have discussed the analytical and simulation results for outage probability. We have compared our results with the scheme in [9], where they randomly pick an antenna at ST for transmission. Fig 2 shows the simulation model of the proposed scheme, in which for the ease of analysis all nodes are assumed to be collinear. The value of d (distance between PT-ST) is considered to be 0.5 and 0.8. The target rate chosen for primary and secondary system is 1 i.e. $R_{pt} = R_{st} = 1$, and we have considered $\frac{P_p}{\sigma^2} = 5$dB.

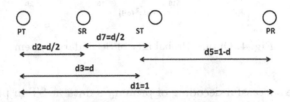

Fig. 2. System Model

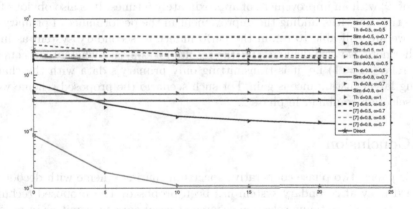

Fig. 3. Outage Probability of Primary System

Fig. 3 and Fig. 4 shows the outage probability of primary and secondary system respectively with respect to $\frac{P_s}{\sigma^2}$. From the plots it is quite obvious, that the outage probability of both primary as well as secondary system is continuously decreasing with the increase in power at secondary transmitter. However this decrement gradually reduces after 10dB because the outage probability also

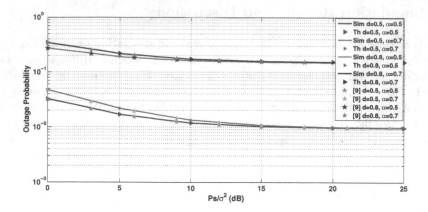

Fig. 4. Outage Probability of Secondary System

depends on the successful decoding of primary's data at ST in phase 1 (from (8 and 21)). The results are shown for two different values of α i.e. 0.5 and 0.7. By transmitting x_s from channel having less instantaneous gain, interference level at PR get reduced which results in considerable improvement in the performance of primary system (approximately 10 times at d = 0.5 and $\alpha = 0.7$ for $\frac{P_s}{\sigma^2} = 5dB$) compared to [9]. Even when half of the power of ST ($\alpha = 0.5$) is allocated to secondary signal, the performance of proposed scheme is still far better than that of [9] with an improvement of approximately 5 times. It is also obvious from Fig. 4 that notwithstanding the improvement in the performance of primary system, we are still able to retain the performance of secondary system as in [9]. Furthermore, we also demonstrate the results for the case wherein ST acts as a pure relay ($\alpha = 1$) i.e. it is transmitting only primary's data with the channel having larger instantaneous gain. For such scenario the proposed scheme works as a selection combiner in phase 2.

4   Conclusion

In this paper, two phase cooperative spectrum sharing scheme with decode and forward relay at secondary system has been proposed. The proposed technique utilizes transmit antenna selection scheme at secondary transmitter in order to reduce interference at primary receiver due to presence of secondary signal. The perfect agreement between the simulated results and the analytically obtained closed form expression for outage probability validated theoretical analysis presented in the paper.

Acknowledgments. Authors would like to thank Dr. Sanjit Kaul for helping us in deriving closed form expression for outage probability of primary system.

References

1. Haykin, S.: Cognitive radio: brain-empowered wireless communications. IEEE Journal on Selected Areas in Communications **23**(2), 201–220 (2005)
2. Akyildiz, I.F., Lee, W.-Y., Vuran, M.C., Mohanty, S.: Next generation/dynamic spectrum access/cognitive radio wireless networks: A survey. Computer Network Journal (ELSEVIER) **50**, 2127–2159 (2006)
3. Zou, Y., Yao, Y.-D., Zheng, B.: Cognitive transmissions with multiple relays in cognitive radio networks. IEEE Transactions on Wireless Communications **10**(2), 648–659 (2011)
4. Tachwali, Y., Basma, F., Refai, H.: Cognitive radio architecture for rapidly deployable heterogeneous wireless networks. IEEE Transactions on Consumer Electronics **56**(3), 1426–1432 (2010)
5. Nosratinia, A., Hunter, T., Hedayat, A.: Cooperative communication in wireless networks. IEEE Communications Magazine **42**(10), 74–80 (2004)
6. Han, Y., Pandharipande, A., Ting, S.H.: Cooperative decode-and-forward relaying for secondary spectrum access. IEEE Transactions on Wireless Communications **8**(10), 4945–4950 (2009)
7. Bohara, V.A., Ting, S.H.: Measurement results for cognitive spectrum sharing based on cooperative relaying. IEEE Transactions on Wireless Communications **10**(7), 2052–2057 (2011)
8. Manna, R., Louie, R.H., Li, Y., Vucetic, B.: Cooperative spectrum sharing in cognitive radio networks with multiple antennas. IEEE Transactions on Signal Processing **59**(11), 5509–5522 (2011)
9. Vashistha, A., Sharma, S., Bohara, V.: Outage & diversity analysis of cooperative spectrum sharing protocol with decode-and-forward relaying. In: 2015 7th International Conference on Communication Systems and Networks (COMSNETS), pp. 1–7, January 2015
10. Chen, Z., Yuan, J., Vucetic, B.: Analysis of transmit antenna selection/maximal-ratio combining in rayleigh fading channels. IEEE Transactions on Vehicular Technology **54**(4), 1312–1321 (2005)
11. Prakash, S., McLoughlin, I.: Predictive transmit antenna selection with maximal ratio combining. In: Global Telecommunications Conference, GLOBECOM 2009, pp. 1–6. IEEE, November 2009
12. Molisch, A.,Win, M.,Winters, J.: Capacity of mimo systems with antenna selection. In: IEEE International Conference on Communications, ICC 2001, vol. 2, pp. 570–574 (2001)
13. Chen, Y., Tellambura, C.: Distribution functions of selection combiner output in equally correlated rayleigh, rician, and nakagami-m fading channels. IEEE Transactions on Communications **52**(11), 1948–1956 (2004)
14. Yates, R., Goodman, D.: Probability and stochastic processes: a friendly introduction for electrical and computer engineers, p. 519 (2005)

A Survey of Machine Learning Algorithms and Their Applications in Cognitive Radio

Mustafa Alshawaqfeh[1], Xu Wang[1], Ali Rıza Ekti[1]([✉]), Muhammad Zeeshan Shakir[2], Khalid Qaraqe[2], and Erchin Serpedin[1]

[1] Department of Electrical and Computer Engineering, Texas A&M University, College Station, TX, USA
{mustafa.shawaqfeh,xu.wang,arekti}@tamu.edu, serpedin@ece.tamu.edu
[2] Department of Electrical and Computer Engineering, Texas A&M University at Qatar, Doha, Qatar
{muhammad.shakir,khalid.qaraqe}@qatar.tamu.edu

Abstract. Cognitive radio (CR) technology is a promising candidate for next generation intelligent wireless networks. The cognitive engine plays the role of the brain for the CR and the learning engine is its core. In order to fully exploit the features of CRs, the learning engine should be improved. Therefore, in this study, we discuss several machine learning algorithms and their applications for CRs in terms of spectrum sensing, modulation classification and power allocation.

Keywords: Cognitive radio · Machine learning · Learning engine · Spectrum sensing · Modulation classification

1 Introduction

The evolution of wireless communications systems and many other devices is continuously subject to two major development trends: a) improvement of existing capabilities, and b) extension and insertion of new features into the existing structures. In what concerns the first trend, one can notice that insertion of new features arises from the fact wireless systems progress very fast in accordance with the market demands. Therefore, wireless systems always require new services and applications. One of the most striking examples for such situations is cell phones. Earlier cell phones were used only for voice transmissions along with limited text messaging applications however contemporary cell phones are capable of transmitting multimedia along with an operating system running on. In what concerns the second trend, a continuous improvement of existing capabilities is a necessity since incorporating new features adds new dimensions that help improve the existing capabilities.

The above mentioned considerations suggest that adaptation and optimization should always be employed as key enabling technologies for the continuous update of communication systems to dynamically changing conditions.

© Institute for Computer Sciences, Social Informatics and Telecommunications Engineering 2015
M. Weichold et al. (Eds.): CROWNCOM 2015, LNICST 156, pp. 790–801, 2015.
DOI: 10.1007/978-3-319-24540-9_66

In this regard, the purpose of this study is to provide a conceptual description of machine learning algorithms used in the design of wireless communication systems in the light of a recently emerging technology called cognitive radio (CR) [1–5]. The idea of CR was first presented by Joseph Mitola III. and Gerald Q. Maguire, Jr. in [3] "The point in which wireless personal digital assistants and the related networks are sufficiently and computationally intelligent about radio resources and related computer-to-computer communication to detect user communications needs as a function of use context, and to provide radio resources and wireless services most appropriate to those needs" [1]. There are many advantages offered by CRs in wireless communications. A CR is basically an intelligent wireless device which is aware of the environment and spectrum and is able to adapt/optimize itself easily to the characteristics of the communication channel to satisfy the user needs. The environment of a CR may include radio frequency (RF) spectrum, user behavior, transmission characteristics and parameters, multi-access interference, localization and data rates of users. The key strengths of machine learning algorithms are their adaptive nature with respect to the dynamic changes of the channel and communication system parameters. In addition, the ability to work without prior knowledge about the communication environment represents another important feature of CRs. These considerations recommend machine learning as a promising technology for CRs.

In this paper, applications of machine learning for *learning engine, spectrum sensing, modulation classification* and *power allocation* in CRs are studied along with currently available methods and approaches to better adapt and optimize the overall system performance. The rest of the paper is organized as follows. The learning engine is presented in Section 2. An overview of key machine learning techniques that can be implemented into the learning engine is presented in Section 3. A review of machine learning applications in spectrum sensing, modulation classification and power allocation for CRs are presented in Section 4, Section 5 and Section 6, respectively. Concluding remarks are provided in Section 7.

2 Learning Engine

The cognitive engine is the brain of a CR system and it enables the system to react intelligently to changes in the environment. Basically, as shown in Figure 1, the CR extends a software-defined radio by adding an independent cognitive engine, which consists of a learning engine and reasoning engine [6]. The learning engine lies in the core of the cognitive engine and it aims to build a model or an objective function based on the inputs that are to be used in taking the right decisions and making the correct predictions.

In the context of CRs, no simple relationship between the system inputs and the objective function is available due to the high complexity and degree of freedom of the software-defined radio (SDR). In this case, several channel statistics, such as transmit power, modulation scheme and sensing scheme, need to be adjusted simultaneously [7]. In such scenarios, adopting a policy-based

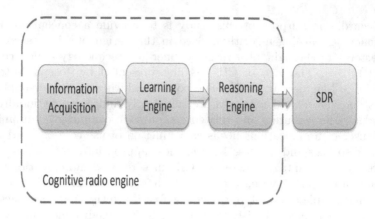

Fig. 1. The structure of cognitive radio engine.

decision making strategy is infeasible due to the large number of states that the cognitive radio networks (CRNs) and its radio frequency (RF) environment assume. In addition, even if the resources are available, considering all the possible states and actions is impossible given the dynamic and random nature of CRNs. Thus, the learning engine is crucial in the operation of the CR engine. A learning engine is adopted to estimate the channel statistics. The results are incorporated into a predictive calculus-based reasoning engine to make decisions and achieve certain objectives.

Several learning algorithms can be used to implement the learning engine. For the sake of brevity, Table 1 lists some of recent works involving the applications of machine learning algorithms in CR. The recent literature shows extensive use of different learning algorithms in CRs which will be discussed in Section 3. However, several factors influence the selection of the learning algorithm to implement the learning engine. For example, one important factor is the availability of prior knowledge about the environment. Supervised learning methods are applicable only if prior information about the environment is known to train the agent. On the other hand, unsupervised learning methods are appealing for scenarios with lack of prior information. The computational complexity of the algorithm is the main limiting factor especially for CRs with limited resources. In general, CRNs and their RF environment exhibit the following characteristics [7]: (*i*) incomplete observation information about the state variable, (*ii*) incorporation of CRs into CRNs and (*iii*) unknown RF environment. Consequently, the learning engine must be designed by taking into account the above characteristics such that the learning method efficiently and optimally adapt to the changes and the incompleteness of the observed information and RF environment.

Table 1. Classification of Papers Exploiting Machine Learning Algorithms

	Supervised Learning		Unsupervised Learning	Reinforcement Learning
	SVM	KNN		
Spectrum Sensing	[8,9]	[8,9]	[8,10]	[11-13]
Modulation Classification	[14,15]	[16]		
Power Allocation				[17-19]

3 Machine Learning

In literature, machine learning techniques can be categorized into three different types, namely, supervised learning, unsupervised learning and reinforcement learning (RL).

3.1 Supervised Learning

Supervised learning is a machine learning approach that infers an objective function from a labeled training data. Thus, this method requires prior information about the environment. The training data consists of input-output pairs. An inferred function is derived based on the samples to map the future input. For instance, the training samples (x_i, y_i) are given and it is assumed that (x_i, y_i) are drawn from some distribution $P(\mathbf{x})$. Classification is the main function for supervised learning and its goal is to find a classifier function f such that it fits and characterizes the training examples. The classifier is used to map and classify the newcoming data. One well-known example of supervised learning methods is referred to as the support vector machine (SVM) and it was first developed in [20]. The original SVM approach builds a linear classifier that maps the input vectors to a high-dimensional space. A nonlinear SVM classification method was proposed by Boser et al [21] using the kernel trick. SVM is exploited in a wide range of machine learning applications due to its accurate predictions, fast evaluation of the targeted function and the robustness against noise and errors. For more information about SVM, the reader is referred to [22,23].

3.2 Unsupervised Learning

In contrast to the supervised learning, the unsupervised learning applies to an environment in which the prior knowledge is unknown. Specifically, the unsupervised learning extracts hidden features from the unlabeled data. Since the samples from unsupervised learning are unlabeled, unsupervised learning receives neither targeted outputs nor environmental rewards. This fact distinguishes the unsupervised learning from the supervised learning and the reinforcement learning. The main functions for unsupervised learning are clustering, dimensionality reduction and blind signal separation [24,25]. In principle, a clustering algorithm aims to group objects into clusters such that the elements in the same cluster are similar to each other and different from the elements placed in any other clusters. There are several clustering algorithms such as K-means or centroid-based clustering [26,27] and mixture models.

3.3 Reinforcement Learning

Reinforcement learning is an online learning method which lies in the middle between supervised and unsupervised learning. The general idea behind the reinforcement learning is to maximize a specific reward function. According to [28], the reinforcement learning consists of three main components: a *policy*, a *reward function* and *value function*. Let S be the set of all possible states of the environment, and A be the set of all possible actions and n denote the time index. A policy $\pi : S \times A \to S$ is the rule that defines the selection of next state s_{n+1} based on the current state-action pair (s_n, a_n). The policy can be deterministic or stochastic. In a deterministic policy, the agent selects the actions in a deterministic fashion based on the current state. The reward function $r_n : A \times S \to \Re$ is a scalar function that maps each state-action pair (s_n, a_n) into a single real number, *reward*, that indicates the reward obtained by selecting the action a_n at state s_n to move into state s_{n+1}. According to the knowledge of the reward function, reinforcement learning is classified into a model-based learning if the reward is known and a model-free learning otherwise. Generally, the reward functions may be stochastic. The reward function determines the immediate or short term reward of an action. However, the agent is interested in the long-run total reward which is defined by the *value function* or *return*. Starting from state s_n, the *return* is the random variable R_n defined as:

$$R_n = \begin{cases} \sum_{k=0}^{\infty} \gamma^k r_{n+k+1} : \text{non-episodic model} \\ \sum_{k=0}^{N} r_{n+k+1} \quad : \text{episodic model}, \end{cases} \tag{1}$$

where $\gamma \in [0,1]$ is the *discount factor*. The goal of the reinforcement algorithm is to find a policy that maximizes R_n. In principle, the optimal policy can be found by exhaustive search of the policy space. This solution is computationally infeasible due to the large (or even infinite) number of policies to be checked. Hence, the core of reinforcement learning algorithms is to find an efficient method to calculate or approximate the function value.

One appealing method is to estimate the function value. Estimation of function values in more details is commonly carried out within a Markov Decision Process (MDP), which represents a general framework for reinforcement learning. MDP is a reinforcement learning environment in which states satisfy Markov property. Markov property means that deciding the next state s_{n+1} depends only on the current state s_n and action a_n. In other words, the current state and actions contain all the required information about future state. Mathematically, this condition can be expressed as follows:

$$Pr\{s_{n+1}, r_{n+1}|s_n, a_n, r_n, s_{n-1}, \ldots, s_0, a_0, r_0\}$$
$$= Pr\{s_{n+1} = s, r_{n+1} = r|s_n, a_n, r_n\}. \tag{2}$$

The Markovian assumption simplifies the analysis by allowing prediction of future rewards based only on the current state and action. A finite MDP means that state and action spaces are finite. A natural way to estimate the value function is to take the sample mean of the received rewards. Since the rewards depend on the selected action, the estimated value function depends on the selected policy. Define the *state-value function for π policy* (V^π) as the expected value of return given that agent is in the s_n state and follows the π policy. For MDP, $V^\pi(s_n)$ is defined as:

$$V^\pi(s_n) = E_\pi[R_n|s_n, \pi]. \tag{3}$$

Similarly, the *action-value function for π policy*, $Q^\pi(s_n, a_n)$, is defined as the expected return starting from state s_n and taking the action a_n and following the policy π. In MDP, $Q^\pi(s_n, a_n)$ can be defined as:

$$Q^\pi(s_n, a_n) = E_\pi[R_n|s_n, a_n, \pi]. \tag{4}$$

It is shown in [28] that the optimal action-value function $Q^*(s_n, a_n)$ satisfies:

$$Q^*(s_n, a_n) = \max_\pi Q^\pi(s_n, a_n)\cdot$$

$$\sum_{s_{n+1}\in S} \left(Pr[s_{n+1}|s_n, s_n] \left[r_n + \gamma \max_{a_{n+1}\in A} Q^*(s_{n+1}, a_{n+1}) \right] \right). \tag{5}$$

One way to maximize the action-value functions is the Q-learning algorithm [29]. Q-learning follows a fixed state transition and does not require prior information about the environment. The update for the one-step version is given by:

$$Q_{n+1}(s_n, a_n) = Q_n(s_n, a_n) + \alpha[r_{n+1} +$$

$$\gamma \max_{a_{n+1}\in A} Q_n(s_{n+1}, a_{n+1}) - Q_n(s_n, a_n)]. \tag{6}$$

The reinforcement learning is subject to a trade-off between exploration and exploitation. This trade-off manifests through the fact that at each stage, the agent has to decide whether to exploit the current highest reward action or to explore new actions for higher rewards. Two action selection methods for controlling the trade off between exploration and exploitation are the ϵ-*greedy* and *softmax* action [28,30]. In ϵ-greedy, the next action is selected either at random with uniform probability ϵ or by selecting the optimal action $a^* = \max_a Q(a, s)$ with probability $1 - \epsilon$. In the softmax method, the action a is selected with probability

$$\frac{exp\{Q(s_n, a_n)/\tau\}}{\sum_{a_{n+1}\in A} exp\{Q(s_n, a_n)/\tau\}}, \tag{7}$$

where τ is a positive weight factor for each action and is referred to as the temperature factor.

Reinforcement learning algorithms differ by how they efficiently compute the value function. Reinforcement learning algorithms can be also divided into *single agent reinforcement learning* (SARL) and *multiple agent reinforcement learning* (MARL). In SARL, the learning process is local at each agent in the sense that rewards for each agent does not depend on the other agents. In MARL, the reward depends on both, the environment and all agent policies and actions. This dependence on other agents' policies complicates the learning process. The interested reader is referred to [28,31,32] for detailed information about the reinforcement learning.

4   Machine Learning for Spectrum Sensing

The main challenge of CRNs is to opportunistically utilize the unused spectrum of the primary system. Also, the CR should be designed in a way to protect the primary users from any interference or quality of service (QoS) degradation. To achieve this goal, the CR must present the ability to detect the occupancy of RF transmission activities in the primary system.

Various methods have been proposed for spectrum sensing [33] such as *matched filter, energy detection* and *cyclostationary detection*. The matched filter [34] is known to be optimal for detecting deterministic unknown signals in additive white Gaussian noise (AWGN). However, the matched filter approach is a coherent method, and impractical for scenarios where the CR compete for large number of bands. Implementation of a matched filter for such scenarios requires to equip the CR device with a large number of synchronization circuits to match the different bands. However, such an approach is not efficient. The basic idea of energy detector [35, 36] is to measure the energy of the received signal, and then to compare it to a threshold to decide the occupancy of the sensed primary band. The main advantages of an energy detector are its simplicity, low cost and the ability to work without any prior knowledge about the waveform of the primary system. However, an energy detector is very sensitive to channel impairments since it is unable to distinguish between the primary signal and noise or any type of interference. Cyclostationary detection is based on the fact that many digital and analog modulated signals have special statistical features because of the inherent periodicity of these signals statistics [37]. In contrast, the noise does not present in general such features. One way to exploit the cyclostationary features is to use the spectral-correlation density (SCD) function. A cyclostationary statistics based detection approach is more immune to stationary noise and interferences. Moreover, cyclostationary provides inherent signal identification since different signals differ in their SCD function. However, these benefits come at the cost of more complexity.

Assessing the RF-spectrum is a high dimensional complex problem due to the large number of parameters involved. Using dynamic programming methods is computationally infeasible especially if the CR devices present power limitations. Machine learning provides an asymptotically close-to-optimal and computationally efficient alternative [30]. Therefore, many papers propose machine learning-based techniques for spectrum sensing.

Spectrum sensing is a typical classification (or clustering) problem in the sense that it is required to identify whether the sensed band belongs to the available class (or cluster). A misdetection occurs if the selected channel is considered to be idle, while it is in reality used by the primary system. Hence, the primary system will be subject to an interference and a collision may occur. On the other hand, a false alarm occurs if the channel is available for the CR but the classifier decides that it is used. Consequently, a degradation of spectrum utilization occurs.

The authors in [8] implemented cooperative spectrum sensing (CSS) using several machine learning techniques. These techniques are the K-means clustering and Gaussian mixture model (GMM) from the unsupervised learning category and the support vector machine (SVM) and the weighted K-nearest-neighbor (KNN) from the supervised learning category. The CSS considered here is a centralized based cooperative sensing. All the energy levels estimated at the CR devices are collected at a CR device (e.g., the central device). This vector of energy levels acts as a feature input for the classification and clustering algorithm to decide whether the channel is available for the CR or not. The channel is idle if it is not utilized by any primary user. Similarly, the CSS scenario is studied in [9] using SVM and KNN.

The authors in [38] proposed a centralized CSS method in which each CR reports its measurements to a central node (another CR device). Then, the linear fusion rule is used to decide the availability of the channel. To enhance the sensing performance, the topology of the CRN is taken into account because measurements carried out by CRs closer to the primary users are more reliable than far away transceivers. The impact of the location information is reflected into the values of the linear coefficients which are determined by the Fisher linear discriminant analysis.

In many cases, the spectrum of interest is very wide and/or non-contiguous. Hence, a single CR device may not sense the whole the spectrum at once. An alternative solution is to assign a subset of k CRs to sense each subband [39]. One issue with the fixed number assignment is that monitoring some subbands with k CRs is more than what is needed to achieve the sensing requirements. And hence, more power consumption is required for the CRs. In [11], a reinforcement learning method with ϵ-greedy action selection is employed to optimize the multiband spectrum sensing and reduce the energy consumption in the CRN. This is achieved by exploiting the occupancy statistic of each subband and then assigning the minimum number of CRs that achieves the required misdetection probability.

5 Machine Learning for Modulation Classification

In general, modulation classification algorithms assume two steps. The first step performs the feature extraction. Examples of features are spectral correlation and cumulants. The second step carries out the classification task (via Naive Bayes, SVM) or clustering task (via KNN, mixture models).

The authors of [16] proposed a two stage classification algorithm using Genetic Programming (GP) and K-Nearest Neighbor (KNN) approach. The proposed algorithm can identify BPSK, QPSK, 16QAM and 64QAM modulation schemes, and exploit the forth and sixth order cumulants of the received signals as features. The first stage divides the signal into three classes: BPSK, QPSK and QAM (both 16 and 64). To differentiate between 16QAM and 64QAM, the third class output is fed into the second stage classifier that distinguishes between 16QAM and 64 QAM.

In [14], two modulation classification approaches are presented. Both of them exploit the SVM classifier. However, they differ in the selection of the feature vector and modulation schemes. The first approach aims to distinguish among 16 QAM, 32 QAM and 64 QAM and uses the demodulation error (i.e, the distance between the received symbol and its nearest neighbor in each constellation) as a feature vector. The second approach aims to distinguish among AM, BPSK, QPSK and BFSK and uses the cyclic spectral correlation as a feature.

The previous works classify only digital modulated signals. In [15], a SVM classification method is proposed to classify two analog modulated signals (AM and FM) in addition to five digitally modulated ones (BPSK, QPSK, GMSK, 16-QAM and 64QAM). The authors use a combination of spectral and higher order cumulants as features. Then, these features are fed into a SVM classifier to identify the modulation scheme.

6 Machine Learning for Power Allocation

As mentioned in Section 3, Q-learning is a simple and efficient way to implement reinforcement learning. The operation of Q-learning requires the definition of a reward function. In power allocation problem, defining the reward function can be easily done in terms of the transmission powers and channel gains. The authors in [17] use centralized Q-learning to address the channel and power allocation problem in CRNs. They consider a scenario where all the transmissions of the CRs are controlled by a cognitive base station. Therefore, the cognitive base station is the learning agent and provides channel and power allocation services to the CRs. In this work, the number of transmission activities of the CRs is modeled as a Poisson process. The state is defined as

$$\mathbf{s}_n = [\text{incoming user index, user(s) on transmission,}$$
$$\text{received power on each channel}]^T$$

and the reward function is defined by:

$$r_n = \sum_{i=1}^{N} \log_2 \left(1 + \frac{P_i f(i) h_i(f(i))}{N_0 + \sum_{j \neq i} P_j f(j) h_j(f(j)) \psi(i,j)} \right)$$

where $f(i)$ and P_i are the channel and power level used by the i'th user, respectively. N_0 denotes the noise power and N stands for the number of users. Function $\psi(i,j)$ is determined by:

$$\phi(i,j) = \begin{cases} 1 & , f(i) = f(j) \\ 0 & , else. \end{cases}$$

A decentralized Q-learning algorithm for power allocation is considered in [18]. The reward criterion is defined by:

$$r_n = \sum_{i=1}^{N} (SINR_i^s - SINR_{Th}^s)^2,$$

where $SINR_i^p$ is the primary network $SINR$ at the I_i'th cell, $SINR_i^s$ denotes the secondary network $SINR$ at the I_i'th cell and N stands for the number of cells.

In [19], a decentralized MARL is considered to control the transmit power and spectrum used by CRs in order to reduce the interference at the primary users. In order to overcome the increased computational complexity of the function value in reinforcement learning for large CRNs, the authors apply an approximation to the value function using a Kanerva-based approximation function. In this paper, the environment state at time index n is defined as $\mathbf{s}_n = [\mathbf{sp}_n, \mathbf{pw}_n]^T$ where $\mathbf{sp}_n$ denotes the vector of spectra and $\mathbf{pw}_n$ stands for a vector of power values across all agents.

7   Conclusion

There is a growing interest in machine learning techniques in assessing the features of CRNs. Therefore, in this study, we investigated the usefulness of machine learning techniques for spectrum sensing, modulation classification and power allocation in CRNs.

Acknowledgments. This publication was made possible by NPRP grant 4-1293-2-513 from the Qatar National Research Fund (a member of Qatar Foundation). The statements made herein are solely the responsibility of the authors.

References

1. Mitola, J.: Cognitive Radio–An Integrated Agent Architecture for Software Defined Radio. Royal Institute of Technology (KTH) (2000)
2. Chen, K.-C., Prasad, R.: Cognitive Radio Networks. John Wiley & Sons, June 2009
3. Mitola, J., Maguire, G.: Cognitive Radio: Making software radios more personal. IEEE Personal Communs. **6**(4), 13–18 (1999)
4. Wang, J., Ghosh, M., Challapali, K.: Emerging cognitive radio applications: A survey. IEEE Communs. Magazine **49**(3), 74–81 (2011)
5. Wang, B., Liu, K.R.: Advances in cognitive radio networks: A survey. IEEE J. Selected Topics Signal Process **5**(1), 5–23 (2011)
6. Clancy, C., Hecker, J., Stuntebeck, E., O'Shea, T.: Applications of machine learning to cognitive radio networks. IEEE Wireless Communs. **14**(4), 47–52 (2007)
7. Bkassiny, M., Li, Y., Jayaweera, S.K.: A survey on machine-learning techniques in cognitive radios. IEEE Commun. Surveys & Tuts. **15**(3), 1136–1159 (2013)
8. Thilina, K.M., Choi, K.W., Saquib, N., Hossain, E.: Machine learning techniques for cooperative spectrum sensing in cognitive radio networks. IEEE Journal on Selected Areas in Communications **31**(11), 2209–2221 (2013)
9. Thilina, K.M., Choi, K.W., Saquib, N., Hossain, E.: Pattern classification techniques for cooperative spectrum sensing in cognitive radio networks: SVM and W-KNN approaches. In: 2012 IEEE Global Communications Conference (GLOBE-COM), pp. 1260–1265 (2012)

800   M. Alshawaqfeh et al.

10. Ding, G., Wu, Q., Song, F., Wang, J.: Decentralized sensor selection for cooperative spectrum sensing based on unsupervised learning. In: 2012 IEEE International Conference on Communications (ICC), pp. 1576–1580, June 2012
11. Oksanen, J., Lundén, J., Koivunen, V.: Reinforcement learning method for energy efficient cooperative multiband spectrum sensing. In: 2010 IEEE International Workshop on Machine Learning for Signal Processing (MLSP), pp. 59–64 (2010)
12. Lo, B.F., Akyildiz, I.F.: Reinforcement learning-based cooperative sensing in cognitive radio ad hoc networks. In: 2010 IEEE 21st International Symposium on Personal Indoor and Mobile Radio Communications (PIMRC), pp. 2244–2249 (2010)
13. Di Felice, M., Chowdhury, K.R., Kassler, A., Bononi, L.: Adaptive sensing scheduling and spectrum selection in cognitive wireless mesh networks. In: 2011 Proceedings of 20th International Conference on Computer Communications and Networks (ICCCN), pp. 1–6 (2011)
14. Freitas, L.C., Cardoso, C., Muller, F.C., Costa, J.W., Klautau, A.: Automatic modulation classification for cognitive radio systems: results for the symbol and waveform domains. In: IEEE Latin-American Conference on Communications, LATINCOM 2009, pp. 1–6 (2009)
15. Petrova, M., Mähönen, P., Osuna, A.: Multi-class classification of analog and digital signals in cognitive radios using support vector machines. In: 7th International Symposium on Wireless Communication Systems (ISWCS), pp. 986–990 (2010)
16. Aslam, M.W., Zhu, Z., Nandi, A.K.: Automatic modulation classification using combination of genetic programming and KNN. IEEE Transactions on Wireless Communications 11(8), 2742–2750 (2012)
17. Yao, Y., Feng, Z.: Centralized channel and power allocation for cognitive radio networks: a Q-learning solution. In: IEEE Future Network and Mobile Summit, pp. 1–8
18. van den Biggelaar, O., Dricot, J., De Doncker, P., Horlin, F.: Power allocation in cognitive radio networks using distributed machine learning. In: IEEE 23rd International Symposium on Personal Indoor and Mobile Radio Communications (PIMRC), pp. 826–831 (2012)
19. Wu, C., Chowdhury, K., Di Felice, M., Meleis, W.: Spectrum management of cognitive radio using multi-agent reinforcement learning. In: Proceedings of the 9th International Conference on Autonomous Agents and Multiagent Systems: Industry track, pp. 1705–1712. International Foundation for Autonomous Agents and Multiagent Systems (2010)
20. Vapnik, V.N.: Statistical Learning Theory. Wiley, New York (1998)
21. Boser, B.E., Guyon, I.M., Vapnik, V.N.: A training algorithm for optimal margin classifiers. In: Proc. ACM Fifth Annual Workshop Computational Learning Theory, Pittsburgh, PA, USA, July, 1992, pp. 144–152 (1992)
22. Byun, H., Lee, S.-W.: Applications of support vector machines for pattern recognition: a survey. In: Lee, S.-W., Verri, A. (eds.) SVM 2002. LNCS, vol. 2388, pp. 213–236. Springer, Heidelberg (2002)
23. Wang, G.: Applications of support vector machines for pattern recognition: a survey. In: Proc. IEEE Fourth International Conference on Networked Computing and Advanced Information Management, pp. 123–128 (2008)
24. Hastie, T., Tibshirani, R., Friedman, J., Franklin, J.: The elements of statistical learning: Data mining, inference and prediction. The Mathematical Intelligencer 27(2), 83–85 (2005)
25. Qiu, R.C., Hu, Z., Li, H., Wicks, M.C.: Cognitive Radio Communication and Networking: Principles and Practice. John Wiley & Sons (2012)

26. Hartigan, J.A., Wong, M.A.: Algorithm as 136: A k-means clustering algorithm. Applied Statistics, 100–108 (1979)
27. Kanungo, T., Mount, D.M., Netanyahu, N.S., Piatko, C.D., Silverman, R., Wu, A.Y.: An efficient k-means clustering algorithm: Analysis and implementation. IEEE Transactions on Pattern Analysis and Machine Intelligence 24(7), 881–892 (2002)
28. Sutton, R.S., Barto, A.G.: Introduction to Reinforcement Learning. MIT Press (1998)
29. Watkins, C.J., Dayan, P.: Q-learning. Machine Learning 8(3–4), 279–292 (1992)
30. Biglieri, E.: Principles of Cognitive Radio. Cambridge University Press (2012)
31. Busoniu, L., Babuska, R., De Schutter, B.: A comprehensive survey of multiagent reinforcement learning. IEEE Transactions on Systems, Man, and Cybernetics, Part C: Applications and Reviews 38(2), 156–172 (2008)
32. Kaelbling, L.P., Littman, M.L., Moore, A.W.: Reinforcement learning: A survey. Journal of Artificial Intelligence Research, 237–285 (1996)
33. Zeng, Y., Liang, Y.-C., Hoang, A.T., Zhang, R.: A review on spectrum sensing for cognitive radio: challenges and solutions. EURASIP Journal on Advances in Signal Processing (2010)
34. Cabric, D., Mishra, S.M., Brodersen, R.W.: Implementation issues in spectrum sensing for cognitive radios. In: Conference Record of the Thirty-Eighth Asilomar Conference on Signals, Systems and Computers, vol. 1, pp. 772–776. IEEE (2004)
35. Sonnenschein, A., Fishman, P.M.: Radiometric detection of spread-spectrum signals in noise of uncertain power. IEEE Transactions on Aerospace and Electronic Systems 28(3), 654–660 (1992)
36. Tandra, R., Sahai, A.: Fundamental limits on detection in low snr under noise uncertainty. In: 2005 International Conference on Wireless Networks, Communications and Mobile Computing, vol. 1, pp. 464–469. IEEE (2005)
37. Kim, K., Akbar, I., Bae, K., Um, J.-S., Spooner, C., Reed, J.: Cyclostationary approaches to signal detection and classification in cognitive radio. In: 2nd IEEE International Symposium on New Frontiers in Dynamic Spectrum Access Networks, DySPAN 2007, pp. 212–215 (2007)
38. Choi, K.W., Hossain, E., Kim, D.I.: Cooperative spectrum sensing under a random geometric primary user network model. IEEE Transactions on Wireless Communications 10(6), 1932–1944 (2011)
39. Oksanen, J., Koivunen, V., Lundén, J., Huttunen, A.: Diversity-based spectrum sensing policy for detecting primary signals over multiple frequency bands. In: 2010 IEEE International Conference on Acoustics Speech and Signal Processing (ICASSP), pp. 3130–3133 (2010)

Author Index

Printed in the United States
By Bookmasters

Printed in the United States
By Bookmasters